Solid State Materials Chemistry

This book explores the fascinating world of functional materials from the perspective of those who are tasked with inventing them, solid state chemists. Written in a clear and accessible style, this book provides a modern-day treatment of solid state materials chemistry for graduate and advanced undergraduate level courses. With over 330 problems and 400 original figures, this essential reference covers a wide range of materials in a holistic manner, including inorganic and organic, crystalline and amorphous, bulk and nanocrystals.

The introductory chapters cover topics such as crystal structures, defects, diffusion in solids, chemical bonding, and electronic band structure. Later chapters focus on important classes of functional materials including pigments, phosphors, dielectric materials, magnets, metals, semiconductors, superconductors, nonlinear optical materials, battery materials, zeolites, metal–organic framework materials, and glasses. The technological applications and synthesis methods used to prepare the materials that drive modern society are highlighted throughout.

Patrick M. Woodward is a Professor in the Department of Chemistry and Biochemistry and holds a courtesy appointment in the Department of Physics at Ohio State University. He is best known for his studies of the structures and properties of perovskite-related materials. He has served as chair of the Solid State Chemistry Gordon Conference (2018), Associate Editor of the *Journal of Solid State Chemistry* (2006–2011), and Vice President of the Neutron Scattering Society of America (2014–2018). He is co-author of the widely used general chemistry textbook, *Chemistry: The Central Science* (Pearson Education Limited, 2018). Patrick is a recipient of an NSF Career Award (2001), a Sloan Research Fellowship (2004), a Leverhulme Visiting Professorship (2017), and is a Fellow of the American Chemical Society (2020).

Pavel Karen is a Professor in the Department of Chemistry at the University of Oslo. His interests include inorganic reaction chemistry, solid state synthesis methods, crystallography, phase relations and thermodynamics, and point-defect chemistry; all components of his teaching portfolio. He is interested in the relationship between structure and properties of less common inorganic solids, such as mixed-valence oxides. Crystal structures are studied by X-ray and neutron diffraction, local structures by Mössbauer spectroscopy, and valence-mixing by calorimetry. He is co-author of the chapter Phase Diagrams and Thermodynamic Properties in the *Handbook on the Physics and Chemistry of the Rare Earths, Volume 30, High-Temperature Superconductors* (Elsevier, 2000). Pavel is a member of the American Chemical Society, the American Crystallographic Association, and of the International

Union of Pure and Applied Chemistry's Division II and the Interdivisional Committee on Terminology, Nomenclature and Symbols.

John S. O. Evans is a Chemistry Professor at Durham University where he served as Head of Chemistry from 2009 to 2014. His research interests are in the synthesis and properties of (mainly) inorganic materials, their structural chemistry, and their real-world applications. In recent years he has worked, inter alia, on negative thermal expansion, symmetry properties of phase transitions, new oxide-chalcogenides and energy-related materials. He has a long-standing interest in developing powder diffraction methods and is co-author of *Rietveld Refinement: Practical Powder Diffraction Pattern Analysis using TOPAS* (De Gruyter, 2019). John was awarded the 1997 Meldola prize of the Royal Society of Chemistry, and was co-awarded the 2015 Royal Society of Chemistry Teamwork in Innovation Award for work with industry.

Thomas Vogt is the Educational Foundation Endowed Chair in the Department of Chemistry and Biochemistry, Director of the NanoCenter and adjunct Professor in the Department of Philosophy at the University of South Carolina. His work focuses on establishing structure–property relationships of solid state materials using X-ray and neutron scattering and electron microscopy. He is recognized as Fellow of the American Physical Society, the American Association for the Advancement of Science, the Institute of Advanced Study at Durham University, and the Neutron Scattering Society of America. Thomas received the Carolina Trustee Professorship of the Board of Trustees in 2018 as well as the University of South Carolina's Educational Foundation Award for Research in Science, Mathematics, and Engineering in 2019.

Solid State Materials Chemistry

Patrick M. Woodward
Ohio State University

Pavel Karen
Universitetet i Oslo

John S. O. Evans
Durham University

Thomas Vogt
University of South Carolina

CAMBRIDGE
UNIVERSITY PRESS

CAMBRIDGE
UNIVERSITY PRESS

University Printing House, Cambridge CB2 8BS, United Kingdom

One Liberty Plaza, 20th Floor, New York, NY 10006, USA

477 Williamstown Road, Port Melbourne, VIC 3207, Australia

314-321, 3rd Floor, Plot 3, Splendor Forum, Jasola District Centre, New Delhi - 110025, India

103 Penang Road, #05-06/07, Visioncrest Commercial, Singapore 238467

Cambridge University Press is part of the University of Cambridge.

It furthers the University's mission by disseminating knowledge in the pursuit of education, learning and research at the highest international levels of excellence.

www.cambridge.org
Information on this title: www.cambridge.org/9780521873253
DOI: 10.1017/9781139025348

First published 2021

A catalogue record for this publication is available from the British Library

ISBN 978-0-521-87325-3 Hardback

Contents

Preface   *page* xvii

Acknowledgments   xix

1 Structures of Crystalline Materials   1

  1.1 Symmetry   1

    1.1.1 Translational Symmetry   2

    1.1.2 Rotational Symmetry   3

    1.1.3 Crystallographic Point Groups and Crystal Systems   5

    1.1.4 Bravais Lattices   5

    1.1.5 Introduction to Space Groups   8

    1.1.6 Symmetry Elements That Combine Rotation and Translation   9

    1.1.7 Space-Group Symbols   11

    1.1.8 Description of a Crystal Structure   12

  1.2 Databases   13

  1.3 Composition   14

    1.3.1 Coordination, Stoichiometry, and Connectivity   15

    1.3.2 The Generalized 8−N Rule   17

  1.4 Structural Principles   18

    1.4.1 Packing of Spheres   19

    1.4.2 Filling Holes   22

    1.4.3 Network Structures   28

    1.4.4 Polyhedral Structures   32

  1.5 Structures of Selected Materials   38

    1.5.1 The Spinel Structure   38

    1.5.2 The Garnet Structure   39

    1.5.3 Perovskite Structures   40

    1.5.4 Silicates   44

    1.5.5 Zeolites   46

    1.5.6 Zintl Phases   47

  1.6 Problems   48

  1.7 Further Reading   51

  1.8 References   51

2 Defects and More Complex Structures   54

   2.1 Point Defects in Crystalline Elemental Solids   54
   2.2 Intrinsic Point Defects in Compounds   55
   2.3 Thermodynamics of Vacancy Formation   58
   2.4 Extrinsic Defects   61
   2.5 Solid Solutions and Vegard's Law   63
   2.6 Kröger–Vink Notation   65
   2.7 Line Defects in Metals   66
      2.7.1 Edge Dislocations   66
      2.7.2 Screw Dislocations   66
   2.8 Planar Defects in Materials   67
      2.8.1 Stacking Faults   67
      2.8.2 Twinning   68
      2.8.3 Antiphase Boundaries   72
      2.8.4 Crystallographic Shear Structures   74
   2.9 Gross Nonstoichiometry and Defect Ordering   75
  2.10 Incommensurate Structures   78
  2.11 Infinitely Adaptive Structures   80
  2.12 Problems   81
  2.13 Further Reading   85
  2.14 References   85

3 Defect Chemistry and Nonstoichiometry   87

   3.1 Narrow Nonstoichiometry in Oxides   87
      3.1.1 Point Defects in a Pure Stoichiometric Oxide   87
      3.1.2 Point Defects upon Oxidation/Reduction of the Stoichiometric
           Oxide   88
      3.1.3 Equilibrium Equations for Oxidative and Reductive
           Nonstoichiometry   89
      3.1.4 Defect Equilibria for Schottky-Type Redox Compensation   90
      3.1.5 Acceptor-Doped Oxides   93
      3.1.6 Donor-Doped Oxides   94
      3.1.7 Solid Solubility of Dopants   94
      3.1.8 Cautionary Note on Defect Models in Pure Oxides   96
   3.2 Wide Nonstoichiometry in Oxides   98
   3.3 Point Defects and Diffusion   99
      3.3.1 Point-Defect Movements   101
      3.3.2 Random Hopping   103
      3.3.3 Hopping Under a Driving Force   104

3.3.4   Hopping Under a Concentration Gradient   105
3.3.5   Hopping Under an Electric Field   106
3.3.6   Relationship between Conductivity and Diffusivity   107
3.3.7   Ambipolar Diffusion   108
3.3.8   Temperature Dependence of Diffusivity   111
3.3.9   Diffusivity and Redox Defect Equilibria   111
3.3.10 Outline of Non-Steady-State Diffusion   112
3.3.11 Cautionary Note on Diffusion in Real Materials   114
3.4 Problems   115
3.5 Further Reading   118
3.6 References   118

4 **Phase Diagrams and Phase Transitions**   120

4.1 Phase Diagrams   120
4.2 Two-Component Phase Diagrams   123
     4.2.1   Without Compound Formation   123
     4.2.2   With Compound Formation   125
     4.2.3   Solid-Solution Formation   128
4.3 Three-Component Phase Diagrams   131
4.4 Structural Phase Transitions   135
     4.4.1   Classification of Phase Transitions   136
     4.4.2   Symmetry and Order Parameters   137
     4.4.3   Introduction to Landau Theory   140
     4.4.4   Second-Order Transitions   141
     4.4.5   First-Order and Tricritical Transitions   144
     4.4.6   Phonons, Soft Modes, and Displacive Transitions   147
4.5 Problems   150
4.6 Further Reading   152
4.7 References   153

5 **Chemical Bonding**   154

5.1 Ionic Bonding   154
     5.1.1   Coulombic Potential Energy   154
     5.1.2   Lattice Energy and the Born–Mayer Equation   156
     5.1.3   Experimental versus Calculated Lattice-Formation Energies 158
5.2 Atomic Orbitals   161
     5.2.1   Energies of Atomic Orbitals   166
     5.2.2   Sizes of Atomic Orbitals   168
5.3 Molecular-Orbital Theory   169
     5.3.1   Homonuclear Diatomics: H_2^+ and H_2 169
     5.3.2   The Heteronuclear Diatomic Case: HHe   173

5.3.3   Orbital Overlap and Symmetry   174
5.3.4   Combination of σ and π Bonding: O_2 175
5.3.5   Symmetry-Adapted Linear Combinations (SALCs)   177
5.3.6   Simple Polyatomic Molecules: BeH_2 and CH_4 179
5.3.7   Conjugated π Bonding: C_6H_6 181
5.3.8   Transition-Metal Complexes: $[CrCl_6]^{3-}$ and $[CoCl_4]^{2-}$ 183
5.3.9   High- and Low-Spin Configurations   186
5.3.10 Jahn–Teller Distortions   188
5.4 Bond Valences   190
5.5 Problems   195
5.6 Further Reading   198
5.7 References   199

6 Electronic Band Structure   200

6.1 The Band Structure of a Hydrogen-Atom Chain   200
6.1.1   The Electronic Structures of Cyclic H_N Molecules   201
6.1.2   Translational Symmetry and the Bloch Function   202
6.1.3   The Quantum Number k 203
6.1.4   Visualizing Crystal Orbitals   204
6.1.5   Band-Structure Diagrams   207
6.1.6   Density-of-States (DOS) Plots   209
6.2 The Band Structure of a Chain of H_2 Molecules   210
6.3 Electrical and Optical Properties   213
6.3.1   Metals, Semiconductors, and Insulators   213
6.3.2   Direct- versus Indirect-Gap Semiconductors   214
6.4 Representing Band Structures in Higher Dimensions   215
6.4.1   Crystal Orbitals in Two Dimensions   215
6.4.2   Crystal Orbitals in Three Dimensions   219
6.5 Band Structures of Two-Dimensional Materials   220
6.5.1   Graphene   221
6.5.2 CuO_2^{2-} Square Lattice   223
6.6 Band Structures of Three-Dimensional Materials   227
6.6.1 α-Polonium   227
6.6.2   Diamond   228
6.6.3   Elemental Semiconductors   230
6.6.4   Rhenium Trioxide   231
6.6.5   Perovskites   233
6.7 Problems   237
6.8 Further Reading   241
6.9 References   242

7 Optical Materials   243

7.1   Light, Color, and Electronic Excitations   243
7.2   Pigments, Dyes, and Gemstones   245
7.3   Transitions between d Orbitals (d-to-d Excitations)   246
    7.3.1   Ligand- and Crystal-Field Theory   246
    7.3.2   Absorption Spectra and Spectroscopic Terms   248
    7.3.3   Correlation Diagrams   252
    7.3.4   Selection Rules and Absorption Intensity   255
7.4   Charge-Transfer Excitations   258
    7.4.1   Ligand-to-Metal Charge Transfer   259
    7.4.2   Metal-to-Metal Charge Transfer   260
7.5   Compound Semiconductors   261
    7.5.1   Optical Absorbance, Band Gap, and Color   262
    7.5.2   Electronegativity, Orbital Overlap, and Band Gap   263
7.6   Conjugated Organic Molecules   265
7.7   Luminescence   267
7.8   Photoluminescence   268
    7.8.1   Components of a Phosphor   268
    7.8.2   Radiative Return to the Ground State   270
    7.8.3   Thermal Quenching   272
    7.8.4   Lanthanoid Activators   274
    7.8.5   Non-Lanthanoid Activators   279
    7.8.6   Energy Transfer   281
    7.8.7   Sensitizers   283
    7.8.8   Concentration Quenching and Cross Relaxation   284
    7.8.9   Up-Conversion Photoluminescence   285
7.9   Electroluminescence   287
    7.9.1   Inorganic Light-Emitting Diodes (LEDs)   287
    7.9.2   Organic Light-Emitting Diodes (OLEDs)   289
7.10   Materials for Lighting   291
    7.10.1   Fluorescent Lamp Phosphors   292
    7.10.2   Phosphor-Converted LEDs for White Light   293
7.11   Problems   294
7.12   Further Reading   298
7.13   References   299

8 Dielectrics and Nonlinear Optical Materials   301

8.1   Dielectric Properties   301
    8.1.1   Dielectric Permittivity and Susceptibility   302

	8.1.2 Polarization and the Clausius–Mossotti Equation	303
	8.1.3 Microscopic Mechanisms of Polarizability	305
	8.1.4 Frequency Dependence of the Dielectric Response	306
	8.1.5 Dielectric Loss	308
8.2	Dielectric Polarizabilities and the Additivity Rule	309
8.3	Crystallographic Symmetry and Dielectric Properties	313
8.4	Pyroelectricity and Ferroelectricity	314
	8.4.1 Ferroelectricity in $BaTiO_3$	314
	8.4.2 Antiferroelectricity	319
8.5	Piezoelectricity	321
8.6	Local Bonding Considerations in Non-Centrosymmetric Materials	324
	8.6.1 Second-Order Jahn–Teller Distortions with d^0 Cations	325
	8.6.2 Second-Order Jahn–Teller Distortions with s^2p^0 Cations	327
8.7	Nonlinear Optical Materials	330
8.8	Nonlinear Susceptibility and Phase Matching	331
8.9	Important SHG Materials	334
	8.9.1 KH_2PO_4	336
	8.9.2 $KTiOPO_4$	336
	8.9.3 Niobates and Tantalates	338
	8.9.4 Organic and Polymer NLO Materials	339
	8.9.5 Borates	340
8.10	Problems	343
8.11	Further Reading	346
8.12	References	346
9	**Magnetic Materials**	**349**
9.1	Magnetic Materials and Their Applications	349
9.2	Physics of Magnetism	349
	9.2.1 Bar Magnets and Atomic Magnets	349
	9.2.2 Magnetic Intensity, Induction, Energy, Susceptibility, and Permeability	352
	9.2.3 Unit Systems in Magnetism	355
9.3	Types of Magnetic Materials	356
9.4	Atomic Origins of Magnetism	357
	9.4.1 Electron Movements Contributing to Magnetism and Their Quantization	357
	9.4.2 Atomic Magnetic Moments	359
	9.4.3 Magnetic Moments for $3d$ Ions in Compounds	363
	9.4.4 Magnetic Moments for $4f$ Ions in Compounds	366
	9.4.5 Note on Magnetic Moments of $4d$ and $5d$ Metals in Compounds	366

9.5   Diamagnetism   367
9.6   Paramagnetism   367
       9.6.1   Curie and Curie–Weiss Paramagnetism   368
       9.6.2   Pauli Paramagnetism   371
9.7   Antiferromagnetism   372
9.8   Superexchange Interactions   374
9.9   Ferromagnetism   377
       9.9.1   Ferromagnetic Insulators and Half-Metals   381
       9.9.2   Ferromagnetic Metals   382
       9.9.3   Superferromagnets   384
9.10   Ferrimagnetism   385
9.11   Frustrated Systems and Spin Glasses   387
9.12   Magnetoelectric Multiferroics   388
9.13   Molecular and Organic Magnets   389
9.14   Problems   391
9.15   Further Reading   394
9.16   References   394

10   Conducting Materials   396

10.1   Conducting Materials   396
10.2   Metals   398
       10.2.1   Drude Model   398
       10.2.2   Free-Electron Model   402
       10.2.3   Fermi–Dirac Distribution   403
       10.2.4   Carrier Concentration   405
       10.2.5   Carrier Mobility and Effective Mass   406
       10.2.6   Fermi Velocity   407
       10.2.7   Scattering Mechanisms   409
       10.2.8   Band Structure and Conductivity of Aluminum   411
       10.2.9   Band Structures and Conductivity of Transition Metals 412
10.3   Semiconductors   414
       10.3.1   Carrier Concentrations in Intrinsic Semiconductors   414
       10.3.2   Doping   416
       10.3.3   Carrier Concentrations and Fermi Energies in Doped
                Semiconductors   419
       10.3.4   Conductivity   421
       10.3.5   p–n Junctions   422
       10.3.6   Light-Emitting Diodes and Photovoltaic Cells   425
       10.3.7   Transistors   426

10.4   Transition-Metal Compounds   428
       10.4.1 Electron Repulsion: The Hubbard Model   428
       10.4.2 Transition-Metal Compounds with the NaCl-Type Structure   431
       10.4.3 Transition-Metal Compounds with the Perovskite Structure   434
10.5   Organic Conductors   437
       10.5.1 Conducting Polymers   438
       10.5.2 Polycyclic Aromatic Hydrocarbons   441
       10.5.3 Charge-Transfer Salts   443
10.6   Carbon   445
       10.6.1 Graphene   445
       10.6.2 Carbon Nanotubes   447
10.7   Problems   451
10.8   Further Reading   454
10.9   References   455

11   Magnetotransport Materials   457

11.1   Magnetotransport and Its Applications   457
11.2   Charge, Orbital, and Spin Ordering in Iron Oxides   458
       11.2.1 The Verwey Transition in Magnetite, Fe_3O_4 458
       11.2.2 Double-Cell Perovskite, $YBaFe_2O_5$ 460
       11.2.3 $CaFeO_3$ and $SrFeO_3$ 462
11.3   Charge and Orbital Ordering in Perovskite-Type Manganites   465
       11.3.1 Spin and Orbital Ordering in $CaMnO_3$ and $LaMnO_3$ 465
       11.3.2 The $La_{1-x}Ca_xMnO_3$ Phase Diagram   468
       11.3.3 Tuning the Colossal Magnetoresistance   470
11.4   Half-Metals and Spin-Polarized Transport   472
       11.4.1 Magnetoresistant Properties of Half-Metals   472
       11.4.2 CrO_2 476
       11.4.3 Heusler Alloys   477
       11.4.4 Half-Metals with Valence-Mixing Itinerant Electrons   480
11.5   Problems   481
11.6   Further Reading   483
11.7   References   483

12   Superconductivity   486

12.1   Overview of Superconductivity   486
12.2   Properties of Superconductors   488
12.3   Origins of Superconductivity and BCS Theory   492
12.4 C_{60}-Derived Superconductors   500
12.5   Molecular Superconductors   505

12.6   BaBiO$_3$ Perovskite Superconductors   509
12.7   Cuprate Superconductors   511
       12.7.1   La$_2$CuO$_4$ "214" Materials   512
       12.7.2   YBa$_2$Cu$_3$O$_{7-\delta}$ "YBCO" or "123" Materials   513
       12.7.3   Other Cuprates   516
       12.7.4   Electronic Properties of Cuprates   517
12.8   Iron Pnictides and Related Superconductors   521
12.9   Problems   523
12.10 Further Reading   526
12.11 References   526

13   Energy Materials: Ionic Conductors, Mixed Conductors, and Intercalation
     Chemistry   529
13.1   Electrochemical Cells and Batteries   529
13.2   Fuel Cells   532
13.3   Conductivity in Ionic Compounds   533
13.4   Superionic Conductors   536
       13.4.1   AgI: A Cation Superionic Conductor   536
       13.4.2   PbF$_2$: An Anionic Superionic Conductor   539
13.5   Cation Conductors   540
       13.5.1   Sodium β-alumina   540
       13.5.2   Other Ceramic Cation Conductors   542
       13.5.3   Polymeric Cation Conductors   543
13.6   Proton Conductors   545
       13.6.1   Water-Containing Proton Conductors   546
       13.6.2   Acid Salts   547
       13.6.3   Perovskite Proton Conductors   548
13.7   Oxide-Ion Conductors   549
       13.7.1   Fluorite-Type Oxide-Ion Conductors   552
       13.7.2   Perovskite, Aurivillius, Brownmillerite, and Other Oxide
                Conductors   553
       13.7.3   SOFC Electrode Materials and Mixed Conductors   555
13.8   Intercalation Chemistry and Its Applications   555
       13.8.1   Graphite Intercalation Chemistry   556
       13.8.2   Lithium Intercalation Chemistry and Battery Electrodes   559
       13.8.3   Lithium-Ion Batteries with Oxide Cathodes   561
       13.8.4   Electrochemical Characteristics of Lithium Batteries   568
       13.8.5   Other Lithium Battery Electrode Materials   569
13.9   Problems   573
13.10 Further Reading   576
13.11 References   576

14 Zeolites and Other Porous Materials   579

    14.1 Zeolites   579
        14.1.1 Representative Structures of Zeolites   581
        14.1.2 Roles of Template Molecules in Zeolite Synthesis   586
        14.1.3 Zeolites in Catalysis   588
        14.1.4 Ion-Exchange Properties   593
        14.1.5 Drying Agents, Molecular Sieving, and Sorption   595
        14.1.6 AlPOs and Related Materials   596
    14.2 Mesoporous Aluminosilicates   597
    14.3 Other Porous Oxide Materials   600
    14.4 Metal–Organic Frameworks (MOFs)   605
        14.4.1 MOF Structures   605
        14.4.2 Some Applications of MOFs   608
    14.5 Problems   612
    14.6 Further Reading   615
    14.7 References   616

15 Amorphous and Disordered Materials   619

    15.1 The Atomic Structure of Glasses   620
    15.2 Topology and the Structure of Glasses   622
    15.3 Oxide Glasses   625
    15.4 Optical Properties and Refractive Index   625
    15.5 Optical Fibers   631
    15.6 Nucleation and Growth   633
    15.7 The Glass Transition   634
    15.8 Strong and Fragile Behavior of Liquids and Melts   639
    15.9 Low-Temperature Dynamics of Amorphous Materials   642
    15.10 Electronic Properties: Anderson Localization   644
    15.11 Metallic Glasses   647
    15.12 Problems   651
    15.13 Further Reading   652
    15.14 References   652

Appendix A: Crystallographic Point Groups in Schönflies Symbolism   655
Appendix B: International Tables for Crystallography   656
Appendix C: Nomenclature of Silicates   661
Appendix D: Bond-Valence Parameters in Solids   662
Appendix E: The Effect of a Magnetic Field on a Moving Charge   663
Appendix F: Coupling j–j 664

Appendix G: The Langevin Function   665
Appendix H: The Brillouin Function   666
 Appendix I: Measuring and Analyzing Magnetic Properties   670
 Appendix J: Fundamental Constants of Exact Value   672
References for Appendices   673
Index   674

Preface

Functional materials are an integral part of daily life. As an example, consider the materials that underpin smartphone technology. The integrated circuitry is made from complex patterns of semiconductors, metallic conductors, and insulators. Organic light-emitting diodes convert electrical signals from the processor into a vibrant high-resolution color display. The display is protected by a screen made from tough but lightweight Gorilla® glass, which is coated with a transparent conducting oxide to make the screen responsive to the touch of a finger. Magnetic materials are used in the speakers, a lithium-ion battery powers the device, specific dielectric materials are used to receive and isolate a call once the signal reaches a base station, and the list goes on.

This book explores the fascinating world of functional materials from the perspective of those who are tasked with inventing them, solid state chemists. We therefore adopt the chemist's definition of a material as a substance whose structure and properties are controlled at the atomic level to produce a specific function. Returning to our example, a modern smartphone contains over half of the non-radioactive elements on the periodic table. A few are used in their elemental form, but in most cases the desired function can only be achieved by combining elements to form compounds. With the periodic table as a palette, how does the chemist design and synthesize the mind-boggling variety of functional materials that future technologies depend upon? That question is the topic this book explores.

The book is written specifically with teaching in mind and is intended primarily for use in upper-level undergraduate or graduate level courses. While our perspective is that of a chemist, the book is accessible to physicists and engineers as well. Mathematical details are given where they add deeper understanding, but the focus is always on relating the properties of a material to the characteristics of the atoms and molecules from which it is built.

The first six chapters cover the fundamentals of extended solids: crystal structures, defects, reactivity, phase diagrams, phase transitions, chemical bonding, and band structure. The remaining chapters, each of which is organized around a specific property or class of materials, show how the properties of modern functional materials can be understood from these fundamental concepts. Recognizing that the field of solid state chemistry is much more expansive than can be covered in a single course, the later chapters are designed to be largely independent of each other. This organization provides the instructor freedom to tailor a course to cover those materials that are most relevant for their students.

Coverage of inorganic and organic materials is interwoven throughout the book to place the emphasis on properties. To keep the scope at a manageable level, neither synthesis nor characterization are covered in detail. Instead, boxes on synthetic methods and characterization methods are placed throughout the book to highlight specific examples. In a similar vein, boxes are used to describe how the properties of nanoscale solids differ from bulk materials (Nanoscale Concepts), and to highlight important technological applications of materials (Materials Spotlight). Students learn by practice, and, in this spirit, we have included dozens of problems at the end of each chapter to allow students to test their understanding of the concepts covered in the chapter. Instructors can obtain a full set of worked solutions on request.

We hope that this book will be a valuable source of learning for the next generations of solid state scientists and engineers and a resource for those who already work in this fascinating field.

Patrick Woodward
Pavel Karen
John Evans
Thomas Vogt

Acknowledgments

We are indebted to several organizations and countless people for their support and encouragement. PMW would like to acknowledge the Leverhulme Foundation for supporting his stay at Durham University as a Visiting Professor during the 2017–2018 academic year. TV spent the beginning of 2018 as a Fellow at Durham University's Institute of Advanced Study. He is grateful to Linda Crowe and the rest of the team at the institute, as well as David Wilkinson, Principal at St. John's College, for their warm hospitality and for providing an environment so conducive to scholarly work. These overlapping stays in Durham were instrumental in making the final push to finish this book. PMW is grateful for many years of support from the Solid State Materials Chemistry program of the National Science Foundation. PMW and JSOE thank Arthur Sleight for inspiration and early career mentoring. In these data-dominated times, PK is grateful to the now Professor Jiří Hanika for teaching Fortran programming in the 1970/1 course Computational Technology at the VŠCHT in Prague, and to the now Ing. František Hovorka, CSc, for the idea of taking an external typing course during the sophomore year 1967/8 at the SPŠCH Praha. We all would like to thank both colleagues and students who provided key feedback on early versions of the chapters. Finally, we thank our families for their patience and support over this long journey.

1   Structures of Crystalline Materials

This book is about functional materials—those that perform a task or a technological operation. By the end of the book, we'll see that all useful properties can ultimately be traced back to structure and dynamics at the atomic level of materials. Understanding structure is therefore of crucial importance.

In this chapter we'll investigate the structures of crystalline materials—those in which atomic arrangements are repeated periodically in three-dimensional (3D) space. Non-crystalline materials are covered in Chapter 15. In the first section, we will discuss the symmetry and crystallography concepts that are important for the description of crystalline substances. A brief introduction to structure databases will follow. In the third section, we'll cover the nomenclature and electron counting rules needed to understand the composition of solids; before learning in the fourth section how structures are built up by packing spheres, connecting coordination polyhedra, or via networks. In the fifth and last section, we'll discuss some structure types encountered later in the book. In addition to the figures and descriptions given in the chapter, readers might find it useful to draw models of important structures with the included structural coordinates.

1.1   Symmetry

In the first section of this chapter, we'll develop the language required to describe the structures of crystalline compounds. We know from everyday life that such materials frequently display an amazing regularity and symmetry on the macroscopic scale—salt crystals can grow as "perfect" cubes and many minerals and gemstones display wonderfully symmetric facets. The origin of this macroscopic symmetry can ultimately be traced back to the symmetry that's present at the atomic scale (Å or 10^{-10} m). This local symmetry is replicated millions of times by translational symmetry to produce symmetric macroscopic objects. It's perhaps worth stating at the outset that there is nothing "magical" about the high-symmetry structures materials adopt. As we will see in Chapter 5, local bonding interactions have inherent symmetry, and dense packing of such units is favorable energetically.

1.1.1   Translational Symmetry

To describe a crystal structure, it is useful to introduce the concept of **lattice**; a spatial pattern of points of equal and equally oriented surroundings. We then define a **motif**, which might be a small group of atoms, a molecule, or a collection of several molecules. If we associate this motif with each of the lattice points, a crystal structure is built as shown in Figure 1.1. We can think of this "association" as re-drawing the motif at a constant displacement from each lattice point. The lattice is an operator of **translational symmetry** of crystal structures, of their periodicity. We can see in Figure 1.1 that the translational symmetry defined by the lattice produces a structure in which the individual atoms in the motif achieve a sensible bonding environment.

The small spatial segment that fully represents the entire structure upon periodic repetition is called the **unit cell**.[1] We can use the analogy of tiles (2D) or bricks (3D) being stacked side by side. In 3D, the unit cell is a parallelepiped, a box whose sides are parallelograms. The size and shape of the unit cell is described with three lattice vectors a, b, c of lengths a, b, c, and angles α, β, γ. The angle α is between b and c, β between a and c, and γ between a and b. Together, a, b, c, α, β, γ are called **lattice parameters** or **unit-cell parameters**.

Positions of atoms inside the unit cell are expressed with **relative or fractional coordinates** x,y,z in terms of fractions of the lattice vectors that define the unit cell. Fractional coordinates define the position (or radius) vector r from the unit-cell origin to the atomic position as $r = xa + yb + zc$. They can always be expressed on a 0 to 1 scale. Because of translational symmetry, a coordinate of 1.2 is equivalent to 0.2, or a coordinate −0.2 is equivalent to −0.2 + 1 = 0.8.

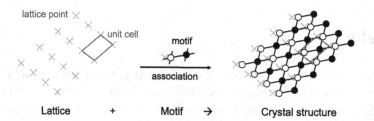

Figure 1.1 Association of an atomic motif with a lattice produces a crystal structure.

---

[1] The following rules apply for choosing the unit cell: (a) its rotational symmetry is the same as that of the lattice, (b) the edges and angles are made as similar to each other as possible, (c) the number of right angles is maximized, and (d) the volume is minimized. Where applicable, the origin coincides with the inversion center of symmetry (Section 1.1.2). In some cases (Section 1.1.4), the cell contains more than one lattice point.

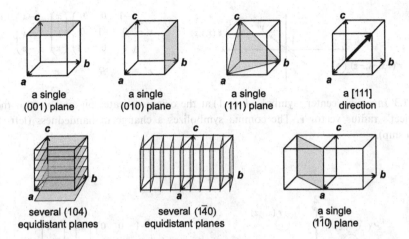

Figure 1.2 Description by indices of lattice planes and lattice directions with respect to the unit cell. An extra cell is drawn in order to show cases where planes facing the origin intercept the lattice vector at negative values.

Directions in the lattice are given with a [uvw] notation. When a line is drawn from the origin, parallel to the desired direction, then u,v,w are simply the relative/fractional coordinates of any point that line goes through, multiplied to give integer values. If it goes through ½,1,½, the direction is [121]. Symbols [242] and [484] would represent the same direction. A set of symmetrically equivalent lattice directions, such as [100], [010], [001], [−100], [0−10], [00−1] in a cubic lattice, is collectively referred to using angle brackets, ⟨100⟩.

It's often useful to define a **set of parallel equidistant planes** in a lattice. These are represented by an hkl notation. Starting with the plane that contains the origin, an hkl set of equidistant planes divides the unit-cell vector **a** into h sections, **b** into k sections and **c** into l sections (Figure 1.2). Dividing into 0 sections is possible and means that the set of planes is parallel with that axis. If the plane that faces the origin crosses **a**, **b**, or **c** at negative values, the appropriate h, k, or l of that set has negative sign (often put above the index, like $\bar{h}$). An (hkl) symbol refers to a plane or to a crystal face (Miller indices). A set of their symmetry-equivalent orientations is denoted in curly brackets, {hkl}.

1.1.2 Rotational Symmetry

A **point group**[2] is a set of symmetry operations that fulfill the mathematical requirements of being a group[3] and act on an isolated geometrical object. The number of

---

[2] At least one point of the object remains unshifted under point-group symmetry operations. Elementary knowledge of point-group symmetry may be an advantage for the reader; see Further Reading.

[3] A group must have a closure (combination of two elements yields an element of the group), fulfill the mathematic associative law (the result of combination of the elements is independent of the order they are applied), have an identity (an element that converts other elements into themselves), and have an inversion (every element has an inverse element; when combined together, they yield the identity element).

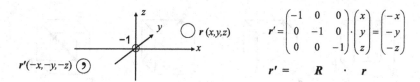

Figure 1.3 Inversion center (symbol −1 or $\bar{1}$) at the origin operates on an object at the endpoint of the object's radius vector *r*. The comma symbolizes a change in handedness (left- to right-hand relationship).

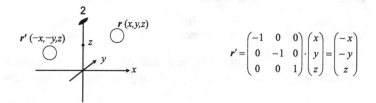

Figure 1.4 Operation of the twofold rotation axis (of symbols 2 in text and ◖ in graphics) on an object at the point *r*.

symmetry operations defines the **order of the group**; the higher the order, the higher the symmetry. The symmetry operations are performed by the **elements of point symmetry**; the identity, inversion center, mirror plane, rotation axis, and rotoinversion axis. All objects possess the **identity**; other symmetry elements may or may not be present.

Let's start with the **inversion center**. Figure 1.3 shows a point x,y,z represented by its radius vector *r* and the effect on that point of an inversion center at the origin of the coordinate system (shown by a small circle). Inversion moves x,y,z to $-x,-y,-z$. Mathematically, we can describe this transformation of *r* to *r′* with the equation $r' = R \cdot r$, where *R* is the matrix of the point-symmetry element, describing its operation.[4] Inversion is given the symbol −1, often typeset as $\bar{1}$.

Symmetry elements that unify a point in lattice space with another one by rotating it in steps of $1/n$ ($n = 1, 2, 3, 4, 6$) of the full circle are called *n*-fold **rotation axes**. The full-circle rotation ($n = 1$) is the identity, a twofold axis ($n = 2$) rotates by ½ of the full circle and has symbol 2, etc. As can be seen in Figure 1.4, twofold rotation around the z axis moves a point of fractional coordinates x,y,z to $-x,-y,z$. A **rotoinversion axis** is a single element, the operation of which combines rotation and inversion. However, the twofold rotoinversion axis −2, shown in Figure 1.5, operates like a mirror plane, *m*, as can be also demonstrated by

---

[4] The three columns in this matrix are the products of the particular symmetry operation on the end-points of the respective unit-cell vectors (1,0,0), (0,1,0), and (0,0,1). This is conveniently used to set up the matrix.

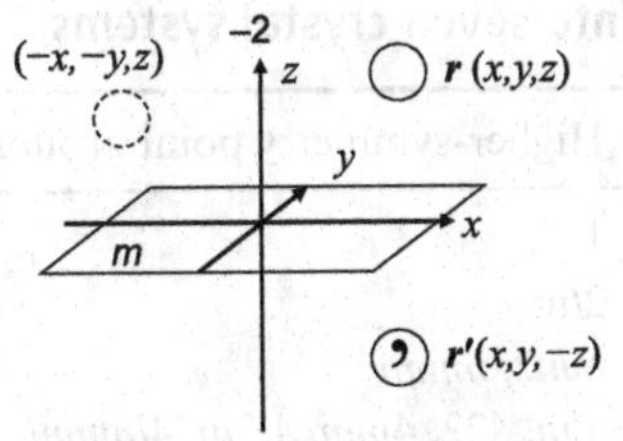

twofold axis operation $\begin{pmatrix} -1 & 0 & 0 \\ 0 & -1 & 0 \\ 0 & 0 & 1 \end{pmatrix} \cdot \begin{pmatrix} x \\ y \\ z \end{pmatrix} = \begin{pmatrix} -x \\ -y \\ z \end{pmatrix}$

inversion operation $r' = \begin{pmatrix} -1 & 0 & 0 \\ 0 & -1 & 0 \\ 0 & 0 & -1 \end{pmatrix} \cdot \begin{pmatrix} -x \\ -y \\ z \end{pmatrix} = \begin{pmatrix} x \\ y \\ -z \end{pmatrix}$

Figure 1.5 Rotoinversion axis -2 is identical with mirror m.

multiplying together the matrices representing 2 and -1. Since $m \equiv -2$, all point-symmetry elements are in fact **elements of rotational symmetry**; the rotational axes 1, 2, 3, 4, 6 and the rotoinversion axes $-1, -2, -3, -4, -6$. Accordingly, the operation matrix of each of them has the symbol $\boldsymbol{R}$.

1.1.3 Crystallographic Point Groups and Crystal Systems

Due to the infinite number of rotation axes, infinitely many point groups are possible for *isolated objects*. However, in crystal structures, the translational symmetry of space filling is only compatible with a small number of rotation axes. Consider that we can tile a plane perfectly with identical rectangular tiles (twofold axis present), isosceles triangles (threefold), squares (fourfold), or hexagons (sixfold), but we can't with pentagons (fivefold) or heptagons (sevenfold), etc. This argument extends to the 3D space filled by the "bricks" of unit cells (Figure 1.1). The point groups with symmetry elements $1, 2, 3, 4, 6, \bar{1}, m \,(\equiv \bar{2}), \bar{3}, \bar{4}, \bar{6}$, which are compatible with translational symmetry, are called **crystallographic point groups**, also known as **crystal classes**.

There are 32 crystallographic point groups and they are classified into seven **crystal systems**; cubic, tetragonal, hexagonal, trigonal, orthorhombic, monoclinic, and triclinic. Each crystal system is defined by its minimum point-group symmetry (Table 1.1, see also Appendix A). If we take the cubic system as an example, the minimum-symmetry point group has symbol 23. It means that at each lattice point, the twofold axes along the x-, y-, and z-coordinate-system axes repeat the threefold axis of the symbol along all body diagonals of the adjacent cells. These threefold axes are easier to visualize and remember as the symmetry condition for the cubic crystal system. So if the actual crystal structure has four intersecting threefold axes, it is cubic. If you do not see intersecting threefold axes, the structure cannot be cubic even if the unit cell has right angles and equal edges.

1.1.4 Bravais Lattices

As noted earlier, a lattice is a collection of points with identical surroundings. Having the highest rotational symmetry of each crystal system, 14 types of **Bravais lattices** are possible

Table 1.1 Sorting 32 crystallographic point groups into seven crystal systems.

Crystal system	Minimum symmetry	Higher-symmetry point groups
Triclinic	1	$\bar{1}$
Monoclinic	2, m	2/m
Orthorhombic	222	$mm2$, mmm
Tetragonal	4, $\bar{4}$	4/m, 422, 4mm, $\bar{4}\,2m$, 4/mmm
Hexagonal	6, $\bar{6}$	6/m, 622, 6mm, $\bar{6}m2$, 6/mmm
Trigonal	3, $\bar{3}$	32, 3m, $\bar{3}m$
Cubic	23	$m\bar{3}$, $\bar{4}\,3m$, 432, $m\bar{3}m$

(Figure 1.6), which represent 14 types of translational symmetry in 3D lattice space[5] of relative coordinates (Section 1.1.1). Some unit cells have lattice points only at the corners, and their lattices are called **primitive Bravais lattices**, labeled with symbol P. Others have lattice points located also at the centers of some or all unit-cell faces or at the unit-cell center, and these are called **centered Bravais lattices**. The body-centered lattice has symbol I (the cell has an additional lattice point at $\frac{1}{2}a + \frac{1}{2}b + \frac{1}{2}c$ relative to a P lattice). The face-centered lattices have symbol F when all unit-cell faces are centered. Symbol A, B, or C is used when just two opposite unit-cell sides are centered, along one direction, a, b, or c. Thus, C-**centering**[6] adds an additional lattice point at $\frac{1}{2}a + \frac{1}{2}b$ relative to a P lattice. A special type of centering, R, occurs in the hexagonal lattice, Figure 1.6. This R-centered lattice is equivalent to a P lattice with a *rhombohedral*[7] unit cell, see Table 1.2. The rhombohedral lattice occurs only in those structures of the trigonal crystal system that carry the symbol R in their symmetry description. The remaining trigonal structures are described with a primitive hexagonal cell P. It is often convenient to express also the R structures on a hexagonal cell (not just the P). Having triple the volume of the rhombohedral cell, the hexagonal cell contains three rhombohedral lattice points: 0,0,0 and $\frac{2}{3},\frac{1}{3},\frac{1}{3}$ with $\frac{1}{3},\frac{2}{3},\frac{2}{3}$, shown in Figure 1.6 labelled as hR.

Figure 1.7 gives an idea why only certain types of centering are possible for certain crystal systems. For example, a C-centered tetragonal cell could always be described with a smaller primitive cell, an F-centered tetragonal with a smaller I-centered cell. Similarly, a monoclinic B cell becomes a smaller P cell, monoclinic F cell becomes a smaller C cell, and a monoclinic I cell is equivalent to a C cell via an A cell rotated around b. However, there are cases where it is useful to choose a non-standard Bravais cell; for example in order to illustrate similarity between two structures.

---

[5] The lattice space defines the orientation and angles of the coordinate-system axes applied to each unit cell of this space.

[6] While International Tables for Crystallography use the British-English forms "centre", "centring", and "centred", the alternative spellings "center", "centering", and "centered" prevail in the USA.

[7] Terms like *trigonal*, *tetragonal*, and *hexagonal* originate in the rotational symmetry; the term *rhombohedral* implies that the Bravais cell is a rhombohedron, hence it refers to the lattice.

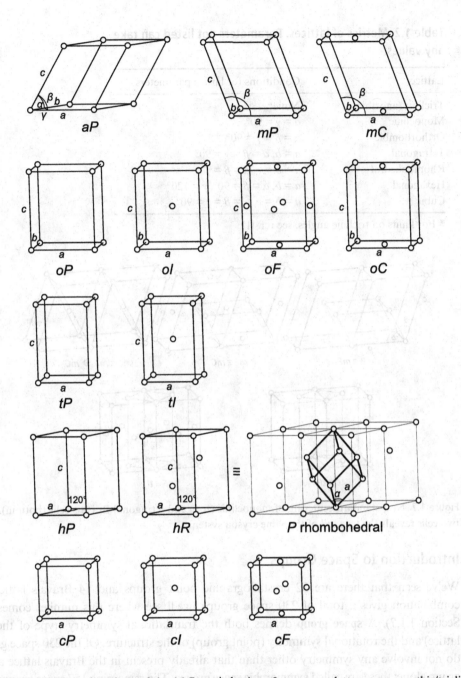

Figure 1.6 Standard settings of the 14 Bravais lattices. Lower-case letters: a = anorthic/triclinic, m = monoclinic, o = orthorhombic, t = tetragonal, h = hexagonal and c = cubic. Upper-case letters refer to centering. The primitive rhombohedral lattice is often described on its equivalent hR cell.

Table 1.2 Metrics of lattices. Parameters not listed can take any values.

Lattice	Conditions for lattice parameters
Triclinic (anorthic)	None*
Monoclinic	$\alpha = \gamma = 90°$
Orthorhombic	$\alpha = \beta = \gamma = 90°$
Tetragonal	$a = b, \alpha = \beta = \gamma = 90°$
Rhombohedral	$a = b = c, \alpha = \beta = \gamma$
Hexagonal	$a = b, \alpha = \beta = 90°, \gamma = 120°$
Cubic	$a = b = c, \alpha = \beta = \gamma = 90°$

* For limits on triclinic angles, see ref. [1].

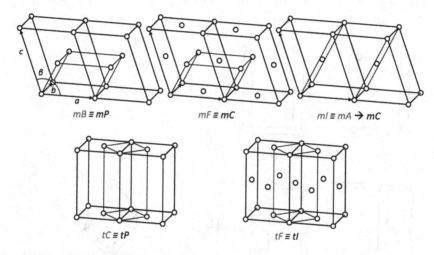

$mB \equiv mP$ $mF \equiv mC$ $mI \equiv mA \rightarrow mC$

$tC \equiv tP$ $tF \equiv tI$

Figure 1.7 Non-standard centering of monoclinic (top) and tetragonal Bravais cells (bottom). Drawing two cells reveals the true cell of the same crystal system.

1.1.5 Introduction to Space Groups

We've seen that there are 32 crystallographic point groups and 14 Bravais lattices. Their combination gives a total of 230 space groups (we'll see where this number comes from in Section 1.1.7). A **space group** defines both the translational symmetry (type of the Bravais lattice) and the rotational symmetry (point group) of the structure. Of the 230 space groups, 73 do not involve any symmetry other than that already present in the Bravais lattice and point group alone; these are called **symmorphic space groups**. The remaining 157 space groups are **non-symmorphic** and possess translations by suitable fractions of lattice vectors, brought about by screw axes or glide planes. Before we explain these two terms, a note on symmetry operators.

When discussing point groups, it was convenient to introduce the matrix operator $\boldsymbol{R}$ that acts on a point r of the relative-coordinate vector r such that $r' = \boldsymbol{R} \cdot r$. Because space groups may include the above-mentioned additional fractional translations, this symbolic language is extended into a **Seitz operator**, $(R \mid t) = R \cdot r + t$, which combines rotations (matrix R) and the possible translations (vector t). If we consider a symmetry element that involves no translations, such as the inversion $\bar{1}$, the Seitz symbol is $(R \mid 0)$ hence $(\bar{1} \mid 0)$. A rotational axis has a direction $[uvw]$ that must be included. The Seitz symbol $(2[001] \mid 0)$ then refers to a twofold rotation around the z axis. A plane has a direction as well—the direction of its normal vector that is oriented perpendicular to the plane. As an example, the mirror in the xy plane of Figure 1.5 has the Seitz symbol $(m[001] \mid 0)$. The Seitz operators that include the fractional translations are explained in the subsection below.

1.1.6 Symmetry Elements That Combine Rotation and Translation

The periodicity of crystal structures (their translational symmetry) means that the set of rotational symmetry elements repeats at each lattice point. This creates additional symmetry elements in between lattice points and may give rise to two types of symmetry elements that aren't present in isolated molecules,[8] screw axes and glide planes. A **screw axis** combines rotation with translation along the axis of rotation. An N_M screw axis (M < N) rotates anticlockwise by 360/N degrees while shifting the image by a distance equal to an M/N fraction of the lattice periodicity along that axis. As an example, the twofold screw axis 2_1 in Figure 1.8 rotates by 180° and shifts by ½ of the vector c (the axis is along z). A 6_3 axis rotates by increments of 60° and each time shifts by ³⁄₆ = ½ of c.

Figure 1.9 shows the symmetry operations of axes 6_2 and 6_4, illustrating that screw axes N_M and $N_{(N-M)}$ produce mirror images of each other. The "tailed hexagon" is the graphic symbol of the sixfold screw axes.

A **glide plane** operates as a mirror plus a plane-parallel shift by half a vector length between two lattice points. When the shift is half of one lattice vector (either a, b, or c), the glide is called an **axial glide plane** a, or b, or c. The operation of the, say, c glide plane is a reflection

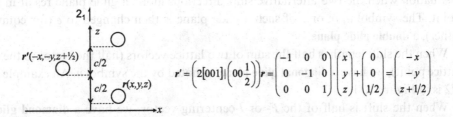

Figure 1.8 The screw axis $2_1[001]$ rotates 180° and shifts by ½ of the vector c. The graphical symbol of 2_1 parallel to the drawing plane is a half-arrow.

---

[8] In the point groups used for isolated molecules, all symmetry elements intersect at a point; in space groups this is no longer true.

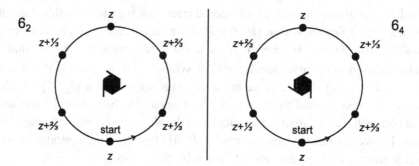

Figure 1.9 Screw axes rotate anticlockwise. Axes 6_2 and 6_4 produce mirror images of each other. The 6_2 axis rotates by 60° and translates by $\frac{2}{6}\,c$, the 6_4 axis rotates by 60° and translates by $\frac{4}{6}$ of the unit-cell length c. Due to translational symmetry, integers are subtracted from fractional coordinates ≥ 1, such as $\frac{3}{3} \equiv 0$ or $\frac{10}{3} \equiv \frac{1}{3}$.

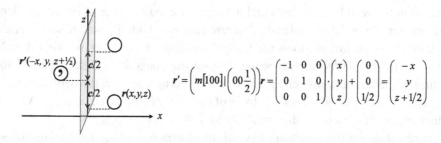

Figure 1.10 The glide plane $c[100]$ creates a mirror image shifted by $c/2$.

followed by shift along c by half the c-length. This is shown in Figure 1.10 with the c-glide plane oriented in the direction of x.[9] Note that an equally oriented plane with a translation along y would be called a b glide. In rare cases (five of the 230 space groups), centering creates a situation when the two alternative shift directions along a glide plane result in the same point. The symbol a, b, or c of such a glide plane is then changed to e (for equivalent or either), a **double glide plane**.

When the shift length is half the sum of two lattice vectors (half the vector to the diagonal lattice point), we have a **diagonal glide plane**, denoted by the symbol n. An example for $a/2 + b/2$ is in Figure 1.11.

When the shift is half of the F- or I-centering vector, we have a **diamond glide plane**, symbol d. This shift can be decomposed into components along lattice vectors $(a/4) + (b/4)$ or $(a/4) + (c/4)$ or $(b/4) + (c/4)$ for the F-centered orthorhombic and cubic lattices, and into $(a/4) + (b/4) + (c/4)$ for the I-centered tetragonal and cubic lattices.

---

[9] As noted above (Section 1.1.5), the direction of a plane is the direction of the plane's normal.

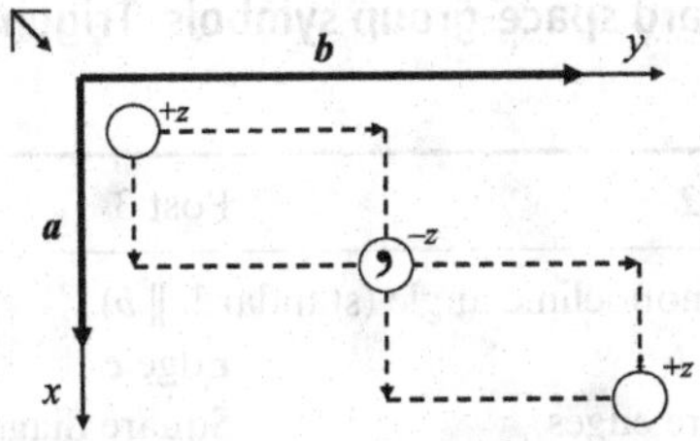

$$\left(m[001]\bigg|\left(\frac{1}{2}\frac{1}{2}0\right)\right)r = \begin{pmatrix} 1 & 0 & 0 \\ 0 & 1 & 0 \\ 0 & 0 & -1 \end{pmatrix}\begin{pmatrix} x \\ y \\ z \end{pmatrix} + \begin{pmatrix} 1/2 \\ 1/2 \\ 0 \end{pmatrix} = \begin{pmatrix} 1/2+x \\ 1/2+y \\ -z \end{pmatrix}$$

Figure 1.11 An orthorhombic diagonal glide $n[001]$ in the xy plane, with its graphical symbol on top left (see also Appendix B). It creates a mirror image shifted by $a/2$ and $b/2$. The subsequent operation of the same diagonal glide then recreates the original point in the next unit cell.

1.1.7 Space-Group Symbols

Shorthand symbols are used to name space groups. These consist of one of the Bravais-lattice symbols, P, F, I, R, A, B, C, followed by the symbol of the crystallographic point group, in which screw axes and glide planes can replace ordinary rotation axes and mirror planes when appropriate, for example $Pnma$. Non-symmorphic space groups are thus recognizable by the presence of symbols such as a, b, c, e, n, d, 2_1, 3_2, 4_2, etc., which indicate the fractional lattice translations. For quick understanding, it is helpful to know the orientation of the rotational-symmetry elements listed in the *standard*[10] space-group symbols (Table 1.3). Consider $I\,4/m\,c\,m$ (with individual posts separated for clarity) as an example. The presence of a fourfold axis, combined with the absence of threefold axes, tells us that this is a tetragonal space group (the I specifies that the Bravais lattice is body-centered). The first post $4/m$ means that both the fourfold axis and the mirror m are in the direction of c, hence 4 is perpendicular to m. The second post tells us that there are glide planes c in the directions of both edges of the square face (edges a and b) of the unit cell. The third post tells us that there are mirror planes in the direction of the diagonal of this square face.

The triclinic crystal system has a P lattice and is compatible with only two symmetry elements; identity and inversion (Table 1.1). There are therefore only two triclinic space groups, those with symbols $P1$ and $P\bar{1}$. As listed in Table 1.1, the monoclinic crystal system allows three types of rotational symmetry; the 2, m, and $2/m$ crystallographic point groups. There are two Bravais lattices (Figure 1.7) available; primitive, P, and base-centered, C. This results in six symmorphic space groups; $P2$, Pm, $P2/m$, $C2$, Cm, $C2/m$. Ten non-symmorphic monoclinic groups would be obtained by replacing 2 with screw axes 2_1 and m with glide planes c; $P2_1$, Pc, $P2_1/m$, $P2/c$, $P2_1/c$, $C2_1$, Cc, $C2_1/m$, $C2/c$, $C2_1/c$. However, not all of these 10 space groups are unique. This is because the combination of C-centering and the rotation axis 2 generates its "own" 2_1 (Figure 1.12). In the standard setting, they appear parallel to b at $x = ¼$ and $¾$. For this reason, $C2_1 \equiv C2$, $C2_1/m \equiv C2/m$, $C2_1/c \equiv C2/c$, and there are not 16

---

[10] Standard orientations are those with unit-cell axes a, b, c chosen by an agreed set of rules, as listed in the International Tables for Crystallography, Volume A. Permutations are possible but non-standard.

Table 1.3 Directions of the three posts in standard space-group symbols. Trigonal directions refer to hexagonal unit cell.

Crystal system	Post 1	Post 2	Post 3
Monoclinic	Perpendicular to the plane of the monoclinic angle (standard, ∥ **b**)		
Orthorhombic	Edge **a**	Edge **b**	Edge **c**
Tetragonal	$4(\bar{4}) \parallel c$	Square edges	Square diagonals
Trigonal	$3(\bar{3}) \parallel c$		
Hexagonal	$6(\bar{6}) \parallel c$	Rhombus edges	Rhombus diagonal, longer
Cubic	Edges	$3(\bar{3})$ in body diagonals	Face diagonals

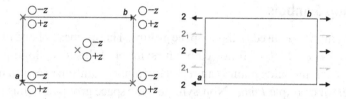

Figure 1.12 Operation of a twofold axis 2[010] combined with monoclinic centering C (lattice points are marked with ×) makes a pattern of points related by screw axes 2_1[010] at $x = \frac{1}{4}$ and $\frac{3}{4}$ (marked with half-arrow heads).

but only 13 unique monoclinic space groups possible. Similar arguments can be developed for all crystal systems leading to a total of 230 unique 3D space groups.

Symmetry information for all space groups is given in the International Tables for Crystallography, Volume A (for more details, see Appendix B of this book). The tables list, inter alia, the positions of all symmetry elements and how an atom at an initial point x,y,z is reproduced in the unit cell by the symmetry elements present.

1.1.8 Description of a Crystal Structure

A crystal structure is defined by its space-group symmetry, unit-cell parameters, and by the coordinates of the atoms in its asymmetric unit. The **asymmetric unit** contains only the atomic coordinates in the unit cell that are crystallographically unique, all other atomic positions are created by the symmetry-element operations. Atoms whose coordinates lie on symmetry elements are said to be on **special positions**, those which don't, are on a **general position** (x,y,z, Appendix B). Different positions are often referred to using **Wyckoff site** labels. The Wyckoff site is labeled with a number and a letter (e.g. 8a). The number is the multiplicity of a site; it gives the number of equivalent points (atoms) generated by the available symmetry operations of that site from any one of them (the "original point"). The letter labels the sites for a given space group. The highest-symmetry site is listed as a and

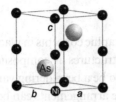

NiAs
$P6_3/mmc$ (#194) hexagonal
$a = b = 3.57$, $c = 5.10$ Å

Atom	Wyckoff site	x	y	z
Ni	2a	0	0	0
As	2c	⅓	⅔	¼

Figure 1.13 Data needed to describe the crystal structure of nickel arsenide.

the lower-symmetry sites as b, c, d... (each with its own set of coordinates for symmetry equivalent points) in the International Tables for Crystallography, Volume A.

For easy drawing, some figures will show auxiliary information that defines the crystal structure; its space group, unit-cell parameters, and fractional coordinates of atoms in the asymmetric unit. For example, in Figure 1.13, a nickel atom is listed in the site $2a$ at 0 0 0 and an arsenic atom in $2c$ at ⅓ ⅔ ¼. Application of symmetry operations in a structure-drawing software (or a look at pages of the space group number 194 in the International Tables, Volume A) yields an additional Ni at 0 0 ½ and an additional As at ⅔ ⅓ ¾, making a total of two formula units of NiAs per unit cell ($Z = 2$).

1.2 Databases

In this chapter, we'll introduce just a few of many structure types behind millions of individual crystal structures. Crystallography information comes in databases. The two with the longest tradition are the Inorganic Crystal Structure Database (ICSD) at the Fachinformationszentrum Karlsruhe in Germany and the Cambridge Structural Database (CSD), curated by the Cambridge Crystallographic Data Center. As of October 2020, the ICSD contained 232012 entries, the CSD 1094733. In addition, National Institute of Standards and Technology (NIST) has developed a wider Inorganic Structure database (NIST ICSD) that combines full structure data (mostly from ICSD) with the unit-cell identification data set curated there as NIST Crystal Data since 1963.

Databases provide powerful software tools for searching known structures, exploring similarities between materials, and for "data mining" to identify important chemical and structural trends. The structural information is exported in a standardized form of the **crystallographic information file** (cif) that is used as input for various crystallographic software packages. Some databases provide calculation of diffraction patterns, useful to identify materials. While the full databases require a license, there are several open-access resources online. Try a search on "icsd demo" or "database of zeolite structures" or "mineralogy database" or the "Crystallography Open Database". Useful demo-version utilities for plotting crystal structures from cif files are searchable as well.

1.3   Composition

With the essential ideas of symmetry and crystallographic concepts in place, we'll spend the rest of this chapter describing principles that rationalize structures as composites of different elements and entities. Materials adopt what at first appears to be a bewildering variety of different atomic arrangements in 3D space. However, many of these arrangements can be readily understood by considering relatively simple structural rules that concern stoichiometry and connectivity patterns in compounds. For those purposes, it will be convenient to introduce the **crystal-chemical formula** [2] that summarizes structural information. These formulas list each *crystallographically non-equivalent* atom[11] in the structure separately and contain information about its coordination to other atoms in superscripted square brackets. The coordination numbers in the brackets follow the same array as the atoms in the formula, and are separated by a comma, while any bonding to atoms of the same element comes after a semicolon. In addition to the coordination number, the bond geometry may be indicated by a lower-case letter (l = line, n = not in a line or plane, t = tetrahedron, o = octahedron, y = pyramid, p = prism, c = cube, co = cuboctahedron, etc.). To keep the formula as simple as possible, it's common to include only connectivities of direct chemical bonds. These simpler formulas will generally be used in this chapter. The nomenclature is best illustrated with examples, such as those given in Table 1.4.

Table 1.4 Examples of crystal-chemical formulas for compounds; full = full* neighborhood, simpler = direct neighborhood (bonding).

Formula	Full crystal-chemical formula Simpler crystal-chemical formula	Brief description of bonding
SiO_2	$Si^{[4t;]}O_2^{[2;]}$ $Si^{[4t]}O_2^{[2]}$	Si tetrahedrally coordinated by 4O while O is coordinated by 2Si
$SrTiO_3$	$Sr^{[8,12co;]}Ti^{[8,6o;]}O_3^{[4,2;]}$ $Sr^{[12co]}Ti^{[6o]}O_3^{[4,2]}$	Sr in cuboctahedron of 12 O and Ti in octahedron of 6O
FeS_2	$Fe^{[6o;]}S_2^{[3;1]}$ $Fe^{[6o]}S_2^{[3;1]}$	Fe in octahedron of 6S; (S–S)$^{2-}$ units present
$MgAl_2O_4$	$Mg^{[12,4t;]}Al_2^{[6,6o;]}O_4^{[(1,3)t;]}$ $Mg^{[4t]}Al_2^{[6o]}O_4^{[1,3]}$	Mg in tetrahedral and Al in octahedral coordination
$Y_3Fe_5O_{12}$	$Y_3^{[4,6,8c;]}Fe_2^{[6,6,6o;]}Fe_3^{[6,4,4t;]}O_{12}^{[(2,1,1)t;]}$ $Y_3^{[8c]}Fe_2^{[6o]}Fe_3^{[4t]}O_{12}^{[2,1,1]}$	Y in a cube of 8O; two Fe sites, one octahedral, one tetrahedral

* Coordination numbers refer to the array of the other atoms in the formula, such as Sr 8-coordinated with Ti and 12-coordinated with a cuboctahedron of O in $SrTiO_3$. Possible bonds to atoms of the same kind follow after the semicolon, such as 1 in FeS_2.

[11] Apart from chemical identity, crystallographically identical atoms have identical environments.

1.3.1 Coordination, Stoichiometry, and Connectivity

If ionic charges are included, the crystal-chemical formula of a binary compound will incorporate three important balances: The electroneutrality balance, the connectivity balance, and the bond-valence balance (Figure 1.14). Let's assume a binary compound C_mA_n of two non-identical elements, *each at one unique site*; one assigned as a "cation" C of charge number c, the other as an "anion" A of charge number a. The crystal-chemical formula is $C_m^{c+}{}^{[N;]}A_n^{a-}{}^{[M;]}$, where N and M are the coordination numbers of C and A, respectively. The **electroneutrality balance** requires that $m \times c = n \times a$. The **connectivity balance** requires that $m \times N = n \times M$ because there must be an equal number of CA and AC connections. Lastly, $c/N = a/M$ is the **bond-valence balance** [3].[12]

Understanding these balances allows us to make simple structural predictions from the chemical formula.[13] Consider SiO_2 as an example. We know from inorganic chemistry that silicon favors tetrahedral coordination, so the connectivity balance $1 \times 4 = 2 \times M$ gives $M = 2$ for the coordination number of oxygen (Table 1.4).

For compounds having more than two sites, balances analogous to those for the two-site formula in Figure 1.14 can be set up. As an example, the mineral hausmannite, Mn_3O_4 (a spinel, Section 1.5.1), has Mn^{2+} in tetrahedral and Mn^{3+} in octahedral coordination. There is one oxygen site. Using this information, we write the formula $Mn^{2+[4t]}Mn^{3+}{}_2^{[6o]}O_4^{[x,y]}$ with the coordination numbers around oxygen as unknowns. Two connectivity balances can be written, $1 \times 4 = 4 \times x$ for Mn^{2+} and $2 \times 6 = 4 \times y$ for Mn^{3+}, and we see that each oxygen is coordinated by one Mn^{2+} and three Mn^{3+}.

For extended structures[14] of ternary and higher-component phases, a **bond graph** is informative as it shows the connectivity visually, see Figure 1.15. The symbolism of a bond graph differs from symbolism of formulas used by molecular chemists, so let's again illustrate it with the simple case of SiO_2. One starts with the crystal-chemical formula (Table 1.4) and writes down one Si symbol surrounded by two O symbols. Then a line is drawn for each bonding connection between two atoms. The lines tell us that each oxygen is coordinated by two silicon atoms and each silicon is coordinated by four oxygens (the two lines between Si and O thus do not mean a double bond). A quick examination of the bond

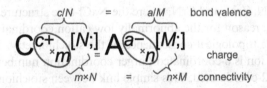

Figure 1.14 Three balances in the crystal-chemical formula of a two-site compound.

---

[12] Termed "the electrostatic valence principle" in ref. [3]. More on bond valences is in Section 5.4.

[13] Such estimates can be useful, see Box 1.1.

[14] Crystal structures with bond networks that extend over the entire volume.

graphs of TiO_2 and CaF_2 shows that as the cation coordination number increases from 4 to 6 to 8, the anion coordination number increases from 2 to 3 to 4, consistent with the connectivity balance.

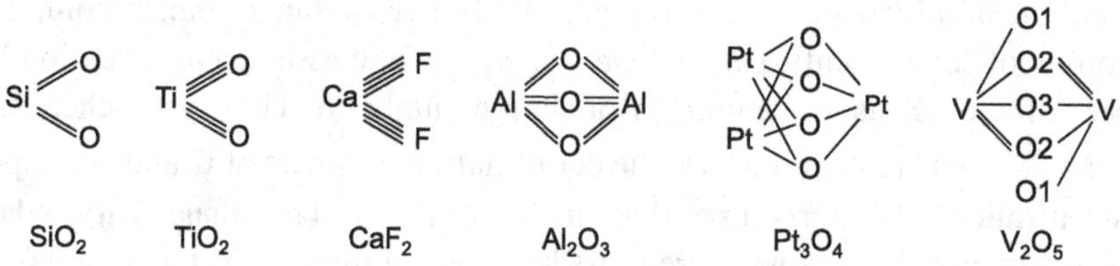

Figure 1.15 Examples of bond graphs. In V_2O_5, five oxygens occupy three crystallographically different sites of differing coordinations to V.

Box 1.1 Synthetic Methods: Preparation of Na₃N

Most binary ionic compounds were discovered long ago. A number of these compounds, such as NaCl and KCl, have played pivotal roles in the development and advancement of civilizations. In principle, such compounds form easily; one brings the two elements together and initiates the reaction. It is surprising to find that, despite repeated attempts, at the end of the twentieth century no one had been able to prepare sodium nitride, Na_3N. This changed in 2002 when Fischer and Jansen reported the first successful synthesis [4]. Like many binary compounds, it was prepared from the elements. However, the preparation conditions were hardly typical. Atomic beams of the two components were generated separately in a microwave plasma and passed into a vacuum chamber where they condensed onto a sapphire substrate cooled to 77 K. At such a low temperature, the elements don't react when deposited onto the substrate, but upon heating to 200 K they begin to form crystalline Na_3N.

This compound adopts the cubic ReO_3 structure (see Table 1.9), with the cation and anion positions reversed. As might be expected for such an elusive compound, Na_3N is not very stable, it decomposes above 360 K.

Why is sodium nitride so difficult to prepare and why is it unstable even after it forms? An important clue comes from the crystal-chemical formula, $Na_3^{[2I]}N^{[6o]}$. Despite its large size, sodium is coordinated by only two nitride ions in a linear geometry. Compare this with stable transition metal nitrides such as ScN, ZrN, and CrN, where the NaCl-type structure gives a cation coordination number of six. The reason for the abnormally low cation coordination number in Na_3N is a simple but unavoidable topological consequence of the stoichiometry. The C_3A stoichiometry means that if the nitride ion is 6-coordinated (larger coordination numbers are rare for N^{3-}) the sodium ion can only be 2-coordinated. This simple link between stoichiometry and coordination number [5] is the fundamental reason why the synthesis of Na_3N has proved so difficult. Typical coordination numbers for Na^+ range from 6 to 12; the coordination number of two is an extreme outlier. Let's also note that the densest-packing principle (Section 1.4.2) supports C_3A of only the smallest cations C occupying voids among the largest anions A, such as Na_3As or Li_3N.

1.3.2 The Generalized 8−N Rule

Atoms bond in patterns that yield stable electron configurations. For an electronegative *sp* atom of *N* valence electrons, the **8−N rule** is valid: *Valence electrons short of 8 are obtained in bonds.* Carbon of 4 valence electrons obtains the missing 4 electrons by forming 4 bonds; phosphorus obtains the missing 3 electrons by forming 3 bonds, etc., as illustrated in Figure 1.16.

In a binary compound, the more electronegative *sp* element, the one that attracts and holds electrons more, will maintain the 8−N rule in order to achieve a stable configuration of the noble gas. Most electropositive *sp* elements will tend to lose their valence electrons to also achieve a noble-gas configuration. The ionic approximation is therefore a convenient way of recognizing the stable configurations in a compound.

Consider our binary solid $C_m A_n$ (Section 1.3.1) of *sp* atoms. The **valence-electron count per anion A**, VEC_A, is calculated from the stoichiometry and from valence-electron numbers e_C and e_A:

$$VEC_A = (m \cdot e_C + n \cdot e_A)/n. \tag{1.1}$$

When $VEC_A = 8$, both atoms achieve the noble-gas configuration as noted above. When $VEC_A > 8$, the excess electrons remain with the cation forming cation–cation bonds or cation-localized electron pairs. When $VEC_A < 8$, the atom A obtains the missing electrons by forming A–A bonds so that each A has an octet. The formal expression of this **generalized 8−N rule** [6, 7] for $C_m A_n$ is:

$$VEC_A = 8 + CC \cdot \frac{m}{n} - AA, \tag{1.2}$$

where the variable *CC* is the number of electrons per "cation" C that form C–C bonds or are localized at the cation as lone pairs, and *AA* is the number of electrons per "anion" A that form A–A bonds. The generalized 8−N rule is useful for analysis of structures of nonmetallic *sp* phases. Let's illustrate this using GaSe and $SnCl_2$ with $VEC_A > 8$ and CdSb and CaC_2 with $VEC_A < 8$.

In GaSe, gallium has three valence electrons, selenium six, $VEC_A = 9$. The excess electron will remain with the Ga cation ($CC = 1$, Equation (1.2) and form single-bonded $(Ga–Ga)^{4+}$ pairs. In

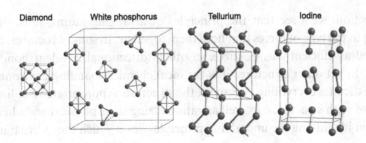

| Diamond | White phosphorus | Tellurium | Iodine |

Figure 1.16 Structures of selected *p* elements drawn to equal scale.

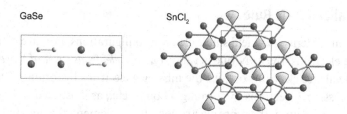

Figure 1.17 Examples of structures with $VEC_A > 8$, in which the excess electrons per anion A are localized at the cation (smaller spheres) as bonds or as lone pairs.

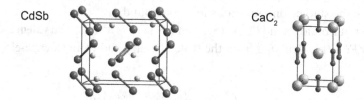

Figure 1.18 Examples of structures with $VEC_A < 8$, when "electrons missing to 8" at the anion A are obtained by sharing in bonds between A atoms.

$SnCl_2$, tin has four valence electrons, chlorine seven, $VEC_A = 9$; one electron in excess. Given two Cl per Sn, two electrons remain at Sn ($CC = 2$) forming the lone electron pair of Sn^{2+}. The structures of GaSe and $SnCl_2$ are shown in Figure 1.17.

In CdSb, the cadmium atom has two valence electrons,[15] antimony five, and $VEC_A = 7$. The missing electron is obtained by sharing between two Sb atoms that form Sb_2^{4-} single-bonded dumbbells ($AA = 1$) isoelectronic with I_2 (Figure 1.18, left). In CaC_2, calcium has two valence electrons, carbon four, and $VEC_A = 5$. The three missing electrons are obtained by sharing in three two-electron bonds ($AA = 3$), and triple-bonded C_2^{2-} pairs isoelectronic with N_2 are present in the crystal structure of CaC_2 (Figure 1.18, right).

1.4   Structural Principles

In this section, we'll see that the principles of arranging atoms, ions, or molecules vary according to the type of forces that hold them together. In some structures, the building units are packed as efficiently as possible, in others directional covalent bonding gives rise to networks. In yet other structures it is the local chemical bonding environment of a central atom that dictates everything else about the structure. Appropriate visualization and understanding of such structures, as well as rationalizing their physical and chemical properties, depends on identifying the underlying principles upon which they were built.

[15] $4d$ orbitals in Cd are sufficiently low in energy that they can be considered part of the core.

1.4.1 Packing of Spheres

The structures of many materials can be understood by exploring how hard spheres pack. If we start in 2D, Figure 1.19 shows two symmetric ways of packing circles and their corresponding unit cells; it's clear that the packing of Figure 1.19b is more space-efficient. The packing efficiency, calculated as the area fraction filled by circles, is 0.785 and 0.907 for the two arrangements.

The densest[16] 3D packing of spheres is based on stacking 2D layers of the type in Figure 1.19b. A second layer of spheres nestles in the dimples of the first layer and is therefore laterally shifted relative to it (Figure 1.20). For the third layer, there are two choices for positioning the spheres. One choice reverses the shift direction so that the spheres in the third layer lie directly above those in the first; the other choice continues shifting in the same direction, in which case the third layer does not lie directly above the first layer. When we view the layers side on (Figure 1.21) the repeat unit is either AB or ABC. The former arrangement is called **hexagonal closest packing (hcp)**, and the latter **cubic closest packing (ccp)**. Both sequences fill space equally efficiently (74.0% of space is occupied; 26.0% is voids between spheres) and would be energetically equivalent for hard spheres. For elements that adopt these arrangements, such as the solid noble gases and many metals, orbital symmetries make one or the other arrangement slightly more stable.

Specific combinations of cubic and hexagonal stacking sequences can be stabilized in some structures; rare-earth metals are typical examples. Figure 1.22 shows how to analyze the sphere packing occurring in α-La. The first step is to orient the unit cell so that we have an on-top view of the densely packed layers (against c in hexagonal structures) and the direction of subsequent shifts is horizontal; as shown in Figure 1.22 bottom. The structure is then rotated 90° so that the normals of the layers point up. We can then identify the closest-packed sphere layers as being A, B, C, mark the subsequent shifts as arrows, and assign letter c to a layer that has local cubic environment and h that has local hexagonal environment according to those shifts. The repetitive sequence of these letters is the basis of the **Jagodzinski–Wyckoff notation** of stacking sequences. The **Ramsdell symbol** gives the number of closest-packed sphere layers per unit cell of a Bravais lattice and a letter to signify the Bravais lattice (H for hexagonal, R for rhombohedral, C for cubic). As an example, the Ramsdell symbol for α-La is 4H.

(a)   (b)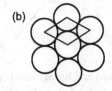

Figure 1.19 Two high-symmetry packings of circles. After Kepler [8].

---

[16] Although these arrangements have long been suggested to be the densest form of packing of equal spheres (J. Kepler's conjecture in ref. [8], apparently responding to the English Admiralty task to determine the most efficient packing of cannonballs on ships), a widely accepted proof was only published in 1998 by Thomas Hales. A panel of referees was 99% certain the proof was correct.

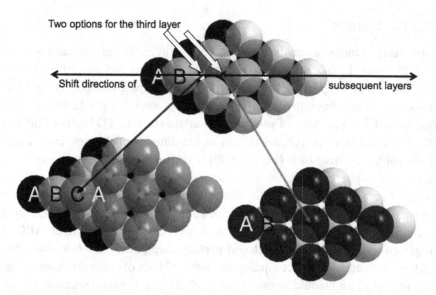

Figure 1.20 Two choices for placement of the third layer of densely packed spheres. One line of the subsequent layer shifts is used to analyze a densest packing.

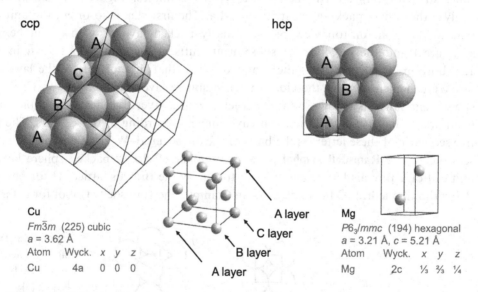

Figure 1.21 Cubic (ccp) and hexagonal (hcp) closest packing of equal spheres. Edges of 8-unit cells are drawn for ccp, with body diagonals perpendicular to the layers. The two unit cells are below; face-centered cubic and primitive hexagonal.

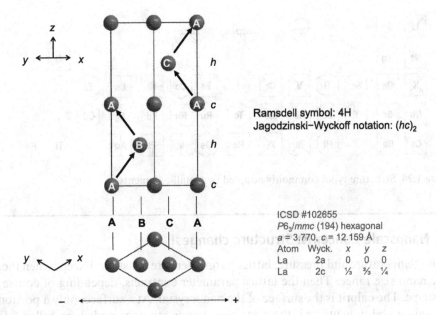

Ramsdell symbol: 4H
Jagodzinski–Wyckoff notation: $(hc)_2$

ICSD #102655
$P6_3/mmc$ (194) hexagonal
$a = 3.770$, $c = 12.159$ Å

Atom	Wyck.	x	y	z
La	2a	0	0	0
La	2c	⅓	⅔	¼

Figure 1.22 How to analyze sphere packing in hexagonal structures (α-La shown).

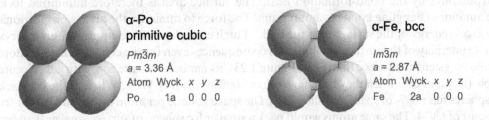

α-Po
primitive cubic

$Pm\bar{3}m$
$a = 3.36$ Å

Atom	Wyck.	x	y	z
Po	1a	0	0	0

α-Fe, bcc

$Im\bar{3}m$
$a = 2.87$ Å

Atom	Wyck.	x	y	z
Fe	2a	0	0	0

Figure 1.23 Spheres in contact in unit cells of primitive and body-centered cubic structures of metals.

In addition to the densest-packed arrangements, two other simple high-symmetry packings are found in metals and are shown in Figure 1.23; the primitive cubic arrangement on the left (adopted by α-Po) and the body-centered cubic (bcc) arrangement on the right. Note that in ccp (Figure 1.21), the spheres are in contact along the face diagonal of the unit cell, in bcc along the body diagonal, and in the primitive cubic packing along the cell edge. The preferred structures for metallic elements are given in Figure 1.24.

Although the densest packing adopted by most elemental metals suggests non-directional bonds, this is true only as a first approximation. It has been shown [9] that bonding electrons may concentrate in certain directions; in aluminum, they prevail in voids formed by every four mutually touching atoms of the cubic closest packing.

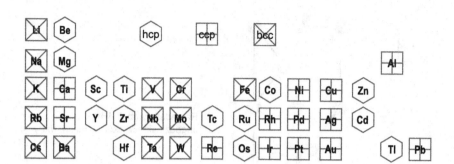

Figure 1.24 Structure types commonly adopted by metallic elements.

Box 1.2 Nanoscale Concepts: Structure changes!

At constant temperature and pressure, lattice parameters are constant. Except when the crystal enters the nano-size range. Then the lattice parameter contracts, depending of course on the crystal's shape. The culprit is the surface of the nanocrystal. At a surface, only a portion of the atoms' bonding ability is utilized, the remaining bonds are open-ended, so-called "dangling bonds". The surface layer is therefore at a substantially higher energy level than the bulk. A system tends to minimize its energy either by increasing the entropy of itself, or of its surroundings by the bond-formation heat. The surface area is therefore minimized to keep the amount of dangling bonds to a minimum. The force to minimize the area is proportional to the excess energy of the surface over the bulk. If high enough, it will compress the entire crystal.

The interplay of the surface and bulk has consequences even for the crystal structure adopted. While tungsten metal is bcc (like Fe in Figure 1.23), its nanoclusters are ccp, with a face-centered cubic (fcc) unit cell, when they have fewer than 7000 atoms [10]. Why? Whereas the bcc packing of spheres fills ~68% of space, fcc fills ~74%. The space volume per atom is then smaller for fcc by a factor of 68/74. The same atoms would pack a smaller fcc sphere, of surface smaller than bcc by a factor of about $(68/74)^{2/3} = 0.945$; exactly $(3/4) \cdot \sqrt[3]{2}$. As the proportion of the surface and of its higher energy content increases with decreasing size of our originally bcc nanoparticle, the energy that can be released by minimizing the surface eventually exceeds the energy needed to form the normally less favored but denser fcc tungsten packing in the bulk, hence the transition.

1.4.2 Filling Holes

As suggested in the previous section, voids are left between spheres of the atomic densest packing. There are two types of voids in hcp and ccp arrangements; **tetrahedral holes** and **octahedral holes** (Figure 1.25). Four spheres surround a tetrahedral hole; three from one layer touching, one from the next. Six spheres surround an octahedral hole; three from one layer and three from the next. One octahedral and two tetrahedral holes are present per each sphere.

Table 1.5 Binary-compound structure types derived by full or fractional filling of octahedral and tetrahedral holes in planes between closest-packed layers of A.

Filling in total of holes	General formula	Structure type hcp of anions	Structure type ccp of anions	The plane-filling sequence
All octahedral	CA	NiAs	NaCl	All full
½ octahedral	CA_2	CdI_2	$CdCl_2$	Empty and full
½ octahedral	CA_2	$CaCl_2$		All ½ full
⅓ octahedral	CA_3	BiI_3	YCl_3	Empty and ⅔ full
⅓ octahedral	CA_3	$RuBr_3$		All ⅓ full
⅔ octahedral	C_2A_3		La_2O_3	Empty, full, full *
⅔ octahedral	C_2A_3	Al_2O_3		All ⅔ full
All tetrahedral	C_2A	**	Li_2O	All full
½ tetrahedral	CA	ZnS wurtzite	ZnS sphalerite	All ½ full
All	C_3A	Na_3As		All full

*Lanthanum fills the octahedral holes in a strongly off-center manner that provides bonding across the empty hole plane with one of the oxygen atoms, by which La achieves coordination number 7. **Repulsion of cations in the two face-sharing tetrahedra prevents this type of hole filling.

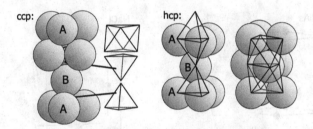

Figure 1.25 Tetrahedral and octahedral holes between layers of ccp and hcp spheres. Note that the label C of the subsequent ccp shifts is partly obscured.

Structures of many simple compounds can be described in terms of smaller atoms (usually cations) occupying the octahedral and/or tetrahedral holes in closest-packed arrangements of larger atoms (usually anions).[17] Table 1.5 summarizes several structure types that can be described in this way. They are discussed in the following paragraphs.

Let's start a more detailed account by analyzing filling of octahedral holes in densest packed arrays. The two-step structure-orientation process is shown in Figure 1.26 for the respective prototypes, NiAs and NaCl. In both, the cation-occupied octahedral holes are

[17] The closest-packed spheres (atoms) begin to touch once the radius of the atom in the octahedral holes becomes less than $\sqrt{2} - 1 \approx 0.414$ of the sphere radius. For tetrahedral holes this occurs when the radius of the smaller atom becomes less than $(\sqrt{3}/\sqrt{2}) - 1 \approx 0.225$ of the closest-packed sphere radius. Hence these radius ratios are the minima for the given coordination to occur, at least in the hard-sphere approximation.

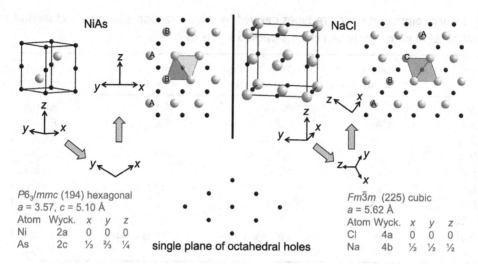

NiAs

P6₃/mmc (194) hexagonal
a = 3.57, c = 5.10 Å

Atom	Wyck.	x	y	z
Ni	2a	0	0	0
As	2c	⅓	⅔	¼

single plane of octahedral holes

NaCl

Fm3̄m (225) cubic
a = 5.62 Å

Atom	Wyck.	x	y	z
Cl	4a	0	0	0
Na	4b	½	½	½

Figure 1.26 The hcp and ccp prototypes NiAs (left) and NaCl (right). Identification of octahedral holes (black) by reorienting the structure from a general view to the top-on view (below) and finally to the side-on view (above) of the closest-packed layers.

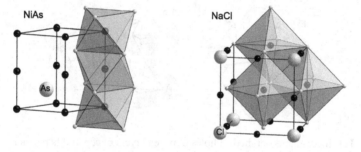

Figure 1.27 Coordination of octahedral holes in NiAs and NaCl structures.

located in the intermediate planes between the closest-packed layers of anion spheres, and the sphere stacking sequence is immediately recognizable.

In NiAs, the closest-packed As atoms have six nearest Ni neighbors forming a trigonal prism (Ni$^{[6o]}$As$^{[6p]}$). In NaCl, both cation and anion arrays have the same arrangement in space and both are octahedrally coordinated; Na$^{[6o]}$Cl$^{[6o]}$. Either of them can therefore be considered closest packed, though this description better suits the larger Cl⁻ anion. Figure 1.27 illustrates how the Ni coordination octahedra in NiAs share opposite faces (in addition to sharing all edges), whereas the octahedra in NaCl do not (only edges are shared). In strongly ionic compounds, the NaCl-type structure is therefore preferred over NiAs in order to avoid the added electrostatic repulsion across the shared octahedral face.

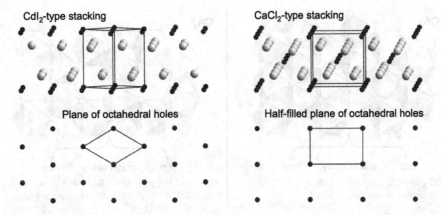

Figure 1.28 Composition CA_2 via filling octahedral holes in hcp: CdI_2 (fully filling every second plane of holes) and $CaCl_2$ (half filling every plane).

Composition CA_2 with "cations" C in half the octahedral holes is achieved in two ways; either by alternating one empty and one filled plane of octahedral holes, or by half filling each plane (Table 1.5). Figure 1.28 compares these two arrangements in the hcp array; the alternately filled arrangement is called the CdI_2 structure type[18] and the homogeneously half-filled one is the $CaCl_2$ structure type.[19] Note that the symmetry of the $CaCl_2$ structure is reduced to orthorhombic by the rectangular pattern of holes.

The alternating empty and filled octahedral planes create a layered structure because the anions face each other across each empty plane of octahedral holes.[20] The polyhedral representation in Figure 1.29 top shows two types of such isolated layers; a CA_2 layer that has all octahedral holes filled and a CA_3 layer with two-thirds of the holes filled. Surprisingly, these two octahedral layers give rise to four structural arrangements. This is because each can stack in either an hcp or ccp array.

The two-thirds filling of every plane of octahedral holes in hcp also occurs in the structure of corundum, $Al_2^{[6o]}O_3^{[4t]}$, an important refractory material. Figure 1.30 shows this two-thirds filling and how it subdivides the NiAs-type infinite columns of octahedra into pairs. The two octahedra share one face and their central Al^{3+} cations repel each other due to the proximity of their ionic charges.

Let's now focus on filling tetrahedral holes. A complete filling of all the tetrahedral holes in a ccp array leads to the composition C_2A of materials such as Li_2O. However, the traditional

---

[18] Perversely, CdI_2 tends to adopt numerous stacking sequences rather than this simple hcp structure.

[19] The structure of rutile, TiO_2, is related, but with anions distorted such that they no longer lie in perfect layers. TiO_2 is then better understood in terms of corner- and edge-sharing octahedra (see Figure 1.45).

[20] The layers are held together by weak electrostatic forces of electric dipoles generated by random fluctuation of the electronic charge in one object, such as the anionic layer, generating a corresponding inverse-charge fluctuation in the other such object (van der Waals forces, see Section 5.1.3).

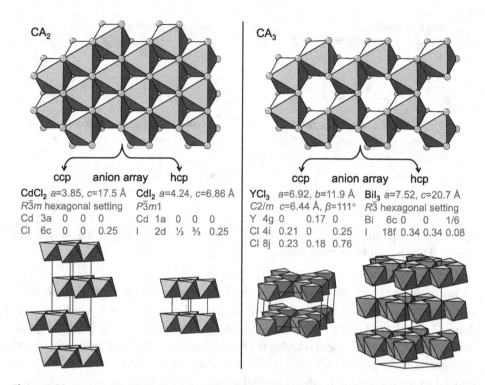

Figure 1.29 Top: Polyhedral representation of a single layer of octahedral holes, fully occupied (left) and two-thirds occupied (right). Bottom: Structure types formed when these layers are packed such that their anions form a ccp or hcp array.

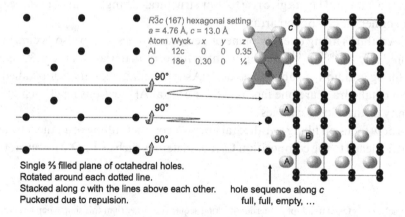

Figure 1.30 Sequence of octahedral-hole filling in hcp oxygens in corundum. The hole plane is two-thirds filled (left; every third hole is empty). All hole sequences in a straight line along c are full, full, empty (side-on view on right). The formed $Al_2^{[6]}O_3^{[4]}$ has pairs of octahedra sharing faces, with repulsion of their central Al^{3+} ions.

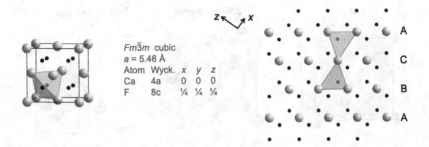

Figure 1.31 CaF$_2$: Identification of tetrahedral holes (black) in the side-on view of the ccp layer stacking of Ca atoms (gray large spheres). Notice a tetrahedral hole both above and below each densest-packed Ca atom, hence CaF$_2$.

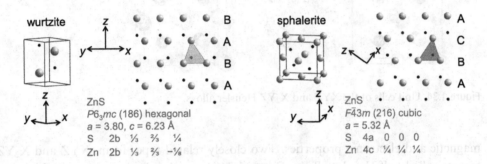

Figure 1.32 Identification of tetrahedral-hole planes (black dots) in the side-on view of the hcp (left) and ccp (right) stacking of two ZnS modifications.

prototype structure used is CaF$_2$,[21] and has a cation ccp array with the fluoride anions in all of the tetrahedral holes. Figure 1.31 illustrates its pattern of densest packing and hole filling.

There are no known structures based on an hcp array with all the tetrahedral holes filled. This can be understood with reference to Figure 1.25 where the face sharing of tetrahedra would have the cations unreasonably close (closer than the cation–anion distance). The prototype structures based on filling half of the tetrahedral holes in hcp and ccp are the wurtzite and sphalerite (zinc blende) polymorphs of ZnS. Identification of their sphere packing and hole filling is illustrated in Figure 1.32.

Polyhedral coordination of the filled tetrahedral holes in unit cells of the prototype ccp and hcp structures is compared in Figure 1.33. More about building up structures from polyhedra is given in Section 1.4.4.

Structures of some ordered metal alloys can be described as densest-packed arrays with filled holes. An example is the family of Heusler alloys that has a variety of important

---

[21] Consequently, Li$_2$O is considered an anti-fluorite type.

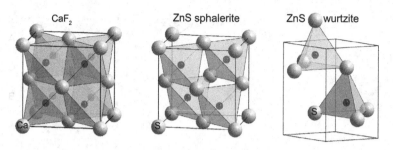

Figure 1.33 Coordination of tetrahedral holes in CaF_2, sphalerite, and wurtzite.

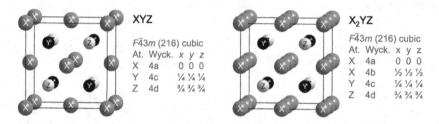

Figure 1.34 Unit cells of the XYZ and X_2YZ Heusler alloys.

magnetic and electronic properties. Two closely related types form, XYZ and X_2YZ, the former called half-Heusler alloys. Figure 1.34 shows that both are derived from a ccp of X atoms in which Y and Z occupy tetrahedral holes in an ordered manner, giving the XYZ structure. In the structure of X_2YZ, the added X fills all octahedral holes.

1.4.3   Network Structures

For many materials, the directional character of the chemical bonding around each atom in the structure is of prime importance. The most natural way to view such structures is to consider them as networks. This often helps to reveal the relationships and similarities among many simple structures. In more complex materials, the field of coordination polymers (Chapter 14) has driven developments in the nomenclature [11, 12], taxonomy [13], and classification [14] of networks. Here, we'll consider only the most elementary terms. Networks with a single type of vertex are **uninodal** networks; all vertices are *N*-connected. An example of a 3-connected uninodal network is graphene (the single sheet of graphite, Figure 1.35), in which carbon forms three sigma bonds and the fourth valence electron is delocalized.

Networks with low coordination numbers may be characterized by a **vertex symbol** [11]. The vertex symbol contains one post for each angle that occurs at the vertex. The post is

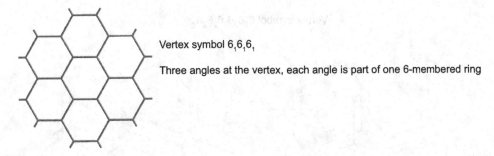

Vertex symbol $6_16_16_1$

Three angles at the vertex, each angle is part of one 6-membered ring

Figure 1.35 A 3-connected uninodal network of graphene.

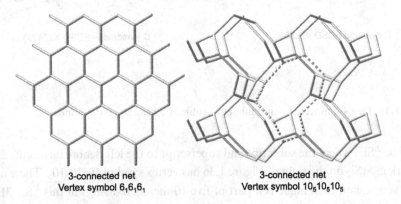

3-connected net
Vertex symbol $6_16_16_1$

3-connected net
Vertex symbol $10_510_510_5$

Figure 1.36 Two 3-connected networks of polysilicide anions in $CaSi_2$ (puckered planar) and $SrSi_2$ (3D). One 10-membered ring is highlighted with a dashed line.

a number with a subscript. The number gives the size of the smallest ring that a particular angle at the vertex is part of. The subscript identifies how many rings are joined at this particular angle. As an example, the vertex symbol for graphene has three posts because there are three angles at the vertex. Each angle forms part of only one smallest, 6-membered, ring so the vertex symbol is $6_16_16_1$.[22]

Let's consider the silicide anions in $CaSi_2$ and $SrSi_2$ as other examples of 3-connected networks. Using the generalized 8−N rule (Section 1.3.2), both compounds have $VEC_A = 5$, and the three missing electrons are obtained by forming three Si–Si single bonds at each Si. The net representing the silicide-anion network on the left of Figure 1.36 has the same vertex symbol $6_16_16_1$ as graphene, but differs in that it is puckered due to accommodation of one nonbonding electron pair at each Si. The standalone crystal-chemical formula for this anion

---

[22] Following Wells, a symbol (6,3) is sometimes used for this honeycomb of 6-membered rings and 3-connected vertices. Do not confuse these Wells symbols with the N,M-connected binodal nets!

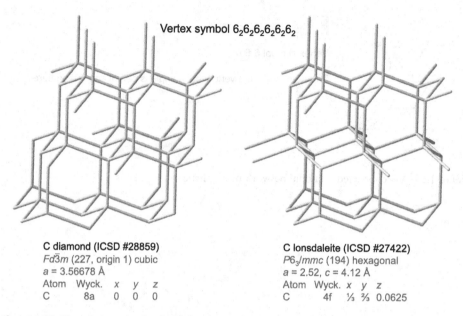

Vertex symbol $6_2 6_2 6_2 6_2 6_2 6_2$

C diamond (ICSD #28859)
$Fd\bar{3}m$ (227, origin 1) cubic
$a = 3.56678$ Å

Atom	Wyck.	x	y	z
C	8a	0	0	0

C lonsdaleite (ICSD #27422)
$P6_3/mmc$ (194) hexagonal
$a = 2.52, c = 4.12$ Å

Atom	Wyck.	x	y	z
C	4f	⅓	⅔	0.0625

Figure 1.37 Two 4-connected uninodal nets—cubic and hexagonal diamond.

would be $^2_\infty \text{Si}^-$, where the subscript and superscript to the left denote its infinite 2D nature. The network in SrSi_2 on the right of Figure 1.36 has vertex symbol $10_5 10_5 10_5$. There are three angles at the vertex, and each angle is a part of five 10-membered rings in this $^3_\infty \text{Si}^-$ 3D network.[23]

Examples of 4-connected nets are shown in Figure 1.37. Both concern elemental carbon forming uninodal nets of the vertex symbol $6_2 6_2 6_2 6_2 6_2 6_2$. The symbol has six posts because there are six angles at each tetrahedral vertex, where each angle is part of two 6-membered rings. For uninodal networks with higher connectivity, we have already encountered in Figure 1.23 the primitive cubic structure of polonium, which is a 6-connected network, and body-centered cubic α-Fe that can be considered as an 8-connected network.

A **binodal** network has two different types of vertices. When characterized using the crystal-chemical formula for a binary phase, $C_m^{c+ \ [N;]} A_n^{a- \ [M;]}$, two different vertices are seen, an N-connected vertex C and an M-connected vertex A. We say that a binodal network has an N,M-connected net. The relationship between uninodal and binodal nets is instructive for understanding similarities among structures.

One type of network-based similarity is **site ordering**.[24] In its simplest case, identical vertices of a uninodal network become occupied by two different atoms in an ordered manner, forming a binodal network. Three examples of this relationship are shown in Figure 1.38. Site ordering removes the equivalence of sites that were symmetry related,

---

[23] The only 3-connected 3D network that has all three bonds and angles around the vertex equal.

[24] One of the homeotypical relationships between two structures defined in ref. [2]. Another such homeotypism is the formation of distortion variants upon decrease in the space-group symmetry.

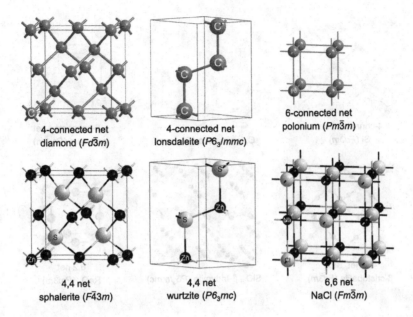

Figure 1.38 Examples of site ordering of uninodal networks (top) to create binodal networks (bottom).

which means that the ordered structure, a **superstructure**, has a lower point symmetry (e.g. $d\bar{3}m$ to $\bar{4}3m$; as in diamond to sphalerite in Figure 1.38), or translational symmetry (the unit cell is multiplied; as in Po to NaCl in Figure 1.38), or both. If the site ordering multiplies the original cell, the new cell is called a **supercell**.

A second important relationship is **network expansion** where a linker is placed between a pair of vertices [11]. For example, we can take the diamond-type net of elemental silicon and put an oxygen atom mid-way between each node, obtaining the high-temperature structure of cristobalite, SiO_2. This and other examples are shown in Figure 1.39 on the same set of uninodal nets as in Figure 1.38. The linker does not have to be a single atom; the primitive cubic 6-connected net expanded with a $-C\equiv N-$ linker is the generic ingredient of the Prussian-blue-type structures.

A final relationship worth mentioning is **vertex decoration**—replacing a vertex with a group of vertices. When this group is a cluster, it is common to describe this as **network augmenting** [11]. An example is the relationship between CaTe and CaB_6 in Figure 1.40. CaTe itself is an example of a material with the **CsCl-type structure**, which in turn can be considered as a site ordering of the α-Fe bcc array of Figure 1.23.

Vertex decoration and network expansion may occur simultaneously. This is a good approach for visualizing open metal–organic frameworks (MOFs). The example in Figure 1.41 starts with oxygen atoms in a cubic Po-type network, which are "decorated" by zinc to $[Zn_4O]^{6+}$ and then linked with the terephthalate anion $[O_2CC_6H_4CO_2]^{2-}$. This forms a ReO_3-type network of these cations and anions.

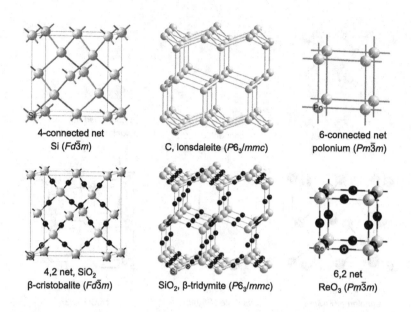

4-connected net
Si (Fd3̄m)

C, lonsdaleite (P6₃/mmc)

6-connected net
polonium (Pm3̄m)

4,2 net, SiO₂
β-cristobalite (Fd3̄m)

SiO₂, β-tridymite (P6₃/mmc)

6,2 net
ReO₃ (Pm3̄m)

Figure 1.39 Examples of network expansion from uninodal to binodal networks.

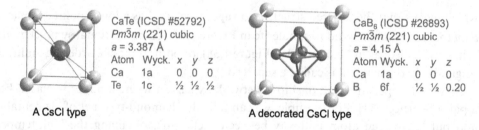

CaTe (ICSD #52792)
$Pm\bar{3}m$ (221) cubic
a = 3.387 Å
Atom Wyck. x y z
Ca   1a   0   0   0
Te   1c   ½   ½   ½

A CsCl type

CaB₆ (ICSD #26893)
$Pm\bar{3}m$ (221) cubic
a = 4.15 Å
Atom Wyck. x y z
Ca   1a   0   0   0
B   6f   ½   ½   0.20

A decorated CsCl type

Figure 1.40 Decoration (augmenting) of anion sites in the CsCl-type structure.

In some cases, two interpenetrating nets run throughout a structure without ever crossing each other. Cu_2O is one such example. In the right-hand side of Figure 1.42, we see two 2,4-connected nets of cristobalite type which never intersect. In each net, Cu^I adopts its preferred coordination number, 2, and oxygen is tetrahedrally coordinated, $Cu_2^{[2t]}O^{[4t]}$.

1.4.4 Polyhedral Structures

In previous sections, we have seen various approaches that help visualize extended structures. In functional materials, however, the local chemistry is often the key to desired properties. Coordination polyhedra are an efficient way to illustrate these local bonding environments, and, often, to visualize the entire structure. We'll find in Chapter 6 that focusing on coordination polyhedra is also a convenient start for developing electronic band-

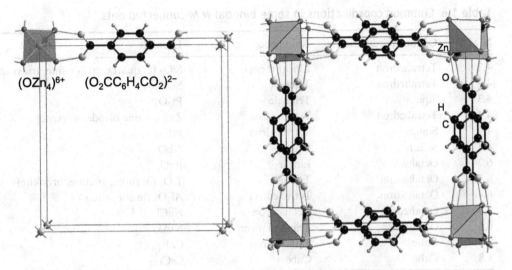

Figure 1.41 Combined site decoration and network expansion in a $Zn_4O(O_2CC_6H_4CO_2)_3$ MOF.

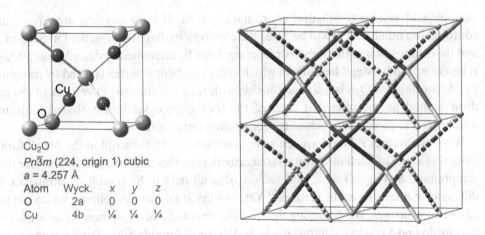

Cu_2O

$Pn\bar{3}m$ (224, origin 1) cubic

$a = 4.257$ Å

Atom	Wyck.	x	y	z
O	2a	0	0	0
Cu	4b	¼	¼	¼

Figure 1.42 Two interpenetrated nets (one solid, one dashed) in cuprite, Cu_2O.

structure ideas and understanding certain properties. In Section 1.5.4 and in Chapter 14, we'll discuss how SiO_4 tetrahedra are the essential building block of many silicate structures.

Building from the ideas in Section 1.4.3, Table 1.6 contains examples of common coordination polyhedra for the "cation" and "anion" in the $C_m^{c+ \, [N;]} A_n^{a- \, [M;]}$ compounds of binodal N,M-connected nets. Many of the structures that were earlier described as based on filling octahedral or tetrahedral holes are also found in this table.

Given the chemical preference to form a certain coordination polyhedron, can we make any generalizations how such equal polyhedra connect into networks? Let's take SiO_2 as an

Table 1.6 Common coordinations in some binodal *N, M*-connected nets.[25]

N,M	Cation coordination	Anion coordination	Example
4,2	Tetrahedron	Linear (bent)	SiO_2 (in quartz, cristobalite, tridymite)
4,3	Tetrahedron	Triangle	Si_3N_4
4,3	Square	Triangle	Pt_3O_4
4,4	Tetrahedron	Tetrahedron	ZnS (in zinc blende, wurtzite)
4,4	Square	Tetrahedron	PtS
4,4	Square	Square	NbO
6,2	Octahedron	Linear	ReO_3
6,3	Octahedron	Triangle	TiO_2 (in rutile, anatase, brookite)
6,4	Octahedron	Tetrahedron	Al_2O_3 (in corundum)
6,6	Octahedron	Octahedron	NaCl
6,6	Octahedron	Trigonal prism	NiAs
8,4	Cube	Tetrahedron	CaF_2
8,8	Cube	Cube	CsCl

example and assume tetrahedral coordination of Si. If both oxygens are equivalent, the coordination number of O will be 2 (see the connectivity-balance argument in Section 1.3.1), and the composition of this polyhedral network can be conveniently described as $SiO_{4/2}$. This is an example of a **Niggli formula**, in which every polyhedral vertex is listed by its connectivity. Accordingly, $SiO_{4/2}$ is a silicon dioxide with four 2-connected O vertices of the polyhedron around Si, meaning that each of the four oxygen vertices is shared with another tetrahedron to form a network of corner-sharing tetrahedra.

When the vertices do not have the same connectivity, the subscript in the Niggli formula is written as a sum of fractions. Various connectivities can therefore be combined to give the same composition. $SiO_{4/2}$, $SiO_{3/3+1/1}$, and $SiO_{2/4+1/2+1/1}$ all refer to SiO_2 built of tetrahedra, but of different connectivities at the vertices.[26] Yet, looking at known CA_2 phases, we do not find the latter two structures. Why? Sharing three or four tetrahedra at a common vertex is not prohibitive; we do find 3-connected tetrahedra in Si_3N_4 (Niggli formula $SiN_{4/3}$) or 4-connected tetrahedra in SiC (Niggli formula $SiC_{4/4}$). Empirically, we can conclude that it is favorable for the oxygen vertices in SiO_2 to have the same bonding environments. When an identical environment is not possible, atoms of the same element usually prefer their environments to be as similar as possible. This preference was postulated by Linus Pauling [3] as the **rule of parsimony**: "The number of essentially different kinds of constituents in a crystal tends to be small."

While this rule aids estimates of polyhedral connectivity in many compounds, does it predict the structure type? In general it does not; several possible spatial arrangements

---

[25] Note that geometries may depart slightly from ideal depending on local site symmetry. Examples do not necessarily represent the only networks that can occur for a given *N, M*-connected net.

[26] $SiO_{4/2}$ means four 2-connected vertices, $SiO_{3/3+1/1}$ three 3-connected and one 1-connected vertex, and $SiO_{2/4+1/2+1/1}$ two 4-connected, one 2-connected, and one 1-connected vertex in the tetrahedron.

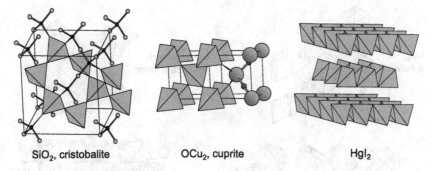

SiO$_2$, cristobalite   OCu$_2$, cuprite   HgI$_2$

Figure 1.43 Structures made of 2-connected tetrahedral vertices.

usually exist. We can see this by considering just a few of the structural variants for the 4,2-connected binodal network of CA$_2$ materials. Both the quartz and cristobalite structures contain corner-sharing tetrahedra but have different 3D arrangements. The OCu$_2$ structure (Figure 1.43) is related to cristobalite, but more complex due to the interpenetration of two networks shown in Figure 1.42. The HgI$_2$ structure also has vertices shared by two tetrahedra but is layered (Figure 1.43).

Let's consider next the structural possibilities for CA$_6$ octahedra sharing vertices, common in functional materials. Starting with the highest n/m composition ratio for C$_m$A$_n$, CA$_6$ in octahedral coordination represents an isolated octahedron. CA$_5$ can be written as CA$_{2/2+4/1}$ where there are two 2-connected and four 1-connected vertices.[27] Such structures contain variously oriented infinite corner-connected octahedral chains. CA$_4$ requires four octahedral vertices shared (CA$_{4/2+2/1}$), and various clustered chains, layers, or networks are possible, as illustrated in Figure 1.44. With all six vertices shared, we have formula CA$_3$ (CA$_{6/2}$) and the octahedral network found in ReO$_3$ and ccp perovskites (Section 1.5.3). Other connectivity patterns are possible if we allow octahedra to share edges instead of just vertices. We've already encountered this in the YCl$_3$/BiI$_3$ structures of Figure 1.29, which contain octahedra that share three edges of six vertices shared by two octahedra each, giving again the Niggli formula CA$_{6/2}$.

As the n/m ratio decreases below three, the average anion coordination number must continue to increase, and shared edges become a necessity. There are several paths to composition CA$_2$. The octahedra must share at least two edges, and the simplest way to do this yields the rutile structure of TiO$_2$ that contains infinite chains of octahedra sharing two opposite edges with the two remaining corners linking the chains together (Figure 1.45). The brookite and anatase polymorphs of TiO$_2$ are other possibilities.

In general, edge and face sharing is more frequent for octahedra than for tetrahedra. This is because shared edges between CA$_4$ tetrahedra lead to very short distances between cations. Face-sharing tetrahedra are not known (we have already encountered the rule that only half

---

[27] Two of six octahedral vertices are shared between two octahedra. Or, stated in the language of coordination chemistry, there are two bridging and four terminal ligand atoms.

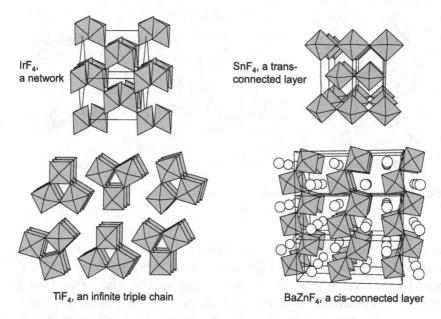

IrF$_4$,
a network

SnF$_4$, a trans-
connected layer

TiF$_4$, an infinite triple chain

BaZnF$_4$, a cis-connected layer

Figure 1.44 Examples of structures with identical octahedra sharing four vertices.

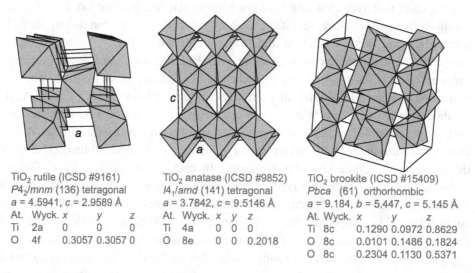

TiO$_2$ rutile (ICSD #9161)
$P4_2/mnm$ (136) tetragonal
$a = 4.5941$, $c = 2.9589$ Å

At.	Wyck.	x	y	z
Ti	2a	0	0	0
O	4f	0.3057	0.3057	0

TiO$_2$ anatase (ICSD #9852)
$I4_1/amd$ (141) tetragonal
$a = 3.7842$, $c = 9.5146$ Å

At.	Wyck.	x	y	z
Ti	4a	0	0	0
O	8e	0	0	0.2018

TiO$_2$ brookite (ICSD #15409)
$Pbca$ (61) orthorhombic
$a = 9.184$, $b = 5.447$, $c = 5.145$ Å

At.	Wyck.	x	y	z
Ti	8c	0.1290	0.0972	0.8629
O	8c	0.0101	0.1486	0.1824
O	8c	0.2304	0.1130	0.5371

Figure 1.45 Edge- and corner-sharing octahedra in TiO$_2$ modifications. Two unit cells are drawn for rutile and anatase.

Box 1.3 Nanoscale Concepts: Polymorphism of TiO_2 nanocrystals

As the proportion of the crystal's surface atoms versus bulk atoms increases towards the ultimate unity for the absolutely smallest cluster, the proportion of the surface's higher energy content also increases. Consider TiO_2 (Figure 1.45). The surface and bulk energies of its three modifications are as follows [15]:

TiO_2	Surface energy (J/m^2)	Relative bulk energy (kJ/mol)
Anatase	0.4	2.6
Brookite	1.0	0.7
Rutile	2.2	0

Rutile has the most stable bulk, yet its surface has the highest energy (sharing fewest octahedral edges in its bulk yields the most dangling bonds at the surface). Below particle sizes of about 35 nm [16], the increasing surface-energy proportion destabilizes the normally stable rutile in favor of brookite. Below 11 nm [16], TiO_2 adopts the anatase structure that shares all octahedral edges and has the fewest dangling bonds at the surface. In the opposite direction, this affects the crystallization of TiO_2, as detailed below.

Since crystals grow from their smallest seed (called the nucleus), TiO_2 obtained by precipitation from an acidic $TiCl_4$ solution is anatase (instantaneous formation, smallest crystallites). This complies with **Ostwald's step rule**—the least stable polymorph *often* crystallizes first [17]. The metastable anatase will turn into the stable rutile only after prolonged aging that allows the crystallites to join and recrystallize. The size and structure of nanoparticles synthesized from solutions depends on the extent of this aggregative growth, called **Ostwald ripening**, in which the tiniest crystals dissolve and regrow on the surface of larger crystals, ultimately minimizing the surface-energy proportion of the system.

---

of tetrahedral holes in an hcp can be filled). In contrast, face-sharing octahedra are found in NiAs (Figure 1.26), corundum (Figure 1.30), the hexagonal perovskites (Figure 1.54), and other structures.

At least one coordination polyhedron exists for every coordination number. As the coordination number increases, several polyhedra become possible; such as the cube, square antiprism, and dodecadeltahedron for coordination number 8, or cuboctahedron, anti-cuboctahedron, and icosahedron for 12. However, upon further increase, the number of known structural examples quickly decreases. One of the highest coordination numbers occurs in the $SmCo_5$-type structure of superferromagnets. Each Sm in this alloy is surrounded by 18 Co atoms that form a 6-capped hexagonal prism. The six capping Co atoms are 3-coordinated, and the twelve prismatic Co atoms are 4-coordinated, giving a Niggli formula $SmCo_{6/3+12/4}$ (Figure 1.46).

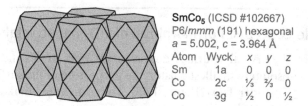

SmCo$_5$ (ICSD #102667)
P6/*mmm* (191) hexagonal
a = 5.002, *c* = 3.964 Å

Atom	Wyck.	x	y	z
Sm	1a	0	0	0
Co	2c	⅓	⅔	0
Co	3g	½	0	½

Figure 1.46 Polyhedral representation of the crystal structure of SmCo$_5$.

1.5   Structures of Selected Materials

1.5.1   The Spinel Structure

The mineral spinel, MgAl$_2$O$_4$, gives its name to a structure type in which two metals of similar atomic sizes are accommodated in tetrahedral and octahedral sites in a ccp array of anions. A simple formula to remember is [tetrahedron]$_1$[octahedron]$_2$O$_4$. A detailed analysis reveals that one-eighth of the tetrahedral and a half of the octahedral ccp holes are occupied in a non-trivial stacking sequence. Table 1.7 illustrates the ubiquity of spinels. Two limiting types of site occupation occur. The first is MgAl$_2$O$_4$ with each ion at its respective site, [Mg^{2+}]$^{[4t]}$[Al^{3+}Al^{3+}]$^{[6o]}$O$_4$, termed a **normal spinel**. In the **inverse spinel**, half the octahedral ions are exchanged with the tetrahedron, as in magnetite [Fe^{3+}]$^{[4t]}$[Fe^{2+}Fe^{3+}]$^{[6o]}$O$_4$. Magnetite forms an inverse spinel due to the ligand-field stabilization energy (LFSE) of the high-spin d^6 Fe^{2+} being larger at octahedral than at tetrahedral sites, while high-spin d^5 Fe^{3+} has no LFSE. LFSE of d^8 Ni^{2+} makes [Ga^{3+}]$^{[4t]}$[Ni^{2+}Ga^{3+}]$^{[6o]}$O$_4$ an inverse spinel, and LFSE of d^4 Mn^{3+} stabilizes normal spinel [Mn^{2+}]$^{[4t]}$[Mn^{3+}Mn^{3+}]$^{[6o]}$O$_4$ with no LFSE from d^5 Mn^{2+}. A spinel of similar ions of no LFSE may show an intersite disorder, such as [Mn^{2+}$_{1-x}$Fe^{3+}$_x$]$^{[4t]}$[Fe^{3+}$_{2-x}$Mn^{2+}$_x$]$^{[6o]}$O$_4$, depending in general on several size- and bonding-related factors.

The connectivity of the cation coordination polyhedra in the spinel structure is hard to visualize, but one key feature is the network of edge-sharing octahedra (Figure 1.47, left). The visualization in Figure 1.47 right is useful for rationalizing the magnetic and electronic properties of spinels that are discussed in more detail in Chapters 9 and 11.

Table 1.7 Oxidation-state combinations for spinels.

Tetrahedron	Octahedron		Chemical formula of an example
1	3	4	LiMn$_2$O$_4$ (a = 8.245 Å)
2	3	3	ZnFe$_2$O$_4$ (a = 8.442 Å)
2	4	2	Fe$_2$TiO$_4$ (a = 8.521 Å)
3	2	3	Fe$_3$O$_4$ (a = 8.394 Å)
4	2	2	Ni$_2$SiO$_4$ (a = 8.045 Å)
5	1	2	LiZnNbO$_4$ (a = 6.082, c = 8.403 Å)
6	1	1	Na$_2$WO$_4$ (a = 9.108 Å)

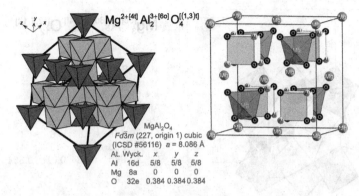

$Mg^{2+[4t]} Al_2^{3+[6o]} O_4^{[(1,3)t]}$

MgAl$_2$O$_4$
$Fd\bar{3}m$ (227, origin 1) cubic
(ICSD #56116) a = 8.086 Å

At.	Wyck.	x	y	z
Al	16d	5/8	5/8	5/8
Mg	8a	0	0	0
O	32e	0.384	0.384	0.384

Figure 1.47 Left: Coordination polyhedra of cations within the unit cell of spinel. Right: The unit-cell content in its simplest representation (the cubes are empty).

1.5.2 The Garnet Structure

The garnet structure type is found in many important magnetic, optical, and magneto-optical materials. Many solid state lasers use yttrium aluminum garnet (YAG) doped with ~1% Nd as the active laser medium. Garnet is a complex, high-symmetry structure, and the simplest formula to remember is [cube]$_3$[octahedron]$_2$[tetrahedron]$_3$O$_{12}$. Table 1.8 shows that a variety of metal atoms can be accommodated at these sites. Intersite disorder is again common, particularly for garnet minerals like pyrope, Mg$_3$Al$_2$Si$_3$O$_{12}$, almandine, Fe$_3$Al$_2$Si$_3$O$_{12}$, and others.

The crystal structure of garnet is not easy to visualize. The crystal-chemical formula together with a polyhedral illustration (Figure 1.48) provides some idea about coordinations and connectivities. The tetrahedra connect to octahedra by corner sharing, whereas three tetrahedral and all six octahedral edges are shared with the rather deformed cubes. As in spinel, all oxygens have a distorted tetrahedral coordination.

Table 1.8 Oxidation states in garnets [cube]$_3$[octahedron]$_2$[tetrahedron]$_3$O$_{12}$.

Cube	Octahedron	Tetrahedron	Examples
2	6	2	Ca$_3$Te$_2$Zn$_3$O$_{12}$ a = 10.930 Å
3	3	3	Y$_3$Fe$_2$Fe$_3$O$_{12}$ a = 12.376 Å
2	3	4	Mg$_3$Al$_2$Si$_3$O$_{12}$ a = 11.459 Å
1	3	5	Na$_3$Sc$_2$V$_3$O$_{12}$ a = 10.913 Å

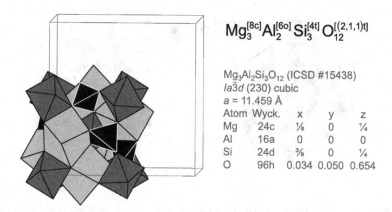

$$Mg_3^{[8c]}Al_2^{[6o]}Si_3^{[4t]}O_{12}^{[(2,1,1)t]}$$

Mg$_3$Al$_2$Si$_3$O$_{12}$ (ICSD #15438)
Ia3d (230) cubic
a = 11.459 Å

Atom	Wyck.	x	y	z
Mg	24c	⅛	0	¼
Al	16a	0	0	0
Si	24d	⅜	0	¼
O	96h	0.034	0.050	0.654

Figure 1.48 Coordination polyhedra in one-eighth of the garnet unit cell, which corresponds to a single formula unit.

1.5.3 Perovskite Structures

The mineral perovskite, $CaTiO_3$, lends its name to a vast family of structures where adjoined $MX_{6/2}$ octahedra form 12-coordinated voids that can be filled by larger atoms A so that the composition becomes AMX_3. Perovskites can also be viewed in terms of closest-packed AX_3 layers in which M occupy one-quarter of the octahedral holes—those that are formed solely by X. Although ccp is most common, perovskites with hcp of AX_3 layers are also known. Figure 1.49 illustrates the ccp case on an ideal cubic perovskite where M fits the octahedral holes exactly.

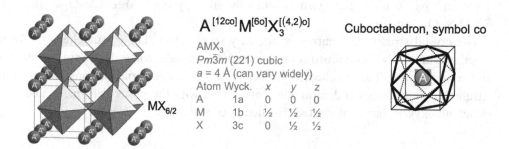

$$A^{[12co]}M^{[6o]}X_3^{[(4,2)o]}$$

AMX_3
Pm$\bar{3}$m (221) cubic
a = 4 Å (can vary widely)

Atom	Wyck.	x	y	z
A	1a	0	0	0
M	1b	½	½	½
X	3c	0	½	½

Cuboctahedron, symbol co

Figure 1.49 Octahedral network in an ideal cubic perovskite (origin at A is chosen).

The ccp array will be more stable than hcp until the increasing size of A prevents the $MX_{6/2}$ octahedra from linking at corners. On the other hand, much smaller A atoms than this maximum can be accommodated while keeping the corners linked. This can be evaluated with the **Goldschmidt tolerance factor** [18] (Figure 1.50) in terms of the ionic radii of A, M, and X. In particular, the Shannon radii [19] for 6-coordinated M, 12-coordinated A, and 2-coordinated O are consistent with Goldschmidt's size considerations: Cubic perovskites, of ideal ratio of the AX and MX bond lengths $AX/(MX\sqrt{2}) = 1$ as in Figure 1.50, top left, are

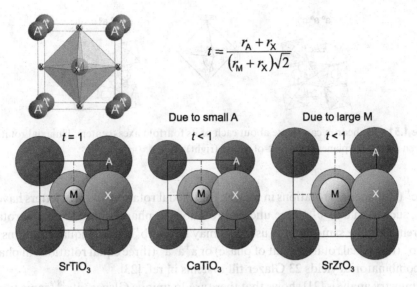

$$t = \frac{r_A + r_X}{(r_M + r_X)\sqrt{2}}$$

Figure 1.50 Tolerance factor *t* and two ways it can become smaller than unity.

stabilized over the *t* range ~1.04 to ~0.98 [20] by minor bonding compromises. When $t \gtrsim 1.04$, the hcp stacking is often observed. In the *t* range between ~0.99 and ~0.83, perovskites are stabilized by structural distortions as discussed below.

Figure 1.50 illustrates two scenarios for *t* < 1: either it is due to M being too large and expanding the MX_6 octahedron, or due to A being too small so that A and X are no longer in contact. The drawing shows that both cases effectively mean that the A atom is smaller than one that would fit exactly. The structure responds to this size mismatch by a coupled rotation of the corner-linked $MX_{6/2}$ octahedra, which expands them in a fixed frame of constant unit cell, accommodating the large M and bringing X closer to A. This **octahedral tilting** can profoundly influence the magnetic and electronic properties of perovskites.

While symmetry analysis [21, 22] is the most rigorous approach to describe the various types of tilting, a simple combinatorics of these coupled rotations is the basis for the **Glazer tilt** classification [23]. It considers a model of linked octahedra and evaluates all combinations of rotations along the three 4-fold axes *of the octahedron* in Figure 1.51. In the network, twisting the central octahedron anticlockwise mechanically requires its in-plane neighbors to twist clockwise in a cooperative manner. The plane of octahedra below the one under consideration, however, is free to rotate in either the same or the opposite sense. Let's start with the simplest case of no rotation at all. It is given a three-letter symbol $a^0 a^0 a^0$ because equal rotations about different axes of the octahedron are symbolized by use of the same letters. A uniaxial rotation is then denoted with "c" as the last letter ($a^0 a^0 c^+$ and $a^0 a^0 c^-$ in Figure 1.51) where superscripts + or − specify whether the octahedra in adjacent layers rotate in the same direction (+, in phase) or in opposite direction (−, out of phase). For biaxial rotation of the octahedron, the letter "b" is added with a superscript sign, such as in

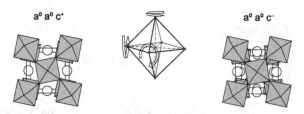

$a^0 a^0 c^+$ $a^0 a^0 c^-$

Figure 1.51 Octahedron can rotate about each of its fourfold axes (center). Uniaxial rotations: In-phase between successive planes (left), out-of-phase (right).

$a^0 b^+ c^+$ (two unequal rotations in phase), for triaxial rotation all three letters have superscript signs, such as in $a^+ b^+ c^+$ (three unequal rotations in phase). Since for equal rotations about different axes the same letter is used, we may have $a^0 b^+ b^+$ (two equal rotations in phase) or $a^- a^- b^+$ (two equal rotations out of phase) or $a^+ a^+ a^+$ (three equal rotations in phase). In total, this combinatorics yields 23 Glazer tilt systems in ref. [23].

Symmetry analysis [21] shows that there are 15 unique Glazer tilts[28] (some tilts also do not form), but three are particularly frequent [24]. One of them is $a^0 a^0 a^0$ for the cubic perovskite of $Pm\bar{3}m$ symmetry. $SrTiO_3$ is one of many examples. Another frequent tilt is $a^- a^- a^-$, which lowers the symmetry to rhombohedral, $R\bar{3}c$. This tilt is equivalent to a single rotation of an octahedron around its threefold axis. Figure 1.52 shows this rotation for $LaNiO_3$ and views of the structure along all three 4-fold axes of the octahedron. These three projections illustrate that all three tilts are equal and out of phase. The high symmetry extends to the coordination polyhedra, fulfilling Pauling's parsimony rule and providing the most favorable ionic bonding of all tilted perovskites. This becomes important as the charge number of the A cation increases. The effect this tilt has on the A-atom coordination is to bring three of the twelve X atoms closer to A while moving three others away, hence approaching the coordination number 9. As the progressing rotation contracts only three of these nine A–X bonds, the range is limited, and the $a^- a^- a^-$ tilting is rare when $t < 0.97$.

The most common tilt is $a^+ b^- b^-$ that lowers the symmetry to orthorhombic $Pnma$. It again accommodates the smaller-than-optimal atom A by bringing some of the twelve X neighbors closer to increase their bond strength more than the rest of them lose it (Section 5.4). An example in Figure 1.53 is $LaFeO_3$, where La has eight close O neighbors and the remaining four at a longer distance. Unlike $a^- a^- a^-$, this tilt allows A to deviate from the center of its coordination polyhedron and is therefore prevalent among perovskites with $t < 0.97$.

---

[28] The reduction from 23 to 15 tilts can be understood by careful consideration of the relationship between symmetry elements and the magnitude of the tilts imposed by Glazer's notation. The number of unique tilts drops from 23 to 17 upon realizing that symmetry cannot restrain tilts of a different sense (in-phase versus out-of-phase) to be of the same magnitude. As an example, $a^+ a^- a^-$ has the same symmetry as $a^+ b^- b^-$, and the latter thus covers both cases. Even if you forced the in-phase tilt to be of equal magnitude to the out-of-phase tilts, there would be no symmetry element that constrained it to be equal, hence, in a real crystal, it would never be more than approximately equal. For a full explanation of how symmetry reduces the number of allowed tilt systems, see ref. [21].

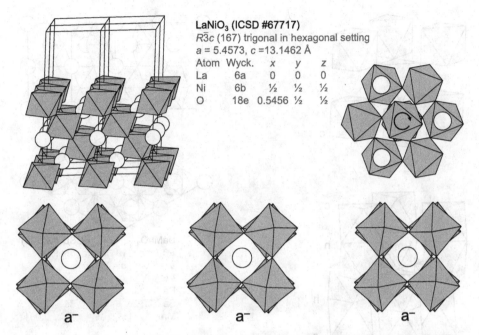

LaNiO₃ (ICSD #67717)
$R\bar{3}c$ (167) trigonal in hexagonal setting
a = 5.4573, c =13.1462 Å

Atom	Wyck.	x	y	z
La	6a	0	0	0
Ni	6b	½	½	½
O	18e	0.5456	½	½

a^- a^- a^-

Figure 1.52 Rotation of Glazer's tilt $a^-a^-a^-$. The actual physical rotation of eight corner-linked octahedra as viewed looking down the threefold axis (top right) and as separated into rotations around the three tilt axes of the octahedron (bottom).

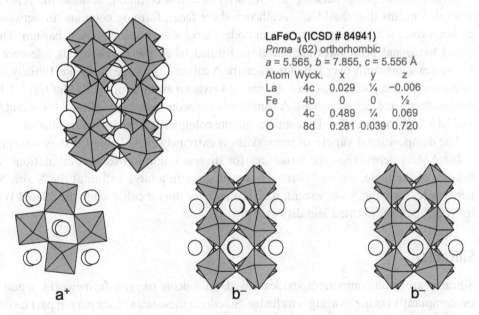

LaFeO₃ (ICSD # 84941)
Pnma (62) orthorhombic
a = 5.565, b = 7.855, c = 5.556 Å

Atom	Wyck.	x	y	z
La	4c	0.029	¼	−0.006
Fe	4b	0	0	½
O	4c	0.489	¼	0.069
O	8d	0.281	0.039	0.720

a^+ b^- b^-

Figure 1.53 Glazer's tilt $a^+b^-b^-$. The octahedron rotates around all three of its axes, in phase around one axis and out of phase around the remaining two.

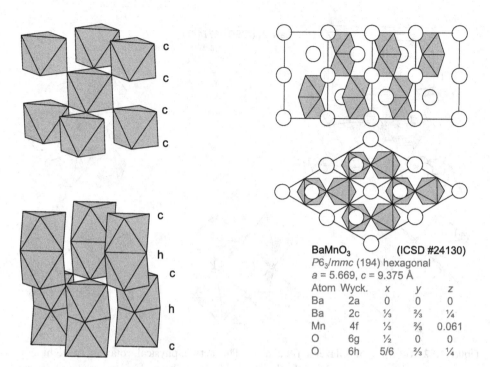

BaMnO$_3$ (ICSD #24130)
$P6_3/mmc$ (194) hexagonal
$a = 5.669, c = 9.375$ Å

Atom	Wyck.	x	y	z
Ba	2a	0	0	0
Ba	2c	⅓	⅔	¼
Mn	4f	⅓	⅔	0.061
O	6g	½	0	0
O	6h	5/6	⅔	¼

Figure 1.54 Left: Comparison of ccp perovskite with a polytype of (hc)$_2$ stacking sequence. Right: The 4H BaMnO$_3$ phase as an example of the latter.

Perovskites with hcp packing of AX$_3$ layers are less common, because the ABAB layer repetition means that the MX$_{6/2}$ octahedra share faces, forming columns. In between these columns there is a lot of space to accommodate large soft A atoms such as barium. These so-called **hexagonal perovskites** are therefore formed when the Goldschmidt tolerance factor t increases significantly over 1, that is, when the A cation is too large for its site. Initially, ordered intergrowths of ccp with hcp occur. Figure 1.54 gives an example of 4H BaMnO$_3$ ($t = 1.11$) that alternates ccp and hcp sequences. A completely hexagonal stacking is found for BaNiO$_3$, with t of 1.13, the structure of which contains infinite columns of face-sharing octahedra.

The compositional variety of perovskites is extremely rich. Table 1.9 shows examples of cubic AMX$_3$ perovskite-type structures for diverse oxidation-state combinations. At the bottom of the table, the anti-perovskites even contain a large cation at the X site. Several perovskites have the A site vacant; ReO$_3$ (our cover star) is cubic while the related WO$_3$ has its WO$_{6/2}$ octahedra tilted and distorted.

1.5.4   Silicates

Silicates are multicomponent oxides based on silicon–oxygen frameworks made up of predominantly corner-sharing tetrahedra. Silicon in these tetrahedra may in part be replaced

Table 1.9 Oxidation-state combinations for AMX$_3$ perovskites.

A	M	X	Examples (cubic or nearly cubic chosen)
+4	+5	−3	ThTaN$_3$ $a = 4.02$ Å
+3	+3	−2	LaAlO$_3$ $a = 3.82$ Å
+2	+4	−2	SrTiO$_3$ $a = 3.90$ Å
+1	+5	−2	CsIO$_3$ $a = 4.67$ Å
	+6	−2	□ReO$_3$ $a = 3.73$ Å
+2	+1	−1	BaLiF$_3$ $a = 4.00$ Å, BaLiH$_3$ $a = 4.02$ Å
+1	+2	−1	KMgF$_3$ $a = 3.97$ Å, (CH$_3$NH$_3$)PbBr$_3$ $a = 5.93$ Å
	+3	−1	□ScF$_3$ $a = 4.00$ Å
−1	−2	+1	AuORb$_3$ $a = 5.50$ Å, ISAg$_3$ $a = 4.90$ Å
−3	−3	+2	SbNCa$_3$ $a = 4.85$ Å, AuNCa$_3$ $a = 4.82$ Å
−4	−2	+2	GeOCa$_3$ $a = 4.73$ Å

by a neighboring element, typically aluminum. Very few silicates contain Si in octahedral coordination,[29] very few feature edge-sharing tetrahedra, and the oxygen vertices are generally 2-coordinated. Silicates are the most common minerals in the Earth's crust and represent about one-third of known inorganic crystal structure types. Although some crystalline silicates, in particular zeolites, have use as functional materials, silicate materials find applications predominantly due to their mechanical properties. The oldest examples of artificial composite materials are silicates such as porcelain, the exquisite properties of which are caused by a texture of fibrous crystals embedded in a glassy matrix.

It's impossible to cover the vast array of silicate structures here; instead we'll focus on simple ideas and nomenclature in common use. We've already seen one way of understanding silicate structures in Section 1.4.3, where we described how the cristobalite modification of SiO$_2$ is derived via expanding the cubic-diamond network of Si and the how the tridymite modification of SiO$_2$ is derived from lonsdaleite. Along this line of thinking, replacement of every second Si in cristobalite with Al and compensation of the thus-formed negative charge by placing Na$^+$ into the cavities in the tetrahedral network leads to carnegieite NaAlSiO$_4$. In a similar way, tridymite can be expanded to the structure of nepheline (Figure 1.55). Such homeotypism occurs also between the quartz modification of SiO$_2$ and β-eucryptite, LiAlSiO$_4$, which is a low-expansion material commonly used as a catalyst support in car exhausts.

The coarsest chemical categorization of silicate anions is based on the network dimensionality. Silicates are classified as **oligo-** and **cyclo-** for finite polyanions, **catena-** for chains, **phyllo-** for layered anions, and the general name of **tectosilicates** is used for 3D networks.[30] Examples of 1D and 2D networks are in Figure 1.56. Their chemical compositions can be determined from

---

[29] Common only at high pressure in phases such as MgSiO$_3$.
[30] From Greek, "oligos" few, "phýllo" leaf, "tekton" builder; and Latin, "catena" chain.

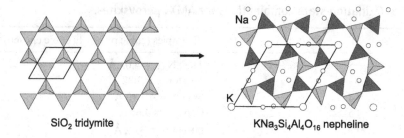

Figure 1.55 Homeotypism between tridymite and nepheline.

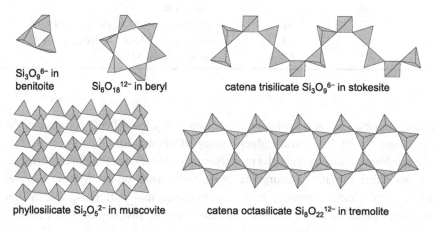

Figure 1.56 Examples of cyclo silicate, catena silicate, and phyllosilicate anions.

the vertex connectivities with Niggli formulas. As an example, all the anions in the top row of Figure 1.56 have the same elemental composition. All have silicon coordination tetrahedra with two terminal and two shared oxygens, $(SiO_{2+2/2})^{2-}$. It is clear that, for infinite polyanions, neither the chemical formula nor the chemical name is very informative of the structural arrangement. A more detailed and informative nomenclature is given in Appendix C.

1.5.5 Zeolites

From the discussion above, it should be clear that the connected tetrahedra in silicates may form an almost infinite number of shapes. One technologically important subgroup of tectosilicates is the zeolites. Zeolites contain interconnected cages made of $SiO_{4/2}$ and $AlO_{4/2}$ tetrahedra, with pores in between them where guest ions, atoms, or molecules can be accommodated. Materials with pores of 2.5–20 Å in diameter are referred to as **microporous**, those with sizes 20–500 Å, **mesoporous**. Owing to the ease with which non-framework chemical species can be exchanged on the large inner surface area, zeolite properties can be

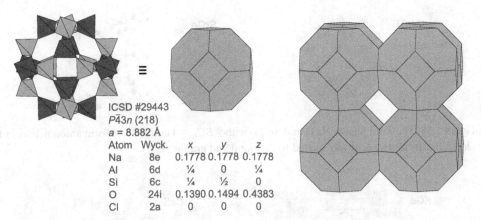

ICSD #29443
$P\bar{4}3n$ (218)
a = 8.882 Å

Atom	Wyck.	x	y	z
Na	8e	0.1778	0.1778	0.1778
Al	6d	¼	0	¼
Si	6c	¼	½	0
O	24i	0.1390	0.1494	0.4383
Cl	2a	0	0	0

Figure 1.57 Graphical representations of the alumosilicate cages in sodalite.

tuned to fit many technological applications (discussed in Chapter 14). One common zeolite cage is the **sodalite cage**, built from 24 tetrahedra as shown in Figure 1.57. The visualization of the interconnected zeolite cages is greatly simplified by omitting the tetrahedra and connecting only their central atoms with a straight line (removing O atoms, the inverse of network expansion).

The connectivity of the cages is accomplished by sharing (Figure 1.57) or by connecting their equivalent faces. When all square faces of the sodalite cage are shared at the Si atom with other sodalite cages, the sodalite structure is formed and contains small pores that do not qualify as micropores. However, when all square faces of sodalite cages are connected via Si–O–Si bonds, forming a neck between the cages, the so-called zeolite A is obtained. When four of the eight hexagonal faces of the sodalite cage are connected to other sodalite cages, zeolite X/Y (faujasite) is formed (Figure 14.2). Owing to important industrial applications, a zeolite nomenclature [25] has been adopted by the International Union of Pure and Applied Chemistry (IUPAC).

1.5.6 Zintl Phases

Zintl phases are compounds of two elements one would consider as metals or semimetals (typically an alkali or alkaline-earth metal and a post-transition metal). Unlike conventional alloys and intermetallics, Zintl phases have fixed stoichiometry and physical properties that are atypical for a metal. They are diamagnetic, brittle, and exhibit low electrical conductivity. Their composition follows the **Zintl–Klemm concept** of the more electropositive metal behaving as a cation that provides electrons for the more electronegative metal to form polyanions in which electrons are shared to complete the octet [26]. The composition can thus be treated by the generalized 8−N rule of Section 1.3.2. As an example, although we normally think of thallium as an electropositive metal, it forms a Zintl phase NaTl with the even more electropositive Na. Because VEC_A = 4, four electrons per Tl must be shared,

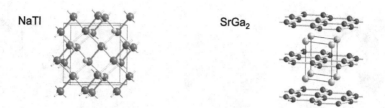

NaTl   SrGa$_2$

Figure 1.58 The Zintl phases NaTl and SrGa$_2$ with $VEC_A = 4$ and $AA = 4$ form anion networks that are electronically and structurally related to diamond and graphite.

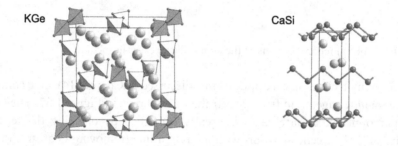

KGe   CaSi

Figure 1.59 Zintl phases KGe and CaSi with anions isoelectronic to elemental nonmetals.

forming four two-electron Tl–Tl bonds at each Tl. Indeed, the crystal structure in Figure 1.58 has a polyanion network $^3_\infty$Tl$^-$ analogous to diamond. Similarly, for SrGa$_2$ (Figure 1.58), $VEC_A = 4$ and four electrons per Ga are shared—three in single bonds, one via conjugation as in graphite. The crystal-chemical formula of the anion is $^2_\infty$Ga$^-$.

In KGe, $VEC_A = 5$, and three two-electron bonds per Ge are formed (Figure 1.59). In the crystal structure of KGe, germanium occurs in Ge$_4^{4-}$ cages isoelectronic with molecules of white phosphorus. Another example is CaSi, where $VEC_A = 6$, suggesting two two-electron bonds per Si (Figure 1.59). In the crystal structure of CaSi, silicon forms polyanion chains $^1_\infty$Si^{2-}, isoelectronic with the fibrous modification of elemental sulfur, and similar to the structure of tellurium shown in Figure 1.16.

1.6 Problems

(Note that some require a structure-drawing program.)

1.1 Write down the $\langle 111 \rangle$ set of symmetry-equivalent directions in a cubic lattice.

1.2 By analogy with Table 1.1, determine the number and type of 2D crystal systems via considering their possible minimum symmetry elements and sketching their Bravais lattices.

1.3 Write down indices for the following sets of equidistant planes:

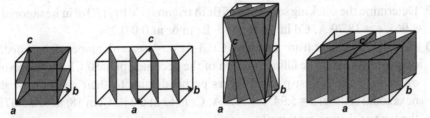

1.4 Sketch a set of equidistant 113 planes in a cubic unit cell.

1.5 Sketch a set of equidistant $1\bar{1}3$ planes in a cubic unit cell.

1.6 State the Bravais lattice and write down the crystallographic point-group symbol for a structure of space-group symbol: (a) $C2/m$, (b) $Fmm2$, (c) $I4/mmm$, (d) $P312$, (e) $R\bar{3}m$, (f) $P\bar{6}m2$, (g) $F23$, (h) $P2_13$, (i) $Ia\bar{3}d$.

1.7 Is it possible for a c glide plane to have the direction of: (a) a axis, (b) b axis, (c) c axis? If not, why is this not allowed?

1.8 Write down crystal-chemical formulas for two-site binary compounds CrN, Cr_2O_3, and CrO_2 with octahedrally coordinated chromium.

1.9 Write down the three balances expressed in the crystal-chemical formulas of the phases in the previous problem.

1.10 Draw the bond graph for the mineral spinel $MgAl_2O_4$ (Figure 1.47).

1.11 Convert the bond graphs in Figure 1.15 into crystal-chemical formulas.

1.12 Describe or sketch a structure for layered V_2O_5 that is consistent with the bond-graph representation in Figure 1.15.

1.13 Use the generalized 8−N rule to identify whether anion–anion or cation–cation bonds are present for the following compounds (assume no cation-localized nonbonding electron pairs): (a) Na_2Tl, (b) $SrSb_2$, (c) $BaTe_2$, (d) InSe.

1.14 Suggest the anion bonding that might occur in MgB_2 and MgC_2.

1.15 Calculate the percentage of available space that's taken up by touching spheres in primitive and body-centered cubic arrangements.

1.16 Imagine a coordination polyhedron with a cation at the center. Now treat the ions as hard spheres and reduce the size of the cation until the anions just touch. What is the radius r of the cation for anions of unit radius in the following coordinations: (a) cube, (b) octahedron, and (c) tetrahedron. Hint: Body diagonal is $\sqrt{3}$ and face diagonal $\sqrt{2}$ times the cube edge.

1.17 From unit-cell parameters in figures in this chapter, calculate the following shortest distances: (a) Na–Cl and Na–Na in NaCl, (b) Ni–Ni in NiAs, (c) Ca–F in CaF_2,

(d) C–C in diamond, (e) Ti–Ti in TiO_2 (rutile), and (f) Ti–O, Sr–O, and O–O in the cubic perovskite $SrTiO_3$ (a = 3.90 Å).

1.18 With a structure-drawing program, determine the stacking sequence in Tb (data for plotting, see Section 1.1.8: $P6_3/mmc$, a = 3.068, c = 14.87 Å, Tb in $2b$ at 0 0 ¼ and $4f$ at ⅓ ⅔ 0.083). Suggest the Ramsdell symbol and the Jagodzinski–Wyckoff notation.

1.19 Determine the stacking sequence of Br in trigonal $CdBr_2$ ($R\bar{3}m$ in hexagonal setting; a = 3.965, c = 18.70 Å, Cd in $3a$ at 0 0 0, Br in $6c$ at 0 0 0.25).

1.20 In the low-temperature form of $CrCl_3$, every second plane of octahedral holes is occupied. What is the filling fraction of the occupied planes? Check the result by a side-on view of the closest-packed layers perpendicular to their subsequent shifts ($R\bar{3}$ in hexagonal setting; a = 5.94, c = 17.3 Å, Cr in $6c$ at 0 0 ⅓, Cl in $18f$ at ⅔ 0 0.0757). What is the type of densest packing?

1.21 La in solid $LaBr_3$ has a tricapped trigonal-prismatic coordination. What is the coordination number of Br? Is it possible that the tricapped trigonal prisms share only corners, or would you expect sharing of edges and/or faces? Check the result by viewing the structure and constructing polyhedra around La ($P6_3/m$, a = 7.971, c = 4.522 Å. La in $2c$ at ⅓ ⅔ ¼, Br in $6h$ at 0.3849 0.2988 ¼).

1.22 Write the Niggli formula and the simple crystal-chemical formula for the CrO_3 structure that contains chains of corner-sharing chromium-centered tetrahedra.

1.23 Construct a bond graph or Niggli formula to determine if it is possible for all anions to be equivalent in a structure of tetrahedrally coordinated cations and stoichiometry of C_2A_3? Which alternative Niggli formula complies best with the rule of parsimony?

1.24 Using the Niggli formula and the rule of parsimony, determine the stoichiometry that results from sharing (a) all corners, (b) all edges, and (c) all faces of a cation-centered cube of anions. Note the structure prototype where you recognize it.

1.25 Write the Niggli formula for C_3N_4 made of identical CN_4 tetrahedra. How many different types of nitrogen vertices are there? What is the coordination number of each?

1.26 In β-Li_3N, nitrogen is 11-coordinated. Write down the Niggli formula of the NLi_3 polyhedron.

1.27 ReO_3 has a 3D network of octahedrally coordinated rhenium. Determine the N, M-connectivity for this binodal network

1.28 Identify the type of derivative network relationship between: (a) CaO ($Fm\bar{3}m$, a = 4.778 Å, O in $4a$ at 0 0 0, Ca in $4b$ at ½ ½ ½) and CaO_2 ($I4/mmm$, a = 3.56 Å, c = 5.95 Å, Ca in $2a$ at 0 0 0, O in $4e$ at 0 0 0.394), (b) $SrMoO_3$ ($Pm\bar{3}m$, a = 3.965 Å, Sr in $1b$ at ½ ½ ½, Mo in $1a$ at 0 0 0, O in $3d$ at ½ 0 0) and the high-temperature Sr_2FeMoO_6 phase ($Fm\bar{3}m$, a = 7.93 Å, Sr in $8c$ at ¼ ¼ ¼, Mo in $4b$ at ½ ½ ½, Fe in $4a$ at 0 0 0, O in $24e$ at 0.253 0 0).

1.29 Which type of similarity do you see between the Laves phase $MgCu_2$ ($Fd\bar{3}m$, a = 7.034 Å, Mg in $8a$ at 0 0 0, Cu in $16d$ at ⅝ ⅝ ⅝) and the spinel in Figure 1.47?

1.30 Rewrite the chemical formulas in Table 1.7 into simplified crystal-chemical formulas of spinel.

1.31 Analyze hole filling in the spinel structure of ccp oxygens. (a) Given the general formula [tetrahedron]$_1$[octahedron]$_2$O$_4$, calculate the total fractions of tetrahedral and octahedral holes filled. (b) Use a structure-drawing program to orient the structure and identify the type of filled holes between layers of densely packed oxygen atoms. (c) Determine the sequence and the fraction of each hole filling.

1.32 With a structure-drawing program, determine which of the three most common tilts the perovskite prototype CaTiO$_3$ (ICSD 62149) adopts.

1.33 Give the Niggli formula for a cyclosilicate anion containing 3 Si.

1.34 What is the formula of the infinite alumosilicate anion in sodalite?

1.35 Use VEC_A modified for the stable 18-electron configuration of Kr to justify the network of Cu tetrahedra in MgCu$_2$.

1.36 Suggest the bonding present in the Zintl phase LiAs.

1.37 Suggest the bonding present in the Zintl phase KIn.

1.38 Suggest the bonding present in the Zintl phase CaIn$_2$. Verify your suggestions with the ICSD and a structure-drawing program.

1.7 Further Reading

A.F. Wells, *"Structural Inorganic Chemistry"* (1984) Oxford University Press.

B.G. Hyde, S. Andersson, *"Inorganic Crystal Structures"* (1989) Wiley.

F.A. Cotton, *"Chemical Applications of Group Theory"* (1990) Wiley.

G. Burns, A.M. Glazer, *"Space Groups for Solid State Scientists"* 2nd edition (1990) Academic Press.

M. O'Keeffe, B.G. Hyde, *"Crystal Structures I: Patterns and Symmetry"* (1996) Mineralogical Society of America.

A. Vincent, *"Molecular Symmetry and Group Theory"* (2001) Wiley.

R. Tilley, *"Crystals and Crystal Structures"* (2006) Wiley.

U. Müller, *"Inorganic Structural Chemistry"* 2nd edition (2007) Wiley.

M. de Graef, M.E. McHenry, *"Structure of Materials: An Introduction to Crystallography, Diffraction and Symmetry"* (2007) Cambridge University Press.

C. Giacovazzo, H.L. Monaco, G. Artioli, D. Viterbo, M. Milaneso, G. Ferraris, G. Gilli, P. Gilli, G. Zanotti, M. Catti, *"Fundamentals of Crystallography"* 3rd edition (2011) IUCr/Oxford University Press.

U. Müller, *"Symmetry Relationships between Crystal Structures"* (2013) IUCr/Oxford University Press.

1.8 References

[1] J. Foadi, G. Evans, "On the allowed values for the triclinic unit-cell angles" *Acta Crystallogr. Sect. A* **67** (2011), 93–95.

[2] J. Lima de Faria, E. Hellner, F. Liebau, E. Makovický, E. Parthé, "Nomenclature of inorganic structure types" *Acta Crystallogr. Sect. A* **46** (1990), 1–11.

[3] L. Pauling, "The principles determining the structure of complex ionic crystals" *J. Am. Chem. Soc.* **51** (1929), 1010–1026.

[4] D. Fisher, M. Jansen, "Synthesis and structure of Na_3N" *Angew. Chem. Int. Ed. Engl.* **41** (2002), 1755–1756.

[5] M. O'Keeffe, B.G. Hyde, "Stoichiometry and the structure and stability of inorganic solids" *Nature* **309** (1984), 411–414.

[6] A. Kjekshus, "The general (8-N) rule and its relationship to the octet rule" *Acta Chem. Scand.* **18** (1964), 2379–2384, and references therein.

[7] E. Parthé, "Valence-electron concentration rules and diagrams for diamagnetic, nonmetallic ion covalent compounds with tetrahedrally coordinated anions" *Acta Crystallogr.* **29** (1973), 2808–2815.

[8] J. Kepler, "*Strena seu de nive sexangula*" Godfried Tambach, Frankfurt am Main (1611).

[9] P.N.H. Nakashima, A.E. Smith, J. Etheridge, B.C. Muddle, "The bonding electron density in aluminum" *Science* **331** (2011), 1583–1586.

[10] H.K. Kim, S.H. Huh, J.W. Park, J.W. Jeong, G.H. Lee, "The cluster size dependence of thermal stabilities of both molybdenum and tungsten nanoclusters" *Chem. Phys. Lett.* **354** (2002), 165–172.

[11] M. O'Keeffe, M. Eddaoudi, H. Li, T. Reineke, O.M. Yaghi, "Frameworks for extended solids: geometrical design principles" *J. Solid State Chem.* **152** (2000), 3–20.

[12] V.A. Blatov, M. O'Keeffe, D.M. Proserpio, "Vertex-, face-, Schläfli-, and Delaney symbols in nets, polyhedra and tilings: recommended terminology" *CrystEngComm* **12** (2010), 44–48.

[13] O. Delgado-Friedrichs, M. O'Keeffe, O.M. Yaghi, "Taxonomy of periodic nets and the design of materials" *Phys. Chem. Chem. Phys.* **9** (2007), 1035–1043.

[14] O. Delgado-Friedrichs, M.D. Foster, M. O'Keeffe, D.M. Proserpio, M.M.J. Treacy, O.M. Yaghi, "What do we know about three-periodic nets?" *J. Solid State Chem.* **178** (2005), 2533–2554.

[15] M.R. Ranade, A. Nawrotsky, H.Z. Zhang, J.F. Banfield, S.H. Elder, A. Zaban, P.H. Borse, S.K. Kulkarni, G.S. Doran, H.J. Whitfield, "Energetics of nanocrystalline TiO_2" *Proc. Natl. Acad. Sci. USA* **99** (2002), 6476–6481.

[16] H. Zhang, J.F. Banfield, "Understanding polymorphic phase transformation behavior during growth of nanocrystalline aggregates: insights from TiO_2" *J. Phys. Chem. B* **104** (2000), 3481–3487.

[17] W. Ostwald, "Studien über die Bildung und Umwandlung fester Körper. 1. Abhandlung: Übersättigung und Überkaltung" *Z. Physik. Chem.* **22** (1897), 289–330.

[18] V.M. Goldschmidt, "Laws of crystal chemistry" *Naturvissenschaften* **14** (1926), 477–485.

[19] R.D. Shannon, "Revised effective ionic radii and systematic studies of interatomic distances in halides and chalcogenides" *Acta Crystallogr. Sect. A* **32** (1976), 751–767.

[20] M.W. Lufaso, P.M. Woodward, "Predictions of the crystal structures of perovskites using the software program SPuDS" *Acta Crystallogr. Sect. B* **57** (2001), 725–738.

[21] C.J. Howard, H.T. Stokes, "Group theoretical analysis of octahedral tilting in perovskites" *Acta Crystallogr. Sect. B* **54** (1998), 782–789.

[22] D. Wang, R.J. Angel, "Octahedral tilts, symmetry-adapted displacive modes and polyhedral volume ratios in perovskite structures" *Acta Crystallogr. Sect. B* **67** (2011), 302–314.

[23] A.M. Glazer, "Classification of tilted octahedra in perovskites" *Acta Crystallogr. Sect. B* **28** (1972), 3384–3392.

[24] P.M. Woodward, "Octahedral tilting in perovskites. II. Structure stabilizing forces" *Acta Crystallogr. Sect. B* **53** (1997), 44–66.

[25] L.B. McCusker, F. Liebau, G. Engelhardt, "Nomenclature of structural and compositional characteristics of ordered microporous and mesoporous materials with inorganic hosts" *Pure Appl. Chem.* **73** (2001), 381–394.

[26] E. Zintl, "Intermetallische Verbindungen" *Angew. Chem.* **52** (1939), 1–6.

2 Defects and More Complex Structures

We have seen in Chapter 1 that the solid state world is dominated by long-range order and beauty. Crystalline materials contain highly symmetric arrangements of atoms that are regularly repeated over millions of unit cells. In this chapter, we will question how realistic this picture is.

In reality, there are a number of ways in which crystalline materials deviate from perfect long-range order and contain imperfections or disorder. This can occur via "mistakes" in the atomic arrangement of a pure material or via the introduction of impurity atoms giving rise to chemical disorder. These defects can occur locally or extend over lines, planes, or 3D volumes of materials. Such effects, even when they occur at very low levels, are vitally important to the chemical and physical properties of materials. They turn low-value minerals into precious gemstones; soft iron into strong and corrosion-resistant stainless steel; and they control the semiconducting properties of silicon in the transistors powering modern electronics.

This chapter also introduces a variety of ways in which materials can deviate from having simple stoichiometric formulae. This can occur either via the presence of defects or chemical substitutions in a material or can have a variety of more complex structural origins. In later chapters, we will see how these various effects influence many of the important properties of functional materials.

2.1 Point Defects in Crystalline Elemental Solids

We have seen that the structures of many elements can be described in terms of regular arrays of spherical atoms. At the local level, this order can be perturbed by three different types of **point defects**; vacancies, interstitials, and substitutional disorder. These are shown schematically in Figure 2.1.

A **vacancy** occurs when an atom is missing from a site in the structure as shown in Figure 2.1, left. An **interstitial** defect occurs when an extra atom sits in a site that would not normally be occupied. An interstitial site can be occupied either by an atom of the same type that

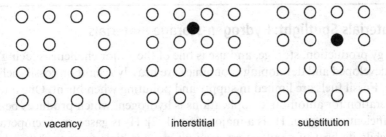

Figure 2.1 Different types of point defects.

makes up the structure, or by an impurity atom. It is common, for example, for small non-metallic elements such as C, N, O, and H to occupy a fraction of interstitial positions in the structures of transition metals. Such materials can be of great technological importance: C interstitials in iron greatly increase its mechanical strength; and Pd can store around 0.6% by mass of H interstitials, which is of interest for hydrogen storage (see Box 2.1).

The third common type of defect is **substitutional disorder** where a foreign atom adopts a site in the structure of a pure element. At low levels, foreign aliovalent[1] atoms are frequently referred to as dopants and can significantly alter chemical (Chapter 3), electronic (Chapters 6 and 10), and optical (Chapter 7) properties of a material. Doping silicon with low levels of Al or P, for example, leads to the formation of p- or n-type semiconductors, respectively. At higher levels of doping, one typically refers to solid-solution or alloy formation. This process can again be used to tailor a material's properties. Real materials exist with any one of these basic types of defects or with various combinations of them.

2.2 Intrinsic Point Defects in Compounds

Similar defects to those depicted in Figure 2.1 for elements can occur in ionic compounds,[2] though with the additional constraint that one must maintain overall electrical neutrality in the crystal. If, for example, one simply removed an Na^+ ion from a compound such as NaCl to create a defect, the crystal would end up with a negative charge. Defects in ionic compounds can only occur in ways that avoid such charge build-up. Defects that can occur in pure compounds are called **intrinsic defects**.

A **Schottky defect** is a cation vacancy compensated by an appropriate number of anion vacancies (Figure 2.2), i.e. it is a missing formula unit. For a binary phase MX, the number of vacancies on cation and anion sites will be equal; for MX_2, two X^- anion vacancies would be needed for every M^{2+} vacancy. In terms of ionic charges, removing M^+

---

[1] Of different valence or oxidation state to the atom it replaces; isovalent means of the same valence.

[2] Our discussion in this chapter will largely be in terms of formal ionic charges in materials. This need not, of course, imply that ionic bonding is actually dominant.

Box 2.1 Materials Spotlight: Hydrogen-storage materials

Efficient energy production, storage, and use is one of the major challenges facing the modern world. Both developed and developing economies are heavily reliant on fossil fuels in all areas of energy use. Fossil fuels are limited in supply and polluting when burnt. One alternative fuel under consideration for automotive applications is hydrogen, which produces benign H_2O.

Safe and efficient storage of H_2 is a major issue [1, 2]. H_2 is gaseous at temperatures above 20.3 K and while its heat of combustion per unit mass is high compared to other fuels (H_2 120 MJ/kg, gasoline 44.5 MJ/kg) the value per unit volume is low: at standard conditions, hydrogen gas has $\Delta H_{comb} = 9.6$ kJ/L and even as a liquid $\Delta H_{comb} = 8.4$ MJ/L compared to 31.2 MJ/L for gasoline. For automotive use, the US Department of Energy state specific targets of a gravimetric energy density of 7.9 MJ/kg (2.2 kW h), corresponding to 6.6 wt% H_2, and a volumetric energy density of 6.1 MJ/L at the time of writing. In addition, the fuel storage in a car needs to be reversible (~1000 times) and allow refueling in 3 minutes—significant challenges!

One possibility is to store H_2 as a high-pressure gas or as a liquid at 20 K. It costs about 15% of the energy content to pressurize H_2 to 700 bar or about 20% to liquefy it (plus around 2% per day to keep it cold). For these and other technological reasons, high-pressure seems the more viable storage solution.

Chemical solutions to the H_2 storage problem can be divided into two main categories—those that rely on chemisorption and formation of chemical bonds, and those that rely on physisorption of H_2. In the chemisorption case, the stability of the compounds formed means that a key challenge is to provide materials that decompose to release H_2 at a low enough temperature (below approximately 90 °C) to be compatible with, for example, polymer electrolyte membranes in fuel cells. For physisorption, the challenge is the opposite: to provide systems that still hold H_2 at room temperature and ambient pressure. The adsorption of H_2 by carbon nanotubes, graphene, polymeric materials, and the metal–organic frameworks of Chapter 14 has received significant attention [3]. Currently, no materials come close to targets under ambient conditions and any adsorption-based storage systems will rely on liquid-N_2 temperatures and the use of several tens of bar pressures.

The alloy $LaNi_5$ is one potential storage material as the number of H atoms that can be stored in interstitial sites (forming $LaNi_5H_6$) per unit volume is around double that of H_2 liquid at its boiling point. The reversibility of H_2 uptake is good, and the H_2 pressure at room temperature is around 2 bar. Unfortunately, the mass content of hydrogen is too low at ~1.4 wt%, and La is expensive. Similar problems face $CoNi_5H_4$ (1.1 wt%) and $PdH_{0.6}$ (0.6 wt%). The latter decomposes close to room temperature and shows excellent reversibility, but suffers from the high cost of Pd metal.

The only way to achieve a higher mass content of hydrogen in a storage compound is to focus on hydrides from the top rows of the periodic table. MgH_2 is one material of interest and can practically store around 7.6 wt% H_2, though its decomposition temperature is rather high (~330 °C) and reversibility is poor. Other materials under investigation include complex hydrides such as $NaAlH_4$. When small amounts of Ti or other metals are included as a catalyst, this material can reversibly produce H_2 by two processes:

Box 2.1 (cont.)

$$NaAlH_4 \rightarrow \tfrac{1}{3}Na_3AlH_6 + \tfrac{2}{3}Al + H_2 \text{ (3.7 wt\% H}_2, > 33\text{ °C)}$$

$$Na_3AlH_6 \rightarrow 3NaH + Al + \tfrac{3}{2}H_2 \text{ (1.8 wt\% H}_2, > 110\text{ °C)}$$

Although $NaAlH_4$ could theoretically release 5.5 wt% H_2, in practice only around 4% can be stored reversibly. In addition, the kinetics are slow.

An interesting strategy for reducing the decomposition temperature of a hydrogen-storage material is to either destabilize the hydrogenated form or stabilize the dehydrogenated form. For example, $LiBH_4$ releases ~13.6 wt% H_2 in an endothermic reaction, $LiBH_4 = LiH + B + \tfrac{3}{2}H_2$ ($\Delta H = 67$ kJ/mol H_2). If one mixes it with MgH_2, the mass hydrogen content is reduced to 11.4 % but the exothermic formation of MgB_2 in the reaction, $LiBH_4 + \tfrac{1}{2}MgH_2 \rightarrow LiH + \tfrac{1}{2}MgB_2 + 2H_2$, makes the total reaction less endothermic ($\Delta H = 25$ kJ mol/H_2), and the release temperature lower.

Other materials under investigation include the lithium amide–imide–nitride system ($LiNH_2$–Li_2NH–Li_3N, which stores 10.4 wt% H_2 and in which nonstoichiometric materials are believed to play a key role [4]), ammonia borane (BH_3NH_3, ~14 wt% release from 100–180 °C) and related lithium/sodium amidoboranes ($LiNH_2BH_3$/$NaNH_2BH_3$, ~11/~7.5 wt% at ~90 °C), though reversibility of all these systems remains an issue.

Even if the chemical challenges can be met, there are still significant engineering challenges for practical H_2-fueled cars. For example, there is significant heat produced when either the endothermic chemical processes described above are reversed on refueling, or the heat of adsorption for cryo-adsorption systems is released. Heavy heat exchangers capable of handling hundreds of kilowatts would be required for chemical systems capable of acceptable refueling rates, reducing the gravimetric storage capacity of the system. More information on hydrogen-storage materials can be found via www.doe.gov.

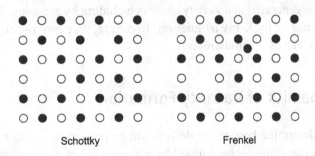

Schottky   Frenkel

Figure 2.2 Schottky and Frenkel defects. Black and white circles represent cations and anions.

from its structural site will perturb the local electroneutrality—the negative charge on surrounding X^- ions is no longer cancelled by the charge of the cation. The uncompensated $1-$ charge is therefore formally assigned to the cation vacancy. The anion vacancy created elsewhere in the crystal at the same time will carry an analogous $1+$ charge. These ideas are formalized in the Kröger–Vink notation outlined in Section 2.6. In a Schottky defect, the negative and positive charges will attract each other, which can lead to **vacancy clustering** in the structure (see Section 2.9). One might also expect that on creation of a defect, the surrounding structure may relax or distort. This is indeed the case but may be difficult to detect using conventional structural characterization techniques (which tend to probe average structures rather than local structures), particularly at low defect concentrations. In Section 2.9, we will see how substantial long-range structural changes can occur at larger defect concentrations.

A **Frenkel defect** consists of a vacancy and a corresponding interstitial ion (of the same charge) to maintain charge balance. Frenkel defects can occur for cations or for anions. AgI commonly shows cation-Frenkel defects, CaF_2 anion-Frenkel defects. The number of Frenkel defects increases with the temperature, and materials with a high concentration of Frenkel defects can have significant ionic mobility at high temperatures. We'll see in Chapter 13 that this defect type gives rise to cationic conductivity in AgI and anionic conductivity in CaF_2-type materials.

We'll mention other types of disorder/defects possible in compounds only briefly. For many materials, intersite disorder is possible, for example the $(Mn^{2+}_{1-x}Fe^{3+}_x)(Fe^{3+}_{1-x/2}Mn^{2+}_{x/2})_2O_4$ spinels we discussed in Section 1.5.1, or disordered alloys such as FePt (see Section 4.4.2). In Chapter 13, we'll also encounter ions that are rotationally disordered in the solid state.

The final intrinsic point defect we'll discuss is a **color center**. It can be described as electrons trapped on vacant anion sites by the positive charge of the vacancy. Isolated trapped electrons are known as F centers. More complex color centers consisting of pairs (M centers) or triplets (R centers) of electrons or trapped holes are also known. The shape of the potential holding the electron in an F center is complex but it can give rise to a localized s-like state as well as a more extended p-like shape. Excitation from one state to the other frequently leads to light absorption in the visible region giving rise to characteristic colors. F centers can be generated in a variety of ways including by exposing alkali halides to excess alkali-metal vapor or by X-ray irradiation. In nature, they give rise to blue forms of calcite and feldspar as well as green diamonds.

2.3   Thermodynamics of Vacancy Formation

What factors determine how many defects will be present in a material? Let's consider an ideal crystal in two dimensions, formed by a single type of atom, such as the one drawn in Figure 2.3. There are N_0 atomic sites in the entire crystal. We can consider the formation of a vacancy as the process of moving one atom from its regular site to a new site on the surface of the crystal. The number of sites becomes $N_0 + 1$. What is the Gibbs energy of creating n such isolated, non-interacting, vacancies?

The main driving force behind vacancy creation is the increase in configurational entropy: there are many possible ways to position the vacancy. The configurational entropy

Figure 2.3 Formation of a vacancy in an ideal crystal.

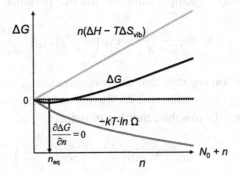

Figure 2.4 Gibbs energy of formation of n vacancies in an ideal crystal with N_0 sites.

contribution to ΔG is $-kT \cdot \ln\Omega$ per n vacancies formed. In this expression, Ω is the number of ways n vacancies can be arranged over the $N_0 + n$ sites, which is given by $\Omega = (N_0 + n)!/(N_0!n!)$ and k is the Boltzmann constant.[3] Note that $-kT \cdot \ln\Omega$ (Figure 2.4) is not a linear function of the number of vacancies. There is also a smaller vibrational contribution to the entropy, ΔS_{vib}, which arises on vacancy formation. The overall favorable increase in entropy is counteracted by the positive enthalpy of vacancy formation. The enthalpy term is positive as the chemical bonds broken on vacancy formation are only partially compensated by the formation of new bonds at the surface site.

At low vacancy concentrations, the ΔH and ΔS_{vib} terms depend simply on the number of vacancies n. Since $\Delta G = \Delta H - T\Delta S$, we can express the total change in Gibbs energy for formation of n vacancies as $\Delta G = n(\Delta H - T\Delta S_{vib}) - kT \cdot \ln\Omega$. This expression is plotted as a function of n in Figure 2.4. The nonlinear nature of the configurational-entropy term means that the minimum in ΔG always occurs at a non-zero value of n; that is, a small equilibrium number of vacancies, n_{eq}, will always be formed. The value of n_{eq} can be calculated by finding the minimum in ΔG by differentiation:

$$\frac{d}{dn}\left(n\Delta H - nT\Delta S_{vib} - kT \cdot \ln\frac{(N_0 + n)!}{N_0!n!}\right) = 0 \tag{2.1}$$

[3] Boltzmann constant $k = 1.380649 \times 10^{-23}$ J/K.

The factorial term can be treated by applying Stirling's formula, $\ln(x!) \approx x\ln x - x$ for large x, which on differentiation leads to:

$$\Delta H - T\Delta S_{\text{vib}} - kT \cdot \ln\left(\frac{N_0 + n_{\text{eq}}}{n_{\text{eq}}}\right) = 0 \tag{2.2}$$

Note that the term in brackets involving n_{eq} is simply the inverse of the fractional concentration of vacancies formed, x_v. Rearrangement shows that the fractional concentration of vacancies depends solely on temperature and the two parameters, ΔH and ΔS_{vib}:

$$x_v = \frac{n_{\text{eq}}}{N_0 + n_{\text{eq}}} = \exp\left(\frac{\Delta S_{\text{vib}}}{k}\right)\exp\left(\frac{-\Delta H}{kT}\right) \tag{2.3}$$

The form of this equation suggests that the concentration of non-interacting point defects at equilibrium[4] follows the mass-action law. We'll investigate the use of such equilibria in Chapter 3.

We can approximate the absolute number of defects by assuming $N_0 \gg n_{\text{eq}}$, which gives:

$$n_{\text{eq}} \approx N_0\exp\left(\frac{\Delta S_{\text{vib}}}{k}\right)\exp\left(\frac{-\Delta H}{kT}\right) \tag{2.4}$$

A similar expression would hold for the formation of isolated interstitials. We can see from this expression that either a low ΔH or high T will favor the formation of defects.

For Schottky defects in an ionic material, one must consider the number of ways of arranging both cation and anion vacancies. The configurational entropy term will be $-2kT \cdot \ln \Omega$. A similar derivation gives the equilibrium number of Schottky defect pairs as:

$$n_{\text{eq,Schottky}} \approx N_0\exp\left(\frac{\Delta S_{\text{vib}}}{2k}\right)\exp\left(\frac{-\Delta H_S}{2kT}\right) \tag{2.5}$$

where ΔH_S is the formation enthalpy of a Schottky defect pair.

For Frenkel defects, one generates n vacancies on the N_0 sites and places them on n interstitial sites out of a possible number of N_I. There will be a configurational entropy term of the form $N_{0/I}!/(N_{0/I} - n)!n!$ for both vacancies and interstitials. One can then show that:

$$n_{\text{eq,Frenkel}} \approx (N_0 N_1)^{\frac{1}{2}}\exp\left(\frac{\Delta S_{\text{vib}}}{2k}\right)\exp\left(\frac{-\Delta H_F}{2kT}\right) \tag{2.6}$$

Table 2.1 gives molar enthalpies and entropies of defect formation for various compounds from which the number of defects can be readily estimated.[5] In Cu metal at 1000 °C, for example,

---

[4] Reaching such equilibrium in solids usually requires high temperatures, it's easy to "trap" non-equilibrium numbers of defects on cooling.

[5] Note the need to convert between the Boltzmann constant k, with units J/K, and the molar gas constant R, with units J/(K mol), when using molar quantities; $N_A k = R$.

Table 2.1 ΔH_f and ΔS_f for defect formation from ref. [5]. Schottky values concern two vacancies such that the energy per vacancy is similar for all three categories.

Defect type	Material	ΔH_f, kJ/mol	ΔS_f, J/(K mol)
Vacancies	Cu	123	21
	Ag	105	12
	Au	91	8.3
	Al	72	18
Schottky	NaCl	235	81
	NaI	193	63
	KCl	245	75
	KI	244	86
Frenkel	AgCl	140–150	45–101
	AgBr	109–124	55–101

one would estimate x_v as around 10^{-4}. If one considers that a 1 cm^3 block of Cu contains around 8.5×10^{22} atoms, the total number of defects ($\sim 10^{19}$) is considerable. It's worth emphasizing that thermodynamics gives the equilibrium number of defects ignoring kinetics—if a real sample has been rapidly cooled from high temperatures, the actual defect concentration might be much higher than the equilibrium value calculated at the temperature of the cooled sample. Defect concentration calculations are illustrated and expanded in the end-of-chapter problems.

2.4 Extrinsic Defects

Defects involving new chemical species are called **extrinsic defects**. For elemental solids, we've seen that foreign atoms can be substituted[6] for atoms of the host material to produce what can be described as doped systems at low levels of incorporation or alloys at higher levels. This is called substitutional disorder. Substitutional disorder is also common, and often deliberately targeted, in ionic compounds. As with intrinsic defects, it is important to consider charge balance. In an ionic oxide MO, one can envisage replacing M^{2+} with another 2+ metal to give a solid solution $M_{1-x}M'_xO$ (so-called **isovalent substitution**); a typical example might be $Co_{1-x}Mn_xO$ that is known to form a solid solution for all values of x. A common and technologically important group of materials based on isovalent cation substitution are luminescent phosphors that rely on doping optically active rare-earth atoms such as Nd and Eu into optically inactive hosts such as La_2O_2S. More examples of this phenomenon are given in Chapter 7.

---

[6] Loose language by soccer commentators means that there's often confusion about usage of the verb to substitute; indeed Fowler [Modern English Usage (1964) Oxford] amusingly describes the verb as a "treacherously double edged sword"! The correct usage is that if we "substitute A for B", B is removed and A is put in its place. "Beckham is substituted for Rooney", means Rooney leaves the pitch. If in doubt use the verb "replace".

In an **aliovalent substitution,** an atom is replaced by one with a different oxidation state. An example might be the substitution of Ca^{2+} for Zr^{4+} in ZrO_2. The introduction of a lower-valent ion means that the substituted site has a formal negative charge that must be compensated (Ca^{2+} has insufficient charge to compensate the negative charge of O^{2-} ions surrounding the former Zr^{4+} site). In this case, charge balance is predominantly achieved through the introduction of vacancies on the anion lattice. For every Zr^{4+} that is replaced by Ca^{2+} (charge difference of 2 units), a corresponding O^{2-} vacancy is required giving rise to a material with formula $Zr_{1-x}Ca_xO_{2-x}$. If one substituted Y^{3+} for Zr^{4+} (charge difference of 1), one oxide-ion vacancy for every two Y^{3+} ions would be required; $Zr_{1-x}Y_xO_{2-x/2}$. There is a variety of possible charge compensations involving substitution, vacancies, or interstitials; these are summarized in Table 2.2.

In semiconducting materials such as Si, dopants with either more or fewer valence electrons can be introduced: a phosphorus atom is able to donate an electron to the conduction band and is said to be a **donor**; an aluminum atom accepts electrons from the valence band (or equivalently adds holes to the valence band) and is said to be an **acceptor**. We'll explore the case of Li^+ acceptor and Al^{3+} donor doping in NiO in detail in Chapter 3.

In cases where the elements present have chemically accessible lower or higher oxidation states, it is possible to achieve charge balance via **redox compensation** without the introduction of vacancies. If we take the example of La_2CuO_4, which contains Cu in the +2 oxidation state, it is possible to replace 7.5% of the available La^{3+} sites with Sr^{2+} to produce $La_{1.85}Sr_{0.15}CuO_4$. We can understand the charge balance in this material in terms of copper changing its oxidation state from +2 to +2.15 on average. A similar Cu oxidation can also be achieved by incorporating extra oxygen in interstitial sites of La_2CuO_4, which produces $La_2CuO_{4.075}$. Again, we can understand the charge balance by considering oxidation of Cu from +2 to +2.15 manifested as a mixture of $0.85Cu^{2+}$ and $0.15Cu^{3+}$ on the Cu site. Both processes convert insulating antiferromagnetic La_2CuO_4 into a superconductor at low temperatures (see Chapter 12).

The range of different types of cation substitutions, combined with the range of different ways in which charge can be compensated, leads to the development of remarkably complex materials from chemically simple starting points. Even for a binary oxide such as NiO, where oxidation/reduction can lead to a nonstoichiometric formula, there are many possibilities. On oxidation, one can incorporate excess oxygen or create Ni vacancies; on reduction, create

Table 2.2 Charge compensation mechanisms in cation-doped materials with fixed integer oxidation states. Symbol $\square$ represents a vacant site.

Dopant charge	Compensated by	Host	Dopant	Substituted material
Higher positive	Cation vacancy	NaCl	Ca^{2+}	$Na_{1-2x}Ca_x\square_xCl$
Higher positive	Anion addition	CaF_2	Y^{3+}	$Ca_{1-x}Y_xF_{2+x}$
Lower positive	Cation addition	SiO_2	Al^{3+}	$Li_xSi_{1-x}Al_xO_2$
Lower positive	Anion vacancy	ZrO_2	Ca^{2+}	$Zr_{1-x}Ca_xO_{2-x}\square_x$
Any	Double substitution	$CaAl_2Si_2O_8$	Na^+	$(Ca_{1-x}Na_x)(Al_{2-x}Si_{2+x})O_8$

Table 2.3 Redox compensations in materials (oxidation states can change).

Redox-active metal	Defect or dopant	Examples	Importance
Oxidized	Cation vacancy	$Fe_{1-x}O$ $LiCoO_2 \rightarrow Li_{1-x}CoO_2$	Vacancy clustering Battery material
Reduced	Cation interstitial	$TiS_2 \rightarrow Li_xTiS_2$ $WO_3 \rightarrow Na_xWO_3$	Battery material Electrochromic
Oxidized	Anion interstitial	$La_2CuO_4 \rightarrow La_2CuO_{4+x}$ $UO_2 \rightarrow UO_{2+x}$	Superconductor Structural evolution
Reduced	Anion vacancy	$YBa_2Cu_3O_7 \rightarrow YBa_2Cu_3O_{7-x}$ $WO_3 \rightarrow WO_{3-x}$	Superconductor Crystallographic shear
Oxidized	Lower-valent cation	$La_2CuO_4 \rightarrow La_{2-x}Sr_xCuO_4$ $LaMnO_3 \rightarrow La_{1-x}Ca_xMnO_3$	Superconductor Magnetoresistance
Reduced	Higher-valent cation	$Nd_2CuO_4 \rightarrow Nd_{2-x}Ce_xCuO_4$ $CaMnO_3 \rightarrow Ca_{1-x}La_xMnO_3$	Superconductor Magnetoresistance

oxygen vacancies or incorporate excess Ni. The ways how to treat these possibilities *in combination* are explored quantitatively in Chapter 3.1. Table 2.3 summarizes some of the possible substitutional mechanisms and gives examples of materials in each category that will be discussed in later chapters.

Just as for cations, substitutions for anions are possible, and there is a similar range of charge-compensation mechanisms. Solid solutions based on anion substitution are generally less common than those based on cation substitution. This is principally because the range of anions of similar size/charge is smaller than the range of metals. However, in later chapters we will meet important materials that can be understood in terms of substitution of O^{2-} for F^-, N^{3-}, or S^{2-} ions.

2.5   Solid Solutions and Vegard's Law

At low levels of substitution and for certain element combinations, substitutions will occur randomly throughout a crystal structure leading to a **solid solution**. In many cases, the properties of solid solutions evolve smoothly as one changes the degree of substitution. This enables fine tuning of important parameters. As an example, the unit-cell parameters and volume of a material often vary smoothly from that of the host (e.g. AY) towards those of the substitutent (BY), as the degree of substitution x in $A_{1-x}B_xY$ is increased. Such a material is said to follow **Vegard's law** (Figure 2.5); for the unit-cell parameter a as an example:

$$a(x) = xa_{BY} + (1-x)a_{AY} \tag{2.7}$$

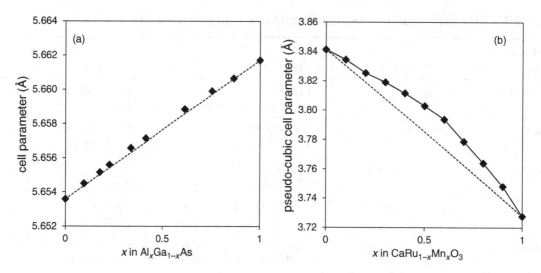

Figure 2.5 (a) Unit-cell parameter of the $Al_xGa_{1-x}As$ thin films [6] and (b) pseudo-cubic cell parameter of $CaMn_xRu_{1-x}O_3$ [7]. Dotted lines show the Vegard's-law prediction.

One can then estimate the degree of substitution directly by determining the cell parameters and rearranging this expression to give $x = [a(x) - a_{AY}]/(a_{BY} - a_{AY})$.

Control of cell parameters via Vegard's law can be particularly important when trying to grow an epitaxial (lattice matching) layer of a material on top of a substrate. For example, a semiconductor laser emitting between 1.2 and 1.65 μm (i.e. within the transparency window of optic fibers) can be prepared by sandwiching $In_{1-x}Ga_xP_{1-y}As_y$ between n- and p-type InP. By adjusting x and y, it's possible to exactly match the cell parameter of the sandwich layer to that of InP. In addition, the band gaps of many semiconductors can also show a Vegard's-law dependence. While maintaining the same cell parameter, different combinations of x and y may have different band gaps allowing control over the device's properties. For a system such as this, our simple form of Vegard's law must take account of each component present and becomes:

$$a(x,y) = xy\, a_{GaAs} + x(1 - y)\, a_{GaP} + (1 - x)y\, a_{InAs} + (1 - x)(1 - y)\, a_{InP} \qquad (2.8)$$

If one introduces appropriate cell parameters ($a_{GaAs} = 5.65$ Å, $a_{GaP} = 5.45$ Å, $a_{InAs} = 6.06$ Å, $a_{InP} = 5.87$ Å), the relationship between x and y to achieve lattice matching simplifies to:

$$x = \frac{y}{2.21 - 0.053y} \qquad (2.9)$$

In practice, Vegard's law is not always followed precisely. For some systems, intermediate members of a solid solution have smaller unit cells than predicted—a **negative deviation** from Vegard's law. Others have larger unit cells than predicted and show a **positive deviation**. This departure is due to the fact that atoms/ions can only be approximately treated as hard

Table 2.4 Examples of Kröger–Vink notation of point defects.

Symbol	Example	Charge number (effective charge)
v_{Na}'	Na^+ vacancy in $Na_{1-2x}Ca_xCl$	−1
$Ca_{Na}^{\bullet}$	Ca^{2+} on a Na^+ site in $Na_{1-2x}Ca_xCl$	+1
F_i'	F^- interstitial in $Ca_{1-x}Y_xF_{2+x}$	−1
$v_O^{\bullet\bullet}$	O^{2-} vacancy in $Zr_{1-x}Ca_xO_{2-x}$	+2

spheres and specific bonding interactions cause small deviations in volume. Even for $Al_xGa_{1-x}As$, the high-precision data of Figure 2.5a show a small but significant positive deviation from Vegard's law; the departure is such that a Vegard's-law derived Al content would be ~3% in error at $x = 0.5$. The data for $CaMn_xRu_{1-x}O_3$ in Figure 2.5b show a much more marked deviation. The authors have interpreted this as being due to the presence of Mn^{3+}/Ru^{5+} (as opposed to the expected isovalent Mn^{4+}/Ru^{4+} substitution) at intermediate compositions.[7] Departure from Vegard's law can therefore give a useful indication of a change in relevant properties of a material under investigation (see also the unit-cell parameter discontinuity in $YBa_2Cu_3O_{7-x}$ upon entering the superconducting regime in Chapter 12). Many materials also show a maximum range of solid solution, beyond which it is impossible to perform further substitution, a **solid-solubility limit**. Departure from Vegard's law can reveal when this limit is reached.

Finally, as we will again see in Chapters 3 and 4, the ease of forming solid solutions is often temperature-dependent, and, as a material is cooled, a solid solution may separate (phase segregate) into its component phases. This can lead to islands of one phase surrounded by a matrix of the second.

2.6 Kröger–Vink Notation

When discussing defects in crystals it is often useful to adopt a shorthand notation for the various types of defects encountered. The most common is the Kröger–Vink notation in which a symbol of the form A_C^B is used. Position A is the chemical symbol of the atom concerned or v for vacancy. Superscript B indicates the effective charge of the defect (remember that a cation vacancy leaves a local excess of negative charge at the defect site; an anion vacancy an excess of positive charge). A dot is used for a positively charged defect (i.e. $A_C^{\bullet}$) and a prime for a negatively charged defect (i.e. A_C'). Finally, the subscript C identifies the site in the crystal on which the defect occurs via the symbol of the host atom or a symbol i if the site is interstitial.[8] Typical examples are given in Table 2.4 using materials introduced in Table 2.2.

---

[7] Note that the y-axis scales on Figures 2.5a/b are very different: the overall percentage change in cell parameter is ~0.1% in (a) and ~3% in (b). In (b) we plot $(volume/4)^{1/3}$ to give a pseudo-cubic-cell parameter related to the simple perovskite cell.

[8] One may also use an s for a surface site (it is often convenient to think of the atoms that are missing in, e.g. a Schottky defect as having migrated to the surface of the crystal).

2.7   Line Defects in Metals

The defects considered up to now have all been randomly distributed, isolated point defects where the probability of finding a defect at a given point in a structure is largely independent of whether nearby sites contain a defect. Energetically favorable defect clustering means, however, that it is common to find aggregated and extended defects in materials. It is this type of defect, which, for example, gives metals their characteristic properties of malleability and ductility and allows them to be worked into useful forms. We'll consider two basic types of **line defects** in metals; edge dislocations and screw dislocations.

2.7.1   Edge Dislocations

An **edge dislocation** can be envisaged as shown in Figure 2.6a illustrating a 2D slice through a crystal structure. An extra plane of atoms (running perpendicular to the plane of the paper) has been inserted into the top half of the crystal. This gives rise to a dislocation line (again perpendicular to the plane of the paper) shown by a star. This defect is similar to the interstitial defect of Figure 2.1b but extends over a 2D plane of the structure.

The existence of edge dislocations helps explain why metals can be relatively easily deformed without cracking or failure. If one applies a force perpendicular to the top half of the crystal of Figure 2.6, then the "extra" plane of atoms can move to alleviate the imposed stress. This can occur in a series of steps in which a single line of bonds breaks and then reforms in the crystal such that the dislocation line moves across the crystal by one lattice spacing at a time. The plane along which the dislocation line moves is called the **slip plane** (shown as a dashed line). In each step only a small number of chemical bonds need to be broken, which helps rationalize the ease with which metals can be deformed. Alternative models of deformation would require the simultaneous breakage of entire planes of metal–metal bonds and would be prohibitively costly in terms of energy. The importance of dislocation motion also helps explain why minor impurities can have a large influence on the mechanical properties of metals. Impurities can trap dislocations at specific locations in a sample and prevent their motion. This process is known as **pinning**.

The description of deformations of materials via the slippage of atomic planes also rationalizes why ceramic materials are usually harder, more brittle, and more prone to cleavage than metals. The delocalized, isotropic nature of metallic bonding allows easy slip-plane formation and slippage as opposed to covalent solids; in an ionic solid slippage of planes by one atomic unit is unfavorable due to the repulsive interactions between ions of like charge.

2.7.2   Screw Dislocations

The second common type of line defect in metals is a so-called **screw dislocation**. One can imagine this as a "spiral staircase" or "corkscrew" arrangement formed by slicing a crystal to

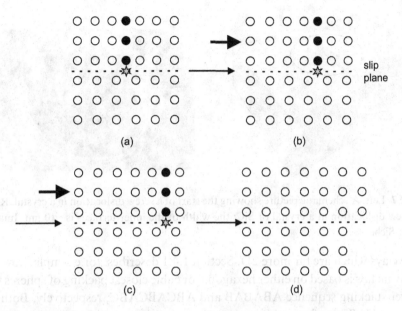

Figure 2.6 An edge dislocation due to an extra plane of atoms (a). Application of a shear stress moves the edge dislocation through the crystal (b, c) eventually forming a step on the crystal surface (d). Such processes occur as metals are worked.

its center and sliding adjacent layers of atoms by one atomic plane as shown schematically in Figure 2.7. As subsequent atoms attach to the growing crystal surface, they will form a spiral arrangement as shown in the experimental image of SiC on the right of the figure. This is a line defect, and the line of the dislocation runs up the center of the spiral. Screw dislocations can move through a crystal under applied stress in a similar manner to edge dislocations.

In practice, the dislocations in metals may appear more complex than simple edge or screw dislocations. However, most common dislocations can be described as a combination of these two basic types.

2.8 Planar Defects in Materials

Many defects in materials involve planes of atoms as opposed to the point and line defects described above. Three categories of **planar defects** found in technologically important materials are stacking faults, twins, and antiphase boundaries.

2.8.1 Stacking Faults

Many materials can be described in terms of layers of atoms or groups of atoms that stack along a certain direction in the structure. These range from metals, which are essentially 3D in their properties, to materials such as graphite, layered transition-metal sulfides, and

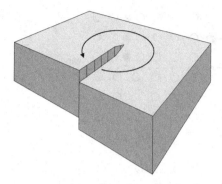

Figure 2.7 Left: A schematic picture showing the start of a screw dislocation in a crystal. Right: An image of a screw dislocation in a crystal of SiC; the width of the inner terraces is ~10 μm. Image courtesy of Dietmar Siche.

silicate clays which are far more 2D. Section 1.4.1 describes, for example, how the structure of many metals is based on either hexagonal or cubic closest packing of spheres that repeat in the layer-stacking sequence ABABAB and ABCABCABC, respectively. Both fill the same percentage (74.05%) of available space and are close in energy. In fact, some materials switch from one sequence to another under different conditions (for example Co from AB to ABC upon heating above 690 K, or Al from ABC to AB under very high pressures). Such materials may also be prone to faults in the stacking sequence. Stacking faults also occur easily in layered materials such as those related to the $CdCl_2$ or CdI_2 structure types (Figure 1.28). In some cases, the forces holding together adjacent layers can be so weak or non-specific that although materials are highly ordered in two dimensions, adjacent layers along the stacking direction rapidly lose registry. This type of disorder, **turbostratic disorder**, is commonly found in graphite, layered clays, and molecular intercalates (Chapter 13).

While stacking faults occur at random in some materials, in others they give rise to ordered structures. Ordered variants differing by their stacking sequences are referred to as **polytypes**, and a schematic example is shown in Figure 2.8. Polytypism (formation of several polytypes) is commonly observed in the layered halides, structurally related transition-metal dichalcogenides, SiC, and many other systems. It is a subset of the more general phenomenon of **polymorphism**: the existence of more than one structural form of a given material.

2.8.2 Twinning

Crystallographic twinning is an extended defect that is important in areas as diverse as the mechanical properties of metals, the optical properties of crystals, and the performance of piezoelectrics and ferroelectrics (Chapter 8). A twinned crystal is defined as an intergrowth of two or more individual crystals of the same species, in which the different portions are related by a symmetry operation that doesn't belong to the point group of the crystal. Each individual is called a **twin component** or a **twin domain** and the symmetry operation relating them a **twin**

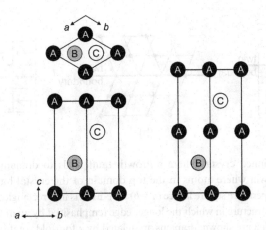

Figure 2.8 Polytypism of an element. Left: ccp, layer sequence ABC. Right: ABAC.

law. The interface between domains is called a **twin boundary** or **domain boundary**, and these can be several atomic layers thick if there is significant structural rearrangement required across the boundary. Figure 2.9 contains some examples. Twinning can be detected in a number of ways. Sometimes it leads to crystals with re-entrant[9] angles between faces. In translucent crystals, it can be seen using cross polarizers in a polarization microscope as different domains show light extinction at different angles. It is also apparent in single-crystal diffraction experiments, since each domain will give rise to its own diffraction pattern; *hkl* reflections from different domains may or may not overlap in reciprocal space depending on the twin law. Twinning can be an unwanted complication in these experiments and is one of the reasons why powder-diffraction methods are often used for structural work.

There are three **twin categories**; growth twins, deformation (glide) twins, and transformation twins. **Growth twins** can arise during the formation of a crystal if it grows from multiple crystallization nuclei. In Figure 2.9a, we can imagine that two domains have grown from different nucleation sites at the twin boundary. **Deformation twins** occur when a shear stress is applied to a crystal and causes each plane of atoms to move by a fraction of a unit cell relative to the layer below it. Major changes in the crystal's macroscopic shape arise from small individual atomic displacements, and the process can occur very rapidly (Figure 2.9b). This type of twinning is important in steel-production, mineralogy, and in the superelastic and shape-memory alloys discussed in Box 2.2.

In functional materials, twinning is most commonly encountered when the shape of the unit cell has a higher point symmetry than the atomic contents.[10] If one imagines producing a crystal by stacking together such unit cells (like building a wall out of parallelepiped bricks), it is easy to make a mistake and start placing bricks upside down without disrupting the overall stacking.

---

[9] A re-entrant angle in an irregular polyhedron is an angle inside that polyhedron, which is greater than 180°, such as on the right-hand side of the crystallite shown in Figure 2.9b.

[10] Or close to higher symmetry. For example, a monoclinic cell with a $\beta = 90.05°$ approximates an orthorhombic cell.

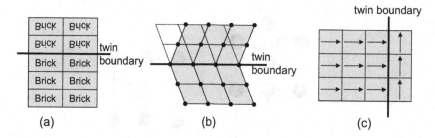

Figure 2.9 (a) A twinned crystal where a growth fault leads to domains related by a mirror plane. (b) A deformation twin where atoms in the top domain of the crystal have been displaced by a shear stress. (c) A high-temperature square lattice ($a = b$) that distorts into rectangles ($a \approx b$) at low temperature will form a twin-domain structure in which the longer edge (emphasized by an arrow) can be aligned in different directions, two of which are shown; domains are related by a fourfold rotation of the square lattice.

Figure 2.9a shows a 2D example where the unit cell is a rectangle and thus has two mirror planes (*mm* symmetry) while the cell contents (here the word "Brick") have lower symmetry (here no symmetry). When a "mistake" is made, a new twin component is formed, related to the first by one of these mirror planes. This process is only likely to occur if the energy penalty for placing the unit cell in an alternative orientation is low. One situation where this happens is when a structure undergoes a phase transition from a high-symmetry to a related low-symmetry form.[11] If only minor structural changes occur, the "lost" symmetry element is likely to act as a twin law giving rise to a **transformation twin**.[12] A common example is when a material is cubic ($a = b = c$) at high temperature but distorts to a lower symmetry tetragonal ($a = b \lesssim c$) structure on cooling. In this process, it is likely that different regions of the crystal will have their longer c unit-cell axis pointed in different directions in 3D space, giving rise to different domains (a 2D simplification is shown in Figure 2.9c where the long cell direction is indicated by an arrow). In fact, doing this will lower the strain across the entire crystal, such that multidomain crystals usually form on cooling.

In a transformation twin, the domains are formally related by a rotational symmetry element that is present in the high-symmetry phase but absent in the low-symmetry phase. They are therefore associated with translationengleiche transitions (see Appendix B) in which the translational symmetry is retained (all lattice points are kept) but a portion of the rotational symmetry is lost. The **number of twin variants** (different twin-domain orientations that will form) can be predicted from the ratio of the number of rotational symmetry operations[13] per lattice point in the two space groups in question. The ratio is called the *index*

---

[11] There are numerous examples discussed in Chapter 4 and throughout this book.

[12] The spatial arrangement of atoms originally related by symmetry is regenerated across the twin boundary when the symmetry operation becomes a twin law.

[13] These symmetry operations are in a numbered list for each space-group entry under the heading "Symmetry operations" in the International Tables for Crystallography, Volume A (see Appendix B), one for each point of the general position. In centered space groups, there is a set associated with each of the centering translations (i.e. with each lattice point per cell). The number of individual domains can be much higher than the number of variants obtained from the symmetry-operations ratio per lattice point.

Box 2.2 Materials Spotlight: Shape-memory alloys

Shape-memory alloys are a remarkable family of structural materials that can be mechanically deformed at low temperature but will regain their original shape on gentle heating. They thus retain a "memory" of their initial shape. What's the origin of this memory effect? On cooling, some materials undergo diffusionless transitions (each atom only moves a small distance relative to its neighbors) which lead to twinning of the type shown in Figure 2.9b. These are commonly called **martensitic transformations** after the well-known example that occurs when austenite (fcc) transforms irreversibly to lower-symmetry martensite (body-centered tetragonal) during the quenching of steel. Shape-memory alloys undergo reversible martensitic transitions in which the details of the low-temperature domain structure control the shape of the crystal. This is shown in the figure below. If a shape-memory alloy is cooled through the phase transition while restricted to a certain physical shape (step 1 in the figure), a twin structure with domains of optimal size and arrangement to fit this particular shape will develop. When this cooled form is mechanically deformed (step 2), the twin-domain boundaries will move into a new interlocked metastable position. In it, the twin domain that best compensates the applied stress has grown at the expense of the others, facilitating the deformation. When that form is heated (step 3) above the temperature of the low- to high-symmetry phase transition, the domain structure, and hence the mechanically deformed shape, will be lost. On subsequent cooling (step 4) the original twin structure "stored" in the sample during its initial manufacture, and hence its shape, is reformed.

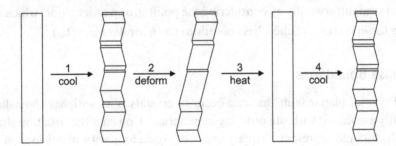

The best-known shape-memory alloy is a 1:1 alloy of Ni and Ti, commonly known as nitinol (<u>ni</u>ckel <u>ti</u>tanium <u>N</u>aval <u>O</u>rdnance <u>L</u>aboratory); other commercial materials are based on Cu-rich CuZnAl and CuNiAl alloys. NiTi is cubic at high temperature but on cooling undergoes a phase transition to structure with a small monoclinic distortion. This martensitic transformation starts at around 60 °C on cooling and 71 °C on warming, but slight changes in composition allow the transition temperature to be controlled between −50 °C and 100 °C. For Cu-based systems, transition temperatures from −180 °C to 200 °C can be achieved. Annealing temperatures of 500–800 °C are typically used to "store" a specific shape in a sample.

Shape-memory alloys have a range of potential applications. Biomedical implants can be prepared, which are inserted into the body in a compressed, deformed state and unfold as they warm to body temperature. Shape-memory fittings to join piping together are available commercially; these are made as tubes, slightly smaller than the pipes, and then deformed at

Box 2.2 (cont.)

low temperature to fit over the two pipes to be joined. On gentle heating, they regain their original dimensions thus forming a tight seal. So-called two-way shape-memory alloys that "remember" the shape of both the low- and high-temperature form can also be prepared. This is achieved by using defects, which create local stresses in a material, to control the places where individual twin domains form. Typing "shape memory movie" into a search engine will provide links to a number of on-line movies that give dramatic illustrations of this effect.

A second property of these alloys—their high **pseudoelasticity** or **superelasticity**—is exploited in eyeglass frames that can withstand distortions without damage. Here, one operates just above the transformation temperature of the material so that the martensitic transition occurs on application of stress, allowing significant mechanical distortion. When the stress is removed, the material reverts to the austenite form and the original shape is regained.

of the translationengleiche subgroup (Appendix B) of the two groups. For example, if a cubic material with space group $Pm\bar{3}m$ (#221) undergoes a phase transition to a tetragonal structure with space group $P4/mmm$ (#123), the former has 48 symmetry operations (or general points), the latter 16. We therefore expect 48/16 = 3 variants. This result can also be obtained straightforwardly by considering the point-group order alone, which again changes from 48 for $m\bar{3}m$ (O_h in Schönfliess notation) to 16 for $4/mmm$ (D_{4h}).[14]

2.8.3 Antiphase Boundaries

A third type of planar fault that can occur in crystals is an **antiphase boundary**. These are frequently associated with site ordering in materials. Consider the situation shown in Figure 2.10, which could represent a square array of oxide ions with metal ions of two different charges (e.g. 1+ and 3+) sitting in the four-coordinate sites. At high temperature, the two types of metal ions are disordered over their common crystallographic site in the structure. On cooling, it may become thermodynamically favorable for them to order in, say, a checkerboard pattern as in the bottom left of Figure 2.10, where cations of low charge are surrounded by high charge and vice versa. Such an arrangement minimizes electrostatic repulsions as it maximizes the separation between the 3+ ions.

The second column of Figure 2.10 represents a situation where an antiphase boundary was formed by two ordered regions meeting out of phase as the crystal cools from its surface. The

---

[14] The point group contains only rotational symmetry operations that leave at least one point unchanged. The order is the number of operations it contains. We include the Schönflies notation of the point groups (Appendix A) as it's familiar to most chemists and most inorganic texts will contain point-group character tables that state the group order.

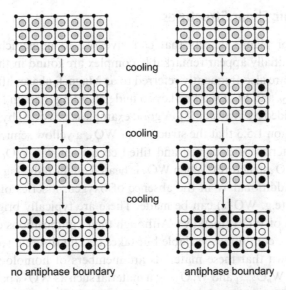

Figure 2.10 Formation of antiphase boundaries. A hypothetical array of oxide anions (small circles) contains an equimolar mixture of 1+ and 3+ metal cations that are disordered at high temperature (gray circles) and order (white/black circles) on cooling from two hypothetical seeds at the opposite crystal surfaces; 50% meet in phase (on the left), 50% out of phase (on the right). The ordered structure here has a quadrupled unit cell reflecting a loss of translational symmetry.

two components are related by a symmetry element present in the high-symmetry structure but missing in the low-symmetry structure. In contrast to twinning, antiphase boundaries form when a subset of translational symmetry elements is lost rather than rotational elements. Antiphase boundaries are therefore associated with klassengleiche subgroups.[15] Clearly, the domain-boundary structure on the bottom right is energetically unfavorable with respect to that in the bottom left, but considerable atomic rearrangement would be required to "heal" the fault. The presence and number of antiphase boundaries will depend strongly on the thermal history of a sample. Real examples where antiphase boundaries occur include M-site ordering in double perovskites such as $A_2MM'O_6$ ($AM_{0.5}M'_{0.5}O_3$) that contain a mixture of differently charged M and M' ions (e.g. 2+/4+ or 1+/5+) on the octahedral site. Similar effects are found in metal alloys such as FePt that has a cubic structure at high temperature, with Fe and Pt disordered over all sites of an fcc lattice, but orders on cooling to a tetragonal structure with alternate layers of Fe and Pt. Faults occur when different ordered domains meet such that two layers of Fe or two layers of Pt are adjacent or the layers grow in different orientations. The ordered material is of interest in that it has high magnetocrystalline anisotropy making it potentially useful for magnetic data storage.

[15] See Appendix B. Klassengleiche subgroups involve a partial loss of translational symmetry (for example, loss of some lattice points due to formation of a supercell or a loss of centering.). The point-group symmetry or crystal class is retained.

2.8.4 Crystallographic Shear Structures

A final family of planar defects that can give rise to nonstoichiometric formulae and structures that initially appear remarkably complex are found in the **crystallographic shear structures**. They are also frequently referred to as **Magnéli phases** after A. Magnéli who first described Mo_9O_{26} and $W_{20}O_{58}$. The key to understanding them in terms of extended planar defects was provided by Wadsley [8]. A good example is provided by the tungsten oxides. We have seen in Section 1.5.3 that the structure of WO_3, a yellow semiconducting material, can be described in terms of distorted and tilted corner-sharing WO_6 octahedra (a distorted variant of the ReO_3 structure). When WO_3 is heated under reducing atmospheres, or when it is reacted with additional W in the absence of oxygen, a series of compounds of general formula WO_{3-x} (e.g. $WO_{2.9}$) can be made. These are typically bright blue in color due to partial reduction of W(VI) to W(V). Although their compositions can be approximated as WO_{3-x}, x is not a continuous variable but takes certain discrete values. Careful structural studies have shown that these materials are members of homologous series with general formulae such as W_nO_{3n-1} and W_nO_{3n-2}; a material such as $WO_{2.9}$ being better formulated as $W_{20}O_{58}$, the $n = 20$ member of the second series.

The structural origin of these general formulae can be understood with respect to Figure 2.11. If one imagines that partial reduction of WO_3 initially removes an entire plane of oxygen atoms, one would generate the structure in the middle of Figure 2.11, in which W atoms on either side of the missing plane are only five-coordinate. If each W in the right-hand five-coordinate plane is shifted by half a unit cell parallel to the two in-plane axes of the original unit-cell edge of pseudo-cubic WO_3, the sixfold coordination of each W is regained. We see a double chain in Figure 2.11 of octahedra that share two edges in addition to sharing their remaining three corners. In the example of Figure 2.11, the octahedra translate solely in the plane of the figure, but we could also imagine octahedra moving out of the plane to share apical edges instead of the equatorial ones.

In reality, many different crystallographic shear planes are possible for WO_3. Figure 2.12 shows some of them. Figure 2.12b shows the (101) fault plane. This corresponds to a twin where all oxygens remain coordinated to two octahedral W atoms, and the formula remains WO_3. As the plane of defects rotates clockwise in Figure 2.12 from (102) to (103) to (001) $\equiv$ (10∞), the 2D projection shows a line of blocks of four edge-sharing octahedra in part (c), a line of blocks of six octahedra in part (d) and a continuous zig-zag belt of octahedra sharing two edges in (e), with a corresponding decrease in the oxygen content as the oxygen connectivity increases. Irregular (102) Wadsley defects have been observed for WO_{3-x} with $x = 0.002$. By $x = 0.05$, these defect planes become ordered, giving a W_nO_{3n-1} family of phases ($n = 20$ for $x = 0.05$) with the integer n decreasing as the defect planes get closer on lowering oxygen content (increasing x). Ordered structures of this type are known down to $n \approx 12$ of $W_{12}O_{35}$ ($WO_{2.92}$ or $x = 0.08$). Beyond $x \approx 0.08$, the spacing of (102) defects becomes too close, and (103) defects are observed instead, leading to W_nO_{3n-2} materials that are well characterized for at

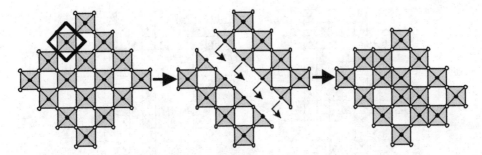

Figure 2.11 A section of the structure of WO_3 showing a projection of corner-sharing octahedra. Removal of a plane of oxygen atoms followed by a displacement along the shear plane recreates octahedral coordination around each W. The pseudo-cubic unit cell of WO_3 is shown as a dark box.

least $n = 26$–18 ($WO_{2.92}$ to $WO_{2.89}$). As might be expected, similar structures are also observed when a small amount of Nb^{5+} is substituted for W^{6+}.

Figure 2.12f shows an extreme case with regularly spaced defects corresponding to a formula of M_2O_5; the ideal[16] structure of V_2O_5. The structure of the mixed-anion compound Nb_3O_7F can similarly be described in terms of crystallographic shear planes of the type shown in Figure 2.12e separated by a single plane of fully corner-shared octahedra. If one allows shear planes in two perpendicular directions, a large family of closely related structures results. An excellent description of the various possibilities is given in the text by Hyde and Andersson (see Further Reading).

A similar phenomenon occurs in the rutile TiO_2 structure leading to a variety of compositions between Ti_3O_5 ($TiO_{1.67}$) and TiO_2. The shear planes are harder to depict as there's no convenient projection to draw, but the resulting structures contain infinite 2D slabs of rutile separated by regions with a portion of face-sharing octahedra. Depending on the shear plane involved, materials with formula Ti_nO_{2n-p} result: $TiO_{1.75}$ to $TiO_{1.89}$ have $4 < n < 9$ and $p = 1$; $TiO_{1.89}$ to $TiO_{1.93}$ have $9 < n/p < 16$ and $p > 1$; $TiO_{1.93}$ to $TiO_{1.98}$ have $16 < n < 40$ and $p = 1$. It's also common (and unsurprising) to find intergrowths of different slabs in electron microscopy images of samples rapidly cooled from high temperature.

2.9 Gross Nonstoichiometry and Defect Ordering

In Section 2.3, we investigated the number of vacancies one might expect to find in a simple ionic material with fixed oxidation states. For compounds with variable metal oxidation states, defect

[16] V_2O_5 is normally described as layers of edge- and corner-sharing VO_5 square pyramids. However, an oxygen from an adjacent layer makes up the sixth coordination site of the V such that it can be described as a highly distorted octahedron.

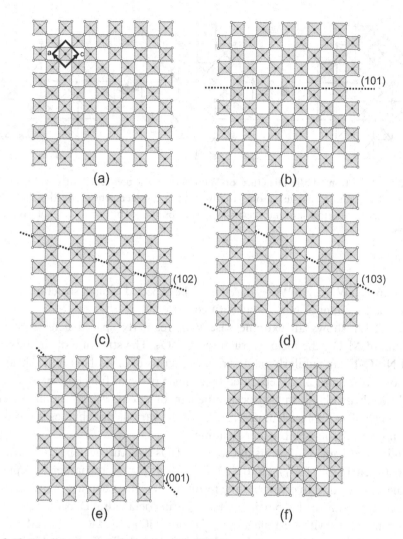

Figure 2.12 Possible planar defects for an ideal WO$_3$-type structure. (a) The WO$_3$ structure viewed down the b axis; the unit cell is bold. (b) A twin boundary. (c–e) Crystallographic shear planes with loss of oxygen from the structure. (f) The idealized structure of V$_2$O$_5$.

levels can be sufficiently high that defect interactions and ordering become important—this can be remarkably complex. The different TiO$_x$ ($0 \leq x \leq 2$) phases that form when Ti is progressively oxidized give several examples [9].

Ti itself is an hcp metal. On oxidation, O^{2-} ions initially adopt interstitial sites in the hcp structure at random before ordering at higher concentrations. At composition Ti$_6$O, oxygen fills one-third of the octahedral holes in half of the hole planes, at Ti$_3$O, two-thirds of them, and at Ti$_2$O all of them, with anti-CdI$_2$-type ordering adopted. The structure and composition probably evolve continuously over much of this range, and Vegard's law is followed up to Ti_3O. Early literature [10] suggested that, around $TiO_{0.67-0.75}$, a structure related to ε-TaN forms but with high O deficiency; this can equally be described as O interstitials in the ω-Ti structure. This same structure type has more recently been reported [11] for a TiO composition synthesized in a Bi flux so there is some doubt about the earlier work. Around $x = 1$, TiO more commonly forms with an NaCl-type structure and a significant number of defects. On the O-deficient side ($TiO_{0.64}$), up to ~36% of oxygen sites are vacant; on the O-rich side ($TiO_{1.26}$), ~23 % of Ti sites and some O sites are vacant. At the 1:1 composition, ~15 % of both sites are vacant such that we might write the formula as $Ti_{0.85}O_{0.85}$. Note that in a case like this, neither elemental analysis nor site occupancies refined from diffraction data (which merely measure relative scattering from cation and anion sites) would reveal the presence of defects; density measurements of the type explored in the end-of-chapter problems are one way of detecting them. At low temperatures, vacancies in TiO order to give a monoclinic structure, in which a regular arrangement of one in six cation and anion sites of the NaCl-type structure is empty. Similarly, Ti_4O_5 ($TiO_{1.2}$) orders at low temperature to a structure in which four out of five cation sites are occupied, and O sites are essentially fully occupied. At higher oxygen contents, one reaches Ti_2O_3 (with the corundum structure), Ti_3O_5 (several polymorphs)[17], Ti_4O_7, and then the crystallographic shear-plane series Ti_nO_{2n-1} (with $n = 5, 6, \ldots$) discussed in Section 2.8.4.

Iron monoxide (wüstite) provides an example where nonstoichiometry and clustering of defects add significant complexity to its ideally NaCl-type structure. In fact, the homogeneity range of FeO does not include the stoichiometric (or integer-valence; see Chapter 3) phase. At 1350 °C, $Fe_{1-x}O$ with $0.06 \leq x \leq 0.16$ can be prepared; at lower temperatures, the range of x is smaller. These materials are thermodynamically stable above ~570 °C (below this Fe and Fe_3O_4 are stable) and there has been considerable investigation into the defect structures of quenched samples. The principal defect present is a vacancy on the regular octahedral Fe^{2+} site (v_{Fe}''; V for brevity in this FeO case), which is charge-compensated by a small number of interstitial Fe^{3+} at nearby tetrahedral holes ($Fe_i^{\cdots}$ or T) like those occupied by Zn in ZnS, sphalerite, Figure 1.32.[18] Due to their effective charge, the v_{Fe}'' and $Fe_i^{\cdots}$ defects attract each other such that locally the Fe^{3+} interstitial is surrounded by four Fe^{2+} vacancies—a V_4T cluster (Figure 2.13). What's more controversial is how these small vacancy clusters are arranged on a longer length scale to form larger clusters.[19]

---

[17] The structural chemistry of Ti_3O_5 is complex, and several polymorphs (α–δ, λ) exist with different arrangements of corner-, edge-, or face-sharing TiO_6 octahedra. The γ polymorph [S.-H. Hong, S. Asbrink, *Acta Crystallogr. Sect. B* **38** (1982), 2570; ICSD 35148] can be viewed as an $n = 3$ rutile shear structure.

[18] We shouldn't be surprised by this as a similar site is occupied by Fe^{3+} in Fe_3O_4 (an inverse spinel structure, i.e. a ccp of O^{2-} with Fe^{3+} in one-eighth of the tetrahedral holes and Fe^{2+}/Fe^{3+} in half the octahedral holes).

[19] If the clusters order in three dimensions, one would see extra superstructure peaks in diffraction patterns due to the larger unit cell involved. Such peaks are indeed observed experimentally, though for some compositions a commensurate superstructure with $a_{sup} = 2.5n \cdot a_{FeO}$ (n is an integer) is formed while at others the superstructure is incommensurate (Section 2.11) with $a_{sup} = (2.51-2.73)n \cdot a_{FeO}$—so-called P$''$ and P$'$ ordering respectively.

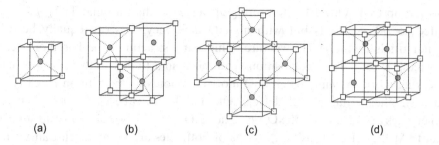

(a)   (b)   (c)   (d)

Figure 2.13 Iron-vacancy clusters proposed for $Fe_{1-x}O$: (a) V_4T cluster, (b) $V_{10}T_4$, (c) $V_{12}T_4$, and (d) $V_{13}T_4$. White squares represent Fe^{2+} vacancies and gray circles interstitial Fe^{3+} sites. The ccp O^{2-} that lie on the unmarked vertices of each cube are omitted for clarity. Note that each cube here has the size of one octant of the NaCl unit cell shown in Figure 1.26.

In one of the early studies on the defect structure of FeO [12], Koch and Cohen proposed the $V_{13}T_4$ cluster shown in Figure 2.13. Other clusters have been suggested, including the $V_{12}T_4$ cluster, which, when regularly spaced at a distance of $2.5a_{FeO}$ (i.e. close to experimentally observed superstructures), would lead to a composition of $Fe_{0.872}O$. Yet other workers have suggested the importance of $V_{10}T_4$ clusters. A single-crystal study [13] investigating both Bragg diffraction and diffuse scattering on a specific $Fe_{0.943}O$ sample suggests that $V_{13}T_4$ and $V_{16}T_5$ clusters are important. The experimental data were interpreted in terms of defect clusters lying on the vertices of a highly distorted cubic lattice with spacing $\sim 2.7a_{FeO}$, with around 50% of these cells containing a defect and 50% being defect-free. The defect and defect-free regions are not homogeneously distributed through the structure. Interestingly, if one considers the size of a $V_{13}T_4$ cluster and the fact that for charge balance (its charge number is $14-$), neighboring octahedral sites must contain a portion of Fe^{3+} ions, close packing of the overall units would require a cell of $\sim 2.5a_{FeO}$. A mixture of larger $V_{16}T_5$ and $V_{13}T_4$ would require a cell of around $2.7a_{FeO}$. These values are consistent with experimental observations and explored in the end-of-chapter problems. The precise structural picture of samples quenched to low temperatures is clearly complex, though it's clear that, at high temperature, isolated vacancies and smaller defect clusters become more important.

2.10 Incommensurate Structures

A number of materials exist that appear to have full occupancy of atomic sites in the structure, yet still possess a nonstoichiometric formula. One case occurs in the so-called **incommensurate structures**, which can't easily be described using the simple 3D concept of one unit cell and space group introduced in Chapter 1. An example is $Sn_{1.17}NbS_{3.17}$; one of a number of $A_{1+x}BX_{3+x}$ compounds that were long thought to have a simple ABX_3 composition. The origin of the structural complexity in this family is at its heart simple. The structure of $Sn_{1.17}NbS_{3.17}$ can be described as alternating layers of NaCl-like SnS and

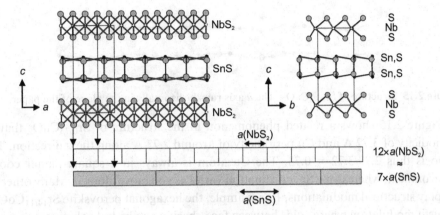

Figure 2.14 The intergrowth structure of $(SnS)_{1.17}NbS_2$. The right-hand view shows that b cell parameters of SnS and NbS_2 segments are identical such that the layers fit together in a simple fashion; the left-hand view shows that the a cell parameters have no simple relationship.

NbS_2 (Figure 2.14).[20] The relative sizes of the ideal unit cells of the SnS and NbS_2 portions of the structure are such that along b the cells fit together but along a they don't. In addition, there is no simple integer ratio of the a cell parameters such that, for example, three units of SnS would match up with two units of NbS_2; the two periodicities do not match.[21] It's therefore not possible to use a simple multiplied unit cell. The situation is rather like tiling a bathroom with rows of tiles of two different lengths—if one starts in one corner of the room and works along the wall, the gaps between the two rows of tiles may never perfectly align.

In fact, for $Sn_{1.17}NbS_{3.17}$, the cell parameters of the two portions ($a_{SnS} = 5.673$ Å; $a_{NbS_2} = 3.321$ Å) are such that the structure approximately matches up after seven SnS and twelve NbS_2 unit cells. The formula of the material can therefore be approximated as $(SnS)_{2\times7}$ $(NbS_2)_{12}$ (2×7 as each SnS cell contains two formula units) or $(SnS)_{1.17}NbS_2$ (equivalent to $Sn_{1.17}NbS_{3.17}$) and could be approximately described on a supercell with $a = 39.8$ Å. This description is, however, rather inelegant, and **incommensurately modulated structures** such as this are better described using the language of superspace groups and modulation functions (structures that are incommensurate in 3D can be conveniently described in a superspace of $(3 + n)D$ with n added modulation periodicities) [14], but this description is beyond the scope of this text. This approach may also be applied to **commensurately modulated structures**, those with periodicities that match a small multiple. The advantage is a low number of structural variables, as opposed to working with a large supercell of many atoms.

---

[20] The NbS_2 layers are similar to those found in CdI_2, but with trigonal prismatic rather than octahedral coordination of Nb.

[21] In the original publication [Meetsma et al., *Acta Crystallogr. Sect. A* **45** (1989), 285–291] the SnS portion of the structure was described with $a = 5.673$ Å, $b = 5.750$ Å, $c = 11.760$ Å in space group *C2mb* with Sn at 0 0.25 0.1335 and S at 0.476 0.25 0.0954 and the NbS_2 portion with $a = 3.321$ Å, $b = 5.752$ Å, $c = 11.763$ Å in space group *Cm2m* with Nb at 0 0 0 and S at 0 0.3335 0.1328. The sets of b and c cell parameters measured differ by less than the experimental uncertainty.

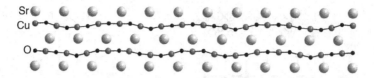

Figure 2.15 Structure of $Sr_{0.73}CuO_2$. The *a* axis runs horizontally in the plane of the paper.

Figure 2.15 shows a related phenomenon in the structure of $Sr_{0.73}CuO_2$ that has a Sr periodicity of 3.72 Å and Cu periodicity of around 2.73 Å along the *a* direction. The Sr:Cu ratio is thus $2.73/3.72 = 0.73$. The Cu atom is always in a square-planar coordination environment, whereas the Sr coordination varies as you move along *a*. Many other materials display structural modulations. For example, the hexagonal perovskite $Sr_{14/11}CoO_3$ [15] has a misfit modulation where voids between face-sharing octahedral columns accommodate an excess of the relatively small Sr ions compensated by a partial reduction of Co^{4+} to Co^{3+}; or metal alloys such as $Zn_{22}Li_6$ [16] show occupational modulation. Incommensurate structures have even been found for simple metals such as Ba and Bi under high pressure.

2.11 Infinitely Adaptive Structures

We'll finish this chapter with a short discussion of other structures that use simple building principles to accommodate nonstoichiometric and continually variable formulae. In fact, we've seen examples of this behavior already in the shear structures of Section 2.8.4, where, as the shear plane or plane spacing changes, a variety of closely related structures with variable composition evolve. Systems where any small compositional change leads to a structure that is unique, even if closely related to those of neighboring compositions, are called **infinitely adaptive**.[22] Another example is the family of Y^{3+} materials of composition $Y(O, F)_{2.13}$ to $Y(O, F)_{2.20}$ that can be formed by the reaction of appropriate quantities of YOF and YF_3. Within this composition range, the excess anions are accommodated in a practically infinite series of very closely related structures, each fluorite-related (as is YOF itself).

We can understand the structures by considering the fluorite MX_2 structure (Figure 1.31) in terms of square grids of X, with a checkerboard arrangement of M above and below the square grid, forming a slab of edge-sharing $XM_{4/4}$ tetrahedra (Figure 2.16, left). Fluorite itself can be built up by alternating this MX slab with square grids of X along a stacking axis, giving $MX + X = MX_2$ overall. In the $Y^{3+}O_{1-m}F_{1+2m}$[23] structures, alternate square grids of X are replaced by a layer of X anions with a triangular grid, which, for an ideal situation, is denser by a factor of $2/\sqrt{3} = 1.155$. The composition of this ideal situation

---

[22] Using the language of Chapter 4, each composition is a single phase. The compositions are so dense that there are effectively no two-phase regions, just one solid-solution range.

[23] The $MX_{2+\delta}$ composition can be expressed in various ways. The general formula can be expressed as $Y_xO_{x-y}F_{x+2y} \equiv YO_{1-m}F_{1+2m} \equiv Y(O_{1-m}F_m)F_{1+m}$; the final representation emphasizes that anion sites in the square grid are fully occupied.

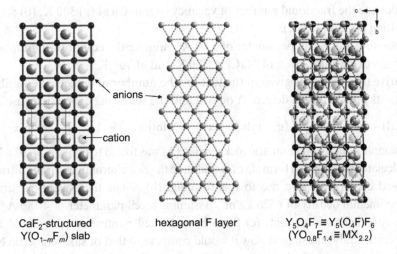

Figure 2.16 The structure of $Y_5O_4F_7$ (right) can be built up from Y(O, F) layers of the fluorite type alternating with hexagonal F anion layers with a denser packing than in the square grid. In the Y(O, F) slab, cation sites below the anion layer are shown in paler gray.

becomes $MX + 1.155X = MX_{2.155}$. Through small adjustments in the size of the triangular anion nets, m in $YO_{1-m}F_{1+2m}$ can be varied from 0.13 to 0.22 ($YO_{0.87}F_{1.26}$ to $YO_{0.78}F_{1.44}$). The way in which the square and triangular grids match up in the plane perpendicular to our imagined stacking direction can lead to large unit cells (analogous to the situation drawn in the lower part of Figure 2.14). In the $m = 0.2$ case of $Y_5O_4F_7$ shown on the right of Figure 2.16, six triangles line up with five squares; in $m = 0.167$ $Y_6O_5F_8$, seven triangles line up with six squares. These ideas are explored in Problem 2.27. In these phases, the Y sites have local coordination numbers varying between eight (like fluorite) and six, and this type of behavior is therefore most likely in materials where cations have flexible coordination environments.

Similar features occur in a range of structures built from different slabs or columns. Traditionally, the structures have been described using large unit cells, but the continuous range of structures within a given system means that they are again often better described as incommensurately modulated structures [17]. While we've only touched on a couple of examples, the complexity and range of structures possible in real materials should be apparent.

2.12 Problems

2.1 Given that Cu adopts a ccp structure with a cubic cell parameter of 3.615 Å, confirm that the equilibrium number of vacancies in a 1 cm^3 sample at 1000 °C is around 10^{19}. Note the need to convert between k, in J/K, and R, in J/(K mol), when using molar quantities: $N_A k = R$.

2.2 Calculate the fractional number of vacancy sites in Cu at (a) 300 K, (b) 800 K, and (c) its melting point (1357 K).

2.3 Assuming a unit-cell parameter of 5.62 Å, estimate the equilibrium number of Schottky defects in a 1 mm³ grain of NaCl at 300 K and at 700 K.

2.4 Derive the equations given in the text for the number of defects at equilibrium, n_{eq}, for Schottky and Frenkel defects. You will find the following expressions useful: for large x,

$$\ln(x!) \approx x\ln(x) - x, \frac{d}{dx}\left((c-x)\ln(c-x)\right) = -\ln(c-x) - 1, \text{ and } \frac{d}{dx}\left(x\ln(x)\right) = \ln(x) + 1.$$

2.5 A sample of nonstoichiometric nickel oxide (**A**) was found to contain 77.70% Ni by mass. (a) Calculate the empirical formula of **A** and state the two alternatives for the intrinsic defect that would on its own give rise to this formula. (b) **A** has the NaCl-type structure and an experimental density of 6526 kg/m³. Assuming a cell parameter of 4.180 Å, state which of the two defects is present. (c) State how the cell parameter of **A** could be determined experimentally and suggest how it would compare to that of stoichiometric NiO.

2.6 A brown sample of zinc oxide was found to have the hexagonal wurtzite structure with $a = b = 3.2495$ Å, $c = 5.2069$ Å ($\alpha = \beta = 90°$; $\gamma = 120°$). Chemical analysis gave 80.765% Zn by mass. Density measurements gave 5810 kg/m³. Determine the formula of the material and state whether it contains oxygen vacancies or interstitial metal atoms.

2.7 Suggest oxidation states for the metal ions in each of the following materials: (a) FeO, $Fe_{0.872}O$, Fe_3O_4, FeS_2; (b) FeTe, $Fe_{1.1}Te$; (c) LaOFeAs, $LaO_{0.9}F_{0.1}FeAs$; and (d) $YBaFe_2O_5$, $NdBaFe_2O_{5.5}$, and $NdBaCo_2O_6$.

2.8 Suggest oxidation states for the metal ions in each of the following materials: (a) TiS_2, $Li_{0.7}TiS_2$; (b) $LaMnO_3$, $La_{0.8}Sr_{0.2}MnO_3$, $La_{0.5}Ca_{0.5}MnO_3$; (c) La_2CuO_4, $La_{1.85}Ba_{0.15}CuO_4$, $La_2CuO_{4.075}$; and (d) $BaPbO_3$, $BaBiO_3$, $Ba_{0.6}K_{0.4}BiO_3$.

2.9 TiO with a 1:1 ratio of Ti:O was synthesized and found to have a NaCl-related structure with $a = 4.1831$ Å and an experimental density of 4927 kg/m³. Comment on these values.

2.10 The table below gives cell parameters and densities for a range of Ti_xO_y materials. Determine the defects present in each. From a graph of your results, estimate x for a 1:1 stoichiometric sample Ti_xO_x.

z in TiO_z	Cell (Å)	Density (g/cm³)
1.32	4.1608	4.713
1.12	4.1755	4.867
0.69	4.2212	4.992

2.11 NbO has an NaCl-related structure, a cell parameter of 4.21 Å and a density of 7.27 g/cm³. Calculate the percentage of vacant sites in the material. Draw a sketch of how the vacancies can be arranged in an ordered way so as to give square-planar coordination of Nb. What is the O coordination?

2.12 GaAs$_{1-x}$P$_x$ has a unit-cell parameter of 5.59 Å. Calculate x and estimate the band gap (E_g) of the material, given a_{GaAs} = 5.65 Å, E_g = 1.42 eV; a_{GaP} = 5.45 Å, E_g = 2.24 eV.

2.13 Using the cell parameters quoted in Section 2.5, confirm the form of Equation (2.9).

2.14 Using the cell-parameter information stated in Section 2.5, calculate the cell parameter expected for In$_{0.76}$Ga$_{0.24}$P$_{0.47}$As$_{0.53}$. Would this composition be lattice matched to any of the four possible end members: InP, GaP, InAs, or GaAs?

2.15 State the Kröger–Vink notation for the predominant defects in each of the materials in Table 2.2.

2.16 Suggest what type of twinning might occur in: (a) an orthorhombic structure with two cell edges approximately equal; (b) a structure with a monoclinc cell with β = 90.1°; (c) an orthorhombic structure with cell parameters a = 3.92 Å, b = 11.21 Å, c = 7.88 Å; and (d) a structure with a conventional primitive monoclinic unit cell with $a \approx c$.

2.17 At high temperatures, BaTiO$_3$ has the cubic perovskite structure. On cooling, it undergoes a series of phase transitions in which the Ti atom moves away from the center of the TiO$_6$ octahedron, and there are changes in the space group and unit-cell parameters causing the ferroelectric behavior discussed in Chapter 8. Determine the number of twin-domain variants (orientations) that could form in the first step when the cubic ($a \approx 4$ Å, space group $Pm\bar{3}m$) structure undergoes a phase transition to a tetragonal structure with space group $P4mm$ and cell parameters $a = b \approx c \approx 4$ Å, $\alpha = \beta = \gamma = 90°$.

2.18 At high temperature, Cu$_3$Au has a disordered fcc structure with space group $Fm\bar{3}m$. On cooling, an ordering transition occurs (similar to that in Figure 4.14) and the low-temperature structure has space group $Pm\bar{3}m$ with Au at 0 0 0 (Wyckoff site 1a) and Cu at ½ ½ 0 (3c). (a) Sketch the low- and high-temperature structures. (b) Is this a translationengleiche or klassengleiche transition? (b) State the number of domain variants in the low-temperature structure and whether they will be related by twin or antiphase boundaries.

2.19 Like tungsten oxides, molybdenum oxides show a variety of crystallographic shear structures. The unit cell for one of them is shown here:

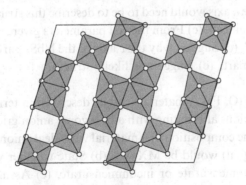

The octahedra also share corners with octahedra in identical layers above and below the plane of the paper. What is the composition of this compound? What is the oxidation state of Mo?

2.20 A mixture of WO_3 and MoO_3 was heated with a small excess of W and Mo metals to give a blue single-phase product containing 25.09% O by mass and a 1:1 W:Mo ratio. Determine the empirical formula of the product and suggest its structure type.

2.21 Assuming the presence of a $V_{12}T_4$ defect cluster of the type shown in Figure 2.13, it is possible to build an ordered superstructure for $Fe_{1-x}O$ using a $5a \times 5a \times 10a$ unit cell. This supercell would contain 16 vacancy clusters. Calculate the composition.

2.22 Assuming that the unit-cell parameter of $Fe_{1-x}O$ is given by Vegard's law as $a = 4.3325 - 0.4103x$ Å, calculate the experimental density for (a) the hypothetical stoichiometric FeO and (b) an Fe-deficient material with an Fe:O ratio of 1.075.

2.23 A sample of $Fe_{1-x}O$ has a measured density of 5491 kg/m^3 and cell parameter of 4.281 Å. Estimate the composition based on these two observations.

2.24 Sketch a V_4T cluster for the $Fe_{1-x}O$ structure showing both O and vacant Fe sites. Indicate the nearest shell of occupied octahedral Fe sites relative to the vacant site. State how many of these sites must on average be occupied by Fe^{3+} to achieve electroneutrality. Write down the Kröger–Vink notation of each defect present.

2.25 Assume that the defect structure of a crystal of $Fe_{1-x}O$ can be described in terms of $V_{13}T_4$ clusters arranged such that they have a cubic cell with $a = 2.7 \times a_{FeO}$, and that on average 50% of these cells contain a defect cluster. Calculate the value of x.

2.26 A layered material with approximate composition $PbNbS_3$ was found to have a very similar X-ray-diffraction pattern to that of $Sn_{1.17}NbS_{3.17}$. Single-crystal studies revealed that the PbS portion could be described with cell parameters of $a = 5.834$ Å, $b = 5.801$ Å, $c = 11.90$ Å and the NbS_2 part with cell parameters of $a = 3.313$ Å, $b = 5.801$ Å, $c = 2 \times 11.90$ Å (ref. [18]). (a) Give a brief description of the likely structure of "$PbNbS_3$". (b) State how large the a axis would need to be to describe this structure using a conventional crystallographic unit cell. (c) From the cell parameters given, calculate the true composition of "$PbNbS_3$". (d) Suggest why the c axis of the NbS_2 part of the structure is double that of the PbS part. (e) Suggest a likely structure for a material of composition $Pb_{1.14}Nb_2S_{5.14}$.

2.27 The structure of $Y(O, F)_{2+\delta}$ materials can be described in terms of a YX slab based on a square grid of anions alternating with a hexagonal anion grid as shown in Figure 2.16. (a) Confirm that the composition of a material with ideal anion arrays (all nearest anion–anion distances equal) would be $MX_{2.155}$. (b) State whether you'd expect this $MX_{2.155}$ material to be commensurate or incommensurate. (c) Assume that the triangles are compressed slightly in the b direction such that horizontal rows of anions align after five squares and six triangles as in Figure 2.16. Calculate the composition of this material. Repeat your calculation for a system in which anions align after six squares and after seven triangles. (d) State the relationship you'd expect between the square and triangular anion grid in $Y_7O_6F_9$.

2.13 Further Reading

Defects: F.A. Kröger, *"The Chemistry of Imperfect Crystals"* (1964) North-Holland; K. Kosuge, *"Chemistry of Nonstoichiometric Compounds"* (1994) Oxford Science Publications; O. Toft Sorensen, *"Nonstoichiometric Oxides"* (1982) Materials Science Series, Academic Press; R.J. D. Tilley, *"Principles and Applications of Chemical Defects"* (1998) CRC Press.

Twinning: A. Putnis, *"Introduction to Mineral Sciences"* (1992) Cambridge University Press; U. Müller *"Symmetry Relationships between Crystal Structures"* (2013) Oxford University Press.

Crystallographic shear: B.G. Hyde, S. Andersson, *"Inorganic Crystal Structures"* (1989) Wiley.

Modulated structures: S. van Smaalen, *"Incommensurate Crystallography"* (2007) Oxford University Press.

2.14 References

[1] A.C. van den Berg, C.O. Aréan, "Materials for hydrogen storage: Current research trends and perspectives" *Chem. Commun.* (2008), 668–681.

[2] U. Eberle, M. Felderhoff, F. Schüth, "Chemical and physical solutions for hydrogen storage" *Angew. Chem. Int. Ed. Engl.* **48** (2009), 6608–6630.

[3] L.J. Murray, M. Dinca, J.R. Long, "Hydrogen storage in metal–organic frameworks" *Chem. Soc. Rev.* **38** (2009), 1294–1314.

[4] W.I.F. David, M.O. Jones, D.H. Gregory, C.M. Jewell, S.R. Johnson, A. Walton, P.P. Edwards, "A mechanism for non-stoichiometry in the lithium amide/lithium imide hydrogen storage reaction" *J. Am. Chem. Soc.* **129** (2007), 1594–1601.

[5] A.R. Allnatt, A.B. Lidiard, *"Atomic Transport in Solids"* (1993) Cambridge University Press.

[6] S. Gehrsitz, H. Sigg, N. Herres, K. Bachem, K. Kohler, F.K. Reinhart, "Compositional dependence of the elastic constants and the lattice parameter of $Al_xGa_{1-x}As$" *Phys. Rev. B* **60** (1999), 11601–11609.

[7] T. Taniguchi, S. Mizusaki, N. Okada, Y. Nagata, S.H. Lai, M.D. Lan, N. Hiraoka, M. Itou, Y. Sakurai, T.C. Ozawa, Y. Noro, H. Samata, "Crystallographic and magnetic properties of the mixed valence oxides $CaRu_{1-x}Mn_xO_3$" *Phys. Rev. B* **77** (2008), 014406/1–7.

[8] A.D. Wadsley, "Nonstoichiometric metal oxides" *Advances in Chemistry Series* **39** (1963), 23–36.

[9] J.L. Murray, H.A. Wriedt, "The O–Ti (Oxygen–Titanium) system" *Bull. Alloy Phase Diagrams* **8** (1987), 148–165.

[10] B.G. Hyde, S. Andersson, *"Inorganic Crystal Structures"* (1987) J. Wiley and Sons.

[11] A. Shinsaku, D. Bogdanovski, H. Yamane, M. Terauchi, R. Dronskowski, "ε-TiO, a novel stable polymorph of titanium monoxide" *Angew. Chem. Int. Ed. Engl.* **55** (2016), 1652–1657.

[12] F. Koch, J.B. Cohen, "The defect structure of $Fe_{1-x}O$" *Acta Crystallogr. Sect. B* **25** (1969), 275–287.

[13] T.R. Welberry, A.G. Christy, "Defect distribution and the diffuse X-ray diffraction pattern of wustite $Fe_{1-x}O$" *Phys. Chem. Miner.* **24** (1997), 24–28.

[14] T. Wagner, A. Schönleber, "A non-mathematical introduction to the superspace description of modulated structures" *Acta Crystallogr. Sect. B* **65** (2009), 249–268.

[15] O. Gourdon, V. Petricek, M. Dusek, P. Bezdicka, S. Durovic, D. Gyepesova, M. Evaina, "Determination of the modulated structure of $Sr_{14/11}CoO_3$ through a (3 + 1)-dimensional space description and using non-harmonic ADPs" *Acta Crystallogr. Sect. B* **55** (1999), 841–848.

[16] V. Pavlyuk, I. Chumak, L. Akselrud, S.Lidin, H. Ehrenberg, "$LiZn_{4-x}$ ($x = 0.825$) as a (3 + 1)-dimensional modulated derivative of hexagonal close packing" *Acta Crystallogr. Sect. B* **70** (2014), 212–217.

[17] S. Schmid, "The yttrium oxide fluoride solid solution described as a composite modulated structure" *Acta Crystallogr. Sect. B* **54** (1998), 391–398.

[18] G.A. Wiegers, A. Meetsma, R.J. Haange, S. Van Smaalen, J.L. De Boer, A. Meerschaut, P. Rabu, J. Rouxel, "The incommensurate misfit layer structure of $(PbS)_{1.14}NbS_2$, 'PbNbS$_3$', and $(LaS)_{1.14}NbS_2$, 'LaNbS$_3$': An X-ray diffraction study" *Acta Cryst. Sect. B* **46** (1990), 324–332.

3 Defect Chemistry and Nonstoichiometry

One key aspect of materials chemistry is the ability to prepare materials with precisely controlled composition. We'll see countless times in later chapters that ever minor changes in chemical composition can hugely influence a material's properties. Defects of the type we've met in Chapter 2 play a key role in both synthesis and composition control. In the first sections of this chapter, we'll investigate how simple ideas of chemical equilibria can give us qualitative and quantitative insights into defect formation. In the second half of the chapter, we'll look at the diffusion of different types of defects, which controls the reactivity of solids and the properties of some functional materials.

3.1 Narrow Nonstoichiometry in Oxides

We learned in Chapter 2 that entropy favors a certain small number of defects in all extended solids. In a metal oxide, defects such as metal or oxygen vacancies and interstitials give rise to a narrow range of oxygen nonstoichiometry around the integer oxidation state of the metal.

3.1.1 Point Defects in a Pure Stoichiometric Oxide

Let's use NiO as an example. Formation of vacancies (Figure 2.4) is one of several possible ways for intrinsic ionic defects to occur in the NaCl-type structure of NiO. In Equations (2.3) to (2.6) we saw that the mass-action law applies to defect formation as if it were a chemical reaction. Using the Kröger–Vink notation (Section 2.6), we can write "chemical" equations for the formation of all possible intrinsic-defect pairs in stoichiometric (1:1) NiO. Table 3.1 shows there are four such pairs:

The term "nil" denotes the value of 0 obtained after crossing out the regular structure sites on both sides of the equation. The first four equations describe structural defects already introduced in Chapter 2. The last equation describes electronic defects. We symbolize them as electrons and holes, but in a redox-prone oxide such as NiO, the electron e' would behave as Ni^+ (an aliovalent

Table 3.1 Formation equations for the four alternative intrinsic ionic-defect pairs in NiO and for intrinsic ionization.

Process	Reaction
Schottky:	$nil = v_{Ni}'' + v_O^{\bullet\bullet}$
Anti-Schottky:	$Ni_{Ni}^{\times} + O_O^{\times} = Ni_i^{\bullet\bullet} + O_i''$
Cation-Frenkel:	$Ni_{Ni}^{\times} = v_{Ni}'' + Ni_i^{\bullet\bullet}$
Anion-Frenkel:	$O_O^{\times} = v_O^{\bullet\bullet} + O_i''$
Intrinsic ionization:	$nil = e' + h^{\bullet}$ $(2Ni_{Ni}^{\times} = Ni_{Ni}' + Ni_{Ni}^{\bullet})$

defect Ni_{Ni}' in the Kröger–Vink notation), whereas the hole $h^{\bullet}$ would represent the oxidized state Ni^{3+} ($Ni_{Ni}^{\bullet}$). The last reaction in Table 3.1 is in principle a disproportionation of divalent nickel into mono- and trivalent defects.[1]

3.1.2   Point Defects upon Oxidation/Reduction of the Stoichiometric Oxide

The possible changes in the numbers of intrinsic defects on oxidation and reduction of the stoichiometric NiO are in Figure 3.1. When NiO is reduced within its homogeneity range,[2] either an O deficit appears at its regular structural sites or an excess of Ni at interstitial sites, yielding $NiO_{1-\delta}$ or $Ni_{1+\delta}O$, respectively. Upon oxidation, either an O excess or a Ni deficit may occur as $NiO_{1+\delta}$ or $Ni_{1-\delta}O$. These alternative responses are based on the four possible intrinsic point-defect reactions in Table 3.1. Which of them actually dominates in a particular material can only be answered by a rather involved experimental study.

One useful rule is that large closest-packed atoms form defects less readily than smaller atoms located in the holes. In nickel oxide and other $3d$ monoxides, oxidation (adding oxygen) creates metal vacancies v_{Ni}'' rather than inserting bulky oxygen interstitials. Likewise, v_{Cr}''' forms in Cr_2O_3 and other corundum-type $3d$ oxides. Reduction (oxygen removal) may form interstitials such as $Zn_i^{\bullet\bullet}$ in ZnO. However, if the size difference is less pronounced, this approximation fails: CdO with the NaCl-type structure, for example, forms oxygen vacancies upon reduction. While the defect type follows from the structure, the propensity for a dominant oxidative or reductive nonstoichiometry depends on the chemistry. A significant nonstoichiometry often develops towards another stable oxidation state of the metal. For example, FeO tends to be oxidized towards Fe^{III}.

---

[1] Chemically, $2Ni^{2+} = Ni^+ + Ni^{3+}$. We will see in Chapter 10 that $nil = e' + h^{\bullet}$ represents a thermal excitation of an electron from a valence band to a conduction band, which leaves a hole behind in the valence band. In a redox oxide such as NiO, the electron becomes trapped as Ni^+ close to the conduction band, whereas the hole is trapped as Ni^{3+} close to the valence band. In Chapter 7, we'll see that optical excitation also creates an electron–hole pair.

[2] By "reduction" or "oxidation" in the context of defects, we'll mean a trace reduction or trace oxidation that creates nonstoichiometry but does not decompose the phase.

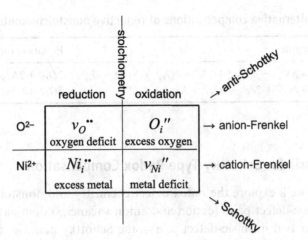

Figure 3.1 The four alternative point-defect compensations for reduction and oxidation of stoichiometric (1:1) NiO are the four alternative intrinsic-defect pairs named on the right.

3.1.3   Equilibrium Equations for Oxidative and Reductive Nonstoichiometry

Just as we did for intrinsic defects in Section 3.1.1, we can write equilibrium equations for the point-defect changes that occur on oxidation or reduction of an oxide. Let's assume O_2 to be the only gaseous reactant[3] in the intrinsic redox reactions of NiO at high temperature.[4] For clarity, we'll proceed via an auxiliary equation of a small amount of oxygen, δO, reacting with NiO. As shown in the middle column of Table 3.2, oxidation forms holes compensated by the intrinsic negative defect v_{Ni}'' or O_i'' (Figure 3.1) in an equation balanced in terms of charges, sites (denoted in subscripts), and chemical elements. Analogously so for reduction and electrons in Table 3.3. Next, the auxiliary reactions are simplified into the defect-formation equation on the right of Table 3.2 and Table 3.3 by crossing out species occurring on both sides, changing to O_2, and removing fractions.

Table 3.2 Two alternative compensations of oxidative nonstoichiometry in NiO.

	Auxiliary scheme	Equation for oxidation
v_{Ni}''	$NiO + \delta O \rightarrow \delta v_{Ni}'' + 2\delta h^\bullet + \delta O_O^x + Ni_{Ni}^x + O_O^x$	$O_{2(g)} \rightleftarrows 2v_{Ni}'' + 4h^\bullet + 2O_O^x$
O_i''	$NiO + \delta O \rightarrow \delta O_i'' + 2\delta h^\bullet + Ni_{Ni}^x + O_O^x$	$O_{2(g)} \rightleftarrows 2O_i'' + 4h^\bullet$

---

[3] An analogous set of equations would have to be set up for every other reacting gas species such as atomic oxygen, ozone, or nickel vapor.

[4] Temperature T provides the needed activation energy (for dissociation of O_2 and diffusion of defects) in the form of thermal energy kT (k is the Boltzmann constant, 1.380649×10^{-23} J/K, see Appendix J).

Table 3.3 Two alternative compensations of reductive nonstoichiometry in NiO.

	Auxiliary scheme	Equation for reduction
$Ni_i^{\bullet\bullet}$	$NiO - \delta O \rightarrow \delta Ni_i^{\bullet\bullet} - \delta Ni_{Ni}^{\times} + 2\delta e' - \delta O_O^{\times} + Ni_{Ni}^{\times} + O_O^{\times}$	$2O_O^{\times} + 2Ni_{Ni}^{\times} \rightleftarrows 2Ni_i^{\bullet\bullet} + 4e' + O_{2(g)}$
$v_O^{\bullet\bullet}$	$NiO - \delta O \rightarrow \delta v_O^{\bullet\bullet} + 2\delta e' - \delta O_O^{\times} + Ni_{Ni}^{\times} + O_O^{\times}$	$2O_O^{\times} \rightleftarrows 2v_O^{\bullet\bullet} + 4e' + O_{2(g)}$

3.1.4 Defect Equilibria for Schottky-Type Redox Compensation

In this section, we'll explore the redox defect chemistry and nonstoichiometry of NiO if Schottky intrinsic-defect pairs (cation and anion vacancies) dominate. We will therefore need to consider two intrinsic-defect pairs—the Schottky pair $v_O^{\bullet\bullet}$ and v_{Ni}'' is one, the other is e' and $h^{\bullet}$. There will be two intrinsic-pair formation reactions and two[5] redox-defect reactions, one for oxidation one for reduction; the equilibria[6] are summarized in Table 3.4:

Table 3.4 Schottky-type* redox compensation in NiO.

Process	Reaction equation	Reaction quotient
Schottky	$nil \rightleftarrows v_{Ni}'' + v_O^{\bullet\bullet}$	$K_S = [v_{Ni}''][v_O^{\bullet\bullet}]$
Ionization	$nil \rightleftarrows e' + h^{\bullet}$	$K_i = [e'][h^{\bullet}]$
Oxidation	$O_{2(g)} \rightleftarrows 2v_{Ni}'' + 4h^{\bullet} + 2O_O^{\times}$	$K_{ox} = [v_{Ni}'']^2[h^{\bullet}]^4 \cdot p_{O_2}^{-1}$
Reduction	$2O_O^{\times} \rightleftarrows 2v_O^{\bullet\bullet} + 4e' + O_{2(g)}$	$K_{red} = [v_O^{\bullet\bullet}]^2[e']^4 \cdot p_{O_2}$

* Reaction quotients (but not the full reaction equations) for anti-Schottky, Frenkel, and anti-Frenkel intrinsic defect pairs are obtained upon appropriate substitution of defect symbols: $Ni_i^{\bullet\bullet} \equiv v_O^{\bullet\bullet}$ and $O_i'' \equiv v_{Ni}''$.

Only three of the four reaction quotients (mass-action terms) are independent, as the equilibrium constants combine to $K_S^2 \cdot K_i^4 = K_{ox} \cdot K_{red}$. We have four unknowns (the fraction of holes, electrons, nickel vacancies, and oxygen vacancies per NiO formula of regular sites), and we need one more equation. This comes from the **electroneutrality condition** requiring equal amounts of positive and negative charges:

$$2[v_{Ni}''] + [e'] = [h^{\bullet}] + 2[v_O^{\bullet\bullet}] \tag{3.1}$$

---

[5] Both involve O_2 gas. If nickel vapor were present as well, two additional redox equations would follow—one for oxidation of the vapor by holes into regular nickel sites, $Ni_{(g)} + v_{Ni}'' + 2h^{\bullet} = Ni_{Ni}^{\times}$, and one for reduction of $Ni_{Ni}^{\times}$ to nickel vapor, $Ni_{Ni}^{\times} + v_O^{\bullet\bullet} + 2e' = 2Ni_{(g)}$. One of their two reaction quotients would be independent of all the others, which allows us to solve one more variable, the Ni vapor pressure.

[6] The unexpressed fractional concentrations of atoms at their regular sites are set to unity due to the low concentration of defects.

The four independent equations can then be chosen and rearranged such that each of them expresses the fraction of only one type of defect as a function of the partial pressure of oxygen, p_{O_2}, having three equilibrium constants as parameters. The equations[7] can be solved either analytically or numerically to give the fractions of defects. Once this is done, the oxygen nonstoichiometry $[v_{Ni}''] - [v_O^{\bullet\bullet}]$ can be evaluated[8] as a function of p_{O_2}.

Figure 3.2 shows logarithmic plots of defect fractions versus p_{O_2} for two limiting cases of dominating intrinsic-defect pair; ionic (left), electronic (right).[9] The approximate oxygen nonstoichiometry is given in the upper plots. High oxygen pressures lead to excess oxygen in the oxide; low pressures to oxygen deficiency. The defect-concentrations in the lower plots illustrate what causes this simple composition change—changing fractions of electronic defects $[e']$ and $[h^\bullet]$, and of ionic defects $[v_{Ni}'']$ and $[v_O^{\bullet\bullet}]$. High $[v_{Ni}'']$ and $[h^\bullet]$ emerge upon

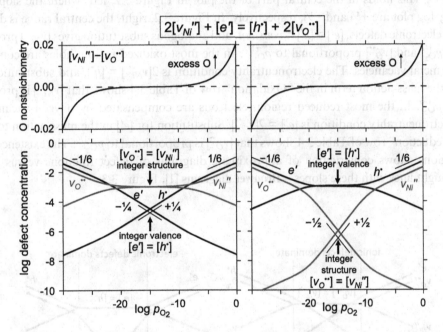

Figure 3.2 Oxygen nonstoichiometry (top) and defect fractions per formula (bottom) in NiO with dominant ionic defects (left; $K_S = 10^{-6}$, $K_i = 10^{-9}$, $K_{ox} = 10^{-9}$) and with dominant electronic defects (right; $K_S = 10^{-12}$, $K_i = 10^{-6}$, $K_{ox} = 10^{-9}$).

[7] $2K_S^{3/2}K_{ox}^{1/4} + K_iK_Sp_{O_2}^{-1/4}[v_O^{\bullet\bullet}]^{1/2} - K_{ox}^{1/2}p_{O_2}^{1/4}[v_O^{\bullet\bullet}]^{3/2} - 2K_{ox}^{1/4}K_S^{1/2}[v_O^{\bullet\bullet}]^2 = 0$

$-2K_SK_{ox}^{1/4} - K_{ox}^{1/2}p_{O_2}^{1/4}[v_{Ni}'']^{1/2} + K_ip_{O_2}^{-1/4}[v_{Ni}'']^{3/2} + 2K_{ox}^{1/4}[v_{Ni}'']^2 = 0$

$2K_{ox}p_{O_2} + K_i(K_{ox}p_{O_2})^{1/2}[h^\bullet] - (K_{ox}p_{O_2})^{1/2}[h^\bullet]^3 - 2K_S[h^\bullet]^4 = 0$

$-2K_SK_i^2 - K_i(K_{ox}p_{O_2})^{1/2}[e'] + (K_{ox}p_{O_2})^{1/2}[e']^3 + 2(K_{ox}p_{O_2}/K_i^2)[e']^4 = 0$.
The unit of concentration is the same as in the equilibrium constants.

[8] The maximum error due to this simplification of the precise $\delta = ([v_{Ni}''] - [v_O^{\bullet\bullet}])/(1 - [v_{Ni}''])$ in NiO$_{1+\delta}$ is 1% of the nonstoichiometry value at the left edge of the graph in Figure 3.2.

[9] PbO is an example of predominant ionic and CuO of predominant electronic defects.

oxidation in oxygen-rich atmospheres, high $[v_O^{\bullet\bullet}]$ and $[e']$ upon reduction in oxygen-poor atmospheres.[10] In between, there are two important points: Zero nonstoichiometry is associated with the **point of integer structure** where occupied metal and oxygen sites have the stoichiometric ratio 1:1 and the intrinsic structural defects compensate each other, $[v_{Ni}'']$ = $[v_O^{\bullet\bullet}]$. The point where the electronic defects compensate each other, $[e'] = [h^\bullet]$, is the **point of integer valence**.[11] For a pure binary oxide, these two points coincide on the p_{O_2} scale.

In several regions of Figure 3.2, the defect fractions have essentially linear variation on the log–log scale. This occurs in ranges where a pair of mutually compensating defects dominates. When Schottky vacancies dominate, the electroneutrality condition in Equation (3.1) simplifies to $[v_{Ni}''] = [v_O^{\bullet\bullet}] = constant$. The mass-action equations for oxidation and reduction in the last two rows in Table 3.4 then show that $[h^\bullet]$ is proportional to $p_{O_2}^{1/4}$ and $[e']$ is proportional to $p_{O_2}^{-1/4}$. This holds in the central part of the plot in Figure 3.2, left, where the slopes on the log–log plot are +¼ and −¼, respectively. In Figure 3.2, right, the central range is dominated by electronic defects, $[e'] = [h^\bullet] = constant$, and analogous substitution gives $[v_{Ni}'']$ proportional to $p_{O_2}^{1/2}$ and $[v_O^{\bullet\bullet}]$ proportional to $p_{O_2}^{-1/2}$. In the most oxidized region, holes are compensated by metal vacancies. The electroneutrality condition is $2[v_{Ni}''] = [h^\bullet]$, and substitution for $[h^\bullet]$ in the mass-action term in the "Oxidation" row of Table 3.4 shows that $[v_{Ni}'']$ is proportional to $p_{O_2}^{1/6}$. In the most reduced region, electrons are compensated by oxygen vacancies. The electroneutrality condition is $[e'] = 2[v_O^{\bullet\bullet}]$. Substitution for $[e']$ in the mass-action term in the "Reduction" row of Table 3.4 shows that $[v_O^{\bullet\bullet}]$ is proportional to $p_{O_2}^{-1/6}$. The existence of linear regions allows construction of approximate diagrams of defect fractions versus p_{O_2} using straight lines with these slopes—**Brouwer diagrams** [1], Figure 3.3.

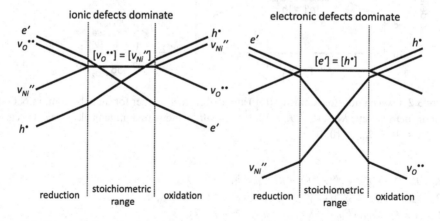

Figure 3.3 Brouwer-diagram sketches corresponding to the previous figure.

---

[10] Recall that $h^\bullet$ is equivalent to Ni^{3+}, e' to Ni^+.   [11] Or of integer oxidation state.

3.1.5 Acceptor-Doped Oxides

How do things change when we replace a small portion of Ni^{2+} with a cation of a lower and fixed oxidation state (such as Li^+), which is called **acceptor**[12] **doping**? All the mass-action equations from Table 3.4 remain the same. The only difference appears in the electroneutrality condition that now includes the additional defect Li_{Ni}':

$$2[v_{Ni}''] + [e'] + [Li_{Ni}'] = [h^\bullet] + 2[v_O^{\bullet\bullet}] \qquad (3.2)$$

Expressing defect fractions as functions of p_{O_2} can be done in the same way as the pure-oxide case, except that we now have four parameters that control the defect equilibria; the three equilibrium constants and the fixed fraction of the acceptor defect, $[Li_{Ni}']$.

Figure 3.4 shows how the defect concentrations of pure NiO in Figure 3.2 change when $[Li_{Ni}'] = 0.02$. The Li^+ acceptor moves the point of integer valence, $[e'] = [h^\bullet]$, to lower p_{O_2} because Ni^{2+} is now more easily oxidized in order to keep the charges balanced. The point of integer structure, $[v_{Ni}''] = [v_O^{\bullet\bullet}]$, moves towards higher p_{O_2} because oxygen vacancies are proportional to $p_{O_2}^{-1}$ and $[v_{Ni}'']$ is proportional to p_{O_2}. Brouwer diagrams are explored in the end-of-chapter problems.

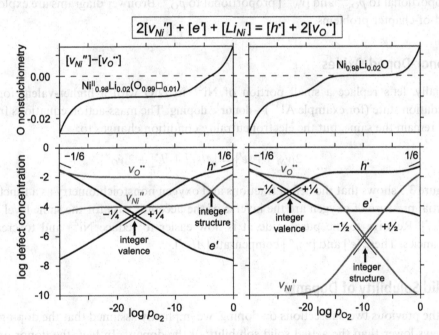

Figure 3.4 Defect fractions per formula in Li-doped NiO, $[Li_{Ni}'] = 0.02$, for dominant ionic (left) and electronic (right) defects with equilibrium constants as in Figure 3.2. The square in the chemical formula in the top left stands for vacancies.

---

[12] In electronics, an acceptor is a neutral atom with fewer valence electrons than the regular atom. In our context, it is an atom of a fixed oxidation state lower than the matrix atom it replaces.

more easily formed after $LiO_{1/2}$ has dissolved in NiO and a higher O_2 pressure is then needed to fill them. Of the two readily formed defects $h^\bullet$ and $v_O^{\bullet\bullet}$ that compensate the Li_{Ni}' acceptor, holes $h^\bullet$ dominate around the point of integer structure and vacancies $v_O^{\bullet\bullet}$ around the point of integer valence.

As before, some regions of Figure 3.4 have a practically linear dependence on $\log p_{O_2}$. The reasons for linearity in the extreme oxidized and reduced regions are the same as for the undoped oxide. The integer-valence point and the integer-structure point split due to the doping, and we get approximately linear ranges around both. Because vacancies $v_O^{\bullet\bullet}$ compensate the acceptor around the point of integer valence, the electroneutrality condition in Equation (3.2) simplifies there to $[Li_{Ni}'] = 2[v_O^{\bullet\bullet}] = constant$. As this also keeps $[v_{Ni}'']$ constant, mass-action equations for oxidation and reduction in the last two rows in Table 3.4 give $[h^\bullet]$ proportional to $p_{O_2}^{1/4}$ and $[e']$ proportional to $p_{O_2}^{-1/4}$. Holes $h^\bullet$ compensate the acceptor around the point of integer structure, hence the electroneutrality condition in Equation (3.2) simplifies to $[Li_{Ni}'] = [h^\bullet] = constant$. Because this also keeps $[e']$ constant, mass-action equations for oxidation and reduction in the last two rows in Table 3.4 give $[v_O^{\bullet\bullet}]$ proportional to $p_{O_2}^{-1/2}$ and $[v_{Ni}'']$ proportional to $p_{O_2}^{1/2}$. Brouwer diagrams are explored in the end-of-chapter problems.

3.1.6   Donor-Doped Oxides

Finally, let's replace a small portion of Ni^{2+} in NiO with a higher-valent ion of fixed oxidation state (for example Al^{3+})—**donor**[13] **doping**. The mass-action equations from Table 3.4 remain the same, but the electroneutrality condition changes to:

$$2[v_{Ni}''] + [e'] = [Al_{Ni}^\bullet] + [h^\bullet] + 2[v_O^{\bullet\bullet}] \tag{3.3}$$

Figure 3.5 shows that the defect fractions and oxygen nonstoichiometry as a function of the partial pressure of oxygen are the inverse of the acceptor case for the same level of doping $[Al_{Ni}^\bullet]$. Relative to the pure oxide, it is now easier to reduce Ni^{2+} and to create nickel vacancies. These $[e']$ and $[v_{Ni}'']$ compensate $[Al_{Ni}^\bullet]$.

3.1.7   Solid Solubility of Dopants

In the previous two subsections on doping, we implicitly assumed that the doping level was always lower than the actual solid solubility of the dopant. In fact, the donor or acceptor solubility in oxides depends on the temperature and partial pressure of oxygen. Before we make some qualitative considerations on the effects of these two variables, let's consider a simpler case of isovalent substitution.

For isovalent substitutions, the effect of temperature on solubility is straightforward. A solution has higher entropy than the sum of its pure components, and increasing

---

[13] In our context, a donor has a fixed oxidation state higher than the matrix atom it replaces.

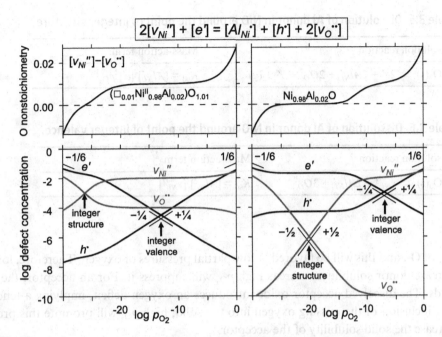

Figure 3.5 Defect fractions per formula in Al-doped NiO, $[Al_{Ni}^{\cdot}] = 0.02$, for dominant ionic (left) and electronic (right) defects with equilibrium constants as in Figure 3.2.

temperature will increase the entropy term $T\Delta S$ that drives the dissolution, thus increasing the solubility. We know empirically that chemically similar isostructural crystals of close enough[14] atomic radii mix completely at high temperatures, but calculations [2] do show that they would eventually demix at low temperatures.

For aliovalent defects, oxidation and reduction make the picture more complex. Let's consider first the effect of temperature. In an oxide[15], increased temperature will favor loss of oxygen (chemical reduction) because the released gas has high entropy. Donor doping makes reduction easier because we can think of the donor oxide (like $AlO_{3/2}$ dissolved in NiO) as bringing excess oxygen into the structure. Increased temperature will therefore favor dissolution of the donor also by promoting reduction (not only by increased $T\Delta S$ of mixing). The situation is different for an acceptor. Since an acceptor introduces an oxygen deficit into the structure, the further oxygen loss at high temperature will disfavor this dissolution, but the general entropic driving force of dissolution will still be present. The effect of increasing temperature on the solid solubility of an acceptor therefore can't be predicted.

The effect of the partial pressure of oxygen can be estimated from the Le Chatelier principle[16]. Dissolution of a donor oxide brings excess oxygen, thus a tendency for easy

---

[14] Size differences less than 15% typically allow complete solubility at high temperatures.

[15] In particular of a redox-active element such as Ni.

[16] *Equilibrium will adjust to minimize the effects of external changes.*

Table 3.5 Dissolution of Al donor in NiO around the point of integer structure.

Dissolution reaction	Mass-action term
$Al_2O_3(s) \rightleftarrows 2e' + 2Al_{Ni}^{\cdot} + 2O_O^{\times} + \frac{1}{2}O_2(g)$	$K_{sd} = [Al_{Ni}^{\cdot}]^2[e']^2 p_{O_2}^{1/2}$

Table 3.6 Dissolution of Al donor in NiO around the point of integer valence.

Dissolution reaction	Mass-action term
$Al_2O_3(s) \rightleftarrows v_{Ni}'' + 2Al_{Ni}^{\cdot} + 3O_O^{\times}$	$K_{sd} = [Al_{Ni}^{\cdot}]^2[v_{Ni}'']$

loss of O_2, and this will be favored at low partial pressures of oxygen. Therefore, low p_{O_2} will increase donor solubility, whereas high p_{O_2} will suppress it. For an acceptor, the opposite holds. The dissolved acceptor oxide introduces an oxygen deficit, implying a tendency for easy inclusion of the missing oxygen into the solid; high p_{O_2} will promote this process and increase the solid solubility of the acceptor.

The combination of these temperature- and p_{O_2} effects is unambiguous only for dissolution of the donor oxide—high temperature and low p_{O_2} will increase the donor's solid solubility. To evaluate this quantitatively, we must include the dissolution reaction of the donor oxide in the set of equilibrium defect-reaction equations. Let's consider the dissolution of the trivalent donor Al in the integer structure of NiO. The defect predominantly compensating the donor will be e' (see Figure 3.5 right), and the dissolution reaction will maintain the 1:1 ratio of the anion and cation sites as shown in Table 3.5.

At equilibrium with excess Al_2O_3, $[Al_{Ni}^{\cdot}]$ is the solid-solubility fraction x of the donor, and we see from the equilibrium equation that $[Al_{Ni}^{\cdot}] = [e']$. Substitution into the mass-action term in Table 3.5 gives $x = K_{sd}^{1/4} p_{O_2}^{-1/8}$, and we observe that the solid solubility of the donor increases when p_{O_2} decreases.

Similarly, we can evaluate the donor solubility about the point of integer valence. The predominant defects compensating the donor will be v_{Ni}'' (see Figure 3.5 left). Table 3.6 above shows that the donor solubility is independent of the partial pressure of oxygen when the defect that predominantly compensates the donor is a structural point defect.

3.1.8 Cautionary Note on Defect Models in Pure Oxides

A pitfall of defect modeling in "pure" oxides is that they are never truly pure. In any oxide, there will be impurities influencing the defect equilibria at some level. Furthermore, some deceptively simple oxides, such as the NaCl-type wüstite of ideal composition FeO, are grossly nonstoichiometric and exhibit clustering of defects (Section 2.9). Real materials

Box 3.1 Synthetic Methods: Oxygen-nonstoichiometry control in oxides

A precise and homogeneous (non)stoichiometry is important for many functional oxide materials. All methods to achieve this start with equilibrium of the oxygen exchange between the point defects in the solid and the surrounding gas atmosphere. Two processes occur. The first is the surface reaction where O_2 splits to, or forms from, two oxide anions and four holes. This is a redox reaction since the holes represent the oxidized state of the metal. The subsequent process is a diffusion-driven homogenization, during which the surface-oxygen excess (or deficit) homogenizes throughout the bulk. The oxygen nonstoichiometry can be controlled by the following parameters, depending on whether the system is closed or open:

Control parameters	Redox reagent	System
p_{O_2}, T	Flowing gas of given p_{O_2}	Open
T	p_{O_2} buffer	Closed
Mass	Oxygen getter or source	Closed

The open systems use hot flowing reaction atmospheres with defined partial pressures of oxygen. These can be mixtures of Ar and O_2 down to $p_{O_2} = 10^{-4}$ bar; mixtures of Ar, H_2, and H_2O (via $H_2O \rightleftarrows H_2 + \frac{1}{2}O_2$) below $p_{O_2} = 10^{-10}$ bar; and CO/CO_2 (via $CO_2 \rightleftarrows CO + \frac{1}{2}O_2$) in the intermediate range. After isothermal equilibration, samples are quenched to low temperature because otherwise they would oxidize during the cool-down.

The closed systems are usually set up in sealed ampoules. The p_{O_2} buffer is a solid redox-couple mixture that maintains constant p_{O_2} when it is in excess of the nonstoichiometric sample and not in contact with it. Like in the previous technique, the constant p_{O_2} value is fixed by the temperature. Gibbs phase rule (Chapter 4) states that $F = C + 2 - P$, where F is the degree of freedom, C is the number of components, and P is the number of phases in equilibrium. As long as the buffer, say, a homogeneous mixture of Ni and NiO, contains Ni and $NiO_{1-\delta}$ at equilibrium, the p_{O_2} above them remains constant at a given temperature: 2 components + 2 intensive variables (temperature + pressure) − 3 phases (Ni, NiO, O_2) = 1 variable that can be varied independently in the current phase system without a phase disappearing. The temperature will therefore fix the p_{O_2}. Almost any p_{O_2} can be achieved by a good choice of the redox couple in equilibrium with O_2, from low (Ni–NiO) to very high p_{O_2} (Ag_2O_2–Ag_2O in gold wraps in anvil cells).

The oxygen-getter/oxygen-source technique is similar, but here it is the *amount* of the added substance that controls the oxygen taken up or released by the sample. Zirconium metal is a good getter that completely and rapidly oxidizes into ZrO_2 but only if relatively high p_{O_2} levels are generated by the sample. When very low p_{O_2} is needed to reduce the oxide, Mg, Zn, or Fe can be used. A common oxygen source is Ag_2O, which releases all O_2 above 450 °C.

A versatile modification of the getter/source technique is solid state coulometry, where specific amounts of oxygen are dosed to or from an enclosed sample electrochemically, via a window made of cubic stabilized zirconia; an excellent conductor of oxide ions (Chapter 13) yet a poor electronic conductor. When appropriate electric charge is supplied from a Pt electrode, the corresponding amount of oxide anions moves through the zirconia window.

range from nearly ideal oxides with randomly distributed dilute point defects, such as Cu_2O, NiO, ZnO, and Cr_2O_3, to quasi-random distributions (CoO), to defect clustering (FeO), to formation of defect-ordered superstructures (CeO_2, PrO_2), or infinitely adaptive defect-ordered structures (e.g. Magnéli phases). Simple clustering, such as dimerization or donor–vacancy association, is common for impurities (dopants) even in oxides that exhibit a random distribution of point defects. In general, we must take extreme caution. On the other hand, the next section shows that some complex oxides can have surprisingly simple defect equilibria.

3.2 Wide Nonstoichiometry in Oxides

Wide nonstoichiometry ranges can occur between two oxidation states of an atom in two closely related integer structures. An example is the $YBa_2Cu_3O_{7-\delta}$ high-T_c superconductor ($0 < \delta < 1$; Figure 3.6), in which one entire oxygen atom per formula can be removed. In $YBa_2Cu_3O_6$, one copper atom has a linear coordination typical of Cu^+ and the other two are Cu^{2+}. In $YBa_2Cu_3O_7$, the linear coordination becomes square planar, consistent with Cu^{3+}.

   This type of nonstoichiometry can also be treated with defect equilibria. We only have to decide which of the limiting structures (Figure 3.6) contains the integer valence that corresponds to the intrinsic situation $[e'] = [h^\cdot]$, where, as Figure 3.2 suggests, properties related to the concentration of these charge carriers should achieve minimum values. Electrical conductivity and oxygen diffusivity decrease towards $YBa_2Cu_3O_6$ (as discussed in ref. [3]), suggesting it is the integer-valence point. The other limit, $YBa_2Cu_3O_7$, is then defined as the integer-structure point by formally considering $YBa_2Cu_3O_{7-\delta}$ as an acceptor-doped $Y_3Cu_3O_7$ with $2Ba_Y'$ per formula.

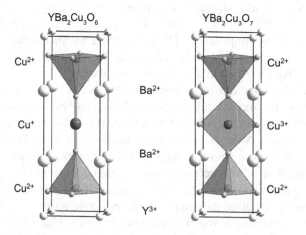

Figure 3.6 Two limiting structures of $YBa_2Cu_3O_{7-\delta}$.

The intrinsic ionic defects in the $YBa_2Cu_3O_7$ integer structure are of the anion-Frenkel type; oxygen vacancies and interstitials. We can then set up the redox compensations in analogy with Table 3.4 and solve the mass-action equations together with the electroneutrality condition. The Frenkel-process reaction is $O_O^x = v_O^{\cdot\cdot} + O_i''$, the oxidation reaction is $\frac{1}{2}O_{2(g)} = 2h^{\cdot} + O_i''$, the reduction reaction is $O_O^x = 2e' + v_O^{\cdot\cdot} + \frac{1}{2}O_{2(g)}$. A profound simplification can be achieved by subtracting the Frenkel reaction from the oxidation, which yields the intuitive reaction of vacancy filling and oxidation:

$$\frac{1}{2}O_{2(g)} + v_O^{\cdot\cdot} = O_O^x + 2h^{\cdot} \tag{3.4}$$

Equation (3.4) does not include the e' and O_i'' defects important for reduction and oxidation beyond the $YBa_2Cu_3O_6$ and $YBa_2Cu_3O_7$ limits, respectively, but it is reasonably valid for $YBa_2Cu_3O_{7-\delta}$ within this range. With vacancies counted from the point of integer structure of the acceptor-doped model, $[v_O^{\cdot\cdot}] \approx \delta$. With holes counted from the point of integer valence, $[h^{\cdot}] \approx 2(1 - \delta)$. The mass-action term for Equation (3.4) is then:

$$K_{vox} = p_{O_2}^{-1/2} 4(1 - \delta)^2 / \delta \tag{3.5}$$

Equation (3.5) has a solution for δ as a function of p_{O_2}:

$$\delta = 1 - \frac{K_{vox}}{8}\left(\sqrt{p_{O_2} + \frac{16}{K_{vox}}\sqrt{p_{O_2}}} - \sqrt{p_{O_2}}\right) \tag{3.6}$$

which describes the entire nonstoichiometry range well for temperatures below 600 °C. As an example, $K_{vox} = 55$ bar$^{-1/2}$ at 500 °C, and $\delta = 0.12$ ($YBa_2Cu_3O_{6.88}$) is calculated. That matches both experiment [4] and the full defect-model approach within their uncertainties. Above 600 °C, a progressive deviation from the experimental reality appears for the highest and lowest values of δ calculated by Equation (3.6).

Further implications of the $YBa_2Cu_3O_{7-\delta}$ nonstoichiometry are in the end-of-chapter problems and in Chapter 12. This example illustrates that in some cases a simple approximation works for wide nonstoichiometry.

3.3 Point Defects and Diffusion

In Section 3.2, we discussed point defects taking part in the reaction of O_2 with a crystalline solid. How can oxygen move inside a compact crystal? How long does it take? In fact, the oxygen speed in $YBa_2Cu_3O_{7-\delta}$ is amazing; one day at 400 °C in O_2 is enough to fully oxidize a pellet of 1 cm in diameter. We can understand this by realizing that point defects are mobile and disperse homogeneously in a material in order to maximize entropy. This is one of many examples of a process called **diffusion**. Solid state diffusion is behind major industrial processes, such as hardening of steel, fabrication of doped semiconductors, or "filtering" oxygen or hydrogen via permeable membranes. It also underpins some important technology, such as the transport of Li^+ in lithium batteries or of O^{2-}/H^+ in fuel cells.

Equations describing diffusion are deceptively simple. **Fick's first law** states that

$$J = D\left(-\frac{dc}{dy}\right) \tag{3.7}$$

when limited to one[17] direction, y. It tells us that the **flux J** (the flow density of the diffusing particle in the matrix or substrate[18]) is directly proportional to the negative gradient of the concentration c of the particle along the flux direction.[19] The minus sign is there because the flow in positive direction occurs along a gradient of higher to lower concentrations, down a negative slope, so we have to turn this negative gradient into a positive number. The proportionality constant D is the **diffusivity** or **diffusion coefficient**.[20] The diffusivity D is characteristic for a particle–matrix pair at a given temperature and pressure; it is not a constant. When the diffusion concerns changes in chemical composition, D is called the **chemical diffusion coefficient** and is denoted as D_A formally implying diffusion of a neutral atom A.

Since time does not appear in Equation (3.7), Fick's first law is valid for **steady-state diffusion**, in which the particle flow into a system equals the flow out. An example is permeation of CO_2 (or heat) through a house wall into an open space. The CO_2 (heat) flux is constant in the steady state, and its concentration profile across the wall is a straight line of negative slope.

Non-steady-state diffusion concerns the spread of diffusing particles as a function of time. The time, τ, is introduced via a continuity consideration—the concentration increase per time equals the net incoming flux per unit length (the negatively taken negative gradient of the particle flux). At every instant, $\partial c/\partial \tau$ in m^{-3}/s equals $-\partial J/\partial y$ in $(s^{-1}\, m^{-2})/m$. Plugging in J from Equation (3.7) leads to **Fick's second law**:

$$\frac{\partial c}{\partial \tau} = \frac{\partial}{\partial y}\left(D\frac{\partial c}{\partial y}\right) \tag{3.8}$$

If D is constant and independent of c, it comes before the differential, $\partial c/\partial \tau = D(\partial^2 c/\partial y^2)$. This equation states that the change of concentration with time is proportional to the curvature of

---

[17] It can be changed to three dimensions by adding terms for x and z, and treating D as a tensor if the material is anisotropic. In our treatment, we will consider diffusion along one direction in an isotropic matrix of a 3D structure, for which D behaves as a scalar.

[18] In a broad sense, the substrate or matrix is everything except the particle whose diffusion we follow.

[19] The flux is the number of atoms $[J] = s^{-1}\, m^{-2}$, or moles, $[J] = mol\, s^{-1}\, m^{-2}$, or kilograms, $[J] = kg\, s^{-1}\, m^{-2}$ passing per unit time through a unit cross-sectional area perpendicular to the flux. The respective concentrations are in m^{-3}, or $mol\, m^{-3}$, or $kg\, m^{-3}$.

[20] A simple consideration for the units of D and, say, the mole flux: Particles that flow at a rate of 1 mol/s perpendicularly through a 1 m^2 window under a concentration gradient of 1 $(mol/m^3)/m$ have diffusivity $D = 1\, m^2/s$. Substitute into Equation (3.7).

the concentration gradient. In a curved gradient, fluxes at any two points are different, Equation (3.7), and will tend to equalize the concentration. When the diffusion proceeds into an enclosed space, the curvature of the gradient disappears once the concentration c becomes equal throughout the bulk. However, if the particle can leave the bulk, a steady state is eventually obtained.

Let's conclude by noting that both of Fick's laws belong to a family of equivalent laws that include thermal and electrical conduction.[21] In the following sections, we shall consider transport along a field gradient in a crystal, termed **lattice (bulk) diffusion**.[22] Uncharged species will move down the chemical-potential slope, whereas charged defects will also move down an electric-potential slope.

3.3.1 Point-Defect Movements

Vacancies and interstitials are the transport vehicles in crystal structures. Atoms at regular sites jump into neighboring vacancies, whereas interstitial atoms jump between interstitial sites (Figure 3.7).[23] Why do atoms jump?

The free energy of an atom is at a minimum at its regular site. Jumps occur because the atom's thermal vibration[24] of frequency v (in s^{-1}) provides a chance of passing the Gibbs-energy barrier $\Delta^{\ddagger}G_m$ for moving to the neighboring vacant site (see Figure 3.9 later on). The likelihood that a vibration becomes a successful jump follows the Boltzmann probability distribution,

$$p_B = \exp(-\Delta^{\ddagger}G_m/kT) \tag{3.9}$$

where kT is the thermal energy. If p_B is, say, 0.0002, on average there will be a jump into a new position every 5000 vibrations. Consider the 2D interstitial in Figure 3.7. It will jump with frequency $p_B v$, but only one-quarter of these jumps will move it forward by the length

---

[21] The thermal-conduction flux is in $J\,s^{-1}\,m^{-2}$, the electrical-conduction flux in $C\,s^{-1}\,m^{-2}$ (where $J\,s^{-1}$ = W and $C\,s^{-1}$ = A). The thermal diffusivity in units of $J\,s^{-1}\,m^{-1}\,K^{-1}$ (under gradient of thermal potential in K per length of 1 m) is normally called thermal conductivity. The electrical diffusivity (under gradient of electric potential in V per length of 1 m, called electric field) in units $C\,s^{-1}\,m^{-1}\,V^{-1}$ (= $S\,m^{-1}$, siemens per meter) is normally called electrical conductivity. The mass-, thermal- or electrical-conduction flux is generally a vector (in contrast to electric- or magnetic-induction fluxes treated later in the book, which are scalars, being an integral amount of a vector quantity over a finite area; a dot product of these two vectors, a scalar).

[22] This is in order to distinguish it from diffusion via extended defects and grain boundaries, which is usually faster than bulk diffusion. However, for high concentrations of point defects, as in many functional materials, bulk diffusion may dominate.

[23] There are also other, more complicated, mechanisms of atom movements.

[24] The vibration frequency v is about $10^{13}\,s^{-1}$. In the Debye model of atomic oscillators connected by elastic springs of chemical bonds in a periodic solid, $\hbar v = k\Theta_D$, where $\hbar = h/2\pi$ and h is the Planck constant ($6.62607015 \times 10^{-34}\,J\,s$), k is the Boltzmann constant ($1.380649 \times 10^{-23}\,J/K$) and Θ_D is the Debye temperature of the solid, typically a few hundreds of kelvins. The $k\Theta_D$ term is the maximum energy (highest frequency) of a sound wave propagating in the solid. Θ_D is proportional to the square root of the "spring constant" (the chemical bond strength).

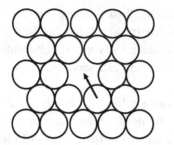 

Figure 3.7 Self-diffusion of an atom via vacancy (left) or interstitial (gray, right).

a along the chosen direction of the diffusion marked with the arrow. The progression rate (in s^{-1}) of that jump in 2D is $r_{\text{progression}} = \frac{1}{4}\,p_B v$.

In general, the progression rate of a jump in the direction of the flux is obtained by multiplying the jump frequency $p_B v$ with the probability p_{avail} that there is a site to jump to, with the probability p_{dir} that such a site moves the atom forwards by a certain distance along the direction of the flux, and with a correlation factor f_c (described below) between the atom that has jumped and the space it left behind:

$$r_{\text{progression}} = f_c\, p_{\text{dir}}\, p_{\text{avail}}\, p_B\, v \qquad (3.10)$$

All three factors, p_{avail}, p_{dir}, and f_c, depend on the crystal structure and the diffusion mechanism. Let's consider a body-centered cubic crystal ($N_n = 8$ nearest neighbors, Figure 1.23) of a metal M with very dilute[25] vacancies as the sole defects. Self-diffusion of the atoms via vacancies will have $p_{\text{avail}} = 8[v_M]$, which is a probability that a vacancy occurs around a selected metal atom when the fraction of vacant sites is $[v_M]$.[26] The atom will then jump into this vacancy. The probability p_{dir} that this particular vacancy will move the jumping atom along the positive direction of the unit-cell edge is $\frac{1}{2}$. This is because 4 of the 8 nearest-neighbor sites advance the atom along this direction (all by $a/2$). Thus, $p_{\text{avail}}\,p_{\text{dir}} = 4[v_M]$ in this example. Now, it is relatively unlikely that there will be another vacancy waiting for our atom to jump further. Why doesn't it jump straight back into the vacancy it left behind? The reason is that it needs to overcome the same energy barrier to jump back. While it waits for this (for a time dictated by the probability), other atoms may fill that vacancy. The probability of the $N_n - 1 = 7$ neighbors not jumping into the vacancy before our atom is just $\frac{1}{2}$. If they don't, our atom jumps back, eliminating its 2 jump chances, and the progression rate of the

---

[25] Vacancies may tend to cluster at temperatures close to the melting point of the solid and speed up the diffusion. In real materials, line and plane defects, as well as grain boundaries, also contribute to diffusion.

[26] Do not confuse the Greek lowercase v symbol for the frequency with the typographically similar italicized Latin lowercase v in the symbol v_M for the vacancy.

original jump is affected by a factor of $1-\frac{2}{7}=0.714$. This rough estimate, $f_c = 1-2/(N_n-1)$, of the correlation factor in Equation (3.10) agrees reasonably with the actual $f_c = 0.727$ obtained by exact and involved derivation for $N_n = 8$ (see Mehrer in Further Reading). This approximation works well for $N_n > 4$.

Instead of the metal atom M, let's consider the vacancy jump. While the atom M jumping via the vacancy mechanism has $p_{avail} = N_n [v_M]$, the vacancy is jumped into by surrounding M atoms, the availability of which at a selected neighbor site is $p_{avail} = 1-[v_M]$ while p_{dir} and f_c remain the same. Now, what is the jump frequency of this vacancy? If it has N_n neighbors, it will get jumped into N_n times more often than by just one atom; hence the jump frequency of the vacancy is $N_n p_B v$ instead of $p_B v$ for the atom M in Equation (3.10). While $r_{progression}(M) = f_c p_{dir} N_n [v_M] p_B v$, we find that $r_{progression}(v_M) = f_c p_{dir} N_n (1-[v_M]) p_B v$. The ratio of these two jump progression rates is $[v_M]/(1-[v_M])$.

In more general terms, the ratio of the atom- and vacancy-progression rates is a consequence of the **jump balance**: as 1 vacancy jumps 1000 times, 1000 atoms jump once. We'll see soon that $r_{progression}$ is directly proportional to the diffusion coefficient D, and we anticipate:

$$\frac{D_{atom}}{D_{vacancy}} = \frac{[v_{atom}]}{1 - [v_{atom}]} \tag{3.11}$$

For low defect fractions, this simplifies to $D_{atom} = D_{vacancy} [v_{atom}]$.

3.3.2 Random Hopping

In the previous section, we realized that atoms in solids jump and move even without an external driving force. In isotropic solids, the movement has equal probability in all directions. The eventual probability distribution follows from a 2D thought experiment called a **random walk**: If you tag an atom on a plane by placing the coordinate cross at it, and let it "walk" n steps, it will end up some distance from the origin. Do the same with another atom, and it will end up somewhere else. If you mark the ends of very many such random walks, there will be no directional preference, only a radial distribution, in 1D.

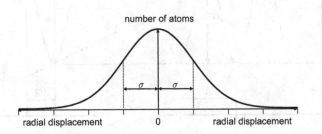

Figure 3.8 Radial distribution of atoms after many long random walks from zero.

The radial distribution is binomial, and becomes normally distributed for an infinite number of steps. It can be shown that the mean square displacement σ_n^2 after n jumps of equal length λ is equal to the sum of the squares of the individual jump vectors, $\sigma_n^2 = n\lambda^2$. Hence after n steps of random walk, the root mean square displacement from the origin is $\sigma = \lambda\sqrt{n}$. The distribution in Figure 3.8 shows that σ is a convenient measure of the progression of the atoms, only a tiny fraction of which will reach the maximum distance λn in the given time $\tau = n/r_{jump}$. Problem 3.20 explores these ideas further.

3.3.3 Hopping Under a Driving Force

At thermal equilibrium, atoms and defects adopt local minima of the Gibbs energy. A typical energy profile between two interstitial sites is shown in Figure 3.9. In the absence of any external force, jumps between the sites will occur at random. If, however, an external field exerting a force F is applied, the jumps will develop a preferred direction. Such a field may be due to a chemical-potential or an electric-potential gradient. In the direction of the force, the Gibbs-energy barrier for migration, $\Delta^{\ddagger}G_m$, will be lowered (Figure 3.9, right) by an energy E_m, while it will be increased in the opposite direction by the same amount. E_m is equivalent to the work done by the force F along the corresponding path $\lambda/2$:

$$E_m \equiv \tfrac{1}{2}\lambda F \tag{3.12}$$

For a jump from site 1 to site 2, the barrier height becomes $(\Delta^{\ddagger}G_m - E_m)$, and the progression rate r_{12} of this jump is $r_{12} = p_{dir}\,p_{avail}\,\nu\,\exp\{(-\Delta^{\ddagger}G_m + E_m)/kT\}$ in analogy with Equation (3.10), or $r_{12} = r_{progression}\exp(E_m/kT)$, where $r_{progression}$ refers to the progression rate under

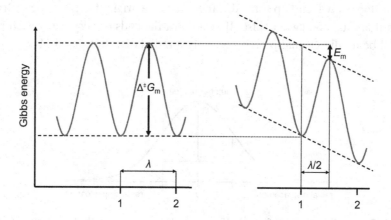

Figure 3.9 Gibbs-energy profile between two interstitial sites. Left: No external force. Right: Under an external force, the forward barrier is lower by E_m.

random movement. When the force F is small, the exponential function can be approximated by a Taylor series with only the linear term[27] retained, $\exp(E_m/kT) \approx 1 + (E_m/kT)$. The frequency of this jump then becomes $r_{12} \approx r_{progression}\{1 + (E_m/kT)\}$. For jumps in the opposite direction, the Gibbs-energy barrier is higher by E_m and accordingly $r_{21} \approx r_{progression}\{1 - (E_m/kT)\}$. Combining these terms gives the net rate $r_{net} = r_{12} - r_{21} = r_{progression}(2E_m/kT)$. Substitution for E_m from Equation (3.12) yields $r_{net} = r_{progression}(\lambda/kT)F$. The net flux J (in $s^{-1}\,m^{-2}$) in a given direction is obtained by multiplying r_{net} (in s^{-1}) by the distance λ (in m) between the two sites and by the local volume concentration c of the atoms (in m^{-3}):

$$J = r_{progression}\frac{\lambda^2 c}{kT} \cdot F \tag{3.13}$$

This result provides two important conclusions: (1) The flux depends on the probability of the random progression in the given direction modified by the effect of the external force. (2) The flux is linearly proportional to the external force.[28] In the following sections, we'll consider two such driving forces—the entropy increase upon homogeneous distribution of defects and the drift of charged defects in an electric field.

3.3.4 Hopping Under a Concentration Gradient

When defects move in a concentration gradient to homogenize their distribution in a sample, the associated entropy increase decreases the Gibbs energy of the system. The driving force F_i per atom i migrating in direction y is given by the gradient of the partial molar Gibbs energy of that atom (gradient of that atom's chemical potential μ_i):

$$F_i = -\frac{d\mu_i}{dy} \tag{3.14}$$

The negative sign in Equation (3.14) means that the transport proceeds from higher values of the potential to lower values, i.e. down the (negative) slope of the potential. The chemical potential per atom i is defined[29] as:

$$\mu_i = \mu_{0i} + kT\ln a_i \tag{3.15}$$

For a dilute ideal solution of the defect i, the activity a_i can be replaced with the volume concentration c_i of the defects (in m^{-3}).[30] Considering that $d\mu_i/dy = (d\mu_i/dc_i)(dc_i/dy)$, the force F_i per atom is:

---

[27] $f(x) = f(0) + f'(0)x$. For $x = 0.1$ in e^x, the error is 0.47%, for $x = 0.5$ the error is 9%.

[28] As long as F is sufficiently small so that the linear term of the Taylor-series approximation is sufficient.

[29] See any textbook on physical chemistry.

[30] If the concentration is expressed in mol/m^3, the Boltzmann constant k is replaced with the molar gas constant $R = N_A k$, where N_A is Avogadro's number ($6.02214076 \times 10^{23}\,mol^{-1}$ exactly).

$$F_i = \frac{1}{c_i} \cdot kT\left(-\frac{dc_i}{dy}\right) \qquad (3.16)$$

Substituting for F in Equation (3.13) gives:

$$J = r_{\text{progression}} \frac{\lambda^2 c_i}{kT} \cdot \frac{1}{c_i} kT\left(-\frac{dc_i}{dy}\right) = r_{\text{progression}} \lambda^2\left(-\frac{dc_i}{dy}\right) \qquad (3.17)$$

This reproduces the macroscopic Fick's first law, Equation (3.7), via microscopic atomic jumps. We see that the diffusivity D_i (in m^2 s^{-1}) of the atom (defect) i in an ideal and dilute solution[31] in the solid is

$$D_i = r_{\text{progression}} \lambda^2 \qquad (3.18)$$

where $r_{\text{progression}}$ (in s^{-1}) is the progression rate at a given site, Equation (3.10), under random movement, and λ is the jump length in the direction of the progression (in m).[32] This is an important result and shows that diffusion under an applied force is controlled by the atom's self-diffusion coefficient.

3.3.5 Hopping Under an Electric Field

From electrostatics we know that the force F_i acting on charge q_i in an electric field of intensity E is $F_i = q_i E$.[33] When the charged defect i moves down the gradient of the electric potential V (in volts),[34] it experiences the electric-field intensity $E = -dV/dy$ and:

$$F_i = q_i\left(-\frac{dV}{dy}\right) \qquad (3.19)$$

Substituting for F in Equation (3.13) and converting the particle flux J_i into the charge flux j_i by multiplying it with the charge (in C) per particle q_i (hence $j_i = q_i J_i$), we obtain

---

[31] If the solution of the defect in the solid were ideal but not dilute, the backward flux of the solvent (i.e. the solid host matrix) would have to be included with its own (intrinsic) diffusion coefficient of the participating fluxes. The combined diffusivity is often called interdiffusivity, denoted as $\tilde{D}$.

[32] For isotropic cubic structures, the probability p_{dir} that a vacancy jump moves a selected atom forward along the direction of the flux, when multiplied by the square of the length λ of the projection of the jump onto that direction, equals $\ell^2/6$, where ℓ is the actual length of the jump. Furthermore, $\ell^2/6$ equals a^2/N_n, where a is the unit-cell parameter and N_n is the number of nearest neighbors of the vacancy. These formulae can be conveniently used instead of $p_{\text{dir}}\lambda^2$ when D is calculated from the combined Equations (3.10) and (3.18).

[33] The unit of the electric-field intensity is V/m. One V/m is the intensity of an electric field where a point charge of $1\,C\,(C \equiv A\,s)$ experiences a force of $1N\,(N \equiv J/m \equiv Ws/m \equiv VA\,s/m)$. 1 N of force per 1 C of charge then equals 1 V/m electric field.

[34] Note that in this context we use the symbol V for the electric potential. Do not confuse with volume.

$$j_i = \sigma_i \left(-\frac{dV}{dy} \right) \qquad (3.20)$$

in which σ_i is the **electrical conductivity** due to defect i,

$$\sigma_i = r_{progression} \frac{\lambda^2 c_i q_i^2}{kT} \qquad (3.21)$$

where $r_{progression}$ is again the progression rate of Equation (3.10) under random movement.

Electrical conductivity is an intrinsic material property and will be treated in more detail in Chapter 10. The conductivity σ_i represents a fraction t_i of the total electrical conductivity σ of the solid due to defects, and t_i is termed the **transference number** of the point defect i. For flux of elementary charges under *constant* electric potential (a steady state), Equation (3.20) adopts the form $j = \sigma E$ that we'll recognize in Chapter 10, Equation (10.2), as one possible expression of Ohm's law.

3.3.6 Relationship between Conductivity and Diffusivity

Whether the driving force is electrical or chemical, it produces a flux. The flux J_i of the defect i at a concentration c_i under the general driving force F_i, Equation (3.13), can be expressed using the experimentally available[35] random diffusivity D_i from Equation (3.18):

$$J_i = \frac{D_i c_i}{kT} \cdot F_i \qquad (3.22)$$

It can also be expressed using the experimentally available[36] electrical conductivity σ_i. This makes use of the **Nernst–Einstein relation** between σ_i (in S/m, 1 S $\equiv$ 1 A/V) and D_i (in m^2/s),[37] obtained when Equation (3.21) is expressed with D_i of Equation (3.18) for a defect i of charge q_i (in C $\equiv$ A s) and carrier density c_i (in[38] 1/m^3):

---

[35] D_i is obtained by tracer-diffusion techniques with isotopes. An isotope is traceable either by mass spectrometry after sectioning the sample, or by measuring radioactivity along the diffusion path if a radioactive isotope is used. As noted earlier, for vacancy mechanisms the tracer-diffusion coefficient is smaller than a purely random diffusion model by the correlation factor f_c. For interstitial diffusion of dilute defects, $f_c = 1$.

[36] σ_i is obtained by measuring electrical conductivity with electrodes that conduct specifically via the defect i. As an example, ZrO_2 (Chapter 13) blocks electronic carriers while allowing oxide anions through.

[37] The D_i thus obtained from conductivity ($D_{conductivity}$) will not necessarily be the same as the tracer-diffusion coefficient for the same atom/defect. The ratio $D_{tracer}/D_{conductivity}$ is then evaluated as the Haven ratio and reflects various movement correlations for the charged defects.

[38] When the amount of charge-carrying defects is expressed in moles, hence the carrier density is in mol/m^3, we replace kT with RT, and the defect's electric charge $q_i = z_i e$ with $z_i F$. Here F is the Faraday constant (1 mole of elementary charges: $N_A e = 6.02214076 \times 10^{23} \times 1.602176634 \times 10^{-19} \approx 96485.3$ C) and z_i is the defect's charge number (the integer charge in units of elementary charge e).

$$\frac{D_i}{kT} = \frac{\sigma_i}{q_i^2 c_i} \qquad (3.23)$$

Substituting from here for D_i in Equation (3.22) yields the flux in terms of σ_i:

$$J_i = \frac{\sigma_i}{q_i^2} \cdot F_i \qquad (3.24)$$

Returning to the Nernst–Einstein relation, we can perform one last step that will lead to the useful concept of **mobility** that we'll encounter in Chapters 10 and 13. Mobility, μ_i, is the speed of a species i per unit intensity of the field that moves it. It is obtained by multiplying Equation (3.23) with the charge q_i in coulombs. Both sides of the equation then have units of m/s per V/m; the left-hand side $\mu_i = D_i q_i / kT$ expresses the mobility of the charged defect via its random diffusivity and the right-hand side $\mu_i = \sigma_i / q_i c_i$ via its electrical conductivity. Finally, we should note that transport of a single type of charged particle is rare. It is usually accompanied by transport of a compensating counter charge, forming two fluxes that together appear as a total flux of neutral atoms. This is explored in the following section.

3.3.7 Ambipolar Diffusion

Transport of charged defects is not limited to situations where an electric field is applied. In many cases, transport under a concentration gradient proceeds via charged defects. Clearly, the transport of a charged defect in the absence of an applied electric field must maintain bulk neutrality. Consider the oxidation of $YBa_2Cu_3O_6$ by O_2 as discussed in Section 3.2. Chemically, a copper cation is oxidized while $\frac{1}{2}O_2$ is reduced to O^{2-}. As O^{2-} migrates into the center of the sample, the positive cation charges (holes) follow its motion; we thus have an influx of both negative (O^{2-}) and positive species ($2h^{\cdot}$), which together represent the flow of oxygen as a neutral atom (O). This type of polar transport of nonpolar species is called **ambipolar diffusion**.

Equation (3.4) describes oxygen transport in $YBa_2Cu_3O_{7-\delta}$ in terms of defects. At the surface, O_2 oxidizes copper (creates $h^{\cdot}$) and is reduced to O^{2-} ions that fill $v_O^{\cdot\cdot}$ coming from the interior. The holes migrate inwards. Because these defects are charged, any non-uniform distribution not only creates a concentration gradient but also a local electric-potential gradient. A flux under the combined chemical and electric potentials therefore requires a formal summation of the two forces F_i used separately in Equation (3.14) and Equation (3.19). The flux of Equation (3.24) then becomes:

$$J_i = \frac{\sigma_i}{q_i^2} \cdot \left(-\frac{\partial \mu_i}{\partial y} - q_i \frac{\partial V}{\partial y} \right) \qquad (3.25)$$

As stated above, there are two such fluxes, an outward flux of $v_O^{\bullet\bullet}$ and a charge-compensating inward flux of $h^{\bullet}$. The charge of one mole of oxygen vacancies is $+2F$, the charge of one mole of holes is $+1F$. The participating mole fluxes are then:

$$J_{v_O^{\bullet\bullet}} = \frac{\sigma_{v_O^{\bullet\bullet}}}{4F^2} \cdot \left(-\frac{\partial \mu_{v_O^{\bullet\bullet}}}{\partial y} - 2F\frac{\partial V}{\partial y}\right) \text{ and } J_{h^{\bullet}} = \frac{\sigma_{h^{\bullet}}}{F^2} \cdot \left(-\frac{\partial \mu_{h^{\bullet}}}{\partial y} - F\frac{\partial V}{\partial y}\right) \tag{3.26}$$

Electroneutrality requires the sum of charged fluxes to be zero. The flux of holes must therefore be twice the flux of oxygen vacancies and run in the opposite direction:

$$J_{h^{\bullet}} = -2J_{v_O^{\bullet\bullet}} \tag{3.27}$$

We cannot establish the actual electric potential created inside the solid, so a good way to proceed is to eliminate the $\partial V/\partial y$ terms from Equations (3.26).[39] By taking the actual inward O flux (see above) as the negative of the oxygen vacancy flux, we obtain:

$$J_O = -J_{v_O^{\bullet\bullet}} = \frac{1}{4F^2} \frac{1}{\dfrac{1}{\sigma_{v_O^{\bullet\bullet}}} + \dfrac{1}{\sigma_{h^{\bullet}}}} \cdot \left(\frac{\partial \mu_{v_O^{\bullet\bullet}}}{\partial y} - 2\frac{\partial \mu_{h^{\bullet}}}{\partial y}\right) \tag{3.28}$$

The form of this equation suggests that the flux of O behaves as if driven by its own chemical-potential gradient—a sum of the gradients of the ionic- and electronic-defect fluxes (of $v_O^{\bullet\bullet}$ and $h^{\bullet}$, respectively):

$$-\frac{\partial \mu_O}{\partial y} = \frac{\partial \mu_{v_O^{\bullet\bullet}}}{\partial y} - 2\frac{\partial \mu_{h^{\bullet}}}{\partial y} \tag{3.29}$$

Note that the combination of the ionic- and electronic-defect conductivities in Equation (3.28) appears in a mathematical form as if ionic and electronic *resistivities*[40] were summed and then inverted into the total ambipolar conductivity σ_O. This can be interpreted as though the ionic and electronic resistors are connected in series in a circuit.

When the O-flux Equation (3.28) is formally rewritten in terms of $\partial \mu_O/\partial y$ and σ_O,

$$J_O = \frac{\sigma_O}{4F^2} \cdot \left(-\frac{\partial \mu_O}{\partial y}\right) \tag{3.30}$$

it becomes obvious that a diffusion coefficient for the neutral species O (the chemical diffusion coefficient D_O) can also be defined. At this point, we need to remember that diffusivities are normally expressed via gradients of concentrations c and not of chemical potentials (Section 3.3.4). Therefore a conversion $\partial \mu_O/\partial y = (\partial \mu_O/\partial c_O)(\partial c_O/\partial y)$ is performed

---

[39] For example by expressing the coulombic term $\partial V/\partial y$ from the equation for holes, substituting it into the equation for vacancies, substituting for the hole flux from Equation (3.27), and rearranging.

[40] Resistivity is the inverse of conductivity.

on Equation (3.30). In the result, we identify D_O as the proportionality parameter between the flux J_O and the concentration gradient $-\partial c_O/\partial y$:

$$D_O = \frac{\sigma_O}{4F^2} \cdot \left(\frac{\partial \mu_O}{\partial c_O}\right) \tag{3.31}$$

Having defined the chemical diffusion coefficient D_O, we can return to the individual point defects and their concentrations. From Equation (3.29), we see that $\partial \mu_O = -\partial \mu_{v_O^{\bullet\bullet}} + 2\partial \mu_{h^\bullet}$, and electro-neutrality dictates that $\partial c_O = -\partial c_{v_O^{\bullet\bullet}} = \partial c_{h^\bullet}/2$ (there are two holes per vacancy). This gives:

$$D_O = \frac{\sigma_O}{4F^2} \cdot \left(\frac{\partial \mu_{v_O^{\bullet\bullet}}}{\partial c_{v_O^{\bullet\bullet}}} + 4\frac{\partial \mu_{h^\bullet}}{\partial c_{h^\bullet}}\right) \tag{3.32}$$

For low concentrations of defects, activity depends on concentration via $\mu = \mu_0 + RT\ln c$; hence $\partial \mu/\partial c \approx RT/c$, where c is the equilibrium steady-state concentration in mol m^{-3}.[41] Converting σ_O back to the defect conductivities, we obtain:

$$D_O = \frac{RT}{F^2} \cdot \frac{1}{\dfrac{1}{\sigma_{v_O^{\bullet\bullet}}} + \dfrac{1}{\sigma_{h^\bullet}}} \cdot \left(\frac{1}{4c_{v_O^{\bullet\bullet}}} + \frac{1}{c_{h^\bullet}}\right) \tag{3.33}$$

What remains is to express the separate diffusion coefficients for the ionic- and electronic-defect components. This is done by substituting for each of the two defect conductivities from the Nernst–Einstein relation of Equation (3.23), which after rearrangement gives:

$$D_O = \frac{1}{\dfrac{1}{D_{v_O^{\bullet\bullet}} \cdot 4c_{v_O^{\bullet\bullet}}} + \dfrac{1}{D_{h^\bullet} \cdot c_{h^\bullet}}} \cdot \left(\frac{1}{4c_{v_O^{\bullet\bullet}}} + \frac{1}{c_{h^\bullet}}\right) \tag{3.34}$$

We see that the diffusion coefficient of the chemical element transported depends on the participating electronic and ionic defects; their individual diffusivities/conductivities, their charge numbers, and their equilibrium concentrations. The diffusion of the neutral chemical element is largely controlled by the slower species,[42] but the slower flux is somewhat enhanced by the faster flux. As an example, the enhancement factor for Equation (3.34) approaches a maximum of $D_O/D_{v_O^{\bullet\bullet}} = 1 + (4c_{v_O^{\bullet\bullet}}/c_{h^\bullet}) = 3$ if $D_{h^\bullet}c_{h^\bullet} \gg D_{v_O^{\bullet\bullet}}4c_{v_O^{\bullet\bullet}}$.

---

[41] For concentrations given in number of particles per m^3, R is replaced by k, and F by e.

[42] Diffusion of neutral O through a purely oxide-ion conductor, such as Ca- or Y-doped zirconia, is limited by the material's low electronic conductivity. It speeds up enormously when a metallic conductor is added to transport the electrons. That can be a Pt powder mixed with the doped-zirconia powder prior to sintering into the device form, or an external wire connecting porous Pt electrodes applied onto the two opposite surfaces along the diffusion path. In contrast, diffusion of O through a good electronic conductor, such as YBa$_2$Cu$_3$O$_{7-\delta}$, is controlled by the oxygen-ion conductivity of the material.

3.3.8 Temperature Dependence of Diffusivity

As we have just seen, point defects are the vehicles of mass transport in crystalline solids. Diffusivity depends on temperature via the energetics of their movement. The energetics may have two contributions—the cost of making the vehicle and the cost of running it. Both costs can be expressed in terms of the free-energy change ΔG (the ability of a system to do work), and the common approximation is to consider the enthalpy and entropy terms in $\Delta G = \Delta H - T\Delta S$ to be independent of temperature.

The making costs are the total Gibbs free energy of formation for the defects relevant in the particular transport. This is derived in Section 2.3, where Equation (2.3) gives the equilibrium fractional concentration of the defect as a function of the changes in atomic vibration entropy (ΔS_{vib}) and enthalpy (ΔH_f) upon defect formation. The running costs stem from Equation (3.9), which represents the increasing Boltzmann probability that at higher temperatures the atom will be more likely to overcome the Gibbs-energy barrier for migration: $\Delta^{\ddagger}G_m = \Delta^{\ddagger}H_m - T\Delta^{\ddagger}S_m$.

Let's consider a metal atom M diffusing by a vacancy mechanism in its own cubic structure where it has N_n nearest neighbors. Substituting $\Delta^{\ddagger}G_m$ in Equation (3.9) with $\Delta^{\ddagger}H_m - T\Delta^{\ddagger}S_m$, considering that $p_{avail} = N_n[v_M]$ in Equation (3.10), expressing $[v_M]$ with Equation (2.3), replacing k with R so that the enthalpy and entropy are per mole, and plugging the result into Equation (3.18), yields:

$$D_M = \lambda^2 f_c\, p_{dir}\, N_n\, \nu \exp(\Delta^{\ddagger}S_m/R + \Delta S_{vib}/R) \cdot \exp(-\Delta^{\ddagger}H_m/RT - \Delta H_f/RT) \qquad (3.35)$$

Here D_M has the temperature dependence of the **Arrhenius equation**, $D = D_0\exp(-E_A/RT)$, in which the activation energy E_A combines the enthalpy contributions, and the pre-exponential term D_0 includes everything else. E_A can therefore be obtained from experimental data by least-squares fitting of the linear slope of $\ln D$ versus $1/T$.

In some cases, the making costs do not apply and only running costs need be considered. This is the case for the diffusion of foreign or extrinsic interstitial atoms in low concentrations, for example of solid-solution carbon atoms in iron. Here the vehicle is simply the extrinsic defect itself, $p_{avail} \approx 1$, and the equivalent expression becomes:

$$D_{interstitial,extrinsic} = \lambda^2 f_c\, p_{dir}\, \nu \exp(\Delta^{\ddagger}S_m/R) \cdot \exp(-\Delta^{\ddagger}H_m/RT) \qquad (3.36)$$

3.3.9 Diffusivity and Redox Defect Equilibria

Let's now consider diffusion in a metal oxide via a metal- or oxygen-vacancy mechanism. Here, we'll have to take into account how the vacancy fraction depends on the equilibria we explored in Section 3.1. The constituent atom i (metal or oxygen) at its regular lattice site will not migrate unless it has an adjacent vacancy. The probability of this is $p_{avail} = N_n[v_i]$, and we modify the previous result (take $f_c = 1$ for simplicity) to

$$D_i = N_n[v_i]\lambda^2 p_{dir} v \exp(\Delta^\ddagger S_m/R) \cdot \exp(-\Delta^\ddagger H_m/RT) \tag{3.37}$$

The vehicle production costs not-yet-included concern the vacancy fraction $[v_i]$ and follow from energetics of the point-defect equilibria the vacancy is involved in.

Let's consider three cases where this $[v_i] = f(T)$ function is straightforward. The first is when the high-temperature vacancy fraction $[v_i]$ in the material becomes "frozen" at lower temperatures when kinetic factors prevent further redox exchange with the reaction atmosphere to establish a new equilibrium. The second is when $[v_i]$ is independent of temperature because it is fixed by aliovalent dopants around the point of integer valence (oxygen vacancies in Figure 3.4 and nickel vacancies in Figure 3.5). The third case is when the vacancy is the dominant defect and $[v_i]$ can be expressed via the temperature dependence of the equilibrium constant K for the defect-formation reaction in terms of the reaction-enthalpy and -entropy change as $K = \exp[(\Delta S/R) - \Delta H/RT]$.[43] As an example, when NiO is oxidized, holes are compensated by nickel vacancies (Figure 3.2). The electroneutrality condition simplifies to $2[v_{Ni}''] = [h^\bullet]$, and from Table 3.4 we see that $[v_{Ni}''] = (p_{O_2}K_{ox}/16)^{1/6}$. Substituting into Equation (3.37) and rearrangement gives the random diffusion coefficient of Ni in oxidized nonstoichiometric NiO via a vacancy mechanism as

$$D_{Ni} = \left(\frac{p_{O_2}}{16}\right)^{\frac{1}{6}} N_n \lambda^2 p_{dir} v \exp\left(\frac{\Delta S_{ox} + 6\Delta^\ddagger S_m}{6R}\right) \cdot \exp\left(-\frac{\Delta H_{ox} + 6\Delta^\ddagger H_m}{6RT}\right) \tag{3.38}$$

where both the Arrhenius activation enthalpy and the pre-exponential factor have two components—one from the defect formation and one from the defect mobility. As stated earlier, v is the (essentially constant) vibration frequency of the atom and p_{dir} is the probability that the jump occurs in the flux direction, advancing the atom by the distance λ. See Footnote 32 in this chapter for alternative expressions of $\lambda^2 p_{dir}$.

3.3.10 Outline of Non-Steady-State Diffusion

Fick's second law, Equation (3.8), expanded into three dimensions, has analytical solutions for time-dependent concentrations of diffusing species across simple shapes such as a plate of infinite thickness (termed a half-space; diffusion enters from one side only), a plate of finite thickness (diffusion enters from both sides), a parallelepiped, a cylinder, or a sphere, all provided D is constant. The principal solutions are in the literature on either particle diffusion [like Crank or Mehrer in Further Reading] or heat conduction. When D depends on the concentration c, this must be properly included in the treatment.

As an example, we'll calculate oxygen-content profiles in a half-space of $YBa_2Cu_3O_6$ exposed to O_2 at 350 °C (Figure 3.10). In this arrangement, the oxygen flux has one direction. Assuming that the surface reduction of O_2 is instantaneous and D is independent of

---

[43] Since $\Delta G = -RT \ln K$.

Figure 3.10 Set-up sketch of oxygen diffusion into half-space of $YBa_2Cu_3O_6$.

concentration, Equation (3.8) is simplified, and the solution starts by introducing the dimensionless (half) length,

$$u = \frac{y/2}{\sqrt{D\tau}} ,$$

(3.39)

as a new variable.[44] This replaces the two variables (τ and y) in Fick's second law $\frac{\partial c}{\partial \tau} = D\left(\frac{\partial^2 c}{\partial y^2}\right)$ with one. Given the new variable u, $\frac{\partial c}{\partial \tau} = \frac{\partial c}{\partial u} \cdot \frac{\partial u}{\partial \tau}$ and $\left(\frac{\partial^2 c}{\partial y^2}\right) = \frac{\partial^2 c}{\partial u^2} \cdot \left(\frac{\partial u}{\partial y}\right)^2$. In the latter two expressions, $\frac{\partial u}{\partial \tau} = -\frac{u}{2\tau}$ and $\frac{\partial u}{\partial y} = \frac{1}{\sqrt{4D\tau}}$ are obtained by differentiating u.

This yields the differential equation $\frac{d^2 c}{du^2} + 2u\frac{dc}{du} = 0$. Substituting $z = dc/du$ simplifies it to $\frac{dz}{du} + 2uz = 0$. Separation of variables gives $\frac{1}{z}dz = -2udu$ that is integrated to $\ln z = -u^2 + const$, from which $z = C \cdot \exp(-u^2)$. Putting back $z = dc/du$ yields $dc = C \cdot \exp(-u^2)du$. The integral of $\exp(-u^2)du$ equals $\frac{\sqrt{\pi}}{2} \cdot erf(u)$, expressed with the Gaussian error function (erf).[45] Given the boundary limits of erf(0) at the concentration $c = 1$ at $u = 0$ at any time and erf(∞) = 1 when $c = 0$ at $u = \infty$ at any time, $C = -2/\sqrt{\pi}$. Integration of the left-hand side of the differential equation gives $c = -erf(u) + K$. The initial condition of $c = 1$ at $u = 0$ yields $K = 1$ and the relative concentration change c_r:

$$c_r = 1 - erf(u)$$

(3.40)

In Figure 3.11 (left), our solution of Equation (3.40) is illustrated for $D_O = 7.144 \times 10^{-13}$ m^2/s at 350 °C. The concentration profile is linear close to the interface as erf(u) $\approx u$ for low u. Even at $u = 0.5$, erf(u) is 0.5205, which happens at the distance $\sqrt{D\tau}$; see Equation (3.39). This distance is called the **penetration depth** (Figure 3.11). It is a characteristic value indicating the extent of penetration of the diffusant. For diffusion solutions based on the erf function, it is close to the diffusion half length; a length where a 50% concentration change occurs in

---

[44] This is possible only if both initial and boundary conditions are functions of u only.

[45] The erf(u) is the integral of the normal probability distribution from its center at 0 to u. That distribution is the limiting large-numbers envelope for the discrete binomial distribution, Figure 3.8.

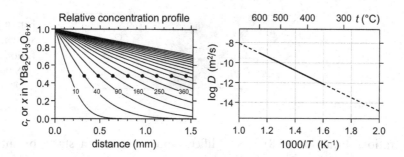

Figure 3.11 Left: Profiles of oxygen-content increase in a thick layer of $YBa_2Cu_3O_6$ exposed to O_2 at 350 °C, after 10, 40, 90, etc. hours. Penetration depths $\sqrt{D\tau}$ are marked by dots. Right: Arrhenius-type temperature dependence of the diffusion coefficient over temperature range 350 to 600 °C.

a given time. Note also that the penetration depth doubles in 2^2 times that time, triples in 3^2 times that time, etc.

Equation (3.40) is derived for a plate of infinite thickness. It can also be used for the initial stages of diffusion into a finite or thin plate, but only until the concentration starts to change significantly at half the plate thickness. The D values for temperatures of interest are easily evaluated from published Arrhenius-type temperature dependencies. The D dependence for our example is plotted in Figure 3.11 (right) and has an activation energy of 129.2 kJ/mol and D_0 of 4.808×10^{-2} m²/s.

3.3.11 Cautionary Note on Diffusion in Real Materials

Caution needs to be applied when investigating diffusion in real oxides and other binary compounds with high ionicity. In addition to possible computational difficulties, problems may be encountered either due to the diffusion model or due to the sample. The model-related problems concern the assumption of single point defects being the transport vehicles. In many oxides, these are in reality aggregated defect clusters, the simplest of which are dimers of interstitials or vacancies. Other model-related problems include disregarding ambipolar diffusion, and, in variable-temperature studies, the fact that D may depend on some defect concentration that is chemically variable due to reaction with the measurement atmosphere. There are also the practical difficulties of arranging the experiment in a manner that correctly approximates the initial and boundary conditions of the model. Sample perfection and morphology is also of concern—in single-crystal measurements, line defects can lead to short-circuit diffusion; in polycrystalline materials, a similar rapid transport can occur at grain boundaries. Last, but not least, omnipresent aliovalent impurities will increase diffusivity by creating point defects as vehicles for the mass transport.

3.4   Problems

3.1   Write Kröger–Vink symbols for the following fully charged point defects in NiO: metal vacancy, oxygen vacancy, lithium acceptor, aluminum donor.

3.2   Name the following defects in TiO_2 and write their Kröger–Vink symbols: Al and Nb dopants at the Ti site; F and N dopants at the O site; oxygen and titanium atoms out of their regular sites.

3.3   State the type of intrinsic defects would you predict in fluorite-type CeO_2 and UO_2.

3.4   Recast Table 3.4 into one valid for a hypothetical MO oxide with cation-Frenkel intrinsic defects.

3.5   Use chemistry or physics to suggest whether ionic or electronic defects will dominate in pure ZrO_2 at high temperatures.

3.6   Consider pure PbO with anion-Frenkel compensation and dominant ionic defects. What are the slopes of the essentially linear dependences in the three limiting regions of the plot of defect fractions versus p_{O_2}? Sketch the Brouwer diagram.

3.7   In fact, pure PbO may show an anti-Schottky disorder. What differences does this bring to the results of Problem 3.6?

3.8   Consider an idealized pure CuO with anion-Frenkel compensation and dominant electronic defects. What are the slopes of the essentially linear functions in the three limiting regions of the plot of defect fractions versus p_{O_2}? Sketch the Brouwer diagram.

3.9   State the limiting slope of the metal-vacancy fraction versus p_{O_2} in the oxidative-nonstoichiometry range for (a) Cu_2O and (b) Cr_2O_3, both of Schottky intrinsic defects.

3.10   Consider a stoichiometric metal oxide of dominant Schottky defects in equilibrium. Which of these two defects will increase its fraction upon either an acceptor or donor doping under the same conditions?

3.11   Consider a stoichiometric metal oxide in equilibrium with its dominant electronic defects. Which of these two defects will increase its fraction upon an acceptor or donor doping under the same conditions?

3.12   Sketch the Brouwer diagram for Li-doped NiO with dominant electronic defects.

3.13   Sketch the Brouwer diagram for Al-doped NiO with dominant ionic defects.

3.14   Write down the dissolution reaction of ZrO_2 in Cr_2O_3 and state how the solid solubility will depend on p_{O_2}: (a) about the point of integer structure, (b) about the point of integer valence.

3.15   The equilibrium constant in Equation (3.5) for oxidation of the oxygen vacancy in $YBa_2Cu_3O_{7-\delta}$ at 500 °C is $K_{vox} = 55$ bar$^{-\frac{1}{2}}$. (a) State the equilibrium composition of $YBa_2Cu_3O_{7-\delta}$ at 500 °C in a flow of Ar gas containing 100 ppm O_2 and in air, both at 1 bar. (b) State the p_{O_2} at which you would anneal $YBa_2Cu_3O_{7-\delta}$ in order to obtain an average oxidation state of Cu^{II}.

3.16 Derive the simplified expression for gross oxygen nonstoichiometry in $NdBaFe_2O_{5+\delta}$ as a function of p_{O_2} assuming that Fe^{3+} defines the point of integer valence.

3.17 Water flows from a $0.2\ cm^2$ hose at 6 kg per minute. State the mass, volume, mole, and molecule flux of H_2O in SI units.

3.18 Calculate the mass flux of carbon under steady-state diffusion through a steel plate separating carbon-rich and carbon-poor gases at 700 °C if subsequent analysis gives carbon concentrations of $1\ kg/m^3$ and $0.5\ kg/m^3$ at the respective depths of 0.5 mm and 1 mm below the surface. Assume $10^{-11}\ m^2/s$ as the diffusion coefficient of C in the iron matrix at this temperature.

3.19 An oxygen-permeable $YBa_2Cu_3O_{7-\delta}$ membrane 1 cm thick operates at 700 °C between pressurized air at 11 bar and pure O_2 at 1 bar. The steady-state O_2 gas production is 0.84 mL at atmospheric pressure and 20 °C (molar volume 24 L/mol) per square meter every second. Calculate the O_2 flux, O atom flux, and, finally, D_O with $K_{vox} = 4.3\ bar^{-\frac{1}{2}}$ (to obtain the O gradient across the membrane) and with molar $YBa_2Cu_3O_{7-\delta}$ volume of $10^{-4}\ m^3/mol$.

3.20 Consider a random 1D walk of n equal steps of length λ originating at zero. (a) What is the probability of deviating by 4λ from the origin in 4 steps? (b) Construct a table of probabilities of reaching points $-n\lambda \ldots + n\lambda$ for up to $n = 4$. (c) Calculate the variance (i.e. the mean squared deviation from the mean) and its root after 4 steps of *unit* length. (d) Derive a formula for the variance σ_n^2 after n steps of length λ and for its root σ_n.

3.21 Calculate the jump frequency of interstitial carbon in bcc iron at 800 °C, assuming a vibration frequency $v = 10^{13}\ s^{-1}$ and an activation energy for hopping $\Delta^{\ddagger}G_m = 62\ kJ/mol$. On average, every x-th carbon vibration overcomes the jump barrier; determine x.

3.22 Assume that bcc iron at 1800 K has a fraction of vacant Fe sites of 0.0001, an Fe atom vibration frequency $v = 10^{13}\ s^{-1}$ and an activation energy for hopping $\Delta^{\ddagger}G_m = 29\ kJ/mol$. What is the jump frequency of Fe atoms? On average, how many vibrations are there between jumps? What is the Fe progression rate at each site?

3.23 Calculate the self-diffusion coefficient of Fe in Problem 3.22, given the unit-cell parameter $a = 2.87$ Å.

3.24 Given the diffusivity $D_C = 10^{-10}\ m^2/s$ for interstitial carbon in bcc iron at 800 °C of $a = 2.87$ Å, estimate the activation energy $\Delta^{\ddagger}G_m$ for hopping of the C atoms ($v = 10^{13}\ s^{-1}$) via interstitial sites.

3.25 The self-diffusion coefficient in Al (fcc; $a = 4.05$ Å) at 600 K is $D_{Al} = 2 \times 10^{-16}\ m^2/s$. Assuming an atomic vibration frequency $v = 4 \times 10^{13}\ s^{-1}$ and an activation energy for hopping $\Delta^{\ddagger}G_m = 58\ kJ/mol$, calculate the site fraction of Al vacancies.

3.26 Verify the statement in this chapter's Footnote 32 that $p_{dir}\lambda^2 = \ell^2/6 = a^2/N_n$, in which the distance a is the unit-cell edge, ℓ the actual jump length, λ its projection onto the unit-cell edge direction, and N_n is the number of nearest neighbors of the vacancy, for (a) primitive cubic, (b) bcc, and (c) fcc packing of spheres.

3.27 Calculate the self-diffusion coefficient in the primitive cubic α-Po (Figure 1.23; $a = 3.36$ Å) at 500 K via vacancy mechanism, assuming 0.001 vacant sites, $v = 10^{13}$ s^{-1}, $\Delta^{\ddagger} G_m = 36$ kJ/mol. Estimate the diffusivity of the vacancy.

3.28 Calculate the self-diffusion coefficient of α-Po along the direction of the cell edge (Figure 1.23; $a = 3.36$ Å) at 500 K via an interstitial mechanism with full availability of interstitial sites, $f_c = 1$, $v = 10^{13}$ s^{-1}, and $\Delta^{\ddagger} G_m = 40$ kJ/mol. Verify the result by orienting the cube to yield three equivalent jumps for positive progression direction and three for negative progression direction.

3.29 A single crystal of fcc iron ($a = 3.6468$ Å) having a few ppm of interstitial carbon is kept at 1000 °C and isolated from its surroundings. Assume $D_C = 2.217 \times 10^{-11}$ m^2/s for the self-diffusion coefficient of interstitial carbon along a via octahedral holes, site availability $p_{avail} = 1$, correlation factor $f_c = 1$ for the interstitial, and atomic vibration frequency $v = 10^{13}$ s^{-1}. (a) State the probability of a vibration of C becoming a jump. (b) State the total length of the path a carbon atom travels in one minute. (c) State that atom's root mean square displacement from its position a minute earlier. (d) Calculate the hopping activation Gibbs energy per mole of interstitial carbon defects.

3.30 Confirm by dimensional analysis in SI that multiplying Equation (3.23) with the charge q_i yields mobility.

3.31 A rectangle of calcium-stabilized zirconia $Zr_{0.85}Ca_{0.15}O_{1.85}$ ($a = 5.13$ Å) is covered with porous platinum electrodes on its opposite faces, heated to 1100 °C in air, and an electrical conductivity of 6 S/m is measured. Assuming $t(v_O^{\bullet\bullet}) = 1$, calculate the diffusivity $D(v_O^{\bullet\bullet})$ and mobility $\mu(v_O^{\bullet\bullet})$ of the oxygen vacancies in the bulk of the sample, neglecting kinetics of surface recombination.

3.32 Assuming that oxide-ion conductivity of $YBa_2Cu_3O_{7-\delta}$ at 500 °C in air is $\sigma(v_O^{\bullet\bullet}) = 10^{-2}$ S/m and the ionic transference number $t(v_O^{\bullet\bullet}) = 10^{-6}$, what is the total electrical conductivity?

3.33 Calculate the enhancement factor $D_O/D_h^{\bullet}$ for Equation (3.34).

3.34 Use Equation (3.34) to generalize the value of the limiting ambipolar enhancement for D_A of a neutral atom A, the flux of which consists of a flux of a z-charged ionic defect and a much faster flux of singly charged electronic defects.

3.35 Figure 3.2 suggests that oxidative nonstoichiometry of NiO at high temperature is achieved via formation of nickel vacancies and holes. The following data have been measured [5, 6] at 1100 °C in O_2: coefficient of (tracer) random diffusion for nickel atoms $D_{Ni} = 10^{-15}$ m^2/s, total electrical conductivity $\sigma = 65.5$ S/m, site fraction 0.0001 of nickel vacancies. Assuming that Ni diffuses with a vacancy mechanism via nickel sites (never at O site), and given $a = 4.20$ Å for the NaCl-type cell, calculate for each of the two majority defects: (a) Concentration per m^3, (b) diffusivity, (c) ionic electrical conductivity and ionic transference coefficient in order to determine whether NiO is an ionic or electronic conductor, (d) mobility.

3.36 Set up the equation for the temperature dependence of D_{Cr} in oxidized Cr_2O_3 where chromium vacancies dominate.

3.37 Check that $u = (y/2)/\sqrt{D\tau}$ of Equation (3.39) is dimensionless.

3.38 A steel blade has been nitridized by exposure to flowing NH_3 at 700 °C for 1 hour. The hardened layer, estimated from the penetration depth $\sqrt{D\tau}$ of the nitrogen atoms, is 20 μm. In what time would the penetration depth reach 0.1 mm?

3.39 Nitridized steel is hard yet brittle, and this means that the bulk of a steel object must remain free of nitrogen. Assuming $D = 10^{-9}$ cm²/s for interstitial diffusion of nitrogen in steel at 700 °C, estimate the minimum thickness of the above steel blade such that the nitrogen concentration at its center would increase by no more than 0.005 of the surface change under 1 (alternatively 25) hour(s) of nitridization. Assume for simplicity that the concentration at the center is twice the concentration that would be caused by nitridization from one side only (in reality it will be less than twice due to decreasing gradient after the two fluxes penetrate each other).

3.40 At 700 °C, an $YBa_2Cu_3O_{7-\delta}$ sphere of 2 cm radius is abruptly exposed to 1 bar O_2 gas. Assuming $D_O = 10^{-5}$ cm²/s and diffusion as the rate-controlling process, calculate how long it will take before 90% of the total oxidation change occurs 1 mm below the surface. Under these conditions, the sphere center will oxidize by less than 1%, so that the sphere can be approximated as an infinite half-space.

3.5   Further Reading

P. Kofstad, "*Non-Stoichiometry, Diffusion and Electrical Conductivity in Binary Metal Oxides*" (1972) Wiley.

J. Crank, "*The Mathematics of Diffusion*" 2nd edition (1975) Clarendon Press.

S. Mrowec (translated to English by S. Marcinkiewicz), "*Defects and Diffusion in Solids*" (1980) Elsevier.

D.M. Smyth, "*The Defect Chemistry of Metal Oxides*" (2000) Oxford University Press.

D.S. Wilkinson, "*Mass Transport in Solids and Fluids*" (2000) Cambridge University Press.

J. Maier, "*Physical Chemistry of Ionic Materials*" (2004) Wiley.

H. Mehrer, "*Diffusion in Solids*" (2007) Springer.

J.-M. Missiaen, "Solid-state spreading and sintering of multiphase materials" *Mater. Sci. Eng. A* **475** (2008), 2–11.

P. Heitjans, J. Kärger (editors), "*Diffusion in Condensed Matter, Methods, Materials, Models*" 3rd edition (2011) Springer.

3.6   References

[1] G. Brouwer, "A general asymptotic solution of reaction equations common in solid-state chemistry" *Philips Res. Rep.* **9** (1954), 366–376.

[2] J.C. Schön, I.V. Pentin, M. Jansen, "Ab initio computation of low-temperature phase diagrams exhibiting miscibility gaps" *Phys. Chem. Chem. Phys.* **8** (2006), 1778–1784.

[3] P. Karen, "Nonstoichiometry in oxides and its control" *J. Solid State Chem.* **179** (2006), 3167–3183 (and references therein).

[4] P. Schleger, W.N. Hardy, B.X. Yang, "Thermodynamics of oxygen in $YBa_2Cu_3O_x$ between 450 °C and 650 °C" *Physica C* **176** (1991), 261–273.

[5] W.C. Tripp, N.M. Tallan, "Gravimetric determination of defect concentrations in NiO" *J. Am. Ceram. Soc.* **53** (1970), 531–533.

[6] M.L. Volpe, J. Reddy, "Cation self-diffusion and semiconductivity in NiO" *J. Chem. Phys.* **53** (1970), 1117–1125.

4   Phase Diagrams and Phase Transitions

In this chapter we will consider two separate but related topics; phase diagrams and phase transitions. Phase diagrams are used throughout materials chemistry to guide the synthesis of known and new materials. Electronic and magnetic phase diagrams are used to help summarize the properties of functional materials under different conditions. We'll see throughout the later chapters of this book that many functional materials undergo structural phase transitions that are intimately linked to their properties. The second half of this chapter will introduce the fundamental concepts needed to understand these.

4.1   Phase Diagrams

Phase diagrams provide a graphical summary of the behavior of chemical systems at equilibrium as a function of external variables. In materials chemistry, these variables are usually composition, temperature, and pressure, but could also include effects such as electric or magnetic field. These diagrams are based on the **phase rule** first proposed by Gibbs, which states that for a system in equilibrium:

$$P + F = C + 2 \tag{4.1}$$

where P is the number of **phases**, F the **degrees of freedom** or **variance**, and C the number of **components**.[1] We can define the **system** as being the part of the universe that we're interested

---

[1] The derivation of this rule is relatively straightforward and is covered in many physical chemistry texts. Briefly, to describe the state of a system consisting of P phases, you need temperature, pressure (two variables), and $(P \cdot C) - P$ mole fractions for P phases of C components each ($-P$ appears because the sum of mole fractions of each phase is one, hence one fraction per phase is redundant). There are then a total of $2 + (P \cdot C) - P$ variables. For phases at equilibrium, chemical potentials for each component must be equal. This gives $C(P - 1)$ independent equations (conditions). The number of variables that still remain free to vary is $F = 2 + (P \cdot C) - P - C(P - 1)$, which can be rearranged to the equation given.

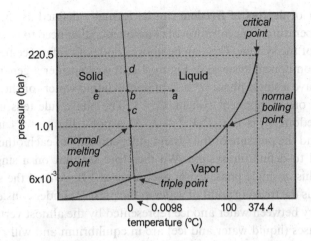

Figure 4.1 Schematic phase diagram of water.

in, anything else is the surroundings. It is easiest to understand the various terms in the phase rule through examples.

Gibbs defined a **phase** as a state of matter that is uniform throughout in terms of both its chemical composition and its physical state. It thus represents a homogeneous part of the chemical system that is bounded by a surface such that it can, at least in principle, be separated from other parts of the system. To take the simple example of H_2O (Figure 4.1) ice, water, and steam are individual phases—each is homogeneous and each could potentially be separated from the other; crystals of ice dispersed in liquid water is therefore a two-phase system.[2] If we consider a mixture of two metals A and B where A forms a dispersion of droplets within B, it is a two-phase system. If A and B form a **solid solution** in which atoms are homogeneously mixed on an atomic length scale, it is a one-phase system. For a material such as SiO_2, which displays polymorphism and has more than one structural form in the solid state (e.g. the quartz, tridymite and cristobalite modifications), each form is a different phase.

We can define the **components** of a system as the minimum number of *independently variable* chemical constituents that we need to define the overall composition. The H_2O phase system, for example, contains a single component because H_2O has a fixed chemical formula. Our two metals A and B would be a two-component system. Al_2O_3 and SiO_2 can also be considered a two-component system if we take Al_2O_3 and SiO_2 as having fixed composition (no nonstoichiometry).[3] This particular two-component system contains three phases: Al_2O_3, SiO_2, and the compound $Al_6Si_2O_{13}$ (often expressed as $3Al_2O_3 \cdot 2SiO_2$ on phase diagrams).

[2] Throughout this chapter, for reasons of pedagogical clarity, single-phase areas in phase diagrams are shaded in gray.
[3] With nonstoichiometry considered, Al_2O_3 and SiO_2 would become part of a three-component system; Si, Al, O.

The number of **degrees of freedom** can be formally defined as the number of variables (pressure, temperature, concentration of components) that need to be specified to fully define the condition of the system. This can be understood with reference to the phase diagram of water (a one-component system) in Figure 4.1. If we consider a general point on the phase diagram, we have a single phase ($P = 1$) present—liquid water for the point labeled a, ice at point e. For a one-component system, $C = 1$. The phase rule tells us that the number of degrees of freedom, F, is $C - P + 2 = 2$. At a point like a, we can thus vary both the temperature and the pressure of the system independently of each other, meaning both have to be specified to define the system. We therefore see that on a single-component phase diagram like this, a single phase exists over an area.[4] If we cool the system until we reach point b, which is the freezing point of water at the pressure under consideration, we reach the phase boundary between water and ice represented by the almost vertical solid line. At this point, two phases (liquid water and ice) are in equilibrium and will remain so indefinitely. The phase rule says that $F = 1 + 2 - 2 = 1$, and we therefore need to specify only one variable to define the system. We can change pressure and temperature and retain two phases but we can't change them independently without destroying one phase or the other. If we increase T, we must decrease p to compensate (i.e. move to point c); if we decrease T, we must increase p (i.e. move to point d). We can see from the phase rule why, when water freezes to ice at constant pressure, the temperature of the system remains constant as long as both ice and water are present and in equilibrium.[5]

At the so-called **triple point**, one has three phases (ice, liquid water, and water vapor) in equilibrium, and $F = 1 - 3 + 2 = 0$. If one changes either p or T, one phase will disappear, so no variables have to be specified to define the system. This is an example of an **invariant point**; one with no degrees of freedom. A line on the H_2O phase diagram thus represents the presence of two phases, and a point where three lines intersect represents three phases. The final point to mention on the water phase diagram is the **critical point**, which occurs at the **critical temperature** and **critical pressure**. Beyond this point, liquid and vapor have the same density and can't be distinguished—a **supercritical fluid** is the only phase present. A practical consequence is that above the critical temperature, gas can't be liquefied by application of pressure alone.

As the focus of this text is on solid state materials, the majority of the chemical systems that we'll consider will be condensed systems, i.e. ones in which the vapor pressures of the substances involved are negligible compared to atmospheric pressure. The pressure can therefore be considered as constant, and one degree of freedom is removed from the system. The phase rule for a condensed system becomes:

---

[4] Note that this is not the case in a phase diagram with two or more components. In a two-component diagram, a single solid phase is represented by a line and in a three-component diagram by a point.

[5] In practice, remaining at equilibrium means using a slow cooling rate. With more rapid cooling, one may not allow time for crystallites of ice to nucleate and thus obtains a supercooled state. This, however, is not an equilibrium state of the system.

$$P + F = C + 1 \tag{4.2}$$

This is the form of the phase rule that will apply for the rest of this chapter.

4.2 Two-Component Phase Diagrams

4.2.1 Without Compound Formation

Figure 4.2 shows one of the simplest phase diagrams for a two-component (A and B) or **binary** condensed system, a system in which A and B form no compounds A_mB_n. Components A and B form no solid solutions (x is either zero or one in $A_{1-x}B_x$) but are completely miscible when molten. The y axis represents the temperature of the system and the x axis the composition. Composition throughout this chapter is expressed in mole fraction (0 to 1), though mole percents are also in common use.[6] The left-hand y axis of Figure 4.2 then corresponds to pure A, and we can read from this axis that A melts at T_4; the right-hand axis represents pure B, which melts at T_5.

There are four main regions on this phase diagram. At low temperature, as A and B form no compounds and no solid solution, one has a two-phase region consisting of a mixture of pure A and pure B solids. When the solid two-phase mixture is heated to temperatures just above T_1, one obtains a mixture of either solid A and liquid, or solid B and liquid, depending

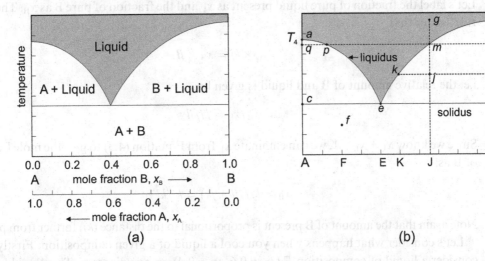

Figure 4.2 (a) The phase diagram for a binary system of A and B that form no compounds and no solid solution. (b) The same phase diagram but with labels for specific points discussed in the text. Single-phase areas are shaded, two-phase areas unshaded.

---

[6] Mole percents are just mole fractions ×100. In some fields (particularly phase diagrams used industrially), the composition axis may be expressed in weight percent.

on whether the composition lies to the left or right of composition E. At composition E, the solid mixture melts directly to form a liquid at point *e*, called the **eutectic point**, which represents the lowest melting temperature in the system. Line *cd* (Figure 4.2b) where the first liquid appears is called the **solidus**. At temperatures above those defined by the line *aeb* (called the **liquidus**), one enters a region where the system is fully molten into a single-phase liquid.

When interpreting such phase diagrams, it is crucial to realize the difference between a two-phase and a one-phase region. For a one-phase region (shaded), each point on the diagram represents a certain state of the phase in terms of composition and temperature. For example, one can form a liquid with any desired composition at any temperature above the liquidus. For two-phase regions the situation is very different. An arbitrary point in a two-phase region does *not* represent an actual state of a phase, but corresponds to the overall composition of two phases in equilibrium. At point *f* in Figure 4.2b (whose composition is marked as F on the composition axis), one has a two-phase mixture of pure A and pure B. The relative amount of each phase present can be read directly from the composition axis. Equivalently (and to prepare ourselves for later diagrams), we can see that the mole fraction of A present, x_A, is given by (distance F to B)/(distance A to B) (i.e. FB/AB) and x_B as AF/AB. Similarly, point *j* on the diagram contains a mixture of pure B and a liquid. The composition of the liquid present is given by point *k* on the diagram (i.e. $x_A \approx x_B \approx 0.5$). The solid present is pure B and represented by point *l*. The relative amounts of liquid of composition K and pure B present can be determined straightforwardly by the **lever rule**. Let's label the fraction of pure liquid present as x_L and the fraction of pure B as x_B. The lever rule[7] states that:

$$x_L \times kj = x_B \times jl \tag{4.3}$$

i.e. the relative amount of B and liquid is given by:

$$x_B/x_L = kj/jl \tag{4.4}$$

Since we know $x_L + x_B = 1$, we can eliminate x_L from Equation (4.3) to give the mole fraction of B as:

$$x_B = kj/(jl + jk) = kj/kl \tag{4.5}$$

Note again that the amount of B present is proportional to the distance (*kj*) further from pure B.

Let's consider what happens when you cool a liquid of a given composition. Firstly, let's consider a liquid of composition E ($x_A \approx 0.6$, $x_B \approx 0.4$) on our diagram. This liquid has the special property of transforming directly to two solids on cooling at the eutectic point *e*. No region of mixed solid and liquid phases exists. Since three phases (A, B, and liquid) are

---

[7] This can be likened to balancing two masses at different distances from a pivot. At balance, $m_1 d_1 = m_2 d_2$ where m_1 and m_2 are the two masses and d_1 and d_2 the distances to the pivot.

present at e, the number of degrees of freedom is zero ($F = C + 1 - P = 2 + 1 - 3 = 0$), and this tells us that the eutectic point e is **invariant**.[8]

The behavior of other compositions is more complex. Consider the composition J ($x_A \approx 0.35$, $x_B \approx 0.65$) initially at point g. When the liquid is cooled to T_3, one meets the liquidus curve at point m where an infinitesimally small amount of pure B will form. Solid B will be in equilibrium with liquid of composition J at this point. On further cooling to T_2 we reach point j. One still has a liquid in equilibrium with pure B, but the composition of the liquid has changed to K—it has become richer in A. The relative amounts of solid and liquid present are given by the lever rule as kj/jl as discussed above. Note that although the liquid and solid have different compositions, the overall composition of the system does not change. As one continues to cool, the liquid composition follows the line ke until at temperature T_1 the final liquid of composition E crystallizes. Below T_1 one forms a solid mixture of A and B of the original overall composition. The system will follow this pathway if cooled under equilibrium conditions. If the system is rapidly cooled, it may follow a different pathway, but the final phases present would be the same.

We can also use Figure 4.2 to understand the effect that solid impurities have on melting points. A sample of pure B at temperature T_3 will be a solid (B doesn't melt until the higher temperature T_5). However, if a small amount of A is added, the solid will be in equilibrium with a small amount of liquid at point m (composition J, $x_A \approx 0.35$, $x_B \approx 0.65$); i.e. a small portion of B will have dissolved in the liquid. If sufficient A is added so that the overall composition becomes J, then all of B will dissolve. The impurity A has lowered the melting point from T_5 to T_3. Since at this point B has just dissolved, we have a saturated solution of B in A. As we continue to add A at this temperature, we retain a single-phase liquid until we reach point p on the liquidus. At this point, a small amount of solid A will form. As more A is added, the amount of solid will increase and liquid decrease until at point q we have pure solid A present. The point p represents a saturated solution of A in B. Since ae and eb represent saturated solutions of A and B, respectively, point e represents a liquid saturated in both solids. The use of one solid to lower the melting point of another is the reason why salt is spread on roads to prevent ice formation. It's also of great practical use in the growth of crystals from a **flux** at lower temperatures than would otherwise be required. Fluxes are chosen to be easily separable from the crystals of interest, for example by dissolution in water in the case of NaCl/KCl fluxes.

4.2.2 With Compound Formation

Most of the systems we'll meet in materials chemistry will be more complex than Figure 4.2—components that undergo reactions in the solid state leading to new materials are far more interesting and exploitable than those that don't! Figure 4.3 shows three possible phase

---

[8] Terms univariant and bivariant are used for $F = 1$ and $F = 2$.

diagrams in which A and B can react to form an intermediate phase AB_2. This phase occurs as a vertical solid line on the phase diagram at $x_A = \frac{1}{3}$, $x_B = \frac{2}{3}$. Note that for composition calculations it is often more convenient to express the formula of AB_2 as $A_{1/3}B_{2/3}$ (a normalized or pseudoatom formula).

Figure 4.3a shows the most straightforward system. Each of the solid phases A, AB_2, and B melts to form a liquid without decomposition (at temperatures T_2, T_1, and T_3, respectively, on the diagram); they are said to have a **congruent melting point**. The phase diagram is then very similar to that of the simple eutectic system in Figure 4.2. In fact, it can be considered as two eutectic systems placed side by side. Figure 4.3b shows a situation in which the compound AB_2 is only stable up to a temperature T_1 above which it decomposes without melting to solid A and B. Above this temperature, the diagram is again essentially identical to Figure 4.2. Other systems show the reverse behavior, i.e. compounds are only stable above a certain temperature. ZrW_2O_8 of Figure 4.8 is one such example (see later).

Figure 4.3c represents a more complex situation in which AB_2 has an **incongruent melting point**, that is, it melts to give a solid and a liquid of different chemical compositions. On heating AB_2 to T_2, it decomposes to give a liquid of composition corresponding to point p and pure B; the relative amounts are given by the lever rule. On further heating, B will gradually dissolve in the melt and the liquid will move in composition from p to r. At point r, all the solid dissolves/melts and a single liquid phase results. At point p, three phases are in equilibrium (liquid, AB_2, and B), $F = C - P + 1 = 0$, and p is therefore an invariant point. As can be seen from Figure 4.3c, the composition of point p lies outside the composition range of AB_2 to B. The composition of the liquid phase therefore can't be expressed in terms of positive quantities of the solid phases with which it is in equilibrium. Such a point is called a **peritectic point**. At a peritectic point there is no minimum in the melting curve as there is at the eutectic point, only a kink.

Systems in which a compound melts incongruently follow relatively complex pathways on cooling. Consider a liquid at point s on Figure 4.3c. On cooling, the liquidus is met at t, and liquid and solid B will be in equilibrium. As cooling continues towards T_2, the liquid will move in composition from t to p. At temperatures just below the peritectic point p, the stable solid phase becomes AB_2. To form this solid, a **peritectic reaction** must occur, in which all the solid B present must react with the liquid to form solid AB_2. A drastic change in the composition of solids present will therefore occur for a very small temperature change. The amount of liquid present also changes dramatically at T_2. Just above T_2 the lever rule shows the system will contain largely liquid ($x_L \approx 0.86$ given by uB/pB) and just below T_2 far less liquid (~ 0.36, uAB_2/pAB_2). On further cooling, the liquid composition follows curve pe, and, at temperatures below T_1, solid A forms along with solid AB_2.

From the complexity of this behavior, it should not be surprising that systems containing phases that melt incongruently frequently show non-equilibrium behavior. If liquid at point s is cooled rapidly and/or if solid B falls to the bottom of the crucible during cooling such that it is essentially removed from the system, the peritectic reaction may not have time to occur completely and one would therefore observe a non-equilibrium mixture of A, AB_2, and B at

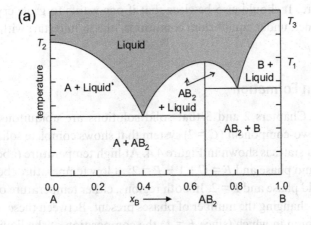

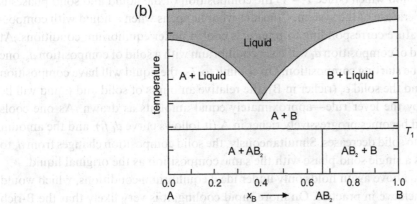

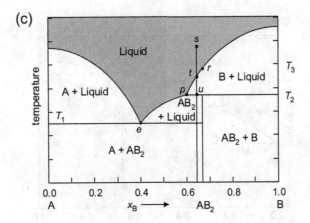

Figure 4.3 Two-component phase diagrams containing an intermediate phase AB_2: (a) AB_2 stable to its melting point. (b) AB_2 decomposes to solid A and B before melting. (c) AB_2 decomposes on melting.

low temperature. It should also be clear that if one wants to grow crystals of composition AB_2 from the melt under equilibrium conditions, one should start with a liquid composition between e and p.

4.2.3 Solid-Solution Formation

We learned in Chapters 2 and 3 that solid solutions are ubiquitous. A schematic phase diagram for a two-component ($C = 2$) system that shows complete solid solution in both the liquid and solid states is shown in Figure 4.4. At high temperature (above the liquidus), one has a single liquid phase and $F = C + 1 - P = 2$; at low temperature (below the solidus), one has a single solid phase and $F = 2$. In both regions, either temperature or composition can be varied without changing the number of phases present. Between these two extremes, one has a two-phase region in which (since $F = 1$) the composition of the liquid and solid phases is fixed by the temperature of the system. Consider what happens when a liquid with composition and temperature corresponding to point c is cooled under equilibrium conditions. At point d_ℓ, the liquid of composition d_ℓ will be in equilibrium with a solid of composition d_s, one richer in B than the starting composition. On cooling to T_3, the liquid will have composition e_ℓ (richer in A) and the solid e_s (richer in B); the relative amounts of solid and liquid will be given as $e_\ell e/e_s e$ by the lever rule—approximately equal amounts as drawn. As one cools further, the liquid becomes progressively richer in A (it follows curve $d_\ell f_\ell$), and the amount of liquid relative to solid decreases. Simultaneously, the solid composition changes from d_s to f_s. At T_1, one has a single solid phase with the same composition as the original liquid.

The description above again holds only under ideal equilibrium conditions, which would be very hard to achieve in practice. On more rapid cooling, it is very likely that the B-rich

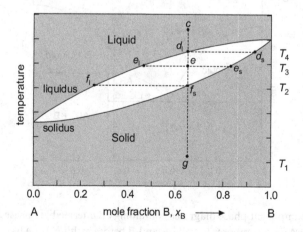

Figure 4.4 Phase diagram for a two-component system showing complete solubility in both solid and liquid states.

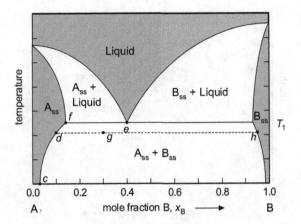

Figure 4.5 Phase diagram for a two-component system in which a limited range of solid solutions is formed.

crystals formed initially wouldn't have sufficient time to fully equilibrate with surrounding liquid. One would then expect crystals whose core is rich in B (compositions as high as d_s) and outer shell rich in A (compositions as A-rich as f_ℓ). Such effects are frequently observed in mineralogy. For many applications, the range of compositions from the core to the outside of the crystal may give rise to deleterious properties, and extended annealing periods may be required to homogenize samples.

A full range of solid solutions as shown in Figure 4.4 would only be expected for components that are chemically very similar, and few phase couples have full thermo-dynamic miscibility at very low temperatures.[9] In most systems at equilibrium, one observes partial solid solution; there is a limit of solubility at a given temperature. Figure 4.5 shows a typical phase diagram for a two-component system in which no compounds form but there is solid solution at either end of the composition range. The diagram is directly related to Figure 4.2a. At the left-hand side of the diagram, we have a region in which A will dissolve B to form a single-phase solid solution $A_{1-x}B_x$ (labeled A_{ss}). The range of solubility generally increases with temperature (curve cdf) and reaches a maximum at the solidus temperature T_1. A similar region exists at the B-rich side of the diagram (B_{ss}). At temperatures below T_1, these regions are separated (the boundary is called the **solvus**) by a two-phase region containing a mixture of A_{ss} and B_{ss}. The composition corresponding to any point on this diagram can be found using the same rules we've used above. Point g, for example, has an overall composition of $x_A \approx 0.7$, $x_B \approx 0.3$; there will be two phases present, A_{ss} and B_{ss}. As drawn, A_{ss} will have composition d ($A_{0.9}B_{0.1}$) and B_{ss} composition h ($A_{0.05}B_{0.95}$). The mole fractions of each of the two phases as given by the lever rule are gh/dh for A_{ss} and dg/dh for B_{ss}

---

[9] Many solid solutions can be quenched from high temperatures and remain stable at low temperatures because of slow kinetics of demixing.

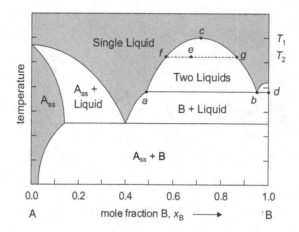

Figure 4.6 Phase diagram for a two-component system with limited solid-solution range for A and liquids rich in A and B immiscible under certain conditions.

(~0.765 $A_{0.9}B_{0.1}$ and ~0.235 $A_{0.05}B_{0.95}$). Note that $(0.765 \times 0.9) + (0.235 \times 0.05) = 0.7$ and $(0.765 \times 0.1) + (0.235 \times 0.95) = 0.3$, as expected for the overall composition.

In the two examples described above, the components have been fully miscible when molten and therefore form a single liquid phase. This is not, however, always the case, and we're familiar in everyday life with oil and water forming two-phase liquid systems. Liquid immiscibility is also encountered in ceramic phase diagrams. The example in Figure 4.6 shows an **immiscibility dome** inside which two liquids exist as separate phases. Point e, for example, will correspond to a mixture of two liquids of compositions corresponding to f and g. Temperature T_1 is called the **upper consolute temperature**. Along the line abd, two liquids (compositions at a and b) and a solid (pure B) are in equilibrium. Point b is an invariant point (three phases are present) called a **monotectic point**. It is similar to the eutectic point, except that one of the phases that forms is a solid (here pure B) and the other a liquid (composition at a).

When we discussed the formation of compounds in Section 4.2.2, we moved from simple eutectic phase diagrams such as Figure 4.2 to the more complex diagrams of Figure 4.3. Each of these diagrams can be readily extended to allow for the formation of solid solutions in the same manner as described above. For example, the incongruently melting system of Figure 4.3c becomes Figure 4.7. While they appear complex at first glance, we can use the same ideas as described above to read such phase diagrams.

A real-world A–B binary system that combines many of the ideas discussed above is shown in Figure 4.8 for ZrO_2–WO_3. The system exhibits two structural forms of ZrO_2 (monoclinic and tetragonal structures) over the temperature range depicted, each of which shows a small range of solid solution. There is a single AB_2 compound ZrW_2O_8 that is only thermodynamically stable above 1105 °C, melts incongruently, and has a peritectic point at 1257 °C. This system is discussed further in the end-of-chapter problems.

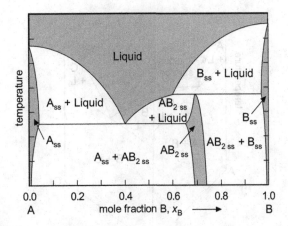

Figure 4.7 Phase diagram for a simple binary system with one intermediate phase AB_2 and a range of solid solutions for each solid phase.

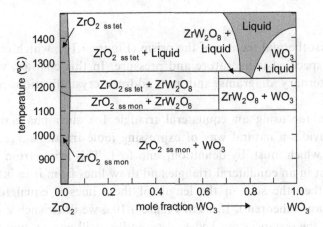

Figure 4.8 Partial phase diagram for the ZrO_2–WO_3 system at ambient pressure, after [1].

4.3   Three-Component Phase Diagrams

For a three-component or **ternary** system, we would need five axes (one for each composition and one each for p and T) to draw a general phase diagram. In practice, an equilateral triangle is used to represent the composition of the system, and temperature is represented by an axis normal to this triangle, forming a prismatic diagram (Figure 4.9). One such prism is required for each pressure of interest. The side faces of the prism represent the binary phase diagrams we've already discussed. Due to the complexity of drawing and reading such diagrams, it is far more common

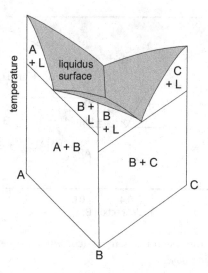

Figure 4.9 A phase diagram for a three-component eutectic system.

to depict an isothermal section of the prism (Figure 4.10), which represents the phases present at a specified temperature and pressure. In this chapter, we will only discuss sub-solidus ternary diagrams and not address crystallization pathways for these systems.

The reason for using an equilateral triangle for such phase diagrams is that its geometry provides a natural way of expressing mole fractions x_A, x_B, x_C of the three components, which must, by definition, sum to 1. This arises from the fact that if you take any point in an equilateral triangle and draw lines from it to intercept each edge at right angles then the sum of the length of these lines is equal to the height of the triangle (Viviani's theorem). If, as in Figure 4.10a, we draw such a triangle and call its height 1 then the distances marked x_A, x_B, and x_C will always sum to 1. One corner of the triangle represents pure A (the others are B and C), and lines drawn parallel to the opposite edge moving towards that corner represent increasing mole fractions of A. This is shown in Figure 4.10a where each of the dashed lines represents a constant mole fraction of A, with the amount of A increasing as one moves from edge CB towards corner A. If we consider point d, it lies on the line representing $x_A = 0.6$. If we were to draw lines parallel to the other edges, d would lie on lines corresponding to $x_B = 0.3$ (lines parallel to the AC edge) and $x_C = 0.1$ (lines parallel to AB).

A second way of determining composition is to use the **triangle rule**, which is the equivalent of the lever rule we used for binary systems. In Figure 4.10b we draw lines from each corner of the triangle through d to the opposite edge. The amount of each component is given by:
$x_A = q/(p+q); x_B = s/(r+s) = 0.3; x_C = u/(t+u) = 0.1.$

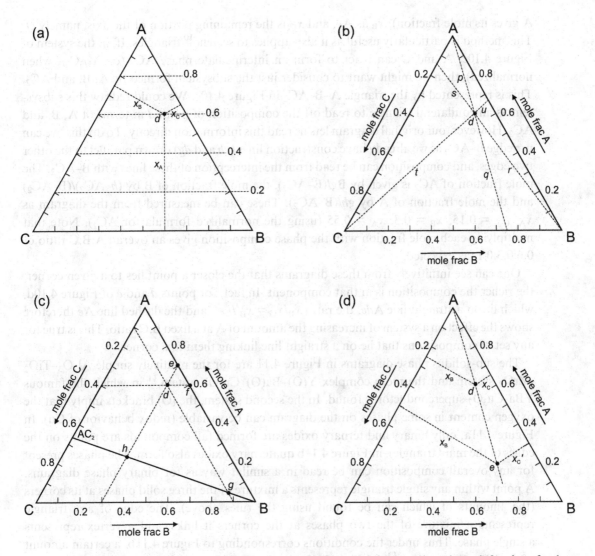

Figure 4.10 Isothermal sections for a three-component system; (a) shows how the mole fraction of a phase increases as one moves towards the corresponding corner, (b) illustrates the triangle rule for determining composition, (c) illustrates how composition can be read from a single axis, and (d) shows a line corresponding to a constant B:C ratio.

A third way of reading compositions, which is often the most useful in practice, is to read the amount of each component directly from one axis. To do this, we take point d in Figure 4.10c and draw dashed lines through it parallel to each edge of the triangle and look where these lines intercept any edge. If we consider edge AB, then the mole fraction of A is given by fB (as in two-component phase diagrams, the length "furthest" from pure

A gives its mole fraction), x_B as Ae, and x_C is the remaining portion of the axis, namely ef. This method is particularly useful as it also applies to scalene[10] triangles. If, in the system of Figure 4.10c, A and C can react to form an intermediate phase AC_2 (or $A_{1/3}C_{2/3}$ when normalized), then we might want to consider just the subsystem formed by A, B, and AC_2. This is represented by the triangle A–B–AC_2 in Figure 4.10c. We could redraw this subsystem as an equilateral triangle to read off the composition of point d in terms of A, B, and AC_2. However, our original diagram lets us read this information directly. To do this, we can use edge B–AC_2 as we already have construction lines (eh and dg) drawn parallel to the other two edges, and composition can be read from the interception of these lines with B–AC_2. The mole fraction of AC_2 is given by Bg/(B–AC_2), the mole fraction of B by (h–AC_2)/(B–AC_2) and the mole fraction of A by gh/(B–AC_2). These can be measured from the diagram as $x_{A_{1/3}C_{2/3}} = 0.15$, $x_B = 0.3$, $x_A = 0.55$ (using the normalized formula for AC_2). Note that multiplying each mole fraction with the phase composition gives an overall A:B:C ratio of 0.6:0.3:0.1, as expected.

One can see intuitively from these diagrams that the closer a point lies to a given corner, the richer the composition is in that component. In fact, for points d and e of Figure 4.10d, which lie on a straight line Ade, the ratio $x_B/x_C = x_B'/x_C'$, and the dashed line Ae therefore shows the effect on a system of increasing the amount of A at a fixed B:C ratio. This is true for any set of compositions that lie on a straight line linking them to a corner.

The sub-solidus phase diagrams in Figure 4.11 are for the relatively simple Al_2O_3–TiO_2 –ZrO_2 system and the more complex Y(O)–Ba(O)–Cu(O) system,[11] in which the famous $YBa_2Cu_3O_7$ superconductor is found. In the second system, the (O) brackets imply that the oxygen content in some phases on the diagram can be variable (redox behavior of Cu). In Figure 4.11a, only binary and ternary oxides are formed (all compounds are points on the edges of the main triangle); in Figure 4.11b quaternary oxides also form. The phases present for any overall composition can be read in a similar way as for binary phase diagrams. A point within any single triangle represents a mixture of the three solid phases at its corners (the amounts of which can be found using the rules above). The edge of any triangle represents a mixture of the two phases at the corners it links, and a vertex represents a single phase. Thus under the conditions corresponding to Figure 4.11b, a certain amount of the $YBa_2Cu_3O_7$ phase[12] (labeled 123 for the Y:Ba:Cu ratio) will be formed from starting compositions anywhere within the triangle enclosed by points labeled CuO, 211, and $BaCuO_2$. Pure material will only be formed using the correct 1:2:3 Y:Ba:Cu ratio. In practice, as we discussed for binary phase diagrams, many systems form solid solutions. These would be represented by an area located around an ideal stoichiometric point.

---

[10] A scalene triangle is a triangle with no equal sides.

[11] Note that in Figure 4.11a the compositions are in mole fractions of Al_2O_3, whereas in Figure 4.11b in mole fractions of $YO_{1.5}$. Al_2TiO_5 (2:1 Al:Ti) therefore lies halfway along the edge joining Al_2O_3 and TiO_2 in Figure 4.11a, whereas $Y_2Cu_2O_5$ (1:1 Y:Cu) lies halfway along the edge joining Y(O) and Cu(O) in Figure 4.11b.

[12] Though the oxygen content may not be exactly 7; see Chapters 3 and 12.

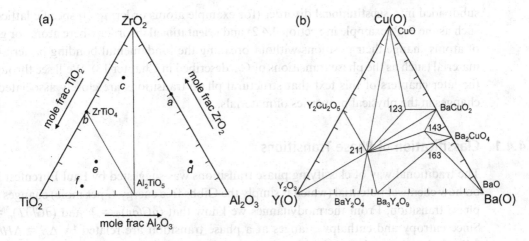

Figure 4.11 Isothermal sub-solidus phase diagrams (neglecting minor solid solubilities) for (a) TiO_2–Al_2O_3–ZrO_2 below 1580 °C [2] and (b) Y(O)–Ba(O)–Cu(O) at 900 °C in O_2 [3].

4.4 Structural Phase Transitions

In the first half of this chapter, we have explored the phase diagrams of a number of simple systems. The concept of **phase transitions** was implicit to this discussion. In Figure 4.1, for example, as we heat from point e to point a, there is a phase transition from solid ice to liquid water. Many materials undergo phase transitions in the solid state where their structure changes. For example, when α-quartz[13] is heated, it undergoes a series of phase transitions. At 573 °C (the **transition temperature** or **critical temperature**, T_c), a relatively subtle transition to β-quartz occurs, in which the connectivity of the $SiO_{4/2}$ tetrahedra is unchanged (i.e. no chemical bonds are broken or formed), but the tilting pattern of tetrahedra changes and the symmetry increases from trigonal to hexagonal. On further heating, more drastic modifications occur in which the bonding pattern changes; at 870 °C, a transition to β-tridymite and, at 1470 °C, to β-cristobalite. The α- to β-quartz transition can be classified as a **displacive** phase transition, because no bonds are broken or formed. The higher temperature transitions are classified as **reconstructive** because they entail a major reorganization of the structure involving bond breaking and formation.[14] The final category of phase transition we'll encounter is **order–disorder** transitions. Order–disorder transitions can be further

---

[13] Polymorphs are often distinguished by Greek letters α, β, γ, etc.

[14] These different classifications aren't used uniformly in the literature. For example, we describe a transition as displacive if atoms only move a short distance, no chemical bonds need to be broken to achieve the conversion, and there is a group–subgroup relationship (Appendix B) between the two space groups, regardless of whether the process occurs in an abrupt or continuous fashion. Others choose to restrict the term displacive to continuous phase transitions.

subdivided into **substitutional** disorder (for example atoms ordering on specific lattice sites such as the FePt example in Section 4.4.2) and **orientational** disorder where atoms or groups of atoms change their positions without breaking the fundamental bonding pattern of the material (such as the phase transitions of C_{60} described in Chapter 13). We'll see throughout the later chapters of this text that structural phase transitions are closely associated with changes in the physical properties of materials.

4.4.1   Classification of Phase Transitions

The traditional way of classifying phase transitions was suggested by Paul Ehrenfest based on how chemical potential (which is simply the Gibbs free energy G per mole) changes at the phase transition. From thermodynamics we know that $(\partial G/\partial p)_T = V$ and $(\partial G/\partial T)_p = -S$. Since entropy and enthalpy changes at a phase transition are related by $\Delta S = \Delta H/T_c$, it follows that for a phase transition such as melting or boiling, in which there are abrupt changes in both volume and entropy/enthalpy, there will be a discontinuity in the first derivative (or slope) of the free energy with respect to temperature (and pressure). Such a transition is therefore called a **first-order** transition. Since constant-pressure heat capacity, C_p, is defined as $(\partial H/\partial T)_p$, it will be infinite at the phase transition.

The existence of a significant latent heat (ΔH) means that first-order transitions display a **hysteresis**, a difference in phase-transition temperature on warming and cooling. We can understand this through a simple thought experiment: if a sample is warmed to T_c and held at precisely this temperature, the phase transition will not initially occur as there is no temperature gradient between the surroundings and sample to allow the flow of latent heat. It is only when the surrounding temperature is raised above T_c that latent heat flows and nuclei of the new phase begin to form. A first-order solid state transition will therefore lag behind the temperature change causing it in any real experiment. The sample remains at temperature T_c during the transformation while the heat supplied feeds the higher entropy of the product (drives the transition). This means that the presence of two phases in coexistence is a criterion for the first-order phase transition. To obtain any hysteresis intrinsically associated with a first-order transition, this kinetic factor has to be eliminated by extrapolating temperature-dependent measurements to the zero rate of temperature change.

A **second-order** phase transition is one in which there is a discontinuity in the second derivative of the free energy with respect to temperature or pressure. The volume and the entropy (and therefore the enthalpy) do not change abruptly at the phase transition, but quantities such as the volumetric coefficient of thermal expansion or heat capacity do.[15] The dependence on temperature of various thermodynamic quantities for classical first- and second-order transitions is shown in Figure 4.12.

---

[15] The volumetric coefficient of thermal expansion, a_V, as $a_V = (1/V)\partial V/\partial T = (1/V)\partial^2 G/\partial P\partial T$ and C_p as $C_p/T = (1/T)\partial H/\partial T = \partial S/\partial T = -\partial^2 G/\partial T^2$.

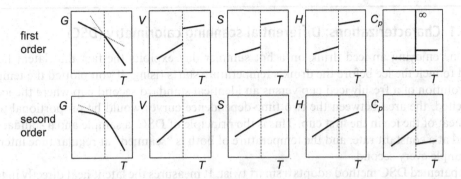

Figure 4.12 Variation in thermodynamic quantities at first- and second-order Ehrenfest phase transitions. The vertical dashed line corresponds to the transition temperature.

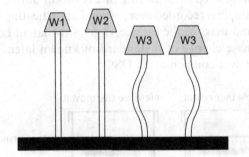

Figure 4.13 The Euler strut. When the load on the strut is increased (weights W3 > W2 > W1), the system will eventually buckle causing a reduction in symmetry.

It is now more common to categorize phase transitions as **discontinuous** or **continuous** with reference to the behavior of entropy or another order parameter (see below) at T_c. At a discontinuous transition, the order parameter changes abruptly. All discontinuous phase transitions are first-order in the Ehrenfest classification. During a continuous phase transition, the order parameter changes in infinitesimally small steps, as in a second-order transition. These concepts are developed further in the following sections.

4.4.2 Symmetry and Order Parameters

We can define many of the phase transitions we'll encounter in terms of the changes in symmetry that occur. Figure 4.13 illustrates the relevance of symmetry using a simple mechanical model, the Euler strut. If we place increasing weights on top of a flexible plastic rod, the rod will buckle or bend at some point. When this occurs, the symmetry of the system is **lowered** or **broken**. We can see from the figure that the system could choose to buckle to the right or the left. Even if the mass was loaded perfectly centrally on the rod, the system would "choose" one direction or the other—ultimately decided in this ideal scenario by random thermal displacements of the system. An analogy to the Euler strut example in materials

Box 4.1 Characterizations: Differential scanning calorimetry (DSC)

Everyone enjoying an iced drink on a hot summer day exploits the fact that latent heat is needed to melt the ice before the drink's temperature starts rising. If you plotted the temperature evolution of a freshly iced cup versus an identical standard second cup where the ice has just melted, the area between the two time-dependence curves would be proportional to the latent heat of the ice in the first cup. This is the principle of DSC: a sample and a standard are warmed at a constant rate, and the temperature of both is "scanned" at regular time intervals, for example every second.

One patented DSC method adopts a smart twist. It measures the latent heat directly in terms of the electrical power to supply it. The sample in a tiny aluminum pan, and an equivalent empty pan for comparison, are each placed in their own tiny thermostat, each having a miniature hot plate and a temperature sensor; the total of about 1 gram weight. The electronics is wired up to keep their heating rate constant during the "scan". The extra power (J/s) needed for the sample is recorded every second upon heating through the phase transition. When summed per unit mass of the sample, it gives the latent heat. The apparatus is usually calibrated using a phase change of a standard of known latent heat and weight. This DSC method is termed a power-compensation DSC.

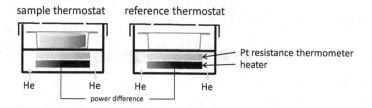

The advantage of the method is a quick thermal equilibration that gives a fast response needed for high rates of heating or cooling, as well as sharp peaks that make it possible to separate closely occurring transitions. In addition to the latent heat of a discontinuous transition, it is possible to determine the heat capacity of the individual phases upon heating or cooling beyond the transition peak.

chemistry could be the phase transitions between perovskite tilt systems outlined in Chapter 1. In the ideal perovskite structure, one has 180° M–O–M bond angles (this ideal structure would normally be the high-temperature, low-pressure limit), which can bend away from 180° on cooling, lowering the symmetry of the system.

As a second model of broken symmetry, consider a simple single-headed arrow that can fluctuate between pointing to the left and pointing to the right. If it switches between these two orientations very rapidly, it would appear as a double-headed arrow to an observer. If, however, its rate of switching was gradually reduced, there would come a point when the

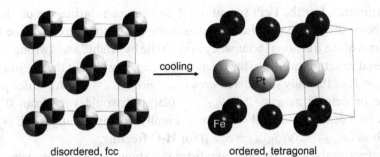

cooling

disordered, fcc   ordered, tetragonal

Pt

Fe

Figure 4.14 Order–disorder transition in FePt alloy.

observer would see it "frozen" into one of two possible orientations. Once the arrow adopts one of the two possible orientations, it changes from appearing double-headed to single-headed and its symmetry is reduced.[16] A structure with an atom dynamically disordered over two closely separated sites at high temperature (for example, an atom in a double-well potential), which chooses one of the two sites and orders on cooling, is one example of this type of behavior. Phase transitions in ferroelectric perovskite materials such as $BaTiO_3$ and $PbTiO_3$, which are discussed in Chapter 8, can also be understood in a similar way. As the material is cooled from high temperature, Ti moves away from a position on a dynamic average at the center of the BO_6 octahedron, freezes in an off-center position, and the material develops a spontaneous polarization. Again, this is associated with a lowering of symmetry at the phase transition on cooling.

Symmetry changes leading to phase transitions can also be brought about by atomic ordering. One important example is found in FePt 1:1 alloys. At high temperatures (above ~1300 °C), FePt has the face-centered cubic structure shown on the left of Figure 4.14, in which Fe and Pt atoms are randomly distributed over all sites of the structure. On cooling, it is thermodynamically favorable to order Fe and Pt atoms in layers perpendicular to the original c axis, and the stable room-temperature structure is tetragonal (this is the so-called $L1_0$ phase). Once again we see that ordering of atoms on cooling lowers the symmetry. This particular transition is of technological relevance as the ordered material has a high magnetocrystalline anisotropy, making it possible to produce magnetically hard nanoparticles with a range of potential applications.

In each of these examples it is useful to introduce an **order parameter**, η, to describe the distortion of the structure relative to the high-symmetry case.[17] This is particularly true when the distortion involves movement of groups of atoms or molecules. In the case of octahedral tilts in perovskites, the tilt angle can be used to define the order parameter. For B cations moving off-center in a BO_6 octahedron, the shift in fractional coordinate of the metal might

---

[16] Note that the point at which the transition is deemed to have occurred might be influenced by the experimental method used to observe it. For example, if one used a photographic technique, a camera with a high shutter speed would record the freezing transition before one with a low shutter speed.

[17] The symbol Q is also commonly used for order parameter.

be the order parameter. For the FePt transition, if we define one lattice site on the right of Figure 4.14 as being the Fe site and the second as being the Pt site, we can define the order parameter in terms of the fractional occupancy of each site by "right" and "wrong" atoms. In a perfectly ordered structure, $occ_{Pt(Fe)}$ (the fractional occupancy of Pt at the Fe site) would be 0, and $occ_{Fe(Fe)} = 1$; in the fully disordered $occ_{Pt(Fe)} = occ_{Fe(Fe)} = 0.5$. An order parameter could therefore be defined as $\eta = 2[occ_{Fe(Fe)} - 0.5]$ and would vary from 0 to 1 for a disordered to fully ordered material on cooling. Finally, entropy change itself is an order parameter, such as $(S_{liquid} - S)/(S_{liquid} - S_{solid})$ for H_2O freezing.

Many readily measurable macroscopic quantities also show a simple dependence on η and can be used to monitor a phase transition. For example the distortion of cell parameters as a perovskite changes from cubic to lower symmetry can be expressed as a spontaneous strain,[18] $\varepsilon = (a - a_{cub})/a_{cub}$, and it typically depends on either η or η^2.[19]

It's worth noting that although the examples described above have an order parameter that could potentially vary continuously (though we'll see that it needn't in practice) from the high- to low-symmetry situation, symmetry itself always changes abruptly—a symmetry element is either present or absent in the structure. A continuous transition concludes when the final infinitesimal heating step causes the order to finally disappear, at which point the new symmetry emerges. For the vast majority of transitions, the high-temperature form has the higher symmetry.

4.4.3   Introduction to Landau Theory

One of the advantages of introducing the order parameter to describe phase transitions in solids is that it allows one to use **Landau theory**[20] to describe their thermodynamics. The basic assumption used in this approach is that the Gibbs free energy of the phase can be approximated as a simple power series in terms of the order parameter η:

---

[18] Strain (ε, the proportional displacement in shape or volume) is the deformation that occurs when a material is subjected to a mechanical stress (σ, a force per unit area). For small deformations where materials behave in an elastic manner, the stress tensor of rank 2 (a 3×3 matrix; diagonalized in engineering to have six non-zero terms) is related to the strain tensor of rank 2 by $\sigma_i = \sum_j c_{ij}\varepsilon_j$ where c_{ij} are elastic stiffness coefficients (originally a tensor of nine 3×3 matrices; in engineering simplified to 36 c_{ij} parameters relating the stress and strain matrices) and subscripts refer to different directional components. For a uniaxial stress and isotropic body, this simplifies to Hooke's law $\sigma = E\varepsilon$ where E is Young's modulus. See, for example, Elliott [*Physics and Chemistry of Solids* (1998), J. Wiley and Sons] for more detail.

[19] Transitions that keep group–subgroup relationships can be categorized according to how the translational symmetry of the lattice changes at the transition: if the lattice centering changes or a superlattice is formed, the transition is called a zone-boundary transition, otherwise it is called a zone-center transition. Here the "zone" refers to the Brillouin zone introduced in Chapter 6 as the volume in reciprocal space that lies closest to each reciprocal lattice point; if lattice points are added or diluted, the zone boundaries change. Zone-boundary transitions (such as the $SrTiO_3$ perovskite tilting discussed in Section 4.4.6) generally have ε proportional to η^2 and zone-center transitions to η.

[20] We call this a phenomenological theory as it explains experimental results mathematically without using a rigorous fundamental physical law.

$$G(\eta) = G_0 + \frac{1}{2}A\eta^2 + \frac{1}{4}B\eta^4 + \frac{1}{6}C\eta^6 + \dots \tag{4.6}$$

where G_0 represents the part of the free energy that does not change at the phase transition. The term $G(\eta) - G_0$ thus represents the *excess* free energy compared to one that the unchanged form would possess if no phase transition occurred. Additional terms can be introduced to this equation if, for example, the order parameter is coupled to properties such as strain. Usually G is independent of the sign of η such that only even powers are needed, as in Equation (4.6), and it's common to adopt the smallest number of terms required to describe a system.

4.4.4 Second-Order Transitions

Let's consider the phase transition shown schematically in Figure 4.15, which involves an oxygen atom being continuously displaced from an ideal site midway between two metal atoms—this could represent a tilting transition of octahedra in a perovskite. We can define an order parameter η in terms of the oxygen displacement from the ideal site. At high temperatures, the oxygen vibrates around its M–O–M midpoint, maintaining high symmetry ($\eta = 0$). At low temperature, the oxygen vibrates around a position displaced to the left or to the right ($\eta \neq 0$). For this situation, oxygen displacements to the left or right ($\pm\eta$) are equivalent, justifying the use of even powers of η in Equation (4.6).

Since the linear M–O–M arrangement is stable at high temperature ($T > T_c$), $G(\eta)$ at $\eta = 0$ must be a minimum. The simplest expression that would produce this has to have a term in Equation (4.6) dependent on η^2 with a positive A coefficient. Below the phase-transition temperature T_c, the linear M–O–M is no longer stable, implying that $G(\eta = 0)$ becomes a local maximum (Figure 4.15). This change requires that A changes sign from positive to negative at the phase transition. The simplest way to express this is to give A a temperature dependence such as:

$$A(T) = a(T - T_c) \tag{4.7}$$

with a positive. This means that $A(T)$ is positive for $T > T_c$ and negative for $T < T_c$. In order to produce minima in our free-energy curve at $\eta \neq 0$ below T_c, it is now necessary to include a term in the $G(\eta)$ series with a positive coefficient, such as one depending on η^4. The Gibbs free-energy expression becomes:

$$G(\eta) = G_0 + \frac{1}{2}a(T - T_c)\eta^2 + \frac{1}{4}B\eta^4 \tag{4.8}$$

We can see from the form of $G(\eta)$ curves in Figure 4.15 that because $A(T)$ changes smoothly with temperature, $G(\eta)$ also changes smoothly, and the phase transition occurs in a smooth or continuous fashion. At the transition temperature T_c, the state of both phases is identical, and

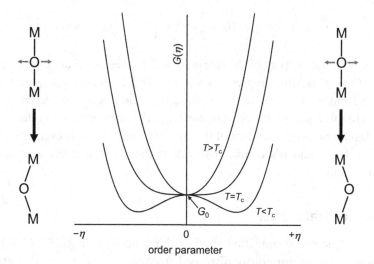

Figure 4.15 Free-energy curves for a second-order displacive phase transition.

the symmetry of the body at this point must contain the symmetry of both phases—the two space groups must be related by a group–subgroup relationship (Appendix B). In general, the high-temperature phase has the higher symmetry, and the low-temperature phase has only a subset of these symmetry elements. This turns out to be a powerful tool when investigating the possible symmetries of the low-temperature structures for a variety of phase transitions. A more detailed discussion of this topic can be found in the literature [4–8].

Equation (4.8) lets us investigate many of the important thermodynamic quantities associated with the phase transition. At any temperature, the equilibrium value of the order parameter is given by the minimum of the $G(\eta)$ curve, i.e. where $(\partial G/\partial\eta) = 0$ and $(\partial^2 G/\partial\eta^2) > 0$. Differentiating Equation (4.8) gives:

$$\frac{\partial G}{\partial \eta} = a(T - T_c)\eta + B\eta^3 = 0 \tag{4.9}$$

The three solutions to this equation are:

$$\eta = 0 \text{ and } \eta = \pm\sqrt{\frac{a}{B}}(T_c - T)^{1/2} \tag{4.10}$$

For $T > T_c$, $\eta = 0$ represents the minimum in Gibbs energy, and there are no other real solutions. For $T < T_c$, this solution ($\eta = 0$) is a local maximum in Gibbs energy, and the other two solutions represent minima.

The way in which η varies with temperature can be expressed even more succinctly than Equation (4.10). If we make $const = \pm\sqrt{aT_c/B}$ then:

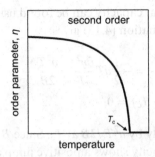
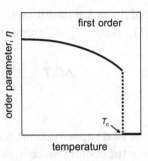

Figure 4.16 Schematic temperature dependence of the order parameter η in the Landau description of a second- and first-order transition.

$$\eta = const[(T_c - T)/T_c]^{\beta} \text{ with } \beta = 1/2. \tag{4.11}$$

The parameter β as a general variable is called the **critical exponent**. Under the Landau approximation, the full structural order of $\eta = 1$ is acquired at $T = 0$ such that $const = 1$ and $\eta = [(T_c - T)/T_c]^{1/2}$. This simple dependence of the order parameter η on temperature (Figure 4.16, left) is followed by many second-order transitions, at least close to T_c.

The transition thermodynamics can be derived as follows. Substitution of the non-zero η from Equation (4.10) into Equation (4.8) gives the excess Gibbs energy acquired by the transition on cooling from T_c to T as:

$$\Delta G = G - G_0 = -\frac{a^2}{4B}(T - T_c)^2 \tag{4.12}$$

Since $G = H - TS$, we can use $S = -dG/dT$ to derive the excess entropy and enthalpy acquired by the phase transition on cooling from T_c to T. The full-transition values can be determined by setting $T = 0$ into the temperature-based terms for ΔS and ΔH below, or by setting $\eta = 1$ with $T_c = B/a$ into their η-based terms:[21]

$$\Delta S = \frac{a^2}{2B}(T - T_c) = -\frac{1}{2}a\eta^2 \tag{4.13}$$

$$\Delta H = \Delta G + T\Delta S = -\frac{a^2}{4B}(T - T_c)^2 + \frac{a^2 T}{2B}(T - T_c) = -\frac{1}{2}aT_c\eta^2 + \frac{1}{4}B\eta^4 \tag{4.14}$$

The derivations are explored in the end-of-chapter problems. Equation (4.14) shows that the excess enthalpy of the system is a double-well function, and the positive sign of a and B in Equation (4.13) means that entropy, as a measure of disorder, decreases as the system distorts (orders) below T_c.

---

[21] Transition entropies and enthalpies are always reported upon heating, hence would have opposite signs to those suggested by Equations (4.13) and (4.14).

The effect of the phase transition on the heat capacity can be found using the relationship $C_p = T(\partial S/\partial T)_p$. Below T_c, differentiating Equation (4.13) gives:

$$\Delta C(T < T_c) = T\frac{\partial \Delta S}{\partial T} = -\frac{1}{2}aT\frac{\partial \eta^2}{\partial T} = \frac{a^2 T}{2B} \tag{4.15}$$

$$\Delta C(T > T_c) = 0 \tag{4.16}$$

We see that the heat capacity changes abruptly by $a^2 T_c/2B$ at T_c. Since B is positive, and we are considering the cooling process, heat capacity shows a positive jump at T_c on cooling to the low-temperature phase (see Figure 4.12), consistent with Ehrenfest's definition of a second-order transition.

Note from this analysis that once a and B are known, all the thermodynamic quantities associated with the phase transition can be calculated. The values of a and B can be obtained by measuring T_c and one of the excess thermodynamic quantities. The way in which an order-parameter approach can be used to describe magnetic transitions is explored in the end-of-chapter problems.

4.4.5 First-Order and Tricritical Transitions

We can apply similar arguments to investigate a discontinuous or first-order transition, provided we can relate the two structures by an order parameter. Here, the high- and low-temperature phases coexist in equilibrium at T_c. This coexistence requires that equal minima in $G(\eta)$ occur for different absolute values of η; one at the order parameter $\eta = 0$ of the disordered phase and two at non-zero $\pm\eta$ of the ordered phase.[22] The simplest Gibbs free-energy function that will allow this coexistence has a fourth-order term negative and a sixth-order term positive:

$$G(\eta) = G_0 + \frac{1}{2}a(T - T_0)\eta^2 - \frac{1}{4}B\eta^4 + \frac{1}{6}C\eta^6 \tag{4.17}$$

where a, B, and C are all positive quantities as written. T_0 is a temperature slightly lower than T_c, and we will derive their relationship shortly. This function has a single minimum at high T, three minima at intermediate temperatures above T_0, and two minima at temperatures below T_0, as depicted in Figure 4.17. As in the second-order case, differentiating Equation (4.17) with respect to η will tell us the order parameter values that give rise to minima in the free energy. Depending on temperature, there are up to three minima at:

---

[22] Note that in the Landau model, the order parameter η is often used in a somewhat relaxed way as a transition parameter and allowed to exceed the value of 1.

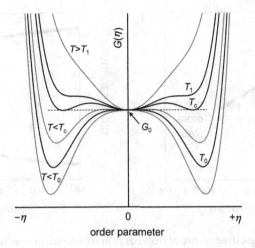

Figure 4.17 Free-energy curves in the Landau model of a first-order phase transition.

$$\eta = 0 \text{ and } \eta = \pm\left(\frac{B + \sqrt{B^2 - 4Ca(T - T_0)}}{2C}\right)^{1/2} \tag{4.18}$$

At high temperatures, we have a single $\eta = 0$ minimum. On cooling, we first reach a temperature $T_1 = T_0 + B^2/4aC$ (higher than T_c or T_0), where two additional local minima appear at $\eta \neq 0$, though they have a higher G than the G at $\eta = 0$.[23] The phase-transition temperature T_c is reached on further cooling, and is defined as the temperature where the three minima in the free-energy curve are all equal to G_0.[24] This can be shown[25] to occur at:

$$T_c = T_0 + \frac{3}{16}\frac{B^2}{aC} \tag{4.19}$$

and T_c lies between T_1 and T_0. Finally, upon cooling through T_0, the local $\eta = 0$ minimum disappears and only the two $\eta \neq 0$ minima remain.

Figure 4.18 shows the temperature dependence of the free energy of each minimum, highlighting how the ordered $\eta \neq 0$ phase becomes thermodynamically stable below T_c. The temperature dependence of the order parameters for first- and second-order transitions are compared schematically in Figure 4.16. Unlike Equation (4.10), the form of Equation (4.18) means that η changes abruptly or discontinuously at a first-order phase transition. The order parameter jumps from 0 to $\pm(3B/4C)^{1/2}$ on cooling through T_c, and, for those phase transitions that are finished at this point, it can be constrained to $\eta \pm 1$ by setting $C = \frac{3}{4}B$. Note that

---

[23] T_1 is the temperature where the expression within the square root $\sqrt{B^2 - 4Ca(T - T_0)}$ in Equation (4.18) equals 0.
[24] This is our equilibrium condition of $\Delta G = 0$ for the phase coexistence at the transition.
[25] At T_c, $G = G_0$ and $dG/dT = 0$ constrain the Landau coefficients. Performing these two operations on Equation (4.17) yields two equations that subtract to give $\eta^2 = 3B/4C$ that is substituted into the $dG/dT = 0$ equation and rearranged to give Equation (4.19).

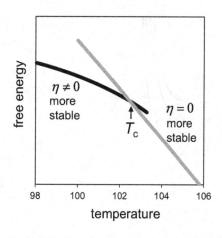 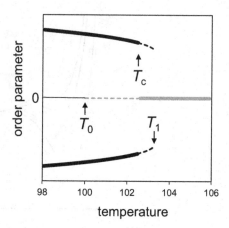

Figure 4.18 Temperature dependence of free energy and order parameter for a first-order transition in the Landau approximation. Dotted regions T_0 to T_c and T_c to T_1 show the temperature ranges, discussed in the text, over which local minima are present in $G(\eta)$.

for the first-order case, the simple jump in η at T_c doesn't imply anything about the pattern of atomic displacements that takes place at the transition. The transition must occur by the nucleation and growth of one phase in the other, as discussed above.

Just as we did for the second-order transition, we can derive other thermodynamic quantities by substituting our expression for η of Equation (4.18) back into Equation (4.17) to express G as a function of T. By differentiating with respect to T (as $S = -dG/dT$), we obtain the entropy, and from $C_p = T(\partial S/\partial T)_p$ the heat capacity.[26] Upon cooling through T_c, the entropy change is $-3aB/8C$ (negative, the system becomes more ordered).[27] The phase coexistence, hence $\Delta G = 0$ at T_c, means that we can use $\Delta H = T\Delta S$ to obtain:

$$\Delta H(T = T_c) = -\frac{3aBT_c}{8C} \tag{4.20}$$

We see that ΔH has a finite negative value for cooling through T_c.[28] Upon heating, this ΔH is positive and is the latent heat supplied to the equilibrium system of two phases at the transition.

We've seen that a differentiator between first- and second-order transitions is the sign of the η^4 contribution to $G(\eta)$. Using the $G(\eta)$ expression of Equation (4.6),[29] a first-order

---

[26] The algebra is tedious but results in $G(T) = G_0 - \dfrac{\left(B^2/4 - aC(T - T_0)\right)^{\frac{3}{2}}}{3C^2} + \dfrac{aB(T - T_0)}{4C} - \dfrac{B^3}{24C^2}$,

$S(T) = S_0 - \dfrac{a}{2C}\left(\dfrac{B}{2} + \sqrt{\dfrac{B^2}{4} - aC(T - T_0)}\right)$, and $C_p(T) = C_{p0} + \dfrac{a^2 T}{4\sqrt{B^2/4 - aC(T - T_0)}}$.

[27] Details explored in Problem 4.16.

[28] Contrast this to Equation (4.14) that shows that $\Delta H = 0$ at $T = T_c$ for a second-order phase transition.

[29] Note that in Equation (4.17) we reversed the sign of the η^4 term to keep all coefficients positive quantities in our main discussion.

transition has $B < 0$ and a second-order, $B > 0$. The case where $B = 0$ is called a **tricritical transition**, and for these $\eta \propto (T_c - T)^{1/4}$ with a critical exponent $\beta = ¼$ (see Problem 4.13).

What can cause the sign of B to become negative so that a phase transition becomes first-order rather than second-order? One opportunity arises when the phase transition distorts the lattice (also termed spontaneous strain). Our free-energy expression should then include additional terms to account for this.[30] As an example, the displacive transition in $SrTiO_3$ can be described by including a quadratic coupling term between strain and order parameter, and the free-energy function can be approximated as:

$$G(\eta) = G_0 + \frac{1}{2}a(T - T_0)\eta^2 + \frac{1}{4}B\eta^4 + \frac{1}{2}\lambda\varepsilon_s\eta^2 + \frac{1}{2}c_{el}\,\varepsilon_s^2 + \ldots \tag{4.21}$$

where ε_s is spontaneous strain, c_{el} an elastic constant, and λ a coupling constant; the final term in this equation is a Hooke's-law-like term with elastic energy proportional to the strain squared. Since the free energy depends on strain, at equilibrium $(\partial G/\partial\varepsilon_s) = 0$, which implies that $\varepsilon_s = -\lambda\eta^2/2c_{el}$. On introducing this to Equation (4.21) we find:

$$G(\eta) = G_0 + \frac{1}{2}a(T - T_0)\eta^2 + \frac{1}{4}\left(B - \frac{\lambda^2}{2c_{el}}\right)\eta^4 + \ldots \tag{4.22}$$

The effect of strain is then to reduce the coefficient of the η^4 term. If the coupling is sufficient to make the overall η^4 coefficient negative, the transition will be first-order.

4.4.6 Phonons, Soft Modes, and Displacive Transitions

The final concept we will discuss for understanding structural distortions at phase transitions is that of the soft mode. In any molecule or material, atoms are never at rest but undergo thermal vibrations. In an isolated molecule, we usually discuss these in terms of normal modes. In H_2O, for example, we can describe the possible vibrations in terms of three normal modes that approximate to a symmetric and an antisymmetric stretch of O–H bonds and a bending of the H–O–H bond angle. If we wanted to describe a hypothetical distorted water molecule with one short and one long O–H bond, one recipe would be to imagine freezing the asymmetric bond stretch at some point along the normal-mode coordinate describing the H displacement.

In an extended solid, the interactions between neighboring atoms mean that they don't vibrate independently. The motion of one atom influences those around it, producing displacement waves that travel through the crystal, termed **lattice modes**. The top panel of Figure 4.19 shows this schematically for an isolated chain of atoms that show transverse

---

[30] In general, excess free energy due to strain will contain terms dependent on $\varepsilon_s\eta$ and $\varepsilon_s\eta^2$ (though higher-order terms can also be included). For zone-center transitions (see Footnote 19) only the linear coupling term is usually required; for zone-boundary transitions only the quadratic term. More detail can be found in the text by Salje [E.K.H. Salje, "*Phase Transitions in Ferroelastic and Co-Elastic Crystals*" (1993) Cambridge University Press].

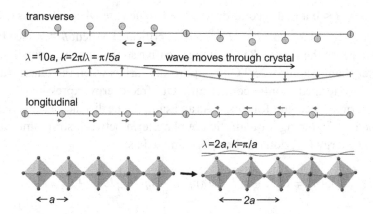

Figure 4.19 Phonons. Transverse and longitudinal phonons with $\lambda = 10a$ in a 1D chain. In the longitudinal case, the wave amplitude represents motion along the chain direction indicated by the small gray arrows. The lower picture shows how a soft mode with $k = \pi/a$ can lead to transition to a structure with doubled unit cell.

displacements from the chain axis with a travelling wave of $\lambda = 10a$. The peaks and troughs of this wave move through the crystal at a velocity (the phase velocity) given by $c = \lambda v$ such that atoms oscillate at frequency v. There are also longitudinal modes, in which the travelling-wave amplitude describes atomic motion to the right (+ amplitude) or left (−) of the time-averaged position along the chain direction. Lattice modes can always be described by waves between $\lambda = \infty$ (all atoms move in the same direction corresponding to a translation of the entire crystal) and $\lambda = 2a$ (atoms in adjacent cells vibrate out of phase). In a normal 3D crystal, we have to consider waves propagating in all directions, and it becomes convenient to label them in reciprocal space using a wave vector $\boldsymbol{k}$ with magnitude $2\pi/\lambda$.[31] The language is analogous to that developed in Chapter 6 to describe band theory and excellent descriptions can be found in texts such as those by Kittel, Ziman, or Dove (see Further Reading).[32] Each lattice mode or wave will affect all the atoms, and the amplitude of the motion depends on its frequency and the temperature. The overall motion of atoms will be a superposition of the motion caused by each of the waves.

To this point we've taken a classical view, but in reality the motion of atoms in solids is determined by quantum mechanics, and the energy of the vibrations is quantized. In the same way that wave-particle duality allows us to describe light waves in terms of particles

[31] As discussed in Chapter 6, the possible values of k (the lengths of $\boldsymbol{k}$) range from 0 to $\pm\pi/a$ with + and − corresponding to waves travelling in opposite directions.

[32] For a crystal with n atoms in the unit cell, there are $3n$ different combinations of motion, each of which has a specific frequency (dependent on interatomic forces) at each value of k; these are called branches. The dependence of v on $\boldsymbol{k}$ (labels direction and λ) for each branch is called a phonon-dispersion curve and is analogous to a band-structure diagram. For a 3D monatomic ($n = 1$) crystal, there will be three branches; one longitudinal and two transverse. At $\boldsymbol{k} = 0$ all atoms move in phase and $v = 0$ for each branch. As $\boldsymbol{k}$ changes, the frequency of the different branches will change.

called photons, vibrational waves can be described as particles called **phonons**. A phonon is the quantum unit of vibrational energy in a crystal and its energy is given by $h\nu$. When vibrational waves propagate heat or sound energy through a crystal, it is carried by the motion of phonons. As temperature is increased, the amplitude of atomic vibrations increases, and this corresponds to an increase in the number of phonons in the crystal. Slightly confusingly, the common usage is that the lattice modes describing the vibrations of atoms are also called phonons.

Anharmonic effects[33] in crystals lead to phonon frequencies varying with temperature. If a material has a relatively low-frequency phonon (or set of related phonons) whose frequency decreases on cooling, we can envisage a situation where the frequency could fall to zero. At this point, the structure becomes unstable with respect to a permanent distortion corresponding to the atomic motion described by the phonon—it becomes **soft** with respect to the distortion. The phonon becomes frozen into the structure and a displacive phase transition occurs to a lower-symmetry structure. The phonon involved is called a **soft mode**. The atoms now vibrate around their new equilibrium positions and the frequency starts to increase again on further cooling.

If we return to our simple example of a H_2O molecule, freezing in different normal modes (bends, stretches) will lead to a different distortion of the molecule with potentially different point-group symmetries. The same is true of soft modes in extended structures, and there is an elegant language that lets you explore the different symmetries of the structures that could form (colloquially called **child structures**) from a high-symmetry (**parent**) **structure**,[34] depending on the symmetry properties (decribed using an irreducible representation or irrep) of the phonon involved. These can be explored through web-based tools such as ISODISTORT (http://stokes.byu.edu/iso/isodistort.php).

As an example, $SrTiO_3$ undergoes a cubic to tetragonal phase transition on cooling through ~110 K due to a soft mode at $k = (\frac{1}{2}, \frac{1}{2}, \frac{1}{2})$. The atomic motions that freeze into the structure below T_c are shown schematically in Figure 4.19 and consist of coupled rotations of TiO_6 octahedra around c, with adjacent octahedra rotated in opposite directions. This is one of the tilted perovskite structures discussed in Section 1.5.3. Since k is non-zero, the unit cell of the low-temperature structure is larger. Similarly, the displacive phase transition between α- and β-quartz can be described by the softening of a ~200 cm^{-1} phonon of β-quartz upon cooling. This phonon is at $k = 0$, so the unit-cell size remains unchanged in α-quartz.[35] In some cases, the wavelength of the soft mode doesn't correspond to an integer number of unit cells of the parent structure. This is another recipe for formation of an incommensurate structure (discussed in Section 2.10), where the structure can't be conveniently described using

---

[33] In the simplest treatment of phonons (the harmonic approximation), it's assumed that the energy of the system depends only on the square of the relative displacements of adjacent atoms. The energy is then the same as that of a set of harmonic oscillators. Higher-order contributions to the energy are called anharmonic terms and are normally treated as a perturbation to the harmonic approximation.

[34] More formally, the highest-symmetry structure is called the aristotype and lower-symmetry structures hettotypes.

[35] $k = 0$ means $\lambda = \infty$ so that all β-quartz cells undergo the same in-phase distortion into α-quartz.

a conventional 3D crystallographic description. In fact, there is a narrow 1.3 K region just below T_c in quartz where the structure is incommensurately modulated for this reason. At lower temperatures, the soft-mode wavelength is said to "lock in" to the lattice and the incommensurate modulation disappears. There is more detail about the intricacies of soft modes and how this simple model applies in real materials in Further Reading.

4.5 Problems

4.1 Give a brief definition of the terms *phase* (*P*), *component* (*C*), and *degrees of freedom* (*F*) in the condensed matter phase rule $P + F = C + 1$.

4.2 Refer to the phase diagram depicted below. (a) State which four phases are stable at 100 °C. (b) What is the name given to the horizontal line separating region 2 from 1 and 3? (c) What are the approximate melting points of A, AB, and B? (d) What happens if you try and melt solid AB$_2$? (e) State what phases are present in each of areas 1–9. (f) Do any of the phases depicted form solid solutions? (g) State the number of phases and degrees of freedom at points *a*, *w*, *x*, *y*, and *z*. (h) Describe what happens when compositions at each of points *a* to *f* are cooled from high temperature. (i) Estimate the relative amounts of solid and liquid when a composition at point *a* ($x_B \approx 0.065$) is cooled to 500 °C, 400 °C, and 300 °C. (j) What might be observed if composition at point *e* is cooled rapidly? (k) State the differences between the peritectic reactions that happen on cooling compositions at points *e* and *f*.

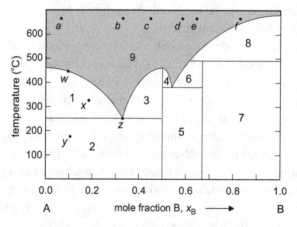

4.3 Using the phase diagram of Figure 4.8: (a) State how you would attempt to prepare a solid polycrystalline sample of ZrW$_2$O$_8$. (b) State how you would attempt to grow single crystals of ZrW$_2$O$_8$.

4.4 In the system Al$_2$O$_3$–BaO, five phases stable above 1300 °C were identified: Al$_2$O$_3$, Al$_{12}$BaO$_{19}$, Al$_2$BaO$_4$, Al$_2$Ba$_3$O$_6$, and BaO. Each was found to melt congruently at 2072 °C, 1900 °C, 1811 °C, 1616 °C, and 1918 °C, respectively. Eutectics form at $x_{BaO} = 0.11, 0.32,$

0.66, and 0.87 (relative to Al_2O_3) with melting points of 1875 °C, 1620 °C, 1480 °C, and 1425 °C. Sketch and fully label the phase diagram of this system.

4.5 Perovskite chemists searching in the CaO–TiO_2 system initially found four phases stable above 1300 °C: CaO, $Ca_3Ti_2O_7$, $CaTiO_3$, and TiO_2. CaO, $CaTiO_3$, and TiO_2 were reported to melt congruently at 2600 °C, 1970 °C, and 1830 °C and $Ca_3Ti_2O_7$ to melt incongruently at 1750 °C. Eutectics were reported at $x_{TiO_2} = 0.29$ and 0.76 with melting points of 1695 °C and 1460 °C. Sketch and fully label a phase diagram for this system.

4.6 Use the phase rule to explain how a mixture of Ni and NiO can be used to provide a controlled low oxygen partial pressure in a closed system.

4.7 If the height of the triangle in Figure 4.10a is 1, prove that the distances x_A, x_B, and x_C must sum to 1. Hint: Write an expression for the total area of triangle ABC in terms of constituent triangles such as ABd.

4.8 Plot the following compositions on a triangular composition diagram. Comment on the compositions of points a, b, and c and of points d, c, and e.

Point	x_A	x_B	x_C
a	0.8	0.1	0.1
b	0.4	0.3	0.3
c	0.2	0.4	0.4
d	0.1	0.7	0.2
e	0.3	0.1	0.6

4.9 State the compounds you would expect to form and their relative phase fractions when oxide mixtures corresponding to points a–f in Figure 4.11 are reacted under equilibrium conditions.

4.10 The following examples of phase transitions are discussed either in this chapter or in other parts of the book. In each case, would you describe the transition as reconstructive, displacive, or order–disorder in nature? State whether or not you would expect to be able to isolate the high-temperature phase at low temperature. (a) The transition from α-quartz to β-quartz at 573 °C. (b) The transitions from β-quartz to tridymite (870 °C) then to β-cristobalite at 1470 °C. (c) The transition at 641 °C of the $a^0b^0c^+$-tilted $NaNbO_3$ perovskite ($P4/mbm$) to the cubic ($Pm\overline{3}m$) perovskite $a^0b^0c^0$. (d)

4.11 State whether symmetry decreases or increases at the water-to-ice phase transition.

4.12 For a ferromagnetic second-order phase transition, assume that the relative magnetization M can be used as the order parameter in $G(M) = G_0 + \frac{1}{2}AM^2 + \frac{1}{4}BM^4$. Sketch the temperature dependence of M on T_c.

4.13 Prove that for a tricritical transition $\eta = [(T_c - T)/T_c]^{1/4}$.

4.14 Show that the expressions given in the text for the critical temperature of a first-order phase transition [Equation (4.19)] and for the abrupt order parameter change of $\eta =$

$\pm(3B/4C)^{1/2}$ at T_c are consistent with the definition that the three minima in the Gibbs energy curve are equal to G_0 at T_c (Figure 4.17).

4.15 Show that the expression relating T_c and T_0 of a first-order phase transition [Equation (4.19)] and the Gibbs energy expression in Footnote 26 are consistent with the definition that the three minima in the Gibbs energy curve are equal to G_0 at T_c (Figure 4.17).

4.16 Show that the entropy and enthalpy changes for a first-order phase transition are given by $\Delta S = -3aB/8C$ and $\Delta H = -3aBT_c/8C$.

4.17 On cooling the cubic perovskite $SrZrO_3$ (space group $Pm\overline{3}m$) from high temperature, it undergoes a phase transition to the tetragonal space group $I4/mcm$. At the phase transition, the c cell parameter doubles and a increases by $\sqrt{2}$. The strain e_t is related to the difference in a and c parameters scaled back to those of the cubic cell. With appropriate normalization, $e_t = (2/\sqrt{3})\left((c-a)/a_{cub}\right)$ where a_{cub} is the cell parameter expected for an undistorted material at that temperature, and is expected to be proportional to η^2. The table below contains unit-cell parameters at various temperatures from ref. [9] that were later analyzed in ref. [10]. Plot the temperature evolution of the cell parameters. By assuming a sensible functional form, predict a_{cub} for each temperature below 1360 K. Comment on your graph. (b) Comment on a plot of $e_t^{0.5}$ versus T. (c) From a suitable plot, show that the data are consistent with the transition being close to tricritical in character and determine the transition temperature T_c.

T (K)	a (Å)	c (Å)	T (K)	a_{cub} (Å)
1160	4.1394	4.1514	1360	4.1536
1180	4.1405	4.1520	1380	4.1546
1200	4.1416	4.1524	1400	4.1555
1220	4.1428	4.1529	1420	4.1565
1240	4.1441	4.1532	1440	4.1575
1260	4.1455	4.1535	1460	4.1585
1280	4.1469	4.1539	1480	4.1594
1300	4.1485	4.1542	1500	4.1604
1320	4.1500	4.1542		
1340	4.1517	4.1537		

4.6   Further Reading

E.M. Levin, C.R. Robbins, H.F. McMurdie, *"Phase Diagrams for Ceramists"* (1964) American Ceramic Society.

H. Okamoto, *"Desk Handbook: Phase Diagrams for Binary Alloys"* (2000) ASM International.

M.T. Dove, *"Introduction to Lattice Dynamics"* (1993) Cambridge University Press.

M.T. Dove, *"Structure and Dynamics"*, Oxford Master Series in Condensed Matter Physics (2003) Oxford University Press.

J.M. Ziman *"Electrons and Phonons: The Theory of Transport Phenomena in Solids"*, Oxford Classic Texts in the Physical Sciences (2001) Oxford University Press.

L.D. Landau, E.M. Lifshitz, *"Statistical Physics, Course of Theoretical Physics, Volume 5"* 3rd edition (2000) Reed Educational and Professional Publishing Ltd.

C. Kittel, *"Introduction to Solid State Physics"* (1996) John Wiley and Sons.

4.7　References

[1] L.L.Y. Chang, M.G. Scroger, B. Phillips, "Condensed phase relations in the systems ZrO_2–WO_2–WO_3 and HfO_2–WO_2–WO_3" *J. Am. Ceram. Soc.* **50** (1967), 211–215.

[2] A.S. Berezhnoi, N.V. Gul'ko, "The system Al_2O_3–TiO_2–ZrO_2" *Dopov. Akad. Nauk Ukr. RSR* **1** (1955), 77–80.

[3] P. Karen, A. Kjekshus, "Phase diagrams and thermodynamic properties" in *"Handbook on the Physics and Chemistry of Rare Earths, Volume 30"*, editors K.A. Gschneidner, Jr., L. Eyring, M. B. Maple (2000) Elsevier, 229–371.

[4] H. Wondratschek, U. Müller (editors), "International Tables for Crystallography, Volume A1: Symmetry relations between space groups" (2011) http://dx.doi.org/10.1107/97809553602 060000110.

[5] L.D. Landau, E.M. Lifshitz, *"Statistical Physics, Course of Theoretical Physics, Volume 5"* 3rd edition (2000) Reed Educational and Professional Publishing Ltd.

[6] B.J. Campbell, H.T. Stokes, D.E. Tanner, D.M. Hatch, "ISODISPLACE. A web-based tool for exploring structural distortions" *J. Appl. Crystallogr.* **39** (2006), 607–614.

[7] H.T. Stokes, D.M. Hatch, "Coupled order parameters in the Landau theory of phase transitions in solids" *Phase Transitions* **34** (1991), 53–67.

[8] E.F. Bertaut, "Representation analysis of magnetic structures" *Acta Crystallogr. Sect. A* **24** (1968), 217–231.

[9] C.J. Howard, K.S. Knight, B.J. Kennedy, E.H. Kisi, "The structural phase transitions in strontium zirconate revisited" *J. Phys. Condens. Matter* **12** (2000), L677–L683.

[10] R.E.A. McKnight, C.J. Howard, M.A. Carpenter, "Elastic anomalies associated with transformation sequences in perovskites: I. Strontium zirconate, $SrZrO_3$" *J. Phys. Condens. Matter* **21** (2009), 015901.

5   Chemical Bonding

Changes in crystal structure invariably lead to changes in physical and/or chemical properties. In some cases, these changes can be dramatic, as illustrated by the contrasting properties of the allotropes of carbon (diamond, graphite, graphene, C_{60}, etc.); in other cases they are subtle but nonetheless important. To understand the relationship between structure and properties, one must first understand chemical bonding.

We begin this chapter with an overview of ionic bonding. From there we move on to the properties of atomic orbitals (AOs) and their interactions to form covalent bonds through the framework of molecular orbital theory. In Chapter 6, we then build upon these principles to describe the formation of bands in extended solids. In this way, covalent and metallic bonding can be understood through a common approach.

5.1   Ionic Bonding

Although there are no compounds where the bonding can be described as purely ionic, the ionic model is a useful approximation for many compounds. We begin our treatment of bonding with a brief overview of the factors that determine the strength of ionic bonding in crystalline solids.

5.1.1   Coulombic Potential Energy

The **coulombic potential energy**, U_C, between two ions of charge numbers z_1 and z_2 separated by a distance d is:

$$U_C = \frac{(z_1 e) \cdot (z_2 e)}{4\pi\varepsilon_0 d} \tag{5.1}$$

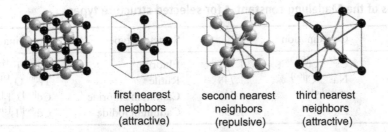

first nearest
neighbors
(attractive)

second nearest
neighbors
(repulsive)

third nearest
neighbors
(attractive)

Figure 5.1 The NaCl structure and the first, second, and third nearest neighbors to a Cl^- ion. The Cl^- ions are gray and the Na^+ ions black.

where e is the elementary charge and ε_0 is the electric constant.[1] To estimate the strength of ionic bonding in a crystal, we treat the ions as point charges and use Equation (5.1) to capture all electrostatic interactions in the crystal, both attractive and repulsive.

To illustrate, consider the electrostatic interactions in the NaCl structure shown in Figure 5.1. We begin with the Cl^- ion in the center of the unit cell and consider the interaction between this ion and all other ions in the crystal. The nearest neighbors are the six Na^+ ions that lie at the center of the faces of the unit cell, at a distance d from the central Cl^- ion. The potential energy of this interaction is negative (an attraction) and amounts to six times Equation (5.1) with z_1 and z_2 equal to -1 and $+1$. The second nearest neighbors are 12 Cl^- ions that lie at the center of each edge of the cubic unit cell, at a distance of $\sqrt{2}d$. The potential energy of this interaction is positive (a repulsion; $z_1 = z_2 = -1$) and is $12/\sqrt{2}$ times Equation (5.1). The third nearest neighbors are eight Na^+ ions that lie at the corners of the unit cell, at a distance of $\sqrt{3}d$. This interaction is attractive ($z_1 = -1$, $z_2 = +1$), and has a magnitude of $8/\sqrt{3}$ times Equation (5.1). This process must be repeated for increasingly distant neighbors leading to an infinite series, of which the first seven terms are as follows:

$$U_C = \frac{(z_1 e) \cdot (z_2 e)}{4\pi\varepsilon_0 d} \left(6 - \frac{12}{\sqrt{2}} + \frac{8}{\sqrt{3}} - \frac{6}{\sqrt{4}} + \frac{24}{\sqrt{5}} - \frac{24}{\sqrt{6}} - \frac{12}{\sqrt{8}} \ldots \right) \tag{5.2}$$

The infinite series expressed in Equation (5.2) does not readily converge because successive shells tend to alternate between those containing anions and those containing cations. Convergence to a value near 1.7476 [times Equation (5.1) for the sole Na^+Cl^- pair] is obtained if the shells are chosen in such a way that the net charge of each shell is nearly neutral [1]. This value, which is the same for all ionic compounds with the NaCl-type structure, is called the **Madelung constant**, A. The net coulombic potential energy per mole of NaCl (z_1, z_2 of opposite signs) is then:

---

[1] U_C is calculated from Coulomb's law, which describes the electrostatic force between two charges, as the work required to separate the positive and negative ions from their initial distance d to infinity. Note also that the electric constant of $\sim 8.8542 \times 10^{-12}$ C/(V m) is often referred to as the permittivity of free space.

Table 5.1 Values of the Madelung constant *A* for selected structure types.

Structure type	Coordination	A	Structure type	Coordination	A
Cesium chloride	$Cs^{[8c]} Cl^{[8c]}$	1.763	Fluorite	$Ca^{[8c]} F_2^{[4t]}$	2.519
Sodium chloride	$Na^{[6o]} Cl^{[6o]}$	1.748	Rutile	$Ti^{[6o]} O_2^{[3]}$	2.408
Wurtzite	$Zn^{[4t]} S^{[4t]}$	1.641	Cadmium chloride	$Cd^{[6o]} Cl_2^{[3n]}$	2.244
Zinc blende	$Zn^{[4t]} S^{[4t]}$	1.638	Cadmium iodide	$Cd^{[6o]} I_2^{[3n]}$	2.192

$$U_C = \frac{z_1 z_2 e^2 N_A}{4\pi\varepsilon_0 d} A \tag{5.3}$$

Similar derivations can be carried out for other structure types. The Madelung constants for some structure types are given in Table 5.1, many others can be found in the literature or calculated using the Ewald method [2]. We see that the Madelung constant increases as the coordination number increases from four for zinc blende/wurtzite to six for sodium chloride to eight for cesium chloride. It is smaller for layered structures with direct anion–anion contacts (CdI_2 or $CdCl_2$) than for structures with the same stoichiometry where the cations are distributed more uniformly (rutile TiO_2). The Madelung constant also increases, almost proportionally, with the number of ions per formula unit.

We can make several generalizations about the factors that optimize electrostatic interactions. The electrostatic attraction holding the ions together increases as the ionic charges increase (MgO will have a more negative U_C than LiF) and as the cation–anion distance *d* decreases (LiF will have a more negative U_C than RbBr), though the latter effect is less dramatic. As a rule, electrostatic interactions are highest in symmetric structures that allow for efficient packing of ions and high coordination numbers.

5.1.2 Lattice Energy and the Born–Mayer Equation

Of the two alternative ways to define **lattice energy** (U_L), we'll use the energy of formation of the ionic crystal at *T* = 0 K:

$$aM^{y+}(g) + bX^{z-}(g) \rightarrow M_a X_b(s) \tag{5.4}$$

This process releases energy of the system into the surroundings, hence U_L is a *negative number* just like the coulombic potential energy U_C introduced in the preceding section.

The lattice energy is an equilibrium of attractive and repulsive interactions. To model it appropriately, we must therefore consider not only the coulombic attraction U_C, but also the repulsive potential energy U_r (Figure 5.2) that arises when electron clouds on the otherwise attracting neighboring ions approach each other too closely. That repulsion can be approximated by a variety of mathematical functions that rise rapidly as the distance d between the two ions decreases. A common choice is the two-parameter exponential function:[2]

$$U_r = Be^{-d/\rho} \qquad (5.5)$$

where B is a magnitude constant that includes N_A, and ρ is an empirical constant typically taken to be 3.45×10^{-11} m (0.345 Å). Combining the coulombic potential energy of the cation and anion charges, Equation (5.3), with the repulsive potential energy arising from interactions between electron clouds on neighboring ions, Equation (5.5), we obtain an expression for the lattice-formation energy U_L of the ionic compound of charge numbers z_1 and z_2:

$$U_L = (U_C + U_r) = \left(\frac{z_1 z_2 e^2 N_A}{4\pi\varepsilon_0 d} A + Be^{-d/\rho} \right) \qquad (5.6)$$

As shown in Figure 5.2, Equation (5.6) has a minimum at the equilibrium interatomic distance[3] d_0 [3] that is calculated[4] as $dU_L/dd = 0$:

$$\frac{dU_L}{dd} = -\frac{z_1 z_2 e^2 N_A}{4\pi\varepsilon_0 d_0^2} A - \frac{Be^{-d_0/\rho}}{\rho} = 0 \qquad (5.7)$$

From here, we can solve Equation (5.7) for B and substitute this expression back into Equation (5.6), giving the **Born–Mayer equation**:

$$U_L = \frac{z_1 z_2 e^2 N_A}{4\pi\varepsilon_0 d_0} A \left(1 - \frac{\rho}{d_0} \right) \qquad (5.8)$$

A generic plot of the coulombic potential energy, the repulsive potential energy, and the lattice-formation energy of an ionic structure as a function of interatomic distance is shown in Figure 5.2. For typical interatomic distances, U_r is about 10–20% of U_C.

---

[2] It is also common to represent repulsive interactions with the expression, $U_r = B/d^n$, where n has a value between 5 and 12 depending upon the electron configuration of the cation and anion. This approach leads to the Born–Landé equation.

[3] This d_0 can be determined either from crystallographic studies or estimated from ionic radii ($d_0 = r_C + r_A$). Ionic radii are derived from observed interatomic distances in a large database of crystal structures. The most extensive and widely used set of radii is that determined by Shannon. These radii are listed in two parallel sets, one called crystal radii (CR) the other ionic radii (IR). Cations have CR that are 0.14 Å larger than their IR, while anions have CR that are smaller than IR by the same amount. The CR set is chosen to approximate ionic sizes determined from minima in the electron density between a given cation and anion and is thought to more accurately represent the true sizes of ions, the IR are chosen to have the same radii for the O^{2-} and F^- ions as initially chosen by Linus Pauling. Both sets give identical estimates of cation–anion distances.

[4] Remember from calculus that the first derivative of a function is equal to zero at either a minimum or a maximum of that function. In this case, $dU_L/dd = 0$ identifies the minimum (most favorable) energy. The fact that it is a minimum rather than a maximum can be shown by taking the second derivative.

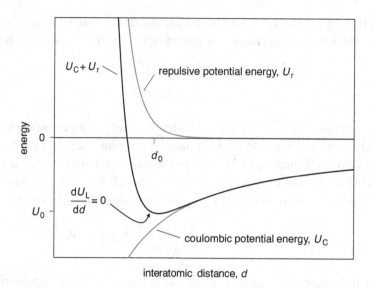

Figure 5.2 The coulombic (U_C) and repulsive (U_r) potential-energy contributions to the lattice-formation energy.

5.1.3 Experimental versus Calculated Lattice-Formation Energies

How does the lattice-formation energy calculated with the Born–Mayer equation compare with experiment? The lattice energy as defined in Equation (5.4) cannot be determined in a single experiment. Instead, the lattice-formation enthalpy, H_L, is determined from several measurements using a thermochemical cycle called a Born–Haber cycle. As the enthalpy change ΔH in the Born–Haber cycle is associated with the same reaction as Equation (5.4), $H_L \approx U_L$, and we will treat the two as numerically equivalent.[5]

As an example, the Born–Haber cycle for KCl is shown in Figure 5.3. In the left branch, energy is supplied to decompose the salt into its constituent elements, atomize molecular chlorine, and evaporate and ionize potassium. In the right branch, energy is released by chlorine atoms acquiring an electron and by the formation of solid KCl from the gaseous ions. The enthalpy of the latter step is the lattice-formation energy we are seeking. For KCl, the value of H_L is found to be −718 kJ/mol using this approach.

It is interesting to note that while we normally think of the electron transfer from metal to nonmetal as a spontaneous process, comparison of the two energies on top of Figure 5.3 shows that for potassium and chlorine this process is endothermic. In fact, this energy is

---

[5] For a process at constant pressure, where the gaseous ions behave ideally, the relationship between lattice-formation enthalpy and energy is given by the equation, $\Delta H_L = \Delta U_L + \Delta nRT$, where Δn is the number of moles of gaseous ions produced (e.g. $\Delta n = 2$ for NaCl, $\Delta n = 3$ for MgF_2). At $T = 0$, the lattice-formation energy and enthalpy are equal, while for finite temperatures the two values differ only slightly ($RT = 2.5$ kJ/mol at $T = 300$ K). If one considers the vibrations of the atoms in the solid and in the gas phase more rigorously, the equations become more complicated. For more details, the interested reader is directed to an article by H.D.B. Jenkins, *J. Chem. Ed.* **82** (2005), 950–952.

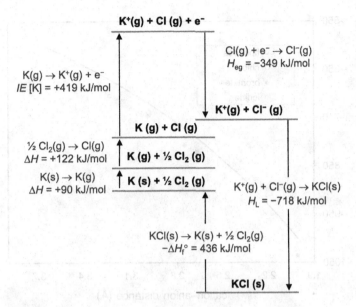

Figure 5.3 The Born–Haber cycle for KCl constructed from enthalpies of sublimation for K, atomization for Cl_2, the ionization energy (*IE*) of K, and the electron-gain enthalpy (H_{eg}) of Cl (note H_{eg} has the same magnitude but opposite sign as electron affinity).

endothermic for any combination of elements. The energy released upon forming an extended ionic solid is the driving force behind the spontaneous and often highly exothermic reactions that occur between metals and nonmetals.

A comparison of experimental lattice-formation enthalpies (H_L) and calculated lattice-formation energies (U_L) for alkali-metal halides with NaCl-type structure is given in Figure 5.4. Given the simplicity of this model, the agreement is surprisingly good. Although the lattice-formation energies are underestimated by 3–6%, the functional dependence is well reproduced in the experimental data.

To obtain more accurate estimates of lattice-formation energy, we must include additional factors, such as van der Waals forces, polarization of ions in low-symmetry environments, and zero-point energy. Each of these effects is briefly discussed below.

Van der Waals forces collectively refer to weak attractive forces arising from induced dipoles in the electron clouds of neighboring atoms and molecules. They are solely responsible for holding atoms together when noble gases solidify, and they are the dominant force when neutral molecules condense to form molecular solids. They tend to favor close packing, which explains why the noble gases crystallize with a ccp structure at low temperatures. London dispersion forces are the most common type of van der Waals forces.[6] They arise from random fluctuations of the electron density around an otherwise spherically symmetric

---

[6] Other types of van der Waals forces include dipole–induced-dipole forces, and interactions between rotating polar molecules.

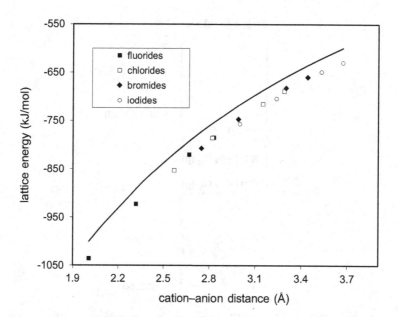

Figure 5.4 Experimental lattice-formation enthalpies H_L for alkali-metal halides with the NaCl-type structure are indicated with symbols. The smooth curve shows the lattice-formation energy U_L as calculated with the Born–Mayer equation for comparison.

atom. The dipolar electrostatic field that is temporarily created in one atom induces a transient dipole of opposite polarity in a neighboring atom. The strength of the van der Waals force increases with increasing atomic number, but in all cases is much weaker than an ionic bond. It also drops off more rapidly with increasing distance, varying inversely with the sixth power of the internuclear separation.

Polarization is a distortion of the electronic charge density of an ion in response to the electric field created by the surrounding ions. The most important type of polarization in ionic crystals is the polarization of large soft anions by adjacent cations that are smaller and often more highly charged. Polarization effects, together with van der Waals forces, are often invoked to help explain the stability of layered structures with direct anion–anion contacts like $CdCl_2$ and CdI_2 [4]. Polarization of the electron distribution of the halide ions reduces the anion–anion repulsions across those ccp or hcp hole planes that do not contain cations (Section 1.4.2) to the point where van der Waals forces are sufficient to stabilize the structure.

The other term that makes a small contribution to the lattice-formation energy is the **zero-point energy** of the crystal. The zero-point energy, a quantum-mechanical concept that comes from the Heisenberg uncertainty principle, is the vibrational energy of the lattice at absolute zero temperature.

How important are these additional contributions to the lattice-formation energy? As a rough guide, van der Waals forces make the lattice-formation energy more negative, and the magnitude of this contribution ranges from a few tenths of a percent to roughly 5% of the total. The zero-point energy is generally less than 1.5% of the total and makes the lattice-formation energy less negative. Because these two corrections are relatively small and opposite in sign, they tend to cancel each other. In systems where these corrections are large, the validity of the ionic model comes into question. More sophisticated approaches are favored when a high degree of accuracy is important.

5.2 **Atomic Orbitals**

The concept of a covalent bond is ubiquitous in chemistry. A variety of approaches spanning a range of complexities can be used to model covalent bonding. Simple approximations such as Lewis structures and valence-shell electron-pair repulsion (VSEPR) provide useful estimates of *molecular* geometry and bonding in many cases. At the other end of the spectrum, advanced computational methods like density-functional theory can be used to make increasingly accurate predictions for crystalline solids. It is, however, not always easy to extract chemical insight from such approaches. Semi-empirical methods, such as extended Hückel theory [5], occupy an intermediate ground and afford insight into the links between local bonding interactions, crystal structures, and physical properties, a useful attribute that merits their inclusion in this book. Models of covalent bonding are typically based on the overlap of AOs. Therefore, we begin our treatment with a closer look at electrons and orbitals.

We start with the simplest possible case, a one-electron atom, and treat its electron as though it were a standing wave rather than a particle. The quantum-mechanical behavior of the electron is described by the partial differential equation first proposed by Erwin Schrödinger[7] in 1926:

$$\hat{H}\,\psi(\boldsymbol{r}) = \left[-\frac{\hbar^2}{2m}\nabla^2 + V(\boldsymbol{r}) \right]\psi(\boldsymbol{r}) = E\psi(\boldsymbol{r}) \tag{5.9}$$

where the function $\psi(\boldsymbol{r})$ describes the electron as a wave, a **wavefunction**. The hamiltonian operator, $\hat{H}$, is a sum of two terms. The first term $-\left(\hbar^2/2m\right)\nabla^2$ represents the kinetic energy of an electron of mass m (the Laplacian ∇^2 is the second partial derivative of the wavefunction), and the second term $V(\boldsymbol{r})$ represents the electron's potential energy as a function of its position vector $\boldsymbol{r}$ from the nucleus. The $\psi(\boldsymbol{r})$ determines the energy E of the electron in Equation (5.9). We will see later how also the size and shape of the electron cloud can be calculated from $\psi(\boldsymbol{r})$. Because each AO has a unique $\psi(\boldsymbol{r})$, we will often refer to $\psi(\boldsymbol{r})$ as the orbital wavefunction, but, more precisely, it is the mathematical description of the standing wave of an electron that occupies a given orbital; an electron wavefunction.

---

[7] This is the non-relativistic, time-independent Schrödinger equation for a single particle moving in an electric field.

That wavefunction is typically expressed in spherical polar coordinates as the product of two functions:

$$\psi_{n,\ell,m_\ell} = R_{n,\ell}(r)Y_{\ell,m_\ell}(\theta, \phi) \qquad (5.10)$$

In these coordinates, the radial part of the wavefunction, R, describes the variation of the electron wave as a function of the distance r from the nucleus (here r is a scalar rather than a vector). The angular part of the wavefunction, Y, describes the shape of the wavefunction in terms of θ and ϕ. Equations for $R(r)$ and $Y(\theta, \phi)$ can be found in many physical chemistry textbooks. We won't go into the detailed mathematics here, but we will qualitatively look at how the size, shape, and energy of the orbital depends upon the quantum numbers n, ℓ, and m_ℓ that determine the wavefunction.

The **principal quantum number**, n, labels the quantized energy and largely determines the size of the orbital. The allowed values of n, which are independent of the other quantum numbers, are positive integers: 1, 2, 3, ... The **orbital angular-momentum quantum number** (sometimes called the azimuthal quantum number), ℓ, labels the quantized orbital angular momentum and dictates the shape of the orbital. The allowed values of ℓ are the integers ranging from 0 to $n - 1$. The values of ℓ are given specific letter designations: $\ell = 0$ corresponds to an s orbital, $\ell = 1$ to a p orbital, $\ell = 2$ to a d orbital, and $\ell = 3$ to an f orbital. Orbitals with the same value of n are said to belong to the same **shell**, and those with the same values of n and ℓ are said to belong to the same **subshell** ($2s$, $3d$, $4p$, etc.).

The **magnetic quantum number**, m_ℓ, labels the quantized orientation of the angular momentum and dictates the orientation of the orbital. The allowed values of m_ℓ are the integers between $-\ell$ and ℓ. This restriction limits the number of orbitals for each subshell to one s orbital ($m_\ell = 0$), three p orbitals ($m_\ell = -1, 0, +1$), five d orbitals ($m_\ell = -2, -1, 0, +1, +2$), and seven f orbitals ($m_\ell = -3, -2, -1, 0, +1, +2, +3$). It is customary to label the orbitals with subscripts that denote the orientation of the orbital with respect to an arbitrary set of Cartesian axes. The p orbitals are labeled p_x, p_y, and p_z, where the subscripts identify the axis along which the lobes of the orbital point, and the d orbitals are labeled d_{xy}, d_{xz}, d_{yz}, $d_{x^2-y^2}$, and d_{z^2}. A fourth quantum number, the **spin quantum number**, m_s, labels the quantized values of spin and can take only two values; $+\frac{1}{2}$ and $-\frac{1}{2}$. The spin quantum number does not enter the wavefunction directly, but through the Pauli exclusion principle[8] it limits the maximum occupancy of an AO to two electrons.

The angular part Y of the wavefunction is a constant when $\ell = 0$, and thus the s orbitals are spherically symmetric. This makes them convenient examples to illustrate how the radial part of the wavefunction changes with increasing n. The wavefunctions for the $1s$, $2s$, and $3s$ orbitals are plotted as a function of r on the left-hand side of Figure 5.5. All three wavefunctions drop off exponentially upon moving away from the nucleus. The $1s$ wavefunction is positive ($\psi > 0$) for all values of r, but the sign of the $2s$ wavefunction

---

[8] The Pauli exclusion principle states that no two electrons in an atom can exist in the same quantum state and therefore cannot have identical quantum numbers.

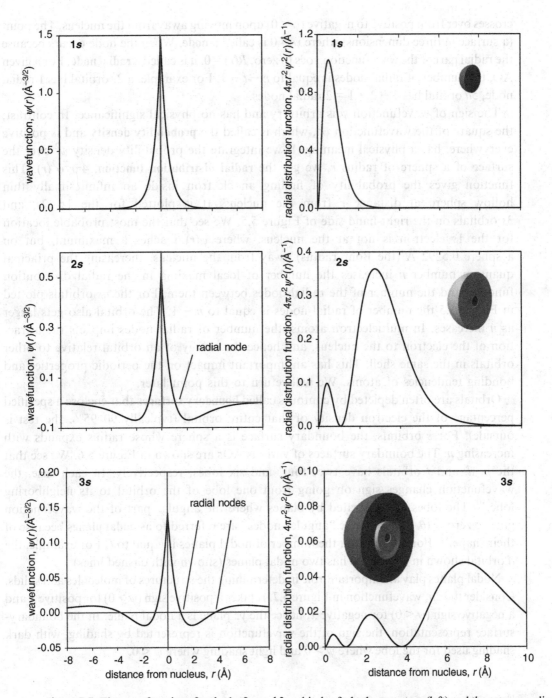

Figure 5.5 The wavefunctions for the 1s, 2s, and 3s orbitals of a hydrogen atom (left) and the corresponding radial distribution functions, $4\pi r^2 \psi^2(r)$ (right). The 1s orbital is the ground state, while the other wavefunctions represent excited states. A cutaway view of each orbital is shown. The dark and light shading indicates a positive ($\psi > 0$) or negative ($\psi < 0$) sign of the wavefunction, respectively.[9]

[9] See also Footnote 15 in this chapter for more discussion on wavefunction sign conventions used in this book.

crosses over from positive to negative ($\psi < 0$) upon moving away from the nucleus. The point (a surface in three dimensions) where $\psi = 0$ is called a **node**. When the node occurs because the radial part of the wavefunction goes to zero, $R(r) = 0$, it is called a **radial node**. For a given AO, the number of radial nodes is equal to $n - \ell - 1$. For example, a $2s$ orbital has 1 radial node, $5d$ orbital has $5 - 2 - 1 = 2$ radial nodes.

The sign of wavefunction ψ is arbitrary and has no physical significance. In contrast, the square of the wavefunction ψ^2, which is called the **probability density** and is positive everywhere, has a physical meaning. If we integrate the probability density ψ^2 over the surface of a sphere of radius r, we get the **radial distribution function**, $4\pi r^2 \psi^2(r)$. This function gives the probability of finding an electron inside an infinitesimally thin hollow sphere at distance r from the nucleus. It is plotted for the $1s$, $2s$, and $3s$ orbitals on the right-hand side of Figure 5.5. We see that the most probable location for the $1s$ electron is not at the nucleus, where $\psi(r)$ reaches a maximum, but on a sphere 0.5292 Å (the Bohr radius) away from the nucleus. Increasing the principal quantum number n increases the number of local maxima in the radial distribution function and the number of the radial nodes between them. For the ns orbitals plotted in Figure 5.5 the number of radial nodes is equal to $n - 1$. The orbital also gets larger as n increases. In multielectron atoms, the number of radial nodes impacts the attraction of the electron to the nucleus, and hence the energy of an orbital relative to other orbitals in the same shell. This has an important impact on the periodic properties and bonding tendencies of atoms. We will return to this point later.

Orbitals are often depicted by contours called **boundary surfaces** that enclose a specified percentage of the electron density of that entire orbital (typically 90–95%, the rest is outside). For s orbitals, the boundary surface is a sphere whose radius expands with increasing n. The boundary surfaces of various AOs are shown in Figure 5.6. We see that the p, d, and f orbitals have two, four, and six lobes, respectively. In each case, the wavefunction changes sign on going from one lobe of the orbital to its neighboring lobe.[10] The lobes are separated by planes where the angular part of the wavefunction goes to zero, $Y(\theta, \phi) = 0$. These "angular nodes" are referred to as **nodal planes** because of their shape.[11] For a given AO, the number of nodal planes is equal to ℓ. For example, the d orbital shown in Figure 5.6 has two nodal planes (shown with dashed lines).

Nodal planes play an important role in determining the structures of molecules and solids. Consider the $2p_x$ wavefunction in Figure 5.7. It takes a positive sign ($\psi > 0$) for positive x and a negative sign ($\psi < 0$) for negative x, hence the yz plane is a nodal plane. In the boundary-surface representation, the sign of the wavefunction is represented by shading, with dark shading used for the lobe where $\psi > 0$ and light shading where $\psi < 0$.

---

[10] Note that the boundary surfaces represent properties of ψ^2, the sign of which is always positive, whereas the shading refers to the sign of the underlying wavefunction ψ.

[11] For some of the d and f orbitals, the surface where the angular part of the wavefunction goes to zero is a cone rather than a plane; the d_{z^2} orbital being one such example. We use the term nodal plane generically to apply to both planes and cones.

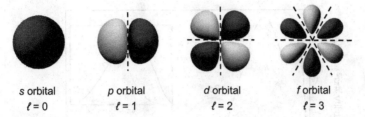

s orbital
$\ell = 0$

p orbital
$\ell = 1$

d orbital
$\ell = 2$

f orbital
$\ell = 3$

Figure 5.6 Boundary surfaces of the angular part of the wavefunction $Y(\theta, \phi)$ for selected orbitals. The light and dark shading represent the sign of the underlying wavefunction. The dashed lines represent the nodal planes.

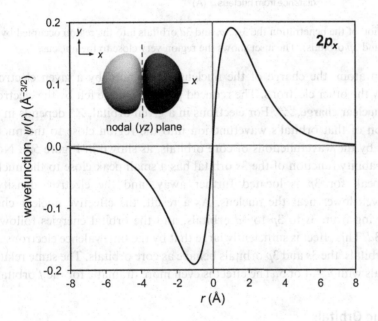

Figure 5.7 The amplitude of $2p_x$ wavefunction plotted along the x-axis line. The contour shape shows the nodal plane at $x = 0$, looking down the z axis.

The size and energy of an orbital depend on the strength of the attraction between the orbital's electron(s) and the nucleus. In one-electron atoms, (H, He$^+$, Li^{2+}, and other so-called hydrogenic atoms), the allowed energy levels are given by:

$$E_n = -\frac{Z^2 hcR_H}{n^2} \tag{5.11}$$

where Z is the charge of the nucleus, h is Planck's constant, c is the speed of light, and R_H is the Rydberg constant. Note that the energy of the orbital only depends upon the nuclear charge Z and the value of n.

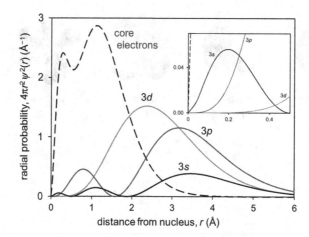

Figure 5.8 An illustration of the penetration the $3s$, $3p$, and $3d$ orbitals into the region occupied by the core electrons ($1s$, $2s$, and $2p$ orbitals). The inset shows the region very close to the nucleus.

In a multielectron atom, the charge of the nucleus experienced by a given electron is partially screened by the other electrons. The reduced positive charge felt by the electron is called the **effective nuclear charge**, Z^*. For electrons in a given orbital, Z^* depends in part upon the penetration of that orbital's wavefunction into the region close to the nucleus, a region dominated by the wavefunctions of core orbitals, as shown in Figure 5.8. Notice how the radial probability function of the $3s$ orbital has a small peak close to the nucleus, whereas the first peak for $3p$ is located further away, and the electron density of the $3d$ orbitals is even lower near the nucleus. As a result, the effective nuclear charge decreases upon moving from $3s$ to $3p$ to $3d$ orbitals, and the orbital energies follow the sequence $3s < 3p < 3d$. This effect is sufficiently large that by the time valence electrons start populating the $3d$ orbitals, the $3s$ and $3p$ orbitals behave as core orbitals. The same relationship holds for orbitals with $n = 4$ or 5. The effect is even more dramatic for the f orbitals.

5.2.1 Energies of Atomic Orbitals

To apply bonding principles in a semiquantitative manner, we need to develop a feel for the properties of AOs. In this subsection and the one that follows, we will examine the periodic trends in the energies and sizes of AOs.

Table 5.2 shows calculated energies (in electronvolts[12], eV) for the valence orbitals across most of the periodic table. Not surprisingly, the general periodic trends in orbital energies are similar to those seen for electronegativity, namely that the valence-orbital energies become increasingly negative upon moving up and to the right. However, there are disruptions in the

---

[12] An electronvolt is the amount of kinetic energy gained by a single electron accelerating from rest through an electric-potential difference of one volt. Accounting for the elementary charge of an electron, 1 eV = $1.602176634 \times 10^{-19}$ J exactly.

Table 5.2 Valence-orbital energies (in eV) of the elements, obtained from relativistic Dirac–Fock calculations [6].*

	H											**He**
1s	−13.5											−24.8
	Li	**Be**					**B**	**C**	**N**	**O**	**F**	**Ne**
2s	−5.3	−8.4					−13.4	−19.2	−26.0	−33.8	−42.6	−52.3
2p							−8.4	−11.0	−13.7	−16.6	−19.7	−23.0
	Na	**Mg**					**Al**	**Si**	**P**	**S**	**Cl**	**Ar**
3s	−4.9	−6.8					−10.7	−14.7	−19.2	−24.0	−29.1	−34.7
3p							−5.7	−7.5	−9.5	−11.5	−13.7	−16.0
	K	**Ca**	**Sc**	**Ti**		**Zn**	**Ga**	**Ge**	**As**	**Se**	**Br**	**Kr**
4s	−4.0	−5.3	−5.7	−6.0		−8.1	−11.7	−15.4	−19.2	−23.3	−27.6	−32.1
4p							−5.6	−7.3	−8.9	−10.6	−12.4	−14.3
3d			−9.1	−10.7		−20.6	−31.5	−43.4	−56.3	−70.3	−85.3	−101
	Rb	**Sr**	**Y**	**Zr**		**Cd**	**In**	**Sn**	**Sb**	**Te**	**I**	**Xe**
5s	−4.0	−4.9	−5.4	−5.8		−7.6	−10.7	−13.8	−16.9	−20.2	−23.7	−27.3
5p							−5.3	−6.7	−8.1	−9.6	−11.1	−12.6
4d			−6.3	−7.9		−19.5	−27.4	−35.5	−44.0	−53.0	−62.3	−72.2
	Cs	**Ba**	**La**	**Hf**		**Hg**	**Tl**	**Pb**	**Bi**	**Po**	**At**	**Rn**
6s	−3.5	−4.4	−4.9	−6.5		−8.9	−12.1	−15.3	−18.5	−21.9	−25.3	−28.9
6p							−5.2	−6.7	−8.1	−9.5	−11.0	−12.5
5d			−6.4	−6.5		−16.5	−23.0	−29.4	−35.9	−42.7	−49.6	−56.8

* The effects of spin–orbit splitting on the energies of the p and d orbitals have been averaged.

smooth periodicity that merit further comment. This is particularly true of the vertical trends in the s-orbital energies of the p-block elements. Firstly, note that upon moving from the second period (B–Ne) to the third period (Al–Ar), the s-orbital energy becomes less negative by a significant amount, while upon moving from the third to the fourth period (Ga–Kr), the s-orbital energy change is smaller, and, in some cases (Ga, Ge), the 4s orbital of the fourth-period element has a lower energy than the 3s orbital of its third-period neighbor. This anomaly can be attributed to the incomplete shielding of the 4s orbital when the d subshell is filled for the first time. A similar discontinuity is seen upon moving from the fifth period (In–Xe) to the sixth period (Tl–Rn), where the unexpectedly deep (i.e. large and negative) orbital energies of s orbitals of the sixth-period elements originate primarily from relativistic effects.[13]

Another important feature to glean from Table 5.2 is how significantly the angular-momentum quantum number ℓ influences the orbital energies. The valence s orbitals are

[13] When the charge of the nucleus is large enough that the velocity of an electron in its vicinity approaches the speed of light, relativistic effects increase the electron's mass. This effect can effectively be neglected for light elements, but it becomes more significant as the charge of the nucleus increases, and is largest for electrons in s orbitals, because they have a non-zero probability at the nucleus.

always much deeper in energy than the p orbitals. This is particularly true for oxygen and fluorine. We can also see a significant change in the energy of the $(n − 1)d$ orbitals relative to the ns orbitals upon moving from left to right across a period. At the beginning of the transition-metal block, the energies of the $(n − 1)d$ and ns orbitals are similar; by the end of the series, the $(n − 1)d$ orbital energy is considerably more negative than the ns orbital energy. As we continue into the p block of the periodic table, the energy of $(n − 1)d$ orbitals becomes so negative that these orbitals effectively act as core orbitals.

5.2.2 Sizes of Atomic Orbitals

The wave nature of electrons makes it impossible to precisely define the size of an orbital. The boundary surfaces shown in Figure 5.6 are one way to approximate the size of an orbital. Another approach is to determine the radius at which the radial distribution function reaches a maximum, r_{max}. Values of r_{max} for valence orbitals across the periodic table are given in Table 5.3. Many of the periodic trends seen for orbital energies also hold for orbital radii. Upon moving up or right, the size of the valence orbitals decreases.

The relative sizes of the valence orbitals for a given atom also contain important information. Notice in Table 5.3 that for each value of the orbital quantum number ℓ, the orbitals belonging to the lowest-energy subshell ($1s$, $2p$, $3d$, and $4f$) are particularly compact. This has some significant implications for bonding. For the second-period elements of the p block (B–Ne), the $2p$ orbitals are comparable in size to the $2s$ orbitals, which is not the case for heavier p-block elements where the p orbitals are substantially larger than the s orbital. As a result, atoms from the second period form short σ bonds with each other, which facilitates formation of π bonds (Section 5.3.4).

Notice that for a given atom the $(n − 1)d$ orbitals are much smaller than the ns and np orbitals, and as a result they interact to a lesser extent with the surrounding atoms. Their small size is one reason why the $(n − 1)d$ orbitals are often found to be partially filled in transition-metal compounds. This leads to several useful properties explored later in the book, including color (Chapter 7), cooperative magnetism (Chapter 9), and metal–insulator transitions driven by changes in external conditions (Chapters 10 and 11). The small size of the $(n − 1)d$ orbitals, with respect to the ns and np orbitals of the same atom, is most pronounced for $3d$ electrons because this is the first d shell to fill and thus it experiences less shielding from the nucleus than electrons populating the $4d$ and $5d$ orbitals.

Although the lanthanoids and actinoids are not shown in Table 5.3, we may conclude that the contracted nature of the $4f$ and $5f$ orbitals is even more dramatic than for the d orbitals. For example, the valence-orbital values of r_{max} for cerium are 2.17 Å for $6s$, 1.12 Å for $5d$, and 0.37 Å for $4f$. Consequently, the $4f$ orbitals have minimal interactions with the orbitals of surrounding atoms and behave more like core orbitals than valence orbitals. The lack of interaction between the $4f$ orbitals and the neighboring atoms is responsible for the sharp lines seen in optical absorption and emission spectra, which make the lanthanoid ions useful as luminescence centers (Chapter 7). This also makes magnetic exchange interactions

Table 5.3 Distances (in Å) at which the radial distribution function reaches a maximum value according to relativistic Dirac–Fock calculations [6].

	H										He
1s	0.53										0.30

	Li	Be				B	C	N	O	F	Ne
2s	1.64	1.09				0.81	0.65	0.54	0.46	0.41	0.36
2p						0.84	0.64	0.52	0.44	0.38	0.34

	Na	Mg				Al	Si	P	S	Cl	Ar
3s	1.79	1.37				1.11	0.95	0.84	0.75	0.68	0.62
3p						1.42	1.15	0.98	0.85	0.76	0.69

	K	Ca	Sc	Ti	Zn	Ga	Ge	As	Se	Br	Kr
4s	2.29	1.83	1.71	1.61	1.18	1.04	0.95	0.87	0.81	0.76	0.72
4p						1.39	1.19	1.06	0.96	0.89	0.82
3d			0.60	0.53	0.30	0.29	0.27	0.25	0.24	0.23	0.22

	Rb	Sr	Y	Zr	Cd	In	Sn	Sb	Te	I	Xe
5s	2.45	2.01	1.85	1.74	1.30	1.24	1.09	1.03	0.97	0.92	0.87
5p						1.56	1.37	1.24	1.15	1.07	1.01
4d			0.96	0.85	0.52	0.51	0.47	0.45	0.43	0.41	0.40

	Cs	Ba	La	Hf	Hg	Tl	Pb	Bi	Po	At	Rn
6s	2.72	2.27	2.11	1.78	1.22	1.13	1.07	1.01	0.97	0.93	0.89
6p						1.59	1.40	1.28	1.20	1.13	1.07
5d			1.19	0.88	0.61	0.59	0.57	0.54	0.53	0.51	0.49

between lanthanoid ions weak, which explains why cooperative magnetic ordering of such ions typically occurs far below room temperature if at all (Chapter 9).

5.3 Molecular-Orbital Theory

Having reviewed the properties of AOs, we now consider what happens when AOs interact to form molecular orbitals. The goal of this treatment is to provide the basic qualitative knowledge needed to interpret and construct simple molecular-orbital (MO) diagrams. The treatment is largely non-mathematical and does not require prior knowledge of group theory. In Chapter 6, we will build on this foundation to model the electronic structures of extended solids.

5.3.1 Homonuclear Diatomics: H_2^+ and H_2

To illustrate the general principles of covalent bonding, we begin by considering the simplest possible molecule, H_2^+. To do so we need to describe the behavior of an electron shared by two nuclei. This is done by defining a wavefunction ψ_{MO} that represents a **molecular orbital** (MO). MOs are similar to the AOs we've already encountered, with the important distinction that they can extend over the entire molecule. One of the most common approaches to defining ψ_{MO} is to treat it as a **linear combination of atomic orbitals** (LCAO):

$$\psi_{MO} = c_1\psi_{AO(1)} + c_2\psi_{AO(2)} \tag{5.12}$$

where $\psi_{AO(1)}$ and $\psi_{AO(2)}$ are AO wavefunctions on atoms 1 and 2, respectively, while c_1 and c_2 are numerical coefficients. In the H_2^+ case, $\psi_{AO(1)}$ is a $1s$ orbital on the first hydrogen atom and $\psi_{AO(2)}$ is a $1s$ orbital on the second atom. The MO wavefunction must meet three criteria; it must be finite everywhere, single valued, and ψ^2 must have an integral. Since the probability of finding an electron when integrated over all space must be unity, the coefficients c_1 and c_2 are chosen so that the MO wavefunction is normalized, which can be expressed mathematically as:

$$\int_0^{2\pi}\int_0^{\pi}\int_0^{\infty} \psi\psi^* dr d\theta d\phi = 1 \tag{5.13}$$

The MO diagram for H_2^+ is shown in Figure 5.9.[14] Each orbital is represented by a horizontal line and each electron by a vertical arrow. The AOs are shown on the sides of the diagram, and the MOs are shown in the center of the diagram. The vertical axis is an energy scale. A basic principle of MO theory is that the number of MOs in a molecule is equal to the number of AOs that combine to form them. In H_2^+, there are two MOs: ψ_+, which results from constructive interference of the two H $1s$ orbitals (their wavefunctions sum upon overlap); and ψ_-, which results from destructive interference of these AOs (their wavefunctions subtract upon overlap). The ψ_+ MO is stabilized with respect to the H $1s$ orbital energy because constructive interference of the two AO wavefunctions increases the electron density between the two positively charged nuclei, leading to an attraction between the two atoms. This orbital is called a **bonding molecular orbital**. Populating this MO stabilizes the molecule with respect to the energy of two isolated atoms. In the **antibonding molecular orbital**, ψ_-, the electron density between the nuclei is lowered, and populating ψ_- destabilizes the molecule.

The AO interactions that make up each MO are represented by sketches where shading gives the sign of the AO wavefunction after it is multiplied by the coefficient that determines its contribution to the MO,[15] and its relative size represents the magnitude of that coefficient. When two otherwise identical H $1s$ AOs have opposite signs (visually represented with opposite shading) in an MO, we say the two orbitals have the opposite phase. The exact

---

[14] The energies and orbital coefficients in this and following figures were calculated using extended Hückel theory as implemented in the Caesar 2.0 software suite.

[15] In these sketches, the shading refers to the sign of the atomic orbital wavefunction at the periphery of the atom (i.e. beyond the outermost radial node) where the interaction with orbitals from neighboring atoms is most significant. For an s orbital, the sign is the same in all directions, while for p, d, and f orbitals the sign alternates from lobe to lobe. The choice of that sign is arbitrary, what matters is the relative sign as we move from orbital lobe to lobe and from atom to atom. This relative sign includes the sign of any numerical coefficient that multiplies the orbital wavefunction.

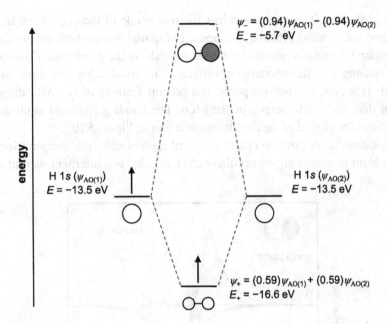

$\psi_- = (0.94)\psi_{AO(1)} - (0.94)\psi_{AO(2)}$
$E_- = -5.7$ eV

H 1s ($\psi_{AO(1)}$)
$E = -13.5$ eV

H 1s ($\psi_{AO(2)}$)
$E = -13.5$ eV

$\psi_+ = (0.59)\psi_{AO(1)} + (0.59)\psi_{AO(2)}$
$E_+ = -16.6$ eV

Figure 5.9 Calculated MO diagram for the H_2^+ molecule. The size of the circle used to represent each AO is proportional to its coefficient. Shading indicates the sign the AO wavefunction acquires from its coefficient in the MO wavefunction; $\psi > 0$ is white and $\psi < 0$ gray. Arrows represent electrons.

values of the orbital energies and coefficients depend upon the level of theory employed, but the general features of the MO diagram will be the same regardless of the sophistication of the theoretical treatment.

The rules for filling MOs with electrons are the same as for AOs. The MOs are filled in order of increasing energy (Aufbau principle) and each MO can hold two electrons of opposite spin (Pauli exclusion principle). When MOs are degenerate (i.e. they have the same energy), the most stable configuration is the one that produces the largest number of parallel electron spins (Hund's first rule).

Analytical expressions for the energy and wavefunction for each MO of H_2^+ are as follows:

$$E_+ = \frac{H_{11} + H_{12}}{1 + S_{12}} \qquad \psi_+ = \frac{1}{\sqrt{2(1 + S_{12})}}(\psi_1 + \psi_2)$$

$$E_- = \frac{H_{11} - H_{12}}{1 - S_{12}} \qquad \psi_- = \frac{1}{\sqrt{2(1 - S_{12})}}(\psi_1 - \psi_2) \qquad (5.14)$$

The wavefunctions and their energies are defined by three parameters: the coulomb integral, H_{11}; the interaction integral, H_{12}; and the overlap integral, S_{12}. The coulomb integral H_{11} is simply the energy of the AO in question, in this case the energy of a hydrogen 1s orbital (-13.5 eV from Table 5.2). The interaction integral H_{12} is a measure of the interaction energy that results from the overlap of AOs. It largely determines the magnitude of the energy splitting between the bonding and antibonding orbitals. The amount of spatial overlap between orbitals is quantified by the overlap integral S_{12}, whose magnitude depends on the interatomic distance as well as the size and symmetry of the overlapping orbitals. Numerical values for these variables can be computed. However, for our purposes, the general features of the MO diagram are more important than the exact energies of the MOs. The bonding (ψ_+) and antibonding (ψ_-) MO wavefunctions are plotted along the internuclear axis in Figure 5.10.

If we add another electron to create a neutral H_2 molecule, the computational complexity of the problem increases due to repulsive electron–electron interactions that are difficult to

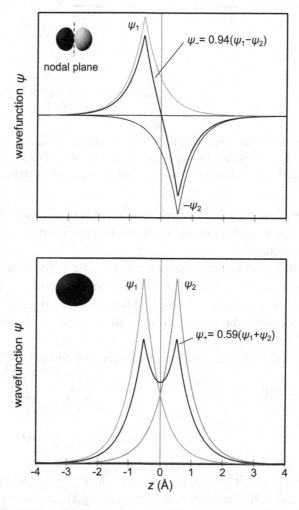

Figure 5.10 Bonding-MO ψ_+ (bottom) and antibonding-MO ψ_- (top) along the internuclear axis (z) for an H_2^+ molecule. The AO wavefunctions ψ_1 and ψ_2 are shown in gray, and the boundary surfaces of the MOs are shown in the upper left corner.

model. We'll return to this point later, but for now we will ignore electron–electron interactions. We can use our simple MO diagram to make some predictions about bonding as electrons are added to the molecule. H_2 has two electrons, and this leads to a doubly occupied bonding orbital ψ_+ and an empty antibonding orbital ψ_- (electron configuration $\psi_+^2\psi_-^0$). This electron count produces a single two-electron bond; the H–H bond order is hence 1.[16]

Increasing the number of electrons in the bonding MO favors a high degree of overlap between the two AOs. For spherically symmetric $1s$ orbitals, this is achieved by reducing the H–H distance. The collapse of one atom on top of the other would give maximal orbital overlap, but the coulombic repulsion between the hydrogen nuclei prevents this from happening.[17] The equilibrium bond distance represents the most favorable balance of these two competing interactions.

If we move to an H_2^- molecule, the electron count becomes three and the electron configuration is $\psi_+^2\psi_-^1$. We have now partially filled the antibonding MO (ψ_-). This weakens the bonding (the bond order is now ½), leading to a reduction in the bond dissociation energy and an increase in the equilibrium bond distance. Intuitively, we would expect H_2^+ and H_2^- to have bonds of equal strength, but this is not quite correct. Because the overlap integral S_{12} is always positive, the energy of the antibonding orbital ψ_- is destabilized more than the bonding orbital ψ_+ is stabilized; see Equation (5.14). This is a general property of MOs that can have implications. In this case it means that the bond in H_2^+ is somewhat stronger than the bond in H_2^-.

5.3.2 The Heteronuclear Diatomic Case: HHe

Let's increase the complexity of the problem by considering the heteronuclear diatomic molecule, HeH. The MO diagram for HeH is shown in Figure 5.11. We now must take into account the different energies of the H $1s$ and He $1s$ orbitals. This impacts the bonding in several ways. Firstly, the AO coefficients in each wavefunction, c_1 and c_2, are no longer equal. For the bonding MO, the He $1s$ orbital coefficient (0.89) is larger than the H $1s$ orbital coefficient (0.24). The opposite relationship holds for the antibonding MO. In each case, the AO closer in energy to the MO makes the larger contribution. This is schematically represented by the relative sizes of the AOs shown in Figure 5.11. The inequality in the coefficients c_1 and c_2 reflects the fact that the electrons are not equally shared, the electron density is shifted toward the more electronegative He.

The energy difference between overlapping orbitals not only affects the wavefunction, it also impacts the MO energies. For the sake of comparison, the same bond distance (1.06 Å) was assumed for the hypothetical HHe molecule in Figure 5.11 as for H_2^+ in Figure 5.9, yet the stabilization of the bonding MO in HHe with respect to the He $1s$ orbital (0.9 eV, see Figure

---

[16] Bond order is calculated as ½[(number of e⁻ in bonding MOs) − (number of e⁻ in antibonding MOs)].

[17] For multielectron atoms, repulsions between core electrons play the primary role in limiting the interatomic distance.

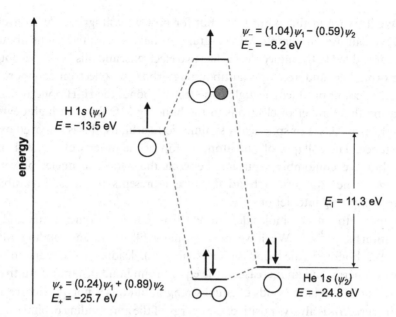

Figure 5.11 Calculated MO diagram for the HeH molecule, Shading represents the sign of the AO wavefunctions when they make up the MOs, $\psi > 0$ (white) and $\psi < 0$ (gray).

5.11), is less than the corresponding stabilization of the bonding MO in H_2^+ (3.1 eV, see Figure 5.9). This illustrates a general feature of MO theory; the stabilization of the bonding MO and destabilization of the antibonding MO depend upon both the *spatial* and *energetic* overlap of parent AOs (on both S_{12} and H_{12}).

Although the covalent stabilization of the bonding MO is smaller in the case of HHe, this does not mean the energy gap between bonding and antibonding orbitals, $\Delta = E_- - E_+$, is smaller. Extended Hückel calculations give Δ values of 10.9 eV for H_2^+ and 17.5 eV for HHe. The energy gap for HHe is larger because the separation between the two MOs contains an ionic contribution. The ionic contribution, E_I, is the difference in orbital energies of the H and He 1s orbitals, $E_I = E_H - E_{He} = -13.5 - (-24.8) = 11.3$ eV, while the covalent contribution accounts for the remaining 6.2 eV of the gap between MOs. In Section 7.5, we will see that both covalent and ionic contributions must be considered to account for changes in the band gaps of compound semiconductors.

5.3.3 Orbital Overlap and Symmetry

For the spherically symmetric s orbitals, the overlap integral S depends only upon the radial portion $R(r)$ of the wavefunction and the interatomic distance. With orbitals that are not spherically symmetric (p, d, and f orbitals), we must also consider the angular geometry. For example, consider the overlap integral between a p orbital (see Figure 5.7) and an s orbital on

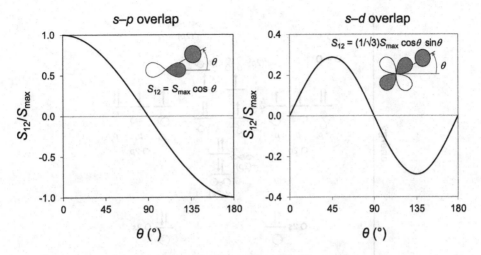

Figure 5.12 Overlap integral S as a function of the angle θ between two orbitals.

a neighboring atom shown in Figure 5.12. In this case, S depends not only upon the interatomic distance, but also upon the angle θ between the two orbitals. When $\theta = 90$, the overlap integral is zero for any value of the interatomic distance because the s orbital overlaps equally with both lobes of the p orbital. The overlap with the lobe of the same sign yields constructive interference, while the overlap with the opposite-sign lobe is equal in magnitude, but destructive. This gives an overlap integral $S = 0$, and the two orbitals are said to be orthogonal. Combinations of other orbitals can give rise to different angular dependencies, such as the overlap between a d orbital (see Figure 5.6) and an s orbital, shown in Figure 5.12. In general, when $S = 0$ we will say that orbital mixing is **symmetry forbidden**. Such geometries can often be identified visually.

5.3.4 Combination of σ and π Bonding: O_2

The final diatomic molecule that we will consider is O_2. We now have five AOs per atom to consider: $1s$, $2s$, $2p_x$, $2p_y$, $2p_z$. The $1s$ orbitals need not be treated explicitly because the large effective nuclear charge they experience confines their wavefunctions to the region close to the nucleus, and their overlap with orbitals on the neighboring atom is negligible. As a rule, we will omit core orbitals when constructing MO diagrams.

The approximate MO diagram for O_2 is shown in Figure 5.13. The diagram is approximate because mixing between MOs formed from the $2s$ and the $2p$ orbitals is neglected. We will return to this point shortly. The vertical separation between the $2s$ and $2p$ orbitals in the MO diagram corresponds to the differences in orbital energy previously discussed (see Table 5.2). The two lowest-energy MOs, $\sigma(2s)$ and $\sigma^*(2s)$, are the respective bonding and antibonding

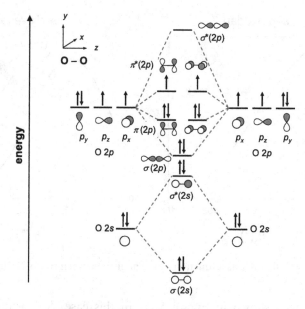

Figure 5.13 MO diagram for the O_2 molecule (along z), with s–p mixing neglected.

orbitals that arise from overlap of the O $2s$ orbitals. They are analogous to the MOs already discussed for H_2. The asterisk as a superscript indicates an antibonding orbital.

The overlap of O $2p$ orbitals depends on the orientation of each orbital with respect to the internuclear axis, which is defined to run parallel to the z direction by convention. The O $2p_z$ orbitals point directly at each other and overlap in a "head-on" fashion forming bonding and antibonding orbitals, $\sigma(2p)$ and $\sigma^*(2p)$. Here, the bonding MO $\sigma(2p)$ is obtained by subtracting wavefunctions of the $2p_z$ orbitals on neighboring atoms, $\sigma(2p) = c_1\psi(2p_z)_1 - c_2\psi(2p_z)_2$, where the coefficients $c_1 = c_2$ are chosen to normalize the wavefunction. The minus reverses the signs of the two lobes of the $2p_z$ wavefunction on atom 2 in such a way that the two AOs overlap constructively between the nuclei. Whenever we encounter orbitals with nodal planes (p, d, and f orbitals), multiplying by a negative coefficient inverts the sign of the AO wavefunction at all points in space, which reverses the shading of each lobe. Whether the orbital has a nodal plane(s) or not, upon multiplying by a negative coefficient we can say that it has the opposite phase of an otherwise identical orbital whose coefficient is positive. The antibonding MO $\sigma^*(2p)$ is formed by adding the two wavefunctions, which leads to destructive interference between the nuclei.

The O $2p_x$ and O $2p_y$ orbitals are aligned perpendicular to the internuclear axis and overlap in a "side-on" fashion. This overlap produces bonding and antibonding orbitals $\pi(2p)$ and $\pi^*(2p)$.[18] Because the overlap of $2p_x$ orbitals is the same as the overlap of $2p_y$ orbitals, these orbitals are doubly degenerate (i.e. two MOs with the same energy). The splitting between the σ

---

[18] The number of nodal planes associated with the bonding MO increases from zero for a σ MO, to one for a π MO, to two for a δ MO (an MO formed by side-on overlap of d orbitals).

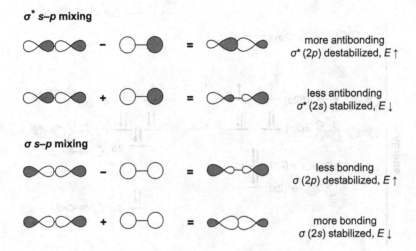

σ* s–p mixing

more antibonding
σ* (2p) destabilized, $E \uparrow$

less antibonding
σ* (2s) stabilized, $E \downarrow$

σ s–p mixing

less bonding
σ (2p) destabilized, $E \uparrow$

more bonding
σ (2s) stabilized, $E \downarrow$

Figure 5.14 MOs of the same symmetry mix: σ(2p) with σ(2s) and σ*(2p) with σ*(2s).

(2p) and σ*(2p) orbitals is larger than the splitting between the π(2p) and π*(2p) because the head-on overlap of $2p_z$ orbitals is larger than the side-on overlap of $2p_x$ ($2p_y$) orbitals; $H_{12}(\sigma) > H_{12}(\pi)$ and $S_{12}(\sigma) > S_{12}(\pi)$.

How does s–p mixing come into the picture? The π(2p) and π*(2p) MOs do not mix with the MOs formed by the 2s orbitals because for every constructive overlap there is an equivalent destructive overlap that cancels it out. However, the σ(2s) and σ(2p) MOs have the same symmetry and can mix, as can the σ*(2s) and σ*(2p) MOs, both interactions are illustrated in Figure 5.14. The s–p mixing stabilizes the lower-energy MOs with predominant 2s character and destabilizes the higher-energy MOs with predominant 2p character. Whenever MOs of the same symmetry mix, the lower-energy MO is stabilized and the higher-energy MO is destabilized.

A more accurate MO diagram for O_2, where s–p mixing has been included, is shown in Figure 5.15. The effect of mixing is most evident near the middle of the MO diagram where the energy of the σ*(2s) MO (now labeled 1σ*) is lowered and that of the σ(2p) MO (now labeled 2σ) is raised. The amount of s–p mixing depends on a number of factors including the energy separation of the 2s and 2p orbitals and the overlap integral. The s–p mixing is sufficiently strong that the 2σ orbital rises above the π-bonding MOs of the diatomic molecules from Li_2 to N_2, while for O_2 and F_2 it remains just below the π MOs.

5.3.5 Symmetry-Adapted Linear Combinations (SALCs)

In the MO diagrams of diatomic molecules, the AOs are placed on the left- and right-hand sides of the diagram while the molecular orbitals are in the middle. This approach does not directly translate to larger polyatomic molecules. However, we can retain this

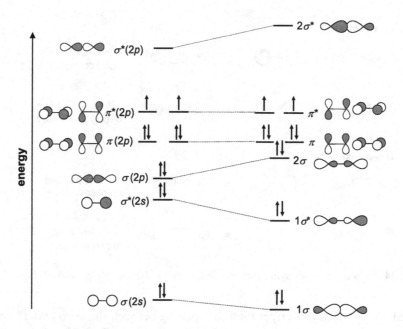

Figure 5.15 MO diagram for O_2 without (left) and with (right) s–p mixing.

approach if we divide the molecule into two fragments and build the MO diagram from them. The most common way to do this is to consider the outer atoms, or ligands, as one fragment and the central atom as the other fragment. The task of combining the AOs of the ligands into a new set of orbitals, the ligand-group orbitals, is based on symmetry. The resulting ligand-group orbitals are called **symmetry-adapted linear combinations** or SALCs for short.

To rigorously construct SALCs requires knowledge of group theory, which is beyond the scope of our treatment. However, once derived, the interactions between ligand-group SALCs and the central atom can be visually estimated from simple symmetry considerations, as demonstrated in Figure 5.16. The ligand SALCs in Figure 5.16 are shown for two, four, and six hydrogens around a central atom possessing s and p valence orbitals, in linear, tetrahedral, and octahedral geometry. In the linear geometry, one SALC can have a net overlap with the s orbital of the central atom, and the other SALC with the p_z orbital. The p_x and p_y orbitals on the central atom are orthogonal to both SALCs and therefore cannot mix to form bonding/antibonding MOs.

Moving from linear to tetrahedral geometry, the number of SALCs follows the number of ligands. The number of orbitals on the central atom and the number of SALCs are now equal, and all orbitals on the central atom have a ligand SALC with the appropriate symmetry for bonding. For an octahedral molecule, there are six SALCs, which a symmetry analysis divides into three groups. The uppermost SALC in Figure 5.16 does not have a nodal plane and has the correct symmetry to interact

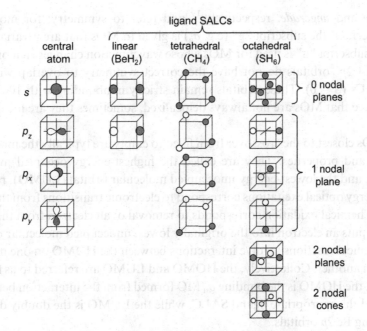

Figure 5.16 Symmetries of s and p orbitals on a central atom (left) and ligand SALCs for linear, tetrahedral, and octahedral arrangements of hydrogen ligands.

with the s orbital of the central atom. The next three SALCs possess a single nodal plane, and each of them has the correct symmetry to interact with one of the p orbitals on the central atom. There are no orbitals on the central atom to interact with the lowest two SALCs. However, as we will see later, when the central atom is a transition metal, these SALCs have the appropriate symmetry to interact with two of the five d orbitals on the central atom.

The SALCs shown in Figure 5.16 are relevant for more than just molecules containing H as a ligand. The symmetries and mixing shown in this figure are representative of σ-bonding interactions between the central atom and ligands in any linear, tetrahedral, or octahedral molecule containing a main-group central atom. For example, the SALCs shown in Figure 5.16 are sufficient to describe the bonding in SF_6. Armed with knowledge of how the SALCs mix with orbitals on the central atom, we will now look at the MO diagrams of some simple polyatomic molecules.

5.3.6 Simple Polyatomic Molecules: BeH_2 and CH_4

The MO diagram for the triatomic molecule BeH_2 is shown in Figure 5.17. As discussed in the preceding section, the Be $2s$ and $2p_z$ orbitals interact with the two ligand SALCs to form two bonding MOs and two antibonding MOs. Their subscripts "g" and "u" are shorthand for *gerade* and *ungerade*, respectively,[19] and refer to symmetry; for molecules with an inversion center, the subscript "g" (e.g. σ_g) is given to MOs that are invariant on inversion, while the subscript "u" is used for MOs whose wavefunction changes sign on inversion. The Be $2p_x$ and $2p_y$ orbitals do not have the correct symmetry to overlap with either of the H $1s$ SALCs ($S = 0$). These orbitals remain strictly nonbonding with 100% Be character. Thus, we see that MOs are not always delocalized, sometimes they are localized on a single atom.

The MOs closest to the crossover from filled to empty are typically the most important for reactivity and properties. They are called the highest-energy occupied molecular orbital (**HOMO**), and the lowest-energy unoccupied molecular orbital (**LUMO**), respectively. The lowest-energy optical excitations correspond to electronic transitions from the HOMO to the LUMO. Chemical oxidation corresponds to removal of an electron from the HOMO, while reduction puts an electron into the originally lowest unoccupied molecular orbital, LUMO. Many chemical reactions involve interactions between the HOMO on one molecule and the LUMO on another. Collectively, the HOMO and LUMO are referred to as **frontier orbitals**. For BeH$_2$, the HOMO is the bonding σ_u MO formed from the interaction between the Be $2p_z$ orbital and the appropriate ligand SALC, while the LUMO is the doubly degenerate set of nonbonding Be $2p$ orbitals.

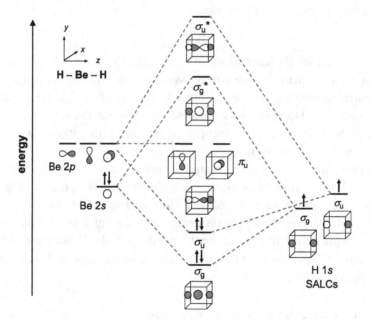

Figure 5.17 MO diagram for BeH$_2$.

---

[19] They originate in group theory. *Gerade* and *ungerade* are German for even and odd.

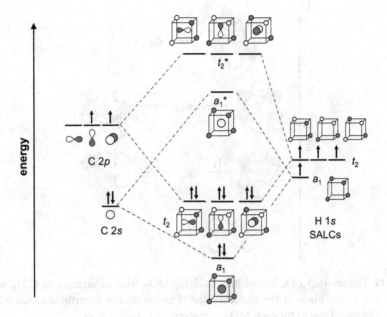

Figure 5.18 MO diagram for CH_4.

The MO diagram for CH_4 is shown in Figure 5.18. The C $2s$ orbital can overlap with the SALC where all four hydrogen AOs have the same phase to form bonding (a_1) and antibonding (a_1*) MOs. Each of the C $2p$ orbitals interacts with one of the SALCs that possess a single nodal plane. These interactions give rise to a triply degenerate set of bonding (t_2) and antibonding (t_2*) MOs.[20] As with BeH_2, there is no mixing of the $2s$ and $2p$ orbitals on the central carbon atom. The well-known valence-bond concept of sp^3 hybridization in tetrahedral molecules is useful in that it tells us that the s and all three p orbitals on the central atom are involved in forming σ bonds. However, the picture of quadruply degenerate bonding orbitals that is sometimes inferred from this description is not consistent with the orbital energies obtained from calculations or seen in photoelectron spectra.

5.3.7 Conjugated π Bonding: C_6H_6

A common feature of many organic functional materials is the presence of a conjugated π-bonding system. The archetypical example is benzene, C_6H_6, a highly symmetric planar molecule possessing 30 valence electrons. The in-plane C $2s$, $2p_x$, $2p_y$, and H $1s$ orbitals

---

[20] We use Mulliken symmetry labels in MO diagrams: Triply degenerate orbitals are labelled t, doubly degenerate orbitals e, and singly degenerate orbitals either a or b, depending upon the symmetry versus the principal rotation axis. A subscript of 1 means the MO is symmetric with respect to a twofold rotation axis perpendicular to the principal axis, while a 2 means it is asymmetric with respect to this axis. A prime symbol signifies an MO that is symmetric with respect to a mirror plane perpendicular to the principal rotation axis, a double prime indicates the MO is asymmetric with respect to such a mirror plane.

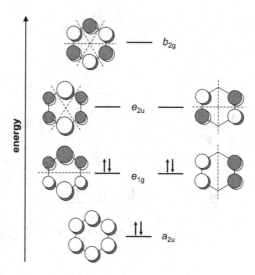

Figure 5.19 The overlap of C $2p_z$ orbitals to form MOs with π character in C_6H_6 as viewed nearly perpendicular to the plane of the molecule (the H atoms do not contribute to the π bonding and are omitted). The nodal planes for each MO are marked with dashed lines.

overlap to form 12 occupied σ-bonding MOs and 12 unoccupied σ-antibonding MOs. The 24 electrons that occupy bonding MOs are responsible for the network of C–C and C–H σ bonds that hold the molecule together. The remaining C $2p_z$ orbitals are oriented perpendicular to the plane of the molecule. These orbitals do not contribute to the σ bonding but they do interact with each other in a π fashion to form an additional six MOs. Just as was the case with the O_2 molecule, the energies of the MOs with π character fall between the σ and σ^* MOs. These π orbitals, which play a major role in determining the chemical and physical properties of benzene, merit a closer look.

The MO diagram for the π interactions in benzene is shown in Figure 5.19. As required, six MOs result from the interaction between six C $2p_z$ orbitals. Not surprisingly, the lowest-energy MO is the one where all six $2p_z$ orbitals have the same phase so that each carbon has bonding π interactions with its neighbors. It should also be intuitive that the highest-energy MO is the one where every $2p_z$ orbital wavefunction is out of phase with its nearest neighbors, leading to antibonding interactions between neighboring carbon atoms. At first glance, the relative energies of the remaining four MOs may be less obvious. However, we can correlate the energies of each MO with the number of nodal planes it possesses. The lowest-energy MO does not have a nodal plane oriented perpendicular to the plane of the molecule.[21] Next come two degenerate MOs, each with one nodal plane, as indicated by the dotted lines in Figure 5.19. The next set of MOs possesses two nodal planes, and the highest-energy MO has three

---

[21] In this discussion, we ignore the nodal plane that coincides with the plane of the molecule, as all six MOs have this nodal plane in common.

nodal planes. The correspondence between the number of nodal planes and the energy should not be surprising. The presence of a nodal plane between two atoms signals destructive interference between AO wavefunctions and is characteristic of an antibonding interaction.

A key feature of the MO diagram for benzene is the delocalized nature of the π bonding. Even in large conjugated molecules, the frontier orbitals span the entire molecule (neglecting the H atoms) as they do in benzene. Consequently, removing (adding) electrons through oxidation (reduction) introduces charge carriers that can move from one end of the molecule to the other. As we will see in later chapters, this means that conjugated organic molecules can exhibit properties where movement (conductivity) of electrons and/or long-range coupling of electron spins (cooperative magnetism) occur.

5.3.8 Transition-Metal Complexes: $[CrCl_6]^{3-}$ and $[CoCl_4]^{2-}$

Many of the functional materials that we will discuss later in the book contain transition metals. Their optical, electrical, and magnetic properties are often dictated by the energy levels and occupation of the five d orbitals, whose orientations are shown in Figure 5.20. Here we focus on the two most common coordination geometries for transition-metal complexes and compounds, the octahedron and the tetrahedron.

Let's begin by considering the octahedral anion $[CrCl_6]^{3-}$. The orbital energies of the chromium $3d$ (-13.5 eV) and $4s$ (-6.6 eV) orbitals are much better matched to the Cl $3p$ orbitals (-13.7 eV) than they are to Cl $3s$ orbitals (-29.1 eV). If we neglect the Cl $3s$ orbitals, there are 18 ligand-group orbitals to consider (six Cl atoms × three $3p$ orbitals per Cl). Using group theory, these ligand-based orbitals can be divided into seven SALCs: one singly degenerate SALC with a_{1g} symmetry, one doubly degenerate SALC with e_g symmetry, and five triply degenerate SALCs with t_{1g}, t_{2g}, t_{2u}, t_{1u}, and t_{1u} symmetry (Figure 5.21). The a_{1g}, t_{1u}, and e_g ligand SALCs are analogous to the SALCs already shown for the octahedral SH_6 molecule in Figure 5.16. They can form σ-bonding/antibonding MOs with Cr orbitals of appropriate symmetry; the a_{1g} SALC with Cr $4s$, the t_{1u} SALC with Cr $4p$, and the e_g SALC with Cr $3d_{x^2-y^2}$ and $3d_{z^2}$ orbitals.

The final three Cr valence orbitals, $3d_{xy}$, $3d_{xz}$, and $3d_{yz}$, have the correct symmetry to form bonding interactions with the t_{2g} Cl SALC. This interaction produces π-bonding and π-

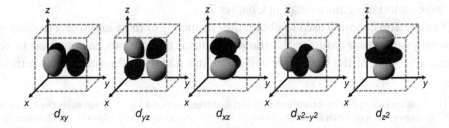

Figure 5.20 The orientations of the d orbitals with respect to a Cartesian coordinate system.

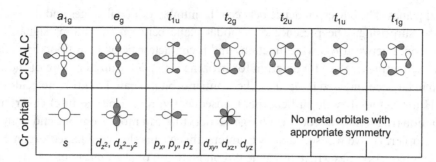

Figure 5.21 Representative Cl $3p$ SALCs and corresponding orbitals on the central Cr atom in $[CrCl_6]^{3-}$ grouped by symmetry. For clarity only the atoms and orbitals in the xy plane are shown.

antibonding MOs. The remaining Cl SALCs are nonbonding ligand-based MOs.[22] Because the chlorine ligands are more electronegative than the metal, the bonding MOs have more ligand character and the antibonding MOs have more metal character.

The MO diagram for $[CrCl_6]^{3-}$ is shown in Figure 5.22. Its complexity may be a bit intimidating at first glance. Fortunately, the chemical and physical properties are dictated largely by the frontier orbitals. The d^3 configuration of chromium(III) means that the half-filled t_{2g} (π^*) orbitals are the HOMO and the empty e_g (σ^*) orbitals are the LUMO. When the d orbitals are partially filled, we need therefore only concern ourselves with the t_{2g} (π^*) and e_g (σ^*) sets of MOs. When the d orbitals are empty, as is the case when the transition metal has a d^0 configuration, the nonbonding ligand-group t_{1g} set becomes the HOMO.

The splitting of the d orbitals into two groups results from the fact that two of them, the Cr $3d_{x^2-y^2}$ and $3d_{z^2}$ orbitals, point directly at the ligands, hence their antibonding (σ^*) inter-action with the e_g Cl SALC has a greater overlap than the π^* interaction between the remaining three Cr orbitals, $3d_{xy}$, $3d_{xz}$, and $3d_{yz}$, and the t_{2g} Cl SALC. Because the σ^* overlap involving the e_g orbitals is more destabilizing than the π^* overlap involving the t_{2g} orbitals, there is a splitting of the antibonding MOs with $3d$-orbital parentage into two sets. The magnitude of the energy separation between the two is referred to as **ligand-field splitting**, Δ. Its value depends on the identity of both the ligands and the transition metal. In many cases, the value of Δ is such that electrons can be excited between d orbitals by visible light, and, as a result, transition-metal compounds are often colored. We will consider colors and bonding of these materials in more detail in Chapter 7.

The electron counting used to obtain the occupancy of these important frontier orbitals is generally straightforward for the transition-metal compounds encountered in solid state chemistry. Most of the ligands we will encounter (chloride, fluoride, oxide, nitride, sulfide,

---

[22] The situation is complicated somewhat by the fact that there are two different sets of triply degenerate SALCs with t_{1u} symmetry, both of which are allowed by symmetry to interact with the Cr $4p$ orbitals. However, the t_{1u} SALC that is shown in the same column as the Cr $4p$ orbital in Figure 5.21 contributes more prominently because it can form both σ and π interactions with each Cr $4p$ orbital. The t_{2u} and t_{1g} SALCs are strictly nonbonding.

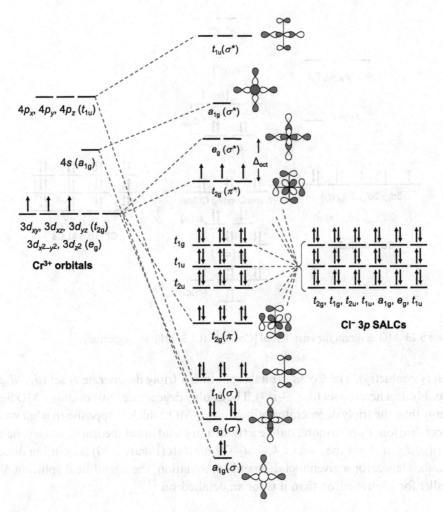

Figure 5.22 MO diagram for the octahedral anion $[CrCl_6]^{3-}$ (Cl $3s$ orbitals neglected). The MOs with t_{1g}, t_{1u}, and t_{2u} symmetry in the middle of the MO diagram are nonbonding Cl $3p$ SALCs.

etc.) are closed-shell ions in the ionic limit. Therefore, the ligand-based SALCs will be completely filled. Any remaining electrons go into the t_{2g} (π^*) and e_g (σ^*) MOs that have predominantly metal d character. Their number is simply the number of the valence electrons of the transition metal minus its oxidation state.

The MO diagram for the tetrahedral transition-metal anion $CoCl_4^{2-}$ is shown in Figure 5.23. As the number of ligands is reduced from six to four, the number of ligand-group orbital SALCs is also reduced, from 18 to 12. Using group theory, the ligand-based orbitals can be grouped into five SALCs; one singly degenerate SALC with a_1 symmetry, one doubly degenerate SALC with e symmetry, and three triply degenerate SALCs (one with t_1, and two

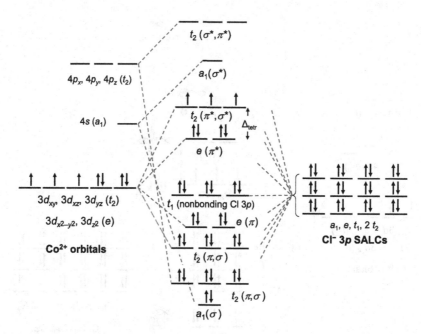

Figure 5.23 MO diagram for tetrahedral $[CoCl_4]^{2-}$ (Cl $3s$ orbitals neglected).

with t_2 symmetry). The Co $3d$ orbitals split into a triply degenerate t_2 set (d_{xy}, d_{xz}, d_{yz}) and a doubly degenerate e set ($d_{x^2-y^2}$, d_{z^2}). The doubly degenerate antibonding e MO lies at lower energy than the triply degenerate antibonding t_2 MO, which is opposite to what we found for an octahedron. Furthermore, unlike a transition metal in octahedral geometry, neither the t_2 set (d_{xy}, d_{xz}, d_{yz}) nor the e set ($d_{x^2-y^2}$, d_{z^2}) of orbitals (Figure 5.20) is pointing directly at the ligands. Hence, for a given metal–ligand combination, the ligand-field splitting Δ is always smaller for a tetrahedron than it is for an octahedron.[23]

5.3.9   High- and Low-Spin Configurations

Up to this point, we have neglected the effects of electron–electron interactions. However, to properly understand the electronic structures and bonding of transition-metal compounds, these effects cannot be ignored. Being negatively charged, electrons repel each other. Due to their close proximity, the repulsive interaction is largest when the two electrons occupy the same orbital. The energy penalty for placing two electrons with antiparallel spins in the same orbital is called the **spin-pairing energy**, P.

In compounds containing first-row transition-metal ions, the ligand-field splitting energy Δ is comparable to the spin-pairing energy P. For many combinations of electron count and

[23] For example, when Co^{2+} ions are doped into MgO, their environment is octahedral and $\Delta_{oct} = 1.19$ eV. When Co^{2+} ions are doped into ZnO, their environment is tetrahedral and $\Delta_{tet} = 0.47$ eV. These values fall surprisingly close to $\Delta_{tet} = (4/9)\Delta_{oct}$ predicted from simple crystal-field theory discussed in Chapter 7.

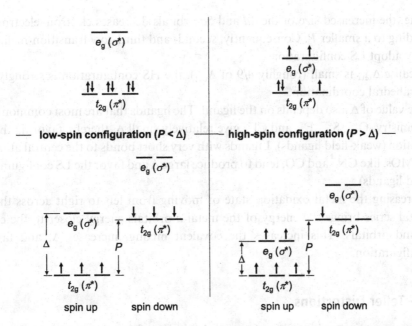

Figure 5.24 Low-spin (left) and high-spin (right) configurations of Co^{3+} in an octahedral environment. Top: Conventional representation without electron–electron interactions included. Bottom: Representation with the pairing energy P shown explicitly.

coordination geometry, the electron configuration depends on which energy is larger, Δ or P. Consider, for example, two possible electron configurations for a d^6 Co^{3+} ion in an octahedral environment (Figure 5.24). These are normally represented with the energy diagrams shown on the top of Figure 5.24. An alternative representation is shown on the bottom of the figure. There, we explicitly see that when $P > \Delta$ (right-hand side) a **high-spin (HS) configuration** is the most stable configuration, whereas when $P < \Delta$ (left-hand side) a **low-spin (LS) configuration** is obtained.

The magnetic, optical, and electrical properties of transition-metal compounds depend upon the spin state (HS or LS) of the transition-metal ion. For a d^6 ion in an octahedral ligand field, the change is particularly dramatic because the HS ion is paramagnetic while the LS ion is diamagnetic. The strength and length of the metal–ligand bond also depend upon the spin state, because populating the σ^* e_g orbitals weakens the bonding more than populating the π^* t_{2g} orbitals. Hence, LS Co^{3+} is smaller (0.685 Å; six-coordinate CR value) than a HS Co^{3+} ion (0.75 Å) [3].

The question of whether the HS or LS state will be more stable depends on the transition-metal ion, the ligands, and the coordination geometry. Some general guidelines are as follows:

- The $4d$ and $5d$ orbitals are larger than the $3d$ orbitals (Table 5.3). As a result, they experience greater overlap with the ligand orbitals, leading to a larger Δ. At the same time, the increased size of the $4d$ and $5d$ orbitals decreases electron–electron repulsions, leading to a smaller P. Consequently, second- and third-row transition-metal ions invariably adopt LS configurations.

- Because Δ_{tetr} is small (roughly 4/9 of Δ_{oct}), the HS configuration is strongly favored for tetrahedral coordination.
- The value of Δ also depends on the ligand. The ligands that are most common in solid state chemistry, O^{2-}, S^{2-}, F^-, and Cl^-, give relatively small Δ, thereby favoring the HS configuration (weak-field ligands). Ligands with very short bonds to the central atom and empty π^* MOs, like CN^- and CO, tend to produce large Δ and favor the LS configuration (strong-field ligands).
- Increasing the metal oxidation state or moving from left to right across the transition-metal series brings the energy of the metal d orbitals energy closer to the energy of the ligand orbitals. This increases the covalent mixing, increases Δ, and favors the LS configuration.

5.3.10 Jahn–Teller Distortions

There are certain combinations of cation electron configurations and ligand geometries that are electronically unstable with respect to a distortion which lowers the symmetry of the molecule (the site symmetry in an extended solid). Jahn–Teller distortions are the most familiar class of electronically driven distortions. The **Jahn–Teller theorem** states that an incompletely filled set of otherwise degenerate MOs will undergo a structural distortion that removes the degeneracy and lowers the energy of these orbitals.

While there are many electron configurations that meet the conditions of the Jahn–Teller theorem, the most important examples occur for octahedral coordination of either a d^9 (e.g. Cu^{2+}) or a HS d^4 (e.g. Mn^{3+}) ion. These two cases undergo Jahn–Teller distortions because the doubly degenerate e_g set is partially filled. The orbital degeneracy can be removed either by lengthening the M–O bonds in the z direction and compressing the bonds in the xy plane or vice versa. The z elongation of the octahedron stabilizes the d_{z^2} orbital (it becomes less antibonding) and destabilizes the $d_{x^2-y^2}$ orbital (it becomes more antibonding) as shown on the right-hand side of Figure 5.25. The xy elongation, shown on the left-hand side of Figure 5.25, does the opposite.

To a first approximation, both types of distortion lower the energy by an equivalent amount, yet the distortion that elongates the octahedron along the z axis occurs almost exclusively. What is the reason for this strong preference? Burdett [7] has shown that the z-elongated octahedron is favored because its partially or fully occupied $3d_{z^2}$ orbital is additionally stabilized over $3d_{x^2-y^2}$ by symmetry-allowed mixing with the empty $4s$ orbital. We will see in later chapters how Jahn–Teller distortions play a key role in the crystal chemistry and physical properties of important classes of materials, such as cuprate superconductors (Chapter 12) and magnetoresistive oxides (Chapter 11).

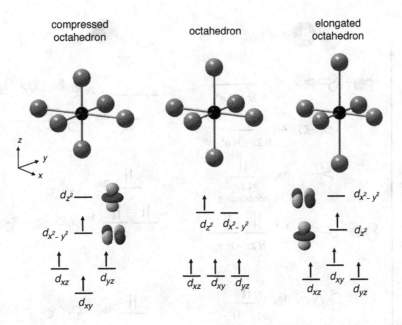

Figure 5.25 Two possible distortions of an octahedron that remove the degeneracy of unequally occupied e_g orbitals for a HS d^4 ion: elongation of the bonds along z (right), and elongation of the bonds in the xy plane (left). In both instances, the remaining bonds contract to maintain the same degree of metal–ligand bonding.

Second-order Jahn–Teller (SOJT) distortions are a related class of electronically driven distortions. SOJT distortions alter the symmetry of the molecule (or crystal) so that two or more MOs that were orthogonal prior to the distortion can interact. The interaction lowers the energy of one or more filled MOs and raises the energy of one or more empty MOs. The driving force for a SOJT distortion is inversely proportional to the energy difference between the two states that are interacting. Consequently, these states are often the HOMO and the LUMO. Both the conventional Jahn–Teller distortion (sometimes called a **first-order Jahn–Teller distortion**) and the second-order Jahn–Teller distortion stabilize the molecule by lowering its symmetry, but for a SOJT the molecule need not possess degenerate, partially occupied MOs as the HOMO.

As an example of the second-order Jahn–Teller distortion, consider the NH_3 molecule. We know that the atoms in NH_3 form a trigonal pyramid and not a triangle. Within the VSEPR framework, the stereochemical influence of the electron lone pair is invoked to explain this, but we can use MO theory to reach the same conclusion. The MOs for NH_3 in both configurations are shown in Figure 5.26. In planar NH_3, the HOMO is a strictly nonbonding N $2p_z$ orbital (labeled $1a_2''$). The distortion to nonplanar NH_3 enables interaction (mixing) between the N $2p_z$ orbital and the N $2s$—H σ^* MO (labeled $2a_1'$) that is symmetry forbidden in the planar geometry. This mixing converts the HOMO from a nonbonding MO to

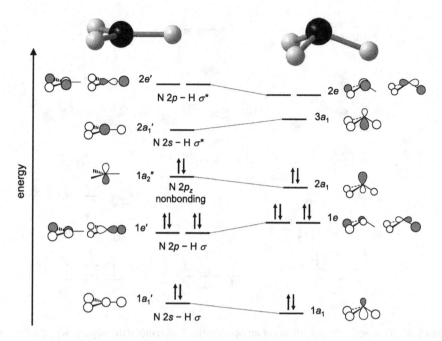

Figure 5.26 MO diagrams for planar (left) and nonplanar (right) NH_3. The SOJT is driven by the interaction between the N $2p_z$ nonbonding and N $2s$—H σ^* MOs that is symmetry forbidden in the planar geometry but allowed in the pyramidal geometry. The notation used to assign MO labels is explained in Footnote 20 in this chapter.

a weakly bonding MO, thereby stabilizing the molecule.[24] We can recognize the HOMO as the lone pair predicted for NH_3 in VSEPR theory. In solid state chemistry, SOJT distortions are often associated either with formation of a stereochemically active electron lone pair or with d^0 ions in octahedral coordination. We will see in Chapter 8 that SOJT distortions play an important role in the crystal chemistry of dielectric and nonlinear optical materials.

5.4　Bond Valences

The bond-valence concept is based on the idea that an atom has a certain bonding power, a valence, distributed over the bonds it forms. This concept takes a particularly simple form in organic chemistry, where each two-electron bond carries a valence of one. It is a powerful predictive concept to know a priori that in stable organic molecules carbon forms four two-electron bonds with its neighbors, oxygen forms two bonds, hydrogen forms one bond, etc.

[24] SOJT distortions lower the energy of some MOs while raising the energies of others, including some filled MOs, which can make it difficult to decide whether a particular distortion should lower the overall energy of the molecule. In most cases, the geometry that achieves the lowest energy for the HOMO also has the lowest total energy.

We saw in Section 1.3.1 that one can also assign bond valences in inorganic solids, but, unlike in most organic compounds, they often take non-integer values. For example, the bond-valence balance (Figure 1.14) gives a valence of ⅙ for an Na–Cl bond in NaCl, and a valence of ⅔ for a Ti–O bond in TiO_2. Recall that these expectation values are calculated by dividing the absolute value of the oxidation state of each atom by the atom's coordination number. In this section, we expand upon this simple concept by relating valences to experimentally observable quantities, such as bond lengths.

We begin with the **valence-sum rule**—the valence v_i of an atom (here meaning the absolute value of its oxidation state) is equal to the sum of bond valences v_{ij} around it,

$$v_i = \sum_j v_{ij} \tag{5.15}$$

where j is the number of bonds formed by atom i, that atom's coordination number. For example, the octahedrally coordinated Ti in TiO_2 has $v_{Ti} = 6 \times v_{Ti-O} = 6 \times ⅔ = 4$, oxygen has $v_O = 3 \times v_{Ti-O} = 3 \times ⅔ = 2$.

To relate these values to experimental data, we need to find a quantitative relationship between bond valence and bond length. While there is more than one function that can be used to approximate this relationship, the most widely used expression is:

$$v_{ij} = \exp\left(\frac{R_{ij}^0 - d_{ij}}{B}\right) \tag{5.16}$$

where v_{ij} and d_{ij} are the valence and length, respectively, of the bond between atoms i and j, R_{ij}^0 is the length of a bond with a valence of one (a single, two-electron bond), and B is a "universal" constant usually taken to be 0.37 Å. The values of R_{ij}^0 are determined empirically from the crystal structures of known compounds. Each cation–anion pair has its own R_{ij}^0 value (see Appendix D for bonds to oxygen).

Owing to the exponential relationship between bond valence and bond length, the bond valence increases faster upon bond contraction than it decreases upon bond expansion. For example, if we start with a bond whose distance $d_{ij} = R_{ij}^0$ and valence $v_{ij} = 1.0$ and shorten it by ~0.256 Å, the valence is doubled (see Figure 5.27). If we lengthen the bond by the same distance, the valence is halved, but this is a smaller change on an absolute scale. A consequence of this asymmetry is the **distortion theorem**—for any ion, lengthening some of its bonds and shortening others, while keeping the bond-valence sum the same, will always increase the average bond length. Consequently, the coordination polyhedron about a cation will have the smallest volume when all anions are equidistant.[25]

---

[25] An important consequence of the distortion theorem is that a too-small cation placed in a symmetric environment of anions will move off center to increase the lengths of some bonds and decrease the lengths of others, thus creating a local dipole moment (Chapter 8).

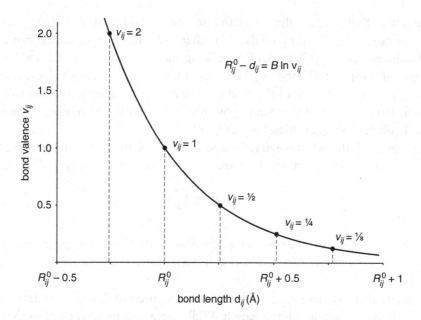

Figure 5.27 The variation in the valence of a bond as a function of bond length.

The **bond-valence method** combines the valence-sum rule in Equation (5.15) with the bond-length bond-valence relation of Equation (5.16). It works forwards or backwards; either to evaluate the bond-valence sum at an atom from experimental bond lengths or to predict bond lengths.

An important application of the former approach is to assess the validity of experimentally determined crystal structures. To check a structure, the valence of each bond is calculated from its bond distance using Equation (5.16), and the valence sum is then evaluated for each crystallographically distinct/inequivalent ion with Equation (5.15). If the bond-valence sum v_i is close to the oxidation state expected for each ion, we can say that the bonding is similar to that seen in the structures that were used to determine the R_{ij}^0 values, which implies a chemically reasonable structure.

Evaluation of the bond-valence sum can also be used to assess atom oxidation states in compounds. As an example, consider the crystal structure of ilmenite (Figure 5.28) which is an ordered variant of corundum containing alternating layers of iron- and titanium-centered octahedra sharing faces. Using the notation introduced in Section 1.3, we express the coordinations as $Fe^{[6o]}Ti^{[6o]}O_3^{[2,2]}$. There are electrostatic repulsions between cations in this structure, which cause the iron and titanium to shift away from each other, resulting in distorted coordination environments for both, as shown in Figure 5.28. The assignment of oxidation states is potentially ambiguous, either $Fe^{2+}Ti^{4+}O_3$ or $Fe^{3+}Ti^{3+}O_3$. Let's see if we can use bond-valence sums to determine the correct values. We begin by assuming $Fe^{2+}Ti^{4+}O_3$ and convert the bond distances Fe–O (2.078 Å, 2.201 Å) and Ti–O (1.874 Å, 2.089 Å) in Figure 5.28 into bond valences:

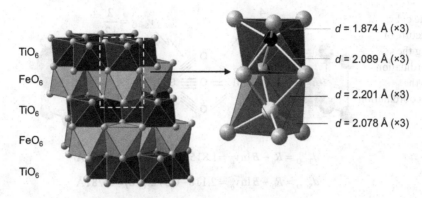

$d = 1.874$ Å (×3)

$d = 2.089$ Å (×3)

$d = 2.201$ Å (×3)

$d = 2.078$ Å (×3)

TiO$_6$

FeO$_6$

TiO$_6$

FeO$_6$

TiO$_6$

Figure 5.28 The crystal structure of ilmenite, FeTiO$_3$.

$$v_{\text{Ti}-\text{O}}(1) = \exp\left(\frac{R^0_{\text{Ti}-\text{O}} - d_{\text{Ti}-\text{O}}}{B}\right) = \exp\left(\frac{1.815 - 1.874}{0.37}\right) = 0.853$$

$$v_{\text{Ti}-\text{O}}(2) = \exp\left(\frac{R^0_{\text{Ti}-\text{O}} - d_{\text{Ti}-\text{O}}}{B}\right) = \exp\left(\frac{1.815 - 2.089}{0.37}\right) = 0.477$$

$$v_{\text{Fe}-\text{O}}(1) = \exp\left(\frac{R^0_{\text{Fe}-\text{O}} - d_{\text{Fe}-\text{O}}}{B}\right) = \exp\left(\frac{1.734 - 2.201}{0.37}\right) = 0.283$$

$$v_{\text{Fe}-\text{O}}(2) = \exp\left(\frac{R^0_{\text{Fe}-\text{O}} - d_{\text{Fe}-\text{O}}}{B}\right) = \exp\left(\frac{1.734 - 2.078}{0.37}\right) = 0.395 \quad (5.17)$$

Summing up the valences about each ion gives $v_{\text{Ti}} = (3 \times 0.853) + (3 \times 0.477) = 3.99$ and $v_{\text{Fe}} = (3 \times 0.283) + (3 \times 0.395) = 2.03$, confirming our guess about the Ti and Fe valences. We can also calculate the bond-valence sum for oxygen; $v_{\text{O}} = 0.853 + 0.477 + 0.283 + 0.395 = 2.008$ that is reassuringly close to the ideal value. Note that if we had initially guessed $Fe^{3+}Ti^{3+}O_3$, the R^0_{ij} values used in our calculations would have been slightly different (1.759 Å for Fe^{3+}–O and 1.791 Å for Ti^{3+}–O), yielding bond-valence sums $v_{\text{Ti}} = 3.84$ and $v_{\text{Fe}} = 2.17$, somewhat farther from the integer values but still leading to the same conclusion.

We can also work in the opposite direction, using the bond-valence method to predict bond lengths. Knowing the crystal-chemical formula of the structure in question, we use the valence-sum rule, Equation (5.15), to convert the oxidation state of an atom into the expected, ideal values of the bond valences. After rearranging Equation (5.16), we use these ideal valences to predict the lengths of bonds in much the same way that ionic radii are used:

$$d_{ij} = R^0_{ij} - B \ln v_{ij} \quad (5.18)$$

Figure 5.29 The cubic perovskite structure of $SrTiO_3$ showing the octahedral coordination of Ti (left) and the cuboctahedral coordination of Sr (right).

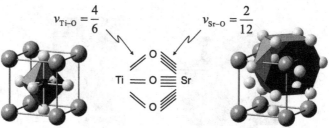

$$d_{Ti-O} = R_0 - B\ln v_{ij} = 1.815 - 0.371\ln(2/3) = 1.965\,\text{Å}$$

$$d_{Sr-O} = R_0 - B\ln v_{ij} = 2.118 - 0.371\ln(1/6) = 2.781\,\text{Å}$$

Let's consider the cubic perovskite $SrTiO_3$ as an illustration of how bond valences can be used to predict distances. The coordination numbers are $Sr^{[12co]}Ti^{[6o]}O_3^{[4,2]}$ as illustrated by the bond graph shown in Figure 5.29. We expect that each of the Ti–O bonds will have a valence of $^4/_6 = ^2/_3$ and each of the Sr–O bonds a valence of $^2/_{12} = ^1/_6$. Given R_{ij}^0 parameters of 1.815 Å and 2.118 Å for the Ti^{4+}–O and Sr^{2+}–O bonds, the expected bond distances are calculated to be 1.965 Å and 2.781 Å, respectively. As discussed in Section 1.5.3, the undistorted cubic perovskite structure will only be stable if the Sr–O and Ti–O lengths are appropriately matched. Our distances can be used to calculate the tolerance factor (see Figure 1.50): $t = d_{A-O}/(d_{M-O}\sqrt{2}) = 2.781/(1.965 \times \sqrt{2}) = 1.001$. Recalling from Chapter 1 that for the cubic perovskite structure to be stable the tolerance factor should be close to 1, we see that the undistorted structure does satisfy the bonding preferences of both Sr^{2+} and Ti^{4+}.

If we replace Sr^{2+} with the smaller Ca^{2+} and repeat the above calculation, we get a smaller tolerance factor, $t = 0.966$, suggesting an octahedral-tilting distortion. At room temperature, $SrTiO_3$ is a cubic perovskite and $CaTiO_3$ a distorted perovskite (exhibiting $a^-a^-b^+$ tilting, Section 1.5.3), in agreement with the bond-valence predictions.

Comparisons of distances obtained from bond valences with those obtained by summing ionic radii show that there is a good agreement between the two approaches. Table 5.4 illustrates this for selected symmetric coordination polyhedra. The agreement should not come as a surprise, as both methods depend on parameters derived from similar sets of structures. Both capture the same trends as the cation coordination number and oxidation-state change. However, the bond-valence approach has advantages. For a given cation–anion pair, only one bond-valence parameter R_{ij}^0 is needed, whereas with ionic radii a different value is associated with each coordination number (both for the cation and the anion). An even more important advantage of the bond-valence approach is the ability to handle distorted environments as easily as symmetric environments. One limitation of bond-valence parameters is that they are not generally tabulated for specific spin states (i.e. HS or LS ions of $3d$ metals), which do affect bond lengths (Section 5.3.9).

Table 5.4 Metal–oxygen bond distances d_{BV} in a polyhedron of coordination number CN, computed from bond-valence parameters R^0 and compared with distances d_{IR} obtained from ionic radii [3]. The four-coordinate radius of oxygen has been arbitrarily used for calculating d_{IR} for all entries.

	R^0 (Å)	CN	v_{ij}	d_{BV} (Å)	d_{IR} (Å)
Ca^{2+}–O	1.967	6	⅓	2.37	2.38
		8	¼	2.48	2.50
		10	⅕	2.56	2.61
Mg^{2+}–O	1.693	4	½	1.95	1.95
		6	⅓	2.10	2.10
		8	¼	2.21	2.25
Zn^{2+}–O	1.704	4	½	1.96	1.98
		6	⅓	2.11	2.12
Al^{3+}–O	1.620	4	¾	1.73	1.77
		6	½	1.88	1.915
Fe^{2+}–O	1.734	4	½	1.99	2.01
		6	⅓	2.14	2.16
Fe^{3+}–O	1.759	4	¾	1.87	1.87
		6	½	2.02	2.025

5.5 Problems

5.1 Consider the infinite series for the Madelung constant of the NaCl-type structure. Its convergence depends on how the successive terms are chosen. As written in Equation (5.2), each successive shell contains ions of the same type (either cations or anions). (a) Calculate the sum of this series for two shells, three shells, etc., up to the full seven shells listed in Equation (5.2). For each successive shell, determine the total number of cations and anions surrounding the central anion. (b) What can you say about the convergence of this series after seven terms? (c) Which of these successive sums is closest to the Madelung constant value of +1.7476? For which sum is the total charge of the cluster closest to zero? (d) What can be done to achieve a more rapid convergence of this series?

5.2 Taking into account both attractive and repulsive interactions, derive an equation analogous to Equation (5.2) for the CsCl structure (Figure 1.40). Include the first four terms in the Madelung series. Hint: You may find this easier to do in terms of the cell edge a, and then convert to the interatomic distance d that is normally used in Madelung formulas.

5.3 CaO adopts the NaCl-type structure with $a = 4.80$ Å. (a) Use the Born–Mayer equation to calculate the lattice-formation energy for CaO. (b) How well does this estimate agree with the value of −3414 kJ/mol determined from the Born–Haber cycle? (c) Calculate the size of the repulsive term as a percentage of the attractive term. (d) Given the fact that SrO has the same structure, would you expect the lattice-formation energy of SrO to be larger or smaller than CaO?

5.4 MgO adopts the NaCl-type structure with $a = 4.22$ Å. (a) Use the Born–Mayer equation to calculate the lattice-formation energy for MgO. (b) Given this estimate of the lattice energy, construct a Born–Haber cycle and estimate the second electron gain enthalpy of oxygen, $O^-(g) + e^- \rightarrow O^{2-}(g)$. Sublimation enthalpy of Mg = +147 kJ/mol, bond dissociation energy of dioxygen = 498 kJ/mol, first ionization energy of Mg = 738 kJ/mol, second ionization energy of Mg = 1451 kJ/mol, first electron gain enthalpy of oxygen = −141 kJ/mol, enthalpy of formation of MgO = −602 kJ/mol.

5.5 Why are there no examples of fluorides with the CdI_2 or $CdCl_2$ structures (Figure 1.28)?

5.6 With the exception of helium, all noble gases solidify at low temperature. The lack of ionic or covalent bonding means that atoms are held together by dispersion forces alone. Given the melting points of the noble gases; Ne = 24 K, Ar = 84 K, Kr = 116 K, Xe = 161 K, what can you say about the strength of the London dispersion forces as the principal quantum number of the outermost shell increases? What is the explanation for this trend?

5.7 Classify each of the following statements about nodes in orbital wavefunctions as true or false: (a) s orbitals have no nodes. (b) The orbital wavefunction always changes sign at a node. (c) The number of nodal planes is determined by the principal quantum number.

5.8 What are the values of the principal and orbital angular-momentum quantum numbers for each of the following orbitals? How many radial nodes and nodal planes does each orbital possess? (a) $4s$ orbital, (b) $5d$ orbital, (c) $4f$ orbital, (d) $2p$ orbital.

5.9 Use the MO diagram of oxygen to determine the oxygen–oxygen bond order in the peroxide ion, O_2^{2-}. Will the O–O distance in peroxide be longer or shorter than in O_2?

5.10 Construct an MO diagram for trigonal-planar BH_3 by analogy with the MOs for trigonal-planar NH_3 in Figure 5.26. Use this diagram to determine the degeneracy and orbital character of the HOMO and the LUMO.

5.11 Consider these six-coordinate ionic radii (IR values from ref. [3]; see also Footnote 3 in this chapter) for divalent, first-row transition-metal ions: $r(Ti^{2+}) = 0.86$ Å, $r(V^{2+}) = 0.79$ Å, $r(Cr^{2+}) = 0.80$ Å, $r(Mn^{2+}) = 0.83$ Å, $r(Fe^{2+}) = 0.78$ Å, $r(Co^{2+}) = 0.745$ Å, $r(Ni^{2+}) = 0.69$ Å. For a fixed oxidation state, the ionic radius normally decreases on moving left to right across the periodic table due to the increasing effective nuclear charge. Why then does the radius increase on moving from V^{2+} to Cr^{2+} to Mn^{2+}?

5.12 For which d-electron counts are there distinct HS and LS states of an octahedrally coordinated transition-metal ion?

5.13 In each of the following pairs, one species contains a transition metal in the HS state and the other in the LS state. Indicate the complex that is most likely to contain the LS ion. (a) $Fe^{[6]}Cl_3$ and $Ru^{[6]}Cl_3$, (b) $[Co^{[6]}(NH_3)_6]^{3+}$ and $[Co^{[4]}Cl_4]^-$.

5.14 NiO adopts the cubic NaCl-type structure while PtO adopts the cooperite structure shown below. (a) What factor do you think is responsible for the differing crystal-chemistry preferences of these two compounds? Hint: Consider the splitting and occupation of the d orbitals for each compound. (b) Which structure type do you think PdO will adopt? Would it be possible to tell from a magnetic measurement?

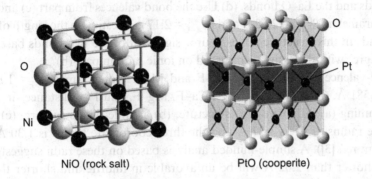

NiO (rock salt)   PtO (cooperite)

5.15 Construct an MO diagram for a linear H_2O molecule by analogy with BeH_2 in Figure 5.17. (a) Determine the degeneracy and orbital character of the HOMO and the LUMO. (b) Identify the orbitals on oxygen that participate in bonding to hydrogen. (c) Now distort the molecule by bending the H–O–H bond and consider how this impacts the MO diagram. How does the orbital character of the HOMO(s) change? (d) Which oxygen orbitals now participate in bonding? (e) Is this distortion an example of a first- or second-order Jahn–Teller distortion?

5.16 MgF_2 adopts the rutile structure (Figure 1.45) with $a = 4.62$ Å and $c = 3.04$ Å. The bond-valence parameter $R^0_{Mg-F} = 1.581$ Å. (a) Use the bond-valence method to predict the length of the Mg–F bonds. (b) Use the Born–Mayer equation to estimate the lattice-formation energy for MgF_2. (c) Comment on the difference between this value and the value of -2978 kJ/mol obtained from a Born–Haber cycle. Is the agreement between calculated and experimental values similar to that observed for the alkali-metal halides discussed in Section 5.1.3? (d) Would you expect the lattice-formation energy of rutile (TiO_2) to be lower (more stable) than MgF_2?

5.17 The structure of ZrV_2O_7 can be derived from the structure of NaCl by replacing Na^+ with Zr^{4+} and Cl^- with $V_2O_7^{4-}$ pyrovanadate groups. The coordination environment of zirconium is octahedral while the local coordination at vanadium is tetrahedral. One of the seven oxygen atoms, O(1), does not bond to Zr, while the other six equivalent oxygens, O(2), bond to both Zr and V. (a) Construct a bond graph (Section 1.3.1) for ZrV_2O_7. (b) What are the idealized valences for the V–O(1), V–O(2), and Zr–O(2) bonds? (c) Given bond-valence parameters $R^0_{Zr-O} = 1.928$ Å and $R^0_{V-O} = 1.803$ Å calculate the expected V–O(1), V–O(2), and Zr–O(2) bond distances.

5.18 In LaOF, which has a structure closely related to fluorite, each lanthanum is surrounded by four oxide and four fluoride ions. Although X-ray-diffraction studies cannot easily distinguish oxygen from fluorine, two bond distances are seen in the crystal structure:

2.42 Å to one anion and 2.60 Å to the other anion. (a) Write the crystal-chemical formula (Section 1.3) for LaOF. (b) Given the eight-coordinate La^{3+} radius of 1.30 Å and the four-coordinate radii for O^{2-} and F^- of 1.24 Å and 1.17 Å, respectively, assign the two observed distances to La–O and La–F bonds. (c) Determine the ideal valences of the La–F bonds and the La–O bonds. (d) Use the bond valences from part (c) and the bond-valence parameters $R^0_{La-F} = 2.02$ Å and $R^0_{La-O} = 2.17$ Å to estimate the length of La–F and La–O bonds in this structure. Does your assignment of the two bonds based on bond valences agree with the assignment based on ionic radii in part (b)?

5.19 The bond-valence parameters for Ca–F and Mg–F bonds are $R^0_{Ca-F} = 1.842$ Å and $R^0_{Mg-F} = 1.581$ Å. Calculate the closest Ca–F, Mg–F, and F–F distances in MgF_2 and CaF_2 assuming (a) a fluorite-type structure, (b) a rutile-type structure. (c) The four-coordinate radius of F^- is 1.31 Å and the three-coordinate radius is 1.30 Å (both IR values from ref. [3]). A simple-minded analysis based on these radii suggests that F–F contacts shorter than 2.62 Å will be unfavorable in fluorite and shorter than 2.60 Å unfavorable in rutile. How do the F–F distances calculated from the cation–anion distances in parts (a) and (b) and simple geometric considerations compare with these limiting distances? (d) The bond-valence parameter for Be–F is $R^0_{Be-F} = 1.28$ Å. Based on this value, do you think BeF_2 would be stable in the rutile structure?

5.20 $MgSiO_3$ is of interest to geologists because it is abundant in the Earth's mantle. Silicon is normally tetrahedrally coordinated by oxygen, but, at the high pressures found in the mantle, silicon becomes octahedrally coordinated and a perovskite structure is formed. (a) Given bond-valence parameters $R^0_{Mg-O} = 1.693$ Å and $R^0_{Si-O} = 1.624$ Å, calculate the expected Mg–O and Si–O distances for a cubic perovskite. (b) Use the distances calculated in part (a) to estimate the tolerance factor t (Section 1.5.3), and state whether you would expect $MgSiO_3$ to form as a cubic or a distorted perovskite. (c) Experiments identify $MgSiO_3$ as a distorted perovskite. The Si–O distances are 1.78 Å (×2), 1.79 Å (×2), and 1.80 Å (×2). How do those compare to your estimate of the Si–O distance in part (a)? (d) The Mg–O distances are 2.00 Å, 2.06 Å (×2), 2.29 Å (×2), 2.41 Å (×2), 2.85 Å, 2.96 Å, and 3.11 Å (×2). Calculate the bond-valence sum for Mg^{2+} as a qualitative estimate of whether this is a reasonable coordination environment for Mg^{2+}.

5.6   Further Reading

T.A. Albright, J.K. Burdett, M.H. Whangbo, *"Orbital Interactions in Chemistry"* (1985) John Wiley and Sons.

J.K. Burdett, *"Chemical Bonding in Solids"* (1995) Oxford University Press.

R. Dronskowski, *"Computational Chemistry of Solid State Materials"* (2005) Wiley–VCH.

D.M.P. Mingos, *"Essential Trends in Inorganic Chemistry"* (1998) Oxford University Press.

Y. Jean, F. Volatron (translated and edited by J.K. Burdett) "*An Introduction to Molecular Orbitals*" (1993) Oxford University Press.

I.D. Brown, "*The Chemical Bond in Inorganic Chemistry: The Bond Valence Model*" (2006) Oxford University Press.

5.7   References

[1] R.P. Grosso Jr., J.T. Fermann, W.J. Vining, "An in-depth look at the Madelung constant for cubic crystal systems" *J. Chem. Educ.* **78** (2001), 1198–1202.

[2] R.A. Jackson, C.R.A. Catlow, "The Madelung constant" *Mol. Simul.* **1** (1988), 207–224.

[3] R.D. Shannon, "Revised effective ionic-radii and systematic studies of interatomic distances in halides and chalcogenides" *Acta Crystallogr. Sect. A* **32** (1976), 751–767.

[4] M. Wilson, P.A. Madden, "Anion polarization and the stability of layered structures in MX_2 systems" *J. Phys. Condens. Matter* **6** (1994), 159–170.

[5] R. Hoffmann, "An extended Hückel theory. I. Hydrocarbons" *J. Chem. Phys.* **39** (1963), 1397–1412.

[6] J.P. Desclaux, "Relativistic Dirac–Fock expectation values for atoms with $Z = 1$ to $Z = 120$" *At. Data Nucl. Data Tables* **12** (1968), 311–406.

[7] J.K. Burdett, "*Chemical Bonding in Solids*" (1995) Oxford University Press, 242–251.

6 Electronic Band Structure

In Chapter 5 we saw how molecular orbital (MO) diagrams can be used to describe the electronic structures of molecules. In this chapter we turn our attention to extended solids, whose electronic structures are represented by band-structure diagrams. The optical, electrical, and magnetic properties of a material are directly linked to its band structure. The chemical reactivity and catalytic properties of a material depend upon the energy levels and symmetry of electronic states near the Fermi level; even dielectric and mechanical properties can be traced to chemical bonding interactions that are intimately linked to the electronic structure. It is therefore essential to develop a working knowledge of the electronic band structures of solids before we can begin to understand the behavior of many functional materials.

In Chapter 5, we learned how MOs can be derived from the overlap of atomic orbitals (AOs) using the linear combination of atomic orbitals (LCAO) approach. In this chapter, we will see how the electronic structures of extended crystalline solids can be built up in the same way. This approach follows directly from MO theory, which makes it particularly intuitive for chemists. It can be applied to solids with electrons that are localized or delocalized. This is an important advantage because many interesting phenomena arise in materials with intermediate degrees of electron delocalization.

To make the visualization and mathematics easier, we begin by considering the electronic structures of 1D systems. The concepts developed in 1D are then extended to describe electronic structures of 2D and 3D crystals.

6.1 The Band Structure of a Hydrogen-Atom Chain

To introduce the concepts associated the electronic structure of an extended solid, we begin with the simplest structure we can imagine; an infinite 1D chain of hydrogen atoms. The following sections develop the band structure of this model system.

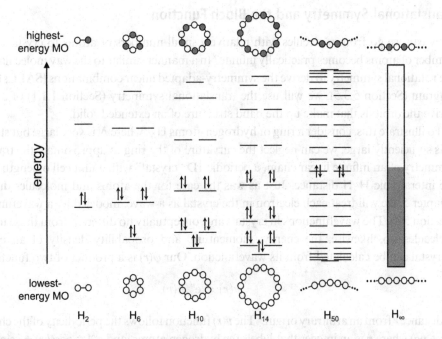

Figure 6.1 MO diagrams for H_2 and several cyclic H_N molecules (N = 6, 10, 14, 50, ∞). For economy of space, only the electrons in the highest-energy occupied MOs are shown for H_{50} and occupied MOs are shaded for H_∞. Shading of the MO images indicates the sign individual H $1s$ wavefunctions acquire in the LCAO formalism; $\psi > 0$ is white and $\psi < 0$ gray.

6.1.1 The Electronic Structures of Cyclic H_N Molecules

Before considering an infinite chain, let's look at the MO diagrams of cyclic all-hydrogen molecules. Because each atom has an identical environment, such molecules are the most appropriate finite-sized approximants to an infinite chain. The MO diagrams for H_2 and cyclic H_N molecules of 6, 10, 14, and 50 atoms are shown in Figure 6.1 together with a molecule extrapolated to H_∞ so that its MOs become continuous. These MO diagrams possess several common features that can be understood from the concepts covered in Chapter 5:

- The number of MOs in the molecule is equal to the number of constituent AOs. For hydrogen, we have one AO per atom: the $1s$ orbital.
- The bonding interactions between nearest-neighbor atoms range from the completely bonding lowest-energy MO to the completely antibonding highest-energy MO.
- As the energies of the MOs increase, the number of nodes increases (see discussion of $2p\,\pi$ MOs of C_6H_6 in Section 5.3.7).
- As the size of the molecule increases, the MOs become more closely spaced in energy, with the highest density occurring near the most bonding and most antibonding orbitals.

6.1.2 Translational Symmetry and the Bloch Function

How can we go from molecules with relatively small numbers of atoms to crystals where the number of atoms becomes practically infinite? In a manner similar to the way molecular chemists use rotational symmetry to derive the symmetry-adapted linear combinations (SALCs) in an MO diagram (Section 5.3.5), we will use the translational symmetry (Section 1.1.1) of a crystal to derive the orbitals that make up the band structure of an extended solid.

To illustrate this, consider a ring of hydrogen atoms H_N, where N is very large but still finite. If N is sufficiently large, we can neglect the curvature of the ring to approximate the translational symmetry of an infinite linear chain, a periodic 1D "crystal" with a unit cell of length a equal to the interatomic H–H distance. Just as was the case for the atoms and molecules discussed in Chapter 5, we will treat each electron in the crystal as a wave, modeled by a wavefunction $\psi(r)$ (Section 5.2). The wavefunctions in crystals are conceptually no different from those in atoms or molecules, so, likewise, the energy, momentum, and probability density of an electron in a crystal can be calculated from its wavefunction. Our $\psi(r)$ is a product of two functions:

$$\psi(r) = e^{ikr}u(r) \tag{6.1}$$

of distance r from an arbitrary origin.[1] The $u(r)$ function follows the periodicity of the chain, $u(r) = u(r + na)$ where n is an integer that labels the hydrogen atoms, and $e^{ikr} = \cos(kr) + i\sin(kr)$. With the parameter k (Section 6.1.3) having units of inverse length, the product kr is dimensionless (the arguments of the cos and sin functions are in radians). The wavefunction $\psi(r)$ of Equation (6.1) is called the **Bloch function** and describes **crystal orbitals**, the infinite-crystal analogs to MOs.

The $u(r)$ in Equation (6.1) is referred to as a **basis set**. Throughout most of this book, we use **tight-binding methods** where AO wavefunctions are the basis set.[2] This is the solid state analogue of MO theory since, in both approaches, each electron wavefunction $\psi(r)$ is a linear combination of atomic orbitals (LCAO). For our linear chain of hydrogen atoms, the basis set is the sum of N hydrogen 1s wavefunctions (ψ_{1s}), each centered at one of the H nuclei along the chain. For each value of k, we combine the H 1s orbitals in a different way to get a unique crystal orbital that can hold two electrons of opposite spin without violating the Pauli exclusion principle (Section 5.2). The manner in which this is done will become clear in Section 6.1.4 where we examine the crystal-orbital wavefunctions of the infinite H-atom chain more closely. In Section 6.2, we will see that when the unit cell contains more than one atom, we can use MOs as the basis set.

The tight-binding model provides a great deal of chemical understanding because the basis set explicitly retains the identities of the atoms that make up the crystal, but the functions used as the basis set need not be AOs or centered on individual atoms.[3] In fact, most modern

---

[1] We can also express the distance from the origin $r = xa$, where x is a fractional coordinate, and the interatomic distance a is the length of the 1D unit cell.

[2] Computational implementations of tight-binding theory typically use mathematical functions that approximate AO wavefunctions, such as Gaussian-type and Slater-type orbitals.

[3] The free-electron model takes the rather drastic approach of assuming the periodic electric-field potential created by the atomic nuclei is uniform throughout the crystal, in which case $u(r) = 1$. This approximation, while

approaches to calculating the electronic structures of crystals rely upon basis sets other than AOs.[4] The details of these methods are beyond the scope of this book, but interested readers are encouraged to consult Dronskowski's *Computational Chemistry of Solid State Materials*, listed in Further Reading.

6.1.3 The Quantum Number *k*

Reducing the electronic structure of an entire crystal down to the contents of a single unit cell greatly simplifies analysis of its electronic structure. This entire simplification is achieved by introducing the parameter k in the Bloch function, Equation (6.1). Although k can take any value, we need only concern ourselves with k values within a finite range. To show this, we return to our H_N ring. If we move a distance Na along the length of the ring, we are back to where we started, and the wavefunction must repeat. Mathematically, this equates to the introduction of a boundary condition that constrains the electronic wavefunction to have the same value at the end of the chain that it has at the beginning:

$$\psi(r) = \psi(r + Na) \tag{6.2}$$

If we apply this boundary condition to the Bloch function, Equation (6.1), recalling that $u(r) = u(r + na)$, we obtain:

$$e^{ikr}u(r) = e^{ik(r+Na)}u(r) = e^{ikNa}e^{ikr}u(r) = [\cos(kNa) + i\sin(kNa)]e^{ikr}u(r) \tag{6.3}$$

This equality only holds if the term in square brackets is unity, which only happens when kNa equals an integer number n of the "sinusoid" periods 2π:

$$kNa = n(2\pi) \tag{6.4}$$

Thus, we see that *for a finite ring* H_N, the allowed values of k are quantized. In analogy to MO theory, a finite ring containing N hydrogen atoms will have one crystal orbital for every atom in the ring (see Figure 6.1), each with a different value of k.[5] If we let n adopt positive and negative integer values ranging from ± 1 to $\pm N/2$,[6] we obtain the following set of values for k:

$$k = \frac{n(2\pi)}{Na} = \pm \frac{2\pi}{Na}, \ \pm \frac{4\pi}{Na}, \ \pm \frac{6\pi}{Na}, \ \dots \ \pm \frac{(N/2)2\pi}{Na} \tag{6.5}$$

computationally simple, is of limited utility (and even more limited accuracy) because it effectively removes chemical bonding from consideration. We will use the free-electron model as an entry point to study the conductivity of simple metals in Chapter 10.

[4] Most AOs possess radial nodes in the core region, and the rapid oscillations of the wavefunction near these nodes makes calculations with an atomic orbital basis set numerically expensive. Most calculations do not include the core electrons and use simpler functions to approximate the wavefunctions of valence electrons in the core region.

[5] In a finite crystal, the number of k values will be equal to the number of unit cells in the crystal.

[6] For convenience, we will assume that N is an even number so that $N/2$ is an integer.

real-space lattice   reciprocal-space lattice

Figure 6.2 The 1D real- and reciprocal-space lattice. In the latter, the interval that is marked corresponds to the first Brillouin zone.

As the number of atoms in the chain N goes to infinity, the difference between successive values of k becomes infinitesimally small, and k becomes continuous. Interestingly, we see that the last term in Equation (6.5), which contains the smallest $(-\pi/a)$ and largest $(+\pi/a)$ value of k, does not depend upon the number of atoms in the chain. This is important because it means that even though k becomes continuous as the chain length goes to infinity, we only need to consider the values of k that fall in the finite range $-\pi/a \leq k \leq +\pi/a$ (of width $2\pi/a$).

To further demonstrate the link between translational symmetry and k, we borrow a concept from X-ray crystallography and transform the real-space lattice into the **reciprocal-space lattice** shown in Figure 6.2. We relate the reciprocal-space lattice parameter a^* (with units of inverse length, just like k) to the real-space lattice parameter a through the relationship $a^* = 2\pi/a$.[7] The range of k values needed to generate all possible crystal orbitals without duplication, $-\pi/a \leq k \leq +\pi/a$, can be expressed in reciprocal-space coordinates as $-a^*/2 \leq k \leq +a^*/2$. This interval, which is the unit cell in 1D reciprocal space, defines the **first Brillouin zone**. Just as knowing the unit cell and its contents in real space defines the crystal structure of a material, knowing the energies and wavefunctions of the crystal orbitals throughout the first Brillouin zone defines its electronic structure.

6.1.4 Visualizing Crystal Orbitals

Let's consider the analogy between MOs and crystal orbitals constructed using the tight-binding approach (Section 6.1.2) more closely. Recall from Equation (5.12) that the MO wavefunctions for the H_2 molecule are of the form

$$\psi_{MO} = c_1 \psi_{AO(1)} + c_2 \psi_{AO(2)} \tag{6.6}$$

where AO wavefunctions $\psi_{AO(1)}$ and $\psi_{AO(2)}$ on atoms 1 and 2 are multiplied by coefficients c_1 and c_2. For a molecule of N constituent AOs, each MO wavefunction can be written: $\psi(r) = \sum_{j=1}^{N} c_j \psi_{AO(j)}$ where $\psi_{AO(j)}$ are individual AO wavefunctions, and the c_j coefficients quantify the contribution of each AO to the MO wavefunction:

---

[7] The relation used in X-ray crystallography would be $a^* = 1/a$ (less frequently $a^* = \lambda/a$ where λ is the wavelength of the incident X-ray radiation). Given the sinusoidal character of the e^{ikr} term of the Bloch function, it is necessary to set $a^* = 2\pi/a$ so the electron density $|\psi(r)|^2$ retains the translational symmetry of the real-space lattice.

Using the tight-binding model for our very large H_N ring, we express the crystal-orbital wavefunctions as:

$$\psi(r) = N^{-1/2} \sum_{n=1}^{N} e^{ikna} \psi_{AO(n)}(r - na) \tag{6.7}$$

where $\psi_{AO(n)}$ is a $1s$ AO wavefunction (ψ_{1s}) on the nth atom, the nucleus of which is located at $r = na$, and the $N^{-1/2}$ term acts to normalize the crystal-orbital wavefunction. We can see that the e^{ikna} terms in the crystal orbital play the same role as the c_j coefficients in an MO. The contribution of each AO to the crystal-orbital wavefunction is weighted by the magnitude and sign of this term. To visualize these crystal orbitals ψ_k for a 1D chain of hydrogen atoms, consider two cases, $k = 0$:

$$\psi_k = N^{-1/2} \left[e^{ik(1a)} \psi_{1s(1)} + e^{ik(2a)} \psi_{1s(2)} + e^{ik(3a)} \psi_{1s(3)} + e^{ik(4a)} \psi_{1s(4)} + e^{ik(5a)} \psi_{1s(5)} + \ldots \right]$$
$$\psi_{k=0} = N^{-1/2} \left[e^0 \psi_{1s(1)} + e^0 \psi_{1s(2)} + e^0 \psi_{1s(3)} + e^0 \psi_{1s(4)} + e^0 \psi_{1s(5)} + \ldots \right] \tag{6.8}$$
$$\psi_{k=0} = N^{-1/2} \left[\psi_{1s(1)} + \psi_{1s(2)} + \psi_{1s(3)} + \psi_{1s(4)} + \psi_{1s(5)} + \ldots \right]$$

and $k = \pi/a$:

$$\psi_{k=\pi/a} = N^{-1/2} \left[e^{i(\pi)} \psi_{1s(1)} + e^{i(2\pi)} \psi_{1s(2)} + e^{i(3\pi)} \psi_{1s(3)} + e^{i(4\pi)} \psi_{1s(4)} + e^{i(5\pi)} \psi_{1s(5)} + \ldots \right]$$
$$\psi_{k=\pi/a} = N^{-1/2} \left[-\psi_{1s(1)} + \psi_{1s(2)} - \psi_{1s(3)} + \psi_{1s(4)} - \psi_{1s(5)} + \ldots \right] \tag{6.9}$$

where $\psi_{1s(n)}$ represents the $1s$ AO on the nth atom.

These two crystal orbitals, sketched in the manner used in Chapter 5, are shown in Figure 6.3, where we can see how the phases of the AOs are altered by the value of k. The two k values chosen, 0 and π/a, have the smallest and largest absolute values of k within the first Brillouin zone, and correspond to the most bonding and most antibonding crystal orbitals, respectively. Orbitals with intermediate values of k will have intermediate energies.

The wavefunctions $\psi(r)$ for these two crystal orbitals and for $k = \pi/2a$ are plotted in Figure 6.4. Notice how the e^{ikr} term imparts a sinusoidal modulation to the wavefunction

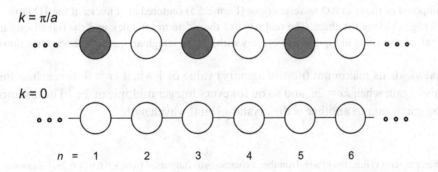

Figure 6.3 The most bonding ($k = 0$) and most antibonding ($k = \pi/a$) crystal orbitals for an infinite chain of hydrogen atoms. The spheres represent individual H $1s$ orbitals, while dark and light shading indicates the sign the AO wavefunction acquires from the e^{ikr} term; $\psi > 0$ is white and $\psi < 0$ gray.

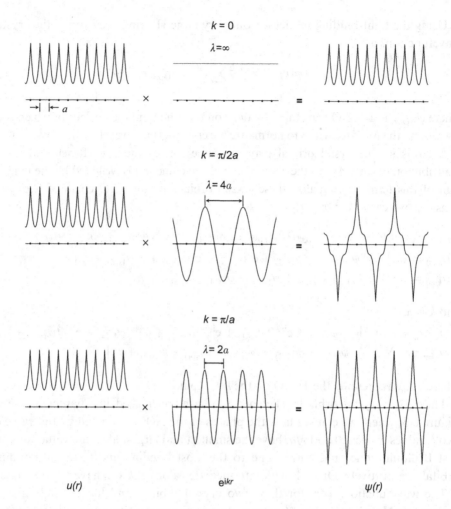

$$u(r) \qquad e^{ikr} \qquad \psi(r)$$

Figure 6.4 Crystal orbitals derived within the tight-binding approximation at $k = 0$ (top), $\pi/2a$ (middle), and π/a (bottom), for an infinite linear chain of hydrogen atoms. The periodic function $u(r)$ (left), is composed of the H $1s$ AO wavefunctions (Figure 5.5) centered at H nuclei at the 1D lattice points (unit-cell edge a) along the chain. The real parts of the e^{ikr} term (middle), and the full Bloch function (right) obtained according to Equation (6.1), vary with k.[8] The imaginary part of the wavefunction is not shown.

that yields its maximum (non-imaginary) value of 1 when $kr = 0$. It reaches this maximum value again when $kr = 2\pi$, and so on for every integer multiple of 2π.[9] The distance $r = \lambda$ from one maximum to another is the crystal-orbital wavelength:

---

[8] Since the $\cos(x)$ function of a real number x completes its full period from $x = 0$ to $x = 2\pi$, it has a wavelength $\lambda = 2\pi$. Consequently, $\cos(2x)$ has $\lambda = \pi$, and $\cos(x/2)$ has $\lambda = 4\pi$.

[9] Because only the real part of $\psi(r)$ is plotted, the function $\psi^2(r)$ that represents the electron density $\rho(r)$ would appear to change from cell to cell, but when the imaginary component is also considered this is not the case.

$$\lambda = 2\pi/k \tag{6.10}$$

For a *free electron*, we can use the de Broglie formula to relate its momentum p and wavelength λ:

$$p = h/\lambda \tag{6.11}$$

If we insert the wavelength of the crystal orbital, $\lambda = 2\pi/k$, into Equation (6.11), we obtain a relationship that expresses the momentum of the electron in a crystal in terms of k:

$$p = hk/2\pi = \hbar k \tag{6.12}$$

Because we are dealing with an electron in a crystal rather than a free electron, $\hbar k$ is termed the **crystal momentum**. Crystal momentum is not quite the same thing as momentum in the classical sense, but, as we will see later, it plays an important role in both the electrical and optical properties of a material.

6.1.5 Band-Structure Diagrams

Each crystal orbital has a specific energy E and crystal momentum $\hbar k$. We can represent the electronic structure of the crystal by plotting the energies of its crystal orbitals as a function of k to create what is called a **band-structure diagram**. The band-structure diagram for our infinite chain of hydrogen atoms is plotted in Figure 6.5. For a chain of finite length (i.e. a large ring), we can imagine the curve in this figure as a series of closely spaced points, each representing a different crystal orbital. There will be one crystal orbital for each hydrogen atom, just as there is one MO per hydrogen in the MO diagrams shown in Figure 6.1. As the number of atoms in the crystal increases, the MO-energy points become more closely spaced. For an infinite crystal, the curve in Figure 6.5 is a continuous function, an infinite set of crystal-orbital points, each with the same $u(r)$ but a different k; a function that defines a **band** of allowed energies. When the unit cell contains more than one atom, the band-structure diagram will have multiple bands, each derived from a different $u(r)$, as we will see in Section 6.2.

Despite its simplicity, there are several things we can learn from Figure 6.5. Firstly, the number of bands is equal to the number of AOs in the unit cell. In this case, we have one atom per unit cell and one AO per atom, which leads to a single band. Secondly, we see that the crystal orbitals in the band run "uphill" in energy from $k = 0$ (bonding) to $k = \pm\pi/a$ (antibonding). The energy span between the top and bottom of a band is the **bandwidth**, W.[10]

In Figure 6.5, the band-structure diagram is plotted for two different H–H distances. Figure 6.5a corresponds to an interatomic distance of 1.0 Å, while Figure 6.5b shows how the

---

[10] As W increases, we also say that the dispersion of the band increases. Wide bands are often called disperse bands.

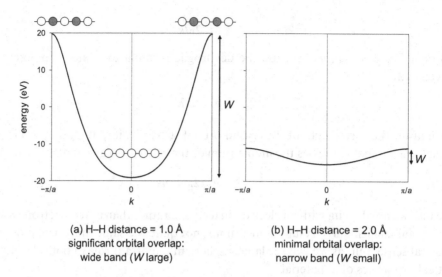

(a) H–H distance = 1.0 Å
significant orbital overlap:
wide band (W large)

(b) H–H distance = 2.0 Å
minimal orbital overlap:
narrow band (W small)

Figure 6.5 The band-structure diagram for an infinite chain of H atoms where the interatomic spacing is (a) 1.0 Å or (b) 2.0 Å. The crystal-orbital fragments shown on the left correspond to $k = 0$ and $k = \pm\,\pi/a$.

band structure changes if the interatomic distance increases to 2.0 Å. In the former case, the hydrogen atoms are close together, resulting in a high degree of orbital overlap. Consequently, the bonding crystal orbital at $k = 0$ is strongly stabilized with respect to an isolated H atom, while the antibonding crystal orbitals at $k = \pm\pi/a$ are highly destabilized, leading to a very wide band of $W \approx 40$ eV. The bandwidth is decreased by a factor of 10 when the interatomic distance becomes 2.0 Å. Narrow bands result when there is little change in the strength of the bonding/antibonding interactions as the value of k changes, either due to poor spatial overlap of orbitals from one real-space unit cell to the next (as is the case for the weakly bonded H-atom chain) or for reasons of symmetry. Electrons that occupy narrow bands are typically highly localized, either on a single atom or a group of atoms within the unit cell. As we will see in Chapter 10, bandwidth is an important parameter in determining the conductivity of a material.

Another point to be made about the band structure of the infinite H-atom chain concerns the shape of the band-structure diagram. The band structure from $k = 0$ to π/a is a mirror image of the band structure from $k = 0$ to $-\pi/a$. This follows from the mathematical properties of the e^{ikr} term in the Bloch function. Because of this symmetry, band structures are typically plotted only for positive values of k. We should also note that the band is not symmetrically distributed about the energy of an isolated H $1s$ orbital ($E = -13.6$ eV), and that the band is flatter at the bottom (near $k = 0$) than at the top (near $k = \pm\pi/a$). This asymmetry, which is more evident for the chain where the atomic spacing is 1 Å (Figure 6.5), is a consequence of the fact that antibonding states are destabilized more than bonding states are stabilized, as previously discussed for the MO diagram of H_2^+ (Section 5.3.1).

6.1.6 Density-of-States (DOS) Plots

A band-structure diagram contains a considerable amount of information; it effectively contains a MO diagram for each point in k space. In a 3D crystal, particularly one with a complicated crystal structure, band-structure diagrams can become quite complex. The additional complexity comes in part because there are more bands (one for each AO in the unit cell), and in part because there are many more values of k to consider in three dimensions than in one dimension.

A mechanism for depicting the electronic structure in a simpler form is to convert the band-structure diagram to a density-of-states (DOS) plot. The DOS plot for the hydrogen chain with an H–H distance of 2.0 Å is shown in Figure 6.6. The vertical axes of the DOS plot and of the band-structure diagram are identical; they give the energies of the electronic states. However, the horizontal axis of a DOS plot is different; it represents the **density of states**, $N(E)$, which is the number of allowed energy levels per unit volume of the solid in the energy range E to $E + dE$, as dE goes to zero.

A peak in the DOS plot indicates a large number of crystal orbitals with similar energies and is therefore related to the slope of the E versus k curve. Recall that for a large but finite chain, the line in Figure 6.6 (left) comprises closely spaced but discrete crystal orbitals. The range where the band flattens out (in this example near $k = 0$ and π/a) will have many crystal orbitals with very similar energies, and the DOS will be high. Where the band is steep (in this example near $k = \pi/2a$), there will be fewer crystal orbitals in the same energy range and the DOS will be lower. We saw a preview of the double-peaked DOS for the infinite H-atom chain in the MO diagrams for cyclic all-hydrogen molecules (Figure 6.1), where the MOs were more closely spaced near the bottom and top of the energy range covered by MOs.

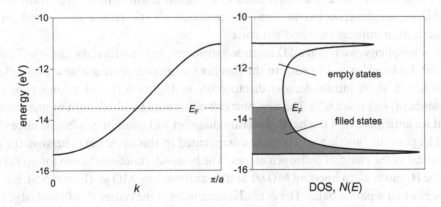

Figure 6.6 The band-structure (left) and density-of-states plot (right) for an infinite H-atom chain of 2.0 Å separation at $T = 0$ K. E_F is the Fermi energy.

DOS plots provide information about energy levels, filling, and widths of bands in a manner that can be assimilated without knowledge of Bloch functions or an understanding of k space. In fact, for depicting the electronic structure of an extended solid, the DOS plot is in many ways the closest equivalent to the MO diagrams discussed in Chapter 5. We will see that it is possible to sketch approximate DOS plots from knowledge of the structure, electron count, and orbital energies in the same way that approximate MO diagrams can be constructed from the same data. Nonetheless, we should not forget that useful information is lost upon transforming a band-structure diagram into a DOS plot.

In band-structure diagrams, just as in MO diagrams, it is important to know which crystal orbitals are occupied by electrons and which ones are empty. The energy level that separates the filled states from the empty states at $T = 0$ K is called the **Fermi energy**, E_F, or **Fermi level**. In Figure 6.6, we see that the Fermi energy in our infinite hydrogen chain cuts the band in such a way that half of the crystal orbitals have energies that fall below E_F and half above it. Such a half-filled band is exactly what we would expect because each H atom contributes one orbital and one electron. Like the orbitals from which they are formed, *each band can hold two electrons (of opposite spin)*. Metallic conductivity is often observed in materials where the Fermi level cuts through a band, leaving it partially occupied. We will return later to the importance of band filling and the Fermi level on the electrical and optical properties of materials.

6.2   The Band Structure of a Chain of H₂ Molecules

You may have been surprised by the result of the preceding section; that an infinite hydrogen chain would exhibit metallic conductivity. Although hydrogen is thought to become metallic at very high pressures, under ambient conditions elemental hydrogen is certainly not metallic. Furthermore, under ambient conditions, a linear chain of hydrogen atoms would not be stable versus dimerization to form H_2 molecules. Let's take a closer look at how this dimerization impacts the band structure.

For simplicity, we retain a 1D extended structure, but this time the chain will comprise H_2 molecules lined up end to end. In the resulting structure, there are now two different H–H contacts, a short intramolecular distance (d_1 in Figure 6.7) and a longer intermolecular distance (d_2 in Figure 6.7). The new unit cell now holds two hydrogen atoms instead of one, and we anticipate that the band-structure diagram will contain two bands instead of one.

The periodic function $u(r)$ is more complicated in this case, but otherwise the analysis is identical to the chain of hydrogen atoms. The basis-set functions to use for $u(r)$ are the MOs of the H_2 molecule; a bonding MO ψ_+ and an antibonding MO ψ_- (Figure 5.9). Each MO will give rise to a separate band. The orbital interactions at the center ($k = 0$) and edge ($k = \pi/a$) of the first Brillouin zone are shown in Figure 6.7. We start by considering the band derived from the bonding MO (ψ_+). At $k = 0$, the intra- as well as intermolecular interactions are

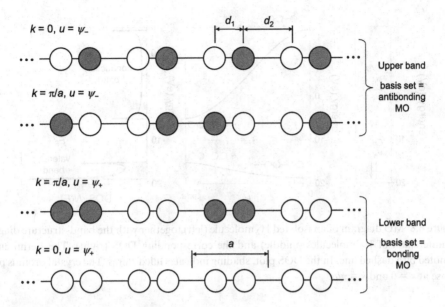

Figure 6.7 Crystal orbitals in an infinite chain of H_2 molecules at $k = 0$ and $k = \pi/a$.

bonding, hence this is the lowest-energy crystal orbital (shown on the bottom in Figure 6.7). Upon moving to $k = \pi/a$, intermolecular interactions become antibonding, whereas intramolecular interactions remain bonding. Consequently, the energy of the bonding band (ψ_+) rises; it runs "uphill" from $k = 0$ to $k = \pi/a$.

Next, we consider the band derived from the antibonding MO (ψ_-). When $k = 0$, both the intra- and intermolecular interactions are antibonding, and, as a result, this is the highest-energy crystal orbital (shown on the top in Figure 6.7). At $k = \pi/a$, the intramolecular interactions are still antibonding, but the intermolecular interactions are now bonding. Hence, this band runs "downhill" in energy from $k = 0$ to $k = \pi/a$. Because the intramolecular H–H distance is shorter than the intermolecular H–H distance ($d_1 < d_2$), the band derived from the bonding MO (ψ_+) will be lower in energy than the band derived from the antibonding MO (ψ_-) throughout the first Brillouin zone (Section 6.1.3).

The calculated band structure of our H_2 molecular chain is shown in Figure 6.8. As predicted, the lower band runs uphill in energy and the upper band runs downhill in energy. Each hydrogen atom has one valence electron so there are two electrons per unit cell, enough to fill the lower band while leaving the upper band empty. Therefore, the Fermi energy lies in the energy gap between bands. Materials where the Fermi level does not cut through a band are semiconductors or insulators, depending on the size of the gap.

In a basic sense, we see that the band-structure diagram of the H_2 chain could have been approximated by broadening each of the two H_2 MOs into a band. This is our first lesson in how to use the MO diagram of a chemical building unit as the starting point for approximating the band structure. The width of the bands depends on the interaction between molecules in

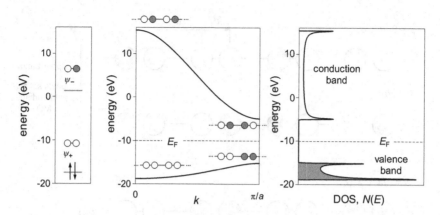

Figure 6.8 MO diagram of an isolated H_2 molecule (left) together with the band-structure diagram for an infinite chain of H_2 molecules (middle) and the corresponding DOS (right). The Fermi energy E_F is denoted by a dashed line. In the DOS plot, shading indicates filled states. The crystal orbitals pictured are those at $k = 0$ and $k = \pi/a$.

neighboring unit cells.[11] If the intermolecular spacing is decreased, the bands will become wider and the energy gap between bands smaller. As the intermolecular H–H distance becomes almost the same as the intramolecular H–H distance, the bonding/antibonding interactions at $k = \pi/a$ in the upper and lower bands will be nearly equivalent, and the energy gap between the bands will become infinitesimally small. When the H–H distances become equal, we revert to a single-atom unit cell and the metallic band structure of the H-atom chain shown in Figure 6.5.[12]

We now see why a structure possessing alternating long and short H–H distances would be more stable than one where the H–H distances are equal; the bond length alternation creates a band gap, thus lowering the energy of the occupied band at the expense of the empty band (Figure 6.8). The H-atom chain is not the only system to undergo this type of distortion. **Peierls' theorem** states that any 1D structure possessing a partially filled band will undergo a distortion that leads to bond-length alternations, splitting the partially filled band to create a gap between filled and empty bands. Distortions of this type are called **Peierls distortions**. A classic real-world example is the Peierls distortion in polyacetylene (Figure 6.9). If all C–C bonds were the same length, the electrons in the π orbitals would be fully delocalized, as shown on the left-

---

[11] The band formed from the antibonding MO (ψ_-) is wider (more disperse) than the band formed from the bonding MO (ψ_+) because antibonding interactions are more destabilizing than bonding interactions are stabilizing.

[12] If we were to artificially use the two-atom unit cell also for the chain where all H–H distances are equal, the real-space unit cell would be twice that of the true one-atom unit cell, and the length of the first Brillouin zone would be halved. The band-structure diagram corresponding to this artificial two-atom unit cell can be derived by drawing a vertical line in the band plot of the H-atom chain (Figure 6.6, left) at $k = \pi/2a$ and folding the plot over that line so that the two wings overlap. The resulting band-structure plot shows two bands widely separated at $k = 0$, but touching at the folding point (the new $k = \pi/a$); the limiting case of the band structure in Figure 6.8. This concept is called band folding.

Figure 6.9 The Peierls distortion in polyacetylene.

hand side of Figure 6.9. However, when the chain distorts to give alternating long and short bonds, the π electrons localize as shown on the right-hand side, and the total electronic energy of the system is lowered. We'll examine the band structure of polyacetylene in more detail in Section 10.5.1.

6.3 Electrical and Optical Properties

The electrical and optical properties of a substance follow directly from its band structure. In this section, we look at some basic connections between band structure and properties. In later chapters, we will revisit these topics in detail.

6.3.1 Metals, Semiconductors, and Insulators

Because the H-atom chain and the H_2 molecular chain have different band structures, the properties of these hypothetical chains will also be different. The electronic conductivity is sensitive to the location of the Fermi level. When E_F cuts through a band, as it does for the H-atom chain, there is almost no energy cost for electrons to move from occupied states just below E_F to empty states just above E_F. As a result, the electrons can move in response to the application of an external driving force, such as an electric field or a temperature gradient. This explains why metals readily conduct electricity and heat. In contrast, when the Fermi energy does not cut through a band, as in the H_2 chain, a non-negligible amount of energy is needed to excite the electron from the filled band to the empty band. In this case, the electrical behavior corresponds to that of a semiconductor or insulator. We will examine these classes of materials more closely in Chapter 10.

Before going further, we need to develop some nomenclature for discussing semiconductors. For a semiconductor, the minimum energy difference between the filled band(s) and the empty band(s) is called the **band gap**, E_g. Although there is not a precise value of E_g that separates semiconductors from insulators, as a rough guideline, materials with $E_g > 3$ eV are typically considered insulators. When discussing semiconductors/insulators, we will refer to the filled bands as **valence bands** and the empty bands as **conduction bands**. Simple DOS sketches of a metal, semiconductor, and insulator are shown in Figure 6.10. These block sketches do not attempt to capture the shape of the various bands as they would be depicted in an actual DOS plot. Instead, each band (or set of bands) is represented by a rectangle and shading is used to show their filling. These crude sketches can be a useful tool for quick approximations of the electronic structure.

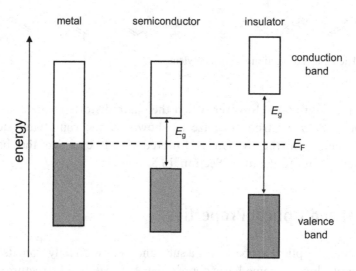

Figure 6.10 Schematic DOS diagrams for a metal (left), a semiconductor (middle), and an insulator (right). Occupied crystal orbitals are shaded.

The shading used in the DOS plots in Figure 6.10 is, strictly speaking, only valid when the temperature is equal to absolute zero ($T = 0$ K). At finite temperatures, the electron distribution smears out as some of the electrons are thermally excited from states below E_F to states above E_F. This effect has important consequences for the conductivity of a material and will be considered in detail in Chapter 10.

6.3.2 Direct- versus Indirect-Gap Semiconductors

In Figure 6.8, it is easy to see that the energy separation between the upper and lower bands changes as a function of k. At $k = 0$ the two bands are separated by ~34 eV, while at $k = \pi/a$, only by ~10 eV. The band gap is defined as the minimum energy separation between the highest energy state in the valence band and the lowest energy state in the conduction band, ~10 eV in this case. When the valence-band maximum and the conduction-band minimum occur at the same value of k, as they do in this example, the material is said to be a **direct-gap semiconductor**. When these two points fall at different values of k, the material is an **indirect-gap semiconductor**. Schematic representations of these two cases are given in Figure 6.11.

To excite an electron from one band to another requires a transfer of energy from the incoming photon to the electron. If k does not change upon moving from one band to another (i.e. the transition is a vertical line in Figure 6.11), the transition only involves transfer of energy. If k changes, however, it is necessary to alter not only the energy but also the momentum of the electron, Equation (6.12). The mechanism for doing this requires transfer of momentum between a lattice vibration (a phonon, see Section 4.4.6) and the electron as the photon is absorbed. Because this is effectively a three-body process, light is absorbed much

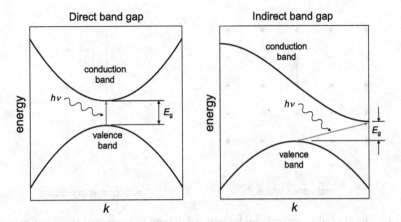

Figure 6.11 The lowest-energy valence to conduction band transition in a direct-gap (left) and indirect-gap (right) semiconductor.

more efficiently in a direct-gap semiconductor than it is in an indirect-gap semiconductor. This has important consequences for the design of optical devices, as we will see in Chapter 7.

6.4 Representing Band Structures in Higher Dimensions

Although there are a few materials whose band structure can be approximated by a 1D model, most materials need to be treated in two or three dimensions. The 2D case is not just a conceptual intermediate on the way to the 3D model, it applies to a number of interesting materials. Surfaces are 2D entities, and their electronic structures are useful constructs for understanding the chemical processes that occur there, such as heterogeneous catalysis. Many compounds are made up of covalently bonded layers that are held together by ionic or dispersion forces. The electronic structures of such materials are largely 2D in character. Other materials are 2D solids in a strict sense, graphene being the best-known example.

The underlying concepts of band theory don't change upon increasing the dimensionality, but the mathematics of the analysis and the representation of the results become more complicated. We will leave the computational details to computers, but it's not possible to discuss 2D and 3D systems without first developing a basic understanding of how they are represented.

6.4.1 Crystal Orbitals in Two Dimensions

Let's start by recalling that the position of any lattice point in a 2D crystal is given by a real-space translation vector, T:

$$T = u\boldsymbol{a} + v\boldsymbol{b}$$

(6.13)

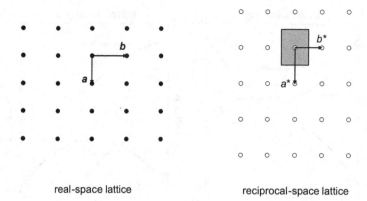

real-space lattice   reciprocal-space lattice

Figure 6.12 The 2D real-space rectangular lattice (left) and its reciprocal-space lattice (right). The shaded rectangle represents the first Brillouin zone. Note that if $b > a$ in real space, $b^* < a^*$ in reciprocal space.

where a and b are the basis vectors of the real-space lattice (Chapter 1.1) while u and v are integers. The real-space unit cell is defined by $u = v = 1$. The corresponding reciprocal-space lattice vector, G, is given by:

$$G = ha^* + kb^* \qquad (6.14)$$

where a^* and b^* are the basis vectors of the reciprocal-space lattice, while h and k are integers. In two dimensions, the reciprocal-space lattice vector a^* must be perpendicular to the real-space vector b, and b^* must be perpendicular to a. The magnitudes of the reciprocal-space lattice vectors are defined so that the dot products $a \cdot a^* = b \cdot b^* = 2\pi$.

For a 2D rectangular lattice, the reciprocal-space lattice vectors a^* of magnitude $2\pi/a$ and b^* of magnitude $2\pi/b$ are shown in Figure 6.12. As in one dimension, the first Brillouin zone contains the complete range of k values needed to generate all possible crystal orbitals without duplication. It is the region in reciprocal space that surrounds $k = 0$ and contains all k points that are closer to this point than to any other reciprocal-space lattice point. To determine the boundaries of the first Brillouin zone, we bisect the reciprocal-space lattice vectors around one of the lattice points by perpendicular lines.

Our next example is the 2D hexagonal lattice shown in Figure 6.13. To construct the first Brillouin zone, we first draw lines from an arbitrary lattice point chosen as the origin to each of its six nearest neighbors. The perpendicular bisectors of these lines define a hexagon that contains all points closer to $k = 0$ than to any other reciprocal-space lattice point. Unlike crystallographic unit cells, the first Brillouin zone is not necessarily a parallelogram in two dimensions (or a parallelepiped in three dimensions), as this example illustrates.[13]

---

[13] In crystallography, a locus of points in space closer to a given lattice point than to any other lattice point is called a Wigner–Seitz cell.

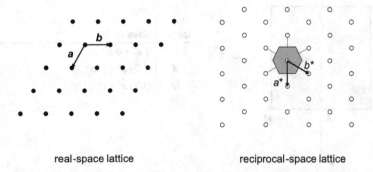

real-space lattice   reciprocal-space lattice

Figure 6.13 The 2D real-space hexagonal lattice (left) and its reciprocal-space lattice (right). On the right-hand side, thin lines connect the reciprocal-space lattice point defined as $k = 0$ with the six neighboring lattice points, and the shaded hexagon represents the first Brillouin zone.

The real-space hexagonal-lattice vectors a and b have same lengths ($|a| = |b|$) and the angle between them is 120°. For calculations in Cartesian coordinate space, we can express them in terms of orthogonal unit-length vectors that are perpendicular ($\hat{x}$) and parallel ($\hat{y}$) to b,[14] as $a = [(|a|\sqrt{3}/2)]\hat{x} - [|a|/2]\hat{y}$ and $b = |b|\,\hat{y}$. Using the relationships $a^* \perp b$ and $b^* \perp a$, one can show that the angle between reciprocal-space lattice vectors is 60° (Figure 6.13). In terms of unit vectors in Cartesian coordinate space, the hexagonal reciprocal-space vectors are $a^* = [4\pi/(|a|\sqrt{3})]\hat{x}$ and $b^* = [2\pi/(|a|\sqrt{3})]\hat{x} + [2\pi/|b|]\hat{y}$.

In a 2D crystal, the electron wavefunction, which was $\psi(r) = e^{ikr}u(r)$ in one dimension (Section 6.1.2), becomes:

$$\psi(\boldsymbol{r}) = e^{i\boldsymbol{k}\cdot\boldsymbol{r}}u(\boldsymbol{r}) \tag{6.15}$$

where $\boldsymbol{r}$ is the position vector or radius vector $\boldsymbol{r} = x\boldsymbol{a} + y\boldsymbol{b}$ (Section 1.1.1) of any point in the crystal in fractional coordinates x and y, and $u(\boldsymbol{r})$ is the lattice-periodic function describing the electric-field potential felt by an electron in the real-space crystal. Whereas k was a scalar in 1D space confined to the interval $-a^*/2 \le k \le +a^*/2$, in 2D space $\boldsymbol{k}$ becomes a vector, but its allowed values are still confined to the first Brillouin-zone interval $-a^*/2, -b^*/2 \le \boldsymbol{k} \le a^*/2, b^*/2$. It can be expressed with fractional coordinates $-\frac{1}{2} < m, n < \frac{1}{2}$ as $\boldsymbol{k} = m\boldsymbol{a}^* + n\boldsymbol{b}^*$ and is referred to as the **wave vector**.[15] It gives not only the wavelength of the crystal orbital, $\lambda = 2\pi/|\boldsymbol{k}|$, but also the direction in which the $e^{i\boldsymbol{k}\cdot\boldsymbol{r}}$ term modulates the wavefunction.[16]

To show the entire electronic structure of a 2D material, we need a 3D plot, two dimensions for $\boldsymbol{k}$ and one for energy. Fortunately, the maxima and minima of many bands are generally found either at the center of the Brillouin zone or at one of its boundaries.

---

[14] Their lengths are $|\hat{x}| = |\hat{y}| = 1$ chosen unit-length in real space or 1 chosen unit-length^{-1} in reciprocal space.

[15] The wave vector is perpendicular to the wave fronts. In two dimensions, the wave fronts are parallel lines that run through points where the crystal-orbital wavefunction takes a constant value.

[16] See Figure 6.17 (right) for a visual illustration of how the sign of the AO wavefunction is modulated by different values of the wave vector $\boldsymbol{k}$.

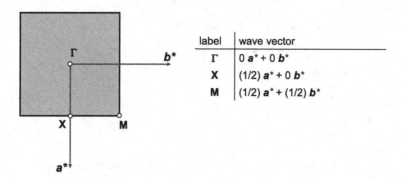

label	wave vector
Γ	0 a^* + 0 b^*
X	(1/2) a^* + 0 b^*
M	(1/2) a^* + (1/2) b^*

Figure 6.14 The first Brillouin zone and special symmetry points for the 2D square lattice. The magnitudes of the reciprocal lattice vectors are $|a^*| = |b^*| = 2\pi/|a|$.

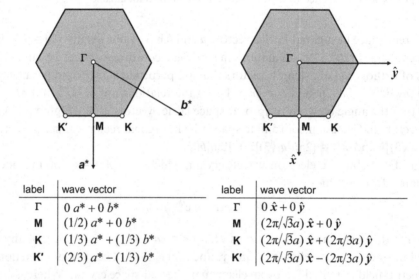

label	wave vector
Γ	0 a^* + 0 b^*
M	(1/2) a^* + 0 b^*
K	(1/3) a^* + (1/3) b^*
K′	(2/3) a^* − (1/3) b^*

label	wave vector
Γ	$0\,\hat{x} + 0\,\hat{y}$
M	$(2\pi/\sqrt{3}a)\,\hat{x} + 0\,\hat{y}$
K	$(2\pi/\sqrt{3}a)\,\hat{x} + (2\pi/3a)\,\hat{y}$
K′	$(2\pi/\sqrt{3}a)\,\hat{x} - (2\pi/3a)\,\hat{y}$

Figure 6.15 The first Brillouin zone and special symmetry points for the 2D hexagonal lattice in terms of the reciprocal-space lattice vectors a^* and b^* (left), and the orthogonal axes (right).

Therefore, we can get a good sense of the band structure by plotting the energy as a function of k along lines that run between various high-symmetry points in the Brillouin zone.[17] For a 2D square lattice, these points are given in Figure 6.14.

The high-symmetry points within the first Brillouin zone of a 2D hexagonal lattice are shown in Figure 6.15. These points are expressed in terms of the reciprocal-space lattice vectors a^* and b^* on the left-hand side of the figure and in terms of the orthogonal unit-length vectors on the right-hand side of the figure.

---

[17] The Γ point has all point-symmetry elements of the lattice, while the other special points retain some but not all symmetry elements.

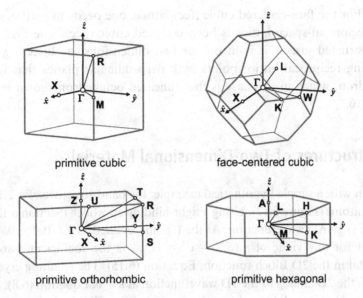

primitive cubic   face-centered cubic

primitive orthorhombic   primitive hexagonal

Figure 6.16 3D Brillouin zones of lattices reciprocal to four selected 3D real-space lattices, with high-symmetry points identified and orthogonal axes drawn.

6.4.2   Crystal Orbitals in Three Dimensions

In three dimensions, the crystal-orbital wavefunctions are completely analogous to the 2D case, Equation (6.15), but the radius vector $r = xa + yb + zc$ and wave vector $k = ma^* + nb^* + oc^*$ are given in three dimensions. The reciprocal-space basis vectors are defined in terms of the real-space lattice vectors by the following relationships:

$$a^* = \frac{2\pi}{V}(b \times c) \qquad b^* = \frac{2\pi}{V}(c \times a) \qquad c^* = \frac{2\pi}{V}(a \times b) \tag{6.16}$$

where V is the real-space unit-cell volume, $V = a \cdot (b \times c)$. From the definition of a vector cross product,[18] a^* is perpendicular to b and c, b^* is perpendicular to a and c, and c^* is perpendicular to a and b. For a cubic lattice with a real-space unit-cell edge a, the reciprocal-space lattice is also cubic with a cell edge of magnitude $a^* = 2\pi/a$. Notice that as the real-space unit cell gets larger, the reciprocal-space cell gets smaller.

We construct the first Brillouin zones of 3D reciprocal-space lattices using the same approach as in two dimensions, but they are now polyhedra instead of polygons. The first Brillouin zones for reciprocal-space lattices that correspond to some common 3D real-space lattices are shown in Figure 6.16, with the high-symmetry points labeled. Those Brillouin zones derived from primitive real-space lattices have simple shapes expected from the 2D

---

[18] The cross product of two vectors (e.g. $b \times c$) is a vector perpendicular to the vectors from which it is formed with a magnitude equal to the area of the parallelogram formed by the original two vectors.

analogy. For the face-centered cubic (fcc) lattice, one needs to realize, or simply accept, that its reciprocal-space lattice is body-centered cubic (bcc). The first Brillouin zone is then constructed just as it is in one or two dimensions, by bisecting lines connecting neighboring reciprocal lattice points with perpendicular planes that form the faces of a polyhedron, which in this case is the truncated octahedron shown in the top right of Figure 6.16.

6.5   Band Structures of Two-Dimensional Materials

Let's begin with a simple hypothetical example: the band structure of a 2D square lattice of hydrogen atoms (Figure 6.17). Using a tight-binding approach (Section 6.1.2), we take $u(r) = \psi_{1s}(r)$, the H $1s$ AO wavefunction. At the Γ point, defined as $k = 0a^* + 0b^*$, the $e^{ik \cdot r}$ term is equal to 1 for any value of r ($e^{ik \cdot r} = e^0 = 1$), as is the coefficient associated with each H $1s$ orbital in the 2D Bloch function, Equation (6.15). The resulting crystal-orbital wavefunction is the 2D analog of the 1D wavefunction at $k = 0$, Equation (6.8). Since the AOs on neighboring H atoms all have the same phase (their coefficients in Equation (6.15) have the same sign), they interfere constructively. This condition maximizes the bonding overlap and therefore corresponds to the lowest-energy point in the first Brillouin zone. At X ($k = \frac{1}{2} a^* + 0 b^*$), the H $1s$ orbital wavefunction contribution alternates between positive and negative (due to the changing sign of the $e^{ik \cdot r}$ term) in the real-space x direction, but remains

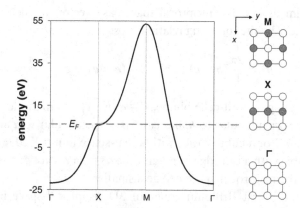

Figure 6.17 The band structure of a 2D square lattice of hydrogen atoms at a 1 Å separation. The crystal orbitals at Γ, X, and M (defined in Figure 6.14) are shown on the right in real-space coordinates. The spheres represent individual H $1s$ orbitals, while dark and light shading indicates the sign the AO wavefunction acquires from the e^{ikr} term; $\psi > 0$ is white and $\psi < 0$ gray.

constant in the real-space y direction.[19] The energy goes up along the Γ to X line because half of the nearest-neighbor H–H interactions change from bonding to antibonding. Finally, the energy reaches a maximum at the M point ($k = \frac{1}{2} a^* + \frac{1}{2} b^*$) where all nearest-neighbor interactions are antibonding. Note that the overall bandwidth W is higher in two dimensions than in one dimension because the number of bonding/antibonding interactions per H atom has increased.

6.5.1 Graphene

Graphite is one of the most familiar layered materials. It contains planar sheets of sp^2 hybridized carbon atoms arranged in a 2D honeycomb pattern (or, if you prefer, chicken wire). In graphite, the layers are stacked in an ordered way to form a 3D structure; however, when a single layer of carbon atoms from graphite is isolated, the 2D entity is called graphene. While it is easy to imagine a graphene sheet, it is quite challenging to isolate layers that are only a single atom thick. It was not until 2004 that researchers found a route to single graphene layers [1]. The discovery of graphene created tremendous excitement, leading to a surge in the study of 2D materials.

The structure of graphene is shown in Figure 6.18. There are two atoms in the hexagonal unit cell. Each carbon has four valence orbitals ($2s$, $2p_x$, $2p_y$, $2p_z$), so we expect a total of eight bands in the band structure. Our earlier look at the bonding in benzene (Section 5.3.7) is instructive. Just as in benzene, the $2s$, $2p_x$, and $2p_y$ orbitals in graphene overlap to form σ

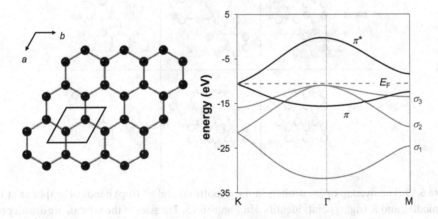

Figure 6.18 Left: A segment of a graphene sheet with the hexagonal unit-cell rhombus. Right: The band structure of graphene with σ bands in gray and π and π^* bands in black. The three empty σ^* bands located at higher energy are not shown.

---

[19] Imagine the 2D wave at X like a corrugated roof with peaks and troughs parallel with y.

bands while the $2p_z$ orbitals overlap to form π bands. By analogy with the MO diagram of benzene, we would expect the π band to be the highest-energy valence band and the π^* band to be the lowest-energy conduction band—the respective solid state equivalents of the highest occupied MO (HOMO) and lowest unoccupied MO (LUMO). Hence, we expect the properties of graphene to be dominated by these bands.

The 2D band structure of graphene is shown on the right-hand side of Figure 6.18. We see that over much of the first Brillouin zone, the σ bands are more stable than the π and π^* bands, as expected.[20] To understand the orbital overlap that gives rise to the π bands, we need only consider a single $2p_z$ orbital per atom. Because there are two atoms per unit cell, there are two bands that originate from the $2p_z$ orbitals; one of them from the π-bonding MO and one from the π-antibonding MO of the two-atom motif (the atom or group of atoms associated with each lattice point, Figure 1.1).

The interaction (bonding, antibonding, nonbonding) between carbon atoms in neighboring unit cells is dictated by the value of k, as shown in Figure 6.19. At Γ, $e^{ik\cdot r} = 1$, hence all nearest-neighbor interactions in the crystal orbital derived from the π MO are bonding, while in the corresponding π^* MO they are antibonding (Figure 6.19, left). This combination maximizes the energy separation between the π and π^* bands. At M, there are two bonding and one antibonding nearest-neighbor interactions for the π band and vice versa for the π^* band (Figure 6.19, middle), reducing the energy separation between the two bands (Figure 6.18).

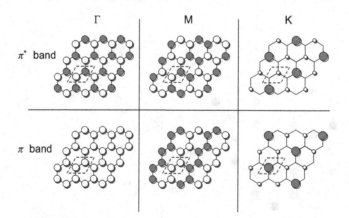

Figure 6.19 The overlap of $2p_z$ orbitals in the π (bottom) and π^* (top) bands of graphene at the Γ (left), M (middle), and K (right) points identified in Figure 6.15. The sizes of the orbitals are drawn proportional to the magnitude of the crystal-orbital coefficients $e^{ik\cdot r}$. For the crystal orbitals at K, only the real part of the wavefunction is shown.

[20] This condition is not met near Γ where the energies of the σ_2 and σ_3 bands reach a maximum. At Γ, mixing (i.e. any bonding or antibonding interaction) between $2s$ orbitals and $2p_x/2p_y$ orbitals is symmetry forbidden (see Section 6.5.2), thus the σ_1 band has pure C $2s$ character while the σ_2 and σ_3 bands have pure C $2p_x/2p_y$ character.

Inspection of the band structure (Figure 6.18) shows that the filled π and empty π^* bands become degenerate at K. A more detailed analysis reveals that the crystal orbitals for both bands are nonbonding at K. The Fermi level is located at the energy where these bands touch. This leads to an unusual electronic structure where the valence bands and conduction bands touch but do not overlap. Materials that possess such an electronic structure are said to be **semimetals**.[21] We will explore the properties of graphene in more detail in Section 10.6.1.

6.5.2 CuO_2^{2-} Square Lattice

The CuO_2^{2-} square lattice (Figure 6.20) is another illustrative and interesting 2D example, as this layer is present in nearly all high-T_c cuprate superconductors. Within each layer, the copper atoms are surrounded by four oxygen atoms forming squares that share corners, and each oxygen atom is linearly coordinated by copper; the Niggli formula is $CuO_{4/2}^{2-}$.

We can approximate the DOS plot from the MO diagram of square-planar CuO_4^{6-} shown on the left-hand side of Figure 6.21. The five highest-energy MOs originate from antibonding Cu $3d$–O $2p$ interactions. Highest is the antibonding interaction between Cu $3d_{x^2-y^2}$ orbitals and O $2p$ orbitals that point directly at each other. The other four antibonding Cu $3d$-based MOs are ~2 eV lower in energy.[22] The remaining MOs, located between −14 eV and −16 eV, are either Cu $3d$–O $2p$ bonding MOs or O $2p$ nonbonding MOs. To facilitate the discussion, we will group the MOs into three separate energy regions, marked I, II, and III in Figure 6.21.

The contribution of each Cu $3d$ orbital to the total DOS, the so-called partial density of states (PDOS), is plotted on the right-hand side of Figure 6.21. We can infer from the small

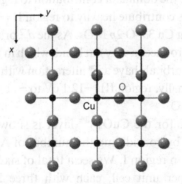

Figure 6.20 A segment of the 2D $CuO_{4/2}^{2-}$ sheet. The unit cell is shown in black.

---

21 In a strict sense, a semimetal is a material whose electronic density of states goes to zero at E_F but is non-zero for any finite energy above or below E_F. This occurs whenever the conduction and valence bands exactly touch, as they do in graphene. The term semimetal also applies in a more general manner to systems where the valence and conduction bands overlap by a small amount, generally at different k points. In graphite, interactions between layers cause the π and π^* bands to overlap slightly.

22 The MO diagram of square-planar CuO_4^{6-} exhibits d-orbital energy levels similar to those seen for a transition-metal ion in an elongated octahedral environment (Section 5.3.10).

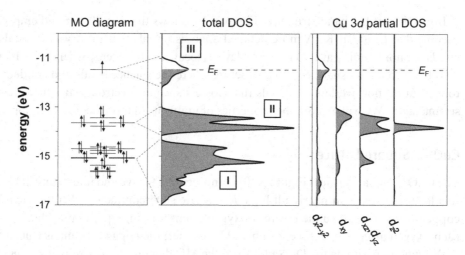

Figure 6.21 The MO diagram of a square-planar CuO_4^{6-} unit (left), the total-DOS plot (middle), and a partial-DOS plot for the infinite $CuO_{4/2}^{2-}$ layer showing the contributions of the individual Cu $3d$ orbitals (right).

degree of Cu character that region I, between -17.0 eV and -14.3 eV, is dominated by oxygen $2p$ orbital contributions. The Cu contributions in the crystal orbitals of the lower and middle sections of region I have Cu $3d$–O $2p$ bonding character. The electronic states near the top of this region, where the Cu contribution is minimal, are best described as O $2p$ nonbonding states.

The Cu $3d$ orbitals make the dominant contribution to regions II and III of the DOS plot. The d_{z^2}, d_{xz}, d_{yz} and d_{xy} orbitals contribute heavily to region II (-14.2 eV to -13.0 eV). These bands originate from antibonding Cu $3d$–O $2p$ MOs. As the d_{z^2} orbital has a rather weak overlap with oxygen in the xy plane, it produces a very flat band, which in turn gives rise to a sharp peak in the DOS. The d_{xz}, d_{yz}, and d_{xy} orbitals have a π^* interaction with the oxygen $2p$ orbitals, the strength of which depends on k. Finally, region III (-12.1 eV to -11.0 eV) originates from the antibonding Cu $3d_{x^2-y^2}$–O $2p$ σ^* MO.

The full band structure for the $CuO_{4/2}^{2-}$ layer is shown in Figure 6.22. The number of bands can be determined by considering the number of AOs in the motif. The O $2p$-orbital contribution is dominant in region I. We see a total of six bands in this region because there are two oxygen atoms per unit cell, each with three $2p$ orbitals.[23] The Cu $3d$-orbital contribution is dominant in regions II and III. Hence, we see four bands in region II (one for each of the doubly occupied Cu $3d$ orbitals d_{z^2}, d_{xz}, d_{yz}, and d_{xy}), and a single half-filled band associated with the Cu $3d_{x^2-y^2}$ orbital in region III.

---

[23] The square-planar CuO_4^{6-} unit contains four oxygen atoms and hence there are $4 \times 3 = 12$ MOs in region I, whereas the band structure is for a CuO_2^{2-} sheet which contains two oxygen atoms per unit cell. Hence there $2 \times 3 = 6$ bands in region I.

Once again, we can understand the shape of the bands in Figure 6.22 by considering the orbital overlap at various points in the Brillouin zone. As this is our first example with more than one element, let's take a closer look at how the mixing of copper and oxygen orbitals varies with changes in k. Consider in Figure 6.23 the band that originates from the antibonding interaction between Cu $3d_{xy}$ and O $2p$ orbitals (this band spans -14.2 eV to -13.0 eV in Figure 6.22). Because this is a π^* band, we choose an antibonding interaction between the Cu $3d_{xy}$ orbital and two O $2p$ orbitals within the CuO$_2$ motif (the orbitals that contribute to the motif are marked with dashed ovals).

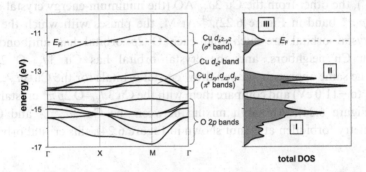

Figure 6.22 The electronic band structure (left) and the DOS plot (right) of a CuO$_{4/2}^{2-}$ layer.

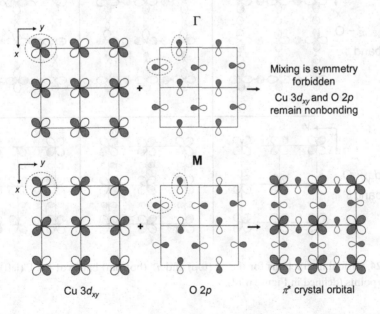

Figure 6.23 The mixing (or lack thereof) of the Cu $3d_{xy}$ and O $2p$ orbitals at Γ and M (defined in Figure 6.14). The orbitals associated with a single CuO$_2$ motif are enclosed in dashed ovals.

Once we have chosen the appropriate orbitals for the motif, the $u(x,y)$ basis set is fixed for all crystal orbitals that belong to this band. The next step is to determine the phases with which the AOs contribute to the crystal orbital (as dictated by the $e^{ik(x,y)}$ term of the Bloch function) at a few high-symmetry values of k. At Γ, the phases of the AOs of the motif are invariant on moving from one unit cell to the next (see Figure 6.23), and we see that the Cu $3d_{xy}$ and O $2p$ orbitals have an equal amount of bonding and antibonding overlap (along both directions). When this happens, orbital mixing (any bonding or antibonding inter-action) is said to be (translational-) symmetry forbidden, and two nonbonding crystal orbitals result. One is derived from the O $2p$ AO (in Figure 6.22, region I, amongst the "O $2p$ bands"), the other from the Cu $3d_{xy}$ AO (the minimum-energy crystal orbital of the Cu $3d_{xy}$–O $2p$ π^* band in Figure 6.22).[24] At M, the phases with which the AOs contribute to the crystal orbital are such that each O $2p$ orbital has antibonding interactions with both Cu neighbors, and the crystal orbital has Cu $3d_{xy}$–O $2p$ π^* character.

Using the same approach, we can draw crystal orbitals for the Cu $3d_{x^2-y^2}$–O $2p$ σ^* band (-12.1 eV to -11.0 eV) and compare them with the Cu $3d_{xy}$–O $2p$ π^* crystal orbitals at Γ, M, and X (Figure 6.24). Although mixing between the Cu $3d_{x^2-y^2}$ and O $2p$ orbitals is also symmetry forbidden at Γ (not shown in Figure 6.24), this crystal orbital is not strictly

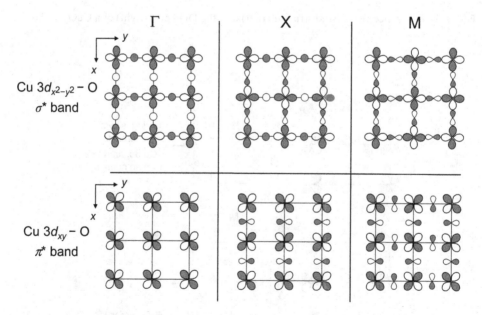

Figure 6.24 Orbital interactions for the σ^* (top) and π^* (bottom) bands at the Γ (left), X (middle), and M (right) points (defined in Figure 6.14).

---

[24] The nonbonding O $2p$ crystal orbital is lower in energy because an isolated O $2p$ orbital has a lower energy than an isolated Cu $3d$ orbital, see Section 5.2.1.

Cu $3d_{x^2-y^2}$ nonbonding because the O $2s$ orbitals we have not discussed up to this point have the proper symmetry to interact with Cu $3d_{x^2-y^2}$. This interaction (see upper left panel of Figure 6.24) is weaker than the Cu $3d$–O $2p$ interaction due to the poor energetic overlap of the Cu $3d$ (-14 eV) and O $2s$ orbitals (-34 eV), yet strong enough to affect the width of this band. Specifically, the σ^* band would become nonbonding at Γ (Figure 6.22), lowering its energy and increasing bandwidth, were it not for this antibonding $3d_{x^2-y^2}$–O $2s$ interaction.

At X, the interactions of Cu $3d_{xy}$ and Cu $3d_{x^2-y^2}$ orbitals with the O $2p$ orbitals become antibonding in the x direction (Figure 6.24, middle), while remaining symmetry forbidden (nonbonding) in the y direction. At M, the Cu $3d$–O $2p$ interactions become antibonding in both directions (Figure 6.24, right). Because the energy of a band increases as the number of antibonding interactions increases, these Cu $3d_{xy}$ π^* and Cu $3d_{x^2-y^2}$ σ^* bands run uphill from Γ to X to M as shown in Figure 6.22.[25]

Since the optical, magnetic and electrical properties are most sensitive to the bands near the Fermi level, the filling and width of the Cu $3d_{x^2-y^2}$ band is critical. Given the d^9 electron configuration of Cu^{2+}, it should not come as a surprise that the Fermi level cuts the Cu $3d_{x^2-y^2}$ band in half (Figure 6.22). As discussed in Section 6.3.1, the presence of a partially filled band should lead to metallic conductivity. However, this picture changes once electron–electron interactions are considered, as we will learn in Section 10.4. These interactions stabilize an antiferromagnetic, insulating ground state in compounds containing CuO_2^{2-} layers, such as La_2CuO_4. Interestingly, when electrons are added to or taken away from the layers, the properties can change dramatically, as we will see when we take up the topic of superconductivity in Chapter 12.

6.6 Band Structures of Three-Dimensional Materials

Finally, we are ready to tackle the band structures of 3D materials.[26] In this section, we will consider several materials that build on the principles already established for lower-dimensional systems, most of which represent important classes of functional materials.

6.6.1 α-Polonium

Perhaps the simplest 3D structure we can imagine is a primitive cubic lattice with a single atom per unit cell. The only element that crystallizes with this structure is α-Po (Figure 1.23). The band structure for α-Po is shown in Figure 6.25. There are four bands, one for each of the valence orbitals of Po; $6s$, $6p_x$, $6p_y$, and $6p_z$. The orbital character of the lowest-energy band is exclusively Po $6s$ (due in part to relativistic effects mentioned in Section 5.2.1). Its shape,

---

[25] At the M point, the Cu $3d_{xy}$ π^* band has the highest energy of the four bands that make up region II.

[26] In the band-structure diagrams for all 3D materials, the energy scale is arbitrarily set so that the highest filled electronic state falls at $E = 0$, as is the customary practice for density-functional theory used to generate these diagrams.

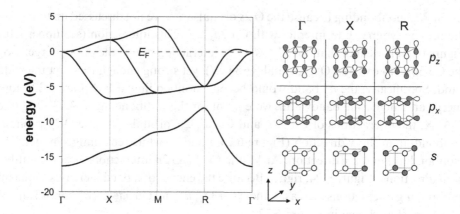

Figure 6.25 The band structure of α-Po, showing the orbital overlap at the Γ, X, and R points (defined in Figure 6.16) for the $6s$, $6p_x$, and $6p_z$ bands. The effects of spin–orbit coupling have not been included in this calculation.

running uphill from Γ to X to M to R, is the 3D analogue of the band structure of a 2D square lattice of hydrogen atoms (Figure 6.17).

The three higher-energy bands originate from Po $6p$ orbitals. These bands are two-thirds filled, and they are therefore cut by the Fermi level. The right-hand side of Figure 6.25 shows the crystal orbitals associated with the $6s$, $6p_x$, and $6p_z$ bands at Γ, X, and R. At Γ, all three $6p$ bands are degenerate; each exhibiting strong σ-antibonding interactions with two neighboring Po atoms and weak π-bonding interactions with their remaining four nearest neighbors. The $6p$ bands are also degenerate at the R point, where the σ interactions have become strongly bonding and the π interactions weakly antibonding. At other points in the first Brillouin zone, the degeneracy of the $6p$ bands is lifted. For example, the $6p_x$ band reaches a minimum at the X point where both σ and π nearest-neighbor interactions are bonding. In contrast, the $6p_y$ and $6p_z$ bands experience strong antibonding nearest-neighbor σ interactions at the X point.

Bismuth lies one element to the left of polonium in the periodic table. The $6p$ orbitals are now half filled. The structure of Bi is related to the primitive cubic structure of α-Po (Figure 6.26), but distorted so that each atom makes three short bonds (d_1 = 3.07 Å) and three long bonds (d_2 = 3.53 Å) to its neighbors. This distortion has a similar origin as the 1D Peierls distortion illustrated in Figure 6.9. An even more pronounced distortion of this type is seen in lighter group-15 elements, such as gray arsenic (d_1 = 2.51 Å, d_2 = 3.15 Å).

6.6.2 Diamond

The crystal structure of diamond has eight atoms in the face-centered cubic unit cell (Figure 1.16 and 1.37), but only two in the equivalent primitive cell (at 0 0 0 and ¼ ¼ ¼ of the rhombohedron representing the primitive cell) normally used for band-structure

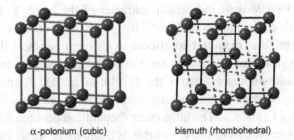

Figure 6.26 The structure of α-polonium (left) showing eight unit cells compared with an analogous fragment of the bismuth structure (right). The short and long bonds in bismuth are marked with solid and dashed lines, respectively.

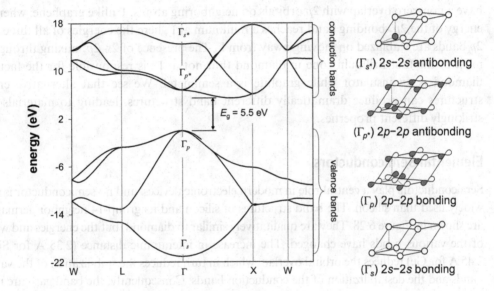

Figure 6.27 The band structure of diamond. Right: Crystal orbitals at the Γ point (only one of the three degenerate crystal orbitals present at Γ_p and Γ_{p*} is shown).

calculations. Using the smaller primitive cell, eight bands result from overlap of the 2s and 2p valence orbitals on carbon (Figure 6.27). The eight valence electrons (four per atom) completely fill the lower four bands. Because the interatomic distances are short (1.545 Å), there is a large degree of orbital overlap stabilizing the filled valence bands, which are bonding, and destabilizing the empty conduction bands, which are antibonding. The net result is a large band gap, 5.5 eV. Diamond has an indirect band gap as the valence-band maximum (at Γ) and the conduction-band minimum (near X) occur at different values of **k**.

Let's take a closer look at the orbital character of the bands at the Γ point. Just as the $2s$ and $2p$ orbitals contribute to different MOs in the MO diagram of CH_4 (Section 5.3.6), the $2s$ and $2p$ orbitals in diamond contribute to different bands at the Γ point. This lack of $2s$ and $2p$ mixing yields just four energies at Γ: singly degenerate crystal orbitals corresponding to bonding and antibonding overlap of the $2s$ orbitals, labeled Γ_s and Γ_{s*} in Figure 6.27, and triply degenerate crystal orbitals arising from bonding and antibonding overlap of the $2p$ orbitals, labeled Γ_p and Γ_{p*}. The large energy separation between the bonding $2s$ (Γ_s) and $2p$ (Γ_p) states results in part from the C $2s$ orbitals being more stable than the C $2p$ orbitals by ~8 eV (see Table 5.2) and in part from the fact that the $2s$ orbital overlap is bonding in all directions at Γ. In fact, the $2s$ orbital overlap at Γ is so large that the antibonding $2s$ crystal orbital (Γ_{s*}) is higher in energy than the antibonding $2p$ crystal orbital (Γ_{p*}).

Moving away from the Γ point, $2s$–$2p$ mixing occurs because the non-zero k alters the phases of the AO contributions to the crystal orbital in a manner that allows the $2s$ orbitals to have a non-zero overlap with $2p$ orbitals on neighboring atoms. Unlike graphene, where the energy of the $2p_z$-bonding band reaches a minimum at Γ, here the energies of all three filled $2p$ bands are stabilized on moving away from Γ. The presence of $2s$–$2p_z$ mixing throughout most of the first Brillouin zone in diamond (but not at Γ) is responsible for the fact that diamond is an insulator while graphite is a semimetal. We see that alternative crystal structures can produce dramatically different band structures, leading to materials with strikingly different properties.

6.6.3 Elemental Semiconductors

Semiconductors play a central role in modern electronic devices, and no semiconductor is more widely used than silicon. The band structures of silicon and its group-14 neighbor germanium are shown in Figure 6.28. They are qualitatively similar to diamond, but the energies and widths of the various bands have changed. The increase in interatomic distance (2.35 Å for Si and 2.45 Å for Ge) reduces the orbital overlap, which in turn reduces the stabilization of the valence bands and the destabilization of the conduction bands. Consequently, the band gaps are much smaller than that of diamond, 1.1 eV for Si and 0.7 eV for Ge. There are also changes in the relative energies of the antibonding crystal orbitals at Γ. The decrease in orbital overlap reduces the splitting between bonding and antibonding states, lowering the energy of Γ_{s*} with respect to Γ_{p*}.[27] As a result, by the time we reach germanium, Γ_{s*} lies below Γ_{p*}.

The orbital overlap decreases further upon moving to α-Sn ($d_{Sn-Sn} = 2.81$ Å), where Γ_{s*} is only marginally higher in energy than the valence-band maximum Γ_p. Not only does this further reduce the band gap ($E_g = 0.1$ eV), it reduces the overall stability of the structure. In fact, tin has two polymorphs; gray tin (α-Sn) is a semiconductor with the diamond structure, whereas white tin (β-Sn) is a metal with a body-centered tetragonal structure. Moving to the

---

[27] Orbital overlap is the most important factor in elemental semiconductors for dictating the relative energy of the bands at Γ because the difference in the energies of the s and p orbitals is almost constant from C to Sn (see Table 5.2).

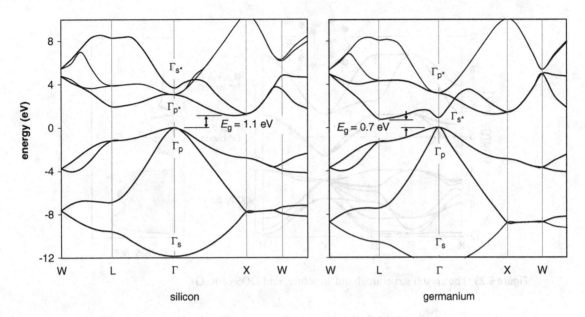

Figure 6.28 The band structures of silicon (left) and germanium (right).

sixth period, we come to Pb where this trend is further exacerbated by relativistic effects (Section 5.2.1) which contract the 6s orbitals and lower their energy with respect to the 6p orbitals. If Pb were to adopt the diamond structure, the antibonding 6s states would be populated before the bonding 6p states are full, destabilizing the structure. Consequently, Pb adopts a cubic closest-packed structure in preference to the diamond structure.

6.6.4   Rhenium Trioxide

The 3D analogue of the square CuO_2^{2-} lattice is the cubic ReO_3 structure (Figure 1.39), a network of corner-sharing octahedra, with a Niggli formula of $ReO_{6/2}$. ReO_3 is not just a model case, it is one of the most highly conducting oxides known. Its conductivity of 1.1×10^7 S/m at room temperature is higher than the conductivity of several elemental metals (Section 10.1).

The electronic band structure of ReO_3 is shown in Figure 6.29. Based on the MO diagram of an octahedrally coordinated transition-metal ion (Section 5.3.8), we expect the Re 5d and O 2p orbitals to make the dominant contributions to the bands near the Fermi level. The octahedral coordination geometry means that the Re 5d orbitals are split into a triply degenerate t_{2g} set (d_{xy}, d_{yz}, d_{xz}) that interacts with the O 2p orbitals in a π fashion, and a doubly degenerate e_g set ($d_{x^2-y^2}$, d_{z^2}) that interacts with the O 2p orbitals in a σ fashion. All Re 5d–O 2p bonding and O 2p nonbonding crystal orbitals are fully occupied. One electron occupies the Re 5d–O 2p π^* set of crystal orbitals, as expected given the $5d^1$ configuration of Re^{6+}.

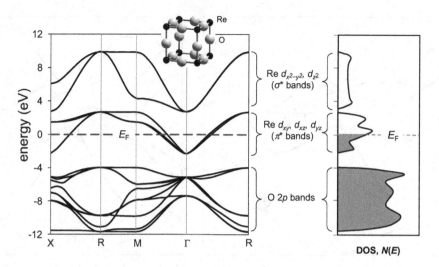

Figure 6.29 The crystal structure, band structure, and DOS of ReO_3.

The primitive unit cell of the ReO_3 structure contains one rhenium atom and three oxygen atoms. Therefore, we expect (3×3) nine bands for the O $2p$ orbitals and (5×1) five bands for the Re $5d$ orbitals. We see the nine O $2p$-based bands between -12 eV and -4 eV in Figure 6.29. Six of them run downhill from Γ, where they are mostly nonbonding O $2p$ crystal orbitals, to R where they have a Re $5d$–O $2p$ and Re $6s$–O $2p$ bonding character. The remaining three are confined to a narrow energy window near the top of the valence band and are largely O $2p$ nonbonding throughout the first Brillouin zone.

The three bands located between -2.3 eV and $+2.6$ eV arise from antibonding π^* interactions between the triply degenerate t_{2g} set of Re $5d$ orbitals and O $2p$ orbitals. The Fermi level cuts through these three bands. The fact that the Fermi level cuts through a reasonably wide band (in this case three bands) is responsible for the high electrical conductivity of ReO_3.

The two highest-energy bands in Figure 6.29 are the Re $5d$–O $2p$ σ^* bands, located between $+2.7$ eV and $+9.9$ eV. The Re $5d$ e_g orbitals make the dominant contribution to these bands. Because the σ interactions lead to a higher degree of spatial overlap than the π interactions, the σ^* bands are wider than the π^* bands, 7.2 eV versus 5.0 eV.

The orbital overlap is shown in Figure 6.30 for three representative bands—a nonbonding O $2p$ band, the Re $5d_{xy}$–O $2p$ π^* band, and the Re $5d_{x^2-y^2}$–O σ^* band—at Γ, X, and M. The picture is very similar to that discussed earlier for the CuO_2^{2-} square lattice (Section 6.5.2). At the Γ point, orbital mixing is symmetry forbidden for Re t_{2g} and O $2p$ orbitals, which makes this set of bands strictly Re $5d$ nonbonding. The σ^* set of bands is also (largely) Re $5d$ nonbonding at Γ. Symmetry allows for Re $5d$–O $2s$ σ^* interactions, but they are weak due to the poor energetic overlap of the O $2s$ and Re $5d$ orbitals. At the X point, all three π^* bands have antibonding interactions with their oxygen neighbors along x, while the interactions in the other two directions remain nonbonding. This leads to a splitting of the three π^* bands,

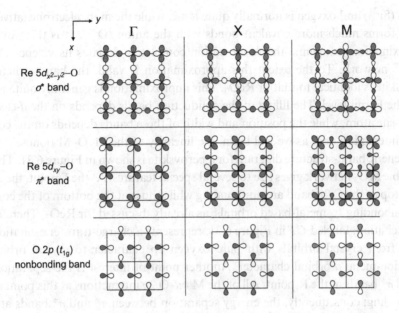

Figure 6.30 The orbital interactions for a nonbonding O $2p$ band (bottom), a Re–O π^* band (middle), and a Re–O σ^* band (top) at the Γ, X, and M points (defined in Figure 6.16).

because the overlap of the Re $5d_{yz}$ orbitals with the x-direction neighbors is negligible, whereas the Re $5d_{xy}$ and $5d_{xz}$ orbitals have significant Re–O interactions along x. Consequently, the latter two bands, being antibonding, have a higher energy than the Re $5d_{yz}$ band at X. For similar reasons, the Re $5d_{x^2-y^2}$ band has a higher energy than the Re $5d_{z^2}$ band at X. At the M point, the Re $5d_{xy}$ and Re $5d_{x^2-y^2}$ bands are completely antibonding, as illustrated on the right-hand side of Figure 6.30. Finally, at the R point all five Re $5d$ bands are completely antibonding (not shown), with each band reaching its highest energy.

The O $2p$ (t_{1g}) band depicted in Figure 6.30 is one of the O $2p$ bands described as nonbonding in the preceding discussion. In a sense, this description is accurate because the band lacks any Re $5d$ character. However, it is not strictly nonbonding if we take O $2p$–O $2p$ interactions into account. If we consider only the nearest-neighbor oxygen interactions, we see a change from nonbonding at Γ to antibonding at M. As a result, the energy of this band increases by a relatively small amount, 1.2 eV, on moving from Γ to M.

6.6.5 Perovskites

Just as the band structure of Si is representative of many semiconductors, the electronic structure of ReO_3 is representative of an important group of materials, the perovskites (Figure 1.49). Throughout the book, we will encounter many functional materials with the perovskite structure. In perovskites such as $SrTiO_3$, the bonding between the larger A-site cation (Sr^{2+}) and oxygen is normally quite ionic, while the more electronegative M-site cation (Ti^{4+}) forms much more covalent bonds with the anion (O^{2-}). It is therefore a reasonable approximation to assume the A-site cation completely donates its valence electrons to the MO_3^{n-} network. To the extent this approximation is valid, the band structure of a cubic perovskite is identical to that of ReO_3. This approximation is generally quite good for bands near the Fermi level. The filling of the conduction bands depends on the d-electron count of the M-site atom, while the position and width of these bands depends on the covalency of the M–O interaction and, as we will learn, the linearity of the M–O–M bonds.

A generic band-structure diagram for a perovskite is shown in Figure 6.31. The symmetry of the cubic MO_3 network gives rise to several special features. At the Γ point, the crystal orbitals at the top of the O $2p$ band are nonbonding while those at the bottom of the conduction band are nonbonding t_{2g} metal-based orbitals, as already discussed for ReO_3. Therefore, the direct-gap excitation, labeled CT in Figure 6.31, represents a charge-transfer excitation in the truest sense; from crystal orbitals with pure oxygen $2p$ character to crystal orbitals with pure transition-metal d-orbital character. Another point of interest is the separation between the π^* and σ^* bands at the R point. All of the M nd–O $2p$ interactions at this point are completely antibonding; consequently, the energy separation between π^* and σ^* bands at R is the solid state equivalent of the ligand-field splitting of an octahedron, Δ.

To examine periodic trends in bonding, let's take a quantitative look at how the band structure evolves as we decrease the electronegativity of the transition metal: $ReO_3 \rightarrow WO_3 \rightarrow KTaO_3 \rightarrow BaHfO_3$. All four compounds have similar band structures, but the exact energies and filling of the bands vary from one compound to the next. The calculated energies of E_g, Δ, CT, and several other parameters are given in Table 6.1.

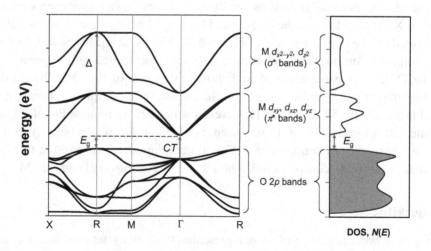

Figure 6.31 A generic band-structure diagram (left) and a schematic DOS plot (right) for an AMO_3 perovskite, where M is a d^0 transition-metal ion. Values of the marked parameters are given in Table 6.1 for several different perovskites.

Table 6.1 Key features from band-structure calculations of cubic ReO$_3$, WO$_3$, KTaO$_3$, BaHfO$_3$, and SrTiO$_3$.

	ReO$_3$	WO$_3$	KTaO$_3$	BaHfO$_3$	SrTiO$_3$
M–O distance (Å)	1.87	1.95	1.99	2.09	1.95
O–O distance (Å)	2.65	2.76	2.81	2.96	2.76
d-electron configuration	d^1	d^0	d^0	d^0	d^0
M–O σ^* bandwidth (eV)	7.2	6.4	6.2	5.1	4.1
M–O π^* bandwidth (eV)	5.0	4.3	4.2	3.5	2.4
Δ (σ^*–π^* at R) (eV)	7.2	6.3	6.1	5.5	3.9
CT (O $2p$–M π^* at Γ) (eV)	2.3	3.4	3.8	5.3	4.1
Calc. band gap, E_g (eV)[†]	Metal	2.4	3.5	5.0	3.4
Exp. band gap, E_g (eV)	Metal	2.4	3.5	5.5	3.1

[†]A scissors correction that increases the band gap by 2.1 eV has been applied to all semiconducting compounds in this table.

On changing from ReO$_3$ to WO$_3$, a dramatic change arises from the fact that the π^* bands in WO$_3$ are empty because W^{6+} has a d^0 configuration. This change in band filling makes WO$_3$ a semiconductor, in sharp contrast to the metallic conductivity of ReO$_3$. As a rule, semiconducting cubic perovskites have indirect band gaps because the valence-band maximum always falls on the line between M and R, while the conduction-band minimum always falls at Γ. The measured band gap of WO$_3$ is 2.4 eV, which is responsible for its yellow color.[28]

Replacing Re with W has other subtle effects on the electronic structure. Because hexavalent tungsten is less electronegative than hexavalent rhenium,[29] it takes more energy to excite an electron from oxygen to the transition metal, hence CT increases from 2.3 eV to 3.4 eV. The widths of the σ^* and π^* bands decrease because the antibonding interactions at the R point, which define the maxima of these bands, are not quite as antibonding, due to longer W–O distances and the fact that the W $5d$ and O $2p$ orbitals have a greater energetic mismatch. The octahedral ligand-field splitting Δ decreases for the same reason. As we continue to decrease the atomic number of the transition metal along WO$_3$ → KTaO$_3$ → BaHfO$_3$, the d^0 electron count is maintained but the nuclear charge steadily decreases. This leads to a decrease in the electronegativity of the transition-metal ion, a decrease in the covalency of the metal–oxygen bonds, and an increase in the bond length. These changes increase CT and E_g and reduce the σ^* and π^* bandwidths.

---

[28] It should be noted that the structure of WO$_3$ is not actually cubic. It is distorted by out-of-center displacements of tungsten atoms and rotations of the octahedra, both of which reduce the bandwidth and increase the band gap. The tungsten displacements are driven by second-order Jahn–Teller distortions.

[29] For a given oxidation state, electronegativities increase and orbital energies become more negative on moving horizontally from left to right across the periodic table (see Table 5.2).

Perovskites containing $3d$ transition metals have narrower σ^* and π^* bands than their $4d$ and $5d$ analogues. Consider $SrTiO_3$ and $BaHfO_3$, where both transition metals come from group 4. Table 6.1 shows that the σ^* and π^* bands are substantially narrower for $SrTiO_3$ than they are for $BaHfO_3$. This effect stems from the fact that $3d$ orbitals are smaller than $5d$ orbitals ($r_{max} = 0.53$ Å for Ti $3d$ versus 0.88 Å for Hf $5d$, Section 5.2.2). The contracted nature of the $3d$ orbitals reduces their overlap with the O $2p$ orbitals, and, as a result, the antibonding $SrTiO_3$ crystal orbitals, particularly the σ^* and π^* crystal orbitals at R, are less destabilized (have a lower energy) than the corresponding $BaHfO_3$ crystal orbitals. The decreased overlap results in the reduced bandwidth for $SrTiO_3$.

Bandwidth in perovskites is an important parameter that can impact the properties, sometimes in a dramatic fashion. Wider bands not only produce smaller band gaps, they are more conducive to electron delocalization. We will see the effect of bandwidth on the electrical conductivity of perovskites in Section 10.4.3. When we encounter such effects, it will be very useful to remember the mechanisms for tuning the bandwidth in perovskites. Table 6.2 contains a summary of those mechanisms.

Changing the identity of the element on the M site is not the only way to control the electronic structure of a perovskite. While the A cation does not usually play a direct role in the electronic structure near the Fermi level, its size controls the linearity of the M–O–M bonds. When the A-site cation is too small for the cubic corner-sharing network, octahedral tilting occurs (Section 1.5.3) leading to bending of the M–O–M bond angles. This introduces some antibonding character into the π^* bands at Γ (remember from Figure 6.30 they were strictly nonbonding at Γ in a cubic perovskite) and reduces the antibonding overlap of the π^* bands at R. The net effect is a reduction in the widths of the π^* bands, which leads to an increase in band gap. Octahedral tilting has a similar effect on the width of the σ^* band, which can have important implications for conductivity and magnetism in some perovskites. Overlap considerations predict that the width of the σ^* bands, W_{σ}^*, should scale proportionally to $\cos \varphi$, where φ is the deviation of the M–O–M bond angle from 180° [2]. As an example, consider what happens when K^+ in $KTaO_3$ is replaced by Na^+. The smaller Na^+ drives an octahedral-tilting distortion that reduces the Ta–O–Ta bond angle from 180° in $KTaO_3$ to 159° (on average) in $NaTaO_3$. This increases the band gap from 3.5 eV in $KTaO_3$ to 4.1 eV in $NaTaO_3$.

Table 6.2 Factors that impact the width of the π^* (W_{π^*}) and σ^* (W_{σ^*}) bands in AMO_3 perovskites.

Change in structure or composition	W_{σ^*} and W_{π^*}
M electronegativity increases	Increase
M–O bond distance increases	Decrease
M oxidation state increases	Increase
$3d$ transition metal substitutes for $4d/5d$ transition metal	Decrease
Octahedral tilting bends M–O–M bond angle	Decrease

6.7 Problems

6.1 MO diagrams for cyclic H_N molecules are shown in Figure 6.1. The AO phases are shown for the lowest- and highest-energy MOs but not for the intermediate MOs. (a) Sketch out the intermediate MOs indicating the phase of each AO for a cyclic H_6 molecule. Determine the number of nodal planes for each of the six MOs. (b) Repeat part (a) for a cyclic H_{10} molecule.

6.2 MO diagrams for cyclic H_N molecules are shown in Figure 6.1. (a) Sketch out the MOs and their relative energies for a square H_4 molecule. Determine the number of nodal planes for each MO. (b) This molecule is prone to a first-order Jahn–Teller distortion. How will the distortion change the shape of the molecule? Redraw the MO diagram after the distortion has taken place.

6.3 Show that the Bloch function given in Equation (6.1) meets the requirement that the electron density, $\rho(r) = \psi^2(r) = \psi^*(r)\psi(r)$, must be periodic according to Equation (6.2).

6.4 The first Brillouin zone of an infinite chain of H atoms has an infinite number of crystal orbitals, each with a different value of k. (a) State the limiting values of k within the first Brillouin zone. (b) The figure below shows the real part of the wavefunction for a crystal orbital $\psi(r)$ with a specific value of k (the positions of the H atoms are denoted by the black dots). Determine the value of k for this crystal orbital. (c) Calculate the momentum of this crystal orbital.

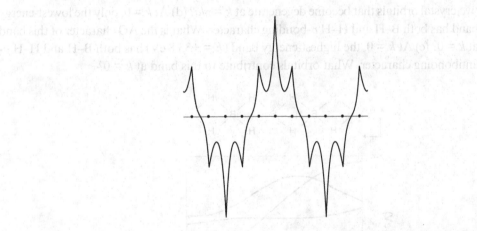

6.5 State whether each of the following statements is true or false. (a) As the magnitude of k increases, the momentum of the electron also increases. (b) As the magnitude of k increases, the wavelength of the corresponding wavefunction also increases. (c) As the size of the unit cell increases, the size of the first Brillouin zone also increases.

6.6 Consider an infinite 1D chain of equally spaced fluorine atoms. (a) Sketch the band structure of this chain. Include all four valence orbitals ($2s$, $2p_x$, $2p_y$, $2p_z$) in the diagram and indicate the position of the Fermi level (assume the chain propagates in the z direction). (b) Sketch a DOS plot. (c) Would you expect this chain to be a metallic conductor?

6.7 Consider an infinite 1D chain of H_2 molecules where the molecular axis is oriented perpendicular to the chain direction, as shown below. (a) How many bands are there in the band structure? (b) Sketch the crystal orbitals for each band at $k = 0$ and $k = \pi/a$. (c) Sketch out the band structure of this chain and include the position of the Fermi level. (d) Sketch a DOS plot. (e) Would you expect this chain to be a metallic conductor? Hint: Develop your answer using the bonding and antibonding MOs of the H_2 molecule.

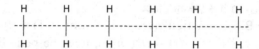

6.8 Consider the infinite 1D chain formed by placing boron atoms between the H_2 molecules from Problem 6.7 to form an infinite chain (shown below) where the B–H distance is 1.27 Å while B–B and H–H are both 1.8 Å. The calculated band structure for this chain is shown below. (a) Which of the boron orbitals can mix (form bonding/antibonding crystal orbitals) with the H_2 bonding orbital, ψ_+, at $k = 0$? Which boron orbital can mix with ψ_+ at $k = \pi/a$? (b) Which of the boron orbitals can mix with the H_2 antibonding orbital, ψ_-, at $k = 0$? Which boron orbital can mix with ψ_- at $k = \pi/a$? (c) What is the AO character of the two crystal orbitals that become degenerate at $k = \pi/a$? (d) At $k = 0$, only the lowest-energy band has both B–H and H–H σ-bonding character. What is the AO character of this band at $k = 0$? (e) At $k = 0$, the highest-energy band ($E = +33.8$ eV) has both B–H and H–H σ-antibonding character. What orbitals contribute to this band at $k = 0$?

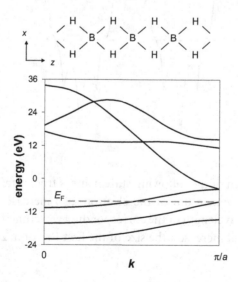

6.9 Consider the band structure for an infinite 1D chain of equally spaced titanium and oxygen atoms shown below, with an overall stoichiometry of TiO. In this calculation, only the Ti $3d$ orbitals and the O $2p$ orbitals have been taken into account. (a) Identify the orbital making the largest contribution to the band that spans the energy range from approximately −12 eV to −6 eV. (b) Identify the orbitals that make the largest contribution to the two flat bands located at roughly −11 eV. (c) Indicate the approximate location of the Fermi level. (d) Would you expect this chain to be a metal or a semiconductor?

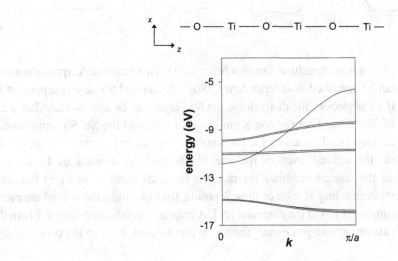

6.10 Consider the electronic structure of α-Po discussed in Section 6.6.1. (a) Sketch the crystal orbitals for the p_y band at Γ, X, and R. (b) Sketch the crystal orbitals for the p_x, p_y, and p_z bands at M. (c) Order the p_x, p_y, and p_z bands from lowest to highest energy at Γ, X, M, and R.

6.11 NbN crystallizes with the NaCl type structure. Sketch out an MO diagram for octahedrally coordinated niobium and use it to sketch a DOS plot for NbN. Indicate the relative area of the σ^*, π^*, and N $2p$ sets of bands, and mark the approximate position of the Fermi level in your DOS sketch. Would you expect NbN to be a metal or a semiconductor?

6.12 GaAs (E_g = 1.4 eV) and ZnSe (E_g = 2.6 eV) are isoelectronic with Ge (Section 6.6.3). Their band structures are shown below. Are they direct- or indirect-gap semiconductors?

GaAs   ZnSe

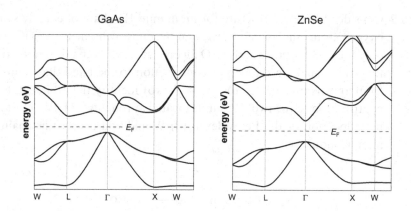

6.13 Consider LaCrSb$_3$, whose structure is shown below (left). To a reasonable approximation, this structure can be described as independent CrSb$_2^{2-}$ layers and Sb$^-$ layers separated by La^{3+} ions [3]. If we neglect subtle distortions, the Sb$^-$ layer can be approximated as a 2D square lattice of Sb$^-$ ions. There is one atom per unit cell, and the Sb–Sb separation is 3.1 Å. The calculated band structure for the idealized Sb$^-$ layer is shown below, on the right. (a) Show the orbital overlap for the Sb $5s$ and $5p$ orbitals at Γ, X, and M. Characterize the nearest-neighbor interactions for each band as (σ or π) bonding, antibonding, or nonbonding at each of these k points. (b) Determine the orbital character of each band, numbered 1–4 in the diagram. (c) LaCrSb$_3$ is a metallic conductor. From the band structure above, would you expect the Sb$^-$ layers to contribute to the conductivity?

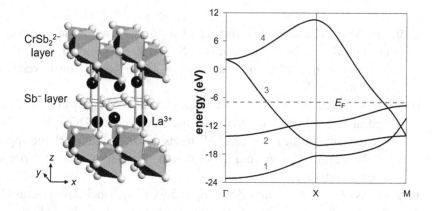

6.14 A 2D square lattice of $3d$ transition-metal atoms 2.3 Å apart has the band structure shown below (the contributions of the $4s$ and $4p$ orbitals have been omitted to simplify the analysis). (a) Considering orbital overlap for each of the five $3d$ orbitals at Γ, X, and M, characterize the nearest-neighbor interactions for each band as (σ, π, or δ) bonding, antibonding, or nonbonding at each of these k points. (b) Use your answers from part (a)

to associate bands numbered 1–3 in the diagram above with a d_{xy}, d_{xz}, or d_{yz} orbital (due to similar behavior at Γ, X, and M, bands 4–5 have contributions from both $d_{x^2-y^2}$ and d_{z^2}). (c) Which d-electron count will provide the strongest metal–metal bonding: d^1, d^4, or d^8?

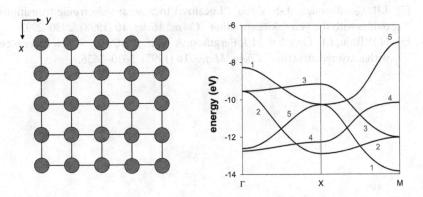

6.15 The perovskite $LaRhO_3$ is a semiconductor. (a) What is the orbital character of the valence band? (b) What is the orbital character of the conduction band? (c) Will a hypothetical cubic $LaRhO_3$ be a direct- or indirect-gap semiconductor?

6.16 $BaZrO_3$ and $CaZrO_3$ are perovskites with similar Zr–O distances. While $BaZrO_3$ is cubic, $CaZrO_3$ is orthorhombically distorted by octahedral tilting. Which of these two compounds will have a larger band gap? Explain your reasoning. Hint: The Zr–O–Zr angle is 180° in $BaZrO_3$ and 146° (on average) in $CaZrO_3$.

6.8 Further Reading

J.K. Burdett, "*Chemical Bonding in Solids*" (1995) Oxford University Press.

P.A. Cox, "*The Electronic Structure and Chemistry of Solids*" (1987) Oxford University Press.

R. Dronskowski, "*Computational Chemistry of Solid State Materials*" (2005) Wiley–VCH, Weinheim.

R. Hoffmann, "*Solids and Surfaces: A Chemist's View of Bonding in Extended Structures*" (1989) Wiley, New York.

E. Kaxiras, "*Atomic and Electronic Structure of Solids*" (2003) Cambridge University Press.

T. Wolfram, S. Ellialtioglu, "*Electronic and Optical Properties of d-Band Perovskites*" (2006) Cambridge University Press.

6.9   References

[1] K.S. Novoselov, A.K. Geim, S.V. Morozov, D. Jiang, Y. Zhang, S.V. Dubonos, V. Grigorieva, A.A. Firsov, "Electric field effect in atomically thin carbon films" *Science* **306** (2004), 666–669.

[2] J.B. Goodenough, J.-S. Zhou, "Localized to itinerant electronic transitions in transition-metal oxides with the perovskite structure" *Chem. Mater.* **10** (1998), 2980–2993.

[3] N.P. Raju, J.E. Greedan, M.J. Ferguson, A. Mar, "LaCrSb$_3$: A new itinerant electron ferromagnet with a layered structure" *Chem. Mater.* **10** (1998), 3630–3635.

7 Optical Materials

We transition into the second part of the book with an in-depth look at materials used for their optical properties. In the first half of the chapter, we consider materials that are valued for the way they absorb light: pigments, dyes, and gemstones. In the second half, we turn our attention to materials that emit light. Materials such as phosphors and light-emitting diodes play a key role in devices that we encounter in our daily lives, including fluorescent and solid state lighting.

7.1 Light, Color, and Electronic Excitations

All forms of electromagnetic radiation travel as self-propagating waves, moving through vacuum at a speed of $c = 2.99792458 \times 10^8$ m/s. The wavelength, λ, and frequency, v, of the wave are related to its speed through the relationship:

$$c = \lambda v \tag{7.1}$$

Electromagnetic radiation can also be treated as a particle called a **photon**. The energy of each photon is determined by its frequency (or wavelength) through Planck's equation:

$$E = hv = hc/\lambda \tag{7.2}$$

where h is the Planck constant of $6.62607015 \times 10^{-34}$ J s for E in joules of the SI system. The photon energy can also be expressed in electron volts (eV), which is the amount of potential energy gained by moving an electron across an electric-potential difference of 1 V (1 eV = $1.602176634 \times 10^{-19}$ J). Instead of wavelength, a spectroscopy unit called wavenumber can be used (in units cm^{-1}, the number of wavelengths that fit in one centimeter). A photon of energy 1 eV represents radiation with a wavenumber of 8065.544 cm^{-1}. As shown in Figure 7.1, visible light makes up only a small slice of the electromagnetic spectrum.

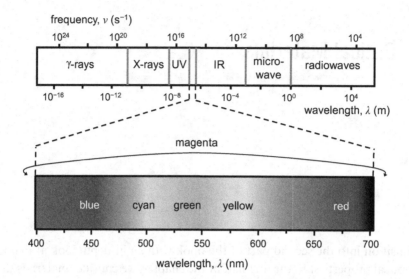

Figure 7.1 The electromagnetic spectrum (upper), and the visible portion of the spectrum (lower), showing the regions corresponding to the primary (RGB) and secondary (CMY) colors.

Colors are perceptions of the human eye. We have three different color receptors with maximum sensitivity at three different wavelengths in the visible spectrum. To understand colors, it is therefore useful to choose three colors of the rainbow as the primary colors of light: red, green, and blue (RGB); and then define three (secondary) colors that arise by mixing neighboring primary colors, as shown in the lower half of Figure 7.1. Mixing blue and green light yields cyan (C), while red plus green gives yellow (Y). The combination of blue plus red we see as a magenta (M). We could of course name many more colors than these six. For example, violet contains more blue light than magenta (i.e. an unequal mixing of blue and red), and orange is more reddish than yellow. Unlike the other primary and secondary colors, magenta cannot be represented by a narrow slice of the visible spectrum; it results from mixing red and blue. Hence, we perceive it as being more reddish than the violet end of the spectrum.

Now let's think about the colors that arise when substances absorb light. When a substance absorbs blue light, both red and green are either transmitted or reflected, and the substance appears yellow to us (remaining R + G = Y, termed the complementary color to B). If a substance absorbs red light, we perceive it as having a cyan color (remaining B + G = C, complementary to R), and one that absorbs green takes on a magenta color (remaining B + R = M, complementary to G). For example, a Cu^{2+}(aq) solution absorbs red light which leads to its familiar cyan color. A substance that absorbs all three RGB primary colors is black, whereas a light source that mixes the three primary colors creates white light. This is why screens of electronic devices are based on pixels, each containing sources that can emit RGB colors, while printers use the CMYK mixing of dyes. The K stands for blacK, formed by mixing C + M + Y of highest saturation. Consider the color of a C + M dye. The C dye absorbs R, while the M dye absorbs G, and the remaining color is B.

At the atomic level, the colors of chemical compounds arise from absorption of a visible light of certain energy hv that leads to an electronic excitation.[1] Such absorptions can be broadly grouped into five categories: (a) d-to-d transitions; (b) charge-transfer transitions; (c) band-to-band transitions in semiconductors; (d) transitions between molecular orbitals (MOs), most commonly π-to-π^* transitions in conjugated organic molecules; (e) f-to-f; and (f) f-to-d transitions.[2] In the following sections, we explore examples of the first four causes of color. Transitions involving f orbitals are explored in Section 7.8 because of their importance in luminescent materials.

7.2 Pigments, Dyes, and Gemstones

A **pigment** is a colored material that is dispersed in a medium in which it is insoluble.[3] The medium might be any number of substances ranging from oil to water to plastic. If a colored material is soluble in the medium in which it is dispersed it is called a **dye**.

Pigments are among the oldest functional materials. The prehistoric artist's palette was confined largely to red, yellow, brown, and black. The reds, yellows, and browns came from iron containing minerals such as Fe_2O_3 (hematite or red ochre) and $FeOOH$ (goethite or yellow ochre), while manganese oxides and carbon were used for black. Reliance on colored materials obtained from natural sources continued until the eighteenth century, when synthetic pigments started to become widely available. A representative list of pigments, dyes, and gemstones, arranged by color, is given in Table 7.1. Minerals that are colored when chemically pure, like eskolaite Cr_2O_3 or malachite $Cu_2CO_3(OH)_2$, are said to be **idiochromatic**. Minerals that are colored due to the presence of dopants, like ruby (Al_2O_3:Cr^{3+}) and spinel ($MgAl_2O_4$:Cr^{3+}) are said to be **allochromatic**. This notation, where the dopant responsible for the color follows a colon after the formula of the undoped host, will be used throughout this chapter. In some instances, the site on which substitutional disorder occurs is obvious. For example, in ruby and spinel the Cr^{3+} substitutes for Al^{3+} as $Al_{2-x}Cr_xO_3$ and $MgAl_{2-x}Cr_xO_4$, respectively. In other cases, such as at low doping levels of aliovalent substitutions, the distribution may be more complex or uncertain.

---

[1] In most substances, the electron's absorbed optical energy is subsequently released in several vibrational steps as thermal energy. Photoluminescent materials are an exception.

[2] One category of electronic transition that doesn't fit neatly into any of these categories are the F centers or color centers discussed in Chapter 2. The violet color of the mineral fluorite, CaF_2, arises from F centers (from *Farbe*, the German word for color).

[3] Pigments can also be black or white.

Table 7.1 Representative pigments, dyes, and gemstones.

Name	Composition	Use	Electronic excitation
Blues			
Azurite	$Cu_3(CO_3)_2(OH)_2$	Pigment	*d*-to-*d* (Cu^{2+})
Cobalt blue	$CoAl_2O_4$	Pigment	*d*-to-*d* (Co^{2+})
Egyptian blue	$CaCuSi_4O_{10}$	Pigment	*d*-to-*d* (Cu^{2+})
Indigo	$C_{15}H_9N_2O$	Dye	MO (π-to-π^*)
Phthalocyanine blue	$Cu(C_{32}N_8H_{16})$	Pigment	MO (π-to-π^*)
Prussian blue	$Fe_4[Fe(CN)_6]_3 \cdot xH_2O$	Pigment	MMCT $Fe^{2+} \rightarrow Fe^{3+}$
Sapphire	$Al_2O_3{:}Fe^{2+}, Ti^{4+}$	Gemstone	MMCT $Fe^{2+} \rightarrow Ti^{4+}$
Ultramarine	$Na_{8-x}[(Si,Al)_{12}O_{24}](S_3,Cl)_{1-2}$	Pigment	MO (S_3^-)
Greens			
Chrome green	Cr_2O_3	Pigment	*d*-to-*d* (Cr^{3+})
Emerald	$Be_3Al_2(SiO_3)_6{:}Cr^{3+}$	Gemstone	*d*-to-*d* (Cr^{3+})
Phthalocyanine green	$Cu(C_{32}N_8Cl_{16-x}H_x)$	Pigment	MO (π-to-π^*)
Malachite	$Cu_2CO_3(OH)_2$	Pigment	*d*-to-*d* (Cu^{2+})
Yellows			
Bismuth vanadate	$BiVO_4$	Pigment	LMCT $O^{2-} \rightarrow V^{5+}$
Cadmium yellow	CdS	Pigment	Band-to-band
Chrome yellow	$PbCrO_4$	Pigment	LMCT $O^{2-} \rightarrow Cr^{6+}$
Orpiment	As_2S_3	Pigment	Band-to-band
Reds			
Alizarin	$C_{14}H_8O_4$	Dye	MO (π-to-π^*)
Cadmium red	$CdS_{1-x}Se_x$	Pigment	Band-to-band
Pyrope	$Mg_3Al_2Si_3O_{12}{:}Cr^{3+}$	Gemstone	*d*-to-*d* (Cr^{3+})
Red lead	Pb_3O_4	Pigment	Band-to-band
Ruby	$Al_2O_3{:}Cr^{3+}$	Gemstone	*d*-to-*d* (Cr^{3+})
Spinel	$MgAl_2O_4{:}Cr^{3+}$	Gemstone	*d*-to-*d* (Cr^{3+})
Vermillion	HgS	Pigment	Band-to-band

7.3   Transitions between *d* Orbitals (*d*-to-*d* Excitations)

Transition-metal compounds make up the largest family of colored substances. In most cases, their color arises from electronic transitions between different *d* orbitals on the same transition-metal ion. In this section, we explore the factors that determine the energy, number, and intensity of the electronic absorption peaks arising from *d*-to-*d* transitions.

7.3.1   Ligand- and Crystal-Field Theory

Understanding how the energetic degeneracy of the *d* orbitals is removed by the surrounding ligands is the first step in understanding the optical properties of transition-metal compounds. **Ligand-field theory** relies on MO calculations (Chapter 5) to establish the *d*-orbital energies. For example, in the MO diagram of an octahedral complex, the $d_z{}^2$ and $d_{x^2-y^2}$ orbitals form σ^* interactions with the ligands, and hence these orbitals are more antibonding than the d_{xy}, d_{xz}, and d_{yz} orbitals that form π^* interactions with the ligands (Section 5.3.8). The σ^* and π^* MOs are usually referred to as e_g and t_{2g} orbitals, respectively, after their symmetry labels. The energy separation between these two sets of orbitals is the **ligand-field splitting**, Δ_{oct}, where the subscript "oct" denotes octahedral coordination.

 Crystal-field theory is a heuristic approach that qualitatively predicts splitting of the *d*-orbital energies due to electrostatic repulsion from the ligands. For example, in an octahedral field, the $d_z{}^2$ and $d_{x^2-y^2}$ orbitals point directly at the ligands and the electrons in them experience more repulsion than those in the d_{xy}, d_{xz}, and d_{yz} orbitals that point between the ligands and thus lie at a lower energy. Both crystal-field and ligand-field theory predict the same patterns of *d*-orbital splitting, which are given for various ligand environments in Figure 7.2. To make quantitative predictions of orbital energies, ligand-field theory is needed, but for deducing the gross features of *d*-orbital splitting in a variety of coordination environments, the simpler crystal-field approach is useful.

 Values of Δ_{oct} for various combinations of transition metal ions and ligands are given in Table 7.2. Examination of these values reveals several trends. Firstly, we see that, for a given *d*-electron count, Δ_{oct} increases as the oxidation state of the metal cation increases, as illustrated by the complex ions containing transition metals with a d^3 configuration: ([Cr(H$_2$O)$_6$]$^{3+}$ > [V(H$_2$O)$_6$]$^{2+}$ and [MnF$_6$]$^{2-}$ > [CrF$_6$]$^{3-}$). This trend results from the trend toward

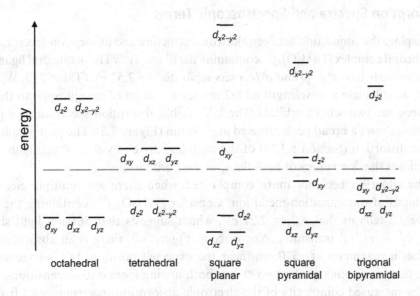

Figure 7.2 Examples of ligand-field splitting diagrams for various coordination environments. The dashed line is the energy of the *d* orbitals in a spherically symmetric environment.

Table 7.2 Ligand-field splitting, Δ_{oct}, for complex ions of various transition metals.

Metal ion	d count	$[MF_6]^{n-}$	$[M(H_2O)_6]^{m+}$
Ti^{3+}	d^1	2.17 eV	2.52 eV
V^{3+}	d^2	2.00 eV	2.21 eV
V^{2+}	d^3	Not formed	1.54 eV
Cr^{3+}	d^3	1.87 eV	2.16 eV
Mn^{4+}	d^3	2.70 eV	Not Formed
Tc^{4+}	d^3	3.52 eV	Not Formed
Re^{4+}	d^3	4.07 eV	Not Formed

shorter metal–ligand bonds as the oxidation state of the metal ion increases. The enhanced overlap raises the energy of the σ^* orbitals more than π^* orbitals, thereby increasing Δ_{oct}. Secondly, we see that Δ_{oct} increases as the valence orbitals on the central metal ion go from $3d$ to $4d$ to $5d$ ($Mn^{4+} \rightarrow Tc^{4+} \rightarrow Re^{4+}$). The larger $4d$ and $5d$ orbitals (Section 5.2.2) experience a stronger overlap with the ligand orbitals, which again leads to greater destabilization of the σ^* orbitals than the π^* orbitals. Finally, for a given ion, Δ_{oct} also depends upon the ligand. For example, the ligand-field splitting increases slightly when fluoride ions are replaced with water molecules.

7.3.2 Absorption Spectra and Spectroscopic Terms

To explore the connection between electronic structure and absorption spectra, consider the octahedral complex $[Ti(H_2O)_6]^{3+}$ containing the d^1 ion Ti^{3+}. The octahedral ligand field splits the d orbitals into $\pi^*(t_{2g})$ and $\sigma^*(e_g)$ sets separated by 2.52 eV (Table 7.2). With Equation (7.2), we calculate a wavelength of 492 nm for a photon of energy equal to the separation between the two sets of orbitals. The UV–visible absorption spectrum of a $[Ti(H_2O)_6]^{3+}$ solution shows a broad peak centered at ~490 nm (Figure 7.3). The peak is split by a Jahn–Teller distortion (Section 5.3.10) of the complex, driven by the $3d^1$ configuration of Ti^{3+}, which lifts the degeneracy of both the e_g and t_{2g} orbitals.

The situation becomes more complicated when there are multiple electrons in the d orbitals of the transition-metal ion. Consider $[V(H_2O)_6]^{3+}$, containing the d^2 ion V^{3+}. Table 7.2 tells us that $\Delta_{oct} = 2.21$ eV, which suggests that 561 nm light should excite a $t_{2g}^2 e_g^0 \rightarrow t_{2g}^1 e_g^1$ transition. As we see in Figure 7.3, there is an absorption peak that reaches a maximum at ~570 nm, near the expected position, but, somewhat unexpectedly, there is another peak at ~390 nm, both arising from d-to-d transitions. The reason for the increased complexity of the electronic absorption spectrum stems from electron–electron interactions.

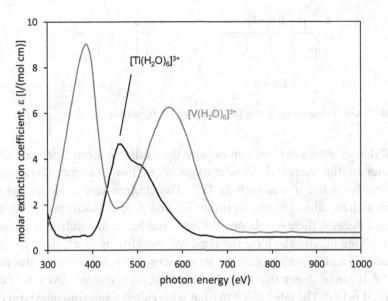

Figure 7.3 The UV–visible absorption spectrum of aqueous solutions of $[Ti(H_2O)_6]^{3+}$ (black) and $[V(H_2O)_6]^{3+}$ (gray).

To understand the $[V(H_2O)_6]^{3+}$ spectrum, we must take a step back and consider the electron–electron interactions in a free V^{3+} ion. Its electron configuration, [Ar] $3d^2$, only gives information on the first two quantum numbers, n and ℓ (Section 5.2). It neither tells us whether the electrons occupy the same *d*-orbital or different *d*-orbitals, nor whether they have the same or opposite spins. Yet these factors have a bearing on the collective energy of the electrons. Coupling between the spin and orbital angular momenta of the electrons also plays a role. To understand absorption spectra of transition-metal ions, we must take these factors into account.

There are two approaches to incorporate electron–electron interactions and spin–orbit coupling into the analysis. Within the **Russell–Saunders coupling** scheme, the spin and orbital momenta are summed separately and then combined. This approach assumes electron–electron repulsions are dominant and treats spin–orbit coupling as a perturbation. Such an approximation is reasonable when the spin–orbit coupling is relatively weak, which is generally applicable for $3d$ transition metals like V^{3+}. The **j–j coupling** scheme represents the opposite approach where spin–orbit coupling is first applied to split the different orbital occupations into states, and then electron–electron repulsions act as a perturbation on those energy levels. For elements with intermediate levels of spin–orbit coupling, like the $4d$ and $5d$ transition-metal ions and the lanthanoids, the coupling is intermediate between these two limiting cases. We will adopt Russell–Saunders coupling throughout this chapter, but the interested reader can find further details on j–j coupling in Appendix F.

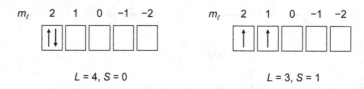

Figure 7.4 Two microstates for a free ion with a d^2 configuration.

The different ways electrons can occupy the available atomic orbitals (AOs) are called **microstates** of the parent electron configuration. For example, there are ten available microstates for a free d^1 ion such as Ti^{3+}. The electron can reside in any one of the five d orbitals with m_ℓ values ranging from 2 to −2, and its spin quantum number m_s can be either $+\frac{1}{2}$ or $-\frac{1}{2}$. Because there is only one electron, and hence no electron–electron repulsions to consider, all ten microstates have the same energy. Moving to d^2, there are now 45 possible microstates, not all of which lie at the same energy. Two such microstates are illustrated in Figure 7.4. It can be shown that these 45 microstates separate into five sets of different energy as discussed below. The label given to each set is called a **spectroscopic term** or **term symbol**.

To derive the term symbols for a d^2 ion, we begin by identifying the values of the total angular-momentum quantum number L and the total spin quantum number S that can arise from the orbital and spin angular momenta of the individual electrons. There are two electrons with spin angular-momentum quantum numbers s_1 and s_2 and orbital angular-momentum quantum numbers ℓ_1 and ℓ_2. According to the **Clebsch–Gordon series**, we find that S and L can take the following values:

$$S = s_1 + s_2, s_1 + s_2 - 1, \ldots, |s_1 - s_2| \tag{7.3}$$

$$L = \ell_1 + \ell_2, \ell_1 + \ell_2 - 1, \ldots, |\ell_1 - \ell_2| \tag{7.4}$$

For a d^2 ion ($\ell_1 = 2$, $s_1 = \frac{1}{2}$; $\ell_2 = 2$, $s_2 = \frac{1}{2}$), the spin angular-momentum quantum number S can have values of 1 and 0, while the orbital angular-momentum quantum number L can have values of 4, 3, 2, 1, and 0. The total orbital angular momentum of a spectroscopic term is denoted by the uppercase letters given in Table 7.3. The total spin is normally reported as the value $2S + 1$, which is called the **multiplicity** of the term. The different values of multiplicity are given the names singlet ($S = 0$ with $2S + 1 = 1$), doublet ($S = \frac{1}{2}$ with $2S + 1 = 2$), triplet ($S = 1$ with $2S + 1 = 3$), and so on.

Table 7.3 Term-symbol notation for the total orbital angular-momentum quantum number L.

$L =$	0	1	2	3	4	5	6	...
	S	P	D	F	G	H	I	..

For a d^2 ion, the 45 microstates divide into the following five sets: nine microstates make up the ^1G ($L = 4$, $S = 0$) term, 21 belong to ^3F ($L = 3$, $S = 1$), five to ^1D ($L = 2$, $S = 0$), nine to ^3P ($L = 1$, $S = 1$), and one to ^1S ($L = 0$, $S = 0$).[4] The electron spins are parallel for the microstates associated with the triplet terms, ^3F and ^3P, and antiparallel for the microstates associated with the singlet terms, ^1G, ^1D, and ^1S. Within the Russell–Saunders scheme, the **total angular-momentum quantum number** J can take values of $J = L + S, L + S - 1, |L - S|$. For the ^1G term, $L = 4$ and $S = 0$, which means there is only a single value of $J = 4$, while for the ^3F term, $L = 3$ and $S = 1$ give $J = 4, 3$, and 2.

Once the allowable values of L and S are known, it's possible to identify the lowest-energy term using Hund's rules. **Hund's first rule** tells us to maximize S as this minimizes electron pairing that costs energy. This rule tells us that the triplet states, ^3F and ^3P, should be lower in energy than the singlet states, ^1G, ^1D, and ^1S. **Hund's second rule** is to maximize L. This rule tells us that the ^3F state is lower in energy than the ^3P state. **Hund's third rule** states that for orbitals which are less than half filled, the ground state J is the smallest possible sum of L and S: $|L - S|$; whereas for orbitals more than half filled, it is the largest possible sum: $|L + S|$. The basis of this rule is in the spin–orbit coupling itself, and it may be violated when the coupling is weak. For the ^3F state, the lowest-energy term has $J = |L - S| = 2$, and the term symbol is $^{2S+1}L_J = {}^3F_2$ ("triplet-F-two"). The ground-state terms for free transition-metal ions with partially filled d-orbitals as predicted by Hund's rules are given in Figure 7.5.

The splitting of the five different sets of degenerate microstates for the V^{3+} ion discussed above is illustrated on a phenomenological (not to scale) energy diagram in Figure 7.6. Note that Hund's rules reliably give the ground-state (lowest-energy) term, but among the excited-state terms the experimentally observed order does not necessarily match the order they predict.

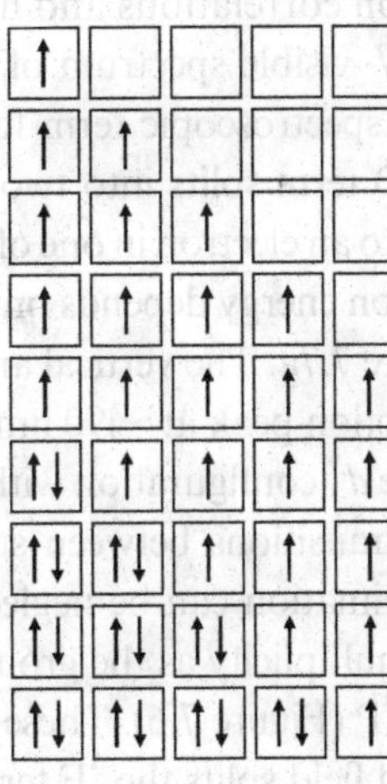

m_ℓ	2	1	0	−1	−2	d^n	L	S	J	term
	↑					d^1	2	½	3/2	$^2D_{3/2}$
	↑	↑				d^2	3	1	2	$^3F_2$
	↑	↑	↑			d^3	3	3/2	3/2	$^4F_{3/2}$
	↑	↑	↑	↑		d^4	2	2	0	$^5D_0$
	↑	↑	↑	↑	↑	d^5	0	5/2	5/2	$^6S_{5/2}$
	↑↓	↑	↑	↑	↑	d^6	2	2	4	$^5D_4$
	↑↓	↑↓	↑	↑	↑	d^7	3	3/2	9/2	$^4F_{9/2}$
	↑↓	↑↓	↑↓	↑	↑	d^8	3	1	4	$^3F_4$
	↑↓	↑↓	↑↓	↑↓	↑	d^9	2	½	5/2	$^2D_{5/2}$

Figure 7.5 The ground-state term symbols for free d^n ions with $n = 1$–9. A representative microstate for each term symbol is shown on the left.

[4] The number of microstates in each term is equal to $(2L + 1)(2S + 1)$.

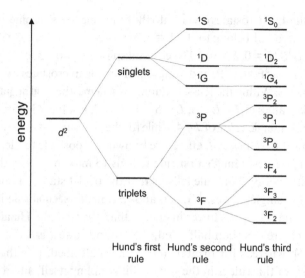

Figure 7.6 Interactions that split the degeneracy of a free-ion d^2 configuration (energies not to scale).

7.3.3 Correlation Diagrams

The relative energies of the spectroscopic terms developed in the preceding section correspond to the free-ion case where all five d orbitals are degenerate. Once ligands are introduced, the individual d orbitals no longer have the same energy, as discussed in Section 7.3.1. Consequently, the free-ion spectroscopic terms discussed above are further split by ligand-field effects, as shown for the octahedral case in Table 7.4. The labels found in the bottom row describe the effects of both electron–electron correlations and ligand-field splitting.

We are now ready to properly explain the UV–visible spectrum of $[Ti(H_2O)_6]^{3+}$ (assuming perfect octahedral symmetry). There is only one spectroscopic term for a d^1 ion, the ^2D term (Figure 7.5).[5] Table 7.4 tells us that a free-ion ^2D term splits into two terms, $^2T_{2g}$ and ^2E_g, in octahedral coordination. These terms correspond to an electron in one of the t_{2g} orbitals or one of the e_g orbitals, respectively. The $T_{2g} \rightarrow E_g$ transition energy depends on the ligand-field splitting Δ_{oct}, as shown in the **correlation diagram**[6] in Figure 7.7a. The vertical arrow in this figure shows the electronic transition responsible for the absorption peak at ~490 nm in Figure 7.3.

The situation becomes more complicated for the d^2 configuration with a free-ion ground-state term of ^3F. As we will see in the next section, transitions between states with different spin multiplicities are very weak and to a first approximation can be neglected. Therefore, we will only consider the terms that have the same spin multiplicity as the ground state. In the d^2 case, the only excited-state term that is also a triplet is ^3P (Figure 7.6). These two states are shown in Figure 7.7b. Introduction of the octahedral ligand field splits the ^3F term into terms $^3T_{1g}$, $^3T_{2g}$, and $^3A_{2g}$, each with a different occupation of the orbitals that are split by the ligand field. The

---

[5] We ignore splitting due to spin–orbit coupling for the sake of simplicity.
[6] Correlation diagrams of this sort are also called Orgel diagrams.

Table 7.4 The splitting of spectroscopic terms of a free ion upon the introduction of an octahedral ligand field.

Free-ion term	S	P	D	F	G
Terms in octahedral field	A_{1g}	T_{1g}	$T_{2g} + E_g$	$T_{1g} + T_{2g} + A_{2g}$	$A_{1g} + E_g + T_{1g} + T_{2g}$

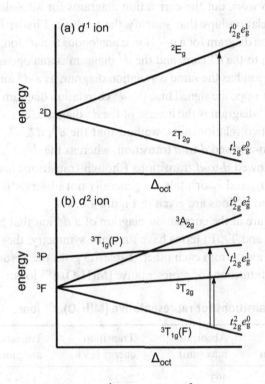

Figure 7.7 Correlation diagram for (a) a d^1 ion and (b) a d^2 ion in octahedral coordination as a function of Δ_{oct}. The vertical arrows show the transitions responsible for the absorption peaks in Figure 7.3. The high-spin d^6 and d^7 correlation diagrams are analogous to the d^1 and d^2 diagrams, respectively, but with different values of S.

$^3T_{1g}$ term corresponds to a configuration with two electrons in the t_{2g} orbitals; the $^3T_{2g}$ term corresponds to one electron in the t_{2g} orbitals and one in the e_g orbitals; and the $^3A_{2g}$ term corresponds to two electrons in the e_g orbitals. The ^3P free-ion term becomes a $^3T_{1g}$ term in an octahedral environment (electron configuration $t_{2g}{}^1 e_g{}^1$) but is not split by the octahedral crystal field. These features are captured in the correlation diagram shown in Figure 7.7b. More comprehensive diagrams called Tanabe–Sugano diagrams must be used to understand transitions between states with different spin multiplicities. We neglect such transitions for now but will return to them in Section 7.8.5 when we consider the luminescence of d^5 ions like Mn^{2+}.

Returning to the spectrum of $[V(H_2O)_6]^{3+}$ (Figure 7.3), we can now assign the absorption peak at ~570 nm to the $^3T_{1g}(F) \rightarrow {}^3T_{2g}(F)$ transition, while the absorption at 390 nm arises from the $^3T_{1g}(F) \rightarrow {}^3T_{1g}(P)$ transition. Both are marked with vertical arrows in Figure 7.7b. The $^3T_{1g}(F) \rightarrow {}^3A_{2g}(F)$ transition not only requires UV photons of an even higher energy, it involves simultaneous excitation of two electrons from the t_{2g} to e_g orbitals and would therefore be exceedingly weak. Experimentally, it is not observed.

It is laborious to work out the correlation diagrams for all d-electron counts. Fortunately, there are several relationships that simplify the analysis. Firstly, for high-spin ions (Section 5.3.9), the correlation diagram for a d^{n+5} ion is analogous to a d^n ion. Thus, the d^6 configuration diagram is identical to the d^1 case, and the d^7 diagram is analogous to d^2. Secondly, it can be shown that a d^{10-n} ion has the same correlation diagram as a d^n ion except that all energies of interaction have the opposite sign. Thus, the d^9 correlation diagram is just the inverse of the d^1 diagram, and the d^8 diagram is the inverse of the d^2 diagram, etc.

Applying these two relationships, we find that the d^1, d^4, d^6, and d^9 configurations all exhibit a single spin-allowed d-to-d transition, whereas the d^2, d^3, d^7, and d^8 configurations have three spin-allowed d-to-d transitions (though transitions involving promotion of two electrons between t_{2g} and e_g orbitals are generally not observed). The correlation diagrams for high-spin d^3 and d^4 ions are given in Figure 7.8.

There is one feature in the correlation diagram of a d^3 ion that has not yet been explained. Because the $T_{1g}(F)$ and $T_{1g}(P)$ terms have the same symmetry, they mix as Δ_{oct} increases, and their energies bend away from each other. Table 7.5 gives the absorption maxima and corresponding electronic transition for representative $[M(H_2O)_6]^{n+}$ ions.

Table 7.5 d-to-d transitions for representative $[M(H_2O)_6]^{n+}$ ions.

	Complex ion	Absorption max (nm)	Transition energy (eV)	Transition assignment*
d^1	$[Ti(H_2O)_6]^{3+}$	492	2.52	$^2T_{2g} \rightarrow {}^2E_g$
d^2	$[V(H_2O)_6]^{3+}$	561	2.21	$^3T_{1g} \rightarrow {}^3T_{2g}$
		389	3.19	$^3T_{1g} \rightarrow {}^3T_{1g}(P)$
d^3	$[Cr(H_2O)_6]^{3+}$	575	2.16	$^4A_{2g} \rightarrow {}^4T_{2g}$
		407	3.05	$^4A_{2g} \rightarrow {}^4T_{1g}(F)$
		265	4.69	$^4A_{2g} \rightarrow {}^4T_{1g}(P)$
d^4	$[Mn(H_2O)_6]^{3+}$	476	2.60	$^5E_g \rightarrow {}^5T_{2g}$
d^6	$[Fe(H_2O)_6]^{2+}$	962	1.29	$^5T_{2g} \rightarrow {}^5E_g$
d^7	$[Co(H_2O)_6]^{2+}$	1230	1.00	$^4T_{1g} \rightarrow {}^4T_{2g}$
		515	2.40	$^4T_{1g} \rightarrow {}^4T_{1g}(P)$
d^8	$[Ni(H_2O)_6]^{2+}$	1180	1.05	$^3A_{2g} \rightarrow {}^3T_{2g}$
		725	1.71	$^3A_{2g} \rightarrow {}^3T_{1g}(F)$
		395	3.14	$^3A_{2g} \rightarrow {}^3T_{1g}(P)$
d^9	$[Cu(H_2O)_6]^{2+}$	794	1.56	$^2E_g \rightarrow {}^2T_{2g}$

*Assuming octahedral coordination. Jahn–Teller distortions are neglected.

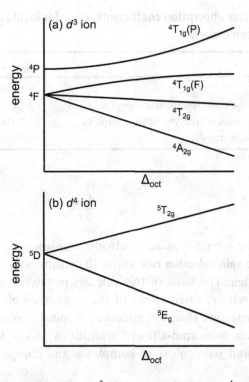

Figure 7.8 Correlation diagram for (a) a d^3 ion and (b) a high-spin d^4 ion in an octahedral environment. The d^8 and d^9 correlation diagrams are analogous to the d^3 and d^4 diagrams, respectively, but with different values of *S*.

Correlation diagrams for tetrahedral coordination can be generated by using the diagram for an octahedrally coordinated d^n ion for a tetrahedrally coordinated d^{10-n} ion. Consequently, the d^1 octahedral diagram is equivalent to the d^9 tetrahedral diagram and vice versa. The d^2 octahedral diagram is equivalent to the d^8 tetrahedral diagram and so on.

7.3.4   Selection Rules and Absorption Intensity

The intensity of an optical absorption is an important factor for applications. The strength with which an electronic transition absorbs light is expressed by its molar extinction coefficient, ε_{max} [in L/(mol cm)],[7] composed of the absorption maximum A_{max} (dimensionless), the concentration *c* of the solution (in mol/dm^3 = mol/L), and the path length of the light, *l* (in cm), according to Beer's law:

---

[7] The molar extinction coefficient is also referred to as molar absorptivity and molar attenuation coefficient, which is the term preferred by IUPAC. Reduced to base SI units, L/(mol cm) is equivalent to m^2/mol.

Table 7.6 Typical molar absorption coefficients, ε_{max}, for localized transitions in transition-metal complexes.

Transition	ε_{max}, L/(mol cm)
Spin-forbidden	<1
Laporte-forbidden, d-to-d centrosymmetric complex	1–20
Laporte-allowed, d-to-d non-centrosymmetric complex	10–1000
Symmetry-allowed charge transfer	1000–50000

$$\varepsilon_{max} = \frac{A_{max}}{cl} \tag{7.5}$$

We can estimate the strength of an electronic transition by considering the following selection rules. The **spin selection rule** states that transitions between states of different spin, S, are forbidden. The basis of this rule lies in the fact that the incident radiation cannot change the relative orientations of the spins of an electron. Spin–orbit coupling can relax the spin selection rule, but nonetheless spin-forbidden transitions ($\Delta S \neq 0$) are always much weaker than spin-allowed transitions ($\Delta S = 0$). This explains the very weak absorptions and pale colors of complexes and compounds containing high-spin d^5 ions.[8]

The **Laporte selection rule** states that in a centrosymmetric environment the only allowed transitions are those between states that have a different parity with respect to the inversion operation.[9] Practically speaking, this means that transitions $s \to s, p \to p, d \to d$, and $f \to f$ are forbidden for ions in a centrosymmetric environment, whereas $s \to p$, $p \to d$, and $d \to f$ transitions are allowed. Although ML_6 octahedral complexes are centrosymmetric, weak d-to-d transitions are observed because vibrations temporarily remove the inversion center and relax the Laporte selection rule. However, these d-to-d transitions are much weaker than fully allowed transitions such as charge-transfer excitations. Approximate values of molar extinction coefficients for several types of transitions are given in Table 7.6. Revisiting Table 7.1, we see that few pigments derive their color from d-to-d transitions because these are relatively inefficient at absorbing light. The few exceptions have transition-metal ions in non-centrosymmetric environments, like tetrahedral and trigonal-bipyramidal geometries, where d-to-d transitions do not violate the Laporte selection rule. For gemstones, it is generally desirable that some light pass through the crystal, hence the light absorption need not be so efficient and d-to-d transitions are often the source of color.

---

[8] You may be wondering how Fe_2O_3 (red ochre) and FeOOH (yellow ochre) could be used as pigments given the d^5 configuration of Fe^{3+}. It is because the color comes from charge-transfer transitions (Section 7.4.1) rather than from d-to-d transitions.

[9] More precisely, transitions between states with a *gerade* (German for even), g, and an *ungerade* (German for odd), u, label are permitted, but g → g and u → u transitions are forbidden.

Box 7.1 Materials Spotlight: Blue pigments through the ages

The evolution of blue pigments provides an interesting illustration of the intersection between art, commerce, and science. Until the eighteenth century, ultramarine was the most highly prized blue pigment. During the Renaissance, it is estimated to have been five times more expensive than gold, which limited its use to artists with wealthy patrons. Michelangelo famously used large quantities of ultramarine in his depiction of the Last Judgment on the altar wall of the Sistine Chapel. Natural ultramarine is obtained by grinding the semi-precious stone, lapis lazuli, of which the dominant component is the mineral lazurite, an aluminosilicate with the sodalite structure (Section 1.5.5). The name ultramarine means "beyond the sea" because the pigment was imported to Europe from mines in what is now Afghanistan. Pure sodalite, which is colorless, has the composition $Na_4Al_3Si_3O_{12}Cl$. In lazurite, the Cl^- ions, which sit near the center of the sodalite cages (see Figure 1.57), are partially replaced by S_3^-, analogous to the ozonide anion O_3^-, together with smaller concentrations of S_2^- and S_4^-. Electronic transitions between MOs belonging to S_3^- give rise to the blue color, while contributions from S_2^- and S_4^- can shift the color towards yellow or red, respectively [1, 2]. The high cost of ultramarine made alternative blue pigments, such as the copper-containing mineral azurite, popular during the Renaissance.

In the early years of the eighteenth century, Prussian blue, $Fe_4[Fe(CN)_6]_3 \cdot xH_2O$, was discovered in Berlin. It can be found in paintings from the Prussian court that date back to 1710. An $Fe^{2+} \rightarrow Fe^{3+}$ charge-transfer excitation gives rise to the intense blue color (Section 7.4.2). Given the scarcity and cost of ultramarine, Prussian blue quickly gained popularity, becoming the first widely used synthetic pigment. A century later, manufacture of cobalt blue, $CoAl_2O_4$, was initiated. Cobalt blue is a normal spinel (Section 1.5.1) that gets its blue color from *d*-to-*d* transitions associated with tetrahedrally coordinated Co^{2+}. Cobalt blue and other synthetic pigments were instrumental to artists from the impressionist era, such as van Gogh, Monet, and Renoir. In modern times, copper pthalocyanine, $Cu(C_{32}N_8H_{16})$, has become the dominant blue pigment. Interestingly, the blue color comes largely from π-to-π^* transitions associated with the pthalocyanine ring rather than from Cu^{2+} *d*-to-*d* transitions.

In 2009, researchers at Oregon State University discovered a new blue pigment with composition $YIn_{1-x}Mn_xO_3$, where the blue color comes from *d*-to-*d* transitions of isolated Mn^{3+} in a trigonal-bipyramidal coordination [3]. The spectra of $CoAl_2O_4$ and $YIn_{0.8}Mn_{0.2}O_3$ (shown in Figure B7.1.1) are similar in the visible range, resulting in a vibrant blue color due to absorbed R + G (note the figure shows reflectance spectra, the opposite of absorbance). In $YIn_{1-x}Mn_xO_3$, the trigonal-bipyramidal coordination splits the Mn^{3+} *d* orbitals into a 2 + 2 + 1 pattern. The visible absorption peak is due to a transition from the half-filled and doubly degenerate $d_{xy}/d_{x^2-y^2}$ orbitals into the empty d_{z^2} orbital. In $CoAl_2O_4$, visible light is absorbed by a $^4A_2 \rightarrow {}^4T_1(P)$ transition on the tetrahedrally coordinated Co^{2+}. Crucially, the transition metal in both materials sits on a site that lacks inversion symmetry, so the transitions are Laporte allowed.

Box 7.1 (cont.)

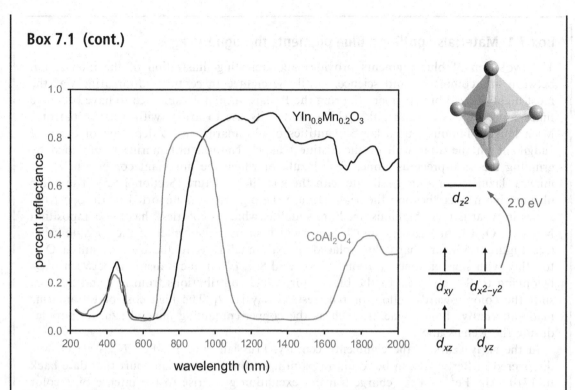

Figure B7.1.1 The reflectance spectra for $YIn_{0.8}Mn_{0.2}O_3$ and $CoAl_2O_4$ and the d-orbital energies and occupancies for the Mn^{3+} ion in a trigonal-bipyramidal coordination. The absorption (where the reflectance is minimal) between 500 and 700 nm in $YIn_{0.8}Mn_{0.2}O_3$ arises from an electronic transition from the $d_{xy}/d_{x^2-y^2}$ orbitals into the empty d_z^2 orbital. Data taken from [4].

In both materials, there are also d-to-d transitions that fall outside the visible range. In cobalt blue there is a broad $^4A_2 \rightarrow {}^4T_1(F)$ transition centered in the infrared (IR) near 1350 nm, while in $YIn_{0.8}Mn_{0.2}O_3$ the d_{xz}/d_{yz} to d_z^2 transition lies in the near UV. While the strong IR absorbance of $CoAl_2O_4$ has little impact on the color, it does lead to absorption of IR light, which can produce unwanted heating. The reduced IR absorbance in $YIn_{0.8}Mn_{0.2}O_3$ is advantageous for use on roofs and other exterior architectural applications. Unfortunately, the high cost of indium makes $YIn_{1-x}Mn_xO_3$ pigments costlier than conventional blue pigments.

7.4   Charge-Transfer Excitations

In a charge-transfer excitation an electron is transferred from an MO whose wavefunction is predominantly associated with a given atom (or group of like atoms) to an MO whose wavefunction is largely associated with a different atom (or group of like atoms). Transitions where an electron is excited from a ligand-based MO to a metal-based MO are called **ligand-to-metal charge transfer** (LMCT) transitions, while a **metal-to-ligand charge transfer** (MLCT) is in the opposite direction. A **metal-to-metal charge transfer** (MMCT) refers to the transfer of an electron from one metal center to a different metal center. Charge-transfer transitions are fully allowed, with molar extinction coefficients several orders of magnitude larger than d-to-d transitions (Table 7.6), an attribute that is ideal for a pigment because bold colors can be realized with relatively small amounts of pigment.

In this section, we take a closer look at LMCT and MMCT transitions. MLCT transitions that fall in the visible region of the spectrum generally require ligands with low-lying unoccupied π^* orbitals, such as bipyridine or phenanthroline. These transitions are rare among pigments and unknown for gemstones, and thus will not be considered here.

7.4.1 Ligand-to-Metal Charge Transfer

A variety of pigments rely on LMCT transitions for their color. Perhaps the best-known examples are salts of the chromate ion, CrO_4^{2-}, such as $PbCrO_4$ (chrome yellow). The MO diagram for a tetrahedral CrO_4^{2-} anion is shown together with the UV–visible absorption spectrum of its aqueous solution in Figure 7.9. A single absorption peak is centered at 375 nm. Even though it reaches a maximum in the UV, it tails into the visible region, absorbing the short-wavelength blue light. The remaining green and red light are transmitted producing the yellow color of chromate (G + R = Y).

The highest occupied molecular orbital (HOMO)-to-lowest unoccupied molecular orbital (LUMO) transition from the triply degenerate t_1 set of orbitals to the doubly

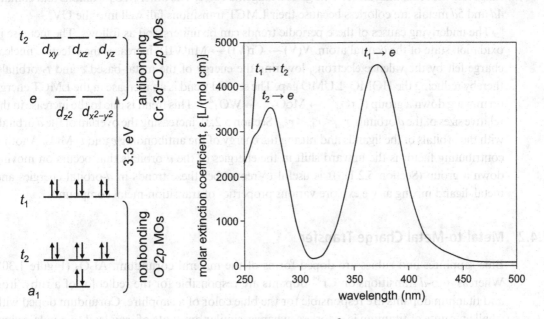

Figure 7.9 A portion of the MO diagram for tetrahedral CrO_4^{2-} (left) along with the UV–visible spectrum of a dilute (0.001 M) aqueous solution of Na_2CrO_4 (right).

Table 7.7 Calculated values of the LMCT gap between the t_1 nonbonding O $2p$ HOMO and the e antibonding LUMO for a series of tetrahedral $MO_4{}^{n-}$ species where M is a d^0 transition-metal ion [6].

Group 5	$t_1 \rightarrow e$ (eV)	Group 6	$t_1 \rightarrow e$ (eV)	Group 7	$t_1 \rightarrow e$ (eV)	Group 8	$t_1 \rightarrow e$ (eV)
$VO_4{}^{3-}$	4.5	$CrO_4{}^{2-}$	3.3	$MnO_4{}^{-}$	2.2		
		$MoO_4{}^{2-}$	5.3	$TcO_4{}^{-}$	4.3	RuO_4	3.1
		$WO_4{}^{2-}$	6.2	$ReO_4{}^{-}$	5.3	OsO_4	4.0

degenerate e set of orbitals is responsible for the 375 nm absorption peak. Recall from our previous discussion of the MO diagram of a tetrahedrally coordinated transition-metal ion (Section 5.3.8), that the unoccupied e orbitals are antibonding MOs with significant Cr $3d_{x^2-y^2}$ and $3d_{z^2}$ orbital character, whereas the t_1 orbitals are nonbonding O $2p$ orbitals. To a first approximation, the transition is a transfer of an electron from oxygen to chromium (the peak at 270 nm is a combination of transitions from lower-lying occupied O $2p$ t_2 MOs into the LUMO and transitions from the t_1 HOMO into the unoccupied Cr-based t_2 orbitals [5]).

The energy of the LMCT transition for a $3d$ tetrahedral $MO_4{}^{n-}$ anion decreases with increasing oxidation number of the transition metal, which lowers the e set of orbitals towards the nonbonding oxygen MOs, reducing the LMCT gap (Table 7.7). The $CrO_4{}^{2-}$ and $RuO_4{}^{2-}$ ions absorb blue and are thus yellow (G + R = Y), whereas $MnO_4{}^{-}$ absorbs strongly in the green range and the remaining B + R = M gives a magenta color. The other $MO_4{}^{n-}$ anions that contain $4d$ and $5d$ metals are colorless because their LMCT transitions fall well into the UV.

The underlying causes of these periodic trends can be understood as follows. The increase in oxidation state of the central atom, V(V) → Cr(VI) → Mn(VII), increases the effective nuclear charge felt by the valence electrons, lowering the energy of the metal-based e and t_2 orbitals, thereby reducing the HOMO–LUMO gap. The second trend is an increase in the LMCT energy on moving down a group ($CrO_4{}^{2-}$ → $MoO_4{}^{2-}$ → $WO_4{}^{2-}$). This trend is due to the increase in the relative sizes of the d orbitals, $r_{3d} < r_{4d} < r_{5d}$ (Section 5.2.3), increasing the overlap of the d orbitals with the orbitals of the ligands and raising the energy of the antibonding e and t_2 MOs. Another contributing factor is the upward shift in the energies of the d orbitals that occurs on moving down a group (Section 5.2.1). It is useful to remember these trends in d-orbital energies and metal–ligand mixing as we explore various properties of transition-metal compounds.

7.4.2 Metal-to-Metal Charge Transfer

Blue sapphires and rubies are doped forms of the mineral corundum, Al_2O_3 (Figure 1.30). Whereas d-to-d transitions on Cr^{3+} dopants are responsible for the red color of a ruby, iron and titanium dopants are responsible for the blue color of a sapphire. Corundum doped with small amounts of titanium is colorless, whereas similar amounts of iron lead to a pale-yellow color. When both are present, the result is the magnificent deep-blue color of a sapphire. In

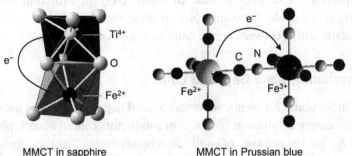

MMCT in sapphire   MMCT in Prussian blue

Figure 7.10 A localized view of the MMCT transitions in sapphire, Al_2O_3:Ti^{4+},Fe^{2+} (left), and Prussian blue, $(Fe^{3+})_4[(Fe^{2+})(CN)_6]_3 \cdot xH_2O$ ($x \approx 14$) (right).

sapphires, co-doping by equal amounts of Ti^{4+} and Fe^{2+} maintains charge balance. When they occupy adjacent six-coordinate sites in the corundum structure (Figure 7.10), it is possible for the absorbed photon to excite an $Fe^{2+} + Ti^{4+} \rightarrow Fe^{3+} + Ti^{3+}$ MMCT transition.

Because of the differences in molar extinction coefficients, doping levels in sapphires and rubies are different. At least 1% chromium must be present in corundum before the deep-ruby-red color appears, whereas the blue color of a sapphire is observed with titanium and iron concentrations as low as 0.01%. In fact, complete substitution leads to the mineral ilmenite, $FeTiO_3$ (Figure 5.28), where the charge transfer band is so intense that it absorbs across the visible spectrum giving $FeTiO_3$ its black color.

Prussian blue is another blue compound whose color originates from an MMCT excitation. It is readily prepared by combining aqueous solutions of Fe^{3+} and ferrocyanide through the following reaction:

$$4[Fe(H_2O)_6]^{3+} + 3[Fe(CN)_6]^{4-} \rightarrow Fe_4[Fe(CN)_6]_3 \cdot xH_2O \qquad (7.6)$$

Prussian blue contains linear Fe^{2+}–$C\equiv N$–Fe^{3+} linkages (see Figure 7.10) that make up an infinite 3D cubic network. The carbon end of the cyanide group coordinates to a low-spin d^6 Fe^{2+} center, while the nitrogen end coordinates to a high-spin d^5 Fe^{3+} center. An intense $Fe^{2+} \rightarrow Fe^{3+}$ MMCT band centered near 705 nm absorbs visible light with $\lambda > 500$ nm. The reflected/transmitted blue and violet light gives the distinctive color of Prussian blue.

7.5 Compound Semiconductors

Thus far, our discussions have been limited to electronic transitions that can be described using a localized picture of bonding. Color can also be realized in systems with delocalized bond networks, such as semiconductors. Historically, many red, orange, and yellow inorganic pigments have been semiconductors with band gaps that selectively absorb a portion of the visible spectrum. Examples include orpiment (As_2S_3), cadmium yellow (CdS), and vermillion (HgS). To understand the color in these compounds, we need to take a closer look at the relationships between color, band gap, and composition.

7.5.1 Optical Absorbance, Band Gap, and Color

As discussed in Section 6.3, in semiconductors a band gap (E_g) separates the occupied valence bands from the empty conduction bands. Semiconductors cannot absorb photons with energies less than E_g, but the presence of continuous bands means that they absorb photons more energetic than E_g over a broad range of wavelengths. Consequently, longer-wavelength light ($hv < E_g$) is either transmitted or reflected, while shorter-wavelength light ($hv > E_g$) is absorbed.

The fact that semiconductors absorb light with photon energies exceeding E_g limits their colors (Figure 7.11). The most energetic visible photons have wavelengths of ~400 nm and energies of 3.1 eV. Semiconductors that have band gaps larger than ~3.1 eV, like ZnS, do not absorb visible light and are white in color. As the band gap decreases, B is absorbed, and the reflected G + R = Y. A vibrant yellow is realized in CdS, where E_g = 2.4 eV. Further reduction of the band gap gradually leads to the absorption of green in addition to blue, and the reflected light changes first to orange and then to red (when both B and G are absorbed equally), as exemplified by HgS with a band gap of 2.0 eV. Further decrease in the band gap darkens the red color until it becomes black for $E_g \le 1.7$ eV, and all visible light is absorbed.

By forming solid solutions, it is possible to precisely control the band gap and tune the color. This strategy has been effectively pursued to make an entire family of pigments amongst solid

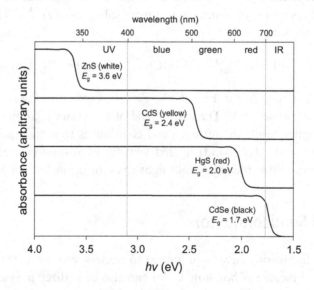

Figure 7.11 Simulated absorbance profiles of four different semiconductors. The curves are offset along the y axis for clarity. The vertical gray lines bracket the range of the visible portion of the spectrum.

solutions formed between CdS (E_g = 2.42 eV) and CdSe (E_g = 1.73 eV). The colors of these pigments range from yellow over orange to red and finally black. Although cadmium-based pigments have many desirable characteristics, their use has declined due to concerns surrounding the toxicity of cadmium. This has spurred efforts to find non-toxic inorganic red, orange, and yellow pigments to replace the lead and cadmium compounds. Candidates that have been proposed include the orange $Ca_{0.5}La_{0.5}TaO_{1.5}N_{1.5}$ [7] and the red Ce_2S_3 [8].

The abruptness of the upturn in absorbance once $hv > E_g$ depends in large part on whether a semiconductor has a direct or an indirect band gap (Section 6.3.2). The absorption coefficient, α, for a direct band-gap semiconductor is proportional to $(hv - E_g)^{1/2}$, whereas for an indirect band-gap semiconductor α is proportional to $(hv - E_g)^2$. Because $hv - E_g \ll 1$ near the band gap, its square root rises sharply upon increasing hv, and a direct band-gap semiconductor has a much sharper absorption edge than a comparable indirect band-gap semiconductor.

7.5.2 Electronegativity, Orbital Overlap, and Band Gap

While silicon remains the dominant material for electronic applications, elemental semiconductors have significant limitations when it comes to optical and optoelectronic applications. In addition to covering a limited range of band-gap energies, silicon and germanium both possess indirect band gaps. Therefore, it is important to understand how the electronic band structures of compound semiconductors differ from those of elemental semiconductors that were previously discussed in Section 6.6.3.

The sphalerite-type (zinc blende) structure (Figure 1.32) of GaAs is an ordered variant of the diamond network of Ge (Figure 1.37). The band structure of GaAs (Figure 7.12) has

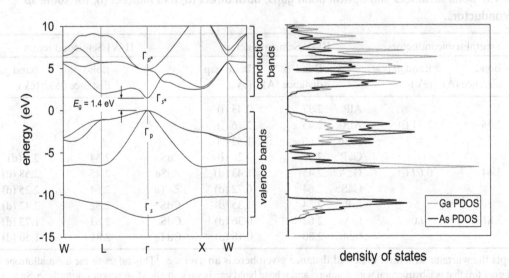

Figure 7.12 The band structure of GaAs (left) together with the partial density-of-states (PDOS) plot showing the individual contributions of Ga and As.

much in common with that of Ge (Figure 6.28), but there are also differences important for applications. GaAs has a larger band gap (1.4 eV versus 0.7 eV), and the conduction-band minimum is located at the Γ point. This means GaAs is a direct-gap semiconductor, unlike Ge.

The increase in band gap comes from the difference in electronegativity between Ga and As, which introduces ionic character into the bonding. This can be seen in the PDOS plot shown on the right-hand side of Figure 7.12. The more electronegative arsenic makes a larger contribution to the valence bands, while gallium makes a larger contribution to the conduction bands. If we further increase the electronegativity difference by going to ZnSe, the band gap increases to 2.6 eV.

The band gaps of several semiconductors are shown in Table 7.8. As we move down the periodic table (e.g. AlP $\rightarrow$ GaAs $\rightarrow$ InSb), the interatomic distance increases and the band gap decreases. This effect arises from decreased orbital overlap, which makes the valence bands less bonding and the conduction bands less antibonding. When we increase the horizontal spacing of the two main-group elements (e.g. Ge $\rightarrow$ GaAs $\rightarrow$ ZnSe), the bond distance remains reasonably constant, but the bond becomes more ionic, and, because the more electronegative element makes a larger contribution to the valence band, its energy is lowered. The opposite occurs for the conduction band, and the band gap increases. This type of manipulation, sometimes referred to as band-gap engineering, plays an important role in designing materials for many electrical and optical devices, including light-emitting diodes (Section 7.9.1).

Table 7.8 Bond distances and optical band gaps, both direct (d) and indirect (i), for some sp^3 semiconductors.

Elemental semiconductors			III–V Semiconductors				II–VI Semiconductors		
	Bond distance (Å)	Band gap (eV)		Bond distance (Å)	Band gap (eV)			Bond distance (Å)	Band gap (eV)
			AlP	2.37	2.43 (i)				
Si	2.35	1.11 (i)	AlAs	2.45	2.16 (i)				
			AlSb	2.66	1.52 (i)				
			GaP	2.36	2.26 (i)	ZnS		2.34	3.6 (d)
Ge	2.44	0.67 (i)	GaAs	2.45	1.43 (d)	ZnSe		2.45	2.58 (d)
			GaSb	2.64	0.72 (d)	ZnTe		2.64	2.25 (d)
			InP	2.54	1.35 (d)	CdS*		2.52	2.42 (d)
Sn†	2.81	~0.0	InAs	2.62	0.36 (d)	CdSe*		2.63	1.73 (d)
			InSb	2.80	0.18 (d)	CdTe		2.81	1.50 (d)

*Adopts the wurtzite structure, the bond distance given here is an average. †This refers to the α-Sn allotrope (also called gray tin) that is isostructural with diamond and whose band gap is very small. Most sources describe α-Sn as a zero band-gap semiconductor (a semimetal) [9].

7.6 Conjugated Organic Molecules

While many organic substances are colorless, those containing a conjugated network of π bonds are an important exception. The π-to-π^* transitions responsible for absorption of visible light are allowed, and their molar absorption coefficients are high. Another attractive feature is the ability to tune the energy of the π-to-π^* transitions by changing the functional groups on the periphery of the molecule. As we move through the book, we will see that conjugated organic molecules make a disproportionately large contribution to the field of functional organic materials.

Two historically important organic molecules used in dyes and pigments are shown in Figure 7.13. Alizarin is the molecule that gives a red color to the dye extracted from the root of a madder plant. This dye was widely used for centuries and is responsible for the color of the "redcoats" worn by British soldiers until the early twentieth century. The color of this dye can be captured in a pigment called madder lake by grinding it with an insoluble inorganic substance, such as alumina. Indigo is another molecular substance from plants that has long been used as a dye. It is most closely associated with the color of blue jeans. Today, synthetically manufactured alizarin and indigo dyes have largely replaced dyes extracted from plants.

Among the simplest conjugated aromatic molecules are the acenes, which are linearly fused benzene rings. Benzene and the first four acenes are shown in Figure 7.14. The electronic structure and MOs associated with the delocalized π network in benzene was described in Section 5.3.7. Consider the six benzene π/π^* MOs in Figure 5.19. As the energies of the MOs increase, there is an increase in the number of nodal planes and a progressive shift from nearest-neighbor interactions that are bonding to those that are antibonding. In benzene and many other conjugated organic molecules, the MOs with net bonding character are filled, while those with net antibonding character are empty.

We can extend the same principles to the larger acenes. As with benzene, each carbon atom contributes one $2p$ orbital to the π network, consequently, the number of π MOs is equal to the number of carbon atoms. The number of nodal planes in the

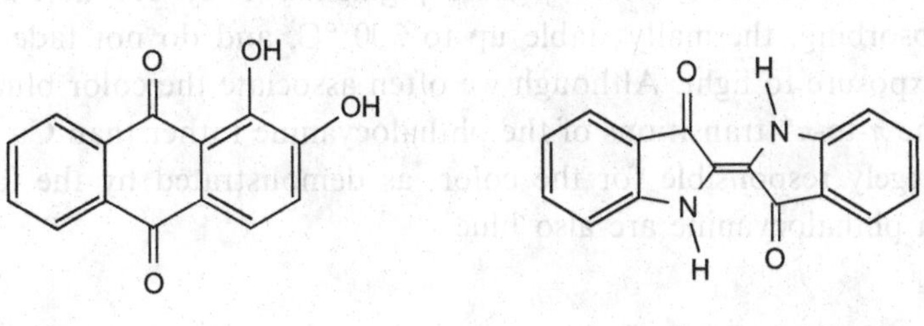

alizarin, $C_{14}H_8O_4$ (red)   indigo, $C_{15}H_9N_2O$ (blue)

Figure 7.13 The molecular structures of alizarin and indigo.

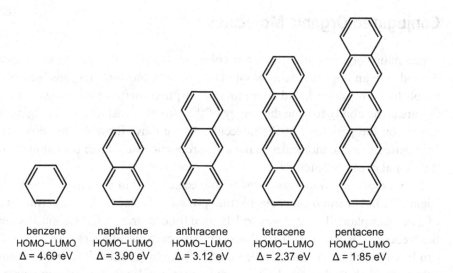

benzene napthalene anthracene tetracene pentacene
HOMO–LUMO HOMO–LUMO HOMO–LUMO HOMO–LUMO HOMO–LUMO
Δ = 4.69 eV Δ = 3.90 eV Δ = 3.12 eV Δ = 2.37 eV Δ = 1.85 eV

Figure 7.14 The HOMO-LUMO gaps, Δ, for benzene and the first four acenes.

HOMO increases from two in benzene (see Figure 5.19), to three in naphthalene, four in anthracene, five in tetracene, and so on.[10] Although benzene is colorless with a large HOMO–LUMO gap (Δ = 4.69 eV), the gap decreases as the size of the π network increases. When we reach tetracene, the lowest-energy π-to-π^* transition has shifted into the visible range, absorbing much of the blue and some green light, leading to an orange color. The HOMO–LUMO gap shifts even further into the visible in pentacene, leading to a dark-red coloration. This trend continues as the size of the π network increases, until we reach the infinite network found in graphene where the band gap goes to zero (Section 6.5.1). While the acenes are not used as pigments or dyes, they form an important class of organic conductors.

The planar porphyrin macrocycle (Figure 7.15, left) is the chromophore responsible for the color of important biological molecules such as chlorophyll. Of the related phthalocyanines (Figure 7.15, right), copper phthalocyanine-based pigments make up the largest class of commercial organic pigments. They are non-toxic, inexpensive, strongly absorbing, thermally stable up to 300 °C, and do not fade appreciably after extended exposure to light. Although we often associate the color blue with Cu^{2+} salts, here it is the π-to-π^* transitions of the phthalocyanine rather than Cu d-to-d transitions that are largely responsible for the color, as demonstrated by the fact that zinc and magnesium phthalocyanine are also blue.

---

[10] Here we include the nodal plane that lies in the plane of the molecule.

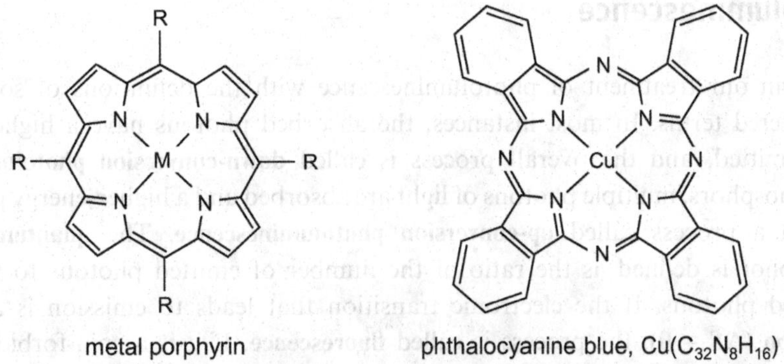

metal porphyrin   phthalocyanine blue, $Cu(C_{32}N_8H_{16})$

Figure 7.15 A generic metal porphyrin molecule (left) and the copper phthalocyanine molecule (right).

7.7 Luminescence

Luminescence describes processes in which materials called **phosphors**[11] are used to convert various forms of energy into electromagnetic radiation, typically in the UV, visible, and IR regions of the spectrum. Luminescent materials appear in a wide range of applications including lighting, display technology, medical imaging, and radiation detection. The most familiar form of luminescence is **photoluminescence**, where electrons are excited by absorption of light that is subsequently reemitted at a different wavelength. Other forms of luminescence exist, depending on the source of energy used to excite electrons. **Electroluminescence** is the direct conversion of electrical energy into light. This type of luminescence, which is the basis for light-emitting diodes, is discussed in Section 7.9. **Cathodoluminescence** occurs when a phosphor is exposed to a beam of electrons accelerated by an electric field. Mechanical energy can also act as the input that leads to luminescence. Examples include: **triboluminescence**, where fracturing materials leads to the emission of light; **piezoluminescence**, which is triggered by the deformation of matter; and **sonoluminescence**, where ultrasonic waves are converted to light. Luminescence can originate from chemical (**chemiluminescence**) or biochemical (**bioluminescence**) reactions when the products formed are in electronically excited states.[12] **Thermoluminescence** occurs when heat activates electrons trapped in excited states, allowing them to relax in a radiative manner to the ground state. **Thermal stimulation of luminescence** is a more accurate description of this process, since the initial excitation and trapping of electrons is caused by interaction with either visible or UV light (afterglow phosphors) or high-energy photons like X-rays and/or γ-rays (storage phosphors).

---

[11] Phosphor is Greek for "light bearer". The element phosphorus shares the same root because white phosphorus is chemiluminescent when slowly oxidized.

[12] One of nature's best-known examples of bioluminescence is the firefly (*Photinus pyralis*), which emits 560 nm light through an enzyme-catalyzed oxidation of luciferin to oxyluciferin.

7.8   Photoluminescence

We begin our treatment of photoluminescence with the definitions of some commonly encountered terms. In most instances, the absorbed photons have a higher energy than those emitted, and the overall process is called **down-conversion photoluminescence**. In some phosphors, multiple photons of light are absorbed and a higher-energy photon emitted through a process called **up-conversion photoluminescence**. The **quantum efficiency** of a phosphor is defined as the ratio of the number of emitted photons to the number of absorbed photons. If the electronic transition that leads to emission is a spin-allowed transition ($\Delta S = 0$), the process is called **fluorescence**. If it is a spin-forbidden transition (typically $\Delta S = 1$), it is called **phosphorescence**. The **decay times**[13] associated with fluorescence range from 10^{-11} s to 10^{-8} s, while phosphorescence has much slower decay times that range from 10^{-6} s to 10^{-2} s. **Persistent phosphors**, such as $SrAl_2O_4$ doped with Eu^{2+} and Dy^{3+}, emit light for hundreds of seconds after the excitation source is turned off, but this does not involve spin-forbidden transitions. Instead, this afterglow is the result of **photoionization**, where incoming photons ionize a site (typically a cation) in the lattice, and the ionized electron is trapped by anion vacancies from which it later escapes via thermal stimulation (i.e. thermoluminescence). Nevertheless, this process is sometimes described in the literature as phosphorescence.

The different electronic transitions that accompany luminescence can be summarized in a Jablonski diagram (Figure 7.16). Following absorption of a photon, an electron is promoted from the singlet ground state to a singlet excited state. Non-radiative **internal conversions** or relaxations into the lowest-energy singlet excited state occur at rates faster than 10^{12} per second. The next step is often emission of fluorescent photons as the electron returns to the ground state at rates of 10^{11} to 10^8 per second. An alternative pathway is non-radiative **intersystem crossing** (ISC) from an excited singlet state to a triplet state. From here, phosphorescence back to the singlet ground state will occur at rates of 10^6 to 10^2 per second, five to six orders of magnitude slower than fluorescence. Yet another pathway, not shown in Figure 7.16, is a non-radiative return to the ground state, an undesirable process that competes with luminescence.

7.8.1   Components of a Phosphor

In the most general sense, a phosphor consists of three components: **sensitizers**, which are sites where incoming photons are absorbed; **activators**, which are sites where photoluminescence occurs through radiative relaxation of electrons; and a **host**, in which both sensitizers and activators are embedded, as illustrated in Figure 7.17. In some phosphors, the sensitizer and the activator are the same ion; when this is not the case, an efficient mechanism for

---

[13] The decay time is the time for a steady-state luminescence intensity to decay to $1/e \approx 36.8\%$ of its original value.

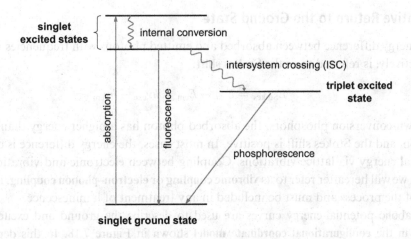

Figure 7.16 A Jablonski diagram illustrating the electronic transitions that can occur in a phosphor after absorption of a photon.

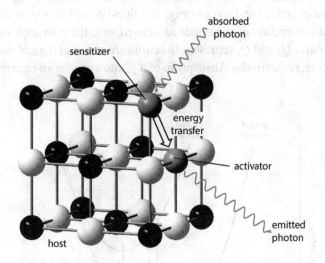

Figure 7.17 The components of a phosphor.

energy transfer from the sensitizer to the activator is needed. Phosphors where the host acts as sensitizer and activator are called **self-activating phosphors**.

The host is important for a variety of reasons. It determines the local coordination environment of the sensitizer and the activator, which play an important role in determining the wavelengths of the absorbed and emitted light. The energy and nature of the vibrations in the host impact the probability that excited-state electrons will return to the ground state radiatively. Finally, the host must possess a band gap large enough to allow the incident light to reach the sensitizers and the emitted light to escape the phosphor.

7.8.2   Radiative Return to the Ground State

The energy difference between absorbed and emitted photons with frequencies ν_{exc} and ν_{em}, respectively, is referred to as the **Stokes shift**:

$$E_{Stokes} = E_{exc} - E_{em} = h\nu_{exc} - h\nu_{em} \tag{7.7}$$

In down-conversion phosphors, the absorbed photon has a higher energy than the emitted photon, and the Stokes shift is positive. In most cases, the energy difference is converted to thermal energy via lattice vibrations. Coupling between electronic and vibrational energy, which we will hereafter refer to as **vibronic coupling** or **electron–phonon coupling**, is an integral part of the process and must be included in any treatment of luminescence.

Parabolic potential-energy curves are used to describe the ground and excited electronic states in the **configurational coordinate model** shown in Figure 7.18. In this depiction, Q is a configurational coordinate used to describe a specific vibrational mode of the luminescent center. In an approximate sense, we can think of Q as representing a bond distance between the activator and the ligands that surround it. The horizontal lines or "rungs" represent vibrational states. They are quantized and span a range of values on the horizontal axis due to the dynamic expansion and contraction of the bonds associated with the relevant vibrational mode.

The coordinates Q_g and Q_e represent the equilibrium metal–ligand distances in the ground and excited state, respectively. Absorption of a photon excites an electron into a higher-lying

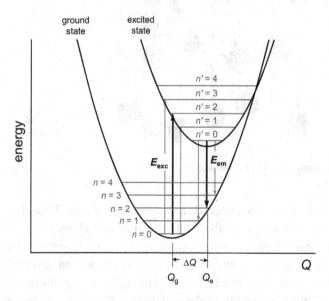

Figure 7.18 Configurational coordinate diagram where Q_g and Q_e represent the equilibrium bond distance of the ground and excited states, respectively. The accessible absorption and emission transitions are shown with vertical arrows, bold arrows signify the most intense transitions.

electronic state that will, in general, have more antibonding character than the ground state, leading to a weakening of the chemical bond. Consequently, the equilibrium bond length of the excited state will be larger than in the ground state ($Q_e > Q_g$). As we will see, the resultant shift in equilibrium bond distance, $\Delta Q = Q_e - Q_g$, is closely coupled to many important characteristics of the luminescence.

If we assume the vibrational motion to be harmonic,[14] the energies of the allowed vibrational states are those of a quantum harmonic oscillator:

$$E_n = \left(n + \frac{1}{2}\right)\hbar\omega \tag{7.8}$$

where $n = 0, 1, 2, 3, \ldots$, and ω is the angular frequency of the oscillator.[15] At room temperature (and below), the lowest-energy vibrational state is typically the most highly populated state and is centered at Q_g (the center of the $n = 0$ rung). Therefore, most electronic excitations originate from this position. The wavefunctions for the excited vibrational states tend to peak at values of Q that are close to where the rung meets the parabola, not unlike the way a pendulum spends more time at the turning points than it does at the bottom of its arc.

According to the **Franck–Condon principle**, electronic excitations occur on timescales much faster than vibrations. Optical transitions are therefore depicted as vertical lines in Figure 7.18, representing no change in bond length upon absorption or emission of a photon. The bold vertical arrow labeled E_{exc} corresponds to the optical excitation that has the maximum intensity. It goes from the center of the $n = 0$ rung on the ground-state parabola, where the ground-state wavefunction has maximum probability, to a point where the $n' = 2$ vibrational-state wavefunction has a high probability (i.e. near the intersection of the $n' = 2$ rung and the excited-state parabola). This transition will be more intense than transitions to other excited vibrational states, because their wavefunctions reach their highest probabilities at more distant values of Q. Although the $n = 0$ to $n' = 2$ transition has maximum intensity, less-intense transitions to other states, shown with gray vertical arrows in Figure 7.18, can also occur.

Once in the excited state, the electron rapidly relaxes to the ground vibrational state ($n' = 0$) via electron–phonon coupling. From this state, it can return to the ground electronic state through emission of a photon, following the same principles that govern absorption. The vertical arrow labeled E_{em} corresponds to the optical emission that has the maximum intensity. Following emission of a photon, the electron returns to the original ground state through further coupling to lattice vibrations. Thus, we see that electron–phonon coupling in both the excited- and ground-state parabolas is the origin of the Stokes shift.

The magnitude of ΔQ in Figure 7.18 gives a measure of the changes in chemical bonding that accompany promotion of an electron into an excited state. This change is captured by

---

[14] The harmonic approximation requires that the restoring force, F, is proportional to displacement, ΔQ; $F = -k_F \Delta Q$, where k_F is a force constant, resulting in a parabolic potential-energy curve, $E = \frac{1}{2}k_F \Delta Q^2$.

[15] The angular frequency of a simple two-body harmonic oscillator is defined as $\omega = (k_F/\mu_r)^{1/2}$ where k_F is the force constant and $\mu_r = (m_1 m_2)/(m_1 + m_2)$ is the reduced mass of the oscillator.

the dimensionless **Huang–Rhys parameter**, S, which is proportional to (ΔQ). Under weak electron–phonon coupling (say, $S < 1$), the shift ΔQ of equilibrium bond distances between the ground and excited states is small. That is the case for electronic transitions between $4f$ orbitals, because the $4f$ orbitals have minimal interactions with the ligands and therefore little impact on the bond distances. In the $\Delta Q = 0$ limit, the ground-state vibrational wavefunctions for both parabolas have maxima at the same value of Q. In this limit, the absorption and emission spectra will consist of single lines called zero-phonon lines, and the Stokes shift will be zero. For small, yet non-zero ΔQ, a small Stokes shift and narrow absorption/emission lines occur. Large ΔQ indicates strong electron–phonon coupling ($S > 5$). This happens when the excited electronic state has very different bonding character than the ground electronic state. Activators from the p block of the periodic table (e.g. Pb^{2+}, Bi^{3+}) or oxyanions (e.g. VO_4^{3-}, WO_4^{2-}) often fall in the strong-coupling regime. Because of the large ΔQ, the ground vibrational state of the lower parabola ($n = 0$) has substantial overlap with several excited vibrational states on the upper parabola, leading to a broad absorption band. For similar reasons, the emission bands will also be broad. Large ΔQ also leads to considerable vibrational relaxation following both absorption and emission, which makes for a large Stokes shift.

Not surprisingly, absorption and emission spectra change with temperature due to changes in the thermal population of different vibrational levels on both parabolas. Whereas at very low temperatures the fine structure of absorption and emission spectra can be resolved, broadening occurs at higher temperatures, which can result in unresolved absorption and emission bands.

7.8.3 Thermal Quenching

We now consider undesirable non-radiative processes that offer the excited electron alternative paths to the ground state. One of the most important is **thermal quenching**, which refers to the process of electrons returning to the ground state by dissipating energy through lattice vibrations. Consider the configurational coordinate diagrams in Figure 7.19, which depict small and large ΔQ. The energy difference between the ground vibrational level in the excited state and the energy where the two parabolas cross is denoted as ΔE. When ΔQ is large, the two parabolas cross at a relatively low energy and ΔE is small. In this case, vibrational levels close to the crossing point will have non-negligible populations at modest temperatures,[16] allowing electrons to cross over from the excited-state parabola to the ground-state parabola where they can return to the ground state non-radiatively through coupling with lattice vibrations. When ΔQ is small, however, the curves cross at much higher energies, and the vibrational states that lie at or above the crossing point only acquire non-negligible

---

[16] The population of vibrational states is given by a Boltzmann distribution, where the probability of being in an excited state whose energy is E above the ground state is proportional to $e^{-E/kT}$, where the thermal energy is given by kT, with $k = 1.380649 \times 10^{-23}$ J/K. For a temperature of 300 K, $kT = 0.0258$ eV.

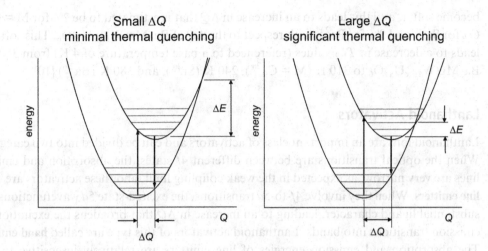

Figure 7.19 Configurational coordinate diagrams for small (left) and large (right) ΔQ. The parameter ΔE represents the energy between the ground vibrational state of the excited electronic state and the energy where the parabolas cross.

populations at high temperatures. In such cases, thermal quenching at room temperature or below tends to be minimal.

Even in cases where $\Delta Q \approx 0$, thermal quenching can still occur if the energy difference between ground- and excited-state parabolas is small. In such cases, the electron can return directly to the ground-state parabola through a process called **multi-phonon emission**, where the lost electronic energy generates several high-energy phonons in the surrounding lattice. As a rule of thumb, multi-phonon emission becomes significant when the energy of the highest-energy phonon mode, $E = h\nu_{max}$, exceeds roughly 20% of the electronic energy difference between the ground vibrational states of the two parabolas. This is a common non-radiative decay pathway for many rare-earth ions.

The **thermal-quenching temperature**, $T_{1/2}$, is the temperature at which a phosphor loses 50% of its emission intensity with respect to an arbitrarily defined base temperature. High values of $T_{1/2}$ are desirable, and in many industrially relevant phosphors it is greater than 100 °C (for base temperature = room temperature). Phosphors with high $T_{1/2}$ usually have stiff hosts to limit expansion of bond distances in the excited state and keep ΔQ small. If the lattice softens, ΔQ increases and ΔE decreases, making non-radiative return to the ground state more probable.

To illustrate the importance of the "stiff" host on thermal-quenching behavior, consider the ordered double perovskites, $Ba_2M(W_{1-x}U_x)O_6$, with M = Mg^{2+}, Ca^{2+}, Sr^{2+}, and Ba^{2+}. The $UO_{6/2}$ entity acts as both sensitizer and activator. An O $2p \rightarrow$ U $6d$ LMCT transition is responsible for light absorption, and ΔQ is large since the uranium-centered octahedra will expand and distort when the excited-state antibonding orbitals are populated. This expansion will compress the M–O bonds of the coordination octahedra that alternate with the $UO_{6/2}$ octahedra. As the size of the M^{2+} cation increases, the M–O bonds lengthen and become softer, and this leads to an increase in ΔQ that is estimated to be 2% for $M = Ca^{2+}$, 6% for Sr^{2+}, and 9% for Ba^{2+}, with respect to the ΔQ of $Ba_2Mg(W_{1-x}U_x)O_6$. This softening leads to a decrease in $T_{1/2}$ values (referenced to a base temperature of 4 K) from 350 K in $Ba_2Mg(W_{1-x}U_x)O_6$ to 310 K ($M = Ca^{2+}$), 240 K (Sr^{2+}), and 180 K (Ba^{2+}) [10].

7.8.4   Lanthanoid Activators

Lanthanoid ions are an important class of activators and can be divided into two categories. When the optical transitions are between different $4f$ states, the absorption and emission lines are very narrow, as expected in the weak coupling limit, and these activators are called **line emitters**. When they involve $4f$-to-$5d$ transitions, the excited-state $5d$ wavefunctions have substantial ligand character, leading to an increase in ΔQ that broadens the excitation and emission transitions into bands. Lanthanoid activators of this type are called **band emitters**. The absorption and emission energies of line emitters are relatively insensitive to their surroundings, while those of band emitters can be altered by modifying the chemical surroundings of the activator ion.

We start with *line emitters*. The lanthanoid elements have a strong preference for the 3+ oxidation state with a $[Xe]4f^n$ electron configuration (Table 7.9), but some can also take either a 2+ or a 4+ oxidation state, particularly when it leads to an empty (Ce^{4+}), half filled (Eu^{2+}, Tb^{4+}), or completely filled (Yb^{2+}) $4f$ subshell. To understand the energy levels of partially

Table 7.9 Ground-state electron configurations of lanthanoid ions.

Z	Element	Ion	Electron configuration	Ground-state term	Ion	Electron configuration	Ground-state term
57	Lanthanum	La^{3+}	$[Xe]4f^0$	$^1S_0$			
58	Cerium	Ce^{3+}	$[Xe]4f^1$	$^2F_{5/2}$	Ce^{4+}	$[Xe]4f^0$	$^1S_0$
59	Praseodymium	Pr^{3+}	$[Xe]4f^2$	$^3H_4$	Pr^{4+}	$[Xe]4f^1$	$^2F_{5/2}$
60	Neodymium	Nd^{3+}	$[Xe]4f^3$	$^4I_{9/2}$			
61	Promethium*	Pm^{3+}	$[Xe]4f^4$	$^5I_4$			
62	Samarium	Sm^{3+}	$[Xe]4f^5$	$^6H_{5/2}$	Sm^{2+}	$[Xe]4f^6$	$^7F_0$
63	Europium	Eu^{3+}	$[Xe]4f^6$	$^7F_0$	Eu^{2+}	$[Xe]4f^7$	$^8S_{7/2}$
64	Gadolinium	Gd^{3+}	$[Xe]4f^7$	$^8S_{7/2}$			
65	Terbium	Tb^{3+}	$[Xe]4f^8$	$^7F_6$	Tb^{4+}	$[Xe]4f^7$	$^8S_{7/2}$
66	Dysprosium	Dy^{3+}	$[Xe]4f^9$	$^6H_{15/2}$			
67	Holmium	Ho^{3+}	$[Xe]4f^{10}$	$^5I_8$			
68	Erbium	Er^{3+}	$[Xe]4f^{11}$	$^4I_{15/2}$			
69	Thulium	Tm^{3+}	$[Xe]4f^{12}$	$^3H_6$			
70	Ytterbium	Yb^{3+}	$[Xe]4f^{13}$	$^2F_{7/2}$	Yb^{2+}	$[Xe]4f^{14}$	$^1S_0$
71	Lutetium	Lu^{3+}	$[Xe]4f^{14}$	$^1S_0$			

*Promethium does not occur naturally. It is radioactive and its longest-lived isotope, ^{145}Pm, has a half-life of 17.7 years.

filled $4f$ orbitals, we need to return to the microstates described by term symbols first introduced in Section 7.3.2. The orbital angular-momentum quantum numbers L can vary from 0 to 6, while the spin angular-momentum quantum numbers can vary from 0 to $7/2$. The possible number of microstates is large and can be calculated with the expression $14!/[n_f!(14-n_f)!]$, with n_f being the number of f-electrons present. Using this expression, we find 3432, 364, and 14 possible microstates for the Gd^{3+} (f^7), Nd^{3+} (f^3), and Ce^{3+} (f^1) ions, respectively.

The interactions that split the $4f$ levels are spin–spin, spin–orbit, and orbit–orbit coupling, which are considerably stronger for the lanthanoids than for the $3d$ transition-metal ions. Conversely, the crystal-field splitting is much smaller than it is for the $3d$ ions due to the limited radial extension of the $4f$ orbitals. We therefore concentrate on the effects of interelectron coupling, and only afterwards allow energies to be shifted by crystal-/ligand-field perturbations. The close similarity of the optical spectra of free lanthanoid ions and those in compounds supports the validity of this approach.

In the 1960s, Dieke and co-workers analyzed optical spectra for Ln^{3+} ions in $LaCl_3$ single crystals and determined energy levels of the various terms, producing what have come to be called Dieke diagrams. The term symbols used in these diagrams are derived using the Russell–Saunders coupling scheme (Section 7.3.2). It is now generally accepted that the coupling is intermediate between the two limiting cases of Russell–Saunders and j–j coupling, and high-level calculations are needed to determine the relative order of the terms [11]. Nonetheless, the Russell–Saunders coupling scheme provides a reasonable approximation of the multielectron energy levels. For example, the ground state of any lanthanoid ion can be correctly predicted using Hund's rules.

As an example, consider the excitation and emission spectra of Eu^{3+}. The electronic ground-state configuration of Eu^{3+} is $[Xe]4f^6$. Applying Hund's rules (Section 7.3.2), we determine that the ground state has $S = 3$ and $L = 3$, which combine into a ^7F term. The possible J values for the ^7F term are the integers between $L + S = 6$ and $|L - S| = 0$, and Hund's third rule tells us that the energy increases as J increases, as shown in Figure 7.20. An f^6 ion has 3003 microstates that can be grouped into 295 distinct $^{2S+1}L_J$ terms, which makes determining the excited-state energies challenging. Fortunately, the photoluminescence of Eu^{3+} activators can be understood from a relatively small subset of the total number of excited states. The relevant transitions are shown on the left-hand side of Figure 7.20. Optical excitation is largely through spin-forbidden transitions from the $^7F_0$ ground state into various low-lying quintuplet states ($S = 2$), as well as LMCT transitions from the surrounding anions to empty Eu $5d$ orbitals, which are typically excited by photons with wavelengths in the 200–300 nm range. The excited-state electrons rapidly relax to the lowest-energy quintuplet state $^5D_0$, then undergo phosphorescence to return to one of the ^7F_J states. Because the spin–orbit coupling is relatively strong, the energies of the various ^7F_J states are well resolved. The energies of the $^5D_0 \rightarrow {}^7F_J$ transitions in Eu^{3+} phosphors are relatively insensitive to their local environment. However, you can see a series of closely spaced sharp lines associated with each $^5D_0 \rightarrow {}^7F_J$ transition caused by subtle crystal-field splitting effects (Figure 7.20).

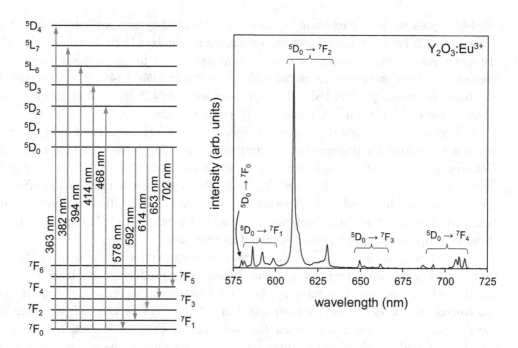

Figure 7.20 The term scheme and separation of energy levels for a free Eu^{3+} ion (left), and the emission spectrum of $Y_2O_3:Eu^{3+}$, a commercial red phosphor used in fluorescent lights (right). Data are taken from ref [12].

The color of Eu^{3+} emission can vary from orange to red, depending upon the host. This happens because the intensities of the different $^5D_0 \rightarrow {}^7F_J$ transitions are highly sensitive to the local symmetry of the Eu^{3+} ion. When Eu^{3+} is located on a site with inversion symmetry, optical transitions other than those where $\Delta J = 0$, ± 1 violate the parity selection rule[17] and are very weak. Only the $^5D_0 \rightarrow {}^7F_1$ (with $\Delta J = 1$) at $\lambda \approx 592$ nm does not violate the parity selection rule, hence the emitted light takes on a reddish-orange color. In hosts where the Eu^{3+} sits on a site without inversion symmetry, such as $Y_2O_3:Eu^{3+}$, crystal-field components mix states of opposite parity into the $4f^n$ configurational levels and the $^5D_0 \rightarrow {}^7F_2$ (~614 nm) emission gains significant intensity (see Figure 7.20).[18] This results in a deeper-red emission, which is generally a desirable attribute in commercial phosphors. This example shows that even though the energies of individual f-to-f transitions are only weakly perturbed by the environment of the activator ion, their relative intensities can be greatly influenced by the local structure imposed by the host.

---

[17] The $J = 0 \rightarrow J = 0$ transition is also forbidden by the parity selection rule even though $\Delta J = 0$.

[18] f-to-f transitions that don't violate the parity selection rule, like $^5D_0 \rightarrow {}^7F_1$, are referred to as magnetic-dipole transitions. Those that do violate this rule, like $^5D_0 \rightarrow {}^7F_2$, are called forced electric-dipole transitions. Y_2O_3 of space group $Ia\bar{3}$, has two different cation sites: $8b$ of site symmetry -3 and $24d$ of site symmetry 2. Forced electric-dipole transitions with significant intensity are only seen for the latter.

Now to *band emitters*. The energies of the empty $5d$ orbitals are sensitive to changes in the chemical environment of the activator ion, because the $5d$ orbitals form antibonding orbitals with anions of the host. The presence of antibonding character in the excited state results in an expansion of the metal–ligand bond lengths with respect to the ground state. The increase in ΔQ leads to a variable Stokes shift, higher rates of thermal quenching, broadening of the absorption and emission bands, and the ability to tune excitation and emission spectra through the appropriate choice of host. The two most important lanthanoid band emitters are Ce^{3+} and Eu^{2+}, which feature $[Xe]4f^1 \leftrightarrow [Xe]4f^0 5d^1$ and $[Xe]4f^7 \leftrightarrow [Xe]4f^6 5d^1$ transitions, respectively. These transitions are fully allowed and give rise to strong absorption and emission bands, which can in turn lead to highly efficient phosphors. Here we concentrate on Ce^{3+} because the $4f^1$ electron configuration simplifies the analysis.

The $4f^1$ ground-state term of Ce^{3+} is split by spin–orbit coupling into two levels, $^2F_{5/2}$ and $^2F_{7/2}$, separated by 0.25 eV. The energy separation between the $4f^1$ ground state and the empty $5d$ orbitals is ~6.3 eV for a free Ce^{3+} ion. This energy separation can be significantly reduced when the Ce^{3+} ion is embedded in phosphor hosts, through two effects—a centroid shift and crystal/ligand-field splitting of the $5d$ orbitals—as illustrated schematically in Figure 7.21. Given the importance of these two effects in designing new phosphors, we examine each separately.

The **centroid shift** is the downward shift in the average energy of all five $5d$ orbitals relative to a free Ce^{3+} ion. The decreased $5d$-to-$4f$ separation is attributed to a reduction in electron–electron repulsions due to delocalization (spreading out) of the excited-state $5d$ Ce^{3+} orbital

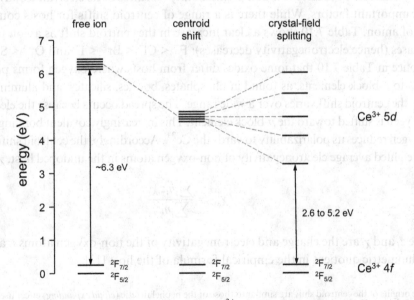

Figure 7.21 Schematic energy-level diagram for Ce^{3+}, showing the combined effects of the centroid shift and crystal/ligand-field splitting in lowering the energy of $4f$-to-$5d$ transition.

Table 7.10 The centroid shift for Ce^{3+} doped into various inorganic hosts grouped according to anions that surround Ce^{3+} in the host. Data taken from ref. [13].

Compound type	Number	Centroid shift (eV)		
		Minimum	Maximum	Median
Fluorides	25	0.54	0.91	0.70
Chlorides	17	1.61	1.89	1.84
Bromides	9	1.92	2.21	2.11
Iodides	4	2.36	2.84	2.67
Oxides (polar covalent)*	75	0.88	1.99	1.30
Oxides (ionic)†	3	1.97	2.59	2.27
Sulfides	4	2.61	2.98	2.80
Selenides	2	3.00	3.21	3.11

*Oxides where the "anionic" portion of the host contains a p-block element that forms polar-covalent bonds to oxygen, such as P, B, Si, or Al. †Oxide hosts with low-electronegativity cations (CaO, La_2O_3, $LaLuO_3$) comparable to Ce^{3+}.

wavefunctions onto the ligand orbitals through its interaction with them.[19] For Ce^{3+}, the magnitude of the centroid shift ranges from approximately 0.5 to 3.2 eV, depending on its neighboring atoms. Dorenbos [13] has examined the centroid shift for Ce^{3+} doped into more than 130 different inorganic hosts, and his results are summarized in Table 7.10. Various factors come into play, including the anion polarizability, the coordination number, and the bond distances. Of these, the anion polarizability (softness of its electron cloud) is one of the most important factors. While there is a range of centroid shifts for hosts containing each type of anion, Table 7.10 shows a clear increase in the centroid shift as anion polarizability increases (hence electronegativity decreases): $F^- < Cl^- < Br^- < I^-$ and $O^{2-} < S^{2-} < Se^{2-}$.

Notice in Table 7.10 that ionic oxides differ from hosts where oxygen forms polar covalent bonds to p-block elements, as found in phosphates, borates, silicates, and aluminates. Among these, the centroid shift varies over a wide range. The spread occurs because the electron density on oxygen is shifted toward the p-block element. This increasingly covalent bonding engagement of oxygen reduces its polarizability towards the Ce^{3+}. Accordingly, the centroid shift decreases as the weighted average electronegativity of non-oxygen atoms in the undoped host, χ_{av}, increases:

$$\chi_{av} = \frac{\sum_i n_i z_i \chi_i}{\sum_i n_i z_i} \tag{7.9}$$

where z_i and χ_i are the charge and electronegativity of the non-oxygen atoms i, and n_i is their stoichiometric quotient in the empirical formula of the host [13].

[19] The origins of the centroid shift are similar to those of the nephelauxetic (*cloud expanding*) effect used to understand the optical spectra of transition-metal complexes. For more details see C.K. Jørgensen, "*Modern Aspects of Ligand Field Theory*" (1971) North-Holland.

Consider the behavior of Ce^{3+} doped into YPO_4 and $YAlO_3$ hosts. In YPO_4, the weighted average Allred–Rochow electronegativity of the non-oxygen atoms is $\chi_{av} = (3\chi_Y + 5\chi_P)/8 = 1.70$, and the centroid shift is 1.19 eV. For $YAlO_3$, $\chi_{av} = (3\chi_Y + 3\chi_{Al})/6 = 1.29$ and the centroid shift is 1.60 eV. The highly covalent P–O bonds of the phosphate groups in YPO_4 reduce the polarizability of the oxygen atoms towards Ce^{3+}, leading to a smaller centroid shift.

Ligand-field splitting of the $5d$ orbitals is the second factor that affects the energies of the $4f$-to-$5d$ transitions. The magnitude of the ligand-field splitting depends primarily upon the coordination geometry (Figure 7.2) and bond distances. Hosts with larger anions tend to have longer bond distances and thus smaller ligand-field splitting.[20] The effects of coordination are illustrated by the differences between a Ce^{3+} ion in an octahedron and a cuboctahedron. Both share the same symmetry ($m\bar{3}m$), and the $5d$ orbitals split into the familiar t_{2g} (d_{xy}, d_{yz}, d_{xz}) and e_g ($d_{z^2}, d_{x^2-y^2}$) sets in both environments. Within the assumptions of crystal-field theory (Section 7.3.1), the t_{2g}–e_g splitting in a cuboctahedron is only 50% of its value in an octahedron, $\Delta_{co} = 0.5\Delta_o$. A larger splitting increases the energy spread of the $5d$ orbitals (Figure 7.21), thereby reducing the energy separation between the $4f$ orbitals and the lowest-energy $5d$ orbital(s). Hence, all other things being equal, an octahedrally coordinated activator will absorb and emit light at longer wavelengths than the same activator in a host where its environment is a cuboctahedron. Empirically, it has been shown [14] that crystal-field splitting for eight-coordinate cube and dodecahedron coordinations is ~80–90% that of an octahedron. For a nine-coordinate tricapped trigonal prism it is much smaller, 40–50% of that seen in an octahedron, similar to the cuboctahedron.

7.8.5   Non-Lanthanoid Activators

In this section, we survey three additional classes of activators: (a) ions of p-block elements with $(n-1)d^{10}ns^2$ configurations, (b) oxoanions of transition metals with a d^0 configuration, and (c) transition-metal ions with partially filled $3d$ orbitals. We begin with the optical transitions of $(n-1)d^{10}ns^2$ activators like Sn^{2+}, Sb^{3+}, Pb^{2+}, and Bi^{3+}. The ns^2 ground state is represented by a singlet $^1S_0$ term, and the ns^1np^1 excited state can be divided into a singlet $^1P_1$ state and three triplet states that are further split by spin–orbit coupling: $^3P_2$, $^3P_1$, and $^3P_0$ (see Figure 7.22). Excitation can occur either through spin-allowed $^1S_0 \rightarrow {}^1P_1$ transitions followed by intersystem crossing into the $^3P_1$ state, or directly from the ground state into the lower-lying $^3P_1$ state. The $^1S_0 \rightarrow {}^3P_J$ transitions are spin-forbidden, but the selection rule is relaxed by spin–orbit coupling. The spin–orbit coupling is high for ions from the fifth period (e.g. Sn^{2+}, Sb^{3+}) and even larger for those from the sixth period (e.g. Pb^{2+}, Bi^{3+}). The $^1S_0 \rightarrow {}^3P_1$ transition is the strongest of the three possible $^1S_0 \rightarrow {}^3P_J$ transitions, because the $^1S_0 \rightarrow {}^3P_0$ and $^1S_0 \rightarrow {}^3P_2$ transitions are forbidden by the parity selection rule. This can be seen in the

[20] As the size of the anion increases, so does its polarizability. Hence, the anions that give the largest centroid shifts tend to give smallest crystal-field splitting.

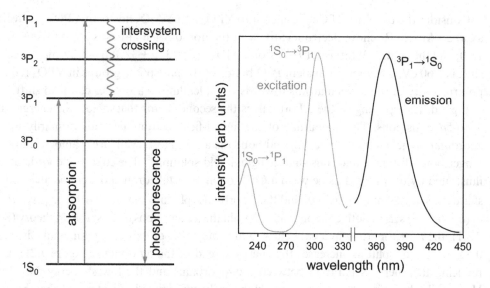

Figure 7.22 The energy-level diagram for a $(n-1)d^{10}ns^2$ ion (left). The excitation and emission spectra of $Ba_2Mg(BO_3)_2{:}Bi^{3+}$ (right). Data taken from ref. [15].

excitation spectrum of $Ba_2Mg(BO_3)_2{:}Bi^{3+}$ in Figure 7.22. Phosphorescence involving a $^3P_1 \rightarrow {}^1S_0$ transition dominates the emission spectrum.[21]

Because the ns^1np^1 excited-state electron configuration is both more antibonding than the ns^2 ground-state configuration and prone to Jahn–Teller distortions (Section 5.3.10), ions like Bi^{3+}, Pb^{2+}, and Sn^{2+} can experience a large reorganization of their coordination environment in the excited state. When this occurs, a large ΔQ, large Stokes shift and broad emission bands are expected. If we start from a symmetric environment like an octahedron, the coordination environment of the ns^1np^1 excited state can undergo a symmetric expansion, a tetragonal distortion (first-order Jahn–Teller distortion), and/or a trigonal distortion (second-order Jahn–Teller distortion) [16]. The tetragonal distortion leads to ligand-field splitting of the $^3P_1$ excited state that can result in splitting of the excitation and emission bands.

The extent of the structural reorganization in the excited state, and hence the size of the Stokes shift, is highly dependent on the structure of the host. When an activator ion is placed on a site that is compressed with respect to its preferred environment, reorganization of the excited state is suppressed, minimizing ΔQ and leading to a small Stokes shift. Conversely, if it is placed on a large site, relaxation of the coordination sphere of the activator ion in the excited state can be extensive, resulting in a large Stokes shift. Stokes shifts for Bi^{3+} activators vary from 0.1 eV in $Cs_2NaYCl_6{:}Bi^{3+}$, where Bi^{3+} substitutes for the smaller octahedrally coordinated Y^{3+}, to 2.5 eV in $Bi_2Ge_3O_9$, where Bi^{3+} has a strong

---

[21] The $^3P_0 \rightarrow {}^1S_0$ tends to make little contribution to absorption or emission spectra. However, for $6s^2$ activators at very low temperatures, the $^3P_0 \rightarrow {}^1S_0$ emission can be observed in some cases.

trigonal distortion in the ground state with three short Bi–O bonds (2.14 Å) on one side and three long Bi–O bonds (2.74 Å) bonds on the other side of the octahedron. In the latter case, Bi^{3+} is thought to adopt a much more symmetric environment in the excited state, and the large ΔQ leads to a large Stokes shift [17].

Oxoanions of d^0 metals, such as WO_4^{2-} and WO_6^{6-}, are further examples of broad-band emitters with large and highly tunable Stokes shifts. $CaWO_4$ is a paradigmatic scintillator material that was used for many decades as an X-ray phosphor. Its luminescence is due to LMCT transitions on the WO_4^{2-} groups, analogous to those discussed for CrO_4^{2-} in Section 7.4.1. Because this excitation promotes electrons from nonbonding O $2p$ orbitals into $5d$ orbitals with significant antibonding character, tungstates exhibit large ΔQ and Stokes shifts that range from 1.2 eV to 2.5 eV. Increasing the coordination number of the central metal decreases the energy of the LMCT transition. This is illustrated by a comparison between $CaWO_4$, where the onset of absorption (the band gap) is ~4.8 eV, and Sr_2MgWO_6, where the onset of absorption for the $WO_{6/2}$ octahedron is ~3.6 eV. Oxoanions of other d^0 transition metals (e.g. VO_4^{3-}, MoO_4^{2-}) can also luminesce. The energies of the excitation and emission bands follow the trends found in Table 7.7.

Transition-metal ions with partially filled d orbitals are the final class of activator that we will consider. Perhaps the best-known example of transition-metal ion luminescence is ruby, where Cr^{3+} is doped into Al_2O_3.[22] Ruby possesses two sharp, closely spaced emission lines near 700 nm. The transition responsible for Cr^{3+} emission is spin-forbidden phosphorescence from a ^2E excited state to the $^4A_2$ ground state. Because this transition is spin-forbidden, the correlation diagrams of Section 7.3.3 are inadequate. Instead, we must turn to the so-called **Tanabe–Sugano diagrams** used to understand both spin-allowed and spin-forbidden transitions. The Tanabe–Sugano diagram for an octahedrally coordinated d^3 ion is shown in Figure 7.23. Because both the excited state (^2E) and ground state ($^4A_2$) have a $t_{2g}^3 e_g^0$ configuration, the energy spacing between them has little dependence on the ligand-field splitting, leading to sharp emission lines and small ΔQ. However, to get the sharp, red emission seen in ruby, it's critical that the ligand-field splitting Δ_o is large enough for the energy of the ^2E level to be lower than that of the $^4T_2$ state whose energy separation from the ground state increases linearly with Δ_o. To meet this criterion, hosts with large Δ_o are needed.

7.8.6 Energy Transfer

So far, we have considered two pathways for excited-state electrons to return to the ground state: radiatively via emission of a photon or non-radiatively through coupling with lattice vibrations. A third possibility is energy transfer, where the excited-state electron returns to the ground state and transfers its energy to a nearby acceptor ion, where an electron is

---

[22] Stimulated emission in ruby was the basis for the first laser, developed in 1960.

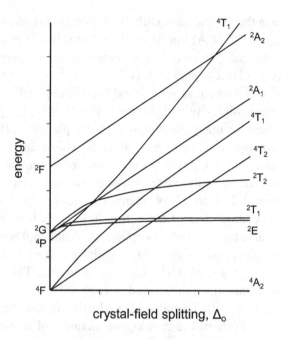

Figure 7.23 A simplified Tanabe–Sugano diagram for an octahedrally coordinated d^3 cation. The spin-forbidden $^2E \rightarrow {}^4A_2$ transition gives rise to luminescence with activators such as Cr^{3+} and Mn^{4+}. The ground-state $^4A_2$ energy is plotted as the x axis.

promoted into an excited state in order to maintain conservation of energy. We can describe this process with the equation:

$$D^* + A \rightarrow D + A^* \tag{7.10}$$

where the ion that is originally in the excited state is called the donor (D), the ion that receives the energy transfer is called the acceptor (A), and an asterisk is used to denote an excited electronic state. Energy transfer underpins the action of sensitizers.

Energy transfer generally occurs by one of two different mechanisms. The first is **Förster resonant-energy transfer (FRET)**, which is based on electromagnetic multipole interactions, predominantly dipole–dipole interactions. In FRET, the electromagnetic field associated with D* interacts with A, leading to a transfer of energy but not electrons between the two sites. The efficiency of FRET scales with d^{-6} for dipole–dipole interactions, where d is the distance between donor and acceptor. The range over which FRET is operative generally doesn't exceed 2–5 nm.[23] To be efficient, the optical transitions should be allowed electric-

---

[23] The strong distance dependence of FRET led to the development of FRET spectroscopy, which allows conformational changes of biomolecules to be followed.

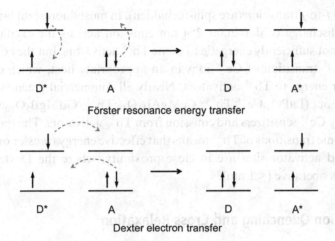

Figure 7.24 FRET and Dexter electron-transfer mechanisms.

dipole transitions, and the emission spectrum of D^* should have significant overlap with the absorption spectrum of A.[24]

An alternative is **Dexter electron transfer** where an excited-state electron is transferred from donor to acceptor while a ground-state electron is transferred in the opposite direction, as shown in the lower half of Figure 7.24. This mechanism does not require overlap between emission and absorption spectra of donor and acceptor, nor does it depend on the selection rules for either transition. It does, however, require a significant overlap of the molecular orbitals on the two sites, which limits its operability to distances ≤ 1 nm.

The dominant type of energy transfer depends on the nature of the donor and acceptor sites. FRET from a broad-band emitter to a line absorber (e.g. $Ce^{3+} \rightarrow Tb^{3+}$) is highly inefficient and energy transfer between these species must rely upon the Dexter mechanism, limiting transfer to near neighbors in the host. The opposite combination, energy transfer from a line emitter to a band absorber, can occur with reasonable efficiency over longer distances via FRET.

7.8.7 Sensitizers

Several otherwise useful activators do not effectively absorb light at practical wavelengths, and therefore can only be used in combination with an appropriate sensitizer. This is particularly true for activators that rely upon spin-forbidden transitions to absorb light, like Tb^{3+} and Mn^{2+}.

Tb^{3+} is an efficient emitter of green light, but it only absorbs strongly at excitation wavelengths smaller than 230 nm, where the $4f^8 \rightarrow 4f^7 5d^1$ transition can be excited (the

---

[24] Though FRET is a non-radiative process that occurs through electric fields, conceptually one can think of the donor emitting a virtual photon that is instantly absorbed by the acceptor.

lower-energy *f*-to-*f* transitions are spin-forbidden). In most fluorescent lamps, a low-pressure Hg-plasma discharge of dominant 254 nm emission acts as the excitation source. These photons are not sufficiently energetic to excite Tb^{3+} activators, but they can effectively excite the $4f^1 \rightarrow 5d^1$ transition of Ce^{3+} ions in an appropriate host, which can then efficiently transfer their energy to Tb^{3+} activators. Nearly all commercial green-emitting fluorescent-lamp phosphors ($LaPO_4$:Ce^{3+},Tb^{3+}; $CeMgAl_{11}O_{19}$:Tb^{3+}; $GdMgB_5O_{10}$:Ce^{3+},Tb^{3+}) rely on absorption by Ce^{3+} sensitizers and emission from Tb^{3+} activators. The spin-forbidden nature of the electronic transitions of Tb^{3+} means that effective energy transfer only occurs when the sensitizer and activator sites are in close proximity, where the Dexter electron-transfer mechanism is operative (< 1 nm).

7.8.8   Concentration Quenching and Cross Relaxation

Energy transfer can also occur between ions of the same type over longer distances, particularly for line emitters whose absorption and emission lines have near-perfect spectral overlap. Multiple transfers are common and can lead to energy migration over significant distances[25] in the host crystal, until an impurity or defect is encountered where non-radiative return to the ground state can occur (so-called killer sites). This type of quenching is called **concentration quenching**. It does not occur at low activator concentrations, where energy transfer is inhibited by the large distances between luminescent centers, but can become significant at higher concentrations. Phosphors that contain Eu^{3+}, Tb^{3+}, and Gd^{3+} activators often show maximum photoluminescence when these ions are present as substitutional dopants in low concentrations. Levels of substitution beyond a few atomic percent lead to a decrease in photoluminescence due to concentration quenching.

In phosphors where the activator has a large Stokes shift, there is minimal overlap between the absorption and emission spectra, limiting energy transfer and the effects of concentration quenching. This explains why self-activating phosphors like $CaWO_4$ and $Bi_4Ge_3O_{12}$, where the activator ion is present as a stoichiometric component of the host, nearly always contain activators that exhibit a large E_{Stokes}.

If the donor only transfers part of its energy, relaxing to a lower energy state but not all the way to the ground state, the energy-transfer process is called **cross-relaxation**. This process can quench certain emission lines while leaving others intact. In the case of Tb^{3+} pairs, the energy difference between $^5D_3$ and $^5D_4$ excited states approximately matches the energy difference between the $^7F_6$ ground state and the higher-lying $^7F_0$ state. At concentrations above 5%, cross-relaxation quenches emission from the $^5D_3$ level in favor of emission from the $^5D_4$ level (see Figure 7.25). The Tb^{3+} ion that is excited into the $^7F_0$ excited state via cross-relaxation can return to the $^7F_6$ ground state non-radiatively through internal conversion. Cross-relaxation is why blue emissions that originate from $^5D_3 \rightarrow {}^7F_J$ transitions are suppressed and green emissions from $^5D_4 \rightarrow {}^7F_J$ transitions are enhanced in phosphors

---

[25] In some instances, the number of energy transfers can exceed 10000 before decay.

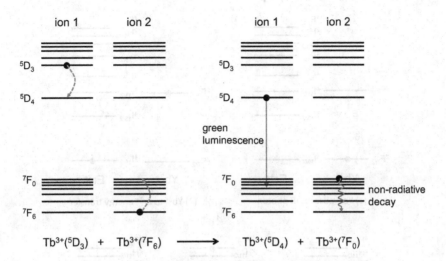

Figure 7.25 Cross-relaxation of Tb^{3+} through an energy transfer from ion 1, the donor, to ion 2, the acceptor. Following the energy transfer, ion 1 emits a green photon while ion 2 undergoes non-radiative relaxation to the ground state.

containing higher Tb^{3+} concentrations. Cross-relaxation is also responsible for quenching blue emissions of Eu^{3+} and the visible emissions of Sm^{3+} and Dy^{3+} at concentrations as low as ~1%.

7.8.9 Up-Conversion Photoluminescence

Up-conversion photoluminescence occurs when the energies of absorbed photons are lower than those of the subsequently emitted photons.[26] The most common up-conversion phosphors convert near-IR radiation to visible light. A typical up-conversion process is represented schematically in Figure 7.26 for $Y_2O_3{:}Er^{3+},Yb^{3+}$. The process begins with the absorption of a near-IR photon (λ = 980 nm) at Yb^{3+}, via the spin- and parity-allowed $^2F_{7/2} \rightarrow {}^2F_{5/2}$ transition. The Yb^{3+} then transfers its energy to Er^{3+} triggering a $^4I_{15/2} \rightarrow {}^4I_{11/2}$ transition. Absorption of a second near-IR photon at Er^{3+} further promotes the electron into the $^4F_{7/2}$ level, from which non-radiative relaxation to the $^2H_{11/2}$, $^4S_{3/2}$, or $^4F_{9/2}$ levels can occur. The final step is radiative return of the Er^{3+} ion to the ground state, leading to emission of 662 nm ($^4F_{9/2} \rightarrow {}^4I_{15/2}$) red light and/or green light with wavelengths of 525 nm ($^2H_{11/2} \rightarrow {}^4I_{15/2}$) and 550 nm ($^4S_{3/2} \rightarrow {}^4I_{15/2}$).

Yb^{3+} is the most widely used sensitizer in up-conversion phosphors because the $^2F_{7/2} \rightarrow {}^2F_{5/2}$ transition has a high absorption cross-section, and there are no accessible higher-energy excited states that permit up-conversion at the Yb^{3+} site. Blue emission can be obtained if Yb^{3+} is paired with a Tm^{3+} activator instead of Er^{3+}. The best host materials have minimal

---

[26] Because $E_{exc} < E_{em}$ up-conversion photoluminescence can be described as an anti-Stokes process.

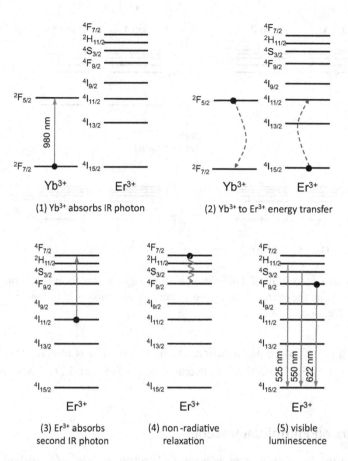

Figure 7.26 The five-step process of up-conversion photoluminescence in Y_2O_3:Yb^{3+},Er^{3+}.

electron–phonon coupling (Section 7.8.2) to reduce thermal relaxation and achieve long-lived excited states. Only then can the excited state persist long enough to allow a second photon to be absorbed before relaxing to the ground state. For example, $NaYF_4$ is a good up-conversion host, in part because it possesses phonon modes whose energies are much smaller than the energy separation of the $^2F_{7/2}$ and $^2F_{5/2}$ states on Yb^{3+}.[27] In hosts with higher-energy phonons, non-radiative decay of the Yb^{3+} $^2F_{5/2}$ excited state competes with absorption of a second photon.

Because up-conversion materials rely on a multiple-photon absorption process rarely found in nature, they are attractive as security markers to protect financial and government documents. Up-conversion photoluminescence is also being explored for applications in

[27] The dominant phonon mode in $NaYF_4$ has an energy of 0.044 eV (~350 cm^{-1}) which is ~29 times smaller than the 1.26 eV separation of the $^2F_{7/2}$ and $^2F_{5/2}$ states.

lasers, next-generation lighting, near-IR photon detectors, nanometer-sized biological labels, and night-vision goggles.

7.9 Electroluminescence

Electroluminescence is the direct conversion of electrical energy into optical energy. The basis of electroluminescence is the radiative recombination of electrons and holes, driven by an electric field. Although electroluminescence can take various forms, here we limit our discussion to inorganic **light-emitting diodes** (LEDs) where electron–hole recombination occurs at the interface between p- and n-type semiconductors, and **organic light-emitting diodes** (OLEDs) where recombination is driven by injecting current into films of organic semiconductors.

7.9.1 Inorganic Light-Emitting Diodes (LEDs)

The electrical properties of semiconductors are highly sensitive to the presence of impurities, particularly aliovalent substitutional impurities (Section 2.4). This topic is covered in detail in Chapter 10, but to understand the operation of LEDs a few basic concepts are touched upon here. If the substitutional impurity has more valence electrons than the atom for which it substitutes, the "extra" electrons are donated to the conduction band, and the semiconductor is said to be doped n-type. If the substitutional impurity has fewer electrons than the atom it replaces, it accepts electrons from the valence band, and the semiconductor is said to be doped p-type. The missing electron in the valence band carries a positive charge and is called a hole.

If p- and n-type semiconductors are joined together, the region where they meet is called a p–n junction. We'll discuss the fabrication, physics, and operation of p–n junctions in Section 10.3.5. For now, we only need to know that when an appropriate voltage is applied to a p–n junction, electrons and holes are driven to the interface and can recombine radiatively to generate light. The host semiconductor is typically the same on either side of the junction, and the energy of the emitted photons is determined by the band gap of the semiconductor. To favor radiative recombination, direct band-gap semiconductors are preferred for use in LEDs, though lower-efficiency LEDs can be made from indirect band-gap materials.

The first practical LEDs were made in the early 1960s from GaAs, which possesses a direct band gap of 1.43 eV, and therefore emits in the near-IR. LEDs made from pure GaAs are still used today as IR sources in fiber-optic communications. Another semiconductor used in LEDs is GaP, which has an indirect band gap of 2.26 eV (see Table 7.8) and emits green light. By forming $GaAs_{1-x}P_x$ solid solutions, it is possible to make LEDs that emit photons with energies intermediate between the two end members. Compositions close to $GaAs_{0.6}P_{0.4}$ are used in red LEDs, while orange and yellow LEDs can be made from compositions with higher GaP content. Unfortunately, the band gap becomes indirect and the efficiency goes down sharply when $x > 0.45$. The emission of GaAs LEDs can also be shifted to shorter wavelengths by forming solid solutions with

AlAs, a semiconductor with almost the same lattice parameter but an indirect band gap of 2.16 eV. The crossover from direct to indirect band gap occurs in the $Ga_{1-x}Al_xAs$ system for compositions with $x > 0.4$. By alloying InP ($E_g = 1.35$ eV, direct), AlP ($E_g = 2.43$ eV, indirect), and GaP it is possible to make $(Ga_{1-x}Al_x)_{1-y}In_yP$ LEDs that emit colors from green to the near-IR. In part because of the absence of arsenic, these have become the preferred semiconductors for yellow, orange, and red LEDs.

Fabricating blue, violet, and UV LEDs challenged researchers for many decades. Solid solutions between ZnS ($E_g = 3.6$ eV) and ZnSe ($E_g = 2.58$ eV) were investigated, but the high concentration of defects and difficulties in obtaining high-quality p–n junctions limited progress. In the late 1980s, Cree introduced a commercial blue LED based on silicon carbide, but the device efficiency was so low that it never gained popularity. In the 1990s, $Ga_{1-x}In_xN$ emerged as the material of choice for blue LEDs. Its wavelength can be tuned from 370 nm (pure GaN) to 470 nm by increasing the indium content. Longer-wavelength emission can be realized, but the efficiency drops as the indium content increases. This is due to compositional segregation upon cooling, caused by the limited solubility of InN in GaN. In Section 7.10.2, we will see that blue LEDs play a key role in modern solid state lighting. While green LEDs can be made in either the $Ga_{1-x}In_xN$ or $(Ga_{1-x}Al_x)_{1-y}In_yP$ systems, a high-efficiency green LED remains elusive. This challenge is sometimes referred to as the "green gap".

Box 7.2 Synthetic Methods: Synthesis and p-doping of GaN

GaN has the hexagonal wurtzite structure and a direct band gap of 3.4 eV. The first demonstration of a GaN LED was in 1972, but the device had an efficiency too low for practical applications. Two major hurdles prevented further progress. Firstly, the lack of a good lattice-matched substrate led to high defect densities in the films from which LEDs are made. Secondly, while n-type samples can be made by replacing some gallium with silicon, it proved difficult to reproducibly prepare p-type GaN.

In the mid 1980s, Akasaki and Amano used metalorganic vapor-phase epitaxy (MOVPE) to grow GaN films on sapphire substrates with lower defect concentrations than achieved previously [18]. Earlier attempts to use sapphire substrates had not produced high-quality films due to the 16% lattice mismatch with GaN. Their breakthrough was the deposition of a 30 nm buffer layer of polycrystalline AlN onto the sapphire substrate at 500 °C, followed by an annealing step at 1000 °C. This approach promotes the growth of small crystallites with preferred orientation upon which the GaN can subsequently nucleate and grow. While the GaN in close proximity to the AlN layer has a high concentration of dislocations, after a few microns the defect concentration is low enough for use in LED applications. Later, Nakamura [19] simplified the process by covering the sapphire at 600 °C with a 20 nm buffer layer of nearly amorphous GaN, before subsequently depositing a highly crystalline GaN film. Without this intermediate buffer layer, hexagonal columns of GaN grow that produce a rough surface and result in poor electrical properties.

Box 7.2 (cont.)

In the late 1980s, Akasaki and Amano observed that Zn-doped GaN emitted more blue light when the device was placed inside a scanning tunneling microscope. Subsequently, it was shown that the p-type behavior of Mg-doped GaN was significantly enhanced when the device was irradiated with low-energy electrons. Nakamura and coworkers [20] showed that these effects were caused by formation of hydride complexes (originating from trimethyl gallium and ammonia used in the epitaxial growth), which passivated the acceptor sites and limited the formation of holes. By irradiating with an electron beam, the unwanted hydrogen is expelled from the sample, activating the acceptors and improving device performance. Annealing at temperatures above 700 °C has a similar effect.

Building on these advances, Nakamura and co-workers produced a blue LED with a quantum efficiency of 2.7% from a $Ga_{1-x}In_xN/Ga_{1-x}Al_xN$ heterostructure in 1994. This demonstration revolutionized the compact-disc industry and triggered a massive surge in research and development activity. Quantum efficiencies have steadily increased over the intervening years and now exceed 80% in state-of-the-art GaN-based LEDs. In 2014, Amano, Akasaki, and Nakamura shared the Nobel Prize in Physics for their work.

7.9.2 Organic Light-Emitting Diodes (OLEDs)

OLEDs convert electrical energy to light through electron–hole recombination. While the overall process has many similarities with the inorganic LEDs just discussed, OLEDs offer distinct advantages. They can be very thin, and the methods of deposition (spin coating, vacuum deposition) are simpler, cheaper, and less energy-intensive than the methods used to deposit films of inorganic semiconductors. They can be made in almost any shape and deposited on flexible materials. Initially, OLED device performance was limited by the poor electrical conductivity of organic materials. However, the emergence of highly conductive polymers such as poly(N-vinylcarbazole) and poly(p-phenylene vinylene) reignited activities in this field. OLEDs now find widespread use in mobile phones, digital cameras, and flat-panel displays, where they compete with liquid-crystal displays (LCDs). Compared with LCDs, OLEDs are thinner, lighter, brighter, produce truer colors, refresh much faster, and consume less power.

A typical OLED is made up of several semiconducting organic materials, sandwiched between two electrodes, one of which must be transparent. A schematic of a relatively simple OLED made of three organic layers is shown in Figure 7.27. The organic semiconducting materials are invariably π-conjugated systems, either small organic molecules or conducting polymers (Section 10.5). When a voltage is applied, electrons are removed at the anode, which is equivalent to injecting holes into the HOMO of the hole-transport layer. At the same time, electrons are injected from the cathode into the LUMO of the electron-transport layer. To facilitate charge injection, the Fermi level of the anode should have a reasonably good

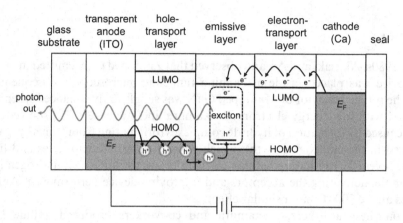

Figure 7.27 A schematic of an OLED with three organic layers: a hole-transport layer, electron-transport layer, and emissive layer.

energetic alignment with the HOMO of the hole-transport layer. This can be achieved by using a transparent conductor, such as $In_{2-x}Sn_xO_3$ (ITO), deposited onto a glass substrate that offers mechanical support and protects the active layers of the OLED from the environment. Similarly, the Fermi level of the cathode should be sufficiently high in energy that it lies close to the LUMO of the electron-transport layer. This necessitates using active metals like Ca, Ba, or alloys like $Mg_{1-x}Ag_x$, which must again be encapsulated due to their moisture sensitivity.

After charge injection, holes and electrons move in opposite directions under the external electrical field, hopping from molecule to molecule. Holes from the hole-transport layer migrate to the emissive layer where they encounter electrons that have migrated from the electron-transport layer. When they meet, the electrostatic attraction between the two oppositely charged particles leads to the formation of a bound electron–hole pair called an exciton. The **exciton** is a neutral quasi-particle that can radiatively decay through electron–hole recombination. The color of the emitted photon is determined by the HOMO–LUMO gap of the emissive layer minus the exciton binding energy, the attractive potential energy that holds the electron and hole together. Exciton binding energies in organic semiconductors are typically on the order of 0.3–0.5 eV.[28]

When an electron and hole meet, they may possess the same or the opposite spin, leading to the formation of both singlet ($S = 0$) and triplet ($S = 1$) excitons. Radiative decay from the triplet state (i.e. phosphorescence) is spin-forbidden, so triplet excitons decay predominantly through non-radiative pathways. Quantum-mechanical momentum conservation tells us that only 25% of all excitons are singlets. This means only one in four excitons decays radiatively, which limits OLED efficiencies. To circumvent this limitation, neutral organo-metallic complexes containing heavy metals like Ir or Pt can be incorporated into the

---

[28] Excitons can also form in inorganic materials, but the binding energies are an order of magnitude smaller than in organic materials.

emissive layer [21]. The presence of a heavy metal leads to strong spin–orbit coupling, which facilitates intersystem crossing and radiative decay of triplet excitons. The triplet state of the organometallic molecule is chosen to lie at a lower energy than that of the semiconducting emissive layer, so that triplet excitons migrate to the organometallic molecules. The incorporation of molecules containing heavy metals dramatically increases the brightness of OLEDs, facilitating their commercialization. In addition to boosting efficiency, the wavelength of photons emitted via triplet-exciton phosphorescence will in general differ from the photon emitted by singlet-exciton fluorescence, because phosphorescence is governed by the HOMO–LUMO gap of the organometallic complex, while fluorescence is predominantly governed by HOMO–LUMO gap of the host organic layer. The flexibility to engineer materials that emit at multiple wavelengths can be useful for applications.

7.10 Materials for Lighting

It's hard to conceive modern life without abundant, inexpensive electric lighting, but until the late nineteenth century cities and homes were still largely illuminated by flame. In the twentieth century, incandescent lighting became ubiquitous, but this revolutionary technology has since largely been replaced by more energy efficient methods of producing white light. Fluorescent lights are still widely used, but they are increasingly being displaced by high-efficiency blue LEDs coupled with down-conversion phosphors to produce white light.

Before discussing the phosphors used in both fluorescent and solid state LED lighting, we must understand the metrics used to evaluate intensity and color. The amount of visible light emitted by a light source is called the **luminous flux**. It is measured in lumens (lm) and is defined as the product of 1 candela (cd) times the solid angle in steradians (sr). A candela is roughly equivalent to the light given off by a single candle. More precisely, the candela is the luminous intensity per unit solid angle weighted by a luminosity function that models the sensitivity of the human eye.[29] **Luminous efficacy** is the ratio of luminous intensity out to electrical power consumed and is measured in units of lumens per watt (W). A typical 100 W incandescent light bulb gives off 15–17 lm/W. By comparison, fluorescent lights produce 50–100 lm/W, and phosphor-converted LEDs can achieve a luminous efficacy of 200 lm/W.

Although hot objects like the filament of an incandescent bulb give off a broad spectrum of light that spans the visible range, it is possible to mimic white light by mixing discrete colors. Mixing blue and yellow light, leads to a "cold" white light, while mixing red, blue, and green can produce a more natural white light, as described in Section 7.1. The color of a white light source is an important parameter and there are various metrics for quantifying color, two of which are touched upon below.

Color temperature is defined by comparing the output of a light source with the light emitted by a black-body radiator, which changes as the temperature increases in the

---

[29] The human eye is most sensitive to green light with a wavelength of 555 nm. A monochromatic source that emits 555 nm light with a radiant intensity of 1/683 W has a luminous intensity of 1 cd.

sequence: red, orange, yellowish-white, white, and ultimately bluish-white. Light sources with color temperatures >5000 K give off bluish-white light and are described as cool, whereas those with color temperatures ranging from 2500 to 3500 K are described as warm.[30] The **color-rendering index** (CRI) is an alternative metric. The CRI is a measure of the ability of a light source to reproduce the colors of various objects faithfully in comparison with an ideal natural light source. The CRI is calculated by comparing the reflection spectra of test colors to the spectra obtained when the same colors are irradiated with a source that simulates sunlight. The CRI for a true black-body radiator, like an incandescent lamp, would be 100. At the other extreme, a white object irradiated with a monochromatic light source, such as a laser, can only reflect a single color (the color of the source) and has a CRI of 0. In general, a CRI in the 70s would be considered acceptable for interior lighting applications; a score in the 80s, good; and a CRI in the 90s, excellent.

7.10.1 Fluorescent-Lamp Phosphors

Fluorescent lights rely upon phosphors to convert the UV light from a mercury discharge lamp to white light. The mercury atoms in the discharge emit about 85% of their radiation at 254 nm, and 12% at 185 nm, so fluorescent-lamp phosphors should absorb efficiently at these wavelengths. Discovery of the halo-apatite phosphor $Ca_5(PO_4)_3(F,Cl):Sb^{3+},Mn^{2+}$ in 1949 was a major turning point for the lighting industry. The Sb^{3+} ions have strong absorption peaks at 255 nm ($^1S_0 \rightarrow {}^3P_1$) and 205 nm ($^1S_0 \rightarrow {}^1P_1$). The Sb^{3+} ions emit a broad band centered in the blue region of the spectrum near 480 nm via a $^3P_1 \rightarrow {}^1S_0$ transition (Section 7.8.5). They also function as a sensitizer for the Mn^{2+} ions, which are not able to efficiently absorb light from the plasma discharge because all electronic transitions in this high-spin d^5 ion are spin-forbidden. Following energy transfer from Sb^{3+}, the Mn^{2+} activators emit a broad band of orange light near 580 nm via a $^4G_{5/2} \rightarrow {}^6S_{5/2}$ transition. The blue emission originating from Sb^{3+} and the orange emission of Mn^{2+} combine to create a "whitish" light. Increasing the Mn^{2+} concentration suppresses the blue emission and enhances the orange emission. Increasing the chloride content shifts the Mn^{2+} emission band to shorter wavelengths. In this way, color temperatures ranging from 2700 to 6500 K can be obtained. Unfortunately, there is a trade-off between optimizing the luminous efficacy and the CRI. If the brightness is high (efficacy ~80 lm/W) the CRI is on the order of 60. It is possible to increase the CRI to 90 through appropriate compositional tuning, but luminous efficacy drops to ~50 lm/W [16].

By combining phosphors that emit in narrow wavelength intervals in the red, green, and blue regions of the spectrum it is possible to achieve high luminous efficacy (~100 lm/W) and good color rendering (CRI ≈ 80). This type of fluorescent light, known as a tricolor lamp, might include the following phosphors: $BaMgAl_{10}O_{17}:Eu^{2+}$ that emits in the blue, near 450 nm; $(Ce_{1-x}Gd_x)MgB_5O_{10}:Tb^{3+}$ or $LaPO_4:Ce^{3+},Tb^{3+}$ that emit in the green, near 540 nm; and $Y_2O_3:Eu^{3+}$ that emits in the red, near 610 nm. All three phosphors have

---

[30] It's somewhat paradoxical that "cool" light has a higher color temperature than "warm" light.

individual quantum efficiencies near 90%. To achieve even better color rendering, needed in museums and store displays, alternative phosphors are chosen that shift the emission maximum of the blue phosphor to longer wavelengths.

7.10.2 Phosphor-Converted LEDs for White Light

Phosphor-converted LEDs are being pursued as *the* light source of the future, due to their combination of high luminous efficacy and long lifetime. In the most common configuration, a $Ga_{1-x}Al_xN$ LED acts as a source of blue light and is combined with phosphors that absorb a fraction of that blue light, then emit the longer-wavelength photons needed for white light. Phosphors for this application should: (a) possess high quantum efficiencies on the order of 90–95%, (b) have excitation maxima that are well matched to the light provided by the LED (emission maxima typically fall between 370 and 470 nm), (c) emit light at wavelengths that provide an optimal CRI, and (d) show minimal loss of efficiency at operating temperatures. To be commercially viable, they should also be non-toxic, inexpensive, and possess excellent thermal, chemical, and photochemical stabilities. Both the temperature and photochemical stability requirements for an LED phosphor are more stringent than those for a fluorescent-lamp phosphor, because the excitation densities of LED-based lighting systems ($\sim$30 W/cm^2) are about three times higher than those in fluorescent lights.

The simplest white LEDs combine blue light with a wavelength of $\sim$450 nm from a $Ga_{1-x}In_xN$ LED and yellow light from a phosphor where Ce^{3+} is doped into an appropriate host, usually $Y_3Al_5O_{12}$ (YAG). The $Y_3Al_5O_{12}:Ce^{3+}$ phosphor has broad absorption and emission bands that peak near 460 nm and 560 nm, respectively (Figure 7.28). By turning to hosts that are complex solid solutions, such as $(Y_{1-x}Gd_x)(Al_{1-y}Ga_y)_5O_{12}:Ce^{3+}$, the emission maximum can be tuned between 510 nm and 580 nm, thereby adjusting color temperature between 3000 K and 8000 K. The CRI of this type of phosphor-converted LED ranges from just below 75 to slightly above 80.

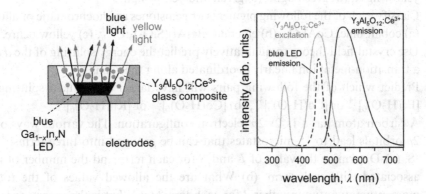

Figure 7.28 A schematic of a phosphor-converted white LED (left). The excitation and emission spectra of a $Y_3Al_5O_{12}:Ce^{3+}$ phosphor superimposed on the emission from a blue $Ga_{1-x}In_xN$ LED (right).

Trichromatic LEDs provide superior CRIs (>90) by combining green and red phosphors with the blue light emitted by the $Ga_{1-x}In_xN$ LED. The phosphors developed for fluorescent lamps are unsuitable for white LEDs because their excitation bands are not well matched to the output of the blue LED. Instead, researchers have largely concentrated on phosphors containing the band emitter Eu^{2+}, particularly for the red phosphor. The $[Xe]4f^65d^1 \rightarrow [Xe]4f^7$ transition of Eu^{2+} usually emits blue or green light when Eu^{2+} is incorporated into an oxide, but in sulfide, nitride, and oxynitride hosts the Eu^{2+} experiences a larger centroid shift, sinking the barycenter of the d-orbital set, and shifting the absorption and emission maxima to longer wavelengths (Figure 7.21).

Binary sulfides with the rock-salt structure like $SrS:Eu^{2+}$ and $CaS:Eu^{2+}$ emit in the red with maxima of 610 nm and 660 nm, respectively. The emission spectrum can be tuned by forming solid solutions or turning to more complex sulfides. Unfortunately, sulfides are prone to corrosion in the presence of minute traces of water, necessitating encapsulation. They also tend to suffer from strong thermal-quenching effects, as illustrated by the green phosphor $SrGa_2S_4:Eu^{2+}$ where the quantum efficiency drops from 75–80% at room temperature to 50% at 170 °C [22]. Nitridosilicates, like $Sr_2Si_5N_8:Eu^{2+}$ (emission maximum 630 nm) and nitridoaluminates, like $SrLiAl_3N_4:Eu^{2+}$ (emission maximum 650 nm), offer much better stability and excellent photoluminescent properties. Further improvements in luminous efficacy can be made by reducing the spectral overlap between the excitation band of the red phosphor and the emission band of the green/yellow phosphor, as well as limiting spillover of the red emission into the near-IR where the human eye cannot detect it [23, 24].

7.11 Problems

7.1 What is the wavelength (in nm), frequency (in s^{-1}), and color of a photon with an energy of 3.60×10^{-19} J? What is its energy in eV?

7.2 What color would a material be if it absorbed (a) red light, (b) blue and green light, (c) yellow light, (d) green light, (e) green and yellow light?

7.3 Classify each of the following pigments or gemstones as idiochromatic or allochromatic: (a) cobalt blue, $CoAl_2O_4$, (b) emerald, $Be_3Al_2(SiO_3)_6:Cr^{3+}$, (c) yellow ochre, $FeOOH$.

7.4 Use crystal-field theory to qualitatively predict the energy splitting of the d orbitals for a transition-metal ion linearly coordinated along the z axis.

7.5 Predict which of the following pairs will have larger ligand-field splitting (Δ_{oct}): (a) $[Fe(H_2O)_6]^{2+}$ or $[Co(H_2O)_6]^{3+}$, (b) $[Co(H_2O)_6]^{3+}$ or $[Rh(H_2O)_6]^{3+}$.

7.6 A carbon atom has a $1s^22s^22p^2$ electron configuration. The various ways of filling the $2p$ orbitals lead to 15 microstates that can be grouped into three terms: ^1D, ^3P, and ^1S. (a) Determine the value of L and S for each term and the number of microstates associated with each term. (b) What are the allowed values of the total angular momentum quantum number J for each term? (c) What is the energy order predicted by Hund's rules for the various $^{(2S+1)}L_J$ terms?

7.7 The UV–visible spectrum of the octahedral complex $[Ni(NH_2CH_2CH_2NH_2)_3]^{2+}$ is shown below. (a) Given that ethylenediamine ligand, $NH_2CH_2CH_2NH_2$, is a neutral bidentate ligand (both nitrogen atoms coordinate to the metal), determine the electron configuration of the nickel ion. (b) Use a correlation diagram to assign each of the three *d*-to-*d* transitions labeled as 1, 2, and 3 below. (c) What color would you predict for a solution of this complex ion from the spectrum below? (d) Compare these transitions to those observed for $[Ni(H_2O)_6]^{2+}$ (Table 7.5) and determine which complex ion has a larger Δ_{oct}.

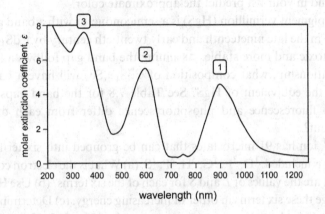

7.8 Which of the two correlation diagrams shown below would be appropriate for tetrahedral $[CoCl_4]^{2-}$? For the applicable correlation diagram, what are the correct term symbols for the lines labeled W, X, Y, and Z?

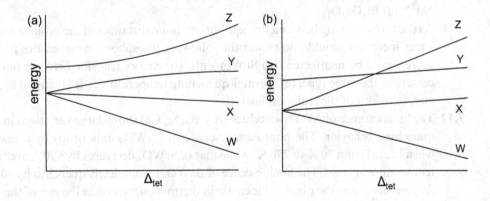

7.9 For which of the following ions is the *d*-to-*d* transition spin-forbidden, Laporte-forbidden, both or neither: (a) tetrahedral Co^{2+}, (b) square-planar Cu^{2+}, (c) octahedral Mn^{2+} (high spin), (d) trigonal-bipyramidal Mn^{3+}?

7.10 Tetrathiomolybdate MoS_4^{2-} is tetrahedral and features a strong $t_1 \rightarrow e$ LMCT absorption at 470 nm. (a) What color would you predict for $(NH_4)_2MoS_4$? (b) What is the energy in eV of the LMCT transition and how does it compare with MoO_4^{2-}? Why does the LMCT energy shift in the way that it does? (c) The compound $Li_3NbS_4 \cdot 2TMEDA$ (TMEDA = tetramethylenediamine) features isolated NbS_4^{3-} tetrahedra. Do you expect the lowest-energy LMCT absorption peak of NbS_4^{3-} to occur at shorter or longer wavelengths than MoS_4^{2-}? Explain your reasoning.

7.11 Which semiconductors in Table 7.8 will have colors other than white or black? For each compound in your list predict the approximate color.

7.12 The red pigment vermillion (HgS) is a semiconductor with a band gap of 2.0 eV. It was replaced in the late nineteenth and early twentieth century by $CdS_{1-x}Se_x$ pigments that are less toxic and more stable. Assuming the band gap follows a linear Vegard's law-type relationship, what composition of $CdS_{1-x}Se_x$ will have a band gap of 2.00 eV, roughly the equivalent of HgS? See Table 7.8 for the band gaps of CdS and CdSe.

7.13 How do fluorescence and phosphorescence differ from each other in theory and experiment?

7.14 The Pr^{3+} ion has 91 microstates that can be grouped into six terms before spin–orbit coupling is included: ^1D, ^3F, ^1G, ^3H, ^3P, ^1S. (a) What is the electron configuration of Pr^{3+}? (b) What are the values of L and S for each of the six terms? (b) Use Hund's first two rules to arrange these six terms in order of increasing energy. (c) Determine the allowed values of the total angular-momentum quantum number J for each of the six terms, and give the lowest-energy $^{(2S+1)}L_J$ term for each.

7.15 For each of the following phosphors, identify the type of electronic transition responsible for luminescence and predict whether it will show weak, moderate, or strong electron–phonon coupling: (a) YVO_4, (b) Y_2O_3:Eu^{3+}, (c) $SrGa_2S_4$:Eu^{2+}, (d) $Y_3Al_5O_{12}$:Nd^{3+}, (e) $Bi_4Ge_3O_{12}$.

7.16 As the difference ΔQ between the equilibrium bond distances of the ground and excited states increases, would you expect the following phosphor characteristics to increase, decrease, or be unaffected: (a) Stokes shift, (b) energy transfer between luminescent centers of the same type, (c) thermal-quenching temperature, (d) Huang–Rhys parameter, (e) width of the emission line(s)?

7.17 The luminescence of AWO_4 scheelites (A = Ba, Sr, Ca) shows large variations in thermal-quenching behavior. The photoluminescence of $CaWO_4$ only drops by a few percent when heated from 90 K to 270 K, while that of $SrWO_4$ decreases by ~90% over the same temperature interval. The luminescence of $BaWO_4$ is completely quenched by 90 K. How do you rationalize the observed increase in thermal quenching as the size of the alkaline-earth ion increases?

7.18 Orange, red, and near-IR LEDs are often made from $GaAs_{1-x}P_x$ solid solutions. For compositions with $x < 0.45$, a direct band gap is observed, while for larger x the band gap is indirect. Assuming a linear Vegard's law relationship between the end members

GaAs (E_g = 1.43 eV) and GaP (E_g = 2.25 eV), determine the shortest wavelength emitted from a composition with a direct band gap. What color of light would such an LED emit?

7.19 The phosphor $BaMgAl_{10}O_{17}$:Eu^{2+},Mn^{2+}, whose excitation and emission spectra are shown below, is of interest as a combined blue and green phosphor in plasma-display panels. The Eu^{2+} ions absorb at 336 nm and emit in the blue with a maximum of 450 nm. There is also energy transfer from Eu^{2+} to Mn^{2+} that leads to the green emission at 512 nm. (a) What are the electronic transitions responsible for emission on Eu^{2+} and Mn^{2+}? (b) What is the Stokes shift for Eu^{2+}? (c) What is the most likely mechanism of energy transfer from Eu^{2+} to Mn^{2+}, Förster (FRET) or Dexter?

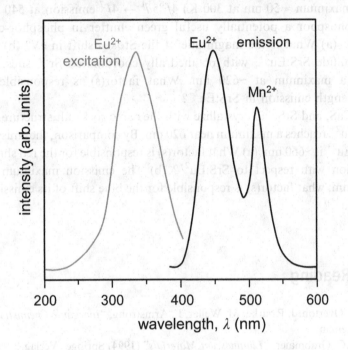

7.20 $CeMgAl_{11}O_{19}$:Tb^{3+} is a green phosphor used in tricolor fluorescent lights. Ce^{3+} ions in the host absorb photons from the Hg-plasma discharge with an excitation maximum near 270 nm and emit via a $[Xe]4f^0 5d^1 \rightarrow [Xe]4f^1$ transition with a Stokes shift of 0.9 eV. The Tb^{3+} dopant emits via *f*-to-*f* transitions from a $^5D_4$ excited state to various ^7F_J levels (see Figure 7.25), with a maximum near 540 nm. (a) At what wavelength will the emission maximum of the Ce^{3+} ion fall? (b) Which lanthanoid ion is responsible for the green light? Why is the other lanthanoid ion needed? (c) What mechanism is responsible for energy transfer between Ce^{3+} and Tb^{3+}?

7.21 When Ce^{3+} ions substitute for La^{3+} in a $LaCl_3$ host, the lowest-energy peak in the photoluminescence excitation spectrum falls at 281 nm, whereas a similar substitution in the double-perovskite host $Cs_2NaLaCl_6$ shifts the lowest-energy peak in the excitation spectrum to 342 nm. The La^{3+} ion in $LaCl_3$ sits in a nine-coordinate tricapped trigonal-prismatic site, while in $Cs_2NaLaCl_6$ the La^{3+} ion is octahedrally coordinated. (a) Would you predict a larger, smaller, or similar centroid shift in $Cs_2NaLaCl_6$? (b) How will differences in the coordination environment shift the lowest-energy Ce^{3+} excitation peak in $Cs_2NaLaCl_6$? (c) Is the observed shift in the Ce^{3+} excitation spectrum due to the centroid shift, the change in coordination, or both?

7.22 In thiogallate phosphor, $SrGa_2S_4$:Eu^{2+}, Eu^{2+} substitutes for Sr^{2+} and is surrounded by a square antiprism of eight sulfide ions. The lowest-energy $5f^7 \rightarrow 4f^65d^1$ excitation is centered at ~480 nm, and the relatively narrow (full width at half maximum ≈ 50 nm at 300 K) $4f^65d^1 \rightarrow 4f^7$ emission at 540 nm, which makes this phosphor a potentially useful green emitter in phosphor-converted tricolor LEDs. (a) What is the magnitude of the Stokes shift in eV? (b) The binary rock-salt sulfide SrS:Eu^{2+}, with octahedrally coordinated Sr^{2+} sites, emits in the red with a maximum at ~620 nm. What factor(s) is responsible for the longer-wavelength emission in SrS:Eu^{2+}?

7.23 SrS, CaS, and $SrSe$ all crystallize with the cubic rock-salt structure. The emission for SrS:Eu^{2+} reaches a maximum near 620 nm. By comparison, the emission maximum for CaS:Eu^{2+} is ~660 nm. (a) What factor(s) is responsible for the red shift of the CaS:Eu^{2+} emission with respect to SrS:Eu^{2+}? (b) The emission maximum for $SrSe$:Eu^{2+} is ~570 nm; what factor(s) is responsible for the blue shift of its emission with respect to SrS:Eu^{2+}?

7.12 Further Reading

P. Atkins, T. Overton, J. Rourke, M. Weller, F. Armstrong, "*Inorganic Chemistry*" 6th edition (2014) W.H. Freeman.

G. Blasse, B.C. Grabmaier, "*Luminescent Materials*" (1994) Springer Verlag.

J.A. Duffy, "*Bonding, Energy Levels and Bands in Inorganic Solids* (1990) Longman Group Essex.

J.E. House, "*Inorganic Chemistry*" 2nd edition (2013) Academic Press.

K. Nassau, "*The Physics and Chemistry of Color: The Fifteen Causes of Color*" 2nd edition (2001) John Wiley and Sons.

J.C. Phillips, J.A. Van Vechten, "Dielectric classification of crystal structures, ionization potentials and band structures" *Phys. Rev. Lett.* **22** (1969), 705–708.

C. Ronda (editor), "*Luminescence: From Theory to Applications*" (2008) Wiley-VCH.

7.13 References

[1] E. Climent-Pascual, J. Romero de Paz, J. Rodriguez-Carvajal, E. Suard, R. Saez-Puche, "Synthesis and characterization of the ultramarine-type analog $Na_{8-x}[Si_6Al_6O_{24}](S_2,S_3,CO_3)_{1-2}$" *Inorg. Chem.* **48** (2009), 6526–6533.

[2] M. Ganio, E.S. Pouyet, S.M. Webb, C.M. Schmidt Patterson, M.S. Walton, "From lapis lazuli to ultramarine blue: Investigating Cennino Cennini's recipe using sulfur K-edge XANES" *Pure Appl. Chem.* **90** (2018), 463–475.

[3] A.E. Smith, H. Mizoguchi, K. Delaney, N.A. Spaldin, A.W. Sleight, M.A. Subramanian, "Mn^{3+} in trigonal bipyramidal coordination: A new blue chromophore" *J. Am. Chem. Soc.* **131** (2009), 17084–17086.

[4] A.E. Smith, M.C. Comstock, M.A. Subramanian, "Spectral properties of the UV absorbing and near-IR reflecting blue pigment $YIn_{1-x}Mn_xO_3$" *Dyes Pigm.* **133** (2016), 214–221.

[5] S. Jitsuhiro, H. Nakai, M. Hada, H. Nakatsuji, "Theoretical study on the ground and excited states of the chromate anion CrO_4^{2-}" *J. Chem. Phys.* **101** (1994), 1029–1036.

[6] A.C. Stückl, C.A. Daul, H.U. Güdel, "Excited-state energies and distortions of d^0 transition metal tetraoxo complexes: A density functional study" *J. Chem. Phys.* **107** (1997), 4606–4617.

[7] M. Jansen, H.P. Letschert, "Inorganic yellow-red pigments without toxic metals" *Nature* **404** (2000), 980–982.

[8] S. Romero, A. Mosset, J.C. Trombe, "Study of some ternary and quaternary systems based on γ-Ce_2S_3 using oxalate complexes: Stabilization and coloration" *J. Alloys Compd.* **269** (1998), 98–106.

[9] S. Kufner, J. Furthmüller, L. Matthes, M. Fitzner, F. Bechstedt, "Structural and electronic properties of α-tin nanocrystals from first principles" *Phys. Rev. B* **87** (2013), 235307.

[10] J.T.W. de Hair, G. Blasse, "Luminescence of octahedral uranate group" *J. Lumin.* **14** (1976), 307–323.

[11] P.S. Peijzel, A. Meijerink, R.T. Wegh, M.F. Reid, G.W. Burdick, "A complete $4f^n$ energy level diagram for all trivalent lanthanide ions" *J. Solid State Chem.* **178** (2005), 448–453.

[12] A.R. Sharits, J.F. Khoury, P.M. Woodward, "Evaluating $NaREMgWO_6$ (RE = La, Gd, Y) doubly ordered double perovskites as Eu^{3+} phosphor hosts" *Inorg. Chem.* **55** (2016), 12383–12390.

[13] P. Dorenbos, "Ce^{3+} 5d-centroid shift and vacuum referred 4f-electron binding energies of all lanthanide impurities in 150 different compounds" *J. Lumin.* **135** (2013), 93–104.

[14] P. Dorenbos, "Crystal field splitting of lanthanide $4f^{n-1}$ 5d levels in inorganic compounds" *J. Alloys Compd.* **341** (2002), 156–159.

[15] N. Lakshminarasimhan, S. Jayakiruba, K. Prabhavathi, "$Ba_2Mg(BO_3)_2:Bi^{3+}$– A new phosphor with ultraviolet light emission" *Solid State Sci.* **72** (2017), 1–4.

[16] G. Blasse, B.C. Grabmaier, "*Luminescent Materials*" (1994) Springer Verlag, Berlin.

[17] C.W.M. Timmermans, O.B. He, G. Blasse, "The luminescence of $Bi_2Ge_3O_9$" *Solid State Commun.* **42** (1982), 505–507.

[18] H. Amano, N. Sawaki, I. Akasaki, Y. Toyoda, "Metalorganic vapor phase epitaxial growth of a high quality GaN film using an AlN buffer layer" *Appl. Phys. Lett.* **48** (1986), 353–355.

[19] S. Nakamura, "GaN growth using GaN buffer layer" *Jpn. J. Appl. Phys.* **30** (1991), L1705–1707.

[20] S. Nakamura, N. Iwasa, M. Senoh, T. Mukai, "Hole compensation mechanism of p-type GaN films" *Jpn. J. Appl. Phys.* **31** (1992), 1258–1266.

[21] H. Sasabe, J. Kido, "Multifunctional materials in high-performance OLEDs: Challenges for solid-state lighting" *Chem. Mater.* **23** (2011), 621–630.

[22] Z. Xinmin, W. Hao, Z. Heping, S. Qiang, "Luminescent properties of $SrGa_2S_4:Eu^{2+}$ and its application in green-LEDs" *J. Rare Earths* **25** (2007), 701–705.

[23] L. Wang, R.-J. Xie, T. Suehiro, T. Takeda, N. Hirosaki, "Down-conversion nitride materials for solid state lighting: Recent advances and perspectives" *Chem. Rev.* **118** (2018), 1951–2009.

[24] Z. Xia, Q. Liu, "Progress in discovery and structural design of color conversion phosphors for LEDs" *Prog. Mater. Sci.* **84** (2016), 59–117.

8 Dielectrics and Nonlinear Optical Materials

Materials can be broadly classified as either conductors or insulators. Conductors can further be divided into electronic and ionic conductors, which are covered in Chapters 10 and 13, respectively. Insulating materials, which are also referred to as dielectric materials, are the subject of this chapter. Dielectric materials find widespread application in electronics. In some cases, the role of the dielectric is simply to insulate active circuit components from each other. In other instances, the dielectric plays an active role as a capacitor, antenna, or filter. In the latter case, the response of the material to external electric fields is of critical importance.

Nonlinear optical (NLO) materials are a subset of the broader class of dielectric materials. When electromagnetic radiation passes through a NLO material new frequencies of radiation are generated. A familiar example is found in green laser pointers, where an NLO crystal is used to convert infrared (IR) light into green light. As we will see NLO effects are only observed in materials that meet specific symmetry criteria. In the later sections of this chapter we'll look at the characteristics of several important NLO materials.

8.1 Dielectric Properties

When an electric field of intensity[1] E is applied to a dielectric material, an electrical polarization develops in response to the applied field. In the following sections, we explore this response, starting with a macroscopic description followed by a closer look at the microscopic origins of the dielectric response.

---

[1] An electric field is a vector field, but when homogeneous or when speaking of a local value of unambiguous direction, only the magnitude of its intensity, E, is sufficient. The unit of E or E is volts per meter. 1 V/m exerts a force of 1 N onto a point charge of 1 C.

8.1.1 Dielectric Permittivity and Susceptibility

A parallel-plate capacitor consists of two conducting plates, each with an area, A, separated by a distance, d, as shown in Figure 8.1. In a vacuum, its capacitance[2] is C_0:

$$C_0 = \varepsilon_0 \frac{A}{d} \tag{8.1}$$

where ε_0 is the electric constant.[3] The quantity of charge, Q, that can be stored[4] is a product of the capacitance and voltage drop across the dielectric, $V = Ed$:

$$Q = CV \tag{8.2}$$

For a parallel-plate capacitor in a vacuum, the stored charge, Q_0, is given by:

$$Q_0 = C_0 V = \left(\frac{\varepsilon_0 A}{d}\right) Ed = \varepsilon_0 A E \tag{8.3}$$

If we place a dielectric material between the plates while maintaining the voltage drop V, the amount of stored charge will increase ε_r times, from Q_0 to Q, due to the polarization of the material.[5] The polarization results in a separation of positive and negative charges, by which the dielectric material acquires an internal field that opposes the applied field E. More charge is then needed to compensate this dielectric polarization while maintaining the

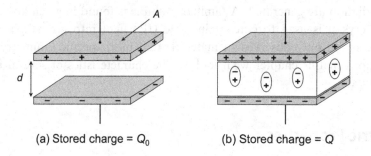

(a) Stored charge = Q_0 (b) Stored charge = Q

Figure 8.1 A parallel-plate capacitor with (a) a vacuum (b) a dielectric material.

---

[2] Its SI unit is one farad, 1 F = 1 C/1 V; it is the charge (in coulombs) a capacitor will accept for the potential across it to change by 1 V.

[3] The electric constant is a fundamental constant whose value (to five significant figures) is 8.8542×10^{-12} F/m. It is also referred to as the vacuum permittivity or the permittivity of free space.

[4] The stored charge is the charge at one of the plates, typically the + charge is considered.

[5] Alternatively, we could also keep the charge constant in our thought experiment. Then the voltage and E would decrease due to the presence of the dielectric material.

Table 8.1 Relative permittivity ε_r for a variety of materials at room temperature.

Substance	ε_r	Substance	ε_r
Air	1.0006	SnO_2	13
Teflon	2	ZrO_2	22
C (diamond)	6	TiO_2	94
Si	12	Perovskite dielectrics	
NaF	5	$BaSnO_3$	18
MgO	10	$BaZrO_3$	43
SiO_2	4	$CaTiO_3$	165
CaF_2	7	$SrTiO_3$	330
PbI_2	21	$KTaO_3$	242
H_2O	80	$Ba_3ZnTa_2O_9$	30

original voltage of the circuit. Consequently, the capacitance will increase ε_r times, from C_0 to C. The factor ε_r is termed the **relative dielectric permittivity**[6]:

$$\varepsilon_r = C/C_0 \tag{8.4}$$

Relative dielectric permittivities of substances vary across a wide range, as shown in Table 8.1, but most insulating solids have $\varepsilon_r < 30$. The reasons for the high permittivity values of water and certain metal oxides, such as the titanates, will become clear later in the chapter. In Section 8.4, we will discuss a special class of dielectric materials called ferroelectrics whose relative dielectric permittivity can be in the tens of thousands.

Upon subtracting 1 from the factor ε_r (i.e. subtracting the contribution of the vacuum from the total capacitance), we obtain a dimensionless value that quantifies the ability of a dielectric material to become polarized in an external field, its **electric susceptibility**, χ_e:

$$\chi_e = \varepsilon_r - 1 \tag{8.5}$$

8.1.2 Polarization and the Clausius–Mossotti Equation

The capacitance of our parallel-plate capacitor increases when a material occupies the space between the plates, because the electric field creates dipoles within the material that oppose the applied field. Thus, polarization is the material property of interest to us. Quantitatively we define bulk **polarization**, P, of a dielectric substance with the following equation (magnitudes-only for simplicity):

---

[6] Terms relative dielectric permittivity and dielectric constant are used interchangeably.

$$P = \varepsilon_0 \chi_e E \qquad (8.6)$$

where the polarization P is expressed in units[7] of F/m. Substituting for χ_e from Equation (8.5), we obtain:

$$P = \varepsilon_r \varepsilon_0 E - \varepsilon_0 E \qquad (8.7)$$

In this expression, we then define the **electric displacement**,[8] which describes the electric field inside the dielectric, as $D = \varepsilon_r \varepsilon_0 E$. We see that D has two components; the electric displacement in a vacuum and the polarization of the material:

$$D = \varepsilon_0 E + P \qquad (8.8)$$

Because (as noted above) D, E, and P are all vectors, the polarization of a crystal in an electric field is anisotropic.

Let's consider polarization from induced local dipole moments within the dielectric. The bulk polarization P [C/m^2] can be obtained by summing up the individual induced dipole moments, p [C m], over the N atoms, ions, or molecules that occupy a cubic meter. For the sake of simplicity, we will speak of polarization of neutral atoms (e.g. Si) for the remainder of this section, but the expressions that are derived here are equally valid for ionic and molecular solids.

The size of the local induced dipole moment p is proportional to the local electric field, E_{loc}, experienced by the atom, with **polarizability**, α, of that atom acting as the proportionality constant:

$$p = \alpha E_{loc} \qquad (8.9)$$

The bulk polarization P (the induced dipole moment per m^3) of a crystal containing N atoms per cubic meter is then:

$$P = N\alpha E_{loc} \qquad (8.10)$$

Unfortunately, E_{loc} differs from the applied external field E because the induced dipole moments of neighboring atoms alter the field strength within the dielectric, hence E_{loc} depends on the atom's location in the crystal. Equation (8.10) is therefore of limited utility as we cannot directly measure E_{loc}. Fortunately, it is possible to derive a relationship between the local and external fields:

---

[7] Because susceptibility is dimensionless, the polarization P has units of C/m^2 since ε_0 [F/m] $\times E$ [V/m] $\times \chi_e = P$ [F V/m^2)] and 1 F = 1 C/V. P is also called the polarization density, because it represents an electric dipole moment per cubic meter [C m/m^3].

[8] Electric displacement is also referred to as electric-flux density or electric induction (the latter in analogy with magnetic induction, see Chapter 9) and is generally a vector.

$$E_{\text{loc}} = \frac{1}{3}(\varepsilon_{\text{r}} + 2)E \qquad (8.11)$$

Substituting Equation (8.11) into Equation (8.10), we obtain an expression for P, which we can equate to polarization as expressed in Equation (8.7). After simplifying this expression, we obtain the **Clausius–Mossotti equation**, which relates the atomic polarizability α, a microscopic quantity, to the dielectric permittivity ε_{r}, a macroscopic quantity:

$$\alpha = \frac{3\varepsilon_0}{N}\left(\frac{\varepsilon_{\text{r}} - 1}{\varepsilon_{\text{r}} + 2}\right) \left[\text{SI}: \ \alpha \text{ in C m}^2/\text{V}, \varepsilon_0 \text{ in C}/(\text{V m}), N \text{ in m}^{-3}\right] \qquad (8.12)$$

Because an average E_{loc} was assumed, Equation (8.12) is strictly speaking only valid for crystal structures of certain symmetry that makes them isotropic with respect to the dielectric-constant tensor. An involved evaluation shows that only cubic crystals meet this criterion, but in practice it works well for other crystal systems if the structure is not too anisotropic.

It is often convenient to work in CGS (centimeter–gram–second) units where $\varepsilon_0 = 1$ and hence α has units of volume. Because $\varepsilon_{\text{r}} = 1 + 4\pi\chi_{\text{e}}$ in the CGSes (electrostatic) system, the Clausius–Mossotti equation takes on a somewhat different form:

$$\alpha = \frac{3}{4\pi}V_{\text{a}}\left(\frac{\varepsilon_{\text{r}} - 1}{\varepsilon_{\text{r}} + 2}\right)[\text{CGSes}: \alpha \text{ in Å}^3, V_a \text{ in Å}^3] \qquad (8.13)$$

where V_{a} is the volume per atom or formula unit, determined by dividing the unit-cell volume by the number of the formula units it contains. For example, the sphalerite form of ZnS ($F\bar{4}3m$, $a = 5.32$ Å, $Z = 4$, Figure 1.33) has a $V_{\text{a}} = (5.32 \text{ Å})^3/4 = 37.6$ Å^3.

8.1.3 Microscopic Mechanisms of Polarizability

Polarizability is a key parameter for understanding dielectric permittivity of materials because it directly relates to properties of the atoms that make up the solid. In an ideal dielectric with no electronic or ionic conductivity,[9] there are three microscopic polarization mechanisms that can contribute to the dielectric response of a material: the **electronic polarizability**, α_{e}; the **ionic polarizability**, α_{i}; and the **dipolar polarizability**, α_{d}. These mechanisms are illustrated in Figure 8.2.

The electronic polarizability α_{e}, which is present in all substances, arises from polarization of the negatively charged electron cloud surrounding the nucleus. The ionic polarizability α_{i} arises from field-induced displacements of cations and anions in opposite directions. In ionic solids, including most technologically important dielectric materials, ionic polarizability is the principal source of polarization. The dipolar polarizability α_{d} arises from reorientations

---

[9] In ionic conductors, very high polarizability can occur due to migration of ions. This type of polarization is called space-charge polarizability.

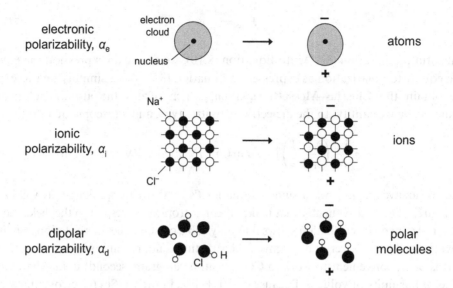

Figure 8.2 Three microscopic polarizability mechanisms. The change from left to right shows the polarization in response to an applied electric field.

of polar molecules in response to the applied field. It is normally only relevant in molecular substances containing polar molecules, such as H_2O and HCl. The α_d contribution tends to be very temperature-dependent because the reorientations tend to freeze out at low temperatures. For most technologically important dielectric materials, the dipolar polarizability is not relevant.

The magnitude of these three contributions typically follows the order $\alpha_e < \alpha_i < \alpha_d$. The relative magnitude of these polarizabilities is reflected in the relative permittivity values given in Table 8.1. In covalent network solids like diamond and silicon, α_e is the only source of polarization and ε_r is fairly small. The ε_r values for "ionic" materials, where α_i plays an important role, vary over a wide range. Large values are often seen in compounds containing transition metals with a d^0 electron configuration. We will explore the reasons for this behavior in Section 8.6. The high permittivity of water stems from its being the only substance in Table 8.1 where α_d contributes.

8.1.4 Frequency Dependence of the Dielectric Response

Polarization involves reorganization of charge. Electronic polarization depends on deformation of electron density, ionic polarization depends on displacements of ions, and dipolar polarization involves reorientations of molecules. All three mechanisms can respond to a static electric field, but when subjected to an alternating field of increasing frequency, v, a limit is eventually reached where the field direction changes faster than the charged entity can follow. When this occurs, a given polarization mechanism no longer contributes to the dielectric permittivity. For dipolar polarization this occurs when $v > 10^9$ Hz (GHz or microwave frequencies), for ionic polarization when $v > 10^{13}$ Hz (THz or IR frequencies), and for electronic polarization when $v > 10^{17}$ Hz (X-rays). This response is illustrated in Figure 8.3, where ε_r is plotted as a function of v for a material where at low frequencies all three polarization mechanisms contribute.[10]

For most useful dielectrics, $\alpha_d = 0$, and the relative permittivity at low frequencies is determined by α_i and α_e. The sum of these two contributions can be determined by measuring the **static dielectric constant**, denoted ε_r ($v \to 0$) or ε_{stat}, via capacitance measurements described in Section 8.1.1. To separate the contributions of α_i and α_e, one then measures the response of the dielectric to visible light where α_i is frozen out. The high-frequency or **optical dielectric constant**, ε_{opt}, is related to the refractive index n (see Section 8.7), through the relationship:

$$\varepsilon_{opt} = n^2 \tag{8.14}$$

For a covalent material like diamond, ε_{stat} and ε_{opt} are very similar ($\varepsilon_{stat} = 5.68$, $\varepsilon_{opt} = 5.66$), which tells us that α_i is essentially zero. For ionic materials like NaCl ($\varepsilon_{stat} = 5.90$, $\varepsilon_{opt} = 2.34$) and LiCl ($\varepsilon_{stat} = 11.95$, $\varepsilon_{opt} = 2.78$), the ionic and electronic contributions are of comparable magnitude. For compounds of unusually high permittivities, like TiO_2 ($\varepsilon_{stat} = 94$, $\varepsilon_{opt} = 7$), the ionic contribution is dominant for reasons that will be discussed in Section 8.6.

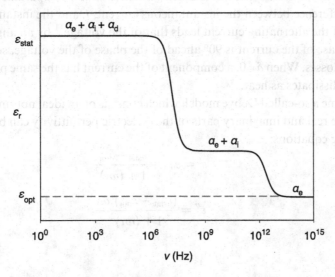

Figure 8.3 Dielectric permittivity ε_r as a function of the frequency v of an alternating applied field for a hypothetical dielectric material where all three polarization mechanisms contribute at low frequencies.

[10] If a material exhibits ionic conductivity, the space-charge polarizability relaxes out when $v > 10^6$ Hz.

8.1.5 Dielectric Loss

An ideal capacitor stores energy without loss or dissipation of energy. In theory, a dielectric material can act as an ideal capacitor, but, in practice, some energy is always lost. Resistive heating due to non-zero conductivity in the dielectric material is one obvious source of loss. However, losses can also occur in an alternating electric field even in the absence of conductivity.

When an alternating field is applied, the capacitor electrodes alternate between positive and negative charges, and the polarization within the capacitor's dielectric tries to follow suit. Some energy is required to drive the movement needed to redistribute charge in the dielectric: molecular reorientations for dipolar polarization, or lattice vibrations (phonons) for ionic polarization. This energy eventually converts to heat and is therefore called **dielectric loss**. In a simplified approach, the dielectric permittivity in an alternating field is a complex number, ε_r^*:

$$\varepsilon_r^* = \varepsilon' + i\varepsilon'' \tag{8.15}$$

where ε' and ε'' are the real and imaginary parts of the dielectric permittivity, respectively. The real part is a measure of the polarization that is in phase with the external field (i.e. it keeps up with the field), while the imaginary part corresponds to polarization that is out of phase with the field (i.e. it lags behind the field). The real component represents the energy that is stored while the imaginary component is a measure of dielectric loss.

Quantitatively, the dielectric loss is defined as $\tan\delta = \varepsilon''/\varepsilon'$. The parameter δ is a measure of the phase difference between the instantaneous current, i, and the instantaneous voltage, v. The phase of the alternating current leads that of the voltage[11] by an angle of $90° - \delta$. When $\delta = 0$, the phase of the current is $90°$ ahead of the phase of the voltage, so that $i \times v = 0$ and there are no losses. When $\delta \neq 0$, a component of the current has the same phase as the voltage, and energy dissipates as heat.

If we assume a so-called Debye model, which consists of an ideal non-interacting system of dipoles,[12] the real and imaginary parts of the dielectric permittivity can be approximated by the following equations:

$$\varepsilon' = \varepsilon_{opt} + \frac{\varepsilon_{stat} - \varepsilon_{opt}}{1 + (\omega\tau)^2} \tag{8.16}$$

$$\varepsilon'' = \frac{(\varepsilon_{stat} - \varepsilon_{opt})\,\omega\tau}{1 + (\omega\tau)^2} \tag{8.17}$$

---

[11] In a capacitor, the current precedes the voltage; current must flow into a capacitor to establish the voltage drop.

[12] The Debye model most accurately approximates dipolar polarization. For ionic and electronic polarization, a harmonic-oscillator model is a better approximation, but leads to a slightly more complicated expression. Qualitatively, both models give a similar dependence on frequency.

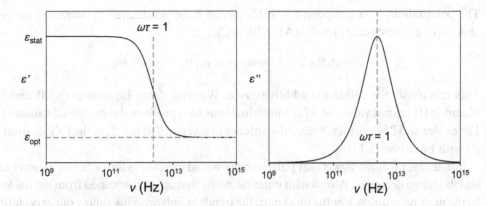

Figure 8.4 The frequency dependence of ε' and ε'' for a material where only ionic and electronic polarization contribute ($\alpha_d = 0$).

where ε_{stat} and ε_{opt} represent the static and optical dielectric constants, the angular frequency, $\omega = 2\pi v$, is calculated from the frequency v of the alternating electric field and from the characteristic relaxation time τ of the polarization. Typical values of τ are on the order of $\sim 10^{-11}$ s for dipolar polarizability and $\sim 10^{-14}$ s for ionic polarizability. The behaviors of ε' and ε'' as a function of frequency are plotted in Figure 8.4.

The imaginary part of the dielectric permittivity ε'' and the dielectric loss, $\tan \delta = (\varepsilon''/\varepsilon')$, reach a maximum when the frequency of the alternating external field matches the resonant frequency of the material, $\omega\tau = 1$. For such frequencies, vibrational modes are excited, and the absorbed energy is dissipated as heat. At frequencies sufficiently low that $\omega\tau \ll 1$, each of the polarization mechanisms saturates before the field reverses; Equation (8.16) reduces to $\varepsilon' \approx \varepsilon_{stat}$ and Equation (8.17) to $\varepsilon'' \approx 0$. At frequencies high enough to make $\omega\tau \gg 1$, the above equations reduce to $\varepsilon' \approx \varepsilon_{opt}$ and $\varepsilon'' \approx 0$. Losses do not occur in the low-frequency limit because there is no coupling between the applied field and lattice vibrations, while in the high-frequency limit the lattice vibrations cannot respond to the rapidly oscillating electric field.

8.2 Dielectric Polarizabilities and the Additivity Rule

From a materials-design perspective, it would be attractive to predict the dielectric permittivity from composition and structure. This is possible when the Clausius–Mossotti equation (8.13) is rearranged to give ε_r as a function of the polarizability α and volume per polarized atom V_a:

$$\varepsilon_r = \frac{V_a + 2\alpha(4\pi/3)}{V_a - \alpha(4\pi/3)} \tag{8.18}$$

The polarizability of a compound $A_xB_yC_d \ldots$ can be approximated by summing the polarizabilities of its constituent "ions" $\alpha(A)$, $\alpha(B)$, $\alpha(C)$, ...:

$$\alpha(A_xB_yC_z\ldots) = x\alpha(A) + y\alpha(B) + z\alpha(C) + \ldots \qquad (8.19)$$

This relationship is called the **additivity rule**. Working from Equations (8.18) and (8.19), Shannon [1] refined values of α for individual ions to reproduce the measured values of ε_r for 129 oxides and 25 fluorides. Selected α values are given in Table 8.2, in the CGSes units Å^3 to go with Equation (8.13).

Intuitively, we might expect that polarizability would increase as the radius of the ion increases and its charge decreases. Atoms with outer electrons that are well screened from the nucleus will be the most polarizable. For the most part, the trends in polarizability follow our expectations by increasing down a group and decreasing across a period. There are two important exceptions, each associated with a specific electron configuration. Firstly, ions with an s^2 lone-pair configuration (Tl^+, Pb^{2+}, Sb^{3+}, Bi^{3+}, Se^{4+}, and Te^{4+}) have larger polarizabilities than expected. Secondly, the polarizabilities of transition-metal "ions" do not continue to decrease when the oxidation state climbs above +3. Both effects result from mixing of empty cation orbitals with occupied anion orbitals and will be examined more closely in Section 8.6.

To illustrate how the ionic polarizabilities and the additivity rule can be used to estimate the dielectric permittivity of a compound, consider the cubic garnet (Section 1.5.2), $Y_3Al_5O_{12}$, with $a = 12.01$ Å. There are eight formula units per unit cell ($Z = 8$), so the volume per formula unit is $(12.01 \text{ Å})^3/8 = 216.4$ Å^3. Ionic polarizabilities from Table 8.2 and Equation (8.19) give the

Table 8.2 Dielectric polarizabilities α (in Å^3) for selected ions from ref. [1].

Li^+	Be^{2+}					B^{3+}			O^{2-}	F^-
1.20	0.19					0.05			2.01	1.62
Na^+	Mg^{2+}					Al^{3+}	Si^{4+}	P^{5+}		
1.80	1.32					0.79	0.87	0.27		
K^+	Ca^{2+}	Sc^{3+}	Ti^{4+}	V^{5+}	Zn^{2+}	Ga^{3+}	Ge^{4+}	As^{5+}		
3.83	3.16	2.81	2.93	2.92	2.04	1.50	1.63	1.72		
Rb^+	Sr^{2+}	Y^{3+}	Zr^{4+}	Nb^{5+}	Cd^{2+}	In^{3+}	Sn^{4+}	Sb^{3+}	Te^{4+}	
5.29	4.24	3.81	3.25	3.97	3.40	2.62	2.83	4.27	5.23	
Cs^+	Ba^{2+}	La^{3+}	Hf^{4+}	Ta^{5+}	Hg^{2+}	Tl^+	Pb^{2+}	Bi^{3+}		
7.43	6.40	6.07	---	4.73	---	7.28	6.58	6.12		

Ce^{4+}	Pr^{3+}	Nd^{3+}	Pm^{3+}	Sm^{3+}	Eu^{3+}	Gd^{3+}	Tb^{3+}	Dy^{3+}	Ho^{3+}	Er^{3+}	Tm^{3+}	Yb^{3+}	Lu^{3+}
3.94	5.32	5.01	---	4.74	4.53	4.37	4.25	4.07	3.97	3.81	3.82	3.58	3.64

polarizability estimate per formula unit: $\alpha(Y_3Al_5O_{12}) = 3\alpha(Y^{3+}) + 5\alpha(Al^{3+}) + 12\alpha(O^{2-})$ = $3(3.81 \text{ Å}^3) + 5(0.79 \text{ Å}^3) + 12(2.01 \text{ Å}^3) = 39.5 \text{ Å}^3$. Plugging $\alpha = 39.5 \text{ Å}^3$ and $V_a = 216.4 \text{ Å}^3$ back into Equation (8.18) yields $\varepsilon_r = 10.7$, in good agreement with the measured value of 10.6.

It is important to note that this approach is only valid for "normal" dielectric materials. Deviations between the calculated and observed values may be attributed to ferroelectricity (Section 8.4), piezoelectricity (Section 8.5), conductivity (ionic or electronic), the presence of rattling ions, compressed ions, and dipolar impurities (such as H_2O).

The polarizabilities in Table 8.2 account for both ionic and electronic contributions to polarization. Shannon and Fischer [2] developed a set of polarizabilities that contain only the electronic contribution by substituting ε_{opt} for ε_r in Equation (8.18). These values are of use for estimating refractive indices, simulating spectra, and various approaches to modeling extended solids. For more details on the relationship between electronic polarizabilities and refractive index, see Section 15.4. As one would expect, the electronic polarizabilities are smaller than the dielectric polarizabilities, because the ionic polarization no longer contributes.

Box 8.1 Materials Spotlight: Microwave dielectrics

Many technologies, including cellular phones, use electromagnetic radiation with wavelengths in the microwave region (frequencies in the GHz range) to transmit information. Several key components in a microwave communication network, such as antennas, transmitters, and filters, are built from dielectric ceramics. One such component is the microwave resonator used in a cell-phone base station, which links the radio signals that cellular phones send and receive with the network switching subsystem. The resonator is typically a hollow ceramic cylinder, called a puck. The dimensions of the puck (on the order of centimeters) are chosen so that it can sustain a standing wave within its body when exposed to microwaves of a specific resonant frequency. This allows it to transmit microwaves that fall within a narrow frequency range and filter out other frequencies.

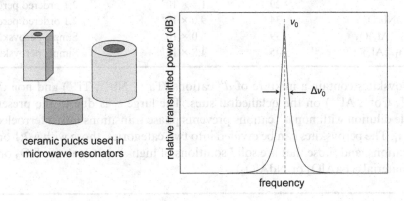

ceramic pucks used in
microwave resonators

Box 8.1 (cont.)

A dielectric material useful for this application should meet three key requirements. Firstly, because the size of the resonator is proportional to $1/\sqrt{\varepsilon_r}$, a high dielectric permittivity is desirable because it allows for a reduction in the size of the resonator. In practice, the optimal range is $20 < \varepsilon_r < 50$. Secondly, the selectivity of the resonator increases as the dielectric loss decreases. The selectivity is defined by a parameter called the **quality factor**, Q, which is approximately equal to $1/(\tan \delta)$. If the transmitted power is plotted as a function of the microwave frequency, a peak is observed at the resonant frequency, as shown above. The quality factor is equal to the resonant frequency, ν_0, divided by the width of the peak, $\Delta\nu_0$, as measured at a transmitted power 3 dB below the peak. Higher Q reduces the crosstalk within the specified frequency range. The value of Q is dependent on the resonant frequency of the puck ν_0, but the product $Q \times \nu_0$ should in theory be constant and is often used as a figure of merit for comparing different materials. For use in cell-phone base stations, $Q \times \nu_0$ should be equal to or larger than 4×10^4. Thirdly, the resonant frequency should be nearly independent of temperature. The temperature dependence is given by the **temperature coefficient of resonant frequency**, τ_f, which quantifies the change in the resonant frequency $\Delta\nu_0$ with a change in temperature ΔT. It is typically expressed in units of ppm/K, $(10^6\Delta\nu_0/\nu_0)/\Delta T$. For use as a microwave resonator, τ_f should be smaller than ± 3 ppm/K.

Most ceramics have ε_r too small for microwave resonators. Ferroelectrics are not suitable because their losses are much too high, which drives down Q. Incipient ferroelectrics like $SrTiO_3$ are not suitable because their permittivity changes too much with temperature. The best microwave resonators are insulating materials with relatively high, temperature-stable, permittivities and very low losses. Many of the best microwave dielectrics come from the perovskite family as seen in the table below. Note the τ_f values for all materials listed below are nearly zero [3].

Material	ε_r	$Q \times \nu_0$ (GHz)	Structure
$Ba_3MgTa_2O_9$	24	2.5×10^5	2:1 ordered perovskite
$Ba_3ZnTa_2O_9$	29	1.5×10^5	2:1 ordered perovskite
$Ba_3(Co_{1-x}Zn_x)Nb_2O_9$	34	9.0×10^4	2:1 ordered perovskite
$(Sr_{1-x}La_x)(Ti_{1-x}Al_x)O_3$	39	6.0×10^4	Simple perovskite
$(Ca_{1-x}Nd_x)(Ti_{1-x}Al_x)O_3$	45	4.8×10^4	Simple perovskite

These perovskites contain a mixture of d^0 cations (Ta^{5+}, Nb^{5+}, Ti^{4+}) and non d^0 cations (Mg^{2+}, Zn^{2+}, Co^{2+}, Al^{3+}) on the octahedral sites. The large ε_r is due to the presence of d^0 cations, while dilution with non d^0 cations prevents phase transitions into a ferroelectric state and reduces τ_f. The perovskites can be divided into two categories, those with a 2:1 ordering of octahedral cations, and those that are solid solutions of high-ε_r materials ($SrTiO_3$ or $CaTiO_3$) and low-ε_r materials ($LaAlO_3$ or $NdAlO_3$).

8.3   Crystallographic Symmetry and Dielectric Properties

All dielectric materials are polarized by their interaction with an external electromagnetic field, but certain properties we encounter later in this chapter can only arise in crystals with specific types of symmetry. In this section, we briefly explore the symmetry restrictions on pyroelectricity, ferroelectricity, piezoelectricity, and second-harmonic generation.

Crystals are classified as either centrosymmetric if they possess an inversion center, or non-centrosymmetric if they do not. Of the 32 crystallographic point groups (Section 1.1.3), 21 are non-centrosymmetric (Table 8.3). Piezoelectricity and second-harmonic generation are permitted in all non-centrosymmetric crystals, except cubic crystals with 432 point-group symmetry, which includes space groups with primitive cubic ($P432$, $P4_132$, $P4_232$, $P4_332$), body-centered cubic ($I432$, $I4_132$), and face-centered cubic ($F432$, $F4_132$) Bravais lattices, illustrating that it is the point-group symmetry, not the Bravais lattice, that dictates which of the above dielectric phenomena are permitted.

Materials that form crystals with a macroscopic electric dipole moment are called **polar materials**.[13] Such a crystal is said to be spontaneously polarized, because the non-zero electrical polarization forms spontaneously, even in the absence of an external electric field. To be polar, a material must not only lack an inversion center, it must also possess a **polar axis**. A rotational axis 1, 2, 3, 4, or 6 is polar if its positive and negative ends are not equivalent, hence the axis *must not* be perpendicular to a twofold axis or a mirror

Table 8.3 Non-centrosymmetric point-group symmetries sorted according to whether they allow piezoelectricity (20 piezoelectric crystal classes), both piezoelectricity and pyroelectricity (10 polar crystal classes), or neither.

| Crystal system | Piezoelectric crystal classes | | Neither |
	Polar crystal classes		
Triclinic	1	$\bar{1}$	–
Monoclinic	2, m	–	–
Orthorhombic	$mm2$	222	–
Trigonal	$3m$, 3	32	–
Tetragonal	$4mm$, 4	422, $42m$, $\bar{4}$	–
Hexagonal	$6mm$, 6	622, $62m$, $\bar{6}$	–
Cubic	–	23, $43m$	432

[13] Somewhat confusingly, in the crystallographic literature the terms polar material and polar crystal class are used to describe any material that crystallizes in a non-centrosymmetric space group. Throughout this book, we use the more restrictive condition that a crystal must contain a polar axis to be considered a polar material.

plane.[14] This narrows the list to 10 polar crystal classes given in Table 8.3.[15] Pyroelectricity and ferroelectricity are only permitted in polar crystals.

8.4   Pyroelectricity and Ferroelectricity

In all polar materials, the magnitude of the bulk polarization changes with temperature; a property referred to as **pyroelectricity**. **Ferroelectrics** are a special class of polar materials where the permanent electric dipole moment of a crystal can be reversed through application of an external electric field. In ferroelectric materials there is a temperature called the **Curie temperature**, T_C, above which thermally induced vibrations of the ions lead to a loss of spontaneous polarization[16], and hence a loss of ferroelectricity. Above T_C, a ferroelectric material enters a **paraelectric** state. Ferroelectric materials are used in capacitors due to their (often) large dielectric permittivity. Because the direction of the electrical polarization switches permanently after application of an electrical field, they also find use in data storage (ferroelectric random-access memory).

Although the term pyroelectric is general, it is often used in a narrower sense to describe those polar materials that are not ferroelectric. Compounds with the wurtzite structure ($P6_3mc$ of crystal class $6mm$, Figure 1.33), such as GaN, provide one such example. Alternating cation and anion layers stack perpendicularly to the c axis in the wurtzite structure, creating a net dipole moment parallel to the c axis. One would have to reverse the layer-stacking sequence from Ga–N–Ga–N– ... to N–Ga–N–Ga– ... to switch the polarity of the crystal. The energy barrier to this reorganization of the crystal is too large to be overcome with an external electric field, hence GaN is a pyroelectric but not a ferroelectric. Gallium nitride and other pyroelectric materials find application as IR (heat) sensors.

8.4.1   Ferroelectricity in BaTiO$_3$

To understand the origins of ferroelectricity, we take a closer look at the archetypal perovskite ferroelectric, $BaTiO_3$. We first relate the dielectric properties of $BaTiO_3$ to changes in its average crystal structure that occur at the Curie temperature, a model that provides a good entry point for understanding structure–property relationships in ferroelectrics. Subsequently, we investigate the phase transitions of $BaTiO_3$ and see how the local structure may differ from the average crystal structure.

Above $T_C \approx 400$ K, $BaTiO_3$ has the cubic perovskite structure (Section 1.5.3). On cooling below T_C, the symmetry lowers from cubic (space group $Fm\bar{3}m$) to tetragonal ($P4mm$). This

---

[14] Neumann's principle states that the physical property of a crystal must have at least the symmetry of the crystal's point group. The polarization arrow does not have a twofold axis or m perpendicular to it.

[15] The oft-used term "unique axis" can be misleading because one may not realize which axis is meant and that it includes the onefold axes in crystal classes 1 and $m \equiv 1m1$ (the extended Hermann–Maugin symbol; see Appendix B).

[16] A crystal that has a net electric dipole moment is said to be spontaneously polarized, because the non-zero electrical polarization persists in the absence of an external electric field.

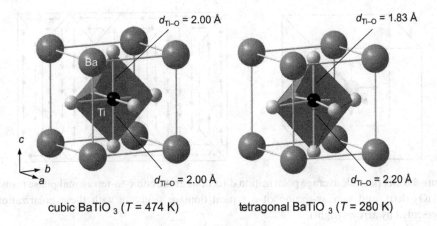

Figure 8.5 The average structure of $BaTiO_3$ above (left) and below (right) T_C. The four equatorial Ti–O bond lengths remain ~2.00 Å in the tetragonal structure.

distortion of the average structure involves a displacement of the titanium ions along the tetragonal fourfold axes, accompanied by a shift of the oxide ions in the opposite direction, resulting in the formation of one short and one long Ti–O bond per octahedron, as shown in Figure 8.5. These cooperative ionic displacements cause the crystal to develop a net polarization, whereby the crystal face the titanium ions approach develops a positive charge and the opposite face a negative charge.

While this simple picture captures the microscopic origins that lead to the formation of local dipole moments in $BaTiO_3$, the macroscopic properties are strongly impacted by the formation of domains. To understand what domains are and how they affect the macroscopic properties of a ferroelectric crystal, we need to investigate the crystal over a much larger length scale.

The high-temperature cubic structure possesses fourfold axes running parallel to each of the three Cartesian directions. Consequently, each titanium atom has six equivalent directions (see Figure 8.6) along which it can displace and still attain the tetragonal symmetry of the average structure below T_C. Over the length scale of hundreds or thousands of unit cells, the average titanium displacements are likely to be in the same direction, but, when viewed over a larger length scale, the crystal actually consists of individual **domains**, each with one of the six different polarization directions. Domains in ferroelectric materials typically have dimensions on the micron length scale, with domain walls a few unit cells thick. The formation of ferroelectric domains helps to minimize the electrostatic and elastic energy of the crystal.[17] If the displacements in neighboring domains are in opposite directions, the boundary is said to be a 180°

[17] Spontaneous polarization within a crystal leads to the formation of a surface charge that creates a depolarizing field oriented oppositely to the bulk polarization. The electrostatic energy associated with the depolarizing field may be minimized if the ferroelectric splits into domains with oppositely oriented polarization. Domain formation also helps to offset strains that arise due to changes in shape of the crystal that occur as it is cooled through the paraelectric–ferroelectric phase transition.

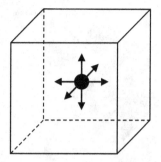 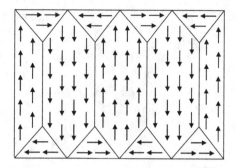

Figure 8.6 Six possible average polarization directions at the cubic-to-tetragonal phase transition in bulk BaTiO$_3$ (left), and a cross-section of a typical domain structure with these polarization directions represented by arrows (right).

domain wall. If the displacements are at right angles to each other, the boundary is a 90° domain wall. Both types of domains are illustrated schematically in Figure 8.6.

The polarization of a ferroelectric crystal as a function of the applied electric field is illustrated in Figure 8.7. This curve is called a **hysteresis loop**, because the polarization that occurs on increasing the field is not reproduced on subsequently decreasing the field. Due to the formation of oppositely polarized domains, the bulk polarization P of a ferroelectric crystal that has never been exposed to an electric field will be nearly zero. As the applied field increases (pathway $1 \rightarrow 2$ in Figure 8.7), the polarization increases until it reaches the **saturation polarization**, $P_{\text{saturation}}$ (point 2 in Figure 8.7). This corresponds to the state where all domains are oriented parallel to the applied field. The saturation polarization of a good ferroelectric typically ranges from 0.1–1 C/m^2. Upon removing the applied field, the polarization decreases, but does not go to zero. The value of the polarization that remains is called the **remanent polarization**, P_{remanent} (point 3 in Figure 8.7). If a field is now applied in the opposite direction, the polarization continues to decrease until it reaches zero. The magnitude of this field is the (negative) **coercive field**, $-E_{\text{coercive}}$ (point 4 in Figure 8.7). Upon increasing this reverse field, the polarization saturates at $-P_{\text{saturation}}$ (under the **switching field**, point 5). From there, the polarization follows the hysteresis loop marked with a solid line in a counterclockwise direction for further changes in applied field. We will see in Chapter 9 that ferromagnets behave similarly, with magnetic dipole moments taking the place of electric dipoles.

The relative permittivity of a BaTiO$_3$ single crystal is plotted as a function of temperature in Figure 8.8. The peak at $T_C \approx 400$ K separates the paraelectric cubic phase from the tetragonal ferroelectric phase. Such a peak at T_C is a characteristic of ferroelectric materials. The peak near 270 K is associated with a transition to an orthorhombic structure, while the peak near 180 K corresponds to a transition to a rhombohedral structure. The tetragonal, orthorhombic, and rhombohedral forms are all ferroelectric.

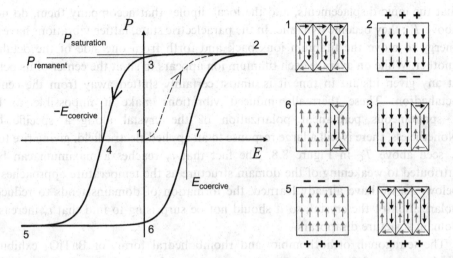

Figure 8.7 Hysteresis loop for a ferroelectric material and schematic evolution of a vertical domain pair (for simplicity) at several points on the loop. The domain structure at point 7 is identical to point 4, but with an oppositely polarized external field.

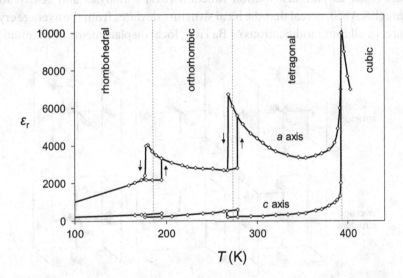

Figure 8.8 The dielectric permittivity ε_r of a single crystal of $BaTiO_3$, measured parallel to the a and c axes at a non-zero electric field, as a function of temperature. Data taken from reference [4].

Naively, one might expect ε_r to be quite small in the cubic structure because there are no permanent local dipole moments, yet for temperatures just above T_C the relative permittivity is higher than it is at most temperatures below T_C. The reason for this behavior is that the ionic displacements, and the local dipoles that accompany them, do not vanish above T_C; they become dynamic. In the paraelectric state, lattice vibrations have sufficient energy to move the titanium ions back and forth from one side of the octahedron to another. While on average each titanium ion appears to sit at the center of its octahedron, at any given instant in time it is almost certainly shifted away from the center of its octahedron. These thermally induced vibrations make it impossible to build up a spontaneous permanent polarization of the crystal along a specific direction. Nonetheless, there is still a large response to an applied electric field, giving rise to the high ε_r seen above T_C in Figure 8.8. The fact that ε_r reaches a maximum can be largely attributed to weakening of the domain structure as the temperature approaches T_C from below. As we have already learned, the formation of domains tends to reduce the net polarization of the crystal, so it should not be surprising to find that ε_r increases as the domain structure disappears.

The tetragonal, orthorhombic, and rhombohedral forms of $BaTiO_3$ exhibit average displacements of Ti towards the corner, edge, and face of the octahedra, respectively, as shown in the lower half of Figure 8.9, each slightly distorting the original cubic cell. Every time the symmetry changes, the domain structure also changes, and this explains the permittivity maximum at each phase transition (Figure 8.8). Interestingly, probes of local structure, such as pair distribution function (PDF) analysis and X-ray absorption fine structure (EXAFS), reveal that the local structure deviates from the average crystallographic structure. In all four modifications of $BaTiO_3$, local displacements of titanium toward a face

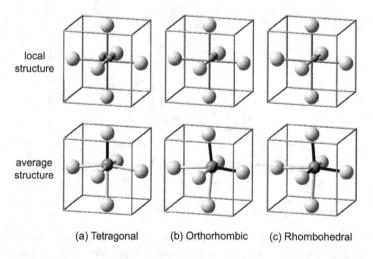

local
structure

average
structure

(a) Tetragonal   (b) Orthorhombic   (c) Rhombohedral

Figure 8.9 The Ti-displacement disorder in $BaTiO_3$ (top), and the resulting average structure (bottom). The small spheres shown in the upper half of the figure represent disordered Ti positions along the body diagonals of the high-temperature cubic unit cell. They are slightly exaggerated for illustrative purposes. The degree of disorder gradually decreases as the symmetry changes from (a) tetragonal, to (b) orthorhombic, to (c) rhombohedral. The darkly shaded bonds are the shortest average Ti–O distances. See ref. [5] for more details.

of its coordination octahedron are observed, giving three short and three long Ti–O bonds [5]. In other words, the local structure around each Ti looks very much like the low-temperature rhombohedral structure shown in Figure 8.9c, even at temperatures where the average structure is tetragonal, orthorhombic, or cubic.

8.4.2 Antiferroelectricity

Not all compounds that have local dipole moments exhibit a spontaneous polarization. **Antiferroelectric** materials possess local dipoles, but their arrangement is such that for every dipole there is an adjacent dipole that is oriented in the opposite direction (antiparallel). Because the dipoles cancel, the antiferroelectric state is nonpolar and does not exhibit a hysteresis loop. Nevertheless, the local dipoles still become dynamic above a certain temperature, called the antiferroelectric Curie temperature, where the material crosses over to a paraelectric state.

Subtle changes in bonding can trigger a crossover between ferroelectric and antiferro-electric behavior. For example, the ferroelectric perovskites $PbTiO_3$ (T_C = 763 K) and $KNbO_3$ (T_C = 707 K) are closely related to antiferroelectric $PbZrO_3$ (T_C = 606 K) and $NaNbO_3$ (T_C = 911 K). In these materials, changes in the perovskite tolerance factor (Section 1.5.3) modify the long-range coupling of the local distortions sufficiently to alter the competition between competing ferroelectric and antiferroelectric ground states. The ferro-electric KH_2PO_4 (T_C = 123 K) and antiferroelectric $(NH_4)H_2PO_4$ (T_C = 148 K) are another such pair; in this instance differences in hydrogen bonding are responsible for their different dielectric properties.

Despite the cancellation of dipoles, antiferroelectrics can exhibit high dielectric permittiv-ity, especially near the Curie temperature. For example, the relative permittivity of $PbZrO_3$ is ~3000 just a few degrees below its Curie temperature. Because the bonding interactions that differentiate ferroelectrics from antiferroelectrics can be subtle, it is sometimes possible to drive a reversible transition from the antiferroelectric state to a ferroelectric state with an applied field, as shown in Figure 8.10. At low fields, the response is linear, but at higher fields a hysteresis loop is observed, as in $PbZrO_3$.

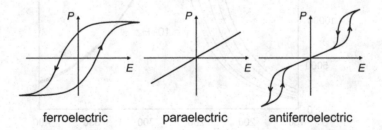

ferroelectric   paraelectric   antiferroelectric

Figure 8.10 Polarization as a function of electric field for a ferroelectric, a paraelectric, and an antiferro-electric material that reversibly transforms to a ferroelectric state when a sufficiently large field is applied.

Box 8.2 Nanoscale Concepts: Relaxor ferroelectrics

Relaxor ferroelectrics, often referred to simply as relaxors, are frustrated ferroelectrics, typically via substitution-induced structural disorder. Many relaxors are perovskites with the generic formula $A(M'_{1-x}M''_x)O_3$ where some degree of disorder exists among the octahedral-site metals, M' and M''. In such compounds it is common that a lone-pair cation, often Pb^{2+}, resides on the A-site and a high-valent d^0 transition metal occupies the M'' site. Importantly, both are prone to off-center displacements that create local dipole moments, as discussed in Section 8.6.

The properties of relaxors and normal ferroelectrics are similar, but the cooperative phase transitions that lead to the formation of large, spontaneously polarized domains are suppressed in relaxors. Like normal ferroelectrics, the dielectric permittivity of a relaxor goes through a maximum as a function of temperature. Unlike normal ferroelectrics where the permittivity peaks sharply near T_C and shows little frequency dependence, relaxors exhibit a broad and frequency-dependent ε_r peak, as shown below for one of the most important relaxors, $Pb(Mg_{1/3}Nb_{2/3})O_3$ (at alternating-current frequencies of 10^{-2}, 10^0, 10^2, 10^4, and 10^5 Hz) [6]. The lack of an abrupt change in crystal structure upon cooling through the permittivity maximum is another feature of relaxors that distinguishes them from normal ferroelectrics. Instead, the polar distortions are confined to nanoscale islands called **polar nanoregions** (PNRs). Whereas in normal ferroelectrics such regions would grow and coalesce upon cooling into the ferroelectric state, in relaxor ferroelectrics the sizes of the PNRs change with temperature yet remain separated from each other by a nonpolar matrix. Although relaxors do not undergo cooperative phase transitions, above the **Burns temperature** the PNRs disappear, and the relaxor enters a paraelectric state. The Burns temperature of $Pb(Mg_{1/3}Nb_{2/3})O_3$ is ~630 K.

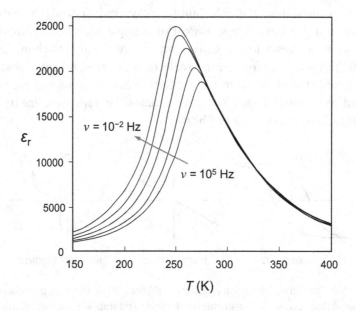

Each PNR possesses a large dipole moment that does not strongly couple to other PNRs.

Box 8.2 (cont.)

The presence of large, uncoupled, dipole moments can produce very large dielectric permittivities, as high as 20000 to 35000 in some relaxors. This makes them attractive for applications in multilayer capacitors. The large permittivities also lead to large electrostriction effects (changes in volume in response to an applied electric field). Furthermore, the absence of a conventional domain structure allows for a fast electrostrictive response to applied fields that makes relaxors ideal for micropositioners in optical devices, low-frequency transducers in sonar systems, and high-frequency transducers in biomedical devices.

8.5 Piezoelectricity

When a suitably oriented mechanical stress is applied to a polar crystal, charges will develop on opposite faces of the crystal. This phenomenon is known as the **piezoelectric effect**. When a suitably oriented electric field is applied to a polar crystal, it will change the dimensions of the crystal, a phenomenon known as the **converse piezoelectric effect**. Both effects are illustrated schematically in Figure 8.11.

The piezoelectric effect originates from distortions of coordination polyhedra that create or modify local dipoles. In polar materials like tetragonal $BaTiO_3$, compression along certain directions amplifies the polarization while in other directions compression diminishes the polarization, as shown in Figure 8.12. Application of an external stress can also induce a net polarization in a nonpolar crystal if its point group belongs to one of the piezoelectric crystal classes listed in Table 8.3. To see how a nonpolar entity with no inversion center can develop a dipole moment, consider the nonpolar trigonal-planar coordination polyhedron. As illustrated in Figure 8.12c, the appropriate external stresses can deform the equilateral triangle in a manner that creates a local dipole.

Given the geometric considerations discussed in the preceding paragraph, it should not come as a surprise that many piezoelectrics are either ferroelectric materials (all ferroelectrics are piezoelectric, but the converse is not true) or structures containing nonpolar building units that lack an inversion center, like trigonal planes or tetrahedra. The applications of piezoelectrics are numerous and amongst the most important of the materials encountered in this chapter. One of the oldest is in sonar, where piezoelectric transducers convert pressure changes associated with sound waves into electrical signals. Piezoelectrics are also used to generate and detect ultrasonic pulses used in medical imaging technologies like ultrasound. Because electrical charges build up when a pressure is applied to a piezoelectric crystal, they can be used to create sparks that ignite flammable gases. On the other hand, the converse

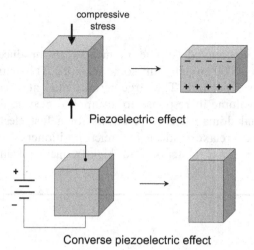

Figure 8.11 The piezoelectric effect where compressive stress leads to polarization of the crystal (top), and the converse piezoelectric effect where an applied electric field changes the shape of the crystal (bottom).

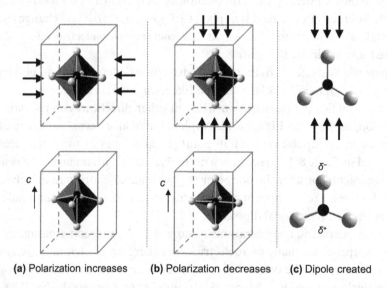

(a) Polarization increases   **(b)** Polarization decreases   **(c)** Dipole created

Figure 8.12 Piezoelectric materials: changes in local polarization in response to stress. (a) A compression in the *ab* plane of tetragonal $BaTiO_3$ amplifies the local polarization, (b) a compression along *c* diminishes it. (c) External stress induces local polarization (fractional charges δ) in a trigonal-planar polyhedron.

piezoelectric effect is used in devices to reproducibly control minute movements by applying a voltage, for example to position the tip of an atomic force microscope.

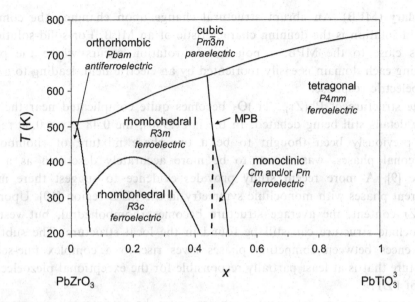

Figure 8.13 The PbZr$_{1-x}$Ti$_x$O$_3$ phase diagram.

While tetrahedral-network solids find use in some applications, such as quartz oscillators for clocks, ferroelectric perovskite oxides dominate the commercial market for piezoelectrics. The most widely used piezoelectrics are based on the Pb(Zr$_{1-x}$Ti$_x$)O$_3$ (PZT) system, the phase diagram of which is shown in Figure 8.13. Pure PbZrO$_3$ becomes antiferroelectric below 506 K. Its structure is distorted from cubic to orthorhombic symmetry by antiparallel displacements of Pb and rotations of the zirconium-centered octahedra (a^{0}b$^-$b$^-$ tilting, Section 1.5.3). Upon introducing a small fraction of Ti at the Zr site, the paraelectric cubic state transforms[18] to a ferroelectric state upon cooling [7]. This phase, labeled rhombohedral I in Figure 8.13, possesses a slight rhombohedral distortion of the cubic cell and space-group symmetry $R3m$. In this structure, Pb^{2+} as well as (Ti/Zr)$^{4+}$ displace along one of the eight ⟨111⟩ directions of the parent cubic cell. A second rhombohedral phase, labeled rhombohedral II in Figure 8.13, appears at low temperatures. This phase of $R3c$ symmetry features a$^-$a$^-$a$^-$ octahedral tilting, in addition to the polar displacements of cations that were already present in the high-temperature $R3m$ phase. For Ti-rich ($x > 0.5$) compositions, Pb(Zr$_{1-x}$Ti$_x$)O$_3$ is isostructural with the average structure of tetragonal BaTiO$_3$ (Figure 8.5) [8].[19]

Near the PbZr$_{0.5}$Ti$_{0.5}$O$_3$ composition, the ferroelectric solid solution changes symmetry from rhombohedral on the Zr-rich side to tetragonal on the Ti-rich side. The solid line separating these two regions in Figure 8.13 is called the **morphotropic phase**

---

[18] This transformation is sluggish and can have a large thermal hysteresis.

[19] The Pb^{2+} displacements in PbTiO$_3$ influence the Ti^{4+} displacements in such a way that both the local and average structure of tetragonal PbTiO$_3$ can be described by displacements along one of the equivalent ⟨001⟩ directions, unlike BaTiO$_3$ where the local displacements are along ⟨111⟩.

boundary (MPB). An abrupt structural change upon changing the composition of a solid solution is the defining characteristic of an MPB. For solid-solution compositions close to the MPB, a polarization rotation occurs where the polar vector defining each domain is easily reoriented by an electric field, leading to a very strong piezoelectric response.

The structure of $Pb(Zr_{1-x}Ti_x)O_3$ becomes quite complicated near the MPB, with some details still being debated. In the late 1990s, the $0.48 \leq x \leq 0.50$ region, which had previously been thought to be a two-phase mixture of rhombohedral and tetragonal phases, was shown to be more accurately described as a monoclinic phase [9]. A more recent study provides evidence to suggest there may be two different phases with monoclinic symmetry in the MPB region [10]. Upon increasing the Zr content, the average structure becomes rhombohedral, but vestiges of the monoclinic structure can still be found in the local structure. The subtle energetic differences between competing phases gives rise to a complex fine-scale domain structure that is at least partially responsible for the exceptional piezoelectric properties of PZT.

In recent years, environmental concerns over the detrimental impacts of lead have motivated a search for Pb-free piezoelectrics. Perovskites such as $(K_{1-x}Na_x)NbO_3$, $Ba(Zr_{1-x}Ti_x)O_3$, and $(Ba_{1-x}Ca_x)TiO_3$, have emerged as possible replacement candidates. Nevertheless, piezoelectrics with properties that match those of PZT remain elusive.

8.6 Local Bonding Considerations in Non-Centrosymmetric Materials

As a general rule, non-centrosymmetric structures are more likely to form when their coordination polyhedra lack an inversion center. Apart from polyhedra that are inherently non-centrosymmetric (triangles, tetrahedra), polyhedra that would otherwise possess an inversion center can become non-centrosymmetric by distorting. For example, when an octahedrally coordinated cation undergoes a second-order Jahn–Teller (SOJT) distortion as in the ferroelectric phases of $BaTiO_3$.

Halasyamani and Poeppelmeier [11] surveyed the ICSD and found ~580 oxides that adopt non-centrosymmetric crystal structures. Among these, two-thirds belonged to polar crystal classes, while the remaining third were nonpolar. Roughly 75% of all entries contain one or more of the following coordination environments: (a) a d^0 cation that undergoes an SOJT distortion (e.g. Ti^{4+}, Nb^{5+}), (b) a p-block cation with an s^2p^0 electron configuration that undergoes an SOJT distortion leading to a stereochemically active lone-pair distortion (e.g. Pb^{2+}, Sb^{3+}), or (c) a tetrahedrally coordinated cation. In the following section, we take a closer look at the bonding that drives SOJT distortions.

8.6.1 Second-Order Jahn–Teller Distortions with d^0 Cations

SOJT distortions were introduced in Section 5.3.10. In extended solids, the most common SOJT distortions are those involving a transition-metal cation with a d^0 configuration and those involving a main-group cation with an s^2p^0 configuration. We consider d^0 cations in this section and s^2p^0 cations in the next.

The molecular-orbital (MO) diagram for an octahedrally coordinated transition-metal atom was considered in detail in Section 5.3.8. When the transition metal has a d^0 configuration, the empty triply degenerate set of MOs with t_{2g} symmetry is the LUMO (lowest unoccupied molecular orbital), while the HOMO (highest occupied molecular orbital) has nonbonding anion character (see Figures 5.21 and 5.22). SOJT distortions are characterized by a shift of the cation out of the center of the octahedron that lowers the symmetry, enabling the unoccupied t_{2g} orbitals to mix with filled anion states, as described below. A thorough treatment of SOJT distortions involving d^0 cations in both molecular and extended solids can be found in the literature [12].

Because perovskite oxides account for a large fraction of ferroelectric materials, let's take a closer look at the bonding that drives SOJT distortions in these materials. Their electronic band structure was covered in detail in Sections 6.6.4 and 6.6.5. For a cubic perovskite with a d^0 cation on the octahedral site (e.g. $SrTiO_3$ or $KTaO_3$), the conduction-band minimum occurs at the Γ point (see Figure 6.31). At Γ, the orbital character of six out of the nine valence bands becomes strictly oxygen $2p$ nonbonding, while the three lowest-energy conduction bands are strictly metal d nonbonding orbitals (d_{xy}, d_{yz}, d_{xz}) as illustrated in Figure 6.30. The key point is that mixing (overlap) of the metal d orbitals and the oxygen $2p$ orbitals to form π-bonding and -antibonding states is symmetry-forbidden at the Γ point.

Now consider how the band structure changes in response to a SOJT distortion. The essential features of this analysis are captured by considering the π interactions and the displacements of the M cation within the xy plane towards the edge of the octahedron. The crystal orbitals at Γ representing the empty d_{xy} band and one of the filled O $2p$ bands with t_{1g} symmetry are shown in the upper half of Figure 8.14. These are representative of the crystal orbitals from conduction and valence bands, respectively, separated by the charge transfer gap CT in Figure 6.31.

If we lower the symmetry by moving the M cations along [110], the symmetry constraints are relaxed, and mixing (overlap) between the two formerly nonbonding crystal orbitals is allowed. Because each oxygen ion now makes one short and one long bond to the neighboring metal atoms, the bonding and antibonding contributions do not cancel out. This introduces π-bonding interactions that stabilize the occupied nonbonding O $2p$ states (see lower middle panel of Figure 8.14), while π^*-antibonding interactions destabilize the empty metal d states (see lower right panel of Figure 8.14). The amount of mixing increases as the energy gap (prior to the distortion) between the O $2p$ (t_{1u}) and M nd (t_{2g}) crystal orbitals decreases. Because the destabilization of

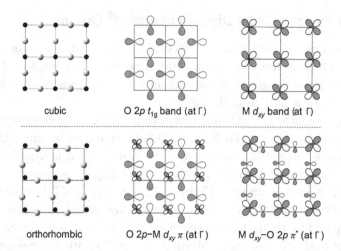

Figure 8.14 Top: Metal–oxygen interactions within an MO_2 plane of a perovskite (left), and crystal orbitals at Γ for the valence-band maximum (middle) and conduction-band minimum (right). Bottom: Changes that occur when the octahedral cation displaces within the plane toward an edge of the octahedron.

the conduction-band states is larger than the stabilization of the valence-band states, the SOJT distortion quickly becomes unfavorable as the d orbitals are populated. This is seen in the crystal chemistry of MO_3 compositions with ReO_3 topologies. SOJT distortions lead to large cation displacements in the d^0 oxides WO_3 and β-MoO_3, whereas the d^1 oxide ReO_3 is cubic.

A survey of AMO_3 perovskites shows that not all compounds that contain a d^0 cation undergo SOJT distortions. SOJT distortions are observed for $KNbO_3$, $NaNbO_3$, and $BaTiO_3$, but not for $KTaO_3$, $NaTaO_3$, $SrTiO_3$, $CaTiO_3$, $AZrO_3$, or $AHfO_3$ (A is Ba, Sr, Ca). This can be understood by realizing that the driving force for the SOJT distortion increases as the energy separation between the metal nd orbitals and the nonbonding oxygen $2p$ states decreases.[20] This energy separation reaches a minimum at the Γ point (Section 6.6.5). SOJT distortions are therefore seen in $KNbO_3$, where the narrow charge-transfer gap facilitates mixing of M nd and O $2p$ orbitals on distortion, but not in $BaZrO_3$ where the gap is larger. SOJT distortions occur in ferroelectric $KNbO_3$ and antiferroelectric $NaNbO_3$, but not in $KTaO_3$ and $NaTaO_3$ for the same reason. Using a variety of approaches, several different studies have reached similar conclusions regarding the tendency for various d^0 "cations" to undergo SOJT distortions [13, 14, 15, 16]: $Hf^{4+} < Zr^{4+} < Ta^{5+} < Ti^{4+} < Nb^{5+} <$

[20] In Figure 6.31, the HOMO–LUMO gap at Γ is labeled a charge-transfer excitation: the smallest direct band gap in a cubic perovskite that allows electromagnetic radiation to excite electrons from nonbonding anion crystal orbitals to empty d orbitals of the central metal atom.

$W^{6+} < V^{5+} < Mo^{6+}$. Notice the similarity of elements that are diagonal neighbors on the periodic table (e.g. Ti^{4+}, Nb^{5+}, W^{6+}).

The charge-transfer gaps of most $A^{2+}TiO_3$ and A^+TaO_3 perovskites make them nearly ferroelectric; close to the border between ferroelectric and paraelectric behavior (except $BaTiO_3$ and $PbTiO_3$, which are ferroelectric). Although $SrTiO_3$ and $KTaO_3$ are cubic and therefore nonpolar at room temperature, they have dielectric permittivities much higher than predicted by the Clausius–Mossotti equation (330 and 242 respectively, see Table 8.1). These permittivities increase further with decreasing temperature as though approaching a transition into a ferroelectric state, but the transition is never realized. Consequently, these phases are referred to as **incipient ferroelectrics**. The close proximity of the ferroelectric state means that only a small perturbation is needed to stabilize ferroelectricity. In the case of $SrTiO_3$, ferroelectricity can be realized by stretching the lattice, which occurs when Sr^{2+} ions are replaced with the larger Ba^{2+} ions to form $BaTiO_3$.[21] In a similar vein, the dielectric properties of thin epitaxial $SrTiO_3$ films are very sensitive to stresses that result from lattice mismatch with the underlying substrate. For example, the dielectric permittivity of $SrTiO_3$ films grown on $DyScO_3$ substrates, which impose a 1% in-plane tensile strain, is increased nearly 20 times compared with bulk $SrTiO_3$ [17]. It has even been shown that isotopic substitution of ^{18}O for ^{16}O can stabilize ferroelectricity in $SrTiO_3$ below 23 K [18].

8.6.2 Second-Order Jahn–Teller Distortions with s^2p^0 Cations

Cations that have an s^2p^0 configuration are also prone to distortions. These distortions are called lone-pair distortions, or more accurately, stereochemically active lone-pair distortions, because their presence is normally inferred from the distortion of the ligands surrounding the s^2p^0 central atom. To understand the driving forces behind these distortions, we again turn to MO theory to show that "lone-pair distortions" are just another type of SOJT distortion as discussed previously in Section 5.3.10.

Consider a main-group cation located on a site with $m\overline{3}m$ (O_h) symmetry. This includes three important coordination environments found in extended solids: the octahedron, the cube, and the cuboctahedron. Once again, the perovskite structure provides us with good examples, as both the smaller six-coordinated cation in the octahedron and the larger 12-coordinated cation in the cuboctahedron have $m\overline{3}m$ site symmetry.

Figure 8.15 shows two perovskites where an s^2p^0 cation drives a SOJT distortion: $CsGeCl_3$ and $BiFeO_3$. The distortion is driven by Ge^{2+} on the six-coordinate site in the former, and by Bi^{3+} on the 12-coordinate site in the latter. In both, the lone-pair cation displaces along

---

[21] According to the distortion theorem (Section 5.4), off-center shifts of an atom within its coordination polyhedron increase the atom's bond-valence sum. In those AMO_3 perovskites where the coordination octahedron defined by AO_3 packing becomes too large (tolerance factor > 1), the M of d^0 configuration and high oxidation state becomes underbonded. The SOJT distortion leads to an increase in the M-atom bond-valence sum, because bonds that shorten gain more bond valence than bonds that lengthen lose.

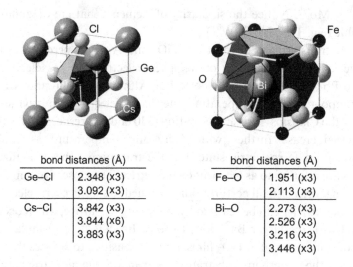

bond distances (Å)			bond distances (Å)	
Ge–Cl	2.348 (x3)		Fe–O	1.951 (x3)
	3.092 (x3)			2.113 (x3)
Cs–Cl	3.842 (x3)		Bi–O	2.273 (x3)
	3.844 (x6)			2.526 (x3)
	3.883 (x3)			3.216 (x3)
				3.446 (x3)

Figure 8.15 The crystal structures of the rhombohedrally distorted perovskites $CsGeCl_3$ (left) and $BiFeO_3$ (right). Only the shortest Ge–Cl and Bi–O bonds are shown.

a threefold axis, lowering the site symmetry to $3m$ and the crystal symmetry to rhombohedral.

Consider the MO diagram for an isolated $GeCl_6^{4-}$ octahedron (Figure 8.16). Because the Ge–Cl π overlap is minimal, we can concentrate on the interactions between the six Cl^- σ donors and the $4s$ and $4p$ valence orbitals of germanium. In a perfect octahedron, these orbitals are orthogonal and do not mix. The HOMO is the strongly antibonding interaction between the Ge $4s$ orbital and the ligand symmetry-adapted linear combination (SALC) with a_{1g} symmetry. The LUMO is the triply degenerate t_{1u} set of orbitals formed from antibonding Ge $4p$–Cl interactions. The relatively small HOMO–LUMO gap provides the necessary driving force for an SOJT distortion.

Figure 8.17 shows the changes to the frontier orbitals (HOMO and LUMO) that result from a shift of Ge^{2+} along the along threefold axis toward the face of the coordination octahedron, as seen in $CsGeCl_3$. In the distorted octahedron with $3m$ point-group symmetry, the Ge $4s$ and $4p_z$ orbitals both have a_1 symmetry and can mix with each other, lowering the energy of the HOMO. In this way, the electron density of the spherically symmetric a_{1g} orbital of the undistorted octahedron is redistributed to a lobe located on the "more open side" of the Ge ion, forming the stereochemically active electron lone pair. At the same time, the energy of the Ge $4p_z$–Cl σ^* MO is raised by the s–p interaction, but, because this orbital is not occupied, there is no energy penalty for this destabilization. Similar symmetry arguments can be made for Bi^{3+} that sits on the 12-coordinate site in $BiFeO_3$.

In oxides where the metal M comes from the fourth period, the M $4s$–O $2p$ interaction is very strong. Consequently, the M $4s$–O $2p$ σ^* level is so antibonding that it is difficult to stabilize the $4s^2 4p^0$ configuration in any geometry. Hence, Ga^+ and Ge^{2+} are rarely observed

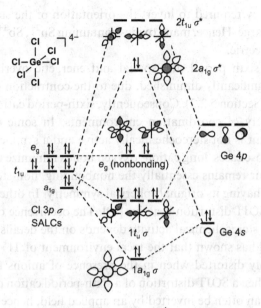

Figure 8.16 The MO diagram for an isolated $GeCl_6^{4-}$ octahedron. Only the valence-shell Ge $4s$ and $4p$ orbitals, and the Cl σ-donor orbitals, are considered.

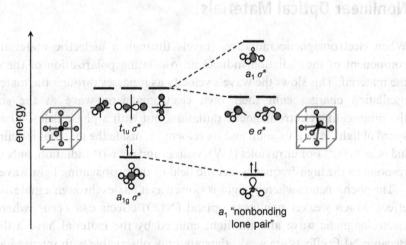

Figure 8.17 HOMO and LUMO of an s^2p^0 cation in a regular octahedral coordination (left) and their response when the s^2p^0 cation shifts toward a face of the octahedron (right) as in $CsGeCl_3$.

in oxides and fluorides. In contrast, many oxides of Sn^{2+}, Sb^{3+}, and Te^{4+} exist because the M $5s$–O $2p$ interaction that gives rise to the $2a_{1g}$ σ^* MO is not as strongly antibonding. However, the interaction is still strong enough that a pronounced stereochemically active electron lone-pair distortion is common. While such distortions produce local dipole moments, the energy required to invert the orientation of the stereochemically active lone pair is often quite large. Hence, many oxides containing Sn^{2+}, Sb^{3+}, and Te^{4+} are pyroelectric but few are ferroelectric.

Moving to the sixth period, the spatial and energetic overlap between metal $6s$ and O $2p$ orbitals is significantly diminished, due to the contraction of the $6s$ orbital driven by relativistic effects (Section 5.2.2). Consequently, sixth-period cations like Tl^{+}, Pb^{2+}, and Bi^{3+} show a wide variety of coordination environments. In some compounds (e.g. $PbWO_4$, $BiVO_4$) the lone pair is not stereochemically active and a symmetric coordination environment results. In those cases, long cation–oxygen bonds minimize the M $6s$–O $2p$ overlap and the $6s^2$ electron pair remains essentially the nonbonding, non-stereochemical, inert pair of the isolated atom, having its original spherical symmetry. In other compounds, like $PbTiO_3$ and $PbZrO_3$, the SOJT distortion is active [19]. The occurrence of the SOJT distortion and the accompanying stereochemical activity depends on the details of the cation–anion interaction. Brown [20] has shown that the local environment of Tl^{+} can vary from completely symmetric to highly distorted when in the presence of anions that require strong, short, bonds with Tl^{+}. When a SOJT distortion of a sixth-period cation does occur, the direction of the displacement can often be inverted by an applied field, hence these cations, particularly Pb^{2+}, are often found in ferroelectrics.

8.7 Nonlinear Optical Materials

When electromagnetic radiation travels through a dielectric material, the electric-field component of the radiation induces an oscillating polarization of the charged species in the material. This slows the wave's velocity as it passes through the material, as though the oscillating charges emit their own electromagnetic wave at the same frequency as the propagating electromagnetic radiation, but with a phase delay. The ratio between the speed of light, c, in a vacuum and its velocity, v, inside the material is defined as its **refractive index**, $n = c/v$.[22] For ultraviolet (UV), visible, and near-IR radiation, only electron clouds can respond to the high-frequency electric field of the propagating light wave.

The phenomenon whereby light is slowed as it passes through a material is a linear optical effect. Much weaker nonlinear optical (NLO) effects can occur, where the propagating electromagnetic wave and the light emitted by the material have a different frequency. Because NLO effects are weak, they are only observable with intense light sources such as lasers. A full treatment of NLO effects and applications is beyond the scope of this text. Instead, we will concentrate on what is arguably the most important NLO effect, **second-harmonic generation (SHG)**.

SHG is a process where two photons with frequency v combine to produce a new photon with twice the frequency, $2v$. Using relationships introduced in Section 7.1, it

---

[22] Be careful not to confuse the italic v for velocity with the Greek letter v for frequency.

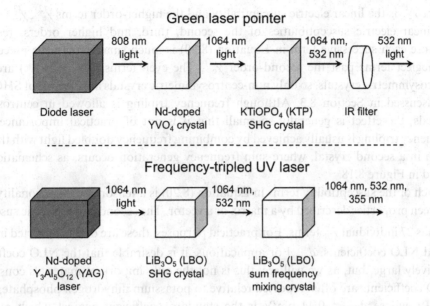

Green laser pointer

808 nm light → Diode laser → 1064 nm light → Nd-doped YVO₄ crystal → 1064 nm, 532 nm → KTiOPO₄ (KTP) SHG crystal → 532 nm light → IR filter

Frequency-tripled UV laser

1064 nm light → Nd-doped Y₃Al₅O₁₂ (YAG) laser → 1064 nm, 532 nm → LiB₃O₅ (LBO) SHG crystal → 1064 nm, 532 nm, 355 nm → LiB₃O₅ (LBO) sum frequency mixing crystal

Figure 8.18 The main optical components of a green laser pointer (top) and a laser that uses two NLO crystals to generate UV light (bottom).

can be shown that the new photon has twice the energy and half the wavelength of the incoming photons. Perhaps the most familiar example is the green laser pointer (Figure 8.18), which uses a $(Y_{1-x}Nd_x)VO_4$ laser crystal to emit IR radiation with a wavelength, $\lambda = 1062$ nm, that is subsequently halved to 532 nm by a $KTiOPO_4$ SHG crystal.[23] SHG crystals are used in many types of lasers to convert IR light to visible and/or UV light.

8.8 Nonlinear Susceptibility and Phase Matching

When polarization was introduced in Section 8.1.2, nonlinear effects were neglected. We can expand Equation (8.6) to take nonlinear effects into account:

$$P = \varepsilon_0 \left(\chi_e^{(1)} E + \chi_e^{(2)} E^2 + \chi_e^{(3)} E^3 + \dots \right) \tag{8.20}$$

---

[23] One drawback of this green laser pointer design is the low SHG efficiency associated with the low power levels employed in a laser pointer. Hundreds of milliwatts of IR light are required for generating the standardized 1 mW of green light. Accordingly, battery life is relatively short and filters must be used to take out the IR light that poses a serious safety hazard to the eye.

where $\chi_e^{(1)}$ is the linear electric susceptibility and the higher-order terms $\chi_e^{(2)}$, $\chi_e^{(3)}$, ... are the nonlinear electric susceptibilities of the second, third, and higher orders, respectively. Because each successive term in Equation (8.20) is much smaller than the preceding one, we neglect terms past the second-order $\chi_e^{(2)}$. The even terms ($\chi_e^{(2)}$, $\chi_e^{(4)}$, ...) are zero for centrosymmetric crystals, so only non-centrosymmetric crystals are capable of SHG activity, as discussed in Section 8.3. Although frequency tripling is allowed in centrosymmetric crystals, the effect is generally so small that it is not of practical importance. Instead, frequency tripling is usually achieved by combining frequency doubled light with the primary beam in a second crystal, where sum frequency generation occurs, as schematically illustrated in Figure 8.18.

Each of the susceptibility terms in Equation (8.20) is a tensor, a proportionality constant between properties described by a matrix or a vector. The second-order electric susceptibility $\chi_e^{(2)}$ has 27 individual χ_{ijk} terms. For practical purposes, these are often converted into the so-called NLO coefficients, d_{ijk}. For applications, it is desirable that the NLO coefficients be relatively large, but, as we will see, this is not the only important materials consideration. NLO coefficients are often reported relative to potassium dihydrogen phosphate, KH_2PO_4 (KDP), whose $d_{321} = 0.44$ pm/V is the standard coefficient against which other NLO materials are measured.[24]

The refractive index of a material is not a constant; it depends on both the frequency of the electromagnetic radiation and the temperature of the material. The frequency dependence is important for SHG because in general the fundamental beam of frequency v and the second-harmonic beam of frequency $2v$ travel at different speeds, which leads to destructive interference that can dramatically reduce SHG efficiency. Fortunately, it is possible to match the refractive indices of the fundamental and second-harmonic beam in birefringent crystals through a technique known as **phase matching**.

To understand phase matching, one must first be familiar with the optical properties of **birefringent crystals**. In such crystals, the refractive index is not isotropic, it varies with the direction and polarization of the light beam in the crystal. In a birefringent crystal that is uniaxial,[25] the component of light that is polarized in the direction of the unique axis is called the **ordinary beam**, while the component polarized orthogonal to the unique axis is called the **extraordinary beam**. The ordinary beam will experience the same refractive index (n_o) regardless of its direction of propagation, whereas the extraordinary beam will experience a refractive index that depends upon the direction it travels through the crystal (n_e). In every

---

[24] The electric field and the polarization response of a crystal are both vectors, and the tensor describing the relationship between the two is a second rank tensor with nine coefficients. Because the SHG process $v + v \rightarrow 2v$ involves three photons, 27 d_{ijk} terms are needed for a general description of second-order NLO effects. When the susceptibilities are independent of frequency, the number of coefficients can be reduced from 27 to 18 terms. For SHG effects, it's common to use a condensed notation because the j and k terms can be permuted, for example, $d_{36} \equiv d_{321} = d_{312}$. For more details, see R. Boyd in Further Reading.

[25] Crystals with a unique axis of symmetry, namely those that belong to the trigonal, tetragonal, or hexagonal crystal systems.

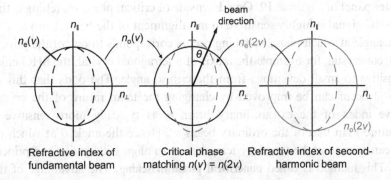

Figure 8.19 Cross-sectional cuts of the optical indicatrix showing the refractive indices of the fundamental (left) and second-harmonic (right) beams in a birefringent crystal. The vertical and horizontal axes represent propagation parallel and orthogonal to the optic axis, respectively. The circle of the refractive index of the ordinary beam, n_o, is solid, the ellipse of the extraordinary beam, n_e, is dashed. The middle panel shows the critical angle, θ, of the propagation direction for which the phase matching occurs.

birefringent crystal, there is at least one direction where $n_o = n_e$, and that direction is defined as the **optic axis**. In uniaxial crystals, the optic axis is parallel with the unique crystallographic axis, whereas biaxial crystals have two optic axes. Double refraction, a process whereby an electromagnetic wave is split into two rays that take slightly different paths because of differences in the refractive indices of the ordinary and extraordinary beams, is a familiar property of birefringent crystals.[26]

The variation of the refractive index as a function of the light-beam direction in a crystal is represented by an ellipsoid, called an optical indicatrix. The shape of the ellipsoid is such that its radius along any given direction is directly proportional to the refractive index in that direction. Because the refractive index of the ordinary beam is isotropic, its optical indicatrix is a sphere. When the beam direction is parallel with the optic axis, the ordinary and extraordinary beams have the same refractive index, as shown on the left- and right-hand sides of Figure 8.19.

The most common method of phase matching, **critical phase matching**, relies on the optical properties of birefringent crystals. The incoming fundamental beam with frequency v is polarized in a plane parallel to the optic axis of the SHG crystal; it acts as an ordinary beam with an index of refraction that is independent of the direction it travels through the crystal. When the second-harmonic beam is generated, its polarization is perpendicular to the fundamental beam and thus it travels through the crystal as an extraordinary beam. Hence, the refractive index of the second-harmonic beam can be tuned by changing the orientation of the propagating beam with respect to the optic axis of the crystal. There is a critical angle, θ, where $n_o(v) = n_e(2v)$, and critical phase matching occurs, as illustrated in

[26] The calcite form of $CaCO_3$ is one of the most familiar examples of a birefringent crystal. The only orientation when it does not split light into two beams is when the light travels parallel to the optic axis.

the center panel of Figure 8.19. One downside of critical phase matching is that the strength of the SHG signal is highly sensitive to misalignment of the beam from θ.

If the angle at which phase matching occurs corresponds to one of the principal axes of the birefringent crystal, for example the a axis of a tetragonal crystal, the SHG efficiency becomes less sensitive to small deviations from the critical angle. The odds that this random coincidence will occur can be improved by changing the temperature of the crystal, because the refractive index of the extraordinary beam n_e is typically more sensitive to changes in temperature than that of the ordinary beam n_o. Hence the angle θ at which phase matching occurs can sometimes be tuned by temperature to align with one of the principle axes of the crystal. This method is called **noncritical phase matching**. The advantage of this approach is decreased sensitivity to small misalignments of the crystal, but it requires precise control of the temperature of the SHG crystal.

A third approach, called **quasi-phase matching**, is used in materials where the two former approaches to phase matching are not practical. The idea is to allow a degree of phase mismatch between the fundamental and second-harmonic beams (due to different refractive indices experienced by the two beams), but before the phase mismatch and destructive interference become too large, a specially engineered domain structure of the crystal inverts the two beams. Every few microns, there is a 180° ferroelectric domain wall that changes the refractive indices in such a way that the relative speeds of the second-harmonic and fundamental beams are inverted. As the fundamental beam is alternatively going faster and slower than the second-harmonic beam, the two never get very far out of phase; hence phase matching.

8.9 Important SHG Materials

While symmetry restrictions narrow the scope of possible SHG materials, there are still thousands of non-centrosymmetric materials from which to choose. What other criteria can be used to select and/or design SHG materials? Becker [21] has proposed seven criteria for selecting good SHG materials: (1) relatively large nonlinear optical coefficients, (2) moderate birefringence, (3) wide transparency range, (4) wide phase-matching range, (5) high light-induced damage threshold (hereafter referred to as simply as damage threshold), (6) good chemical and mechanical stability, and (7) ease of crystal growth.

Some of the most important inorganic SHG materials are listed in Table 8.4. They can be divided into four different families: (a) KH_2PO_4 (KDP) and related compounds, (b) $KTiOPO_4$ (KTP) and related compounds, (c) niobates and tantalates, and (d) borates. KDP was the first commercial SHG material, but it has largely been supplanted as new materials have been discovered and commercialized. KTP and related materials are attractive because of the ease with which very large, high-quality crystals can be grown. The large NLO coefficients of the niobates and tantalates make them attractive with low-power laser sources, but applications are limited by low damage thresholds and

Table 8.4 Properties of commercially important SHG materials.

	KH$_2$PO$_4$ (KDP)	KTiOPO$_4$ (KTP)	LiB$_3$O$_5$	BiB$_3$O$_6$	β-BaB$_2$O$_4$	LiNbO$_3$
Space group	$I\bar{4}2d$	$Pna2_1$	$Pna2_1$	$C2$	$R3c$	$R3c$
Point group	$\bar{4}2m$	$mm2$	$mm2$	2	$3m$	$3m$
Melting point	526 K	1445 K	1107 K	999 K	1368 K	1526 K
*Refractive indices**	$n_o = 1.494$ $n_e = 1.460$	$n_x = 1.738$ $n_y = 1.745$ $n_z = 1.830$	$n_x = 1.566$ $n_y = 1.590$ $n_z = 1.606$	$n_1 = 1.917$ $n_2 = 1.757$ $n_3 = 1.784$	$n_o = 1.655$ $n_e = 1.542$	$n_o = 2.232$ $n_e = 2.156$
Transparency range (nm)	200–1500	350–3500	160–2600	290–2500	190–3500	420–5200
Effective NLO coefficient †	1.0	8.4	2.7	~9	5.3	40
Damage threshold (GW/cm^2) ‡	0.25	1.0	9	**	5	0.3
Comments	Crystals can be grown from solution, hygroscopic	High NLO coefficient, inexpensive compared to borates	Highest damage threshold among commercial materials	Higher NLO coefficients than most borates	Broad transmission and phase matching ranges, high damage threshold	High NLO coefficients, used in electro-optic applications

* As measured at 1064 nm and room temperature. For uniaxial rhombohedral and tetragonal crystals, the refractive index of the ordinary beam, n_o, and the extraordinary beam, n_e, are given. For biaxial orthorhombic and monoclinic crystals, three refractive indices are needed. †The effective NLO coefficients are versus the $d_{321} \equiv d_{36}$ NLO coefficient of KDP. ‡The damage thresholds are for a 10 ns pulse of 1064 nm radiation. **The damage threshold for BiB$_3$O$_6$ is reported to be similar to LiB$_3$O$_5$.

lack of transparency in the UV. The borates are versatile SHG materials due to their broad transparency range, high damage threshold, and ease of phase matching. The structures and properties of the key members of each family are discussed in more detail in the subsequent sections.

8.9.1 KH₂PO₄

Potassium dihydrogen phosphate, KH_2PO_4 (KDP), crystals were among the first to produce useful levels of frequency-doubled light. KDP is tetragonal, and its anionic network, linked by hydrogen bonds along c, is shown in Figure 8.20. Since two of four oxygens of the phosphate anion form covalent bonds to hydrogen, their bonds to P become single bonds, and the phosphorus atom shifts toward the remaining two oxygens to compensate for this loss of bonding. This shift generates a net dipole moment parallel to the c axis (see Figure 8.20). The most attractive aspect of KDP as an NLO material is the possibility to obtain large, high-quality crystals. However, KDP has several limitations, including a relatively small NLO susceptibility, a low damage threshold, and hygroscopicity.

8.9.2 KTiOPO₄

$KTiOPO_4$ (KTP) has several advantages over KDP. Not only does KTP have an NLO coefficient eight times larger than KDP, it has a higher damage threshold, and is not hygroscopic. Good crystals of KTP can be grown either hydrothermally or from a flux.

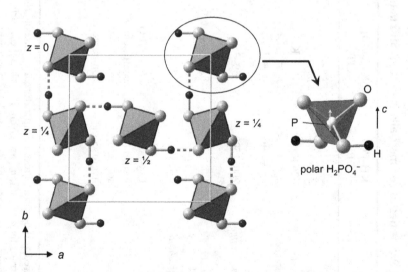

Figure 8.20 The structure of KH_2PO_4 viewed against the c axis. Covalent O–H bonds are solid, hydrogen bonds are dashed. K^+ omitted for clarity. An individual $H_2PO_4^-$ anion is shown rotated on the right. The white arrow indicates the shift of the phosphorus parallel to c.

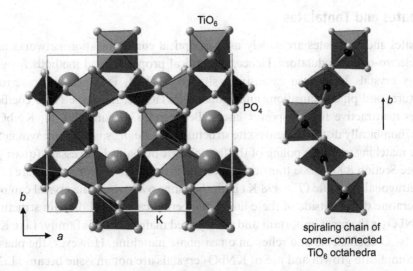

Figure 8.21 The structure of KTiOPO$_4$. A closer view of one of the spiraling corner-connected TiO$_{4+2/2}$ chains is shown on the right. The short Ti–O bonds ($d <$ 1.75 Å) are drawn to indicate the displacements of Ti.

These factors make KTP one of the most popular SHG materials, particularly for low- and medium-power lasers.

While KH$_2$PO$_4$ and KTiOPO$_4$ are both phosphates, the structural origins of the non-centrosymmetry are not the same. In KTiOPO$_4$, the non-centrosymmetry arises from SOJT distortions of Ti ions with a d^0 configuration, which yield larger NLO coefficients than in KDP. The structure of KTiOPO$_4$ can be described as spiraling chains of corner-shared Ti-centered octahedra held together by phosphate groups. The K$^+$ ions sit in the channels within this framework. Four of six oxygens around each titanium come from the tetrahedral phosphate groups, while the other two oxygens connect the octahedra into the spiral chains seen in Figure 8.21.

A closer look at the bonding within the octahedron shows that the bonds are not symmetric. Titanium forms four intermediate-length bonds (1.95–2.05 Å) with oxygens of the surrounding phosphate groups. It completes its octahedron by forming one short bond (1.71–1.74 Å) and one long bond (2.10–2.15 Å) to the bridging oxygens that are not bonded to phosphorus. The net effect is a long–short ... Ti–O–Ti–O– ... bond alternation along the spiraling chain (Figure 8.21), creating a polar axis along the chain direction. Other members of the KTP family include KTiOAsO$_4$ (KTA), RbTiOPO$_4$ (RTP), and RbTiOAsO$_4$ (RTA). By replacing phosphate with arsenate, NLO coefficients increase. More importantly, the long-wavelength transparency increases from 3500 nm to 5000 nm.

8.9.3   Niobates and Tantalates

Niobates and tantalates are widely used in optical communication networks as waveguides and electro-optic modulators. Hence, the optical properties and methods for growing high-quality crystals have been extensively studied. $KNbO_3$ is a perovskite ferroelectric with structures and phase transitions analogous to $BaTiO_3$. It has large NLO coefficients, which makes it attractive for low-power lasers. Between 263 K and 498 K, $KNbO_3$ adopts an orthorhombically distorted perovskite structure with $Amm2$ space-group symmetry. To attain phase matching, periodic poling of the ferroelectric domains is necessary (quasi-phase matching, see Section 8.8). Phase transitions into either the rhombohedral structure ($T < 263$ K) or the tetragonal structure ($T > 498$ K) limit the range over which the crystal can operate. If the temperature drifts outside of these limits, the periodically poled domain structure is lost.

$LiNbO_3$ is the most important and widely used material in this family. Like $KNbO_3$, it has large NLO coefficients and relies on quasi-phase matching. However, the phase transitions that complicate growth and use of $KNbO_3$ crystals are not an issue because $LiNbO_3$ has no phase transitions below its Curie temperature ($T_C = 1483$ K). While the robust polar state in $LiNbO_3$ prevents the switching that would be required for ferroelectric applications, this aspect of its crystal chemistry is an advantage for NLO applications. $LiNbO_3$ is also used as a pyroelectric and a piezoelectric.

The structure of $LiNbO_3$ (Figure 8.22), and the isostructural $LiTaO_3$, can be described as an ordered variant of the corundum structure (Figure 1.30). In Al_2O_3 and other oxides with the corundum structure, electrostatic repulsions lead to a displacement of the cations away from the shared octahedral face. In $LiNbO_3$, the cations are ordered so that each pair of face-sharing octahedra contains one lithium and one niobium. This destroys the inversion center in the

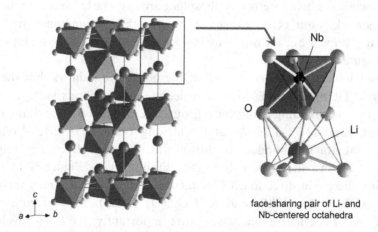

face-sharing pair of Li- and
Nb-centered octahedra

Figure 8.22 Left: The unit cell of $LiNbO_3$ with Nb-centered octahedra shaded. Right: A closer view at the bonding within each pair of face-sharing octahedra.

structure, lowering the space-group symmetry from $R\bar{3}c$ to $R3c$. Just like in corundum, the central atoms (Li^+ and Nb^{5+}) repel each other leading to highly distorted trigonal-antiprismatic environments. The Nb–O distances are 1.88 Å (×3) and 2.13 Å (×3), while the Li–O distances are 2.05 Å (×3) and 2.27 Å (×3). Although Li and Nb move in opposite directions, the niobium ion carries a higher charge so that the distortion creates a local dipole moment. Because all niobium atoms shift in the same direction, a net dipole moment develops parallel to the c axis.[27]

While the niobates and tantalates have the largest NLO coefficients among commercial inorganic materials, their properties are not well suited for many applications. Their low damage threshold rules them out for high-power lasers. Because they have very limited transparency in the UV (the absorption edge for $LiNbO_3$ is 400 nm), they cannot be used if the SHG light is in the UV. Finally, the need to engineer the domain structure to achieve quasi-phase matching adds a level of complexity to the crystal-preparation process.

8.9.4 Organic and Polymer NLO Materials

The NLO materials discussed thus far depend on displacements of ions to form local dipole moments that are responsible for the NLO response. Another approach to NLO materials design is to employ molecular fragments that have highly polarizable bond networks. Molecules that exhibit conjugated π bonding tend to show high polarizability, particularly those with an electron-donating group on one end and an electron acceptor on the other. A prototypical example is 4-(N, N-dimethylamino)-4'-nitrostilbene (Figure 8.23), where two benzene rings, connected by an ethylene group (stilbene), make up the conjugated π system; the dimethylamino group, $-N(CH_3)_2$, acts as the donor, and the nitro group, $-NO_2$, on the opposite end of the molecule acts as the acceptor [22].

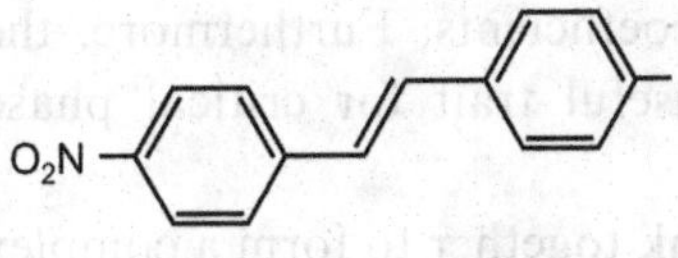

Figure 8.23 The structure of 4-(N, N-dimethylamino)-4'-nitrostilbene.

In such molecules, the electromagnetic field of the propagating light wave can induce significant reorganization of the delocalized π-bonding electron density. The effective NLO coefficients of the best molecular species can be orders of magnitude larger than the classic inorganic NLO materials, but several practical considerations have limited their use. Firstly, it is difficult to control the crystallization to obtain a non-centrosymmetric crystal. This is especially true for conjugated polymers that have attracted interest for NLO applications. Secondly, the optical quality of organic crystals tends to be poor and the damage threshold low. Finally, conjugated organic molecules tend to absorb strongly in the near-UV and visible regions of the spectrum.

---

[27] The $LiNbO_3$ and ilmenite ($FeTiO_3$) structures are closely related ordered variants of the corundum structure, with one important difference. In $LiNbO_3$, the Li–Nb pairs all point in the same direction (e.g. Li up and Nb down), which produces a polar structure, while in ilmenite the orientation of the Fe–Ti pairs alternates, leading to a centrosymmetric structure.

8.9.5   Borates

Conjugated organic molecules are not the only species that possess delocalized π bonds. Trigonal-planar polyatomic anions like NO_3^-, CO_3^{2-}, and BO_3^{3-} all have polarizable delocalized π-bonding MOs. Of these non-centrosymmetric building blocks, borates are by far the most important for NLO applications. About 36% of borates crystallize in non-centrosymmetric space groups, as compared to ~15% of all inorganic solids [23]. Another attractive feature of the borates is a large HOMO–LUMO gap that makes them transparent to near-UV light.

Two important borates are β-BaB_2O_4 and LiB_3O_5. These materials were developed in the 1980s, and they have subsequently become among the most important NLO materials. The appeal of the borates stems from three factors: (a) damage thresholds are much higher than KTP or $LiNbO_3$, (b) many borates are transparent to wavelengths that extend below 200 nm, and (c) relative ease of phase matching.

The NLO coefficients and birefringence of borate crystals largely depend on the concentration and orientation of trigonal-planar BO_3 groups, containing sp^2-hybridized boron. The structure of β-BaB_2O_4 is shown in Figure 8.24. A key feature is the presence of planar $B_3O_6^{3-}$ anions oriented perpendicular to the polar threefold axis. This structural arrangement leads to large NLO coefficients. Furthermore, the coplanar orientation results in high birefringence, a useful trait for critical phase matching.

In LiB_3O_5, triangles and tetrahedra of borate anions link together to form a complex 3D network. There are two important changes with respect to β-BaB_2O_4. Firstly, only

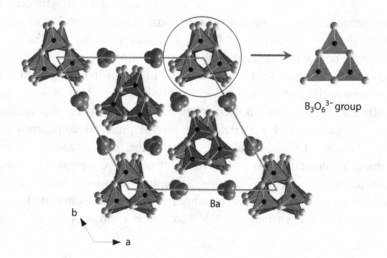

Figure 8.24 The structure of β-BaB_2O_4. The planar $B_3O_6^{3-}$ anion is shown on the right.

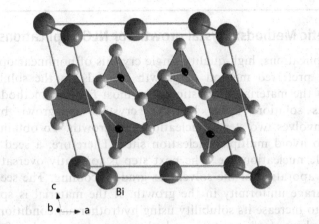

Figure 8.25 The structure of BiB_3O_6 showing the chains of corner-connected BO_4 tetrahedra and BO_3 triangles.

two-thirds of the boron atoms are in a trigonal-planar coordination environment. Secondly, the coplanar alignment of these triangles is lost. Consequently, LiB_3O_5 has both smaller NLO coefficients and a smaller birefringence than β-BaB_2O_4. Its advantage is a higher damage threshold, among the highest for commercial SHG materials. Phase matching with LiB_3O_5 crystals is often achieved via noncritical phase matching ("temperature tuning").

An attractive design strategy is to combine the polarizability of the borate anions with the high nonlinearities that come with SOJT distortions of cations. Unfortunately, compounds that contain borate anions with cations prone to SOJT distortions are relatively rare, and in those that do exist, the orientations of the cations and anions are not well aligned. An exception is BiB_3O_6 (Figure 8.25) where Bi^{3+} with a pronounced stereochemically active lone pair is present. Calculations of the NLO coefficients show that both the borate anions and the Bi^{3+} cations make important contributions to the NLO response, which is much higher than LiB_3O_5, even though both compounds contain similar concentrations of tetrahedral and trigonal-planar boron.

There are many other borate materials that have been investigated for SHG applications. $CsLiB_6O_{10}$ is a congruently melting structural derivative of LiB_3O_5, which makes crystal growth easier. $Sr_2Be_2B_2O_7$ and $KBe_2BO_3F_2$ are transparent down to 150–160 nm and thus are well suited for applications involving short-wavelength UV light. The isostructural compounds $YCa_4(BO_3)_3O$ and $GdCa_4(BO_3)_3O$ can be doped with luminescent rare-earth ions (e.g. Nd^{3+}) to use in self-frequency-doubling lasers, where the same material is used for lasing and nonlinear frequency conversion. This results in compact devices capable of producing UV, blue, and green light.

Box 8.3 Synthetic Methods: Crystal growth for NLO applications

For practical applications, high-quality single crystals of nonlinear optical materials are a necessity. The preferred method of growth depends on the solubility and melting characteristics of the material in question. The most familiar method of crystal growth is from aqueous solution. Large KH_2PO_4 crystals are grown by this technique. Crystallization involves two steps, nucleation and growth. To obtain large single crystals one needs to avoid multiple nucleation sites. Therefore, a seed crystal is used to introduce a single nucleation site. The next step is to gently oversaturate the solution either through evaporation of the solvent or gradual cooling. The seed crystal is slowly rotated to encourage uniformity in the growth. If the material is sparingly soluble, it may be possible to increase its solubility using hydrothermal conditions. This is the case for $KTiOPO_4$, where seed crystals and the source material, often polycrystalline $KTiOPO_4$ in the presence of mineralizers like KH_2PO_4 and KNO_3, are placed in a hydrothermal autoclave [24]. Once the desired high temperature and high pressure have been reached, a temperature gradient is used to grow the seeds at the cooler end into large crystals.

Many commercial SHG materials are not water-soluble, which means that other methods of growth must be employed. The best strategy depends on the melting characteristics of the material. Congruently melting compounds (Section 4.2.2) are often grown by the **Czochralski method**: A seed crystal is placed in contact with the surface of the melt whose temperature is kept slightly above the melting point. The seed is then slowly pulled out of the melt, setting up a temperature gradient that causes the melt to crystallize on the tip of the seed crystal. The seed is usually rotated as it is pulled out to maintain uniformity of the temperature and composition of the melt.

Surprisingly, many commercially important SHG materials do not melt congruently. **Incongruently melting** compounds decompose upon melting, forming a solid and a melt of differing compositions. Their crystal growth requires knowledge of the phase diagram to properly select the composition of the starting mixture such that the melt formed will be in equilibrium with the composition of the desired crystal. Another possibility is a "flux" method. A flux is an additive, typically a low-melting salt, whose molten state acts as a solvent. Growth of NLO crystals from a flux generally involves **top-seeded solution growth**, where a rotating seed crystal is slowly pulled out of the flux, similar to the Czochralski method. β-BaB_2O_4 is often grown from NaF fluxes. $KTiOPO_4$ can be grown from fluxes such as $K_4P_2O_7$ and $K_6P_4O_{13}$. LiB_3O_5 and $LiNbO_3$ are grown by "self fluxing", that is by using an excess of one of the reagents, B_2O_3 and Li_2O, respectively. The disadvantage of a flux is that it can introduce undesired impurities and/or lead to nonstoichiometry.

8.10 Problems

8.1 Derive the Clausius–Mossotti equation for the CGSes system of units.

8.2 Use the Clausius–Mossotti expression in Equation (8.13) and Table 8.2 to estimate permittivities of SnO_2 (rutile-type, $P4_2/mnm$, $Z = 2$, $a = 4.74$ Å, $c = 3.19$ Å), TiO_2 (rutile-type, $a = 4.59$ Å, $c = 2.96$ Å) and ZrO_2 (baddeleyite-type, $P2_1/c$, $Z = 4$, unit-cell volume $= 141$ Å). Does the Clausius–Mossotti equation give a reasonably accurate estimate for each compound when compared to the experimental values of ε_r given in Table 8.1? If not, what is the origin of the discrepancy?

8.3 Why does the polarizability of the lanthanoid ions decrease as the atomic number increases?

8.4 Lanthanoid zirconates $Ln_2Zr_2O_7$ (Ln = La, Pr, Nd, Sm, Eu) adopt the cubic pyrochlore structure ($Fd\bar{3}m$, $Z = 8$), with cubic unit-cell edge $a = 10.80$ Å (La), 10.69 Å (Pr), 10.67 Å (Nd), 10.59 Å (Sm), 10.55 Å (Eu). Use this information, Equation (8.13) and Table 8.2 to estimate the dielectric permittivity of these phases. Does the Clausius–Mossotti equation predict that the dielectric constant will increase or decrease as the radius of the rare-earth ion decreases?

8.5 Use the Clausius–Mossotti expression to estimate the dielectric permittivities of the cubic perovskites $BaZrO_3$ ($a = 4.19$ Å), $KTaO_3$ ($a = 3.99$ Å), and $SrTiO_3$ ($a = 3.90$ Å). Compare your estimates to the observed values in Table 8.1. How does the divergence between calculated and observed values correlate with the tendency for the octahedral cations to undergo SOJT distortions? What does this tell you about compositions for which the Clausius–Mossotti equation can reliably be used?

8.6 Identify three characteristics of a relaxor ferroelectric that distinguish it from a normal ferroelectric.

8.7 Why don't first-order Jahn–Teller distortions, such as those seen with six-coordinate Cu^{2+} or Mn^{3+}, typically lead to polar materials?

8.8 In each of the following pairs, identify the cation that is more likely to undergo an SOJT distortion. Briefly explain your reasoning. (a) Ti^{4+} or V^{4+}, (b) Mn^{3+} or Ta^{5+}, (c) Mo^{6+} or Zr^{4+}, (d) Sn^{4+} or Te^{4+}.

8.9 In a simple cubic perovskite like $SrTiO_3$, all oxygens are equivalent. Each makes two bonds to Ti^{4+} and four to Sr^{2+}, as shown in a bond graph below.

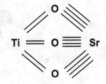

(a) Draw a comparable bond graph for a 1:1 ordered perovskite Ba_2ScTaO_6 that obeys Pauling's rule of parsimony (Section 1.4.4). (b) Use the bond graph to estimate the Ba–O, Sc–O, and Ta–O bond valences. (c) Draw a comparable bond graph for the 1:2 ordered perovskite $Ba_3ZnTa_2O_9$ and determine the minimum number of chemical environments needed for oxygen is this structure. For each such type of oxygen, determine the number of bonds it makes to each cation. (d) What are the bond valences for Ba–O, Zn–O, and Ta–O? What are the bond-valence sums for each type of oxygen? (e) Based on your answers to parts (b) and (d), do the oxygen bond valences suggest a bonding instability that might trigger an SOJT distortion of the Ta^{5+} ions? Explain your reasoning.

8.10 Is it possible for a cubic material structure to be (a) non-centrosymmetric, (b) piezoelectric, (c) pyroelectric?

8.11 The resonant frequency of a microwave resonator can be approximated by the relationship:

$$v_0 \approx \frac{c}{d_{cavity}\varepsilon_r}$$

where c is the speed of light and d_{cavity} and ε_r are the diameter and permittivity of the dielectric puck, respectively. (a) What diameter would give a resonant frequency, v_0, of 850 MHz for a puck made of $Ba_3(Co_{1-x}Zn_x)Nb_2O_9$ ($\varepsilon_r = 34$)? (b) How would the diameter change for a resonator made from $(Ca_{1-x}Nd_x)(Ti_{1-x}Al_x)O_3$ ($\varepsilon_r = 45$)?

8.12 The perovskite $BiInO_3$ can be prepared using high-pressure synthesis. At room temperature, it has $Pna2_1$ space-group symmetry, with an SHG signal 120–140 times that of α-quartz. Upon heating, no phase transitions occur until it decomposes into In_2O_3 and $Bi_{25}InO_{39}$ at 873 K. Based on this information, what can you say about the potential of $BiInO_3$ for application as (a) a pyroelectric, (b) a ferroelectric, (c) a piezoelectric? (d) How would you expect the dielectric constant to change upon heating?

8.13 $BiAlO_3$ and $BiGaO_3$ can be prepared by high-pressure synthesis. The space group of $BiAlO_3$ is $R3c$ while that of $BiGaO_3$ is $Pcca$. Based on the symmetry alone, what can you say about the possibility for (a) ferroelectric, and (b) piezoelectric behavior in these two phases?

8.14 PbO adopts the tetragonal litharge structure that can be described as a distorted variant of the CsCl structure:

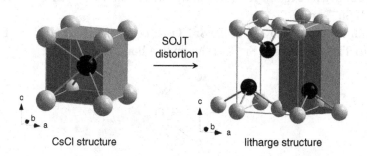

CsCl structure   litharge structure

The SOJT distortion driven by Pb^{2+} leads to a large displacement of the cation toward a square face of the original cubic coordination environment and a corresponding elongation of the c axis of the unit cell. (a) The bond-valence parameters for Pb^{2+}–O bonds are $R^0 = 2.11$ Å and $B = 0.37$ Å. Use Equation (5.18) ($d_{ij} = R_{ij}^0 - B\ln v_{ij}$) to estimate the Pb–O bond length in the hypothetical cubic CsCl-type structure of PbO.

Use this distance to calculate the unit-cell edge and volume. (b) The actual structure of PbO has space group $P4/nmm$ and $Z = 2$, with $a = 3.974$ Å, $c = 5.022$ Å. Calculate the unit-cell volume per formula unit and compare with your prediction for PbO with the CsCl structure. What volume expansion (in %) is thus needed to make room for the stereochemically active electron lone pair? (c) If we take Pb^{2+} and O^{2-} to be equal in size (eight-coordinate radii are 1.43 Å and 1.28 Å, respectively, for Pb^{2+} and O^{2-}), how does the "volume" of a lone pair compare with the "volume" of an oxide ion?

8.15 The black modification of SnO is isostructural with PbO litharge of the previous problem. (a) The bond-valence parameters for Sn^{2+}–O are $R^0 = 1.98$ Å and $B = 0.37$ Å. Use Equation (5.18) ($d_{ij} = R_{ij}^0 - B\ln v_{ij}$) to predict the Sn–O bond length in the hypothetical CsCl-type SnO. Use this distance to calculate the unit-cell edge and volume. (b) The crystal structure of SnO has space group $P4/nmm$ and $Z = 2$, with $a = 3.803$ Å, $c = 4.838$ Å. Calculate the unit-cell volume per formula unit and compare with your prediction of the volume for "cubic" SnO with the CsCl structure. What percent expansion is needed to make room for the stereochemically active electron lone pair?

8.16 We can approximate the Sn^{2+} coordination in the hypothetical CsCl-type SnO of the previous problem by an SnH_8^{6-} cube. The point-group symmetry for a cube is $m\bar{3}m$ (O_h). In SnH_8^{6-}, the Sn $5s$ and $5p$ orbitals have a_{1g} and t_{1u} symmetry, respectively. The hydride ligands form two triply degenerate SALCs with t_{1u} and t_{2g} symmetry, as well as two singly degenerate SALCs with a_{1g} and a_{2u} symmetry. Use this information to construct an approximate MO diagram for a cubic SnH_8^{6-} molecule. What is the degeneracy, bonding character, and symmetry of the HOMO and LUMO?

8.17 The structure of $NH_4H_2PO_4$ is closely related to KH_2PO_4. A projection of the $NH_4H_2PO_4$ structure (comparable to KH_2PO_4 in Figure 8.20) is shown below. The lightly shaded phosphate tetrahedra are at $z = ½$, the darker ones at $z = 0$. The hydrogen atoms are represented by black spheres. Based on the pattern of O–H bonds in this figure, predict the displacements of the phosphorus atoms. Would you expect $NH_4H_2PO_4$ to be an SHG material like KDP?

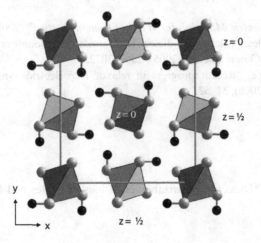

8.18 In 2005, two new polymorphs of BiB_3O_6 (β- and γ-BiB_3O_6) were synthesized using boric acid as a flux [25]. All three polymorphs adopt closely related monoclinic structures. While α-BiB_3O_6 has $C2$ space-group symmetry, β- and γ-BiB_3O_6 both crystallize with the $P2_1/n$ space group. Would you expect β- and γ-BiB_3O_6 to show NLO properties comparable to α-BiB_3O_6?

8.19 Determine the crystal system and point-group symmetry from the space-group symbol for each of the following borates. In each case determine the point group and state whether the symmetry permits SHG activity. (a) $I\bar{4}2d$ for $CsLiB_6O_{10}$, (b) $Ia\bar{3}d$ for $Sr_4Li(BO_3)_3$, (c) $Ama2$ for $Ca_4Na(BO_3)_3$, (d) $P2_12_12_1$ for CsB_3O_5, (e) $P2_1/n$ for $BaLiBO_3$, (f) $C2$ for $CsBe_2BO_3F_2$.

8.20 For borates with similar orientation of the BO_3 groups, the NLO coefficients should roughly scale with the coplanar character and density of the BO_3 groups. The following borates have highly coplanar anions: β-BaB_2O_4 ($R3c$, unit-cell volume 1731 Å^3, $Z = 18$), $Sr_2Be_2(BO_3)_2O$ ($P\bar{6}c2$, unit-cell volume 290.8 Å^3, $Z = 2$), KBe_2BO_3F ($R32$, unit-cell volume 318.0 Å^3, $Z = 3$). (a) Calculate the concentration of BO_3 groups per unit volume for each compound. (b) The largest NLO coefficients for each compound are the $d_{22} = 2.3$ pm/V for β-BaB_2O_4, $d_{11} = 1.52$ pm/V for $Sr_2Be_2(BO_3)_2O$ and $d_{11} = 0.8$ pm/V for KBe_2BO_3F. Do the NLO coefficients scale with the concentration of BO_3 groups? (c) $CsBe_2BO_3F_2$ ($C2$, unit-cell volume 243 Å^3, $Z = 2$) also has nearly coplanar BO_3 groups. How would you expect its NLO coefficients to compare with the other three compounds?

8.11 Further Reading

S. Elliot, *"The Physics and Chemistry of Solids"* (1998) Wiley.

A.R. West, *"Solid State Chemistry and its Applications"* 2nd edition (2014) Wiley.

R.W. Boyd, *"Nonlinear Optics"* 3rd edition (2008) Academic Press.

P.S. Halasyamani, K.R. Poeppelmeier, "Noncentrosymmetric oxides" *Chem. Mater.* **10** (1998), 2753–2769.

M.T. Sebastian, *"Dielectric Materials for Wireless Communication"* (2008) Elsevier.

I.M. Reaney, D. Iddles, "Microwave dielectric ceramics for resonators and filters in mobile phone networks" *J. Am. Ceram. Soc.* **89** (2006), 2063–2072.

A.A. Bokov, Z.G. Ye, "Recent progress in relaxor ferroelectrics with the perovskite structure" *J. Mater. Sci.* **41** (2006), 31–52.

8.12 References

[1] R.D. Shannon, "Dielectric polarizabilities of ions in oxides and fluorides" *J. Appl. Phys.* **73** (1993), 348–366.

[2] R.D. Shannon, R.X. Fischer, "Empirical electronic polarizabilities in oxides, hydroxides, oxy-fluorides, and oxychlorides" *Phys. Rev. B* **73** (2006), 235111.

[3] I.M. Reaney, D. Iddles, "Microwave dielectric ceramics for resonators and filters in mobile phone networks" *J. Am. Ceram. Soc.* **89** (2006), 2063–2072.

[4] W.J. Merz, "The electrical and optical behavior of $BaTiO_3$ single-domain crystals" *Phys. Rev.* **76** (1949), 1221–1225.

[5] M.S. Senn, D.A. Keen, T.C.A. Lucas, J.A. Hriljac, A.L. Goodwin, "Emergence of long-range order in $BaTiO_3$ from local symmetry-breaking distortions" *Phys. Rev. Lett.* **116** (2016), 207602.

[6] A.A. Bokov, Z.G. Ye, "Recent progress in relaxor ferroelectrics with the perovskite structure" *J. Mater. Sci.* **41** (2006), 31–52.

[7] F. Cordero, F. Craciun, F. Trequattrini, C. Galassi, P.A. Thomas, D.S. Keeble, A.M. Glazer, "Splitting of the transition to the antiferroelectric state in $PbZr_{0.95}Ti_{0.05}O_3$ into polar and antiferrodistortive components" *Phys. Rev. B* **88** (2013), 094107.

[8] A. Yoshiasa, T. Nakatani, A. Nakatsuka, M. Okube, K. Sugiyama, T. Mashimoto, "High-temperature single-crystal X-ray diffraction study of tetragonal and cubic perovskite-type $PbTiO_3$ phases" *Acta Crystallogr. Sect. B* **72** (2016), 381–388.

[9] B. Noheda, D.E. Cox, G. Shirane, J.A. Gonzalo, L.E. Cross, S.-E. Park, "A monoclinic ferro-electric phase in the $Pb(Zr_{1-x}Ti_x)O_3$ solid solution" *Appl. Phys. Lett.* **74** (1999), 2059–2061.

[10] N. Zhang, H. Yokota, A.M. Glazer, D.A. Keen, S. Gorfman, P.A. Thomas, W. Rena, Z.G. Ye, "Local-scale structures across the morphotropic phase boundary in $PbZr_{1-x}Ti_xO_3$" *IUCrJ* **5** (2018), 73–81.

[11] P.S. Halasyamani, K.R. Poeppelmeier, "Noncentrosymmetric oxides" *Chem. Mater.* **10** (1998), 2753–2769.

[12] R.A. Wheeler, M.H. Whangbo, T. Hughbanks, R. Hoffmann, J.K. Burdett, T.A. Albright, "Symmetric vs. asymmetric linear M–X–M linkages in molecules, polymers, and extended networks" *J. Am. Chem. Soc.* **108** (1986), 2222–2236.

[13] H.W. Eng, P.W. Barnes, B.M. Auer, P.M. Woodward, "Investigations of the electronic structure of d^0 transition metal oxides belonging to the perovskite family" *J. Solid State Chem.* **96** (2003), 535–546.

[14] F. Cora, C.R.A. Catlow, "QM investigations on perovskite-structured transition metal oxides: Bulk, surfaces and interfaces" *Faraday Trans.* **114** (1999), 421–442.

[15] M. Kunz, I.D. Brown, "Out-of-center distortions around octahedrally coordinated d^0 transition metals" *J. Solid State Chem.* **115** (1995), 395–406.

[16] K.M. Ok, P.S. Halasyamani, C.D. Casanova, M. Llunell, A.P. Alemany, S. Alvarez, "Distortions in octahedrally coordinated d^0 transition metal oxides: A continuous symmetry measures approach" *Chem. Mater.* **14** (2006), 3176–3183.

[17] W. Chang, S.W. Kirchoefer, J.M. Pond, J.A. Belloti, S.B. Qadri, "Room-temperature tunable microwave properties of strained $SrTiO_3$ films" *J. Appl. Phys.* **96** (2004), 6629–6633.

[18] M. Itoh, R. Wang, Y. Inaguma, T. Yamaguchi, Y.J. Shan, T. Nakamura, "Ferroelectricity induced by oxygen isotope exchange in strontium titanate perovskite" *Phys. Rev. Lett.* **82** (1999), 3540–3543.

[19] M.W. Stoltzfus, P.M. Woodward, R. Seshadri, J-H. Klepeis, B.E. Bursten, "Structure and bonding in $SnWO_4$, $PbWO_4$ and $BiVO_4$: Lone pairs vs. inert pairs" *Inorg. Chem.* **46** (2007), 3839–3850.

[20] I.D. Brown, "*The Chemical Bond in Inorganic Chemistry: The Bond Valence Model*" (2002), Chapter 8, Oxford University Press.

[21] P. Becker, "Borate materials in nonlinear optics" *Adv. Mater.* **10** (1998), 979–992.

[22] T. Vijayakumar, I.H. Joe, C.P.R. Nair, V.S. Jayakumar, "Efficient electron delocalization in prospective push–pull non-linear optical chromophore 4-[N, N-dimethylamino]-4'-nitro stilbene (DANS): A vibrational spectroscopic study" *Chem. Phys.* **343** (2008), 83–99.

[23] D. Xue, K. Betzler, H. Hesse, D. Lammers, "Nonlinear optical properties of borate crystals" *Solid State Commun.* **114** (2000), 21–25.

[24] R.A. Laudise, W.A. Sunder, R.F. Belt, G. Gashurov, "Solubility and P–V–T relations and the growth of potassium titanyl phosphate" *J. Cryst. Growth* **102** (1990), 427–433.

[25] L.Y. Li, G.B. Li, Y.X. Wang, F.H. Liao, J.H. Lin, "Bismuth borates: Two new polymorphs of BiB_3O_6" *Inorg. Chem.* **44** (2005), 8243–8248.

9 Magnetic Materials

9.1 Magnetic Materials and Their Applications

While most people are familiar with the concept of a magnet, many do not realize the ubiquity of magnetic materials in our everyday lives. Every electric motor contains a ferro- or ferrimagnet. So does every headphone, loudspeaker, and power-supply transformer. Magnetic card strips contain ferrimagnetic γ-Fe_2O_3. The old technology of tape recording used γ-Fe_2O_3 or ferromagnetic CrO_2. Hard disks in computers have recording platters coated with ferromagnetic alloys patterned on a nanometer scale.

The fundamental origin of magnetism can be traced back to the movement of an electrical charge. This can be the flow of current in an electrical circuit or, as in the solids we'll discuss, due to the quantum-mechanical properties of electrons in atoms. In this chapter we first introduce some of the key physical concepts of magnetism and define some of the quantities involved. We will discuss how to understand concepts such as *diamagnetism* and *paramagnetism* of isolated atoms and their assemblies. We will then move on to study the origins of cooperative phenomena such as *antiferromagnetism*, *ferromagnetism*, and *ferrimagnetism* and how these can be controlled and exploited in functional materials. It is perhaps worth noting at the outset that magnetism is an area where we'll encounter unfamiliar units and where it's often more convenient to work in non-standard units, the so-called CGSem units, than the standard SI system. We'll generally adopt SI but will choose to list the alternative CGSem units in situations where they're most commonly encountered in the literature.

9.2 Physics of Magnetism

9.2.1 Bar Magnets and Atomic Magnets

Most people are familiar with the everyday properties of bar magnets from childhood toys and school science experiments. We know that they send magnetic field as though emanating

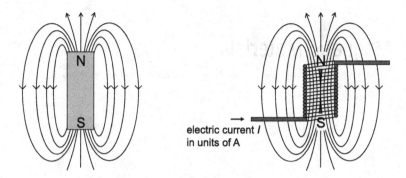

Figure 9.1 Magnetic field generated by a bar magnet and by an electric solenoid in a 2D rendering.

from one end, through the surrounding space, and back at the opposite end, forming a magnetic loop. When suspended, the bar aligns itself with the magnetic field of the Earth. By convention, the end that points towards the Earth's magnetic pole in the north is called the magnet's north pole, N, and the other end the south pole, S. The N and S poles of two bar magnets would attract each other until they unite in a longer bar of just two poles. Accordingly, no division of a bar magnet isolates the poles. However small, a magnet always behaves as an N, S magnetic dipole.

Amazingly, the same magnetic field can be generated by sending a direct electric current through a solenoid (Figure 9.1). This not only facilitates the physical description of magnets via moving electrical charges, but also suggests that magnetism has atomic origins, being caused by the movement of charges in atoms.

Let's extract a single current loop from the solenoid in Figure 9.1. The magnetic field lines around such a circular loop are drawn in Figure 9.2. The product of the loop area A and the loop current I has the magnitude μ of a vector $\boldsymbol{\mu}$, called the **magnetic dipole moment**:[1]

$$\mu = IA \tag{9.1}$$

This defines the SI unit[2] of the magnetic dipole moment as A m^2.

Let's now consider, on Figure 9.3, just one electron orbiting with velocity v at radius r.[3] As a rotating particle, it has an angular momentum that is a vector,[4] as shown in Figure 9.3. As

---

[1] All magnetic moments are magnetic dipole moments, and we typically omit the word "dipole". For convenience, we often choose to deal only with the magnitude of this vector, as will be done for most of this chapter, and omit the "magnitude" by speaking of its absolute value as a "magnetic moment".

[2] The unit A m^2 is equivalent to joule per tesla, J/T = kg m^2 s^{-2}/(kg s^{-2} A^{-1}), in which the joule is the unit of work or energy and (as you'll see shortly) the tesla is the unit of magnetic induction.

[3] Be careful not to confuse the italic v for velocity with the Greek letter ν for frequency.

[4] The direction of the angular-momentum vector follows the right-hand rule of fingers indicating the circular motion *of the particle*.

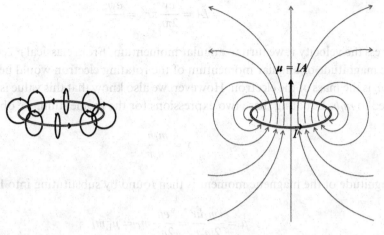

Figure 9.2 Left: Electric current generates a magnetic field with a specific orientation. Right: Magnetic-moment vector μ due to electric current I in a circular loop of area A whose vector is not drawn but is parallel with μ (as it follows the thumb of the right hand with fingers along the *current* direction). The current direction is by definition the opposite of the movement of the electrons.

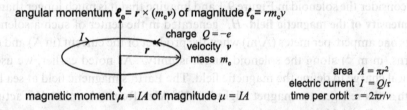

Figure 9.3 Electron as an orbiting and charged particle has an angular momentum and gives rise to a magnetic moment.

a rotating charge, it produces magnetic moment. The current I is the flow of the electron's charge per time $\tau = 2\pi r/v$ to complete one circuit,

$$I = -\frac{e}{\tau} = -\frac{ev}{2\pi r} \tag{9.2}$$

where e is the *elementary charge*, $1.602176634 \times 10^{-19}$ C (Appendix J). Accordingly, the current direction is the opposite of the electron's direction, turning the area vector A as well as the magnetic moment downwards in Figure 9.3, in agreement with the right-hand rule in the caption of Figure 9.2. We see that the magnetic-moment vector has the opposite direction of the angular-momentum vector. From now on, however, let's consider only magnitudes. The magnitude of the magnetic moment is:

$$\mu = IA = \frac{ev}{2\pi r}\pi r^2 = \frac{evr}{2} \tag{9.3}$$

To express the velocity v, we turn to angular momentum. From classical mechanics we know that the magnitude of angular momentum of the rotating electron would be given by $m_{e}vr$, where m_{e} is the mass of the electron. However, we also know that this value is quantized[5] and restricted to $m_{\ell}\hbar$. Equating these two expressions for the momentum, we obtain for v:

$$v = \frac{m_{\ell}\hbar}{m_{e}r} \tag{9.4}$$

The magnitude of the magnetic moment is then found by substituting into Equation (9.3):

$$\mu = \frac{em_{\ell}\hbar r}{2m_{e}r} = \frac{e\hbar}{2m_{e}}m_{\ell} = \mu_{B}m_{\ell} \tag{9.5}$$

Here we have introduced the **Bohr magneton** $\mu_{B} = e\hbar/2m_{e}$ as a convenient unit for the very small atomic magnetic moment, $\mu_{B} = 9.27401 \times 10^{-24}$ A m^2.

9.2.2 Magnetic Intensity, Induction, Energy, Susceptibility, and Permeability

Let's consider the solenoid in Figure 9.1 and imagine that it is much longer than its diameter. The **intensity of the magnetic field**, H,[6] generated in the center of such a solenoid, along its axis, is one ampere per meter (A/m) when the product of the current (in A) and of the density of turns (in m^{-1}) along the solenoid length is unity.[7] As noted earlier, we use the moving electrical charge to define the magnetic field. The Earth's magnetic field at sea level is in tens of A/m and a refrigerator magnet might be ~10^5 A/m. Remember that H is actually a vector, and so are B and M, introduced below.[8]

When a medium is immersed into a magnetic field, it is penetrated by it. The density of the magnetic force lines in the medium is called the **magnetic induction**, B.[9] The magnetic induction is defined as the force that a homogeneous magnetic field exerts on a unit length of a straight wire carrying unity current (Appendix E). The unit of B, one N/(A m) = J/(A m^2), is named the

---

[5] It is quantized with the orbital magnetic quantum number, m_{ℓ} (Chapter 5). For our rotating electron, the quantization follows already from the condition of having an integer number m_{ℓ} of wavelengths $\lambda = h/m_{e}v$ around the orbit when the electron is taken as a standing wave.

[6] Also called magnetic-field strength. When used to magnetize objects, it is termed magnetizing field. The electric analogy of H is the electric-field intensity E introduced in Section 8.1.

[7] A current of 0.001 A in a solenoid of 1000 turns per meter of its length will generate $H = 1$ A/m.

[8] We choose to deal only with the magnitude because magnetic measurements are typically performed on powder samples of isometric shape, where magnetic susceptibility (see next page) is measured as a scalar. In a single crystal, it would generally be a tensor, a correlation scheme between two vectors.

[9] B is therefore also called magnetic-flux density, but physicists often prefer just magnetic field. The electric analogy of B is the electric-flux density D (the electric displacement introduced in Section 8.1.2).

tesla (T). The relationship between the field intensity H and the induction B_0 in a vacuum (absence of any medium, hence the zero) is:

$$B_0 = \mu_0 H, \tag{9.6}$$

where μ_0 ($4\pi \times 10^{-7}$ N/A^2) is the magnetic constant.[10] Inside any other medium, the induction B will be the sum of the contribution from the external field ($\mu_0 H$) and from the medium itself ($\mu_0 M$):

$$B = \mu_0 (H + M) \tag{9.7}$$

where M is the **magnetization** of the medium, its magnetic moment per unit volume (A m^2/m^3 = A/m).[11] The dependence of M on H characterizes a material's magnetic properties at a given temperature. The magnetization M is the key quantity because it represents a material-specific response to an external magnetic field.

Materials that are less penetrable for a magnetic field than a vacuum have $M < 0$ and are said to be **diamagnetic**, while others, such as **paramagnetic** materials, concentrate the magnetic field in their volume.[12] As shown in Figure 9.4, in an uneven magnetic field (under a gradient), paramagnets will move towards regions of highest magnetic intensity (be attracted), whereas diamagnets will move away (be repelled).

Magnetization M is conveniently described relative to the field intensity H that caused it—by **volume (magnetic) susceptibility**, χ_v; the susceptibility per unit volume of the material,

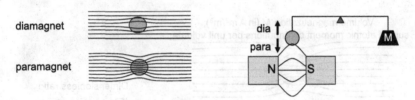

Figure 9.4 A paramagnet is attracted into the magnetic field, diamagnet is repelled; the paramagnet will appear to weigh more in the experiment sketched.

---

[10] Also known as the permeability of free space (or vacuum permeability), it is defined in SI units via the force of 2×10^{-7} N exerted by a current of 1 A flowing in two 1-m-long and 1-m-distant parallel wires in absolute vacuum. The Laplace–Biot–Savart law then gives μ_0. Do not confuse μ_0 with the symbol μ for the magnetic moment and μ_B for Bohr magneton. The electric analogy of μ_0 is the electric constant ε_0 introduced in Section 8.1.1.

[11] The electric polarization P (Section 8.1.2) is not analogous to this magnetization M; it is analogous to the rarely used term magnetic polarization, $J = \mu_0 M$, so that $B = \mu_0 H + J$ corresponds to Equation (8.8) for electric displacement. The use of M and not J in this book follows from the definition of the magnetic moment μ via electric current in a loop, as opposed to the alternative definition via the mechanical force moment of the magnetic pole of a magnet. The pole's exact coordinate is not well defined for an atom, whereas electric charges are separate and can be approximated as points.

[12] As do the cooperative antiferromagnetic, ferromagnetic, and ferrimagnetic materials we'll meet later.

$$\chi_v = \frac{M}{H} \tag{9.8}$$

which is dimensionless in the SI system. Susceptibility is an important parameter of a magnetic material. It's often convenient to divide χ_v with the mass density (in kg/m^3) to obtain the **mass susceptibility**, χ_m (in m^3/kg). Multiplying χ_m with the molar mass (in kg/mol)[13] yields the **molar susceptibility**, χ_{mol} (in m^3/mol).

If we substitute Equation (9.8) for χ_v back into Equation (9.7), we obtain

$$B = \mu_0 H (1 + \chi_v) \tag{9.9}$$

from which we see that diamagnetic materials (for which induction B is less than in a vacuum) have negative χ_v, and paramagnetic materials have positive χ_v. Typical values of χ_{mol} at room temperature might range from -16×10^{-10} m^3/mol for a diamagnetic compound like H_2O to a few 1000×10^{-10} m^3/mol for a paramagnetic transition-metal compound.

Figure 9.5 summarizes the relationships of B, H, and M. It shows that the volume susceptibility arises from introduction of dimensionless variables when Equation (9.7) for magnetization in a medium is divided by Equation (9.6) for magnetization in a vacuum. The second dimensionless variable thus obtained is the **relative magnetic permeability**, μ_r, defined as $\mu_r = B/(\mu_0 H) = \chi_v + 1$. The value of μ_r tells us how many times the magnetic induction is increased by the given material as opposed to a vacuum.[14] Materials with a high induction in their interior have permeability much larger than 1; vacuum has $\mu_r = 1$.

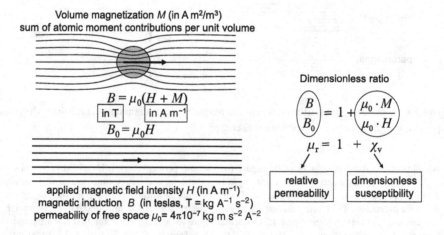

Figure 9.5 Dimensionless magnetic susceptibility in SI units.

[13] Or χ_v with the molar volume in m^3/mol.

[14] Just like the relative dielectric permittivity ε_r in Section 8.1.1 told us how many times the stored electric charge in a capacitor increased upon inserting a material into it instead of a vacuum.

Up to now, we assumed that the moment of the immersed magnetic body is aligned with the external field. If it is not, the magnet will experience a torque to align it. It will have a potential energy, an ability to do work. The **potential energy of a magnetic moment** cross aligned in a homogeneous external field is a dot product of the vectors of the moment μ and of the magnetic induction B. It's a projection of μ on B times $|B|$, a scalar. When the magnetized object is aligned within the applied field, we can consider it to be at the bottom of an energy well of depth μB.

9.2.3 Unit Systems in Magnetism

We've mentioned in the introduction that there's an alternative set of units to SI units in magnetism. For various reasons, this centimeter–gram–second electromagnetic (CGSem) system of units is still sometimes used when working with magnetism. In this system, the μ_0 terms disappear from the equations given above. The variety of units can be very confusing when one starts working in this area. Figure 9.5 is recast into CGSem units in Figure 9.6.

The conversions are not purely decimal, because the two systems build on different system of units and their relations. For example, the law in Appendix E about the force F acting on a unit segment l of a unit-current-carrying wire in a field of unit magnetic induction B is dimensionally correct in CGSem only when the unit of current is 10 A. The CGSem unit for the magnetic moment caused by a loop of electric current, the "electromagnetic moment unit" (emu) is then realized by 10 A cm^2, whereas the SI unit is 1 A m^2. Therefore, 1 emu = 10^{-3} A m^2. As a unit of magnetic moment, the emu also equals erg/G in analogy to J/T = A m^2 in SI. The conversions are summarized in Table 9.1.

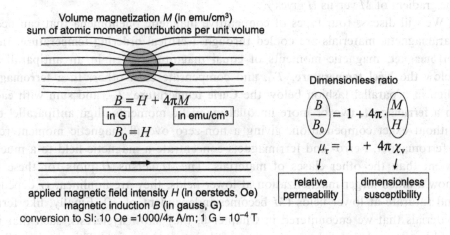

Volume magnetization M (in emu/cm³)
sum of atomic moment contributions per unit volume

$$B = H + 4\pi M$$
in G   in emu/cm³
$$B_0 = H$$

Dimensionless ratio

$$\left(\frac{B}{B_0}\right) = 1 + 4\pi \cdot \left(\frac{M}{H}\right)$$
$$\mu_r = 1 + 4\pi\chi_v$$
relative permeability   dimensionless susceptibility

applied magnetic field intensity H (in oersteds, Oe)
magnetic induction B (in gauss, G)
conversion to SI: 10 Oe =1000/4π A/m; 1 G = 10^{-4} T

Figure 9.6 Dimensionless magnetic susceptibility in CGSem units.

Table 9.1 CGSem–SI conversions for quantities in Figure 9.5 and Figure 9.6.

	Quantity		SI units		CGSem units
H	Field intensity	$1000/4\pi$	A/m	$= 1$	Oe
B	Induction	10^{-4}	T	$= 1$	G
M	Volume magnetization	1000	A/m	$= 1$	emu/cm^3
M_m	Mass magnetization	1	A m^2/kg	$= 1$	emu/g
χ_v	Volume susceptibility	4π	dimensionless	$= 1$	emu/(cm^3 Oe)
χ_m	Mass susceptibility	$4\pi\times10^{-3}$	m^3/kg	$= 1$	emu/(g Oe)
χ_{mol}	Molar susceptibility	$4\pi\times10^{-6}$	m^3/mol	$= 1$	emu/(mol Oe)
μ	Magnetic moment	10^{-3}	A m^2	$= 1$	emu

9.3   Types of Magnetic Materials

When we come to look at collective properties of magnetic moments and the technologically important properties they impart, we'll categorize magnetic materials into six basic types. The magnetic behaviors of the five most important are summarized in a cartoon form in Figure 9.7, together with dependences of key quantities on field and temperature. **Diamagnetic** materials repel a magnetic field (Figure 9.4) and thus have a negative susceptibility χ (a vacuum has $\chi = 0$). Their relative permeability is less than one. **Paramagnetic** materials contain unpaired electrons, have magnetic dipoles, concentrate a magnetic field and thus have a positive χ. The defining property of paramagnets is that their magnetic atoms or ions act independently of each other. In weak magnetic fields, plots of M versus H are linear for diamagnets/paramagnets; having negative/positive slopes, respectively, because the gradient of M versus H gives χ_v.

We will discuss four types of cooperative magnetic phenomena that can occur when paramagnetic materials are cooled through a critical **ordering temperature**. In an **antiferromagnet**, magnetic moments of equal magnitude align in an antiparallel fashion below the **Néel temperature**, T_N, and compensate each other. In a **ferromagnet**, they align in a parallel fashion below the **Curie temperature**, T_C, and sum with each other. In a **ferrimagnet**, two or more unequal magnetic moments align antiparallel below T_C without exact compensation, giving a non-zero overall magnetic moment resembling a ferromagnet. Ferro- and ferrimagnets concentrate a magnetic field to a much greater extent than the other clases of materials. The M versus H plots for these materials show a much larger magnetization at lower fields, they are nonlinear (χ depends on H) and saturate at lower fields (M becomes independent of H). Finally, like ferroelectric materials that we encountered in Chapter 8, they show **hysteresis**, such that when the field is removed they retain a portion of their magnetization. This is what allows them

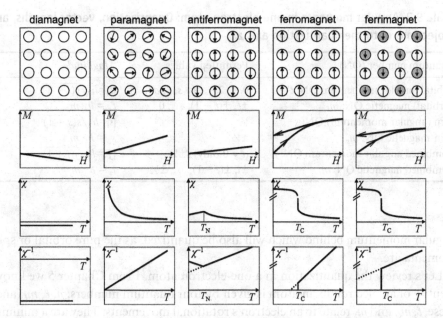

Figure 9.7 Types of magnetic materials and their typical dependences of M versus H and of χ and χ^{-1} versus T.

to be made into permanent magnets. A final category of magnetic materials is the **spin glass** that contains moments that are "frozen" in a disordered arrangement (Section 9.11).[15]

9.4   Atomic Origins of Magnetism

9.4.1   Electron Movements Contributing to Magnetism and Their Quantization

The movement of an electron in an atom can be imagined in terms of the electron orbit around the nucleus (hence having its orbital angular momentum[16]) and of the electron spin around its own axis (hence having its spin angular momentum). Both contribute to magnetism. In Section 9.2.2 we saw how the orbiting electron's angular momentum yields a magnetic moment. The spin angular momentum does not lend itself to a macroscopic interpretation of a "spinning electron", this is a quantum-mechanical effect. When both orbit- and spin-magnetic moments are present, they will combine into a total moment, the

---

[15] Having read "spin glass" and "frozen", one might rename a paramagnet as a "spin fluid" of spin orientations. In ferro-, ferri-, and antiferromagnets, the ordered spin orientations would be a "spin crystal".

[16] For electrons in atoms, the term orbital angular momentum was adopted because electrons occur in orbitals. As the momentum originates from the motion of an electron orbiting the nucleus, the term is not meant to imply that this electron is confined to a single orbital.

Table 9.2 Angular momenta in one-electron atom: Quantization, vector lengths, and projections onto the quantization axis z.

Quantum number (QN)	Permitted values	*Magnitude*
Orbital angular momentum QN: ℓ	$0, 1, 2, \ldots (\equiv s, p, d, \ldots)$	$|\ell| = \hbar\sqrt{\ell(\ell+1)}$
(Orbital) magnetic QN: m_ℓ	$\pm\ell, \pm(\ell-1), \ldots, 0$	$\ell_z = \hbar \cdot m_\ell$
Spin (angular momentum) QN: s	½	$|s| = \hbar\sqrt{s(s+1)}$
Spin-magnetic QN: m_s	$\pm½$	$s_z = \hbar \cdot m_s$
Combined angular momentum QN: j	($\ell \neq 0$ only) $\ell \pm ½$	$|j| = \hbar\sqrt{j(j+1)}$
Combined magnetic QN: m_j	$\pm j, \pm(j-1), \ldots, \pm½$	$j_z = \hbar \cdot m_j$

angular momentum behind which will also be quantized, as the pure orbital or spin angular momenta are.

Let's review the quantization in a one-electron atom. From Chapter 5 we know that the identity of an electron in an atom is given by four quantum numbers; n, ℓ, m_ℓ, and m_s.[17] Of these, ℓ, m_ℓ, and m_s relate to an electron's rotational movements. They are a minimum subset of the in-total six quantum numbers defined in Table 9.2: two for orbit, two for spin, and two for their total. Of these six, ℓ, s, j quantize the respective angular-momentum vectors $\boldsymbol{\ell}$ (orbital), $\boldsymbol{s}$ (spin), and $\boldsymbol{j}$ (orbital and spin combined), and m_ℓ, m_s, m_j quantize the behavior in magnetic field; they represent the respective projections ℓ_z, s_z, j_z of the allowed orientations of those three vectors onto the axis of quantization, which is the axis of the external magnetic field. For each angular-momentum quantum number ℓ, or s, or j, there are $2 \times (\ell$ or s or $j) + 1$ such possible orientations. Each orientation has a different z-projection magnetic moment, μ_z, and the electron's potential energy is therefore split into a multiplet when the external magnetic field is applied—**Zeeman effect** (also termed Zeeman splitting).

The magnitudes μ (see also Footnote 1 in this chapter) of the absolute and z-projection magnetic moments are calculated from the *Magnitude* expressions for angular momenta in Table 9.2 and

$$\mu = g_i\, \mu_B \frac{Magnitude}{\hbar} \tag{9.10}$$

where g_i ($i = \ell$ or s or j) is one of three different g factors: For orbital moment, $g_\ell = 1$ classically. For spin, $g_s = 2.002319$ for a single[18] electron, known as the **electron-spin g factor** (in many-electron systems denoted g_e). For combinations of spin and orbit, the g_j factor must be calculated as shown in Section 9.4.2. The energy split of the multiplet under a magnetic field applied along z is by $U_{m,i}$ (for $i = \ell$ or s or j, see Figure 9.10) calculated from the

---

[17] For example, the value of ℓ tells us what type of orbital the electron occupies ($s, p, d,$ or f) …

[18] For some atoms, the exact value varies slightly with environment.

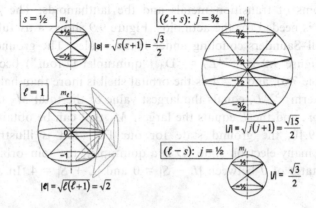

Figure 9.8 Single electron's spin angular momenta (left top) combined with orbital angular momenta (left bottom) yield two choices for the total angular momentum j (right), quantized in units of $\hbar$. The sizes of the two spheres on the right correspond roughly to superposition of the spin cones over the orbital angular momentum vectors at left bottom (precisely in the classical limit of $\ell \to \infty$).

magnitudes of z-projection moments, μ_z, onto the field axis (Table 9.2) and mirrored by $\pm$ values of m_ℓ, m_s, or m_j (Table 9.2) around the initial energy taken as zero:

$$U_{m,i} = B \cdot \mu_z = B \cdot g_i \mu_B m_i \tag{9.11}$$

Only m_ℓ and m_s follow directly from the electron's identity in the atom. The m_j is obtained by **spin–orbit coupling**. The one-electron angular momentum j is a vector sum of the orbital and spin angular momenta, $j = \ell + s$, which are quantized. As shown in Figure 9.8, top left, the spin angular-momentum vector can lie anywhere on the surface of a cone around the field direction and has two possible orientations ($m_s = +\frac{1}{2}, -\frac{1}{2}$). Its components along the field direction are $+\frac{1}{2}\hbar$ and $-\frac{1}{2}\hbar$ (Table 9.2). This spin angular momentum is brought into the picture of the orbital angular momentum shown for a p electron ($\ell = 1$) in Figure 9.8, bottom left, where the angular-momentum vector quantizes in an external magnetic field into three cones with respective projections $-1\hbar$, 0, $+1\hbar$ onto the field direction. Since the spin (vector) has two possible orientations ($m_s = \frac{1}{2}, -\frac{1}{2}$), two spheres containing the m_j-quantized cones are obtained, a smaller one for $j = \ell - \frac{1}{2}$ (Figure 9.8, bottom right) and a larger one for $j = \ell + \frac{1}{2}$ (Figure 9.8, top right). The two j states differ in energy. The difference is a measure of the spin–orbit coupling strength, which is proportional to Z^4 and hence increases rapidly with the atomic number Z.

9.4.2 Atomic Magnetic Moments

When looking at the magnetic properties of a material, we'll typically be interested in the most stable or ground state of its magnetic atoms. The Russell–Saunders coupling under Hund's rules (Section 7.3.2) yields correct ground-state magnetic moments for most free atoms of transition metals and the lanthanoids. The j–j coupling scheme (Appendix F) is needed for the actinoids. Figure 9.9 shows a useful summation scheme for the Russell–Saunders coupling under Hund's rules. The ground state for the Fe^{2+} example in Figure 9.9 is $^{2S+1}L_J$ = $^5D_4$ ("quintuplet-D-four") because $S = 2$, $L = 2$ (hence D; Table 7.3), and $J = 4$ as the orbital shell is more than half filled. The S in the ground-state term $^{2S+1}L_J$ equals the largest value of the total M_S that can be obtained by summing m_s, and the L equals the largest M_L that can be obtained by summing m_ℓ.

In Figure 9.10, the ground state for our Fe^{2+} ion is illustrated on an energy diagram. The many-electron state ^5D is a quintuplet; the spin–orbit coupling produces $2S + 1 = 5$ states of J between $|L - S| = 0$ and $|L + S| = 4$. In an applied magnetic

Electrons:		↑	↑	↑	↑	↑	↓				
First, maximum S	m_s	½	½	½	½	½	−½	−½	−½	−½	−½
Second, maximum L	m_ℓ	2	1	0	−1	−2	2	1	0	−1	−2
Third Hund rule	$J =$			$\|L-S\|$					$\|L+S\|$		

Figure 9.9 Hund's rules in a summation scheme to determine the ground state of a *free* atom or ion with d-valence electrons via Russell–Saunders coupling, illustrated on Fe^{2+} having $^{2S+1}L_J$ = $^5D_4$.

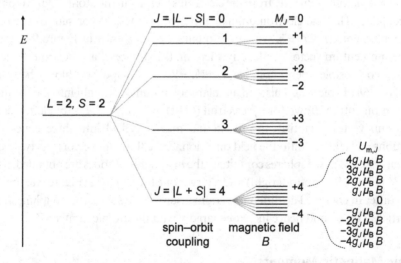

Figure 9.10 Energy diagram for ^5D sub-states of an isolated Fe^{2+} under spin–orbit coupling and a magnetic field. For the ground state $J = 4$, magnetic potential energies $U_{m,J}$ are listed to illustrate the meaning of the split in magnetic field.

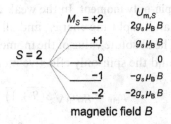

Figure 9.11 Energy diagram for isolated Fe^{2+} under spin-only approximation.

field, each of these states splits in energy into a multiplet of M_J values from $-J$ to $+J$, differing by 1.[19]

Note that the free-ion scheme will not necessarily be valid in a compound of a magnetic atom. We know from Chapter 7 that the ligand field splits the degeneracy of d-electron levels. It is then a question of the mutual symmetry and occupation of the remaining degenerate orbitals whether or not there is a contribution from the orbital angular momentum. If not, as is the case of some $3d$ ions in compounds, we say that the orbital moment is quenched, and the "spin-only" splitting of the ground state in the applied field is much simpler (Figure 9.11). As f electrons are much less sensitive to ligand-field effects, a scheme (see Figure 9.13 later) analogous to Figure 9.9 applies to the magnetic ground state of both free and bonded lanthanoid ions (except for Sm^{3+} and Eu^{3+}). This is discussed in more detail in Section 9.4.3.

Magnetic moments are measured at non-zero fields and non-zero temperatures. The measured value then depends on how the states exemplified in Figure 9.10 are populated. This is controlled by the relative size of the thermal energy kT versus both the spin–orbit J-splitting energy and the M_J-splitting energy that increases with increasing applied field B, Equation (9.11). Two simple cases emerge, depending on the field and temperature:

1. In very strong fields of tens of teslas at very low temperatures of a few kelvins, a **saturation moment**, μ_{sat}, is achieved for paramagnetic materials. It corresponds to the z-projection value of the magnetic moment as calculated with Equation (9.10) and the appropriate z-projection *Magnitude* listed in Table 9.2, in which m_ℓ, m_s, m_j are replaced by M_L, M_S, M_J. In magnetically ordered materials, μ_{sat} is obtained via neutron diffraction.
2. In weak fields at high temperatures, an **effective moment**, μ_{eff}, is measured. It corresponds to the absolute magnetic moment, the value of which is calculated with Equation (9.10) and the appropriate *Magnitude* listed in Table 9.2, in which ℓ, s, j are replaced by L, S, J.

Four cases of these two types of moment are discussed in more detail in the following.

---

[19] It's worth stating that for a full electron shell, L, S, and J must all be zero, so the total angular momentum will be zero and there will be no magnetic moment; the atom will be diamagnetic. This also means that we only need to consider valence electrons when calculating magnetic moments.

The weak-field limit for spin-only moment. In the weak-field limit, the magnetic M_S split is weak, kT is much larger than its splitting energy, and all levels of the S multiplet in Figure 9.11 are equally occupied. The absolute value of the moment in Table 9.2 is applicable, not its projection onto the z axis, and the spin-only effective moment μ_{eff} is:

$$\mu_{\text{eff}} = g_{\text{e}}\mu_{\text{B}}\sqrt{S(S+1)} \tag{9.12}$$

Since $g_{\text{e}} \approx 2$ and $S = \frac{1}{2} n_{\text{u}}$, where n_{u} is the number of unpaired electrons, the approximate value of μ_{eff} in Bohr magnetons is given by $\sqrt{n_{\text{u}}(n_{\text{u}}+2)}$.

The strong-field limit for spin-only moment. Under large B, the difference in the M_S substate energies is large, and, at low temperatures, only the lowest M_S is occupied (for example the $M_S = -2$ level in Figure 9.11). The saturated magnetic moment of the magnetized substance is then calculated from the z-component magnitude in Table 9.2:

$$\mu_{\text{sat}} = g_{\text{e}}\mu_{\text{B}}S \tag{9.13}$$

Setting $g_{\text{e}} \approx 2$ makes μ_{sat} in Bohr magnetons equal to the number of unpaired electrons.

The weak-field limit for spin–orbit coupling. Let's consider the case where kT is less than the J splitting (spin–orbit) but larger than the M_J splitting (magnetic). This means that the lowest J multiplet is rather evenly populated, as exemplified by rare-earth ions at ambient temperatures. The grand total angular momentum J produces the magnetic moment. Because J is composed of spin and orbital contributions, vector summation needs to be applied.

Figure 9.12 shows the summation graphically. Because of the "doubled" contribution from S ($g_{\text{e}} \approx 2$), the instantaneous moment is not aligned with the angular momentum J, but

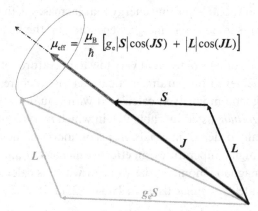

$$\mu_{\text{eff}} = \frac{\mu_{\text{B}}}{\hbar}\left[g_{\text{e}}|S|\cos(JS) + |L|\cos(JL)\right]$$

Figure 9.12 Vector addition of the orbital and spin angular momenta (in black), and of their contributions towards the magnetic moment (in gray), valid in the weak-field case described in text. The effective magnetic moment, μ_{eff}, is proportional to the sum of projections of $g_{\text{e}}S$ and L onto J. For visual clarity of the summation, the angular and magnetic moments are drawn parallel to each other instead of antiparallel.

rapidly sweeps a cone around it (Larmor precession), so that only the projection onto J over time should be considered. The contributions towards the effective magnetic moment μ_{eff} from the two angular momenta will also be projections: $|L|\cos(JL)$ and $g_e|S|\cos(JS)$. The two angles can be expressed from the cosine law, which says that $|L|^2 = |J|^2 + |S|^2 - 2|J||S|\cos(JS)$ and that $|S|^2 = |J|^2 + |L|^2 - 2|J||L|\cos(JL)$ (Figure 9.12). The magnitudes of the angular momentum vectors ($|S|$, $|L|$, and $|J|$) are then expressed in terms of quantum numbers. Hence, for J,

$$\mu_{eff} = g_J \cdot \mu_B \sqrt{J(J+1)}, \tag{9.14}$$

where:

$$g_J = g_e \frac{J(J+1) - L(L+1) + S(S+1)}{2J(J+1)} + \frac{J(J+1) + L(L+1) - S(S+1)}{2J(J+1)} \tag{9.15}$$

For g_e approximated as 2, this simplifies into the **Landé factor**:

$$g_{Land\acute{e}} = 1 + \frac{J(J+1) - L(L+1) + S(S+1)}{2J(J+1)} \approx g_J \tag{9.16}$$

The strong-field limit for spin–orbit coupling. Under large B at low temperatures, only the lowest M_J is occupied, and the saturated magnetic moment of the magnetized substance is then calculated as $\mu_{sat} = g_J \mu_B J$ from the z-component magnitude in Table 9.2 but using g_J of Equation (9.15).

9.4.3 Magnetic Moments for 3d Ions in Compounds

Values of μ_{eff} for 3d transition-metal ions are given in Table 9.3. The spin-only values are much closer to those observed experimentally than to those calculated using the total angular momentum. The reason for this is the well-known ligand-field effect, also discussed in Section 5.3.8 and Section 7.3.1. As an example, a d^1 Ti^{3+} free ion has a 10-fold degenerate ^{2}D ground state (the electron can be spin up or spin down in any one of five d orbitals, the multiplicity $2S + 1 = 2$, Figure 7.5). In an octahedral ligand field, the degeneracy is lifted, resulting in t_{2g} (d_{xy}, d_{xz}, d_{yz}) and e_g (d_{z^2}, $d_{x^2-y^2}$) orbital sets.[20] The e_g orbitals are ~2.5 eV above the t_{2g} and will not be populated under normal conditions, so the Ti^{3+} electron is in t_{2g}. One therefore has a sixfold degenerate ground state. Whether the orbital angular momentum contributes or is quenched then depends on the symmetry of these degenerate orbitals. We can understand this by returning to our picture of the orbit angular momentum resulting

---

[20] The ground-state term under a ligand field is found in correlation diagrams, dealt with in Section 7.3.3. In short, for d^n in an octahedral field, the terms can be A$_{1g}$, A$_{2g}$, E$_g$, T$_{1g}$, or T$_{2g}$, in a tetrahedral field the subscript g falls off. In a weak octahedral/tetrahedral field ($\equiv$ high-spin case), free-ion terms D (d^1, d^9, d^4, d^6) split to E + T$_2$, the F terms (d^2, d^8, d^3, d^7) split to A$_2$ + T$_1$ + T$_2$, the S term (d^5) is not split but renamed A$_1$. Note that the subscripts are not J as in the magnetic-field-induced R–S term, but relate to symmetry.

Table 9.3 Electronic configurations, free-ion terms, configurations, and ground terms for high-spin magnetic moments in Bohr magnetons for selected $3d$ transition-metal salts featuring an octahedral ligand field. Note that different ground terms arise for low-spin cases. Experimental data taken from Nicholls [1].

d^n	Free ion	HS config.	HS ground term	Orbital contribution?	μ_{eff} calculated* $g_J\sqrt{J(J+1)}$	μ_{eff} calculated $g_e\sqrt{S(S+1)}$	μ_{eff} expt 80 K	μ_{eff} expt 300 K	Example
$3d^1$	$^2D_{3/2}$	t_{2g}^1	$^2T_{2g}$	Yes	1.55	1.73	1.4	1.8	Cs_2VCl_6
$3d^2$	$^3F_2$	t_{2g}^2	$^3T_{1g}$	Yes	1.63	2.83	2.7	2.7	$(NH_4)V(SO_4)_2\cdot12H_2O$
$3d^3$	$^4F_{3/2}$	t_{2g}^3	$^4A_{2g}$	No	0.77	3.88	3.8	3.8	$KCr(SO_4)_2\cdot12H_2O$
$3d^4$	$^5D_0$	$t_{2g}^3 e_g^1$	^5E_g	No	0.00	4.90	4.8	4.8	$CrSO_4\cdot6H_2O$
$3d^5$	$^6S_{5/2}$	$t_{2g}^3 e_g^2$	$^6A_{1g}$	No	5.92	5.92	5.9	5.9	$K_2Mn(SO_4)_2\cdot6H_2O$
$3d^6$	$^5D_4$	$t_{2g}^4 e_g^2$	$^5T_{2g}$	Yes	6.70	4.90	5.4	5.5	$(NH_4)_2Fe(SO_4)_2\cdot6H_2O$
$3d^7$	$^4F_{9/2}$	$t_{2g}^5 e_g^2$	$^4T_{1g}$	Yes	6.63	3.88	4.6	5.1	$(NH_4)_2Co(SO_4)_2\cdot6H_2O$
$3d^8$	$^3F_4$	$t_{2g}^6 e_g^2$	$^3A_{2g}$	No	5.59	2.83	3.3	3.3	$(NH_4)_2Ni(SO_4)_2\cdot6H_2O$
$3d^9$	$^2D_{5/2}$	$t_{2g}^6 e_g^3$	^2E_g	No	3.55	1.73	1.9	1.9	$(NH_4)_2Cu(SO_4)_2\cdot6H_2O$

* Notice that, due to Hund's third rule, spin–orbit coupling makes the experimental moment smaller than the spin-only moment when d orbitals are less than half filled and bigger when more than half filled.

from the circular motion of a charged particle around an axis (Figure 9.3). The orbital angular momentum occurs when an electron has *two* orbitals available at about the *same energy level*, that are *rotationally equivalent* around the axis of the electron's conceived circular motion. Around z, such symmetry-related orbitals would be d_{xz} and d_{yz}. Hence, in our example of Ti^{3+} in an octahedral ligand field, an orbital contribution to the magnetic moment is possible. In a d^3 configuration of octahedral Cr^{3+}, on the other hand, the angular momenta of the d_{xy}, d_{xz}, and d_{yz} sphere cancel.

Symmetry being the key, the rule is such that a T ground term[21] allows the orbital angular-momentum contribution (Table 9.3). The overall picture is unfortunately more complex still in that spin–orbit coupling leads to a further splitting of the t_{2g} orbitals of Ti^{3+} even before applying the external magnetic field. The energy separation is of the order of kT, leading to the population, hence magnetic moment, changing with temperature. For d^1 Ti^{3+} in an octahedral ligand field, calculations[22] of thermally induced populations of the microstates predict that μ_{eff} would vary from 0 to $\sqrt{5}$ μ_B as the temperature changes from 0 to ∞ K. The experimental values are far less extreme and somewhat fortuitously end up being close to spin-only predictions at room temperature. The less extreme temperature dependence of such moments is partly due to covalency (electrons shared with ligands) and partly because the true local environment of d^1 Ti^{3+} will not be a regular octahedron but will undergo Jahn–Teller splitting (Section 5.3.10) into levels where the moment would be temperature-independent. Similar arguments hold for the other T ground states, such as those for d^2, d^6, and d^7 in an octahedral field. Compounds of these ions therefore show moments that depart from spin-only values due to spin–orbit coupling, and have contributions from thermally excited states that bring a change with temperature.

Things are more straightforward for ions with ground terms A_{2g} and E_g that allow no orbital contribution. In an octahedral field, such d^3, d^4, d^8, and d^9 ions are found experimentally to have μ_{eff} values close to spin-only predictions, and temperature effects due to thermally excited states[23] are weak. For d^5, spin-only μ_{eff} values are observed.[24]

A peculiarity occurs for the octahedral d^3/d^8 ions (A_{2g} terms), called **zero-field splitting**. Even before magnetic field splits the spins states (into an array from $-M_S$ to $+M_S$, see Figure 9.11), the ligand field splits them a little, yet differently; according to their absolute value, into larger and smaller $|M_S|$. This leads to a paramagnetic anisotropy; the split-state

[21] Adopted in an octahedral field for t_{2g}^1, t_{2g}^2, $t_{2g}^4e_g^2$, and $t_{2g}^5e_g^2$; tetrahedral field $e^2t_2^1$, $e^2t_2^2$, $e^4t_2^4$, and $e^4t_2^5$.

[22] According to Kotani theory.

[23] Minor deviations from spin-only values arise due to mixing with higher-energy T terms of the same multiplicity as the ground term. These are always present for d^3/d^8 A_{2g} and d^4/d^9 E_g terms and can lead to μ_{eff} values lower (for a d shell less than half filled) or higher (for a d shell more than half filled) than expected. In other words, an admixture of excited states introduces some degree of spin–orbit coupling.

[24] There are no higher T terms of the same multiplicity to induce even minor deviations.

occupancies slightly change with the direction of the magnetic field and so do the suscepti-bilities if measured on a single crystal. A weak anisotropy is also caused in the octahedral d^4 and d^9 ions (E_g terms) by Jahn–Teller distortion.

9.4.4 Magnetic Moments for 4*f* Ions in Compounds

For the lanthanoids, the situation is much simpler than above (see Table 9.4). The Russell–Saunders scheme can still be applied, and the contracted nature of the 4f orbitals means that ligand-field effects are far less important. Hund's third rule (Section 9.4.2) is obeyed, and the agreement between μ_{eff}(calc) using J and μ_{eff}(expt) is very good in most cases. For Sm^{3+} and Eu^{3+}, however, the experimental values are much higher than expected for the ground state. This is because there are low-energy excited states partially occupied due to thermal energy, and the moment rapidly falls on cooling.

As an example of the calculation, the Nd^{3+} (f^3) Russell–Saunders $^{2S+1}L_J$ term is $^4I_{9/2}$ where $S = \frac{3}{2}$ yields $2S + 1 = 4$, $L = 6$ yields the I (Table 7.3), and $J = |L-S|$ is $6 - \frac{3}{2} = \frac{9}{2}$. Equation (9.16) then gives $g_{Landé} = 0.727$, with which Equation (9.14) yields $\mu_{eff} = 3.61$ μ_B.

9.4.5 Note on Magnetic Moments of 4*d* and 5*d* Metals in Compounds

When comparing the 3d and 4f ions just discussed, we see that relatively simple formulas can be applied to estimate magnetic moments for most of them (except Sm^{3+} and Eu^{3+}). Estimates became rather complex for 4d and 5d metals since their larger spin–orbit coupling ($\propto Z^4$) is comparable to ligand-field effects, and the situation is intermediate between the relatively clear-cut 3d and 4f cases.

Table 9.4 Calculated μ_{eff} in Bohr magnetons for rare-earth ions.

Ion	f^n	R–S term	μ_{eff}(calc)	μ_{eff}(expt)
Ce^{3+}	$4f^1$	$^2F_{5/2}$	2.53	2.4
Pr^{3+}	$4f^2$	$^3H_4$	3.58	3.5
Nd^{3+}	$4f^3$	$^4I_{9/2}$	3.61	3.5
Pm^{3+}	$4f^4$	$^5I_4$	2.68	not available
Sm^{3+}	$4f^5$	$^6H_{5/2}$	0.84	1.5
Eu^{3+}	$4f^6$	$^7F_0$	0.00	3.4
Gd^{3+}	$4f^7$	$^8S_{7/2}$	7.95	8.0
Tb^{3+}	$4f^8$	$^7F_6$	9.73	9.5
Dy^{3+}	$4f^9$	$^6H_{15/2}$	10.65	10.6
Ho^{3+}	$4f^{10}$	$^5I_8$	10.61	10.4
Er^{3+}	$4f^{11}$	$^4I_{15/2}$	9.58	9.5
Tm^{3+}	$4f^{12}$	$^3H_6$	7.56	7.3
Yb^{3+}	$4f^{13}$	$^2F_{7/2}$	4.54	4.5

Electrons:		↑	↑	↑											
First, maximum S	m_s	½	½	½	½	½	½	½	-½	-½	-½	-½	-½	-½	-½
Second, maximum L	m_ℓ	3	2	1	0	-1	-2	-3	3	2	1	0	-1	-2	-3
Third Hund rule	$J=$				$|L-S|$							$|L+S|$			

Figure 9.13 Hund's rules summation to determine the ground state of a $4f$ atom or ion via Russell–Saunders coupling, illustrated on Nd^{3+} of $^{2S+1}L_J = {}^4I_{9/2}$.

9.5 Diamagnetism

In addition to the orbital and spin contributions to the magnetic moment of an atom with an unpaired electron, there is a diamagnetic contribution. Diamagnetism is caused by the change in the motion of all electrons in an applied field, and that includes the paired electrons that we did not have to consider so far. We can imagine this effect as occurring due to induced currents. **Lenz's law of induction** states that the "magnetic field of an induced loop current is oriented against the applied field". Diamagnetism will therefore make χ negative by expelling the field from the material. In a field with a gradient, the diamagnet will be driven out to regions of weaker field. This is the origin of the famous levitating-frog experiment that abounds on the internet. Diamagnetism occurs in all materials, though its effect is eclipsed by the positive susceptibility arising from any unpaired electrons.[25] Of the elements with no unpaired electrons, Bi is the strongest diamagnet[26] (it has the largest number of electrons and $J = 0$) with $\chi_v = -166 \times 10^{-6}$. In all metals, an additional weak diamagnetic effect due to delocalized electrons occurs, termed **Landau diamagnetism**.[27] Normal diamagnetism is not a desperately interesting phenomenon; however, in superconductors (Chapter 12) the inner currents ideally expel all the magnetic field, and one has a **perfect diamagnet** of $\chi_v = -1$. Such an extreme diamagnetic susceptibility of a superconductor is able to levitate trains.

9.6 Paramagnetism

Paramagnetism is a situation when atomic magnetic moments in the given substance are randomly oriented. Thermal agitation disorders their orientations, whereas an applied magnetic field tends to align them. A paramagnetic material will concentrate the magnetic

---

[25] Nevertheless it's important to correct for diamagnetic contributions when measuring and interpreting properties of paramagnetic substances. Such diamagnetic corrections are often made from tabulated values of Pascal's constants as mentioned in Appendix I.

[26] A compass with a Bi needle points east–west; true to the prefix "dia" meaning across.

[27] Also conjugation electrons of a graphite plane (graphene) cause a strong anisotropic diamagnetism with $\chi = -400 \times 10^{-6}$ for the field perpendicular to the plane. Similarly for aromatic conjugated molecules.

field in its volume (χ is positive) and be attracted to regions of high field. Typical values of χ are 10^{-3} to 10^{-5}. Paramagnets can be split into two categories according to the localized or delocalized nature of the contributing electrons, which in turn determines how susceptibility depends on temperature. Substances with localized electrons are classified as **Curie or Curie–Weiss paramagnets**, while those with delocalized unpaired electrons are called **Pauli paramagnets**.

9.6.1 Curie and Curie–Weiss Paramagnetism

Substances that contain non-interacting paramagnetic moments have a susceptibility that varies inversely with temperature and obeys the **Curie law**,

$$\chi = \frac{C}{T} \tag{9.17}$$

where C is the Curie constant. This hyperbolic temperature dependence arises from the competition between the increasingly stronger thermal agitation of the independent moments and the tendency of the applied field to align them (Figure 9.14). Examples of **Curie paramagnets** include compounds with well-separated transition-metal ions, such as the Tutton's salt $(NH_4)_2Mn(SO_4)_2 \cdot 6H_2O$ and the iron/chromium alums $NH_4M(SO_4)_2 \cdot 12H_2O$ (M = Cr, Fe), rare-earth compounds where the magnetism arises from localized f electrons, and $O_2(g)$. The Curie law is valid unless the applied field is very strong (>1 T) and the temperature very low (<20 K). Deviations from the Curie law can be checked visually by plotting the reciprocal susceptibility versus temperature, which provides a straight line, $\chi^{-1} = C^{-1}T$, when the Curie law is obeyed (Figure 9.14).

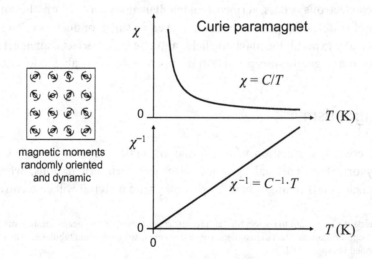

Figure 9.14 Curie paramagnet and Curie law: Temperature dependence of the magnetic susceptibility χ (top), and linearization via its inverse (bottom).

Materials that exhibit interactions between individual magnetic moments will only behave as paramagnets above the temperature at which the thermal energy overcomes those interactions and effectively randomizes the moments. That is above T_C for ferromagnets and above T_N for antiferromagnets. The **Curie–Weiss law** is then valid,

$$\chi = \frac{C}{T - \theta} \tag{9.18}$$

where θ is the **Weiss constant**. Figure 9.15 shows χ and χ^{-1} versus absolute temperature of these two types of Curie–Weiss paramagnets. As magnetization in general aligns magnetic moments in one direction, a negative Weiss constant means a negative propensity to the alignment, as in antiferromagnets, whereas a positive θ means a positive propensity to align the moments, as in ferromagnets. The ferromagnetic order becomes thermally randomized when a positive $\theta \approx T_C$ is reached upon warming, at which point it starts to resemble a Curie paramagnet at 0 K. Since an antiferromagnet's $\theta \approx -T_N$, the antiferromagnet is as weak a magnet at 0 K as a Curie paramagnet would be at a temperature T_N. It should be noted that when $|\theta|$ and the ordering temperature are very different, it indicates some type of frustration working against the order (Section 9.11).

The derivation of the Curie law is in principle quantum mechanical because susceptibility depends on thermally induced populations of the levels that are split from M_J in an applied magnetic field (Figure 9.10). For Curie paramagnets at temperatures larger than ~20 K, in

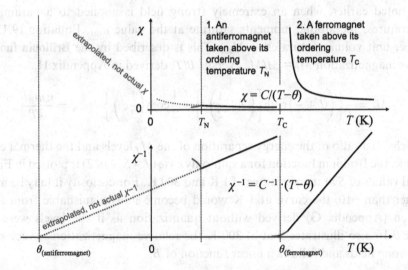

Figure 9.15 Comparison of the two opposite Curie–Weiss paramagnets: A ferromagnet at high temperatures and antiferromagnet at lower temperatures. Their temperature dependence of the magnetic susceptibility χ is on top, and its linearization on the bottom shows the difference in their Weiss constants θ.

fields that are not exceptionally strong, the quantum-mechanical result derived in Appendix H is well approximated by the Curie law,

$$\chi_N = N \frac{\mu_{eff}^2 \mu_0}{3kT} \tag{9.19}$$

where N is the number of magnetic atoms per which the susceptibility is expressed[28] and μ_{eff} is the effective magnetic moment (in A m^2) from which S or J is calculated, leading to the number of unpaired electrons per magnetic atom. As an example, for a spin-only moment, Equation (9.12) is introduced into Equation (9.19) for μ_{eff}. With N_A replacing N, χ_{mol} per mole of magnetic atoms is obtained:

$$\chi_{mol} = \frac{N_A g_e^2 \mu_B^2 \mu_0}{3kT} S(S+1) = 6.3003 \times 10^{-6} \frac{S(S+1)}{T} \text{ m}^3 \text{ mol}^{-1} \tag{9.20}$$

Note that since χ is proportional to μ_{eff}^2 in Equation (9.19), in a compound of two magnetic atoms their μ_{eff} values aren't additive. Susceptibility χ and therefore μ_{eff}^2 values are. For example, consider the perovskite-related LaSrMnMoO$_6$. It contains two atoms with unpaired electrons. High-spin Mn^{2+} has five unpaired electrons hence $S = 2.5$, from which a spin-only $\mu_{eff} = 5.916 \, \mu_B$ is calculated with Equation (9.12). The other magnetic atom is d^1 Mo^{5+} of a spin-only $\mu_{eff} = 1.732 \, \mu_B$ calculated from $S = \frac{1}{2}$. The average moment per magnetic atom is $\mu_{eff} = \sqrt{0.5 \times 5.916^2 + 0.5 \times 1.732^2} = 4.36 \, \mu_B$. It compares well with the average moment of 4.33 μ_B obtained [2] by least-squares fitting with Equation (9.19) of the experimental χ versus $1/T$ data for LaSrMnMoO$_6$.

As noted earlier, when an extremely strong field is applied to a paramagnet at low temperatures, the atomic moments saturate at the value μ_{sat}, Equation (9.13), and their sum per unit volume saturates at M_{sat}. This is described by the **Brillouin function** for the relative magnetization $M_r = M/M_{sat}$ versus B/T, derived in Appendix H,

$$M_r = \frac{1}{J} \left\{ \left(J + \frac{1}{2} \right) \coth \left[\left(J + \frac{1}{2} \right) \xi \right] - \frac{1}{2} \coth \left(\frac{1}{2} \xi \right) \right\} \qquad \xi = \frac{g_J \mu_B B}{kT} \tag{9.21}$$

in which ξ is a ratio of the energy separation of the M_J levels and the thermal energy. As an example, the Brillouin function for a spin-only case ($J = S$, $g_J \approx 2$) is plotted in Figure 9.16 for several values of S at temperatures of 1 K and 300 K. For curiosity it may be noted that for S larger than ~10, the curve at 1 K would become indistinguishable from the Langevin function (Appendix G) derived without quantization as if the levels were continuous. Figure 9.16 also illustrates that at 300 K the relative magnetization is very small even at very strong fields and follows a linear function of B.

---

[28] If N is per unit volume, the dimensionless volume susceptibility χ_v is obtained, if per mol, χ_{mol} is obtained, etc.

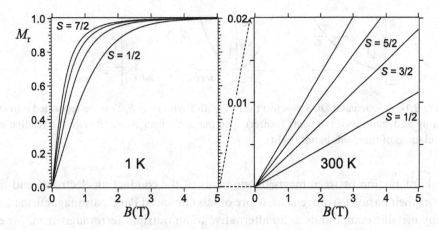

Figure 9.16 Relative magnetization $M_r = M/M_{sat}$ versus an applied field B at 1 K and 300 K, calculated with the Brillouin function for a spin-only case and several values of S.

9.6.2 Pauli Paramagnetism

Curie–Weiss paramagnetism is applicable to substances where the unpaired electrons are localized. When this is not the case, the **temperature-independent Pauli paramagnetism** can be observed. It occurs when the electrons can be reasonably well described as a gas of non-interacting particles. This picture is only appropriate in metallic conductors where the valence electrons are highly delocalized, such as in alkali-metals and Mg, Ca, Sr, Ba, and Al. Other metallic elements are either diamagnetic, like Cu and Bi, or have more complicated magnetism due to interactions between electrons.

We can understand the origins of Pauli paramagnetism through the schematic density-of-states (DOS) picture in Figure 9.17, where we draw separate bands for electrons with moments[29] up ($\uparrow$) and down ($\downarrow$). In the absence of a magnetic field, the energy of electrons in each of these sub-bands will be identical and there will be the same number of $\uparrow$ electrons as $\downarrow$. In the presence of a magnetic field, the energy of the spin moments parallel to it will sink by $\mu_B B$, the energy of the single-electron moment, and the energy of those that are antiparallel will increase by the same amount.

The Pauli paramagnetic moment for the electron-gas model is small, but still three times larger than the opposite Landau diamagnetic moment of those electrons. It is much smaller, though, than the moment of Curie paramagnets.[30] In many transition metals, an **enhanced Pauli paramagnetism** occurs, which varies slightly with temperature. It is associated with

---

[29] The spin magnetic moment is opposite to the spin angular momentum (Figure 9.3).

[30] The reason is as follows: Starting with Equation (9.19) for susceptibility under Curie law, for a free electron gas, one calculates a volume susceptibility with an N_V of the conducting electrons per unit volume, replaces the thermal energy kT with the much higher Fermi energy E_F (a different statistics also applies—one of Fermi–Dirac, not that of Boltzman), and the factor ⅓ with ⅔. The high E_F is also the reason why Pauli paramagnetism is temperature independent; the kT energy is very small compared to E_F.

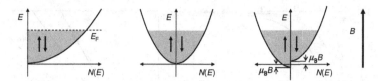

Figure 9.17 A schematic DOS plot (on the left, with Fermi level E_F) can be separated into sub-bands of spin-up and spin-down moments (center). The magnetic field B will change the relative energies and populations of these sub-bands (right).

weak interaction between magnetic moments of the conducting electrons, and this slightly favors their parallel arrangement. More or less enhanced Pauli paramagnetism also occurs in many metallic compounds as an alternative to an outright ferromagnetism, for example in RuO_2.

9.7   Antiferromagnetism

Antiferromagnetic materials obey the Curie–Weiss law, Equation (9.18), at high temperature where they behave as paramagnets with a negative value of θ. On cooling through the **Néel temperature**, T_N, their magnetic moments align in an antiparallel fashion. Because of this interaction among the moments, the effect of an external field is smaller on an antiferromagnet than a paramagnet, and its susceptibility, while positive, is smaller than in the paramagnetic state close to T_N. Plots of χ and χ^{-1} versus temperature are shown in Figure 9.18.

Antiferromagnetic materials also show a significant anisotropy below T_N. The susceptibility when the external field is parallel to the moments ($\chi_\parallel$) is considerably smaller than when the field is perpendicular to them ($\chi_\perp$), Figure 9.18.[31] A powdered sample will show a weighted average $\frac{1}{3}\chi_\parallel + \frac{2}{3}\chi_\perp$ of those two lines. We can understand this anisotropy by considering how the thermal agitation of the spins will be affected when the field is applied either parallel or perpendicular to the antiferromagnetic alignment. Let's start at 0 K, assume no thermal agitation and hence a fixed direction of the opposing moments. If a weak external magnetic field is applied *along* these moments, it cannot change the already saturated and compensating moments, and $\chi_\parallel$ is zero. Above 0 K, thermal agitation decreases the magnetization in both sets of opposing moments equally in zero field. When an external field is applied, it strengthens the set of moments parallel to it (by decreasing the amplitude of their thermal agitation) and weakens the set of antiparallel moments (by increasing the amplitude of their thermal agitation). This means that $\chi_\parallel$ increases with increasing T. If the weak external field is applied at 0 K *across* the antiferromagnetic moments, it tilts both sets of

---

[31] Ferromagnets also possess anisotropy of magnetization, but its manifestation is hidden by ferromagnetic domain formation.

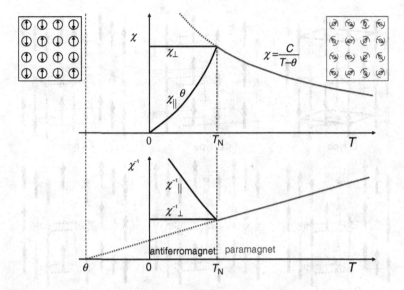

Figure 9.18 Susceptibility of an antiferromagnet as a function of temperature.

moments somewhat into its direction, producing a non-zero $\chi_\perp$. On increasing temperature, the external field influences the increasing thermal amplitudes of both sets equally and the total tilting is not affected. The magnitude of $\chi_\perp$ remains constant up to T_N. The behavior shown in Figure 9.18 is valid for a weak external field. In the presence of a very strong external field, stronger than the antiferromagnetic interactions, the behavior is different. In the perpendicular direction, the tilt increases until all moments are aligned with the field, whereas in the parallel direction, the opposing moments at some point will flip to yield a ferromagnetic state.

There are numerous ways in which magnetic moments can be arranged to give an antiferromagnetic pattern, even for simple structure types [3]. Figure 9.19 shows the possible ordering patterns for a primitive cubic arrangement of magnetic atoms.[32] It's worth noting that ordering of antiferromagnetic moments leads to a magnetic unit cell that is larger than the chemical unit cell. This is probed by neutron diffraction as explained in Box 9.1.

In addition to simple spin arrangements of the type shown in Figure 9.19, more complex possibilities exist. In some materials, spin alignments follow a helical arrangement as one moves through the structure, like the steps of a spiral staircase. The pitch of the helix may be such that the spin arrangement repeats after a few unit cells (e.g. with steps at a 120° angle, the spin would repeat after three chemical unit cells) or it may be that there is no simple registry such that the magnetic structure is incommensurate (Section 2.10) with the chemical structure.

---

[32] We'll for example see in Chapter 11 that $CaMnO_3$ adopts type G while $LaMnO_3$ the type A arrangement, and an important group of materials is obtained when these two form a solid solution.

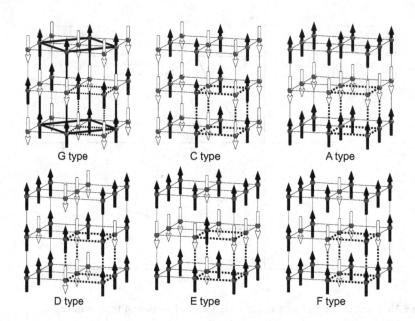

Figure 9.19 Types of antiferromagnetic order on a *P* cubic lattice of magnetic atoms. One of many possible moment directions is drawn; more generally, a + and − would replace the arrows. The chemical cell is in a thick dotted line. The magnetic unit cell is drawn for the G type in bold as an example.

9.8   Superexchange Interactions

Antiferromagnetic ordering turns out to be very common in insulating transition-metal oxides and fluorides. For example, monoxides of Mn to Ni order antiferromagnetically below the Néel temperatures given in Table 9.5. The moments of metal atoms alternate their up/down orientation in all three directions. The magnetic information leading to this order is transmitted by the nonmagnetic linking atoms in a process called **superexchange** [4, 5, 6].

The chemically most intuitive description of superexchange in oxides and fluorides is the partial covalent model. We'll assume initially a *d*-metal M cation (Lewis acid) and a p^6 anion L ("ligands" O^{2-} or F^-, Lewis bases) in an M–L–M bonding segment. We've seen in Chapter 5 that e_g orbitals on M have the correct symmetry for strong σ overlap[33] with a *p* orbital on L. If we consider the straight 180° interaction d^3–L–d^3 of Figure 9.20 (top) we can therefore envisage a partial transfer of one of the ligand *p* electrons to an e_g orbital on the metal on the left (a polar covalent bond). Hund's first rule suggests that the M t_{2g} electrons will have the same spin as that shared with L, here ↑. Coming to the M on the right, the same *p* orbital of

---

[33] For σ bond overlap, the overlapping portion must not change sign when coordinate axes of any of the two involved orbitals are rotated around the line connecting the two atoms. Such σ overlaps are p_z with d_{z^2}, and $d_{x^2-y^2}$ with either p_x or p_y. Such π overlaps along *x* are p_y with d_{xy}, and p_z with d_{xz}. Along *y*, they are p_x with d_{xy}, and p_z with d_{yz}. Along *z*, they are p_x with d_{xz}, and p_y with d_{yz}. Orthogonal, non-overlapping, orbitals are p_x with either d_{xz} or d_{xy} along *x*, p_y with either d_{xy} or d_{yz} along *y*, and p_z with either d_{xz} or d_{yz} along *z*.

Table 9.5 Antiferromagnetic NaCl-type oxides and KMF_3 perovskites of M = Mn–Ni.

Phase	T_N (K)	d_{M-O} (Å)	Phase	T_N (K)	d_{M-F} (Å)
MnO	122	2.22	$KMnF_3$	88	2.09
FeO	198	2.14	$KFeF_3$	115	2.06
CoO	293	2.13	$KCoF_3$	135	2.04
NiO	523	2.09	$KNiF_3$	275	2.01

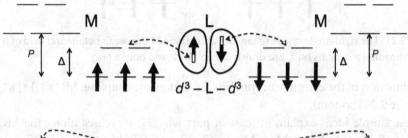

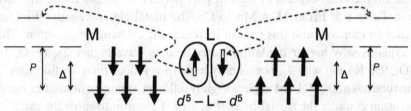

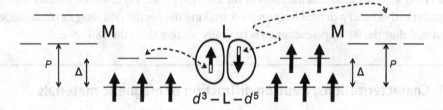

Figure 9.20 Superexchange M–L–M 180° coupling with e_g orbitals of M. L is a p^6 anion, M is a d^n cation in a weak Δ-splitting octahedral L field. The corresponding spin-up and spin-down levels are separated by the spin-pairing energy P. Electron-spin sharing (partially filled arrows on L) corresponds to partial presence of the L p electron at the cation owing to partial covalency of the M–L bond.

L shares its ↓ electron, making the unpaired t_{2g} electrons ↓ on that M. There is thus an overall antiferromagnetic coupling of magnetic moments on the two metal sites, M(↑)(↑)L(↓)(↓)M. Overlap arguments suggest that a decrease in the bond angle from 180° will reduce the strength of this antiferromagnetic superexchange.

A similar situation occurs for a high-spin d^5–L–d^5 superexchange (Figure 9.20, middle) though now the Pauli exclusion principle dictates that the spin on M is opposite to that on L, which leads to M(↓)(↑)L(↓)(↑)M that is again antiferromagnetic. For a d^3–L–d^5 case,

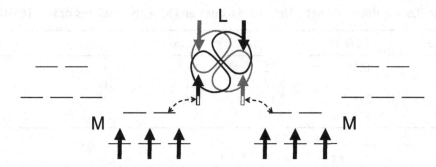

Figure 9.21 The right-angle superexchange of d^3 cations M in a weak octahedral field of p^6 anions L. The two orthogonal p orbitals on L are drawn one in black and one in gray.

a combination of these arguments predicts ferromagnetic coupling, M(↑)(↑)L(↓)(↑)M, as shown in Figure 9.20 (bottom).

These simple ideas explain at least in part why T_N increases along the MO and KMF$_3$ series in Table 9.5. From M = Mn to Co, the metal electronegativity increases, causing covalency to increase, and this in turn increases the exchange interaction. The same argument explains why higher oxidation states tend to have higher T_N (FeO, 198 K; versus α-Fe$_2$O$_3$, 953 K) and why T_N is generally higher for oxides than for fluorides.

When one has a 90° M–L–M angle, a slightly different argument holds. For our d^3–L–d^3 case, we can again consider the partial transfer of an L ↑ electron towards the metal ion on the left (Figure 9.21). However, a *different p* orbital is overlapping with M on the right. We can therefore use Hund's rule to state that the spins of the emerging holes on L will be parallel such that an L ↑ electron in this other p orbital is again used, making the overall coupling ferromagnetic. It should be noted that the 90° superexchange is typically weaker than the 180° one.

Box 9.1 Characterizations: Neutron diffraction of magnetic materials

X-ray diffraction is one of the most powerful probes of crystal structure. When a crystal is illuminated with X-rays of wavelength similar to interatomic separations, interference of X-rays scattered from the electron clouds of atoms leads to radiation being diffracted in certain directions. The direction of the diffracted radiation carries information on the size and shape of the unit cell of the material (via Bragg's law stating that the radiation wavelength $\lambda = 2d_{hkl} \sin\theta$, where θ is the glancing angle between the beam and the atomic plane hkl). Intensities of the diffracted radiation peaks carry information on where atoms lie within that cell. Neutrons can also be produced (either at reactor or spallation sources) with wavelengths comparable to interatomic distances. Neutrons are scattered by atomic nuclei giving rise to diffraction patterns sensitive to isotopes. Since neutrons are fermions, they have half-integer spin and will also be scattered by the spin the atom may have due to its unpaired electrons (paired electrons not contributing). In contrast, diffraction of X-rays as bosons of zero spin is generally insensitive to magnetic moment. For parallel spins of all identical atoms (inside a domain), the diffracted Bragg peak will be a simple sum of the contribution from the so-called "nuclear" and "magnetic" scattering. This is shown in the bottom row of the figure below, where the filled peaks represent nuclear scattering and the open peaks, overall scattering. In the case

Box 9.1 (cont.)

of anti-parallel magnetic moments, neutrons will "see" two different types of atoms, the spin-ups and spin-downs. Diffraction from magnetic unit cells that are multiples of the "chemical cell" (also known as the "nuclear cell") will produce additional peaks solely due to spin ordering. These "magnetic" peaks are marked below by the black *hkl* indices for three of the antiferromagnetic ordering types shown in Figure 9.19 (while the nuclear peaks have gray indices referring to the unit cell drawn). This contribution to the diffraction pattern from magnetic scattering means that neutron diffraction can be used to determine magnetic structures of ordered moments.

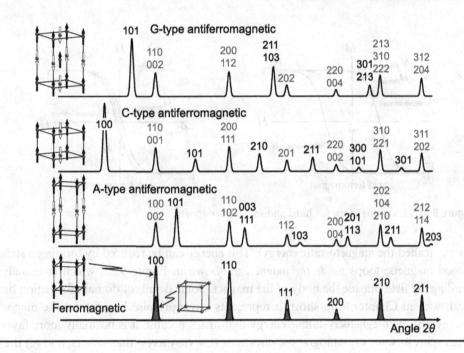

9.9   Ferromagnetism

Ferromagnetism is the magnetic behavior of parallel orientations of atomic magnetic moments (Figure 9.14). Ferromagnets order their moments below a critical temperature called the **Curie temperature**, T_C. The total atomic magnetic moment exhibited by such materials is called **spontaneous magnetization**.

From this description we might expect that all ferromagnetic materials, such as all pieces of iron, would have a permanent magnetization below T_C and attract each other—which we know isn't the case. The reason for this discrepancy lies in the **domain structure** of ferromagnets. If all atomic moments in a material point in the same direction, it behaves like the bar magnet of Figure 9.1 and needs to send a powerful magnetic field through its neighboring space. This costs

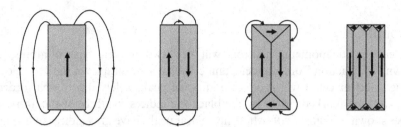

Figure 9.22 The reason why ferromagnetic materials adopt domain structure is to avoid sending the field into the poorly magnetizable neighborhood.

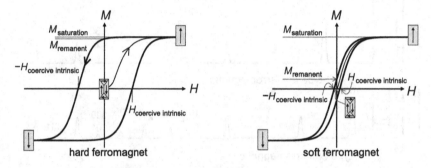

Figure 9.23 Hysteresis loops for hard and soft ferromagnets.

energy (called the **magnetostatic energy**). This energy can be reduced by forming a structure of closed magnetic loops *inside the magnet*, as shown in Figure 9.22, which essentially avoids sending the field outside the body of the magnet. Some details of domain formation have been dealt with in Chapter 8. In short, it represents a compromise between (a) the magnetostatic energy, (b) the **magnetocrystalline energy** that arises because it is normally more favorable to align spins in some crystallographic directions (the **easy axes**) than others, and (c) the **magnetostrictive energy** of strain caused by **magnetostriction**, the contraction along that favorable direction of magnetization where the spins attract themselves. Boundaries between domains are called **Bloch walls**. For Fe at room temperature, these might be around 650 Å thick.

From this description, one may again wonder why the domain-structured ferromagnets can become magnetic at all. The domain structure of a material may not be very strong; it only needs to be more stable than the alternative of sending the field through the material's surroundings. An external magnetic field can therefore reorient domains and change their size. Favorable domains with easy axes of magnetization close to the field direction will therefore grow at the expense of less favorable ones, and then eventually reorient closer to the field direction as the magnetic saturation is approached. Such changes in domain structure are behind the characteristic M versus H hysteresis loops of ferromagnetic materials shown in Figure 9.23.

From the origin at zero for zero H, the magnetization M follows the light curve of Figure 9.23. Saturation is easy and achieved at fields far lower than for an isolated paramagnetic moment. High magnetization occurs already at a low field. Although there is an energetic cost associated with moving domain walls past dislocations or defects in a material, the external field is able to overcome this energy. When the external field is removed, the new domain structure in the material may become partially locked by defects and dislocations rearranged by magnetostriction. On reducing the field to zero, the material then retains **remanent**[34] **magnetization**, $M_{remanent}$.[35] This type of domain realignment is the experiment one performs when "stroking" a needle with a permanent magnet to magnetize it. Permanent magnets are manufactured commercially by cooling a glowing hot material in a strong permanent field that keeps the forming domains aligned. A high $M_{remanent}$ is obtained when the field is removed from the cold product.

Figure 9.23 shows that it then takes an opposite field, the negatively taken intrinsic[36] **coercive field**, H_{ci} [7], to fully demagnetize the ferromagnet so that $M = 0$. Ferromagnets with easily movable domains have a low H_{ci} and are called **soft magnets**. Ferromagnets with a large H_{ci} are called **hard ferromagnets** and have important applications as permanent magnets in motors. For data storage, a high coercive field is not desirable because even higher fields are then needed to rewrite the information. On the other hand, high remanent magnetization is essential. The hysteresis loop for a data-storage material should therefore approach the form of a narrow and tall rectangle. In contrast, good permanent magnets have hysteresis loops that are both tall and wide. One useful way of representing magnetic measurements on such magnets is to plot B versus H. The **maximum energy product**, $(BH)_{max}$ (in units of T A/m = J/m^3), can then be defined, as shown in Figure 9.24, via the point on the demagnetization line where the product of B and H is a maximum.

The temperature dependence of the magnetization of a ferromagnet is shown in Figure 9.25. Below T_C, it resembles the curve for the order–disorder transition[37] of second order (Figure 4.16). Above T_C, the magnetization becomes so low that it would not be seen on the given scale of M, and the plot in Figure 9.25 is continued with reciprocal susceptibility[38] instead, following the Curie–Weiss law of Equation (9.18).

Let's take first the $T > T_C$ side of the plot in Figure 9.25, where the originally ferromagnetic material is thermally disordered into a paramagnet and obeys the Curie–Weiss law. Weiss, in his phenomenological description of ferromagnetism, assumed that the ferromagnetic alignment of atomic moments is transmitted by an **internal field**, H_W, between these moments,[39] the strength of which is proportional to the sample magnetization:

$$H_W = \lambda M \tag{9.22}$$

---

[34] An older version of "remnant". Now used only in this physics meaning. It is best not to confuse the two.

[35] The demagnetizing field (the field lines external to the sample in Figure 9.22) is much weaker than an external field and can't move domain walls past all defects.

[36] The intrinsic coercive field H_{ci} (also called intrinsic coercivity) zeroes the magnetization M, whereas the coercive field, H_c (coercivity), zeroes the magnetic induction B, $B = \mu_0(M + H)$.

[37] This approach to describing $M(T)$ is illustrated in Problem 4.12.

[38] Because $\chi^{-1} = H/M$, Equation (9.8), it is as though M^{-1} were plotted for unity field.

[39] Weiss called this field a molecular field. The field is constant throughout the material; it represents a mean-field approximation of the quantum-mechanical exchange interactions between the atomic moments.

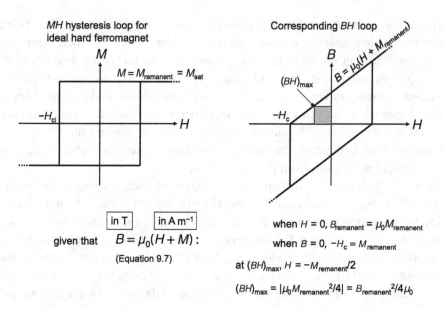

Figure 9.24 Maximum energy product for an idealized ferromagnet in SI units. See Footnote 36 of this chapter for the coercive-field symbols.

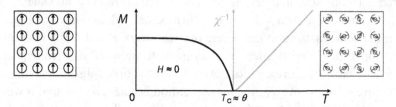

Figure 9.25 Magnetization M of a ferromagnet as a function of temperature. Paramagnetic reciprocal susceptibility is plotted after M drops to very low values at T_C. Note that only when the applied field $H \to$ 0 will M fall sharply to zero at T_C.

with λ being called the internal-field constant. The total field that magnetizes the moments is $H_{\text{total}} = H + H_W$. Let's recollect that the Curie law is derived on the assumption that magnetic moments do not interact with each other. How would the law change when the magnetization is almost entirely due to the field these very moments produce? Using H_{total} in Equation (9.8), we write for the inherent total paramagnetic volume susceptibility under Curie law:

$$\chi_{\text{inherent}} = \frac{M}{H_{\text{total}}} = \frac{M}{H + \lambda M} = \frac{C}{T} \tag{9.23}$$

Then we take the latter identity, and express the *measured* susceptibility $\chi = M/H$,

$$\chi = \frac{M}{H} = \frac{C}{T - C\lambda} \tag{9.24}$$

which has the same form as the Curie–Weiss law in Equation (9.18), only $C\lambda$ replaces the temperature $\theta \approx T_C$ above which a ferromagnetic material magnetizes like a paramagnet. The expression $T_C \approx C\lambda$ means that T_C is a measure of the strength parameter λ of the field that causes the moments to align parallel; it is a measure of the exchange field.

Let's now turn to the $T < T_C$ side of the plot in Figure 9.25 and consider the temperature dependence of the ferromagnetic magnetization. The descending curve originates in the Brillouin function for the relative magnetization M_r, Equation (9.21), of paramagnets. In order to apply it to a ferromagnet, B is replaced with $\mu_0 H_{total}$, where $H_{total} = H + H_W$, as above. However, H_W is a function of M, Equation (9.22), hence of M_r, and it turns out that these two functions for $M_r(T)$ of a ferromagnet (Appendix H) can only be solved numerically. The solution yields the simple curve in Figure 9.25.

We haven't yet addressed the materials-chemistry question of what makes a good ferromagnet; that is, what are the forces that cause magnetic moments to align? Before discussing this, it's useful to separate ferromagnets into three different categories. One way of doing this is via schematic DOS plots of the type we used in Figure 9.17, where we split the DOS to show the two sub-bands for the majority/minority spins with magnetic moments parallel/antiparallel to the field. These are often called **spin-polarized** bands. This viewpoint allows us to split materials into categories of **ferromagnetic metals** (with DOS non-zero at E_F for both majority and minority spins), **ferromagnetic half-metals** (with the majority spin only at E_F, a gap occurs for minority spins), and the rare case of **ferromagnetic insulators** (with no spin at E_F), as shown in Figure 9.26.

9.9.1 Ferromagnetic Insulators and Half-Metals

Ferromagnetic insulators (semiconductors) are rare. Examples are weak ferromagnets of low T_C such as EuO (T_C = 69 K) and EuS (16 K) of rock-salt structure, a $PbCl_2$-type EuH_2 (18 K), $EuLiH_3$ cubic perovskite (37 K), or the transparent K_2NiF_4-type salt Rb_2CrCl_4 (52 K). The nonmetallic ferromagnet $La_2Mn^{IV}Ni^{II}O_6$ (T_C = 280 K [8]) arises from the high-spin d^3–L–d^8 superexchange combination (Section 9.8). An obtuse-angle superexchange leads to weak ferromagnetism in highly tilted perovskites $YTiO_3$ (T_C = 30 K) and $SeCuO_3$

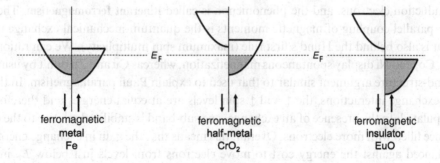

Figure 9.26 Types of ferromagnets according to schematic spin-polarized bands.

(26 K) or in the $CdCr_2Se_4$ spinel (130 K). These materials are insulating or semiconducting above T_C (EuO is yellow) but their conductivity increases below T_C by many orders of magnitude. The reason is the easy hopping of electrons between sites of the same spin.

Half-metals, as illustrated in Figure 9.26 (center), are solids that for one spin orientation have a gap, whereas the other orientation is gapless, like in a metal. Half-metals are conductors of spin-polarized electrons. The spin orientation also means that all half-metals are magnetically ordered. A typical half-metal is CrO_2, with $T_C = 392$ K and an *integer* spin-only ferromagnetic moment of 2.0 μ_B per Cr at 0 K. This saturation moment corresponds to complete polarization of the two nonbonding t_{2g} electrons present in stoichiometric CrO_2, as required by the half-metal definition. More about this in Chapter 11.

There are several *origins of ferromagnetism in nonmetallic materials*. One is the ferromagnetic superexchange detailed in Section 9.8. Another is an absence of antiferromagnetic superexchange; a weak ferromagnetism (of low T_C) may then appear due to direct interaction of the two not-so-distant magnetic atoms along the array M–L–M. Another factor contributing to ferromagnetism (at least locally) is **double exchange** in valence-mixed compounds (Chapter 11), where an electron is present simultaneously at two or more neighboring sites of an element. This requires a parallel orientation of spins at the sites involved. The term double exchange was chosen [9] for cases where the imagined hopping happens over an L atom in between the two sites. A typical example is magnetite, Fe_3O_4 (see also Section 9.10), in which minority-spin t_{2g}^1 electrons of octahedrally coordinated Fe^{2+} hop into the same yet empty orbital of Fe^{3+} at an equivalent octahedral site. Another example is $La_{2/3}Sr_{1/3}MnO_3$, where the e_g^1 electron of Mn^{3+} hops to Mn^{4+}. With close enough energies, even orbitals of two different metals may give rise to double exchange and ferromagnetism. An example is sharing of the t_{2g}^1 Fe^{2+} electron with the d^0 configuration of Mo^{6+} in the half-metallic ferromagnet Sr_2FeMoO_6 of the perovskite-related structure [10]. Because electron hopping is involved in stabilizing ferromagnetic interactions, double-exchange ferromagnets are close to being metallic.

9.9.2 Ferromagnetic Metals

Ferromagnetism well above room temperature is displayed by three elements: Fe, Co, and Ni with $T_C = 1043$ K, 1388 K, and 627 K, respectively. Their magnetic coupling is due to conduction electrons, and the phenomenon is called **itinerant ferromagnetism**. The origin of the parallel coupling of magnetic moments is the quantum-mechanical exchange interaction that is also behind the Hund's first rule (maximum spin multiplicity). We can rationalize why Fe, Co, and Ni display spontaneous magnetization, whereas Cu and Zn don't by using a simple band-structure argument similar to that used to explain Pauli paramagnetism. In the absence of exchange interactions, the ↑ and ↓ spin levels are at equal energies and therefore equally populated. In the presence of an exchange, one sub-band is stabilized relative to the other, and hence filled with more electrons. Overall, this means that the gain in exchange energy must be balanced against the energy cost to move electrons from levels just below E_F in the thus-emerging minority-spin band to the top of the thus-emerging majority-spin band. The question

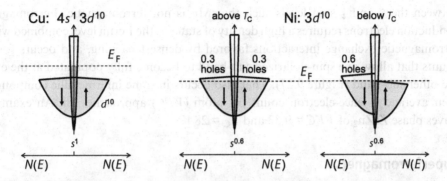

Figure 9.27 Comparison of a spin-polarized DOS plot at non-zero T for diamagnetic Cu (idealized Cu-atom configuration $4s^1 3d^{10}$) with one for ferromagnetic Ni (idealized Ni-atom configuration $3d^{10}$), showing the effect of the exchange interaction below T_C.

of whether a metal will be magnetic depends in large part on this balance. When the number of states $N(E_F)$ at the Fermi level is high, the electrons move to states that are only marginally higher in energy, and the cost is low. This is the case for Co, Fe, and Ni, where the Fermi level cuts through the d bands. When $N(E_F)$ is small, the electrons transferred to majority-spin bands are forced to occupy states that are significantly more antibonding, and the cost is high. This is the case for metals like Cu, where the Fermi level cuts through the s band. By way of example, a schematic DOS plot is shown in Figure 9.27.

For Cu, E_F lies in the broad $4s$ band of 1 electron per atom, where the DOS is low (Figure 9.27, left). The energy cost of flipping that electron and putting it onto the other sub-band is too large to be offset by any exchange energy, and Cu therefore doesn't show ferromagnetism. For Ni, one has a total of 10 valence electrons. Of these, around 9.4 reside in the d band and 0.6 in the s band. The exchange splitting is such that the majority-spin band is completely filled[40] and the 0.6 holes lie entirely in the minority band. The saturation magnetization per Ni atom is therefore 0.6 μ_B. This simple band-structure picture helps rationalize why magnetic moments recorded experimentally don't suggest an integer number of electrons contributing to the moment. Similar arguments hold for Co and Fe, which have fewer valence electrons than Ni, increasing the excess of spin-polarized electrons and increasing M_{sat}. The picture is very similar to that used to explain Pauli paramagnetism, but the origin of the energy splitting of sub-bands is the internal exchange energy rather than an external magnetic field. An element that sits on the fence is Pd. It has $N(E_F)$ nearly high enough for spontaneous magnetization, and this leads to high magnetization in applied fields (an enhanced Pauli paramagnetism). Importantly, since high $N(E_F)$ is associated with narrow bands, a decrease in dimensionality leads to ferromagnetism in Pd films and nanoparticles.

Our simple picture of itinerant ferromagnetism of elements ends around Fe. Mn adopts a different crystal structure, has broad d bands of low $N(E_F)$ that increases the cost of moving

---

[40] Chemists may appreciate that with 9.4 d electrons a sub-band forms with full d^5 configuration.

between the ↑ and ↓ spin levels such that Mn is not ferromagnetic. Ferromagnetism of conduction electrons requires a high density of states at the Fermi level combined with strong ferromagnetic exchange interactions favored by densest packing, and occurs for electron counts that allow one spin-polarized sub-band to become fully populated at the expense of the other sub-band (Figure 9.27). That also occurs in some intermetallic compounds when their average valence-electron count per atom (VEC) approaches 10. An example is the Laves phase $ZrZn_2$ of $VEC = 9.33$ and $T_C = 28$ K.

9.9.3 Superferromagnets

A combination of d-shell and f-shell magnetism in **superferromagnets** $SmCo_5$ and $Nd_2Fe_{14}B$ has produced the strongest permanent magnets so far, i.e. those with the highest maximum-energy products $(BH)_{max}$. Since their discovery in the 1960s, these materials have replaced other permanent magnets in most applications. Table 9.6 shows that their coercivities exceed those of traditional magnets. The strength of superferromagnets may be illustrated by the fact that a cube-shaped magnet of 1 inch edge cannot be pulled straight from an iron plate by hand; a task comparable to lifting a mass of 50 kg. The best ceramic magnet, $BaFe_{12}O_{19}$, of the same size would behave like a 3 kg mass.

What are the origins of such exceptional properties? For $SmCo_5$ they can be traced back to the magnetic structure shown in Figure 9.28. The high remanence is because the Co moment is high, 1.8 μ_B at 4 K. The Sm moment of 0.38 μ_B is parallel but decreases to about zero at room temperature, while the ordered Co moments persist up to the very high T_C of these magnets.[41] One can think of the Sm "stuffing" atom as causing just a slight reduction in T_C

Table 9.6 Magnetic properties of some anonymized commercial polycrystalline hard magnets: $B_{remanent}$ (remanence), H_c (coercivity), energy product $(BH)_{max}$, and T_C.

Composition	$B_{remanent}$ (T)	H_c (kA/m)	$(BH)_{max}$ (kJ/m^3)	T_C (K)
~$SmCo_5$	1.05	730	200	1100
~$Nd_2Fe_{14}B$	1.23	900	280	550
$Al_7Ni_{15}Co_{35}Cu_4Ti_5Fe_{34}$	1.05	62	70	1150
$BaFe_{12}O_{19}$	0.40	190	9	742

[41] The moment magnitudes can be accounted for by considering spin–orbit coupling and localized moments. For Sm^{2+} ($4f^65s^2p^6$), the calculated saturated moment $\mu_{sat} = g_J$ (Table 9.3) is $0\,\mu_B$, based on $S = 3$ and $L = 3$ from the Russell–Saunders scheme in Figure 9.13 and $J = |L - S|$. However, J approaching zero partially uncouples L and S and brings a second-order Zeeman effect (see Orchard in Further Reading) contribution to the magnetic moment, and the moment becomes non-zero (even at low temperatures). Treating $SmCo_5$ as a Zintl phase (Section 1.5.6), one can speculate that formation of Sm^{2+} provides two electrons to five Co. The average saturation moment for the thus formed $3d^{9.4}$ cobaltide configuration in this compound is $g_J = 1.80\,\mu_B$, calculated with $S = 0.3$ and $L = 1.2$, interpolated via the Russell–Saunders scheme of Figure 9.9.

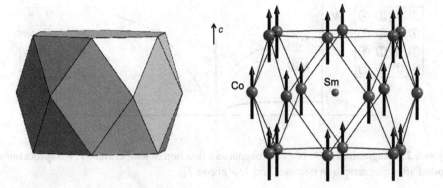

Figure 9.28 Magnetic structure of SmCo$_5$ at room temperature.

relative to the 1388 K of hexagonal close-packed Co. The key role of Sm is in the extremely high magnetocrystalline anisotropy it induces. It triggers a combination of crystal-field effects and spin–orbit coupling that makes magnetization very easy in one direction and very difficult in others (high coercivity). When such magnets are manufactured commercially, intentional phase admixtures and careful processing are used to orient grains, control grain boundary effects, and help pin magnetic domains.

9.10 Ferrimagnetism

In Section 9.8 we've described how superexchange gives rise to antiferromagnetic coupling in insulating materials. In Section 9.9 we learned that ferromagnetic insulators are extremely rare. How then do we produce an insulating material with ferromagnet-like properties? One solution is to use a ferrimagnet (Figure 9.29) where antiferromagnetic coupling of opposing moments of unequal magnitude (e.g. due to two different metals being involved) leads to a net overall magnetization. While bulk properties are similar to a ferromagnet, the atomic-scale interactions are more complex. A ferrimagnet of two magnetic atoms may have up to three types of couplings, one dominant antiferromagnetic and two ferromagnetic.[42] This takes away the simplicity of the Weiss approach, which would have to incorporate three Weiss fields between two moments of unequal magnitude, producing an asymptotic approach to Curie–Weiss behavior at high temperatures. The behavior below T_C is also more complicated, as each of the two relative saturation moments will follow their own temperature dependence.[43] Experimentally, a linear Curie–Weiss plot is obtained in the high-temperature extrapolation, having negative θ indicative of the antiferromagnetic coupling.

---

[42] The white–gray antiferromagnetic interaction and the white–white or gray–gray ferromagnetic interactions of Figure 9.29 (left) cartoon.

[43] At some point, the two moments may even cancel; the point is termed the compensation temperature.

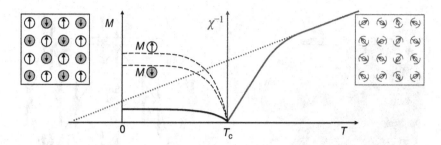

Figure 9.29 Magnetization M of a ferrimagnet as a function of temperature T. Reciprocal susceptibility is plotted after magnetization becomes very low above T_C.

On cooling towards T_C, there is a marked increase in χ, followed by an onset of spontaneous magnetization at T_C.

Ferrimagnetism is common among spinel-type materials; indeed the ferrimagnetic spinel Fe_3O_4 (lodestone ≡ leading stone) with a T_C of 860 K was the technological solution the ancient Chinese either employed directly in compasses or used to magnetize needles. The Fe_3O_4 spinel (Section 1.5.1) has antiferromagnetic coupling between the tetrahedral and octahedral sites, $[Fe^{3+}(\downarrow)]^{tet}[Fe^{3+}Fe^{2+}(\uparrow)]^{oct}O_4$. Since magnetic moments of Fe^{3+} at both sites cancel, the saturation moment is solely due to the 4 unpaired electrons of high-spin d^6 Fe^{2+}, hence $4g_e(\frac{1}{2}) = 4\,\mu_B$. The ability of the spinel structure to accommodate various elements yields a variety of magnetic properties as well as the ability to control the saturation moment via substitutions.

Another group of important ferrimagnetic materials has the garnet-type structure (Section 1.5.2). Its simplest representation is $[\text{cube}]_3[\text{octahedron}]_2[\text{tetrahedron}]_3O_{12}$. Let's choose $Gd_3Fe_5O_{12}$ as one example of many. Its "magnetic" crystal-chemical formula can be summarized as $[Gd^{3+}(\uparrow)]_3^{[8c]}[Fe^{3+}(\uparrow)]_2^{[6o]}[Fe^{3+}(\downarrow)]_3^{[4t]}O_{12}$. Two of the d^5 tetrahedral Fe^{3+} moments are cancelled by the d^5 (both are $S = \frac{5}{2}$) octahedral Fe^{3+} such that the overall saturation moment is given by $3|\mu(Gd^{3+})| - |\mu(Fe^{3+})|$. Garnets can be prepared as large single crystals that are stable, colored, translucent insulators with a bulk ferromagnetic-like behavior. This unique combination of properties is utilized in magneto-optical devices: disk memories, converters of magnetic field into optical images, and sensors of movement or rotation. Garnets are also used in optoelectronics and microwave technology as semi-permeable transmission media called Faraday rotators.[44]

---

[44] The Faraday effect is a usually tiny rotation of the plane of polarized light upon *transmission* through a transparent magnet, the angle of which is proportional to the magnetic-field component in the direction of the light beam. The angle achievable in garnets is rather high, around 45°. An analogous effect occurring upon *reflection* from surfaces of magnetized crystals is called the Kerr magneto-optical effect.

9.11   Frustrated Systems and Spin Glasses

In earlier sections, we've looked at exchange interactions that lead to ferro- or antiferromagnetic ordering. In some situations, it may not be possible to simultaneously satisfy the energetic requirements for all components in the system. This is called **frustration** and can lead to a variety of energetically equivalent ground states being present. One of the best-known magnetic examples of this occurs on the so-called kagome lattice made up of corner-sharing triangles.[45] With antiferromagnetic coupling between moments on two corners of the triangle, there is an immediate problem when placing a moment on the third corner (Figure 9.30). Although the kagome frustration has been treated theoretically in considerable detail, only a few real examples exist. One is the hydronium jarosite, $(H_3O)Fe_3(SO_4)_2(OH)_6$ [11]. A similar frustration occurs for tetrahedral networks. The tetrahedral frustration is often referred to as "spin ice" by topological analogy with the disordered arrangement of protons in hydrogen bonds of $H_2O(s)$ described by Pauling [12]. In general, the amount of frustration is often quantified by the ratio of the Weiss temperature θ obtained from the Curie–Weiss high-temperature behavior and the freezing/ordering temperature such as T_g (freezing to a glass of spins, see below) or T_N (antiferromagnetic ordering). The $|\theta/T_{\text{freezing/ordering}}|$ will be large for a frustrated material (spin-ice freezing) and close to 1 for a perfect antiferromagnet (spin ordering).

A second way to achieve frustration is via random site occupation in alloys. In metallic alloys containing a few percent of a magnetic dopant such as Fe in Au, Mn in Cu, or Er in Y, exchange via the host electrons leads to random interactions between the magnetic dopants.[46] Similar effects occur in inherently disordered alloys such as amorphous $GdAl_2$.

A **spin glass** is a material formed by cooling the multi-degenerate spin-frustrated ground state through a "freezing" transition[47] into a glass-like, fixed, residual spin disorder. The

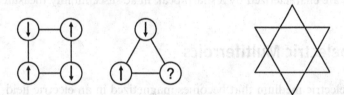

Figure 9.30 Magnetic frustration. While it's possible to have all nearest-neighbor antiferromagnetic interactions satisfied on a square, this is not possible for a triangle. Materials based on the kagome lattice (a segment is shown on right) then show spin-glass behavior.

[45] *Kago-me* = "basket-eye" in Japanese, a light basket of bamboo strains woven such that large hexagonal openings, "eyes", and small triangles emerge.

[46] Below a certain temperature, further cooling begins to decrease the conductivity of the alloy—against what is common for metals—as the increasing strength of the coupling condenses more conducting electrons around the magnetic impurity (the **Kondo effect**).

[47] The term "freezing" is used in analogy with water-ice freezing upon intrinsic disorder of H-atom orientations around O-atom nodes of the tetrahedral network.

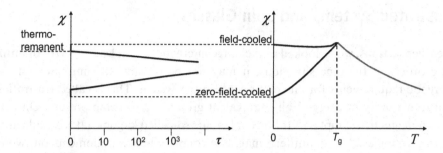

Figure 9.31 Right: Equilibrium magnetic susceptibility of a spin glass upon cooling through the spin-glass temperature T_g with or without an external field. Left: Their isothermal relaxations as a function of time τ after the external field was switched off or on, respectively.

temperature dependence of the susceptibility of a typical spin glass is shown in Figure 9.31 (right). At high temperature, the material is paramagnetic and follows the Curie–Weiss law.[48] On cooling through T_g, the spins interlock into one of many possible ground states of the different frustrated combinations. Below T_g, the susceptibility depends on the field under which the material was cooled. The larger the field in which the material is cooled, the higher the interlocked magnetization. The high-field-cooled susceptibility of the spin glass remains high as a function of temperature. The zero-field-cooled susceptibility is low and decreases with temperature as local antiferromagnetic interactions are less agitated. The conversion between these two extreme interlocked results (the former created in zero field, the latter in a high field) takes a long time (Figure 9.31). The relaxation times also depend on whether or not equilibrium has actually been reached upon the field- and no-field cooling below T_g. Magnetic properties of a spin glass therefore depend strongly on its thermal history and are reminiscent of physical properties of high-viscosity plastics or other amorphous materials (Chapter 15). Spin glasses are characterized by a sharp peak in ac susceptibility measurements close to T_g.

9.12 Magnetoelectric Multiferroics

A magnetoelectric medium that becomes magnetized in an electric field, E, and electrically polarized in magnetic field, H, is a largely unfulfilled idea that dates back to Pierre Curie at the end of the nineteenth century. In particular, ferroelectric ferromagnets[49] are of interest [13]. Potential applications include data-storage devices, magnetic or electric modulation of

---

[48] The paramagnetic state may or may not transform directly into the spin glass. If an intermediate ferromagnetic state is formed upon cooling, the final spin glass at the lowest temperature is called **reentrant** spin glass, in reference to the re-emerging magnetic disorder.

[49] This is the most prominent pair of the so-called multiferroics, materials that combine two types of field-induced, ferroic, alignment (of magnetic dipole moments due to a magnetic field, of electric dipole moments due to an electric field, of strains by a stress field) thus achieving coveted cross-coupling phenomena when one field tunes the alignment pertinent to the other field.

transducers, multifunctional sensors, and actuators. Yet making this happen in a single chemical compound needs careful control of symmetry [14, 15] and is challenging compositionally. Most ferromagnets have unpaired d electrons and are good electrical conductors (we have already noted the rarity of insulating ferromagnets), while ferroelectrics need to be insulators with non-centrosymmetric structures and are typically associated with the d^0 configuration (Chapter 8). One possible solution is to mix d^n and d^0 ions in one network, but this dilutes the ferroelectric and ferromagnetic subsets and weakens both the ferroelectric coupling and magnetic superexchange. Loss of ferroelectricity ensues, with the dilute magnetic arrangement having a low ordering temperature, if any at all.

Another path is to introduce transition metals into the pyro- or ferroelectric salts of some sp anions (typically non-centrosymmetric borates or phosphates). However, only weak magnetic interactions are achieved, for example in ferroelectric, weakly ferromagnetic, and ferroelastic boracites ($M_3B_7O_{13}X$; M = transition metal, X = halogen).

The opposite approach is to stuff potentially ferromagnetic networks with cations that have a stereochemically active lone pair, such as Pb^{2+}, Bi^{3+}, Sn^{2+}, or Sb^{3+}. Indeed, $BiFeO_3$ is a ferroelectric antiferromagnet with a weak ferromagnetic moment due to canting of spins. A more robust ferromagnetism has been obtained by replacing La^{3+} in La_2MnNiO_6 with the lone-pair cation Bi^{3+} to form Bi_2MnNiO_6 [16]. This leads to simultaneous ferroelectricity and ferromagnetism below 140 K. A problem of this approach is that the ferromagnetism, or its coupling with ferroelectricity, still is weak. How does one achieve the desired high magnetoelectric coupling between properties?

A window of opportunity appears in geometrically frustrated networks where the bonding compromise creates off-center positions for atoms in their coordination polyhedra. As an example, in hexagonal $YMnO_3$ (not a perovskite), off-center Y^{3+} cause ferroelectric polarization that combines with a frustrated magnetic order of Mn^{3+} that is easily reoriented [17].

A related way of smuggling ferroelectricity into a magnetic material is to remove the centrosymmetry of suitable valence-mixed magnetically ordered networks by ordering them into integer valences. The loss of inversion symmetry that may occur upon such charge ordering then gives rise to a weak ferroelectricity. The ferrimagnetic spinel magnetite, Fe_3O_4, is reported to be ferroelectric below the Verwey transition at 125 K (see Chapter 11). An analogous symmetry-breaking transition in the $NdBaFe_2O_5$ perovskite ($Pmmm \rightarrow P2_1ma$) removes inversion centers in this *antiferromagnet* [18]. A similar case occurs in $LuFe_2O_4$ [19]. However, a common problem with electrical polarization in these phases is their relatively high electrical conductivity, which is undesirable for ferroelectric applications.

9.13 Molecular and Organic Magnets

The magnetic materials discussed so far were solid oxides or metals of high specific weight, which have to be manufactured by high-temperature processing. The possibility of creating molecular magnets that could be synthesized in solution and readily processed to a useful form is attractive. Unfortunately, only a small number of *sp* molecules have unpaired electrons; examples include NO and O_2 (in its ground state). Liquid oxygen is very strongly attracted to magnetic fields, but as a liquid it cannot acquire any permanently oriented moment on its own. Solid O_2 is antiferromagnetic. Ozonides, such as $[N(CH_3)_4]^+ [O_3]^-$ (stable up to 70 °C in the dark), contain one unpaired spin per anion, but their magnetic properties have not been studied. Superoxides (with an O_2^- anion) also carry an unpaired electron.

Other verified cases of *sp* magnetism involve free-radical species, such as NO_2. Unfortunately, molecules with unpaired electrons have a propensity for dimerizing and thus losing their magnetism. Some organic free radicals are more stable, such as triphenyl-methyl $(C_6H_5)_3C^•$ and its high-spin polymer, but have not been shown to be ferromagnetic [20]. The first organic ferromagnet was reported [21] in the orthorhombic β-modification of *para*-nitrophenyl nitronyl nitroxide (*p*-NPNN) where the unpaired electron is carried by a conjugated moiety (Figure 9.32), but the T_C of 0.6 K is very low.

Some success in terms of bulk ferromagnetism has been achieved with transition-metal-containing molecules [22]. As an example, decamethylferrocene $Fe^{II}[C_5(CH_3)_5]_2$, diamagnetic due to low-spin iron, is oxidized by the electron acceptor tetracyanoethylene (TCNE, Figure 9.33) into $[Fe^{III}(C_5(CH_3)_5)_2]^+$ and $[TCNE]^-$. Here, the d^5 Fe of the cation carries one

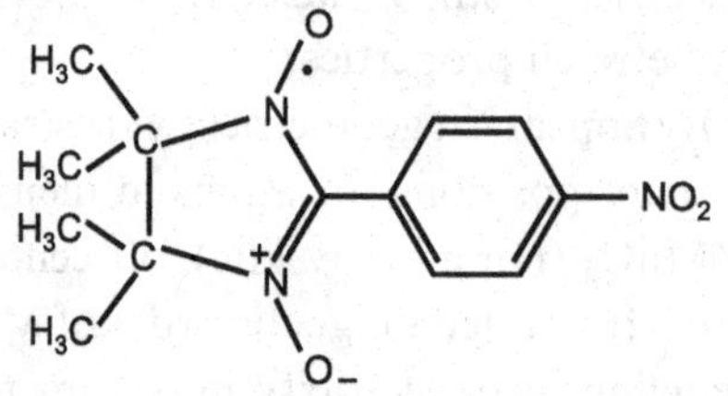

Figure 9.32 The first ferromagnetic radical, *p*-NPNN.

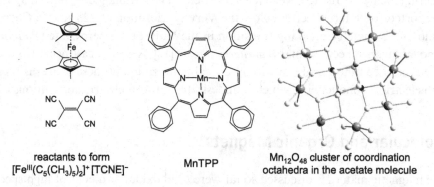

reactants to form
$[Fe^{III}(C_5(CH_3)_5)_2]^+ [TCNE]^-$

MnTPP

$Mn_{12}O_{48}$ cluster of coordination
octahedra in the acetate molecule

Figure 9.33 Molecular structures: $Fe^{II}(C_5(CH_3)_5)_2$ and TCNE (left), MnTPP (middle), $Mn_{12}O_{48}$ cluster of the "acetate-molecule" magnet described in text (right).

unpaired electron, and the crystal becomes ferromagnetic below $T_C = 4.8$ K [23]. It is also an electrical insulator, and its ferromagnetism occurs despite no direct covalent bonding between the metal and TCNE. Higher T_C values do occur with covalent interactions. For example, $[MnTPP]^+ [TCNE]^-$ (MnTPP is a Mn^{3+} complex of tetraphenylporphyrin; Figure 9.33) has $T_C = 16$ K and a structure of planar cations having their Mn centers covalently linked by nitrogens of TCNE into 1D chains, hence it is not a simple isolated "molecule".

There is considerable interest in **single-molecule magnets** (SMMs) containing one or several metal centers with a fixed magnetization direction due to anisotropy within a molecule, the environment of which also shields them from forming intermolecular couplings. One way to achieve the anisotropy is a significant negative zero-field splitting (this occurs for the A_{2g}-term d^3/d^8; Table 9.3) such that the largest M_S values become lowest in energy, and S is as large as possible. Another way is through Jahn–Teller distortion of d^4 or d^9 ions (Section 9.4.3). An example is the $Mn_{12}O_{48}$ cluster of Mn^{4+} and Mn^{3+} coordination octahedra inside the $Mn_{12}O_{12}(CH_3COO)_{16}(H_2O)_4$ "acetate molecule" (Figure 9.33, right), in which 90° O–Mn–O superexchange coupling of the 8 Mn^{3+} ions is ferromagnetic, as is the coupling of 4 Mn^{4+} in the Mn_4O_4 central cube. Their antiparallel orientation gives an $S = 8 \cdot 2 - 4 \cdot 3/2 = 10$ ground-state ferrimagnet.

SMMs with a single metal center would be one more step towards the ultimately smallest unit to store binary data written by an external magnetic field. The task is not trivial, as the examples above suggest that these SMMs would keep their memory only at very low temperatures. One of the most interesting candidates has therefore been based on Dy^{3+} with magnetism of its five unpaired electrons augmented by an orbital-momentum contribution (see Table 9.4). When Dy^{3+} is complexed with the large tris(*tert*-butyl)cyclopentadienyl ligands that, together with bulky $[B(C_6F_5)_4]^-$ anions, prevent intermolecular coupling, the molecule shows magnetic hysteresis up to 60 K [24]. The magical border of the liquid-N_2 temperature was crossed with hexakis(isopropyl)cyclopentadienyl ligands with magnetic hysteresis up to 80 K [25].

9.14 Problems

9.1 Are penguins found at the south pole of the Earth's magnetic field?

9.2 What is the orbital angular momentum for an s electron and why?

9.3 Convert the Earth's magnetic field of 0.7 Oe to SI units.

9.4 What is the volume magnetization M and what is its SI unit?

9.5 Which $3d$ electron configurations (high- and low-spin where applicable) yield in octahedral ligand field an orbital magnetic moment that is quenched or zero? State the reason in each case.

9.6 State whether or not you would expect an orbital contribution to the magnetic moment of Ni^{II} in octahedral or tetrahedral coordination. Calculate the spin-only moment μ_{eff}.

9.7 Determine the spin–orbit coupling ground-state terms $^{2S+1}L_J$ for isolated Mn^{4+}, Mn^{3+}, and Mn^{2+}. Calculate μ_{eff} in Bohr magnetons, with and without an orbital contribution.

9.8 Show that, when the orbital momentum is quenched, Equation (9.14) simplifies to the spin-only model.

9.9 Calculate effective and saturated spin-only magnetic moments (in μ_B) per Cr in $NH_4Cr(SO_4)_2 \cdot 12H_2O$.

9.10 Calculate effective and saturated Russell–Saunders total moments (in μ_B) per high-spin Fe in $NH_4Fe(SO_4)_2 \cdot 12H_2O$.

9.11 For $NH_4Fe(SO_4)_2 \cdot 12H_2O$: (a) Calculate the molar susceptibility at 298 K. (b) Given that the cubic unit cell has $a = 1.243$ nm and contains four formula units, convert the molar susceptibility to dimensionless χ_v. (c) Calculate the relative permeability.

9.12 Estimate the spin-only dimensionless susceptibility of O_2 as an ideal gas at 273.15 K and 100 kPa.

9.13 Superconductors expel a magnetic field from their bulk. In liquid N_2, 102.5 mg of a $YBa_2Cu_3O_{6.9}$ powder of apparent density 3000 kg/m^3 was placed into the center of a miniature induction coil (a pick-up coil), and the induced voltage dropped by 450 mV. At 26 °C, a 38.4 mg standard increased the (empty-coil) voltage by 5.73 V. Assuming a linear voltage response to susceptibility, calculate the fraction of the magnetic field expelled from the inside of the superconductor, given that the standard was a Curie–Weiss paramagnet with $C_m = 9.164 \times 10^{-4}$ m^3 K/kg and $\theta = -24$ K. Hints: The expelled field corresponds to the dimensionless (volume) diamagnetic susceptibility of the superconductor. Because the samples were weighed, mass susceptibility must be calculated first.

9.14 Calculate the effective and saturated, spin-only, $g_e = 2$, paramagnetic moments per high-spin Mn in $La_{0.5}Sr_{0.5}MnO_3$.

9.15 Calculate the effective paramagnetic moment per magnetic atom in $DyCrO_3$.

9.16 Mass susceptibility of Sr_2MnMoO_6 (molar mass 422.114 g/mol) was measured as a function of temperature, from which the Curie constant, C_m, was obtained by least-squares fitting as $C_m = 1.306 \times 10^{-4}$ m^3 K/kg. Determine the oxidation states of Mn and Mo. Hints: Select an expression for the Curie constant C in Equation (9.19). Replace μ_{eff} in it with μ_{eff}/μ_B so that μ_{eff} is now in Bohr magnetons. Assuming two magnetic atoms per formula, Mn and Mo, calculate the number N of these atoms in one kilogram of the sample.

9.17 The molar susceptibility χ_{mol} of the Prussian-blue $CsFe[Cr(CN)_6]$ Curie paramagnet is 0.0163 emu/mol at 300 K and 0.0094 emu/mol at 200 K (CGSem). Using the CGSem formula $\mu_{eff} = 2.828\sqrt{C}$ in Bohr magnetons per formula unit of Appendix I, the Curie law $C = \chi T$, and the spin-only as well as $g_e = 2$ approximation, determine the spin state of Fe^{2+}.

9.18 Calculate the value with which you divide molar magnetization M_{mol} in emu/mol in order to get the amount of Bohr magnetons per mole.

9.19 Reciprocal mass susceptibility χ_m^{-1} for $(NH_4)_2Fe(SO_4)_2 \cdot 6H_2O$ (molar mass 392.139 g/mol) at $T = 100$ K, 200 K, and 300 K was measured in a weak field as 1037000 kg/m^3, 2075000 kg/m^3, and 3112000 kg/m^3, respectively. (a) Plot and fit T versus χ_m^{-1}. What type of magnetism do you see? Is it expected? (b) Assuming a spin-only moment, is iron in a high- or low-spin state?

9.20 Reciprocal mass susceptibilities χ_m^{-1} for CoO at $T = 400$ K, 500 K, and 600 K in a weak field are 2.000×10^6, 2.315×10^6, and 2.635×10^6 kg/m^3, respectively. Plot and fit T versus χ_m^{-1}. Assuming a spin-only moment, is cobalt in a high- or low-spin state? What type of magnetic order would you expect at low temperature from the Weiss constant θ?

9.21 What is the percentage of magnetic saturation in an $S = \frac{1}{2}$ paramagnet at 10 K under a field of magnetic induction 1 T?

9.22 Estimate the magnetic induction B_W and the intensity H_W of the Weiss field for a spin-only material of $S = \frac{1}{2}$ that orders ferromagnetically at 300 K. How does B_W (in teslas) compare with fields generated in typical laboratory magnetometers?

9.23 Predict the type of magnetic coupling for the following materials assuming localized electrons: (a) VO with the NaCl-type structure, (b) LaCrO$_3$ with the perovskite structure, (c) the hypothetical ordered perovskite La$_2$Cr^{III}Fe^{III}O$_6$, (d) cubic CsNi[Cr(CN)$_6$] with Ni and Cr like Na and Cl in rock salt and CN groups (as if one p-group pseudoatom) in between each Ni and Cr.

9.24 Predict the type of cooperative magnetism in CrCl$_3$ having 90° interactions.

9.25 α-MnS and MnO adopt the rock-salt structure. The Néel temperature of 130 K for α-MnS is higher than 117 K observed for MnO despite the fact that the Mn^{2+} ions are much farther apart in α-MnS. Suggest a factor responsible for this behavior.

9.26 Data suggest that the majority-spin 3d sub-band of Co is full. If so, how many electrons are in the delocalized 4s band, given the saturated ferromagnetic moment of 1.751 μ_B per Co and a g factor for Co of 2.17?

9.27 Transport measurements show Fe has 0.95 electrons in the delocalized s band. Given a ferromagnetic moment of 2.22 μ_B per Fe and g factor 2.094, determine the filling in the spin-polarized d sub-bands. Sketch an equivalent picture to Figure 9.27.

9.28 Would you expect Au$_4$V to be ferromagnetic?

9.29 A sample of C$_{60}$ has 1 mass ppm of Fe impurity. When heated at 700 °C under pressure, C$_{60}$ is converted to a carbon polymer and iron to Fe$_3$C. Estimate the mass magnetization in J/(T kg) ($\equiv$ A m^2/kg) of the product first magnetized and then taken to the Earth's field. For Fe$_3$C, assume $M_{remanent} = 700$ kA/m and the density $\rho = 7000$ kg/m^3.

9.30 Derive the formula for the maximum energy product of an ideal ferromagnet in Figure 9.24.

9.31 The spinel MnFe$_2$O$_4 \equiv [(Mn^{2+})_{1-x}(Fe^{3+})_x]^{tet}[(Fe^{3+})_{1-x/2}(Mn^{2+})_{x/2}]_2^{oct}O_4$ has a saturated moment independent of x. Explain why.

9.32 Which of the following garnets is not diamagnetic: $Ca_3Te_2Zn_3O_{12}$, $Y_3Fe_2Fe_3O_{12}$, $Mg_3Al_2Si_3O_{12}$, $NaCa_2Zn_2V_3O_{12}$?

9.33 What is the saturation magnetization, M_{sat}, in Bohr magnetons per formula unit of ferrimagnetic $Y_3Fe_5O_{12}$ and $Gd_3Fe_5O_{12}$ with Gd^{3+} in $4f^7 5s^2 p^6$ configuration?

9.34 The so-called Cu_6 ring (cluster) magnet has six Cu^{2+} coordination octahedra sharing opposite edges of two oxygens. Which type of magnetism would you expect?

9.15 Further Reading

A.F. Orchard, *"Magnetochemistry"* (2003) Oxford University Press.

N.A. Spaldin, *"Magnetic Materials: Fundamentals and Device Applications"* (2003) Cambridge University Press.

S. Blundell, *"Magnetism in Condensed Matter"* (2001) Oxford University Press.

R.L. Carlin, *"Magnetochemistry"* (1986) Springer-Verlag.

B.N. Figgis, M.A. Hitchman, *"Ligand Field Theory and Its Applications"* (2000) Wiley.

A.P. Guimarães, *"Principles of Nanomagnetism"* 2nd edition (2017) Springer.

9.16 References

[1] D. Nicholls, *"Complexes and First-Row Transition Elements"* (1979) Macmillan, p. 106–7.

[2] E.N. Caspi, J.D. Jorgensen, M.V. Lobanov, M. Greenblatt, "Structural disorder and magnetic frustration in $ALaMnMoO_6$ (A = Ba, Sr) double perovskites" *Phys. Rev. B* **67** (2003), 134431/1–11.

[3] E.O. Wollan, W.C. Koehler, "Neutron diffraction study of the magnetic properties of the series of perovskite-type compounds $La_{1-x}Ca_xMnO_3$" *Phys. Rev.* **100** (1955), 545–563.

[4] J.B. Goodenough, "An interpretation of the magnetic properties of the perovskite-type mixed crystals $La_{1-x}Sr_xCoO_{3-\delta}$" *J. Phys. Chem. Solids* **6** (1958), 287–297.

[5] P.W. Anderson, "New approach to the theory of superexchange interactions" *Phys. Rev.* **115** (1959), 2–13.

[6] J. Kanamori, "Superexchange interaction and symmetry properties of electron orbitals" *J. Phys. Chem. Solids* **10** (1959), 87–98.

[7] H.W.F. Sung, C. Rudowicz, "Physics behind the magnetic hysteresis loop: A survey of misconceptions in magnetism literature" *J. Magn. Magn. Mater.* **260** (2003), 250–260.

[8] N.S. Rogado, J. Li, A.W. Sleight, M.A. Subramanian, "Magnetocapacitance and magnetoresistance near room temperature in a ferromagnetic semiconductor La_2NiMnO_6" *Adv. Mater.* **17** (2005), 2225–2227.

[9] C. Zener, "Interaction between the *d* shells in the transition metals" *Phys. Rev.* **81** (1951), 440–444.

[10] J. Lindén, T. Yamamoto, M. Karppinen, H. Yamauchi, T. Pietari, "Evidence for valence fluctuation of Fe in Sr_2FeMoO_{6-w} double perovskite" *Appl. Phys. Lett.* **76** (2000), 2925–2927.

[11] A.S. Wills, G.S. Oakley, D. Visser, J. Frunzke, A. Harrison, K.H. Andersen, "Short-range order in the topological spin glass $(D_3O)Fe_3(SO_4)_2(OD)_6$ using xyz polarized neutron diffraction" *Phys. Rev. B* **64** (2001), 094436/1–8.

[12] L. Pauling, "The structure and entropy of ice and of other crystals with some randomness of atomic arrangement" *J. Am. Chem. Soc.* **57** (1935), 2680–2684.

[13] D.I. Khomskii, "Multiferroics: Different ways to combine magnetism and ferroelectricity" *J. Magn. Magn. Mater.* **306** (2006), 1–8.

[14] H. Schmid, "Some symmetry aspects of ferroics and single phase multiferroics" *J. Phys.: Condens. Matter* **20** (2008), 434201/1–24.

[15] M.S. Senn, N.C. Bristowe, "A group-theoretical approach to enumerating magnetoelectric and multiferroic couplings in perovskites" *Acta Crystallogr. Sect. A* **74** (2018), 308–321.

[16] M. Azuma, K. Takata, T. Saito, S. Ishiwata, Y. Shimakawa, M. Takano, "Designed ferromagnetic, ferroelectric Bi_2NiMnO_6" *J. Am. Chem. Soc.* **127** (2005), 8889–8892.

[17] B.B. Van Aken, T.T.M. Palstra, A. Filippetti, N. Spaldin, "The origin of ferroelectricity in magnetoelectric $YMnO_3$" *Nat. Mater.* **3** (2004), 164–170.

[18] P.M. Woodward, E. Suard, P. Karen, "Structural tuning of charge, orbital, and spin ordering in double-cell perovskite series between $NdBaFe_2O_5$ and $HoBaFe_2O_5$" *J. Am. Chem. Soc.* **125** (2003), 8889–8899.

[19] N. Ikeda, H. Ohsumi, K. Ohwada, K. Ishii, T. Inami, K. Kakurai, Y. Murakami, K. Yoshii, S. Mori, Y. Horibe, H. Kito, "Ferroelectricity from iron valence ordering in the charge-frustrated system $LuFe_2O_4$" *Nature* **436** (2005), 1136–1138.

[20] J.S. Miller, "Organic magnets: A history" *Adv. Mater.* **14** (2002), 1105–1110.

[21] M. Tamura, Y. Nakazawa, D. Shiomi, K. Nozawa, Y. Hosokoshi, M. Ishikawa, M. Takahashi, M. Kinoshita, "Bulk ferromagnetism in the β-phase crystal of the *p*-nitrophenyl nitronyl nitroxide radical" *Chem. Phys. Lett.* **186** (1991), 401–404.

[22] J.S. Miller, "Magnetically ordered molecule-based materials" *MRS Bull.* **32** (2007), 549–555.

[23] J.-H. Her, P.W. Stephens, J. Ribas-Arin, J.J. Novoa, W.W. Shum, J.S. Miller, "Structure and magnetic interactions in the organic-based ferromagnet decamethylferrocenium tetracyanoethenide, $[FeCp*_2]^{+}[TCNE]^{-}$" *Inorg. Chem.* **48** (2009), 3296–3307.

[24] C.A.P. Goodwin, F. Ortu, D. Reta, N.F. Chilton, D.P. Mills, "Molecular magnetic hysteresis at 60 kelvin in dysprosocenium" *Nature* **548** (2017), 439–442.

[25] F.-S. Guo, B.M. Day, Y.-C. Chen, M.-L. Tong, A. Mansikkamäki, R.A. Layfield, "Magnetic hysteresis up to 80 kelvin in a dysprosium metallocene single-molecule magnet" *Science* **362** (2018), 1400–1403.

10   Conducting Materials

10.1 Conducting Materials

An electrical current is a flow of charged particles, and electrical conductivity is a measure of how easily a current can pass through a material. Materials where the current is carried by electrons are called **electronic conductors**, while those where the current is carried by ions are known as **ionic conductors**. In this chapter we take close look at the origins of electronic conductivity in a wide variety of materials. Ionic conductors are covered in Chapter 13.

It is difficult to imagine a material property that spans a greater range than electrical conductivity, nearly 30 orders of magnitude (Table 10.1). To put the conductivities of various materials in context, we need to define the units associated with conductivity. We begin with Ohm's law:

$$I = \frac{V}{R} \tag{10.1}$$

where I is the current in amperes, V is the potential difference in volts, and R is the resistance in ohms. Resistance is the material property of interest to us, but, as an extrinsic property, its value depends on the size and shape of the sample. To express Ohm's law in a form that is independent of the sample dimensions, we replace current in Equation (10.1) with **current density**, $J = I/A$ (current per area),[1] potential difference with **electric-field intensity**, $E = V/L$ (voltage per length), and resistance with **resistivity**, $\rho = RA/L$ (resistance per length and per inverse area). In doing so, Equation (10.1) becomes:

$$J = \frac{V}{AR} = \frac{EL}{AR} = \frac{EL}{A(\rho L/A)} = \frac{E}{\rho} = \sigma E \tag{10.2}$$

---

[1] Because the electric current (in amperes) is the number of flowing charges (in coulombs) per second, the current density J (in A/m^2) is the same as the charge flux J [in [C/(m^2 s)]. We've encountered the flux J already in Chapter 3.

Table 10.1 Electrical conductivity, σ, of selected materials at room temperature.

Substance	σ (S/m)	Substance	σ (S/m)
Ag	6.2×10^7	$Bi_2Ru_2O_7$	2×10^5
Cu	5.9×10^7	$LaNiO_3$	1×10^5
Al	3.8×10^7	Doped polyacetylene	8×10^4
Na	2.1×10^7	Fe_3O_4	2×10^4
ReO_3	1.1×10^7	$YBa_2Cu_3O_7$*	1×10^2
Ti	2.5×10^6	Ge	2×10^0
La	1.6×10^6	Si	10^{-3}
$SrMoO_3$	1.0×10^6	NiO	10^{-8}
Bi	7.7×10^5	Al_2O_3	10^{-12}
Mn	6.2×10^5	S	10^{-15}
NbN	4×10^5	SiO_2 (quartz)	10^{-16}
TiO	3×10^5	Teflon	10^{-22}

*Conductivity in the *ab* plane of the crystal.

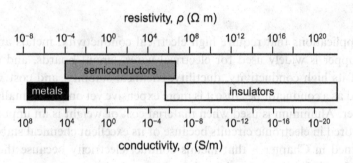

Figure 10.1 Approximate ranges for the conductivity and resistivity of metals, semiconductors, and insulators.

where σ is **conductivity**, the inverse of resistivity, $\sigma = 1/\rho$. Resistivity has units of ohms meter [Ω m = (V/A) m] and conductivity has units of siemens per meter [S/m = (A/V)/m], where $1 \text{ S} = 1 \ \Omega^{-1}$. Conductivity and resistivity are intrinsic properties of materials. Approximate ranges of conductivity and resistivity for metals, semiconductors, and insulators are illustrated in Figure 10.1.

Metals are the most familiar electronic conductors, yet we see from Table 10.1 that not all metals are equally able to conduct electricity. For example, silver and copper have conductivities that are two orders of magnitude greater than manganese and bismuth. As the name suggests, semiconductors, like silicon and germanium, are far less conductive than metals. However, semiconductors differ from metals in ways other than their absolute conductivity.

Semiconductors become more conductive with increasing temperature, whereas temperature has the opposite effect on a metal. The conductivity of a semiconductor can be increased dramatically by adding small amounts of other elements, a process called doping, whereas impurities typically decrease the conductivity of a metal. In this chapter we explore the origins of these differences.

Metallic elements and alloys are not the only materials that are good conductors. Transition-metal oxides and nitrides, like ReO_3 and NbN, have conductivities that are comparable to many metals. However, unlike elemental metals, the electrical conductivities of transition-metal compounds span a very wide range. For example, TiO is only 10 times less conductive than titanium metal, whereas stoichiometric TiO_2 is highly insulating. The differences between TiO and TiO_2 can be attributed to differences in the occupation of the d orbitals, but not all transition-metal oxides with partially filled d orbitals exhibit metallic conductivity. NiO and TiO both crystallize with the NaCl structure and both have ions with partially filled d orbitals, but NiO is less conductive than TiO by 13 orders of magnitude. Clearly, many aspects of conductivity cannot easily be understood without a closer look at the underlying mechanisms of conductivity in different classes of materials.

10.2 Metals

In most applications that require high electrical conductivity, metals are the materials of choice. Copper is widely used for electrical wires, circuit boards, and electrical contacts because of its high conductivity, ductility, chemical stability, and cost. Silver is not commonly used as a conductor because it is more expensive yet only marginally more conductive than copper. Aluminum is used when material cost or weight is an important factor, while gold is favored in electronic circuits because of its excellent chemical stability.

We learned in Chapter 6 that metals conduct electricity because the Fermi level cuts through a band. In such materials, there is very little energy cost for electrons to move from occupied to empty crystal orbitals (the infinite crystal equivalents of molecular orbitals, see Chapter 6), allowing easy movement in response to an external electric field. In fact, a partially filled band is a necessary but not sufficient condition for metallic conductivity. Unfortunately, this simple picture does not give any clues as to why some metals are better conductors than others. Nor does it make any prediction about the how conductivity will change as a function of temperature or defect concentration.

10.2.1 Drude Model

One of the earliest attempts to explain the conductivity of metals was put forward by Paul Drude in 1900. It predates the development of quantum mechanics and is therefore a classical treatment. Although too simplistic in many ways, the Drude model serves as a useful introduction to conduction. Many of the parameters used to describe conduction in

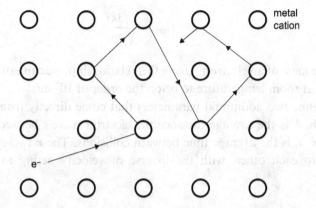

metal
cation

e⁻

Figure 10.2 The trajectory of an electron in the Drude model in the absence of an applied field.

the Drude model, such as carrier concentration and mobility, have been retained in more modern treatments.

We begin by assuming that a metal can be treated as an ordered array of positively charged stationary ions surrounded by a negatively charged sea of delocalized valence electrons. The valence electrons move throughout the crystal and behave as an ideal gas, albeit one where the "gas molecules" have a very small mass. If we were to follow the path of a single electron, we would see that it undergoes frequent collisions with the stationary ions, leading to a random trajectory, as shown schematically in Figure 10.2.

In the kinetic theory of gases, the atoms or molecules of the gas are in constant random motion, and all collisions are assumed to be elastic. Potential energy is neglected, so the energy of each particle is simply its kinetic energy, $E = \frac{1}{2}mv^2$, where m and v are the mass and velocity of particle, respectively. Here we can treat velocity as a scalar (a speed). Not all atoms/ molecules of the gas move with the same speed; at any given moment, some are moving much faster than the ensemble average and others much slower. The spread of their energies follows a Maxwell–Boltzmann probability distribution, with an average kinetic energy of:

$$E = \frac{3}{2}kT \tag{10.3}$$

where $k = 1.380649 \times 10^{-23}$ J/K is the Boltzmann constant and T is the absolute temperature. Since the particles, here electrons, all have the same mass, we can calculate the rms velocity[2] of randomly moving electrons in the ensemble from the definition of kinetic energy and Equation (10.3):

---

[2] This value is called the root mean square (rms) velocity because it is obtained from the average kinetic energy E that is proportional to velocity squared. The average (mean) velocity for the Maxwell–Boltzmann probability distribution of kinetic energies is smaller, 92.1% of the rms velocity.

$$v_{\text{rms}} = \sqrt{\frac{3kT}{m_e}} \tag{10.4}$$

where m_e is the mass of an electron. Using this relationship, we can estimate the rms velocity of an electron at room temperature to be on the order of 10^5 m/s.

Next, we define two additional parameters that come directly from kinetic theory. The **mean free path**, ℓ, is the average distance the electron travels between collisions, and the **relaxation time**, τ, is the average time between collisions. These two quantities are directly proportional to each other, with the inverse of velocity acting as the proportionality constant:

$$\tau = \frac{\ell}{v} \tag{10.5}$$

Because metal structures are closely packed, we might assume that the electrons do not travel far without being deflected by a collision with an ion. The Drude model yields qualitatively reasonable estimates of the conductivity for main-group metals if we choose a mean free path of a few nanometers. This distance amounts to a collision every few unit cells, which does not seem unreasonable. Using a value of 1 nm for the mean free path and taking our earlier velocity estimate of 10^5 m/s, we obtain a relaxation time τ on the order of 10^{-14} s, one collision every 10 femtoseconds.

Even though electrons are moving quickly, in the absence of an electric field the motions are random, and the net flow of charge is zero. The situation changes once an electric field E is applied. The electrons are attracted to the positively charged end of the conductor by a force $F = -eE$, where e is the elementary charge (1.602×10^{-19} C), and the negative sign originates from the two vectors being oppositely aligned.[3] The net velocity of the electron along the direction of the electric field, called the **drift velocity**, v_d, is equal to the product of the acceleration, a, due to the field ($F = ma$) and the time τ that elapses before a collision changes the path of the electron:

$$v_d = a\tau = \left(\frac{F}{m_e}\right)\tau = -\frac{Ee\tau}{m_e} \tag{10.6}$$

If we consider only the magnitudes (absolute values) of the vectors, we can calculate the drift velocity. An electric potential of 1 V applied along a wire 1 cm long generates an electric field of magnitude 100 V/m. Taken together with our earlier estimate of 1×10^{-14} s for τ, we can restate Equation (10.6) with scalars to obtain $v_d = Ee\tau/m_e \approx 0.2$ m/s for the drift velocity. This net velocity of an electron is more than five orders of magnitude smaller than our earlier estimate of its average thermal velocity. The fact that the electrons keep colliding with the

---

[3] The direction of the electric-field vector is defined to be from positive to negative charge. The force attracts the electron towards the positive end.

stationary ions explains why the drift velocity is so much smaller than the thermal velocity. We can think of these collisions as a source of resistance to the electrical current flow.

Before returning to macroscopic quantities, we need to define one more quantity, **electron mobility**, μ, which is the electron's drift velocity divided by the electric-field intensity acting on it, has units of $m^2/(V\,s)$ and is a measure of the ease with which an electron moves through a crystal:

$$\mu = \frac{v_d}{E} \tag{10.7}$$

Combining Equations (10.6) and (10.7), we obtain the following scalar expression for the mobility of an electron:

$$\mu = \frac{e\tau}{m_e} \tag{10.8}$$

As the collisions become less frequent (the relaxation time τ increases), the electric field has more time to accelerate the electron along its path and increase its mobility.

Returning to macroscopic quantities, it can be shown that the current density J (in A/m^2) is equal to the number of conduction electrons per unit volume, n, multiplied by their charge and drift velocity:

$$J = nev_d = ne\mu E \tag{10.9}$$

By comparison with Equation (10.2) ($J = \sigma E$) we see that:

$$\sigma = ne\mu \tag{10.10}$$

This is a key equation for understanding the conductivity of a material. The electrical conductivity of a material depends upon the concentration of charge carriers n and their mobility μ.

It is quite difficult to directly measure the relaxation time in a metal, but, as a crude approximation, we might assume that τ and hence μ do not change substantially from one metal to the next. To the extent that this assumption is valid, the conductivity should scale with n, the number of charge carriers per unit volume. In the Drude model, all valence electrons are free to respond to the applied field and act as charge carriers. To calculate n within this model, one simply needs to multiply the atomic density by the number of valence electrons per atom.

The valence-electron concentration and room-temperature conductivity of several metals are shown in Table 10.2. Although the conductivities of the alkali metals do increase with increasing valence-electron concentration, we see that any relationship between the two variables quickly breaks down as we start moving around the periodic table.[4] To better understand these periodic trends, as well as many other aspects of metals, we need a more realistic treatment of the conduction electrons.

[4] There are many other experimental observations that reveal the shortcomings of the Drude model; the heat capacity of metals and the conductivity of alloys are two examples.

Table 10.2 Conductivity σ does not scale with the valence-electron concentration n.

Metal	n (m^{-3})	σ (S/m)	Metal	n (m^{-3})	σ (S/m)
Rb	1.1×10^{28}	0.8×10^7	Ag*	5.9×10^{28}	6.2×10^7
K	1.3×10^{28}	1.4×10^7	Au*	5.9×10^{28}	4.5×10^7
Na	2.5×10^{28}	2.1×10^7	Cu*	8.4×10^{28}	5.9×10^7
Ca	4.6×10^{28}	2.9×10^7	Zn*	13.1×10^{28}	1.7×10^7
Mg	8.6×10^{28}	2.3×10^7	Sc	12.0×10^{28}	0.18×10^7
Al	18.0×10^{28}	3.8×10^7	Ti	20.5×10^{28}	0.25×10^7

* The d subshells of Cu, Ag, Au, and Zn are assumed to be full and are not counted in the valence-electron count.

10.2.2 Free-Electron Model

The failure of the Drude model to quantitatively explain the conductivities of metals should not come as a surprise; after all, electrons follow the laws of quantum mechanics, where the energy of each electron is dictated by the band it occupies and its wave vector k (Section 6.4.2). The simplest quantum-mechanical description of electrons in a metal is the free-electron model. It assumes that the electric potential felt by an electron is constant and equal to a constant potential V_0 everywhere inside the crystal and effectively infinite outside the crystal (Section 5.2). This potential for a 1D version of this idealized crystal is plotted in Figure 10.3. You might think that such a simple model would provide little insight to real materials, but, with some fine tuning, it works reasonably well for simple main-group metals.

Recall from Section 6.1.2 that in a 1D crystalline solid the electronic wavefunction takes the form $\psi(r) = e^{ikr}u(r)$, where $u(r)$ is a periodic function representing the potential felt by the electron. For a given band, each electron must have a unique value of k. In this sense, k acts as a quantum number specifying the crystal orbital that the electron occupies, but it is more than just a label. The value of k is inversely proportional to the wavelength λ of the crystal-orbital wavefunction $k = 2\pi/\lambda$; Equation (6.10). In a 3D crystal, it becomes a vector k (Section 6.4.2) that is parallel to the propagation direction of the wave (perpendicular to its wave front), the wavelength is $\lambda = 2\pi/|k|$ and the wavefunction becomes $\psi(r) = e^{ik \cdot r}u(r)$.[5]

In the free-electron model, the potential is uniform throughout the crystal, $u(r) = 1$, and the wavefunction simplifies to $\psi(r) = e^{ik \cdot r}$. When this wavefunction is inserted into the time-independent Schrödinger equation, $H\psi(r) = E\psi(r)$, the energies E of the allowed electronic states are:

---

[5] Perhaps the simplest illustration of this is the crystal orbitals for a 2D sheet of hydrogen atoms depicted in Figure 6.17. At the X point ($k = \pi/ak_x + 0k_y$), the wavefunction changes sign (propagates) upon moving through the crystal in the x direction. The vector nature of k is essential to distinguish this crystal orbital from that at the Y point ($k = 0k_x + \pi/ak_y$); both have the same wavelength but the waves propagate in different directions.

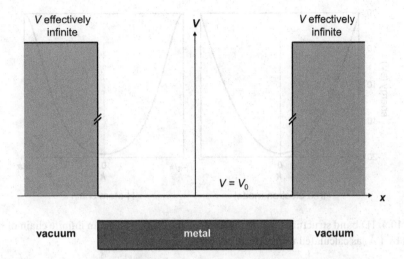

Figure 10.3 The potential V (arbitrary scale, units V) experienced by an electron in the free-electron model in a 1D metal of finite length.

$$E = V_0 + \frac{\hbar^2 k^2}{2m_e} \tag{10.11}$$

To apply the free-electron model to real materials, the electron mass, m_e, is replaced with the **effective mass**, m^*:

$$E = V_0 + \frac{\hbar^2 k^2}{2m^*} \tag{10.12}$$

The effective mass is an adjustable parameter that varies from one material to another. We will return to its significance later.

Using Equation (10.12), we can plot energy as a function of k to obtain a band-structure diagram. The band structure, which is a simple parabola with a minimum at $k = 0$, is plotted in Figure 10.4, where it is compared with the band structure for a 1D chain of hydrogen atoms discussed in Section 6.1.5. Even though they are derived using quite different assumptions, the tight-binding and free-electron band structures are surprisingly similar. To obtain this level of agreement, two parameters in the free-electron model must be adjusted. The constant potential V_0 must be chosen so that the bands have the same energy at $k = 0$, and the effective mass m^* must be adjusted so that the bandwidths are similar.

10.2.3 Fermi–Dirac Distribution

In the electron-gas Drude model, the energy of each electron is taken to be its kinetic energy that follows a Maxwell–Boltzmann distribution, whereby the average electron energy is directly proportional to the absolute temperature, Equation (10.3). In any quantum-mechanical treatment, including the free-electron model, the energy of each electron is

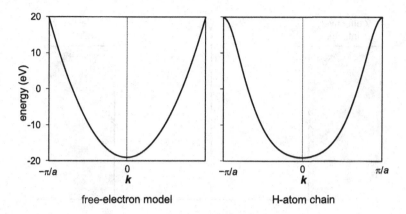

Figure 10.4 1D band structure for the free-electron model (left), and an infinite chain of hydrogen atoms spaced by 1 Å, as calculated using a tight-binding model (right).

determined by the crystal orbital it occupies, which is dictated by its wave vector k. Because the Pauli exclusion principle prevents two electrons of the same spin from occupying the same crystal orbital, the imposition of quantum mechanics leads to a very different distribution of electron energies. That is not to say temperature has no effect in the more realistic quantum-mechanical treatment, as we see below.

At finite temperatures, thermal energy excites some electrons into states with energies greater than the Fermi energy, E_F, while creating vacancies in states with energies less than E_F. The probability that a given electronic state (a crystal orbital) is occupied, $f(E)$, is given by the **Fermi–Dirac distribution function**:

$$f(E) = \frac{1}{1 + \exp[(E - E_F)/kT]} \tag{10.13}$$

where E is the energy[6] of the electronic state in question. For example, at 300 K, $f(E) = 0.02$ for a crystal orbital whose energy is 0.10 eV above E_F, which means there is only a 2% probability that this state will be occupied. The Fermi–Dirac distribution is plotted for a metallic solid at $T = 0$ and 300 K in Figure 10.5. At 300 K, the electron distribution is smeared out by a few tenths of an electronvolt on either side of the Fermi energy E_F.

The imposition of Fermi–Dirac statistics changes our picture of conductivity in an important way. For an electron to move through the crystal in response to an applied electric field, there must be an unoccupied state of similar energy for the electron to move into. Electrons at energies well below the Fermi level, where $f(E) \approx 1$, do not meet this criterion. As a result, only a small percentage of the total number of valence electrons contribute to the conduction process.

---

[6] Do not confuse the symbol E used for energy with the E used for electric-field intensity, or k representing the Boltzmann constant with the symbol used for the wave vector.

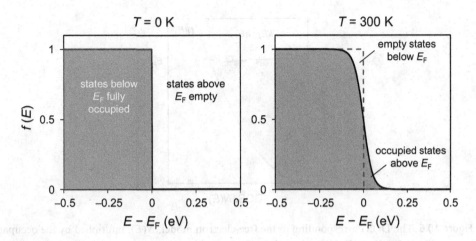

Figure 10.5 Fermi–Dirac distribution function, $f(E)$, in a metal near E_F at $T = 0$ (left) and 300 K (right).

10.2.4 Carrier Concentration

Conductivity is proportional to the carrier concentration n per volume and to the mobility of those carriers μ, Equation (10.10). What do the free-electron model and Fermi–Dirac statistics tell us about how the carrier concentration n changes as the valence-electron concentration (per unit volume) varies from metal to metal? For the parabolic band that comes out of the free-electron model in Section 10.2.2, it can be shown that the density of states (DOS), $N(E)$, is proportional to the square root of the energy, $N(E) \propto E^{1/2}$. This is shown in Figure 10.6, along with the product $f(E) \cdot N(E)$, which emphasizes the occupied states at a given temperature. The electrons that occupy states close to E_F are the ones that contribute to the conductivity, while those well below E_F cannot access empty states needed to move through the crystal. If we increase the concentration of valence electrons, band filling, and the DOS at the Fermi level, $N(E_F)$ will also increase. Hence, the free-electron model predicts that n will increase, even if in a nonlinear fashion, as the concentration of valence electrons increases. This helps to explain why the conductivities of the alkali metals increase as they become smaller (Rb → K → Na), and the valence-electron concentration goes up (Table 10.2).

Closer inspection of Table 10.2 shows that the conductivity does not change uniformly as the valence-electron density increases. The discrepancy between theory and observation stems in part from the inability of the free-electron model to approximate the band structures of most metals. It works best for the alkali metals where the valence electrons occupy one half of a single band originating from overlap of s orbitals. For other elements, the Fermi level cuts through multiple bands, and the free-electron model breaks down. In later sections, we will take a more accurate look at the band structures of real metals.

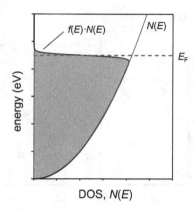

Figure 10.6 The DOS corresponding to the free-electron model, $N(E)$, multiplied by the occupancy probability at 500 K, $N(E){\cdot}f(E)$, (shaded).

10.2.5 Carrier Mobility and Effective Mass

Carrier concentration is not the only variable that determines the conductivity of a substance. Mobility plays an equally important role; but how are we to assess the changes in mobility from one substance to another? Do the conduction electrons in copper have a higher or lower mobility than those in potassium, and what about transition metals like titanium? To answer these questions, we need to take a closer look at the relationship between band structure and electron mobility.

Effective mass is a kind of "fudge factor" that allows many predictions of the simple free-electron model to be applied more generally than would otherwise be justified. As a rule of thumb, when the Fermi level cuts through a wide band, the electrons behave as high-mobility "light" charge carriers with relatively small effective masses. Conversely, when the Fermi level cuts through a narrow band, electrons have low mobility, and m^* is large.[7]

Effective mass can be defined in a more quantitative way by differentiating Equation (10.12) twice with respect to the wave vector k. In real materials, effective mass is a tensor whose magnitude depends upon the directions of the wave vector and the applied electric field. Fortunately, the parabolic band shape of the free-electron model evolves from its minimum at $k = 0$ identically in any direction of k space, which allows us to obtain an expression that gives a scalar value of the effective mass:

$$\frac{d^2E}{dk^2} = \frac{\hbar^2}{m^*} \qquad (10.14)$$

Rearranging to solve for m^* we get:

---

[7] To help remember this guideline think of the bands as the tracks of a rollercoaster and the carriers as the cars on those tracks. The speed of the car is a proxy for mobility.

$$m^* = \frac{\hbar^2}{(\mathrm{d}^2 E/\mathrm{d}k^2)} \qquad (10.15)$$

The effective mass of the conduction electrons therefore depends on the curvature of the bands that cross the Fermi level. In Section 6.1.5, it was stated without full justification that flat (narrow) bands are associated with localized electrons and wide bands with highly delocalized electrons. Now we see the rationale; flat bands have very little curvature, which makes the denominator of Equation (10.15) small and leads to a large effective mass. The opposite relationship holds for wide bands.

Although the qualitative relationship between bandwidth and effective mass is quite general, the quantitative use of Equation (10.15) is much more limited. For a parabolic band, Equation (10.14) yields m^* constant, independent of the wave vector k. For more realistic band structures this is not the case; there are different values of m^* at different values of k. The usefulness of Equation (10.15) is generally limited to substances where the charge carriers reside near the bottom or top of a band whose shape is approximately parabolic. As we will see later, this approximation often works reasonably well for semiconductors.

When Equation (10.15) is used to describe charge carriers near the top of the band, we get an unusual result. Consider the band structure of the 1D hydrogen atom chain shown on the right-hand side of Figure 10.4. Because the band has an inflection point midway between $k = 0$ and $k = \pi/a$, the quantity $\mathrm{d}^2 E/\mathrm{d}k^2$ will be negative when the band is more than half filled. Consequently m^* will also be negative. What does it mean to have a negative effective mass? It means that nearly full bands conduct as though the charge carriers were positively charged. That is to say, the charge carriers move in the opposite direction than expected for electrons. These charge carriers, called **holes**, respond to an electric field just like an electron with a negative mass. We will return to holes when discussing semiconductors where they play an important role.

10.2.6 Fermi Velocity

In the Drude model, a Maxwell–Boltzmann distribution was used to estimate the average energy of the electrons, and, through Equation (10.4), their rms velocity. The free-electron model produces a different distribution of energies, which means our earlier velocity estimate of $\sim 10^5$ m/s from Section 10.2.1 is suspect. Let us revisit the question of carrier velocity, this time within the more realistic quantum-mechanical picture.

The velocity of a particle is directly proportional to its momentum, $p = mv$, and, as we learned in Chapter 6, the crystal momentum of an electron is related to the wave vector k through Equation (6.12), $p = \hbar k$.[8] Combining these expressions, we express the electron's velocity in terms of its wave vector:

---

[8] Though not quite the same as true momentum, crystal momentum must be conserved when electrons are scattered by collisions with lattice vibrations (phonons), impurities, or other electrons. Crystal momentum must also be conserved when electrons absorb or emit photons.

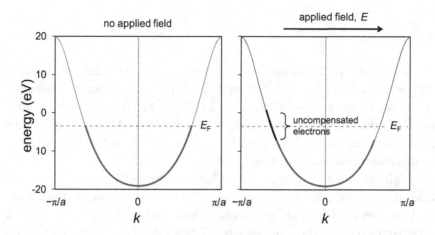

Figure 10.7 The occupied states in a partially filled band of a 1D metal in the absence of an external electric field (left), and once an external electric field is applied (right). Thermal excitations of the carriers have been neglected.

$$v = \frac{\hbar k}{m} \tag{10.16}$$

The band structure of an arbitrary 1D metal with a single band is plotted in Figure 10.7, where a thick line is used to denote occupied states. In the 1D case, we can treat the wave vector k as a scalar. The electrons near the center of the Brillouin zone have low velocities, while those with larger values of k are moving much faster. In the absence of an external field, the distribution is symmetric on either side of $k = 0$, which means that for every electron with a momentum $\hbar k$ there is another electron with the same absolute value of momentum moving in the opposite direction. Consequently, there is no net migration of charge and no electric current flowing. While thermal excitations will smear out the occupied states near E_F slightly (as governed by the Fermi–Dirac function), the distribution will remain symmetric with respect to k regardless of the temperature.

If an electric field is applied in the positive k direction, the states with negative k are stabilized and those with positive k are destabilized. This shifts the distribution of occupied states slightly to the left, as shown in Figure 10.7.[9] Because the displacement is small (the shift has been exaggerated in Figure 10.7), most electrons with wave-vector length $-k$ are still cancelled by an electron with a wave vector of length $+k$. However, near the Fermi level, a few occupied states with wave vectors of $-k$ are not compensated by an electron moving in the opposite direction. It is these uncompensated electrons that are responsible for the flow of current. We should also note that if the band is completely full, as it is for an insulator, the applied field is not able to shift the occupied states and there is no flow of current in response to the field.

[9] The states with negative values of k are stabilized by an electric field because of the negative charge of an electron. Remember that by convention electrons flow in the opposite direction of current.

Because the uncompensated electrons are equally distributed around the Fermi level, their average energy is E_F. By setting E_F equal to the average kinetic energy of the conduction electrons ($\frac{1}{2}mv^2$), we can calculate the rms velocity of the conduction electrons. This value, called the **Fermi velocity**, v_F, is:

$$v_F = \sqrt{\frac{2E_F}{m^*}} \tag{10.17}$$

Using experimentally measured values of E_F and setting $m^* = m_e$, we can estimate v_F. For copper, $E_F = 7.0$ eV and $v_F = 1.6 \times 10^6$ m/s, whereas for aluminum these values are 12 eV and 2.0×10^6 m/s.[10] The Fermi velocities for these metals are more than an order of magnitude larger than the velocity calculated with Equation (10.4). Thus, we see that by treating the electrons classically, the Drude model overestimates the number of carriers, but underestimates their velocity. Because the two estimates err in opposite directions, the Drude estimates of the conductivity are closer to experimental values than one might expect.

10.2.7 Scattering Mechanisms

By expressing the mobility term in Equation (10.10) with Equation (10.8) and replacing m_e with m^*, we can obtain an expression for conductivity in terms of relaxation time:

$$\sigma = ne\mu = \frac{ne^2\tau}{m^*} \tag{10.18}$$

Using Equation (10.5) ($\tau = \ell/v$) we can write a similar expression that relates conductivity to mean free path:

$$\sigma = \frac{ne^2}{m^* v_F}\ell \tag{10.19}$$

These equations make intuitive sense; as the relaxation time and mean free path increase, the electrons are better able to respond to an applied field and the conductivity increases. From conductivities measured at room temperature, Equations (10.17) and (10.19) provide mean free paths of 42 nm and 29 nm for Cu and Al, respectively. This is considerably larger than our earlier estimate of a few nanometers (Section 10.2.1). With mean free paths that are tens of nanometers, the conduction electrons are able to pass by hundreds of atoms between collisions. Given the close packing of atoms in a metal, this behavior is somewhat surprising. It's akin to hitting a golf ball through a dense forest and finding that it travels several

---

[10] Absolute values of E_F can be confusing. In the free-electron model, the bottom of the band defines zero energy, and E_F values are positive numbers that get larger as the band filling increases. That is the formalism used here. In tight-binding theory, the Fermi level is always a negative number determined by orbital overlap and the ionization energies of the atomic orbitals. In other band-structure calculation methods, E_F is often arbitrarily set to zero energy. Note that in Figure 10.4 both plots employ the scale used for tight-binding calculations.

hundred yards without hitting a tree. As we will soon see, the mean free paths at low temperatures can be even longer. How can we understand this behavior?

Two facets of quantum mechanics are responsible for the unexpectedly long mean free path of the electron. Firstly, as the conducting electron approaches an atom, it experiences repulsions with the atom's electrons, which tend to cancel out the Coulomb attraction to the positively charged nucleus. Secondly, wave–particle duality allows the electron to behave as a wave with a wavelength given by the de Broglie relation, $\lambda = h/(m_e v)$. Using $v_F = 2 \times 10^6$ m/s, we obtain a wavelength of 3.6 Å for conduction electrons in aluminum. This value is comparable to the atom spacing in a crystal. By acting as a wave, the electron can pass through the crystal indefinitely without scattering. When an electron wave travels through the crystal, atoms absorb energy from the wave and radiate it back, so that the wave continues with the same direction and intensity, much in the same way a light wave traveling through a crystal would behave.[11]

In a perfect crystal, the conduction electrons would not be scattered by collisions with atoms and there would be no resistance to current flow. In real materials, deviations from perfect periodicity disrupt the transport of conduction electrons and give rise to finite resistance. We can classify the imperfections into two categories: defects and lattice vibrations. Defects, such as substitutional impurities, interstitial atoms, and vacancies, break the periodicity of the lattice and lead to scattering of the electrons. Even if you could create a defect-free crystal, at finite temperatures lattice vibrations break the periodicity of the lattice and result in scattering of the conduction electrons.

The resistivities of high-purity aluminum and copper are plotted as a function of temperature in Figure 10.8. Notice that both metals become more conductive as the temperature decreases. At 1 K, the resistivity of aluminum is 1×10^{-12} Ω m and that of copper 2×10^{-11} Ω m. If we convert these values to conductivities (1×10^{12} and 5×10^{10} S/m, respectively) and compare with the room-temperature values, we see that aluminum and copper are 25000 and 830 times more conductive at 1 K than at room temperature.

What is responsible for the increased conductivity of metals at low temperature? To answer this question, consider Equation (10.19). The Fermi velocity depends on the Fermi energy through Equation (10.17), which is essentially independent of temperature. For a metal, the carrier concentration n does not change dramatically with temperature. If anything, n might decrease subtly at low temperature as the Fermi–Dirac function sharpens. Therefore, we can only explain the increased conductivity by concluding that the mean free path of the conduction electrons has increased significantly at low temperature. By assuming n is constant as T decreases, we can estimate a lower bound on the increase in mean free path. In high-purity aluminum, such an estimate gives a mean free path at 1 K of 0.7 mm, which means the electron is able to pass through more than a million unit cells between collisions!

As seen on the right-hand side of Figure 10.8, a crossover in the temperature dependence of the resistivity happens at low temperature. At low temperatures, the thermal vibrations of

---

[11] The velocity of the electron wave is slowed just like a light wave is slowed when it moves through a crystal, in the latter case by an amount equal to the refractive index.

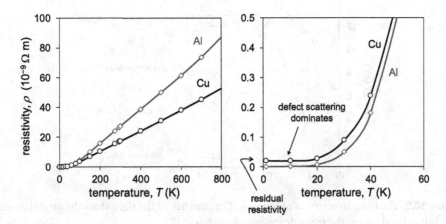

Figure 10.8 The resistivities for high-purity single crystals of aluminum and copper as a function of temperature. The plot on the right-hand side shows an expanded view of the low-temperature region. Data taken from ref. [1].

the atoms become negligible, and scattering due to impurities and defects dominates. This contribution to the scattering is independent of temperature. At sufficiently low temperatures, below roughly 20 K for high-purity Al and Cu, the resistivity reaches a nearly constant value known as the **residual resistivity**. The value of the residual resistivity of a metal is determined largely by the concentration of defects. As the defect concentration decreases, the residual resistivity also goes down. At higher temperatures, scattering due to lattice vibrations dominates, and the resistivity increases linearly as a function of temperature.[12]

10.2.8 Band Structure and Conductivity of Aluminum

The free-electron model can be applied with reasonable success to the alkali metals, but for most other metals we need a more sophisticated model. Let's use the concepts developed in Chapter 6 to discuss the band structure and conductivity of aluminum. Aluminum has a cubic closest-packed structure with a face-centered cubic (fcc) unit cell. Instead of using the fcc cell, we choose to construct the band structure with the smaller primitive unit cell that contains a single atom. Since there are four valence orbitals per atom ($3s$, $3p_x$, $3p_y$, $3p_z$), the band structure contains four bands. The $3s$ and $3p$ orbitals do not mix at Γ, which leads to a large energy separation because the $3s$ orbitals have a bonding overlap at Γ while the $3p$ orbitals overlap in an antibonding manner, as shown on the right-hand side of Figure 10.9. The observation of a minimum for the "$3s$" band at Γ parallels the band structures of the 1D (Figure 6.5) chain and 2D (Figure 6.17) sheet of H atoms. This feature arises because the maximum bonding overlap of a lattice of s orbitals always occurs at Γ, where all atomic orbitals that make up the crystal orbital have the same phase. The lower half of the "$3s$" band has a parabolic shape similar to the free-electron band structure, but agreement

---

[12] More complex behavior is seen in some metallic conductors, as discussed in Section 15.10.

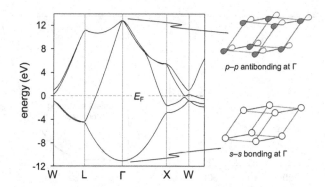

Figure 10.9 The band structure of aluminum. The sketches on the right show the orbital contributions of the 3s band and one of the triply degenerate 3p bands at Γ.

with the free-electron model breaks down as the energy increases. The free-electron model is not a good approximation near the Fermi level. Therefore, we cannot use the free-electron model to estimate the carrier concentration or mobility in aluminum.

It is more complicated to explain the band structure as we move away from Γ because the 3s and 3p orbitals mix, but we can make some observations about the properties from the band structure. There is no gap between bands, which means that, regardless of the electron count, the Fermi level will cut through one or more bands, resulting in metallic conductivity. The bands are wide, with widths that vary from approximately 10 eV to 16 eV. Wide bands lead to high electron mobilities, so it is not surprising that aluminum is a highly conducting metal.

10.2.9 Band Structures and Conductivity of Transition Metals

How do transition metals differ from main-group metals? We must now consider the d orbitals, and how their inclusion impacts the band structure. The room-temperature electrical conductivities of metals in the d block of the periodic table are given in Table 10.3. Within a group, the conductivities are similar, whereas along a period, the changes are quite irregular. One conspicuous feature is the abrupt jump in conductivity between groups 10 and 11, even though elements from both groups adopt fcc structures. What is responsible for the high conductivity of the group 11 elements?

Because we need to include the ns, np, and $(n-1)d$ orbitals, we expect a total of nine bands in the band-structure diagram. Figure 10.10 compares the band-structure diagrams and DOS plots for Ag and Pd. Four wide bands originate from overlap of the 5s and 5p orbitals. They are similar in shape to those seen in aluminum (although not identical due to mixing with the 4d orbitals). Moving toward Γ, the energy of the band with predominant 5s character decreases toward a minimum while the bands with predominant 5p character increase toward a maximum. In addition to these four bands, five much narrower bands arise from the 4d orbitals. The contribution of the 5s, 5p, and 4d bands can be seen more clearly in

Table 10.3 Room-temperature conductivities (in 10^6 S/m) and structure types of the transition metals, arranged as found in the periodic table (group numbers shown at the top).

3	4	5	6	7	8	9	10	11
Sc	Ti	V	Cr	Mn	Fe	Co	Ni	Cu
1.8	2.5	5.0	7.9	0.6	10.0	17	14	59
hcp*	hcp	bcc*	bcc	Other	bcc	hcp	fcc*	fcc
Y	Zr	Nb	Mo	Tc	Ru	Rh	Pd	Ag
1.8	2.4	6.7	20	5.0	14	23	10	62
hcp	hcp	bcc	bcc	hcp	hcp	fcc	fcc	fcc
Lu	Hf	Ta	W	Re	Os	Ir	Pt	Au
1.8	5.3	7.7	20	5.6	12	21	9.4	45
hcp	hcp	bcc	bcc	hcp	hcp	fcc	fcc	fcc

* The abbreviations hcp, bcc, and fcc refer to hexagonal closest-packed, body-centered cubic, and face-centered cubic (cubic closest-packed) structures.

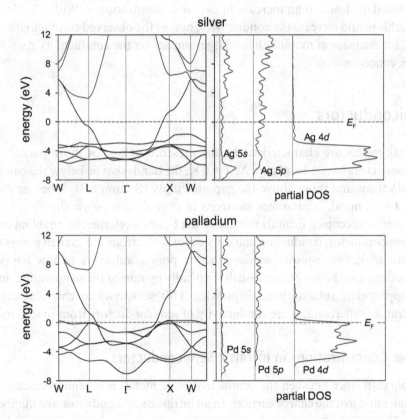

Figure 10.10 The band structure of Ag (top) and Pd (bottom). Fermi energy E_F is set to 0 eV. The partial DOS illustrates the contributions of the 5s, 5p, and 4d orbitals (right).

the partial DOS plots on the right-hand side of Figure 10.10. The $4d$ bands are narrower because the $4d$ orbitals are more contracted than the $5s$ and $5p$ orbitals (Section 5.2.2), limiting overlap with orbitals on neighboring atoms. Hence, we get a band structure that contains wide bands with $5s/5p$ character together with a set of narrow bands of $4d$ parentage.

How do we rationalize the high conductivity of Ag in terms of its band structure? Although palladium and silver have similar band structures, the Fermi level cuts through the bands at different points. In silver, the $4d$ bands are fully occupied, and the Fermi level cuts through a wide sp band, which leads to high-mobility carriers. In palladium, the electron count per atom decreases by one electron and the Fermi level cuts through the upper part of the $4d$ bands. These bands are much narrower, and therefore the electrons that populate these bands have lower mobilities. Consequently, palladium has a six times smaller conductivity than silver, even though they are isostructural and neighbors on the periodic table. In fact, all metals in groups 3–10, where the Fermi level cuts through the d bands, have lower conductivities than Cu, Ag, and Au.

It is worthwhile noting that when E_F cuts through the narrow d bands, the DOS at the Fermi energy is considerably larger than it is in metals where E_F cuts through the much wider sp bands. This leads to an increase in carrier concentration n. With all other things being equal, this would increase the conductivity, but, as the observed conductivities of Pd and Ag show, the decrease in mobility has a larger impact on the conductivity than the increase in carrier concentration.

10.3 Semiconductors

Whereas metals are characterized by the presence of one or more partially filled bands, semiconductors have a band gap. At $T = 0$ K, the bands that lie below the band gap are filled with electrons and those above the gap are empty (Section 6.3). When an external field is applied to a metal, it promotes electrons in orbitals just below the Fermi level to nearly degenerate unoccupied orbitals just above the Fermi level, thereby enabling current to flow. In a semiconductor, conduction can only occur if electrons are excited across the band gap, and this makes high-purity semiconductors poor conductors at low temperatures. The conductivity can, however, increase dramatically by raising the temperature and/or introducing appropriate substitutional impurities. This sensitivity to chemical substitution and temperature differentiates the conductivity of semiconductors from that of metals.

10.3.1 Carrier Concentrations in Intrinsic Semiconductors

One big difference between the conductivities of metals and semiconductors concerns the concentration n of the charge carriers. In an intrinsic semiconductor, the number of electrons thermally excited into the conduction band at any given temperature must be equal to the number of empty states (holes) created in the valence band. If the densities of states in the

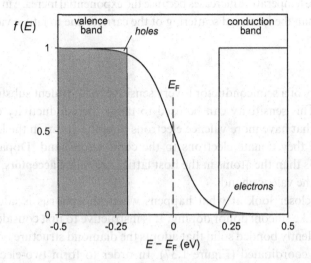

Figure 10.11 The Fermi–Dirac distribution function $f(E)$ near E_F at $T = 300$ K in a semiconductor with a band gap of 0.4 eV. The shading represents occupied states.

valence and conduction bands are equal, this condition requires that the Fermi level be located halfway between the two bands, as shown in Figure 10.11. In practice, the densities of states of the two bands are not exactly equal, and E_F shifts slightly from the halfway point, but the shift is normally quite small.

We see from Figure 10.11 that the concentration of charge carriers is governed by the tails of the Fermi–Dirac distribution. For crystal orbitals that lie within the conduction band, $E - E_F$ is typically much larger than kT, and the distribution in Equation (10.13) simplifies to[13]:

$$f(E) = \frac{1}{1 + \exp[(E - E_F)/kT]} \cong \frac{1}{\exp[(E - E_F)/kT]} = \exp[-(E - E_F)/kT] \qquad (10.20)$$

It can be shown that the carrier concentration n in a semiconductor is proportional to the value of $f(E)$ at the conduction-band edge, where $E - E_F = E_g/2$, thus:

$$n \propto \exp[-E_g/2kT] \qquad (10.21)$$

For silicon whose band gap $E_g = 1.1$ eV, Equation (10.21) gives $n = 1 \times 10^{16}$ m^{-3} at $T = 300$ K, which translates to roughly one charge carrier for every 10^{12} atoms, many orders of magnitude less than in a metal. This difference in carrier concentration helps to explain why high-purity semiconductors have low conductivities (Table 10.1). Equation (10.21) also reveals another important difference between semiconductors and metals. Semiconductors become more

---

[13] Expressed in electronvolts, the Boltzmann constant is $k = 8.617 \times 10^{-5}$ eV/K and $kT = 0.0258$ eV at 300 K.

conducting as the temperature increases because the exponential increase in carrier concentration is much larger than the increased scattering of the carriers due to lattice vibrations.

10.3.2 Doping

The conductivity of a semiconductor is very sensitive to aliovalent substitution (Section 2.4), called **doping**. This sensitivity can be used to tailor the conductivity of a semiconductor. Dopant atoms that have more valence electrons than the atoms in the host lattice are called **donors**, because they donate electrons to the conduction band. Dopants that have fewer valence electrons than the atoms in the host lattice are called **acceptors**, because they accept electrons from the valence band.

Let's take a closer look at what happens when phosphorus is added to an elemental semiconductor like silicon: donor doping. It is instructive to first consider the local bonding. Silicon is a covalently bonded solid that adopts the diamond structure type, where each atom is tetrahedrally coordinated (Figure 1.37). In order to form two-electron bonds with its neighbors, each silicon atom must use all four of its valence electrons. Phosphorus has five valence electrons so that when it substitutes for silicon, there is one extra electron that is not needed to satisfy local bonding requirements. This extra electron can easily be excited into the conduction band where it is free to move through the crystal.

Let's shift from this local picture to the delocalized view represented by a band-structure diagram. In silicon, the valence bands are made up of bonding crystal orbitals and the conduction bands of antibonding crystal orbitals (Section 6.6.3). As the bonding crystal orbitals are fully occupied, the extra electron associated with the phosphorus dopant goes into the conduction band. Because the conduction band is nearly empty, the electron has many nearly degenerate crystal orbitals to choose from and is relatively free to move through the crystal. Given our earlier estimate of one free charge carrier for every 10^{12} atoms in pure Si, we see that even very low levels of doping can dramatically alter the carrier concentration. In theory, even one part in 10^9 of the phosphorus donor would increase the carrier concentration by three orders of magnitude over that of pure silicon![14] Consequently, carrier concentration and conductivity can be controlled by introducing dopants in defined quantities, something that cannot be done with a metal. Such extreme sensitivity to dopants also explains why semiconductor processing takes place in clean rooms where great care is taken to avoid unintentional incorporation of impurities.

Acceptor doping can be illustrated by aluminum substituting for silicon (Kröger–Vink notation Al_{Si}' in Section 2.6). In this case, the dopant atom does not have enough valence electrons to satisfy the local bonding requirements of the matrix. The inability to completely fill the bonding states results in an unfilled crystal orbital at the top of the valence band, which, as we learned earlier, is called a hole. If we apply an electric field, electrons from

---

[14] In real samples, it's quite difficult to reduce the impurity levels to the point where they have no influence on the conductivity. Germanium crystals have been grown with impurity levels as low as 1 part in 10^{10}.

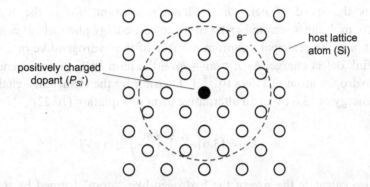

Figure 10.12 Orbit of a conduction-band electron around a (positively charged) donor.

elsewhere in the crystal can move into the hole. It is easier to keep track of the movement of holes (which are few in number) than it is to follow the movement of electrons (which are much more numerous), provided we remember one key distinction—holes are positively charged and move in the opposite direction to electrons when an electric field is applied.[15]

The extra electron that phosphorus donates to the conduction band is not quite as free as the above description would suggest. Phosphorus has one more proton than silicon, which makes the phosphorus site positively charged ($P_{Si}^{\cdot}$). The conduction-band electron is attracted to this positively charged site and will be bound to the impurity much like the electron in a hydrogen atom is bound to the nucleus (see Figure 10.12). In fact, with some modifications, we can estimate the energy that binds the electron to the substitutional impurity just as we would calculate the energy levels in a hydrogen atom.

The differences between the bound states of a substitutional impurity and the energy levels in a hydrogen atom are as follows. Firstly, the charge of the "nucleus" is screened by other electrons and reduced by the relative dielectric permittivity, ε_r, of the host. Secondly, the kinetic energy of the electron is modified because it is moving in the conduction band of the semiconductor rather than in free space. Therefore, we need to use the effective mass m^* in place of the free-electron mass m_e. Finally, the energy zero corresponds not to a vacuum but to the bottom of the conduction band, E_{cb}. Taking these factors into account, it can be shown that the energy required to excite an electron bound to a donor into the conduction band, the **donor ionization energy**, E_d, is equal to:

$$E_d = E_{cb} - E_n = \left(\frac{1}{\varepsilon_r}\right)^2 \left(\frac{m^*}{m_e}\right)\left(\frac{Z^2 hcR_H}{n^2}\right) = \left(\frac{Z}{n\varepsilon_r}\right)^2 \left(\frac{m^*}{m_e}\right) hcR_H \quad \text{(in J)} \qquad (10.22)$$

[15] Consider a mostly filled auditorium with a few empty seats at the front. When the show starts and people are encouraged to move closer to the stage, the people (analogous to electrons) move toward the front while the empty seats (analogous to holes) move toward the back of the auditorium.

where c is the speed of light, h is Planck's constant, R_H is the Rydberg constant $(1.0973\times10^7 \text{ m}^{-1}$, the wavenumber $1/\lambda$ of the lowest-energy photon that can ionize a hydrogen atom), n is the principal quantum number of the hydrogen-like orbital, and Z is the Kröger–Vink defect charge. As $E = h\nu = hc/\lambda$, the term hcR_H gives the energy required to ionize a hydrogen atom, 2.1799×10^{-18} J. Division by the elementary charge, e, gives the Rydberg energy of 13.6 eV in an alternative form of Equation (10.22):

$$E_d = 13.6\left(\frac{Z}{n\varepsilon_r}\right)^2\left(\frac{m^*}{m_e}\right) \quad \text{(in eV)} \tag{10.23}$$

We can also calculate the size of the hydrogen-like "atom" formed by the donor and its electron. In the Bohr model of the hydrogen atom, the radius of an electron in a $1s$ orbital is 52.9 pm (the Bohr radius). In a doped semiconductor, the radius of the donor electron, r_d, is modified by the relative dielectric permittivity and the effective mass of the host:

$$r_d = 52.9\varepsilon_r\left(\frac{m_e}{m^*}\right) \quad \text{(in pm)} \tag{10.24}$$

E_d is the energy required to excite a hydrogenic electron from the ground state (principal quantum number $n = 1$), where it is trapped, to the conduction band ($n = \infty$), where it can move freely through the crystal. It is referred to as the donor ionization energy, as shown in Figure 10.13. Due to the dielectric screening of the semiconductor and the effective-mass correction (in most semiconductors $m^*/m_e < 1$), a substitutional impurity is much easier to ionize than a hydrogen atom. For a group-15 dopant in silicon, the ionization energy is calculated to be 0.03 eV, in reasonable agreement with 0.045 eV measured for phosphorus in silicon. The ionization energies of selected dopants in silicon and germanium are listed in Table 10.4, including acceptors. The ionization energies are in general smaller in germanium, in part because Ge has a higher relative dielectric permittivity ($\varepsilon_r = 16$ versus 12 for Si). We also see that for various dopants in a given host, the ionization energies are quite similar, as

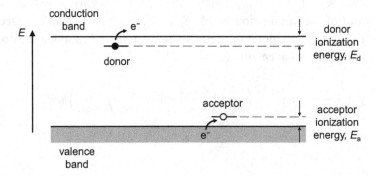

Figure 10.13 Impurity ionization energies with respect to the valence- and conduction-band edges for donor and acceptor doping.

Table 10.4 Ionization energies (in eV) of various donors and acceptors in Si and Ge.

Donors	Si	Ge	Acceptors	Si	Ge
P	0.045	0.013	B	0.045	0.010
As	0.054	0.014	Al	0.057	0.011
Sb	0.042	0.010	Ga	0.065	0.011

we might expect given the fact that the chemical identity of the dopant does not enter into Equation (10.22) (except its charge Z).

Figure 10.13 and Table 10.4 suggest that this treatment of a donor-bound electron trapped beneath the conduction band can be inverted into an entirely analogous case of the acceptor-bound hole trapped above the valence band. The positively charged hole experiences a weak attraction to the negatively charged dopant, forming a hydrogen-like bound state. Figure 10.13 shows how the acceptor states sit just above the valence band so that when an electron jumps into this state, a hole is created in the valence band. The energy required to excite the bound hole into the valence band is called the **acceptor ionization energy**, E_a. We can see from Table 10.4 that E_a and E_d are of similar magnitudes.

10.3.3 Carrier Concentrations and Fermi Energies in Doped Semiconductors

Real semiconductor crystals invariably contain both donors and acceptors. Electrons in the conduction band arise from ionized donor states, whereas holes in the valence band arise from ionized acceptor states. In addition, conducting electrons and holes are created by thermal excitation across the band gap. To further complicate matters, electrons can fall from the donor levels into empty acceptor levels. Deriving expressions for the concentration of charge carriers in a completely general manner is therefore quite complicated. Fortunately, the situation is much simpler in a few important limiting cases of temperature and doping.

Let the total volume concentration of electrons in the semiconductor conduction band be n (just as it is for metals) and the concentration of holes in the valence band p (both in units of m^{-3}). In the absence of doping, we can use Equation (10.21) to estimate n, and because a hole is created for each electron excited into the conduction band, $p = n$.

We will then use the symbols N_d and N_a to denote the concentrations of the respective donor and acceptor impurities (in m^{-3}). When the concentrations of the two types of dopants are not equal, the behavior of the semiconductor is determined largely by the majority dopant. Semiconductors containing more donors than acceptors ($N_d > N_a$) are said to be **n-type semiconductors**. Semiconductors where the acceptor dopants are in the majority ($N_a > N_d$) are said to be **p-type semiconductors**.

The carrier concentration and Fermi level in doped semiconductors show temperature dependences that differ from intrinsic semiconductors. Consider the behavior of an n-type semiconductor. At very low temperatures, there is not enough thermal energy to ionize the donor states and the carrier concentration in the conduction band is very low. As the donor states are almost fully occupied, the Fermi level must lie above the ionization energy E_d of these donor states.[16] This temperature range is called the **freeze-out regime**. In this regime, we can estimate the carrier concentration n by substituting E_d for the band gap in Equation (10.21):

$$n \propto \exp[-E_d/2kT] \tag{10.25}$$

When the temperature is increased above the freeze-out regime, the donor states begin to ionize. As a result, n increases and E_F moves downward. At some higher temperature, nearly all donor states are ionized and carrier concentration is effectively equal to the dopant concentration, $n = N_d$. This is called the **saturation regime** because the carrier concentration remains essentially constant as the temperature continues to climb.

As the temperature continues to increase, the concentration of the intrinsic carriers that are generated via the valence- to conduction-band excitation increases exponentially. At low temperatures, their concentration is much smaller than the concentration of carriers that come from doping, and we can effectively ignore the intrinsic-carrier contribution (at least for the majority carriers, electrons in this case). However, with increasing temperature, the concentration of intrinsic carriers will eventually approach and exceed the now saturated concentration of carriers coming from dopants. At high enough temperatures, the concentration of conduction-band electrons generated by thermal excitation becomes dominant and the carriers coming from doping can be neglected. This is called the **intrinsic regime** because the properties of the semiconductor are essentially the same as an undoped semiconductor of the same material. E_F is then located at the midpoint of the band gap, and $n \approx p$, both of which are governed by Equation (10.21).

The temperature dependence of the carrier concentration for an n-type semiconductor is plotted as a function of temperature in Figure 10.14. The carrier concentration in a p-type semiconductor evolves in a similar manner.

The crossover from the freeze-out regime to the saturation regime depends upon the donor ionization energy. For typical dopants, this occurs well below room temperature. The crossover from the saturation regime to the intrinsic regime depends upon the dopant concentration and the band gap. Apart from very small band-gap semiconductors and/or very low dopant concentrations, the intrinsic regime is only reached at temperatures exceeding room temperature. For devices, it is preferable to operate in the saturation regime where the carrier concentration is not strongly temperature dependent. For most doped semiconductors, this is the situation at room temperature, which means the Fermi level is not too far from the ionization energy of the dominant dopant. Thus, E_F is strongly shifted toward the

---

[16] Remember that $f(E) = 0.5$ when $E = E_F$, so that E_F must lie above the donor ionization energy until more than 50% of the donor states are ionized.

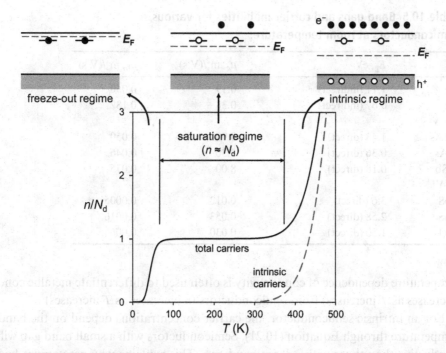

Figure 10.14 Temperature variation of the total carrier concentration per donor, n/N_d, for an n-type semiconductor. In the upper schematic, the filled circles and shaded regions represent electrons and the open circles, holes.

conduction band in an n-type semiconductor, and toward the valence band in a p-type semiconductor.

10.3.4 Conductivity

The electrical conductivity of a semiconductor is governed by the carrier concentration and mobility, just as it is for a metal. Unlike a metal, we now have two types of carriers, electrons and holes, whose currents are additive. We can therefore modify Equation (10.10) to a general expression for the conductivity of a semiconductor:

$$\sigma = \sigma_n + \sigma_p = n\mu_e e + p\mu_h e \tag{10.26}$$

where μ_e and μ_h are the mobilities of the electron and the hole, respectively. Like a metal, the scattering from lattice vibrations increases with temperature (Section 10.2.7), which tends to decrease the mobility and hence the conductivity, but, unlike a metal, the carrier concentration increases exponentially with temperature. The latter term dominates, and, as a result, the temperature dependence of the conductivity shows approximately the same temperature variation as the carrier concentration (Figure 10.14). When new materials are prepared, the

Table 10.5 Band gaps and carrier mobilities for various semiconductors at room temperature.

	E_g, eV	μ_e, m²/(V s)	μ_h, m²/(V s)
Si	1.11 (indirect)	0.19	0.050
Ge	0.67 (indirect)	0.38	0.182
III–V			
GaAs	1.43 (direct)	0.90	0.050
InAs	0.36 (direct)	3.30	0.046
InSb	0.18 (direct)	8.00	0.075
II–VI			
ZnS	3.6 (direct)	0.012	0.0005
ZnSe	2.58 (direct)	0.053	0.0016
CdTe	1.50 (direct)	0.030	0.0065

temperature dependence of conductivity is often used to differentiate metallic conductors (σ decreases as T increases) from semiconductors (σ increases as T increases).

For an intrinsic semiconductor, the carrier concentrations depend on the band gap and temperature through Equation (10.21). Semiconductors with a small band gap will be more conductive than those with a large band gap. This explains why germanium has a higher intrinsic conductivity than silicon (see Table 10.1).[17] The carrier mobilities depend on effective masses (Section 10.2.5) which in turn depend on the curvature of the appropriate bands in the band-structure diagram, Equation (10.15). Because electrons and holes reside in different bands, their mobilities are different, Table 10.5. The observation that electron mobilities are higher than hole mobilities is general to nearly all semiconductors. The largest electron mobilities are found for III–V semiconductors.

10.3.5 p–n Junctions

The p–n junction is an integral part of many electronic devices, including transistors, light-emitting diodes (LEDs), and solar cells. A **p–n junction** is formed when a p-type semiconductor and an n-type semiconductor are brought into intimate contact. Typically, the host semiconductor is the same on either side of the junction, and the junction is formed by an abrupt transition from acceptor to donor doping. In practice, rather than fusing together two separate crystals, dopants of one type are diffused into a layer of a semiconductor that was already doped in the opposite sense. For example, acceptors can be diffused into an n-type semiconductor crystal, or conversely donors can be diffused into a p-type crystal.

---

[17] We need to remember that doping, which is always present to some extent, also plays an important role in determining the carrier concentration.

Setting aside the practicalities of manufacturing a p–n junction, let's consider what would happen if we literally brought crystals of the same host semiconductor, one doped p-type and the other n-type, into intimate contact and let the system come to equilibrium. Initially, the Fermi level in the n-type material is higher than it is on the p-type side (Figure 10.15). This imbalance causes the electrons on the n-type side to flow over to the p-type side and holes to flow in the opposite direction. The electrons and holes recombine, annihilating each other over a fairly narrow region on either side of the junction. When the electrons flow out of the n-type side, they leave behind ionized donors that carry a positive charge (e.g. $P_{Si}^{\cdot}$). In a similar manner, the holes leave behind ionized acceptors that carry a negative charge (e.g. Al_{Si}'). As a result, the junction develops an electric field that opposes further diffusion of electrons from the n-type side and holes from the p-type side into the junction. This process is illustrated in Figure 10.15.

When two solids are in equilibrium with each other, their respective Fermi levels must be at the same energy. For this to happen, while maintaining the p- and n-type character of the two sides far away from the junction, the valence- and conduction-band edges must bend across the junction, as shown on the lower left-hand side of Figure 10.15. What is the physical origin of band bending in a p–n junction? It is a direct consequence of the uncompensated charges of the ionized impurities. As shown on the lower right-hand side of Figure 10.15, a positive charge develops on the n-type side of the junction, and a corresponding negative charge

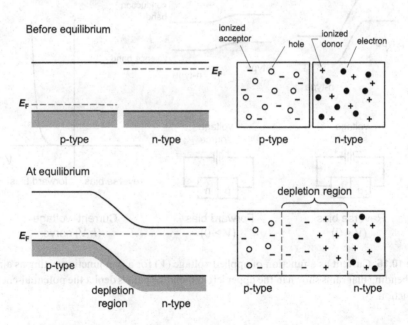

Figure 10.15 A schematic representation of the development of a p–n junction before (upper) and after (lower) equilibrium is established. The energy-level diagram (left) and charge distribution (right) across the junction are shown.

develops on the p-type side. Because electrons are attracted to a positive charge and repelled by a negative charge, the band edges on the n-type side of the junction are lowered and those on the p-type side are raised. The drop in potential energy that occurs upon crossing a p–n junction made from silicon is on the order of 0.5–0.7 eV. Dividing this potential-energy difference in electronvolts (eV) by 1 (electron) we obtain the **built-in potential**, V_{bi}, in volts.[18]

At the center of the junction, $n \approx p$, and the carrier concentration is characteristic of an intrinsic semiconductor. As discussed in Section 10.3.1, carrier concentrations in intrinsic semiconductors are very low unless either E_g is small or the temperature is large. Therefore, the transition region is almost completely depleted of mobile carriers, and for this reason is called the **depletion region**. The depletion region acts as an insulating barrier between n-type and p-type sides of the junction.

The most basic property of a p–n junction is **rectification**, which refers to the phenomenon whereby current can pass much more easily in one direction than in the other. Energy-level diagrams for a p–n junction in both reverse and forward bias are shown on the left-hand side of Figure 10.16. In these diagrams, it is thermodynamically favorable for electrons and holes to occupy the states where they have the lowest potential energy. Electrons in the mostly empty conduction band will therefore flow downhill, while holes inside the valence band move uphill. In the absence of an external electric field, the system is at equilibrium and there

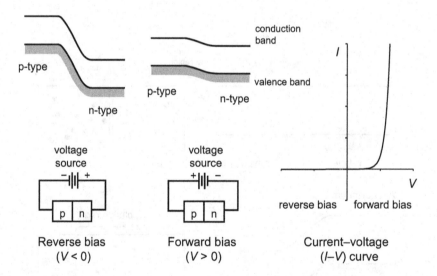

Figure 10.16 Current as a function of applied voltage (V) for a p–n junction acting as a rectifier. The band-bending diagrams shown in the upper left and middle panels depict the potential-energy drop across the junction.

---

[18] For definition of electronvolt, see Footnote 12 in Chapter 5.

is no net flow of current. When the equilibrated junction is connected with its p-side to the negative pole and n-side to the positive pole of an external voltage source, as shown in the left-hand panel of Figure 10.16, the band bending at the junction increases. Since the applied voltage results in practically zero current flow, the junction is said to be in **reverse bias**. If the polarity of the voltage source is reversed, as shown in the middle panel of Figure 10.16, the band bending at the junction is reduced, and current begins to flow across the junction, which is now said to be in **forward bias**. If you put the junction into a circuit of alternating current, only one polarity passes through the junction.[19]

Let's consider the effects of applied biases at the microscopic level. In reverse bias, the positive applied voltage pulls electrons on the n-type side away from the junction, and similarly the negative pole attracts holes on the p-type side. This leads to more ionized donors on both sides of the junction, increasing the width of the depletion region, as well as the magnitude of the potential-energy drop across the junction. As a result, the flow of current across the junction is very small. A forward bias has the opposite effect; once the applied voltage exceeds V_{bi}, the band bending is fully suppressed and the current increases sharply, leading to the asymmetric I–V curve shown in Figure 10.16. This is a feature that is characteristic of a rectifying junction.

10.3.6 Light-Emitting Diodes and Photovoltaic Cells

When a forward bias is applied to a p–n junction, electrons from the n-type side and holes from the p-type side are forced into the depletion region, and this leads to a high rate of recombination. Energy equal to the band gap is released, either radiatively or non-radiatively, as the electrons fall into the holes. If the semiconductor is a direct-gap semiconductor, **radiative-recombination** events are often favored, leading to emission of photons with

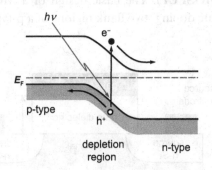

Figure 10.17 Light absorption and carrier separation in a photovoltaic cell.

[19] A rectifier is a device that converts alternating current to direct current.

energies determined by the band gap. This is the principle behind light-emitting diodes (LEDs) that were discussed in Section 7.9. **Non-radiative-recombination** events, where the recombination energy is dissipated by lattice vibrations, reduce the efficiency of LEDs. This is particularly problematic for semiconductors with indirect band gaps.

Photovoltaic cells operate much like LEDs, but the timing of events is opposite, as shown in Figure 10.17. Photons with energies greater than the band gap are absorbed by the semiconductor creating an electron–hole pair. Normally the electron and hole would recombine, but, if the photon is absorbed in the depletion region, band bending drives the electron downhill towards the n-type side and the hole uphill toward the p-type side. The separated charge carriers can then be run through an external circuit to do work.

The voltage that can be obtained from a photovoltaic cell is limited by the extent of band bending at the p–n junction, the upper limit being the band gap of the semiconductor. The current that can be obtained is limited by the number of photons absorbed in the depletion region. This creates a trade-off in terms of the semiconductor band gap: increasing the band gap increases the output voltage yet lowers the current, because photons with energies smaller than E_g are not absorbed. Decreasing the band gap has the opposite effect. For generating energy from sunlight incident upon the Earth's surface, the optimum band gap is ~1.4–1.5 eV.

10.3.7 Transistors

The transistor is a key component in many electronic devices, and its discovery ranks as one of the most important inventions of the twentieth century. The first three-contact transistor was built at Bell Labs in 1947. The invention sparked a huge research effort that eventually led to integrated circuits and computer chips. There are different types of transistors but the most common transistor in today's integrated circuits is the **metal-oxide-semiconductor field-effect transistor (MOSFET)**. The basic design of a MOSFET is shown in Figure 10.18. Through appropriate doping, two flank regions of a p-type semiconductor are made n-type.

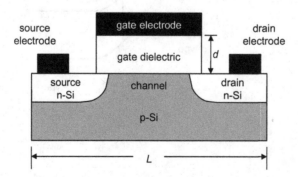

Figure 10.18 A schematic representation of an n-channel MOSFET on a Si wafer.

These two n-type regions are called the source and the drain, respectively, while the p-type region that separates them is called the channel. An insulating-oxide dielectric is deposited on top of the channel. This oxide is called the gate. Finally, metallic contacts are made to the source, drain, and gate. The architecture described here corresponds to an n-channel MOSFET.[20] If the n- and p-type regions are reversed, a p-channel MOSFET is formed.

In the n-channel MOSFET, the source is connected to the negative voltage and the drain to the positive voltage. In the absence of voltage on the gate, the consecutive n–p and p–n junctions at the source–channel and drain–channel interfaces do not allow any current through the transistor, even in the presence of a potential difference between the source and drain. When a positive voltage is applied to the gate, the holes in the p-type channel are repelled, and minority carrier electrons are attracted to the gate. If the gate voltage exceeds a threshold value, this effect is large enough to create a thin inversion layer (n-type > p-type) in the channel. Because the inversion layer acts as an n-type semiconductor, there is now a path for current to flow between the source and the drain. In this way, a MOSFET can act as a switch: "on" when the gate voltage exceeds the threshold value and "off" when it does not. The switching speed of a MOSFET depends in part on the carrier mobility. For this reason, n-channel MOSFETs, where the current is carried by electrons, are generally preferred to p-channel MOSFETs, where the current is carried by lower-mobility holes (see Table 10.5).

Until relatively recently, the materials used in a MOSFET were not very diverse. The source, drain, and channel were made from silicon; the gate dielectric was SiO_2 formed by oxidizing the silicon; and the metallic contact to the gate was heavily doped polycrystalline silicon. The reliance on silicon is based on the need to simplify the fabrication of millions of transistors on a single chip. In recent years, things have become more interesting as the drive to put more and more transistors on a chip has led to a dramatic reduction of the channel length, L.

The gate voltage needed to create an inversion layer depends upon the capacitance of the gate dielectric, which in turn depends upon the dimensions and the dielectric permittivity of the gate dielectric. Recollect from Section 8.1.1 that the capacitance of a parallel-plate capacitor $C = \varepsilon_r\varepsilon_0 A/d$. As the size of the MOSFET decreases, the area A of the gate dielectric decreases and the capacitance goes down. If no other changes are made, a higher voltage is required to create the inversion layer, but this is undesirable because it increases power consumption and generates more heat. For decades, engineers have been compensating for the shrinking gate area by steadily decreasing the thickness d of the SiO_2 gate dielectric. By the time the channel length reached 65 nm in 2006, the SiO_2 layer was only 1.2 nm thick! Further reduction in this thickness was not feasible because leakage, including quantum-mechanical tunneling, across the gate dielectric becomes problematic.

---

[20] When the gate voltage is sufficiently large to turn on the transistor the channel becomes n-type and current flows across the channel, hence the name n-channel MOSFET.

The solution to this engineering dilemma is straightforward in principle: compensate for the decrease in A by increasing ε_r. Since the relative permittivity of amorphous SiO_2 is only 3.9, finding a higher ε_r alternative does not sound too daunting. However, to be practical, the replacement material must meet several criteria. It must be highly insulating, not form compounds with silicon at the gate–channel interface, and have bands that are sufficiently offset from silicon to avoid unwanted leakage current. The material that emerged from an extensive search was amorphous HfO_2, whose permittivity is approximately 30. The seven-fold increase in ε_r allowed engineers to shrink the channel length to 32 nm by 2010, while increasing the thickness of the gate dielectric. At the same time, it was necessary to change the gate-electrode material from heavily doped Si to more conducting materials like TiN. To further increase the density of transistors, manufacturers have since moved away from the planar geometry of a MOSFET to a geometry called a FinFET, where the gate wraps around three sides of the channel. Further efforts to keep pace with Moore's law[21] are likely to involve incorporation of even more exotic materials and unconventional architectures.

10.4 Transition-Metal Compounds

Compounds containing transition-metal ions with partially filled d orbitals are a fascinating and complex class of materials. Their electrical and magnetic properties vary widely and can change dramatically in response to subtle changes in temperature, pressure, and/or chemical substitution. To a first approximation, we would expect most transition-metal compounds to be metallic because the bands originating from d orbitals are partially filled. A glance at the properties of transition-metal compounds quickly reveals behavior that cannot be explained within the conventional band-structure approach. For example, some metal oxides with partially filled d orbitals are metallic (e.g. TiO and $SrFeO_3$), while others are semiconducting (e.g. NiO and $LaFeO_3$). Compounds like VO_2 and $LaTiO_3$ show both behaviors, semiconducting at low temperature and metallic at high temperature. In the sections that follow we explore the reasons behind this behavior.

10.4.1 Electron Repulsion: The Hubbard Model

The fundamental reason transition-metal compounds show complex electronic behavior can be traced back to strong repulsive interactions between electrons. In the band model we have used up to this point, the effects of electron–electron repulsion have been assumed to be uniform throughout the crystal. This approximation tends to work well for metals and semiconductors where the bands are wide and the electron wavefunctions highly delocalized, but there are many instances where this simplifying approximation breaks down. Hund's

---

[21] Moore's law is an empirical prediction that the density of transistors on an integrated circuit will double approximately every two years.

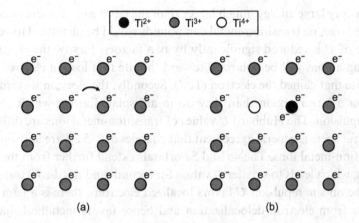

Figure 10.19 (a) A square array of Ti^{3+} ions. (b) The same array after an electron moves from one site to another, initiating conduction.

rules (Section 7.3.2) owe their existence to electron–electron repulsion, and high-spin transition-metal ions (Section 5.3.9) would not exist were it not for electron–electron repulsions. Compounds where the physical properties cannot be understood without taking electron–electron interactions into account are called **strongly correlated materials**. These materials play a prominent role in many functional materials, particularly those discussed in chapters 11 and 12.

Accounting for the effects of electron repulsion in solids is a challenging problem. One approach, the **Hubbard model**, simplifies the problem by assuming that the only electron–electron repulsions of importance are those between valence electrons on the same atom. To understand the Hubbard model, consider the 2D array of Ti^{3+} ions shown in Figure 10.19a. As Ti^{3+} is a d^1 ion, the above-defined repulsions are absent. The situation changes once we allow the electrons to move, as they must for the material to be conducting. Once an electron moves from one titanium atom to another, Figure 10.19b, we create a Ti site containing two electrons and must pay an energy penalty, U, equal to the on-site repulsion between the two electrons.

What is the magnitude of this "Hubbard" U? When the electron moves from one site to another, the net change in the system can be expressed as a disproportionation reaction: $2Ti^{3+} \rightarrow Ti^{4+} + Ti^{2+}$. For titanium ions in the gas phase, we can precisely determine this quantity from the ionization energy, IE, and the electron-gain enthalpy, H_{eg}, of a Ti^{3+} ion[22]:

$$Ti^{3+}(g) \rightarrow Ti^{4+}(g) + e^- \qquad IE \text{ of } Ti^{3+} = 4th \ IE \text{ of } Ti = 43.3 \text{ eV}$$
$$Ti^{3+}(g) + e^- \rightarrow Ti^{2+}(g) \qquad H_{eg} \text{ of } Ti^{3+} = -(3^{rd} \ IE \text{ of } Ti) = -29.3 \text{ eV}$$
$$2Ti^{3+}(g) \rightarrow Ti^{4+}(g) + Ti^{2+}(g) \qquad U = IE + H_{eg} = 14.0 \text{ eV}$$

[22] Remember from Chapter 5 that electron-gain enthalpy H_{eg} has the same magnitude and opposite sign as electron affinity, EA.

This is a very large energy, much larger than the band gap of a semiconductor. If U were always this large, no transition-metal compounds would be metallic. However, in solids, the magnitude of U is reduced significantly by two factors. Firstly, the electron clouds of the surrounding anions will be polarized toward the site that lost an electron (Ti^{4+}) and away from the site that gained the electron (Ti^{2+}). Secondly, the electron wavefunctions in a solid extend over a larger region than they do in an isolated atom, which reduces the on-site electron repulsion. The Hubbard U values of transition-metal ions are difficult to determine precisely, but there is general agreement that energies of 2–5 eV are appropriate for first-row ($3d$) transition-metal ions. The $4d$ and $5d$ orbitals extend further from the nucleus than the $3d$ orbitals, which leads to smaller U values for second- and third-row transition-metal ions.

While the on-site repulsion U favors localized electrons, there is a gain in kinetic energy that results from electron delocalization and hence favors metallic behavior. In strongly correlated materials, the competition between these two effects determines whether the material will be insulating or metallic. For transition-metal compounds, the competing forces tend to be closely balanced, so that either metallic or insulating behavior can be realized depending on several factors.[23]

To understand the behavior of transition-metal compounds, we incorporate the Hubbard model into the band model that we have been using up to this point. To illustrate this integration, let's return to our 2D array of Ti^{3+} ions. To make things a little more realistic, we add nitride ions so that each titanium is surrounded by a square of nitride ions and vice versa (see Figure 10.20). Unlike our earlier array of Ti^{3+} ions, this structure is charge-balanced.

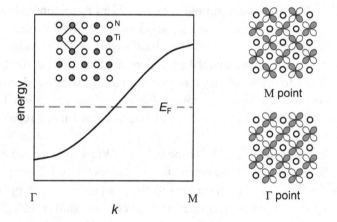

Figure 10.20 The band structure for the Ti $3d_{xy}$ band in the 2D array of TiN discussed in the text, including the crystal orbitals at Γ and M.

[23] For lanthanoid compounds where the electrons reside in highly contracted $4f$ orbitals, the Hubbard U values are large, 5–12 eV, and localized-electron behavior dominates.

The resulting array is not a completely hypothetical construct, it is identical to a (001) layer in the NaCl-type structure of TiN.[24]

To keep things simple, we will only consider the Ti $3d_{xy}$ band. Firstly, we derive the band structure in the absence of electron-repulsion effects. There is one titanium atom per unit cell, so the band structure consists of a single band. One that runs uphill from Γ ($k_x = k_y = 0$), where the Ti–Ti overlap is bonding, to M ($k_x = k_y = \pi/a$), where the Ti–Ti interactions are antibonding (Figure 10.20). At both Γ and M, the N $2p$ orbitals cannot contribute to the crystal orbital for reasons of symmetry (see Section 6.5.2). Each titanium ion has a single d electron, so the d_{xy} band is half filled. The resulting band structure, where the Fermi level cuts through the band at the halfway point, is reminiscent of the band structure of the 1D hydrogen-atom chain encountered in Chapter 6 (Figure 6.5). From this picture, we would expect metallic conductivity.

What happens when we introduce on-site electron repulsion? The Hubbard U splits the single two-electron band into two one-electron bands, which we will refer to as the lower Hubbard band and the upper Hubbard band. The lower band reflects the energy when each titanium ion holds a single electron. The upper band represents the higher energy that would be experienced by an electron that moves to a site where there is already an electron present.

If the titanium atoms are sufficiently far apart, their interaction is negligible. In this scenario, each band will be flat and the energy separation between bands will be equal to the on-site repulsion U of two electrons (the Hubbard U). If we shrink the lattice parameter, the Ti $3d_{xy}$ orbitals begin to overlap as the titanium atoms approach each other and the bands develop a width, W. In the absence of mixing with other bands, the lower and upper Hubbard bands will have the same width. The bandwidth will increase as the Ti–Ti distance decreases, similar to what was demonstrated for the 1D hydrogen chain in Figure 6.5. When the Hubbard U is larger than the bandwidth W, as shown on the left-hand side of Figure 10.21, the lower Hubbard band is filled and separated from the empty upper Hubbard band (remember the Hubbard bands can only hold one electron each). In this scenario, on-site repulsion prevails and leads to localized electrons and semiconducting behavior. Compounds that are semiconducting (insulating) because the on-site electron repulsion is larger than the bandwidth ($U > W$) are called **Mott–Hubbard insulators**. When the bandwidth W exceeds the Hubbard U ($W > U$), as shown on the right-hand side of Figure 10.21, the two bands overlap, and the Fermi level cuts through both bands. In this case, there is no gap between the bands, and metallic behavior is observed.

10.4.2 Transition-Metal Compounds with the NaCl-Type Structure

To see the effects of electron–electron repulsion in real compounds, perhaps the best place to start is with binary oxides and nitrides that crystallize with the NaCl-type structure. In this structure, the coordination octahedra share edges in all three dimensions, as illustrated on

---

[24] The Ti–Ti distance in TiN (3.00 Å) is very similar to the Ti–Ti distance in titanium metal (2.95 Å), so this structure is a reasonable approximation to a hypothetical square array of d^1 ions.

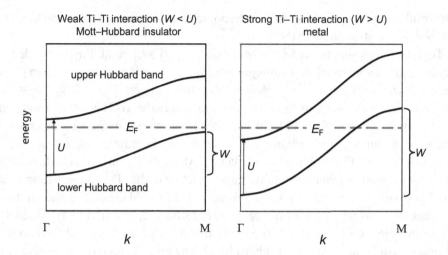

Figure 10.21 The Ti $3d_{xy}$ band in our 2D model of TiN for the respective weak (left) and strong (right) Ti–Ti interactions.

the left-hand side of Figure 10.22 (a 3D analog of the 2D model of TiN discussed in the previous section). Both metallic and insulating behavior can be found in this family, which makes it a good place to study the competing forces at work in transition-metal compounds.

The transition-metal ions are octahedrally coordinated by the anions splitting the d orbitals into t_{2g} (d_{xy}, d_{yz}, d_{xz}) and e_g ($d_{x^2-y^2}$, d_{z^2}) sets (Section 5.3.8). The t_{2g} and e_g orbitals form relatively narrow bands that lie between the filled anion $2p$ bands and the empty cation $4s$ and $4p$ bands, as shown on the right-hand side of Figure 10.22. Because the cations are in low oxidation states (+2 or +3), the metal–anion interactions are fairly ionic, which in turn leads to relatively narrow e_g bands.[25] The bands arising from t_{2g} orbitals are somewhat wider because the d orbitals can overlap directly across the shared edges, as discussed in the previous section and illustrated in Figure 10.20. The widths of the t_{2g} and e_g bands in TiO are estimated to be approximately 6 eV and 4 eV, respectively [2, 3]. The widths of the e_g bands are sufficiently small that $W < U$, and the electrons that go into these orbitals are localized. For the t_{2g} bands, no such broad generalization can be made.

The properties of transition-metal monoxides with the NaCl-type structure are summarized in Table 10.6. TiO is a metal and Pauli paramagnet (of delocalized electrons, Section 9.6.2). VO lies close to the crossover from metallic to semiconducting [3]. The conductivity is sensitive to defects (both vanadium and oxygen vacancies are normally present), and the magnetic susceptibility has a temperature dependence that is intermediate between a Pauli paramagnet and a Curie–Weiss paramagnet (of localized unpaired electrons, Section 9.6.1).

---

[25] The maximum width of a d orbital band is limited by the difference in energy between nonbonding and antibonding crystal orbitals. Therefore, when the metal–anion distances are long and the orbital interactions are weak, as is often the case for larger ions in low oxidation states, we expect narrow bands.

Table 10.6 Structural, electrical, and magnetic properties of transition-metal monoxides with NaCl-type structure.

Compound	M–M distance (Å)	d-orbital r_{max} (Å)*	Electrical properties	Magnetic properties[†]
TiO (d^2)	2.94	0.53	Metallic	Pauli PM
VO (d^3)	2.89	0.48	Semimetallic	PM
MnO (d^5)	3.14	0.41	Semiconductor	AFM, T_N =122 K
FeO (d^6)	3.03	0.38	Semiconductor	AFM, T_N =198 K
CoO (d^7)	3.01	0.36	Semiconductor	AFM, T_N =293 K
NiO (d^8)	2.95	0.34	Semiconductor	AFM, T_N =523 K

*r_{max} is the distance from the nucleus where the radial distribution function of the $3d$ orbitals reaches a maximum value (Section 5.2.2). [†]PM = paramagnetic, AFM = antiferromagnetic.

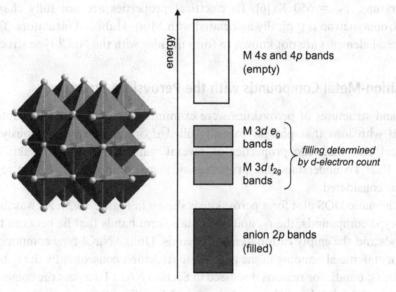

Figure 10.22 The NaCl-type structure (left), along with a schematic DOS diagram for first-row transition-metal nitrides and oxides with this structure (right).

The properties of the remaining compounds, MnO through NiO, are unambiguously characteristic of localized electrons, implying that $W < U$.

The crossover from delocalized to localized electronic behavior is driven by contraction of the d orbitals as we move from early to late transition metals. The calculated values of r_{max} for the $3d$ orbitals in Table 10.6 show that this contraction is significant. The width of the t_{2g} bands decreases because the orbital contraction is not paralleled by a decrease in the metal–metal distance,[26] hence the orbital overlap is diminished and the t_{2g} bands become

---

[26] The expansion of the unit cell and hence the M–M distance on going from VO to MnO results from filling the e_g orbitals, which are σ antibonding. These orbitals are empty in TiO and VO and doubly occupied in MnO, CoO, and NiO.

increasingly narrow. The increase in the Néel temperature, T_N, can be attributed to the increasing electronegativity of the transition-metal ion, which leads to an increase in covalency of the metal–oxygen bonds and hence stronger superexchange interactions that propagate through the M–O–M σ bonds (Section 9.8).

Many early transition-metal nitrides (TiN, ZrN, HfN, NbN, and TaN) are metallic and Pauli paramagnetic, strong evidence that $W > U$. VN is also metallic, but its low-temperature magnetic susceptibility shows a more pronounced temperature dependence than expected for a Pauli paramagnet (Section 9.6.2). This is the first sign that electron repulsions are starting to impact the properties. This becomes more pronounced with CrN, which orders antiferromagnetically with a T_N of 280 K, an indication of significant electron–electron interactions. It is generally accepted that $W \approx U$ for CrN [4, 5]. MnN crystallizes with a tetragonally distorted NaCl-type structure and orders antiferromagnetically at quite high temperature, $T_N = 650$ K [6]. Its electrical properties are not fully characterized, but antiferromagnetism is typically associated with Mott–Hubbard insulators. The later transition-metal elements are not known to form nitrides with the NaCl-type structure.

10.4.3 Transition-Metal Compounds with the Perovskite Structure

The band structures of perovskites were examined in some detail in Chapter 6, but compounds with ions that possess partially filled d orbitals were consciously avoided. The electrical and magnetic properties of several transition-metal perovskites are listed in Table 10.7. To understand the properties of these compounds, electron repulsion effects must be considered.

A schematic DOS plot for a perovskite is shown in Figure 10.23. As was the case with the NaCl-type compounds, the t_{2g} and e_g orbitals form bands that lie between the filled anion $2p$ bands and the empty cation $4s$ and $4p$ bands. Unlike NaCl-type compounds, there is no direct metal–metal bonding in the perovskite structure, consequently the e_g bands are wider than the t_{2g} bands, for reasons discussed in Section 6.6.5. To assess the competition between localized- and delocalized-electron behavior, the width of these bands is a critical parameter. Recall that both sets of bands become wider when the transition-metal oxidation state increases, or when a $3d$ transition-metal ion is replaced by a $4d$ or $5d$ transition-metal ion from the same group. These considerations are important for making sense of the trends in Table 10.7.

The perovskites listed in Table 10.7 can be sorted by their electrical properties as either semiconducting or metallic. The semiconductors can further be sorted into those that are diamagnetic ($LaScO_3$, $SrTiO_3$, $KNbO_3$, $LaCoO_3$, and $LaRhO_3$) and those that order antiferromagnetically. Those that are diamagnetic are sometimes called band insulators/semiconductors because electron-repulsion effects are not needed to explain the presence of a band gap: $LaScO_3$, $SrTiO_3$, and $KNbO_3$ are semiconducting because the d-orbital bands are empty, just like the perovskites discussed in Section 6.6.5. $LaCoO_3$ and $LaRhO_3$ are semiconducting because the transition-metal ions have a low-spin d^6 configuration, resulting

Table 10.7 The electrical and magnetic properties of selected perovskites.

Compound	Electron configuration		Electrical properties	Magnetic properties
LaM^{3+}O$_3$, M = 3d				
LaScO$_3$	d^0	$t_{2g}^0 e_g^0$	Semiconducting	Diamagnetic
LaTiO$_3$	d^1	$t_{2g}^1 e_g^0$	Semiconducting	AFM, T_N = 138 K
LaVO$_3$	d^2	$t_{2g}^2 e_g^0$	Semiconducting	AFM, T_N = 142 K
LaCrO$_3$	d^3	$t_{2g}^3 e_g^0$	Semiconducting	AFM, T_N = 290 K
LaMnO$_3$	d^4	$t_{2g}^3 e_g^1$	Semiconducting	AFM, T_N = 100 K
LaFeO$_3$	d^5	$t_{2g}^3 e_g^2$	Semiconducting	AFM, T_N = 750 K
LaCoO$_3$	d^6	$t_{2g}^6 e_g^0$	Semiconducting*	Diamagnetic*
LaNiO$_3$	d^7	$t_{2g}^6 e_g^1$	Metallic	Pauli PM
SrM^{4+}O$_3$, M = 3d				
SrTiO$_3$	d^0	$t_{2g}^0 e_g^0$	Semiconducting	Diamagnetic
SrVO$_3$	d^1	$t_{2g}^1 e_g^0$	Metallic	Pauli PM
SrCrO$_3$	d^2	$t_{2g}^2 e_g^0$	Metallic	Pauli PM
SrMnO$_3$	d^3	$t_{2g}^3 e_g^0$	Semiconducting	AFM, T_N = 235 K
SrFeO$_3$	d^4	$t_{2g}^3 e_g^1$	Metallic	FM, T_C = 130 K
SrCoO$_3$	d^5	$t_{2g}^4 e_g^1$	Metallic	FM, T_C = 280 K
AMO$_3$, M = 4d				
KNbO$_3$	d^0	$t_{2g}^0 e_g^0$	Semiconducting	Diamagnetic
BaNbO$_3$	d^1	$t_{2g}^1 e_g^0$	Metallic	Pauli PM
SrMoO$_3$	d^2	$t_{2g}^2 e_g^0$	Metallic	Pauli PM
SrRuO$_3$	d^4	$t_{2g}^4 e_g^0$	Metallic	FM, T_C = 165 K
LaRuO$_3$	d^5	$t_{2g}^5 e_g^0$	Metallic	Pauli PM
SrRhO$_3$	d^5	$t_{2g}^5 e_g^0$	Metallic	PM†
LaRhO$_3$	d^6	$t_{2g}^6 e_g^0$	Semiconducting	Diamagnetic
LaPdO$_3$	d^7	$t_{2g}^6 e_g^1$	Metallic	Pauli PM

*Below 100 K, LaCoO$_3$ is a diamagnetic semiconductor with a low-spin $t_{2g}^6 e_g^0$ configuration. At higher temperatures, electrons are excited into the e_g bands leading to paramagnetism and metallic conductivity. See text for more details. †The paramagnetism of SrRhO$_3$ is intermediate between Pauli and Curie–Weiss.

in filled t_{2g} bands and empty e_g bands. In LaCoO$_3$, the gap separating these two bands is very small. As a result, LaCoO$_3$ is semiconducting and diamagnetic only at low temperatures. As the temperature rises, electrons in LaCoO$_3$ are thermally excited into e_g bands, leading paramagetism and eventually to metallic conductivity ($T > 650$ K) [7, 8].

For each of the remaining (non-diamagnetic) semiconducting perovskites, either the t_{2g} bands, the e_g bands, or both, are partially filled. When the bandwidth is smaller than the on-site repulsion ($W < U$), they are Mott–Hubbard insulators, whereas, for the opposite case ($W > U$), they are Pauli-paramagnetic metallic conductors with fully delocalized electrons. Those compounds that are metallic, but do not show Pauli paramagnetism, namely the ferromagnetic SrFeO$_3$, SrCoO$_3$, and SrRuO$_3$ lie in a regime where $W \approx U$. In these compounds, the interactions between electrons are not sufficient to fully localize them, but they do impact the properties in significant ways, similar to the previously discussed compounds VO and CrN.

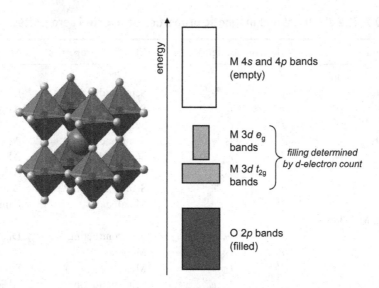

Figure 10.23 The perovskite-type structure (left), along with a schematic DOS diagram for first-row transition-metal oxides with this structure (right).

It is interesting to consider the periodic trends that dictate whether a perovskite will be a Mott–Hubbard insulator ($W < U$) or a Pauli paramagnetic metal ($W > U$). Among the $LaMO_3$ compounds where M is a $3d$ transition-metal ion, the combination of the smaller $3d$ orbitals and the relatively low oxidation state of M (+3) leads to fairly narrow bands. Hence, all $LaMO_3$ perovskites with partially filled $3d$ bands are Mott–Hubbard insulators with the exception of $LaNiO_3$. In $LaNiO_3$, there are two factors that combine to make W larger than the other $LaMO_3$ compounds. Firstly, +3 is a high oxidation state for a late-transition metal like nickel. This results in short, highly covalent Ni–O bonds, which in turn raise the energy of the antibonding crystal orbitals at the R point, leading to wider t_{2g} and e_g bands (see Section 6.6.5). That effect, coupled with the partially filled e_g bands, push this compound toward the $W > U$ category. It should be noted that if La^{3+} is replaced by a smaller lanthanoid ion (e.g. Nd^{3+}), the induced octahedral tilting narrows the e_g bands (Section 6.6.5). This triggers a transition from metallic conductivity at high temperature to semiconducting behavior at low temperature [9]. From this, we can infer that W in $LaNiO_3$ is just barely large enough to stabilize delocalized metallic conductivity to low temperature.

Among $SrMO_3$ perovskites containing $3d$ transition-metal ions, delocalized electrons and metallic conductivity are the norm rather than the exception. At first glance, this may seem surprising because the same M constitutes an antiferromagnetic semiconductor when the A-site cation is La. This change in behavior can largely be attributed to the increased bandwidth that results when the oxidation state of M increases from +3 to +4. The sole exception in Table 10.7 is $SrMnO_3$, where the d^3 configuration leads to a half-filled t_{2g} bands. The on-site repulsion is particularly high for half-filled bands because for a t_{2g} electron to move from one Mn^{4+} to another, it must either go into the higher-energy e_g bands or flip its spin and remain in the t_{2g}-orbital manifold. It should also be noted that W is only slightly larger than the Hubbard U for some of the metallic $SrMO_3$ compounds in Table 10.7. Replacing Sr^{2+} with the smaller Ca^{2+} leads to octahedral tilting distortions that narrow the bandwidth (Section 6.6.5), resulting in semiconducting behavior for both $CaCrO_3$ and $CaFeO_3$. The response of $CaFeO_3$ to electron localization is particularly interesting, and is discussed further in Chapter 11.

When M in $SrMO_3$ is a second-row transition-metal ion with partially filled t_{2g} or e_g orbitals, the larger spatial extent of the $4d$ orbitals leads to wider bands. Hence $W > U$, and metallic conductivity is observed for all of the $4d$ perovskites in Table 10.7, except for the band insulators $KNbO_3$ and $LaRhO_3$.

10.5 Organic Conductors

Organic compounds are not an obvious place to look for conducting materials. Most organic solids consist of small molecules held together by relatively weak intermolecular forces. The lack of chemical bonds between molecules does not favor electronic wavefunctions that are delocalized over the entire crystal, as they must be in highly conducting materials. Until the 1970s, there was little enthusiasm for the idea that organic substances could be useful conductors of electricity, but scientists have subsequently discovered organic substances that are semiconducting, metallic, and superconducting. In fact, the conductivity of organic substances spans a range comparable to that of oxides. Because organic conductors are lightweight and potentially flexible, they lend themselves to novel applications where inorganic conductors cannot be used, such as flexible electronics. They also offer new approaches to processing. For example, some organic conductors can be dissolved in solvents and printed like an ink [10]. In this section, we examine some prototypical organic conductors and take a closer look at what makes these substances different from most organic materials.

A common feature among organic conductors is a conjugated π system: a network of alternating single and double bonds where each atom is sp^2 hybridized and the remaining p orbital participates in a delocalized π-bonding network (Section 5.3.7). Conjugated molecules and polymers possess delocalized molecular orbitals (MOs), which are a necessary criterion for an organic conductor. In conjugated systems, the highest occupied molecular orbitals (HOMOs) have π-bonding character and the lowest unoccupied molecular orbitals (LUMOs) have π-antibonding character. As the size of the conjugated network increases, the HOMO–LUMO gap decreases, but, even for large conjugated systems, a gap generally remains, which makes most organic conductors inherently semiconducting.[27] To attain metallic conductivity, it is necessary to either remove electrons from the π orbitals or add electrons to the π^* orbitals. As we will see, there are various strategies for introducing charge

---

[27] We can think of graphene (Section 6.5.1) as the largest possible 2D conjugated system and in graphene the gap between π and π^* crystal orbitals goes exactly to zero.

carriers. Organic conductors can be broadly subdivided into three categories: conducting polymers, polycyclic aromatic hydrocarbons, and charge-transfer salts.

10.5.1 Conducting Polymers

Polymers are a natural place to look for organic conductors because, like inorganic semiconductors, they contain extended networks of covalently bonded atoms. However, most familiar polymers (polyethylene, polypropylene, polystyrene, polyester, Teflon, etc.) are insulating because their backbones are made up of sp^3-hybridized carbon atoms connected by single bonds. Polyacetylene is the simplest polymer where every atom along the polymer backbone is sp^2-hybridized, and therefore part of a conjugated π network (Figure 10.24).

Let's take a closer look at the electronic structure of *trans*-polyacetylene. As with benzene (Section 5.3.7) and graphene (Section 6.5.1), the electrical and optical properties are determined almost exclusively by the π and π^* orbitals. For *trans*-polyacetylene, there are only two carbon atoms in the repeat unit (Figure 10.25), which means the basis sets of interest are the π and π^* MOs of the C_2H_2 repeat unit.[28] Once the basis sets, $u(x)$, are determined, the task is analogous to our earlier derivation of the band structure of a 1D chain of H_2 molecules (Section 6.2). The crystal orbitals corresponding to these two MOs are shown at $k = 0$ and π/a in Figure 10.25.[29] Immediately we see that at $k = 0$ the band formed from the π MO is completely bonding and will have the lowest possible energy, while the band formed from the π^* MO is completely antibonding and will have the

trans-polyacetylene   *cis*-polyacetylene

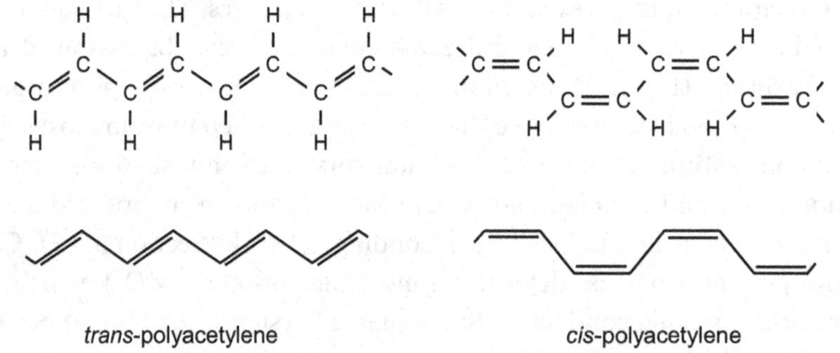

Figure 10.24 The *trans*- and *cis*- forms of polyacetylene.

---

[28] In constructing the π and π^* bands, we can neglect the orbitals that form σ bonds because the σ and π orbitals are orthogonal.

[29] In this treatment, we consider the C_2H_2 repeat unit as a unit cell with 1D translational symmetry. This approximation is a useful construct for visualizing the electronic structure of polyacetylene, but, in practice, long-range crystalline order in polyacetylene varies depending on synthetic conditions.

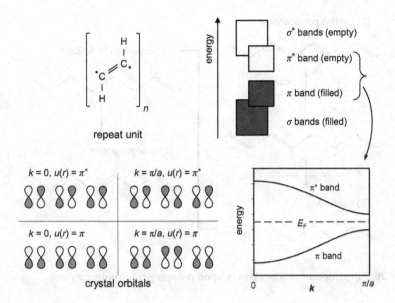

Figure 10.25 The C_2H_2 repeat unit in *trans*-polyacetylene (top left). A block diagram of the electronic structure (top right). The crystal orbitals formed from the carbon $2p_z$ orbitals at $k = 0$ and π/a looking at a side-on view of the chain (bottom left). The band structure, showing only the valence (π) and conduction (π^*) bands (bottom right).

highest possible energy. At $k = \pi/a$, the interaction between carbon atoms in neighboring unit cells changes sign so that the band formed from the π MO runs uphill from $k = 0$ to π/a and the band formed from the π^* MO runs downhill. The individual widths of the π and π^* bands are estimated to be 5–6 eV [10].

If all carbon–carbon distances were identical, the bands would touch at $k = \pi/a$ and *trans*-polyacetylene would be a metal. However, the distances alternate between 1.44 Å (single bond) and 1.36 Å (double bond).[30] This alternation lifts the degeneracy and is responsible for the presence of a band gap of 1.7–1.8 eV. As mentioned in Section 6.2, this is a classic case of a Peierls distortion.

Conjugated polymers are "doped" by a different mechanism than inorganic semiconductors: by oxidation/reduction rather than aliovalent substitution. For example, when polyacetylene is exposed to iodine vapor, I_2 molecules penetrate the substance, partially oxidizing $(CH)_x$ chains and forming dilute cationic charges. In total, one I_2 takes two electrons from a pair of double bonds, turning them into single bonds of one C^+ (where the π electron was taken) and one $C^{\cdot}$ radical (where the π electron remained), as shown in Figure 10.26, left. The reduction product I^- reacts with excess I_2 to form I_3^- counter ions. When the double bond becomes a single bond, a local distortion involving an increase in the C–C distance occurs.

[30] This difference is not quite as pronounced as the difference between carbon–carbon double and single bond lengths typically found in organic molecules: 1.32 Å (double) and 1.54 Å (single).

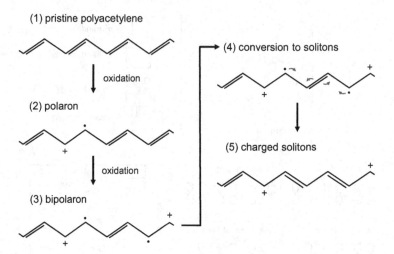

Figure 10.26 Formation of charge carriers upon p-doping *trans*-polyacetylene.

In a conducting polymer like polyacetylene, the cation–radical defect pair can move along the polymer chain, distorting the C–C distances in its vicinity as it moves. A charge carrier that distorts the structure when it moves through a crystal is termed a **polaron**. In polyacetylene, the positive charge of the polaron is not very mobile at low doping levels because it is localized by attraction to the negatively charged counter ion I_3^-.

As the doping level increases, the polarons in a given chain eventually begin to interact, and the potential exerted by negatively charged counter ions becomes more uniform. When two polarons approach each other, they form a **bipolaron** as shown in Figure 10.26, left. While bipolarons are not the important charge carriers in *trans*-polyacetylene, they are in other conducting polymers. In polyacetylene, the unpaired electrons in the bipolaron tend to combine, forming a new double bond and leaving behind two positively charged carbon atoms. At each of these carbocations, shown in the lower right-hand side of Figure 10.26, the pattern of bond alternation is reversed (i.e. the bond patterns on either side of the carbon are mirror images of each other). A carbon atom that forms two single bonds and reverses the pattern of bond alternation is called a **soliton**.[31] In highly conducting samples of *trans*-polyacetylene, positively charged solitons are the dominant charge carriers.

Interactions between chains also play an important role. Polymers contain chains of different lengths, the packing of which invariably introduces some disorder. The interchain bandwidths are at least an order of magnitude smaller than intrachain bandwidths. For these reasons, conductivity in polymers is limited by interchain hopping. The mechanisms of

---

[31] Solitons can also be neutral, in which case it is simply a free radical. When it is charged, it can be either positively charged (carbocation) or negatively charged (carbanion). In all three cases, the pattern of single and double bonds on either side of the soliton are mirror images of each other.

interchain hopping are beyond the scope of this text, but suffice to say that the properties of conducting polymers are sensitive to the distribution of chain lengths and crystallinity, both of which depend on the details of synthesis and processing.

Although *trans*-polyacetylene is the simplest, first, and most highly conducting polymer, its use in practical applications is limited by its instability towards oxidizing agents and moisture. Additional examples of conducting polymers include doped polyaniline, polypyrole, polythiophene, polyethylene dioxythiophene (PEDOT), and numerous others. Many of these are quite stable in air, although their conductivities are not as high as polyacetylene. The interested reader is directed to a review by Heeger [10] for more details about other conducting polymer systems and their applications.

10.5.2 Polycyclic Aromatic Hydrocarbons

Polycyclic aromatic hydrocarbons (PAHs) are planar molecules made up of fused benzene rings. We examined a subset of PAHs, the acenes, in Section 7.6 where we learned that the band gap decreases as the number of the rings increases—from 4.69 eV in benzene to 1.85 eV in pentacene (see Figure 7.14). The latter value is similar to the band gap of *trans*-polyacetylene, which suggests that the larger PAHs might have useful electrical properties.

Box 10.1 Materials Spotlight: The discovery of conducting polymers

In the early 1970s, Hideki Shirakawa found a way to prepare films of polyacetylene via Ziegler–Natta-catalyzed polymerization of gaseous acetylene ($HC \equiv CH$). When the polymerization was carried out at low temperature ($-78\,°C$), a copper-colored film was deposited on the walls of the reaction vessel. This film was shown to be predominantly (>95%) *cis*-polyacetylene. When the polymerization was carried out at higher temperatures ($150\,°C$), a silver-colored film was deposited that was subsequently shown to be *trans*-polyacetylene.

While polyacetylene has a conjugated network, Shirakawa's films were not particularly good conductors for the reasons discussed in the text. The *trans* isomer has a band gap of 1.7–1.8 eV and a conductivity of 10^{-2} S/m to 10^{-3} S/m, while the *cis* isomer has an even larger band gap (~2.2 eV) and a lower conductivity (10^{-7} S/m to 10^{-8} S/m). When Shirakawa told Alan MacDiarmid about his silvery films, MacDiarmid suggested that it should be possible to increase their conductivity through doping. Shirakawa, MacDiarmid. and Alan Heeger showed that when exposed to halogen gas (I_2, Br_2, or Cl_2), polyacetylene is partially oxidized, which results in p-type doping and leads to a dramatic increase in conductivity. By 1977, they were able to raise the conductivity of *trans*-polyacetylene films to 3×10^3 S/m [11]. Subsequently, conductivities as high as ~10^5 S/m have been obtained in doped polyacetylene films. In 2000, the three scientists shared the Nobel Prize in Chemistry for this discovery.

While molecules with a conjugated π system possess delocalized MOs, not all conjugated molecules form conductors when they crystallize. For a molecular crystal to be conducting, there must be communication between MOs on different molecules. The optical properties of PAHs provide a clue to this interaction. For example, the longest-wavelength absorption for an isolated pentacene molecule (in solution or frozen in an argon matrix) is at ~580 nm, corresponding to a HOMO–LUMO gap of 2.1 eV. When pentacene crystallizes, this absorption shifts to ~670 nm, corresponding to a band gap of 1.8 eV [12, 13]. The red shift of the absorption edge implies interactions between discrete MOs to form bands. However, compared with inorganic semiconductors, or even the intrachain π and π^* bands in polyacetylene, the bands in pentacene are very narrow, only a few tenths of an electronvolt wide. Bandwidths of this magnitude are typical of small-molecule organic conductors.

The widths of the bands formed from π and π^* MOs are sensitive to the way molecules pack in the solid state. Linear acenes, like tetracene and pentacene, tend to pack in a "herringbone" arrangement (Figure 10.27) where the long edge of one molecule points toward the face of another. The valence bands[32] in pentacene are formed from the HOMO of a pentacene molecule that has six nodal planes. As shown in Figure 10.27, the highest-energy

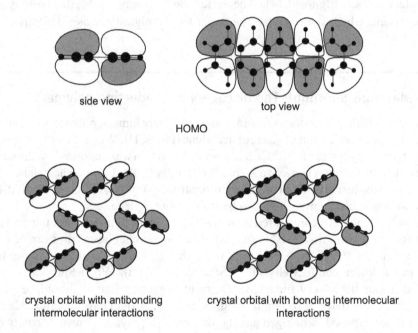

side view   top view

HOMO

crystal orbital with antibonding   crystal orbital with bonding intermolecular
intermolecular interactions   interactions

Figure 10.27 Top: The HOMO of pentacene $C_{22}H_{14}$ with its six nodal planes seen sideways and from top. Bottom left: Antibonding pattern of the "sideways" intermolecular crystal-orbital interactions at the top of the valence band. Bottom right: Analogous pattern of bonding interactions at the bottom of the valence band. After ref. [14].

[32] There are two molecules per unit cell and thus two valence bands formed from the HOMO of pentacene.

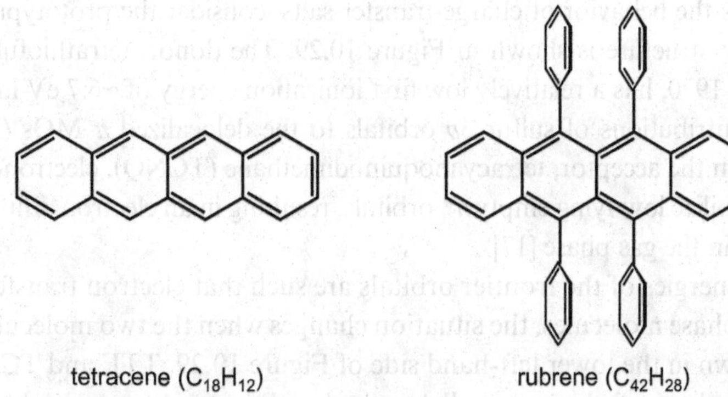

tetracene (C₁₈H₁₂)   rubrene (C₄₂H₂₈)

Figure 10.28 The molecular structure of tetracene (left) and rubrene (right).

point in the valence band belongs to a crystal orbital with antibonding intermolecular interactions, while the lowest-energy crystal orbital in the valence band features bonding interactions between HOMOs on neighboring molecules. The difference in these two energies determines the width of the valence bands and is a measure of the strength of bonding interactions between molecules. Despite the relatively narrow bandwidth, conductivities as high as 1×10^4 S/m have been reported in iodine-doped pentacene films [15].

By functionalizing the conjugated core of molecules like tetracene and pentacene, it is possible to alter the energy levels of the HOMO and LUMO, the crystal packing, and the reactivity. Among such substituted variants, rubrene (5,6,11,12-tetraphenyltetracene, $C_{42}H_{28}$) is one of the most extensively studied PAHs. Rubrene (Figure 10.28) is less reactive and therefore more stable than tetracene or pentacene, which makes it possible to grow high-purity single crystals. Rubrene holds the record mobility for an organic semiconductor, $\mu = 0.0040$ m²/(V s). This value, while much lower than values observed in crystalline inorganic semiconductors (Table 10.5), is approaching values seen in amorphous silicon. Unfortunately, films of rubrene tend to be poorly crystalline. For devices that incorporate thin films of organic conductors, pentacene is often the material of choice.

10.5.3 Charge-Transfer Salts

In the previous section we learned that conjugated polymers and polycyclic aromatic hydrocarbons are intrinsically semiconducting. Doping is required to attain high conductivity, even then the conductivity decreases as the temperature is lowered. To find organic conductors that behave like metals, we turn to charge-transfer salts. These compounds contain two types of molecules whose frontier orbitals are energetically aligned so that electron density is transferred from a donor molecule with a high-lying HOMO to an acceptor molecule with a low-lying LUMO. The energy levels of the two molecules in the salt are such that in the solid state the transfer is only a fraction of an electron per donor–acceptor pair. In this way, partially filled bands can be obtained without doping.

To illustrate the behavior of charge-transfer salts, consider the prototypical example, TTF–TCNQ, whose structure is shown in Figure 10.29. The donor, tetrathiofulvalene (TTF), first synthesized in 1970, has a relatively low first ionization energy of ~6.7 eV in the gas phase [16], due to the contributions of sulfur $3p$ orbitals to the delocalized π MOs (see Footnote 31 in Chapter 12). In the acceptor, tetracyanoquinodimethane (TCNQ), electron-withdrawing cyanide groups stabilize low-lying empty π^* orbitals, resulting in an electron affinity that is estimated to be ~2.8 eV in the gas phase [17].

While the energies of the frontier orbitals are such that electron transfer would not occur between gas-phase molecules, the situation changes when the two molecules form the crystal structure shown in the lower left-hand side of Figure 10.29. TTF and TCNQ form alternating cationic and anionic layers parallel to the bc–plane, each composed of planar molecules that stack tightly along the b axis of the monoclinic unit cell due to dispersion forces between their π orbitals. This widens the bands shown on the right-hand side of Figure 10.29. The two bands that cross E_F and run uphill from Γ to Z, are derived from the π-acceptor orbitals on TCNQ. The nearly degenerate pair of bands that cross E_F and run downhill from Γ to Z are derived predominantly from $3p$ π orbitals of the sulfur atoms in the TTF molecule. The overlap of these two pairs of bands is responsible for the partial electron transfer from TTF to TCNQ. From spectroscopic signatures, such as the C–N stretching frequencies of the TCNQ molecules, the transfer is estimated to be 0.59 e^- per donor–acceptor pair.

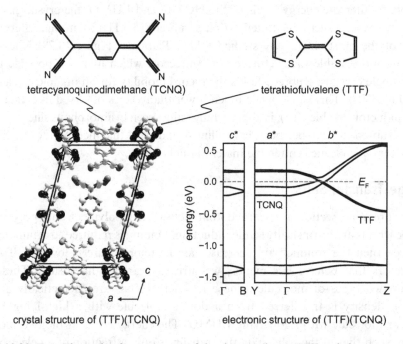

Figure 10.29 Top: The molecular structures of the electron donor TTF and electron acceptor TCNQ. Left: The crystal structure of TTF–TCNQ. Right: The band structure of TTF–TCNQ near the Fermi level. After ref. [18].

The Fermi level cuts through the frontier orbitals of both TTF and TCNQ molecular stacks, leading to metallic conductivity. The room-temperature conductivity is 6.5×10^4 S/m [19]. Upon cooling, conductivity increases as expected for a metal, reaching a value of 1.5×10^6 S/m at 58 K before undergoing a metal–insulator transition upon further cooling [20]. The bandwidth is only substantial along the b^* axis of the reciprocal lattice, which corresponds to the b axis of the real-space lattice, the direction along which the molecules stack. This anisotropy of the electronic structure is confirmed by experimental measurements that reveal conductivity parallel to b three orders of magnitude larger than it is in perpendicular directions. Because conduction occurs predominantly along the stacks of similar molecules, TTF–TCNQ is a quasi-1D metal.

The frontier orbitals on the donor and acceptor molecules must be appropriately aligned to achieve partial charge transfer. For example, if the hydrogens on the central benzene ring of TCNQ are replaced with fluorine (F_4TCNQ), the acceptor becomes more electronegative and a full electron transfer occurs, leading to an insulating behavior. Another approach to stabilize partially filled π bands is to have one singly charged polyatomic anion, such as PF_6^-, AsF_6^-, SbF_6^-, BF_4^-, or NO_3^- gaining its anion charge from two π-donor molecules. On average, two donor molecules give up a total of one e^- leading to p-doping of the donor-based valence band by one-half hole per donor molecule. These compounds, called Bechgaard salts, are known for their superconductivity and are discussed in more detail in Section 12.5.

10.6 Carbon

In the previous section, we explored the electronic conductivity of hydrocarbons with networks of conjugated π bonds. In this section, we focus on two low-dimensional allotropes of carbon with (in principle) infinitely conjugated networks: graphene and carbon nanotubes. We will see that the conductivity of these materials is exceptional in many ways. The fullerenes, yet another conducting allotrope of carbon, are covered in Chapter 12, due to their superconductivity.

10.6.1 Graphene

Graphene is an infinite 2D network of fused hexagonal rings. Each carbon atom is sp^2 hybridized and forms bonds to three equidistant neighbors. As discussed in Section 6.5.1, the electronic structure of graphene near the Fermi level is dominated by the π and π^* bands that originate from overlap of the carbon $2p_z$ orbitals. Using a tight-binding approximation, it is possible to derive an analytical expression for the energies of these two bands as a function of the 2D wave vector, k [21]. A 3D contour plot of the energy throughout the first Brillouin zone using this model is shown in Figure 10.30. Notice how the two bands are widely separated at Γ, where the π interactions between nearest neighbors are bonding in the lower band and antibonding in the upper band. The situation is very different at the K and K' points (Figures 6.18 and 6.19), where the two bands touch. Those six points in

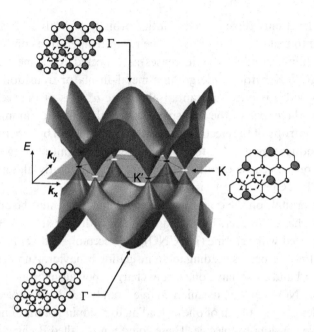

Figure 10.30 The energies of the π and π^* bands in graphene as plotted in 2D reciprocal space (k_x and k_y in Cartesian coordinates, for convenience, rather than the hexagonal reciprocal lattice vectors a^* and b^* shown in Figure 6.13). The real-space orbital-overlap diagrams at Γ and K are taken from Figure 6.19. Image of the 2D band structure reprinted with permission from C.J.M. Beenakker, "Colloquium: Andreev reflection and Klein tunneling in graphene" *Rev. Mod. Phys.* **80**, 1337–1354, 2008. Copyright 2008 by the American Physical Society.

k space where the filled and empty bands touch are called **Dirac points** and the surrounding regions **Dirac cones**. The Fermi level lies at the precise energy where the bands touch, which makes graphene a perfect semimetal.

Because the filled π band touches but does not overlap with the empty π^* band, anything that perturbs the position of the Fermi level will have large impact on the carrier concentration. The 2D structure of graphene makes it is possible to apply an electric field across a graphene sheet, with the electric-field gradient perpendicular to the graphene layer. The electric field shifts the position of the Fermi level and the charge-carrier concentration increases as it shifts away from the Dirac points, leading to an increase in conductivity. This process is called gating. Depending upon the sign of the applied voltage, the Fermi level can be shifted upward, leading to conductivity dominated by electrons, or downward whereby holes become the dominant charge carriers. In both cases, the carrier concentration increases linearly with the voltage for modest fields. Graphene is highly sensitive to molecules adsorbed on the surface. Exposure to electron-donating species induces n-type conduction, whereas exposure to electron-withdrawing species leads to p-type conduction.

The semimetallic nature of graphene is not the only aspect of its electronic structure that sets it apart from most conducting materials. In the free-electron model, the band energy varies with square of the wave vector ($E \propto k^2$), as expressed in Equation (10.12). In real materials the band structure is more complicated, but, in conventional semiconductors, the bands are approximately parabolic ($E \propto k^2$) near the minimum of the conduction band and the maximum of the valence band. In contrast, the band energies in graphene change linearly with the wave vector k in the vicinity of the K and K′ points where the π and π^* bands touch, as given by the relationship:

$$E(k) = \pm v_F \hbar k \tag{10.27}$$

where $v_F \approx 10^6$ m/s is the Fermi velocity in graphene. The Dirac cones and linear relationship between E and the wave vector k arise from the half-filling of the π bands and the honeycomb topology of the crystal structure (see Section 6.5.1). See ref. [22] for more details.

Graphene's electronic structure near the Fermi level is compared with a conventional semiconductor in Figure 10.31. In conventional semiconductors, v_F depends on the position of the Fermi level, Equation (10.17), whereas in graphene it is independent of both E_F and k. Since effective masses are given by the curvature of the bands, Equation (10.15), the carriers in graphene behave as though they are massless.[33] Since mobility is inversely proportional to the effective mass, one might expect charge carriers in graphene to have infinitely large mobilities.[34] While this is not observed in practice, the carrier mobilities in graphene are exceptionally high. They approach 20 m^2/(V s) at low temperatures [23] and can be as high as 2.5 m^2/(V s) at room temperature [24]. While mobilities this high are known for III–V semiconductors like InSb, the exceptional feature of graphene is that its carrier mobility does not decrease as the carrier concentration increases. The high mobility leads to mean free paths on the micron length scale in high-quality graphene, even at room temperature! This attribute makes it possible to construct devices from graphene capable of **ballistic transport**, meaning that the carriers pass from one end of the graphene sheet to the other end without being scattered by a defect or lattice vibration. Another feature that sets the mobilities in graphene apart is the symmetry between electron and hole mobilities; they are the same in graphene but typically very different in conventional semiconductors.

10.6.2 Carbon Nanotubes

Carbon nanotubes are hollow cylinders of graphene sheets, typically a few nanometers in diameter. They come in two varieties: single-walled carbon nanotubes (SWCNTs) and multi-walled carbon nanotubes (MWCNTs). The latter are made up of cylindrical SWCNTs of increasing diameter wrapped around each other, held together by van der Waals forces. Because the electronic and transport properties of MWCNTs tend to be a composite of the

---

[33] The effective mass is only zero when the Fermi level is at the energy where the bands touch. If the Fermi level is shifted, the carriers develop a mass, but it is still quite small compared to conventional materials. For this reason, the charge carriers in graphene are sometimes referred to as massless Dirac fermions.

[34] The charge carriers in graphene obey the Dirac equation, an equation that combines quantum mechanics and general relativity. This gives rise to interesting analogies between the charge carriers in graphene and the world of particle physics.

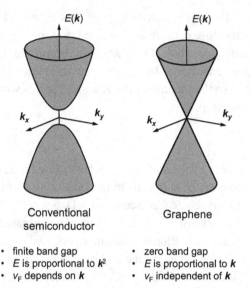

Conventional
semiconductor

Graphene

- finite band gap
- E is proportional to k^2
- v_F depends on k

- zero band gap
- E is proportional to k
- v_F independent of k

Figure 10.31 Differences between the linear dispersion near the Fermi level of graphene and the parabolic dispersion seen in a conventional semiconductor.

individual SWCNT from which they are made, the treatment that follows is limited to SWCNTs.

We will treat carbon nanotubes as 1D crystals that possess translational symmetry along the long axis of the tube. Unlike graphene whose structure is uniquely defined, SWCNTs can adopt an infinite number of structures, each with its own diameter and chirality. Fortunately, there is a relatively simple nomenclature that can be used to define the structure of any carbon nanotube in terms of graphene's lattice vectors.

As illustrated in Figure 10.32, we begin by specifying the circumferential vector, C, which defines the circumference of the tube by connecting two equivalent sites on a graphene sheet.[35] The circumferential vector can be expressed as a linear sum of the a and b lattice vectors of graphene in the general form $C = na + mb$, where n and m are integers. The next step is to define the translational vector, T, that is perpendicular to C and links two equivalent sites. The rectangle formed by C and T defines the unit cell of the carbon nanotube. To complete the process, we imagine excising a strip, of width defined by C and infinite length, from the graphene sheet and rolling it up into a tube by connecting opposite sides of the strip. In this manner, the structure of the carbon nanotube is completely defined by the integers n and m, and is expressed using the notation (n,m). For example, the tube depicted in Figure 10.32 is a (4,2) carbon nanotube.

Once the (n,m) indices of the circumferential vector are known, many useful characteristics of the tube can be calculated. The diameter of the tube d is related to its circumference:

---

[35] The terms circumferential vector and roll-up vector are used interchangeably in the literature.

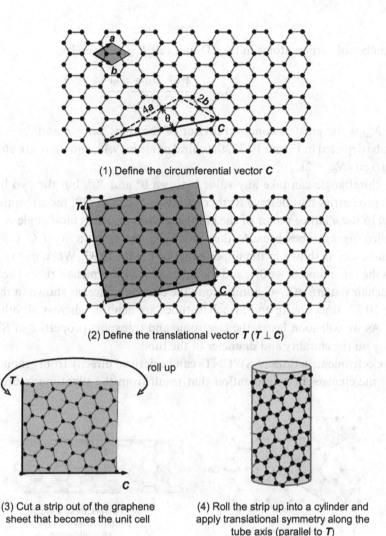

(1) Define the circumferential vector **C**

(2) Define the translational vector **T** (**T** ⊥ **C**)

roll up

(3) Cut a strip out of the graphene
sheet that becomes the unit cell

(4) Roll the strip up into a cylinder and
apply translational symmetry along the
tube axis (parallel to **T**)

Figure 10.32 A step-by-step illustration of how the (n,m) notation used to describe the structure of a carbon nanotube is related to the lattice vectors a and b of a graphene sheet. In this example $n = 4$ and $m = 2$.

$$d = \frac{|C|}{\pi} = \frac{a\sqrt{n^2 + nm + m^2}}{\pi} \qquad (10.28)$$

where a is the graphene unit-cell edge, whose length is 2.46 Å. The so-called chiral angle, θ, which is the angle between the lattice vector $\boldsymbol{a}$ and the circumferential vector $\boldsymbol{C}$ (see the top image in Figure 10.32), is related to n and m through the relationship:

$$\tan\theta = \frac{\sqrt{3}m}{2n + m} \qquad (10.29)$$

The number of carbon atoms in the 1D unit cell N_{cell} is given by:

$$N_{cell} = \frac{4(n^2 + nm + m^2)}{N_{cd}} \tag{10.30}$$

where N_{cd} is the greatest common divisor of the sums $2m + n$ and $2n + m$. For the (4,2) nanotube depicted in Figure 10.32, the chiral angle is 19.1°, and there are 56 carbon atoms in the unit cell ($N_{cd} = 2$).

The chiral angle can take any value between 0° and 30°, but the two limiting cases are special geometries that deserve further mention. When $m = 0$, the circumferential vector is parallel to the *a* lattice vector of the graphene sheet and the chiral angle is zero. Such tubes are called zigzag tubes because they exhibit a zigzag pattern of C–C bonds along the circumference, as shown in the upper panel of Figure 10.33. When $n = m$, the chiral angle attains the other limit ($\theta = 30°$), and the tubes are called armchair tubes because they exhibit an armchair pattern of C–C bonds along the circumference, as shown in the lower panel of Figure 10.33. Both zigzag and armchair tubes are achiral, whereas all other SWCNTs are chiral. As we will soon learn, the electronic and transport properties of SWCNTs depend crucially on the chirality and diameter of the tube.

The electronic structures of SWCNTs can be derived directly from graphene, provided we neglect the changes in hybridization that result from the slight curvature of the tube. As

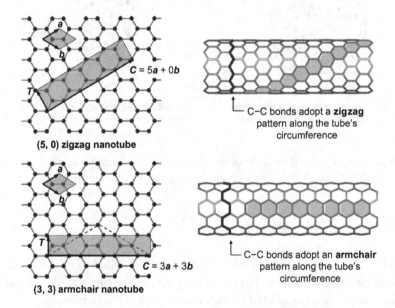

Figure 10.33 A zigzag carbon nanotube ($m = 0$) whose chiral angle $\theta = 0°$ (top), and an armchair carbon nanotube ($n = m$) whose chiral angle $\theta = 30°$ (bottom). Note the rolled-up tubes on the right are illustrative of each type, they do not correspond directly to the tubes shown on the left.

graphene has full translational symmetry in two dimensions, the wave vector k can take any value in the first Brillouin zone, whereas a SWCNT is finite along its circumference, which quantizes the allowed values of k to a series of parallel lines whose spacing is inversely proportional to the diameter of the tube [25]. When the condition $kC = 2\pi q$ is met (q is an integer), the Dirac points fall on a line of allowed k values. This condition is met in SWCNTs where $(n - m)/3$ is an integer, and such nanotubes are metallic. All other SWCNTs are semiconducting with a band gap that is inversely proportional to the diameter. Thus, armchair nanotubes are always metallic, because $n - m = 0$, whereas zigzag and chiral nanotubes can be either metallic or semiconducting, depending upon the values of n and m.

The analysis above assumes all $2p$ π orbitals are locally orthogonal to the network of σ bonds. This assumption is not strictly true due to the curvature of the nanotube, and thus each carbon atom experiences a subtle trigonal distortion from the planar geometry it adopts in graphene. For example, the carbon atoms in a (5,5) SWCNT with a diameter of 6.8 nm are bent ~6° from planar. The degree of such "pyramidalization" is smaller in larger tubes. This distortion perturbs the electronic structure in a couple of important ways. Firstly, it opens a very small band gap ($E_g < 0.1$ eV) in zigzag and chiral nanotubes that would otherwise be metallic because $(n - m)/3$ is an integer. Interestingly, in armchair nanotubes, this distortion does not prevent the allowed wave vectors from running through the Dirac points, and the tubes remain metallic. The band dispersion in metallic nanotubes is nearly linear close to the Fermi level, and the electrical transport properties along the tube axis are similar to graphene.

The ability to tune from metallic to semiconducting with a subtle change in chirality or diameter makes SWCNTs highly unusual materials that are attractive for applications. At the same time, it presents a significant practical barrier to use in electronic devices. Most syntheses produce a mixture of metallic (or nearly so) SWCNTs, where $(n - m)$ is a multiple of three, and semiconducting SWCNTs, in roughly a 1:2 ratio. To make devices reproducibly, the SWCNTs must be separated according to their chirality, not to mention separated from MWCNTs that are also present. Approaches to doing this include chemically digesting the more reactive tubes (i.e. those with larger pyramidalization angles), dielectrophoretic separation, chromatographic methods, and ultracentrifugation [26]. Finding routes to obtain nanotubes with homogeneous properties is essential if they are to find applications that fully exploit their electronic transport properties.

10.7 Problems

10.1 Use the conductivity of copper given in Table 10.1 to calculate the resistance of a copper wire that is 3 mm in diameter and 20 cm long.

10.2 Show by dimensional analysis that Equation (10.9) gives the proper units (A/m^2) for the current density J. Remember 1 A = 1 C/s.

10.3 Confirm Equation (10.10) by dimensional analysis in SI units. Remember that 1 A = 1 C/s.

10.4 (a) Estimate the carrier mobility μ in Na metal at 300 K via Equation (10.10), using the conductivity of Na in Table 10.2, and the Drude-model assumption that all valence electrons can respond to an applied field. (b) Use it to calculate the relaxation time τ. (c) What is the rms velocity of the conduction electrons at 300 K in the Drude model? (d) Use your answers to estimate the mean free path of an electron. How does it compare with the unit-cell parameter of 4.3 Å for Na metal? How does it compare with the estimates of mean free path given in Section 10.2.7? How do you account for the discrepancy between the two estimates?

10.5 The conductivities and crystal structures of the heavier s-block metals are given in the table below. (a) Calculate the valence-electron concentration n in m^{-3} and arrange the elements from lowest to highest n. (b) Use these values and those for lighter s-block elements Na, K, Mg, and Ca in Table 10.2 to construct a plot of electrical conductivity (in S/m) versus valence-electron concentration (in m^{-3}). (c) What can you say about how well these metals follow the predictions of the Drude model?

	Rb	Cs	Sr	Ba
Conductivity (S/m)	8.3×10^6	5.0×10^6	7.7×10^6	2.9×10^6
Crystal structure	bcc	bcc	fcc	bcc
Lattice parameter (Å)	5.58	6.14	6.08	5.03

10.6 Show that at any finite temperature the crystal orbital at the Fermi level will have a 50% probability of being occupied.

10.7 At which energy separation from E_F is the probability $f(E)$ of an electronic state in a metal being occupied equal to 0.01 at (a) 10 K, (b) 100 K, (c) 300 K, and (d) 1000 K.

10.8 (a) Within the assumptions of the Drude model, calculate the rms velocity of a valence electron in sodium metal at $T = 300$ K. (b) Use a quantum-mechanical model to estimate the Fermi velocity of the free carriers (assume $m^* = m_e$) in Na given that $E_F = 3.1$ eV. (c) Which velocity estimate is larger? (d) Do the two models predict different temperature dependences of the free-carrier velocity?

10.9 What is the temperature dependence of the conductivity for (a) a metal, (b) an intrinsic semiconductor?

10.10 Compare electrical properties of copper and nickel. Which has higher carrier concentration? Which has higher carrier mobility? Which has higher conductivity and why?

10.11 Calculate the value of $f(E)$ at the conduction-band edge in intrinsic Ge ($E_g = 0.7$ eV) at 300 K. Assume the Fermi level is located midway between the valence- and conduction-band edges.

10.12 Calculate the donor ionization energy and Bohr radius for a donor impurity in InP (the dielectric permittivity $\varepsilon_r = 12.1$, effective mass $m^* = 0.07 m_e$).

10.13 Will the temperature range over which a semiconductor that is in the saturation regime increase or decrease in response to the following changes: (a) the band gap of the semiconductor increases, (b) the effective mass of the carriers increases, (c) the dielectric constant of the semiconductor increases?

10.14 In compounds containing lanthanoid ions, the $4f$ orbitals are sufficiently localized that interactions with orbitals on neighboring atoms can be neglected, hence $W \approx 0$. In this limit, the Hubbard U can be experimentally determined from photoelectron spectra. (a) Use ionization energies for gas-phase ions to estimate U for Pr^{3+} and Gd^{3+} (the third and fourth ionization energies of Pr are 21.6 eV and 39.0 eV, respectively, those for Gd are 20.6 eV and 44.0 eV). (b) Measured values of U for these two ions in solids are estimated as 6 eV for Pr^{3+} and 12 eV for Gd^{3+}. Why are the estimated values from part (a) higher than the measured values? (c) Why is U so much higher for Gd^{3+} than it is for Pr^{3+}? (d) By analogy, would you expect a higher U for TiO or MnO?

10.15 The perovskite $NaOsO_3$ undergoes a transition from a paramagnetic metal to an antiferromagnetic semiconductor upon cooling below 410 K. (a) What is the electron configuration of osmium in this compound. (b) Most perovskites containing ions with partially filled $5d$ orbitals are metals at all temperatures, what feature of $NaOsO_3$ makes it different?

10.16 Each of the following compounds contains an octahedrally coordinated transition-metal ion and behaves as an insulator/semiconductor. Identify each as either a band insulator or a Mott–Hubbard insulator: (a) Fe_2O_3, (b) $MgCr_2O_4$, (c) Li_2PtO_3.

10.17 The images below show a portion of a chain of the conducting polymer polypyrrole, before (top) and after (bottom) oxidative doping. (a) Is the charge carrier formed in this process a polaron, bipolaron, or soliton? (b) Is the conducing polymer that results doped n- or p-type?

10.18 Consider the electronic structure of graphene. (a) At how many points in the first Brillouin zone do the conduction and valence bands of graphene meet? (b) What is the name given to those points? (c) The bands that meet are derived from which atomic orbitals on carbon: $2s$, $2p_x$, $2p_y$, and/or $2p_z$?

10.19 Name at least three ways in which the electronic structure and transport properties of graphene differ from a conventional semiconductor.

10.20 What percentage of single-walled carbon nanotubes in a typical synthesis batch will be semiconducting? Explain your answer.

10.21 Consider the circumferential and translational vectors of a potential carbon nanotube shown below. (a) What are the *n* and *m* values of the circumferential vector **C**? (b) Is the **T** vector shown below a valid translational vector?

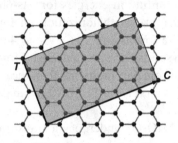

10.22 Consider the unit of a carbon nanotube projected onto a graphene sheet shown below. (a) What are the *n* and *m* values of this carbon nanotube? (b) Is this a metallic or semiconducting nanotube? (c) What is the diameter and chiral angle of this nanotube? (d) How many atoms are contained in the unit cell?

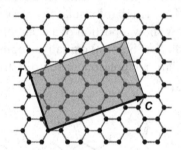

10.8   Further Reading

P.A. Cox, "*The Electronic Structure and Chemistry of Solids*" (1987) Oxford University Press.

S. Elliot, "*The Physics and Chemistry of Solids*" (1998) Wiley.

D.I. Khomskii, "*Transition Metal Compounds*" (2014) Cambridge University Press.

A.J. Heeger, "Semiconducting polymers: The third generation" *Chem. Soc. Rev.* **39** (2010), 2354–2371.

M. Ouyang, J.L. Huang, C.M. Lieber, "Fundamental electronic properties and applications of single-walled carbon nanotubes" *Acc. Chem. Res.* **35** (2002), 1018–1025.

R. Saito, G. Dresselhaus, M.S. Dresselhaus, "*Physical Properties of Carbon Nanotubes*" (1998) Imperial College Press.

A.H. Casto Neto, F. Guinea, N.M.R. Peres, K.S. Novoselov, A.K. Geim, "The electronic properties of graphene" *Rev. Mod. Phys.* **81** (2009), 110–155.

10.9 References

[1] "Electrical resistivity of pure metals," in "*CRC Handbook of Chemistry and Physics*" 98th edition, Editor: J.R. Rumble (Internet Version 2018) CRC Press/Taylor & Francis.

[2] A. Yamasaki, T. Fujiwara, "Electronic structure of the MO oxides (M = Mg, Ca, Ti, V) in the GW approximation" *Phys. Rev. B* **66** (2002), 245108.

[3] F. Rivadulla, J. Fernández-Rossier, M. García-Hernández, M.A. López-Quintela, J. Rivas, J.B. Goodenough, "VO: A strongly correlated metal close to a Mott–Hubbard transition" *Phys. Rev. B* **76** (2007), 205110.

[4] A. Herwadkar, W.R.L. Lambrecht, "Electronic structure of CrN: A borderline Mott insulator" *Phys. Rev. B* **79** (2009), 035125.

[5] R.M. Ibberson, R. Cywinski, "The magnetic and structural transitions in CrN and (Cr, Mo)N" *Physica B* **180** (1992), 329–332.

[6] K. Suzuki, Y. Yamaguchi, T. Kaneko, H. Yoshida, Y. Obi, H. Fujimori, H. Morita, "Neutron diffraction studies of the compounds MnN and FeN" *J. Phys. Soc. Japan* **70** (2001), 1084–1089.

[7] M.A. Señarís-Rodríguez, J.B. Goodenough, "LaCoO$_3$ revisited" *J. Solid State Chem.* **116** (1995), 224–231.

[8] M.W. Haverkort, Z. Hu, J.C. Cezar, T. Burnus, H. Hartmann, M. Reuther, C. Zobel, T. Lorenz, A. Tanaka, N.B. Brookes, H.H. Hsieh, H.J. Lin, C.T. Chen, L.H. Tjeng, "Spin state transition in LaCoO$_3$ studied using soft X-ray absorption spectroscopy and magnetic circular dichroism" *Phys. Rev. Lett.* **97** (2006), 176405.

[9] G. Catalan, "Progress in perovskite nickelate research" *Phase Transitions* **81** (2008), 729–748.

[10] A.J. Heeger, "Semiconducting polymers: The third generation" *Chem. Soc. Rev.* **39** (2010), 2354–2371.

[11] H. Shirakawa, E.J. Louis, A.G. MacDiarmid, C.K. Chiang, A.J. Heeger, "Synthesis of electrically conducting polymers: Halogen derivatives of polyacetylene (CH)x" *J. Chem. Soc., Chem. Commun.* (1977), 578–580.

[12] M. Bendikov, F. Wudl, D.F. Perepichka, "Tetrathiafulvalenes, oligoacenenes, and their buckminsterfullerene derivatives: The brick and mortar of organic electronics" *Chem. Rev.* **104** (2004), 4891–4945.

[13] J.E. Anthony, "The larger acenes: Versatile organic semiconductors" *Angew. Chem. Int. Ed. Engl.* **47** (2008), 452–483.

[14] G.A. de Wijs, C.C. Mattheus, R.A. de Groot, T.T.M. Palstra, "Anisotropy of the mobility of pentacene from frustration" *Synth. Met.* **139** (2003), 109–114.

[15] T. Minakata, I. Nagoya, M. Ozaki, "Highly ordered and conducting thin film of pentacene doped with iodine vapor" *J. Appl. Phys.* **69** (1991), 7354–7356.

[16] D.L. Lichtenberger, R.L. Johnston, K. Hinkelmann, T. Suzuki, F. Wudl, "Relative electron donor strengths of tetrathiafulvene derivatives: Effects of chemical substitutions and the molecular environment from a combined photoelectron and electrochemical study" *J. Am. Chem. Soc.* **112** (1990), 3302–3307.

[17] R.N. Compton, C.D. Cooper, "Negative ion properties of tetracyanoquinodimethane: Electron affinity and compound states" *J. Chem. Phys.* **66** (1977), 4325–4329.

[18] M. Sing, U. Schwingenschlögl, R. Claessen, P. Blaha, J.M.P. Carmelo, L.M. Martelo, P.D. Sacramento, M. Dressel, C.S. Jacobsen, "Electronic structure of the quasi-one-dimensional organic conductor TTF–TCNQ" *Phys. Rev. B* **68** (2003), 125111.

[19] J. Ferraris, D.O. Cowan, V. Walatka, J.H. Perlstein, "Electron transfer in a new highly conductive donor–acceptor complex" *J. Am. Chem. Soc.* **95** (1973), 948–949.

[20] M.J. Cohen, L.B. Coleman, A.F. Garito, A.J. Heeger, "Electrical conductivity of tethrathiofulvalinium tetracyanoquinodimethane (TTF)(TCNQ)" *Phys. Rev. B* **10** (1974), 1298–1307.

[21] A.H. Casto Neto, F. Guinea, N.M.R. Peres, K.S. Novoselov, A.K. Geim, "The electronic properties of graphene" *Rev. Mod. Phys.* **81** (2009), 110–155.

[22] A. Jorio, R. Saito, G. Dresselhaus, M.S. Dresselhaus, Chapter 2 in *"Raman Spectroscopy in Graphene Related Systems"* (2011) Wiley-VCH.

[23] K.I. Bolotin, K.J. Sikes, Z. Jiang, M. Klima, G. Fudenberg, J. Hone, P. Kim, H.L. Stormer, "Ultrahigh electron mobility in suspended graphene" *Solid State Commun.* **146** (2008), 351–355.

[24] M. Chhowalla, D. Jena, H. Zhang, "Two-dimensional semiconductors for transistors" *Nat. Rev. Mater.* **1** (2016), 1–15.

[25] M. Ouyang, J.L. Huang, C.M. Lieber, "Fundamental electronic properties and applications of single-walled carbon nanotubes" *Acc. Chem. Res.* **35** (2002), 1018–1025.

[26] M.C. Hersam, "Progress towards monodisperse single walled carbon nanotubes" *Nat. Nanotechnol.* **3** (2008), 387–394.

11   Magnetotransport Materials

11.1   Magnetotransport and Its Applications

While present-day data processing is based on binary digits of electric charge being on and off, a future spin-electronics paradigm may utilize not only the charge but also electron spin. Materials that only carry electrons with spin of one direction (spin-polarized current) would be essential ingredients of spintronic devices. The quest for materials in which magnetization and magnetic order is correlated with spin-polarized conductivity has therefore been one of the research motors in solid state chemistry.

Magnetoresistance, the change in resistivity upon applying an external magnetic field, has already been widely adopted in applications. Computer hard disks have read/write heads made of materials that rely on this property. Magnetoresistance is normally a weak effect occurring due to the distribution of electron velocities. However, if a junction of two conductors with opposite spins is made, one creates a magnetoresistant device called a spin valve (Figure 11.1). One of the materials is chosen so that it can easily flip its magnetization in an external magnetic field. Without a field, the assembly is insulating because the conduction electrons in the two materials are oppositely polarized. Under a field, their spins align parallel, allowing the spin-polarized portion of the current through. If the magnetic field stems from the pattern of magnetized particles on a spinning hard-disk platter, you have converted the recorded information into a time-resolved pattern of electric current. In some materials, the spin flip needed can be induced by a sufficiently large pulse of spin-polarized current. This is termed spin injection and represents a first step towards the holy grail of spintronics; the spin transistor.

Materials with spin-polarized itinerant electrons are closely associated with ferromagnetism. However, the degree of spin polarization in ferromagnetic metals is not high: Jullière [1] studied the change in conductance across two ferromagnetic films, first oriented with parallel and then with antiparallel magnetizations (upon tunneling through an intermediate 100 Å sheet of semiconducting Ge). The maximum relative change in conductance[1] G (=1/R)

---

[1] A conductance (symbol G) in siemens (S) is the inverse of resistance (symbol R) in ohms (Ω).

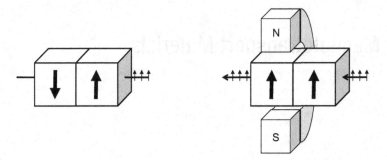

Figure 11.1 Principle of the spin valve with two conductors of strictly opposite spin orientations and an external field that can switch the orientation parallel.

between these orientations, $(G_{par} - G_{apar})/G_{par}$, was a mere 0.14. This gives the spin-polarization fraction $P = 0.27$ with the formula Jullière used,

$$magnetoresistance \equiv \frac{R_{apar} - R_{par}}{R_{apar}} = \frac{G_{par} - G_{apar}}{G_{par}} = \frac{2P^2}{1 + P^2} \qquad (11.1)$$

of **spin polarization** P defined as

$$P = \frac{n^\uparrow(E_F) - n^\downarrow(E_F)}{n^\uparrow(E_F) + n^\downarrow(E_F)} \qquad (11.2)$$

where $n^\uparrow(E_F)$ and $n^\downarrow(E_F)$ are the densities of states at the Fermi level for the respective spin-up and spin-down tunneling electrons. In Section 11.4 we return to this in more detail. Higher degrees of spin polarization may be obtained from itinerant valence-mixing electrons of ferro- or ferrimagnetics such as magnetite Fe_3O_4.

11.2 Charge, Orbital, and Spin Ordering in Iron Oxides

11.2.1 The Verwey Transition in Magnetite, Fe_3O_4

Magnetite, Fe_3O_4, is the oldest known magnetic material, with a long history of use in compasses. It is also one of the earliest examples of a mixed-valence [2] oxide. Fe_3O_4 adopts the inverse spinel structure (more details in Section 1.5.1), $[Fe^{3+}]^{tet}[Fe^{2+}Fe^{3+}]^{oct}O_4$, whereby a mixture of high-spin Fe^{2+} and Fe^{3+} is accommodated on octahedral cation sites. Spectroscopic measurements at ambient conditions show that all iron atoms on the octahedral sites are identical. This is reminiscent of aromatic conjugation in benzene where delocalization of six π electrons over six carbons leads to six equivalent carbon-to-carbon bonds of bond order 1.5. In magnetite, each minority-spin electron of Fe^{2+} is delocalized

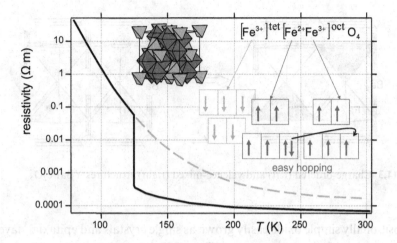

Figure 11.2 Electrical resistivity across the Verwey transition in an Fe_3O_4 single crystal, plotted from data in [4], and extrapolation of the low-temperature trend of relatively easy hopping from Fe^{2+} to Fe^{3+}.

over the octahedral sites. Just as the σ electrons in benzene remain localized, so do the five majority-spin electrons per Fe in Fe_3O_4. Here the analogy ends, though. While conjugation stabilizes benzene by some -150 kJ/mol (released energy), the delocalization in magnetite is weakly endothermic, it consumes about 1 kJ per mole Fe of heat that feeds the increased entropy of the delocalization product when the thermal energy becomes high enough upon warming; just like melting of ice. As a consequence, the delocalized electron localizes upon cooling, and the high-temperature "valence-mixed" state freezes into a low-temperature "charge-ordered" state.

The transition into the valence-mixed[2] state in magnetite is called the **Verwey transition**, after Evert Verwey, who first reported [3] a spectacular drop in electrical resistivity upon heating through the $T = 125$ K transition temperature (Figure 11.2). Magnetite is a good electrical conductor because of its magnetism: The unpaired spins of iron at all octahedral sites are parallel as they point in the opposite direction to the tetrahedral spins, forming a ferrimagnet (Section 9.10). The minority-spin electron of Fe^{2+} can easily hop from one octahedral iron to the other, with no need to flip its spin. On top of this situation, the resistivity suddenly drops upon heating through 125 K when the valence-mixing delocalization sets in.

The valence-mixing electrons are all of the minority spin (Figure 11.2), and this makes the magnetite crystal a source of spin-polarized current; an electrical conductor where the majority of carriers are electrons of one spin only. The fact that magnetite is a

---

[2] For clarity, we shall use "valence mixed" when the "valences" of two oxidation states are thermally mixed via sharing electrons. The low-temperature situation is called "charge ordered". Note that the term "mixed valence" introduced by Robin and Day [2] is general and covers both cases.

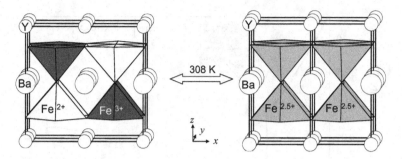

Figure 11.3 Charge-ordered (left) and valence-mixed (right) structures $YBaFe_2O_5$.

compositionally simple oxide easily grown as single crystals and epitaxial[3] layers, with a very high magnetic ordering temperature (T_C = 858 K), makes it an attractive candidate for spintronic applications. Since its crystal structure becomes remarkably complex in the low-temperature charge-ordered state [5], we'll use a perovskite derivative to illustrate a simple valence-mixing process.

11.2.2 Double-Cell Perovskite, $YBaFe_2O_5$

The crystal structure of $YBaFe_2O_5$ [6] is derived from the perovskite aristotype ABO_3 by ordering two types of A atoms, Y and Ba, and removing one of every six oxygens (Figure 11.3). The size difference of Y^{3+} and Ba^{2+} drives their ordering into layers, doubling one of the unit-cell edges. By confining the oxygen vacancies to the yttrium layer, Y^{3+} becomes eight-coordinate, while the larger Ba^{2+} retains a twelve-coordinate environment. More importantly, all iron coordinations become square pyramidal, and the oxidation state of Fe is +2.5 on average.

The charge-ordered $YBaFe_2O_5$ (Figure 11.3, left) undergoes a thermally induced phase transition where (most of) the Fe^{2+} and Fe^{3+} pairs start sharing the minority-spin Fe^{2+} electron upon formation of valence-mixed $Fe^{2.5+}$ (Figure 11.3, right). The two phases coexist while the latent heat is supplied (or removed), like ice and water, marking a first-order transition. The resistivity drops by two orders of magnitude upon heating through the transition, as with magnetite.

The high-temperature structure has all Fe-coordination square pyramids identical, even locally in both space and time. At low temperatures (Figure 11.3, left), chains of Fe^{3+} and chains of Fe^{2+} pyramids pack into the checkerboard arrangement shown. This is surprising because the coulombic minimum for *point charges* would entail Fe^{3+} and Fe^{2+} alternating in all three directions, not just in two. Why does $YBaFe_2O_5$ adopt a structure that is not predicted from electrostatic interactions?

---

[3] An MgO substrate with unit-cell parameter a = 4.21 Å is about a half of a = 8.40 Å for magnetite.

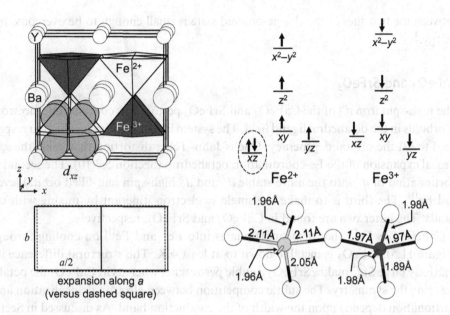

Figure 11.4 Orbital ordering of d_{xz} in $YBaFe_2O_5$ at low temperatures (left). Fe^{2+} and Fe^{3+} coordinations with distances and energies of their d orbitals (right).

Such a violation is indicative of ordering of *spatial charges*, entire orbitals. Expansion along x and contraction along y of the Fe^{2+} square pyramid identify d_{xz} as the ordered orbitals, as illustrated in Figure 11.4, top left. Such a preferential orientation of an orbital throughout the periodic structure is called **orbital ordering**. As the Fe^{2+} d_{xz} orbital is doubly occupied (Figure 11.4, right), the uniaxial expansion is a Jahn–Teller distortion. Because these d_{xz} orbitals are ordered in the periodic structure, we speak of a **cooperative Jahn–Teller distortion**.

Below T_N of about 430 K, the valence-mixed $YBaFe_2O_5$ becomes an antiferromagnet. The antiferromagnetic coupling occurs in all three directions of each double-pyramidal slab of Figure 11.3, right. Only Fe atoms of two different slabs, which face one another across the Y layer, couple ferromagnetically in the valence-mixed state. This parallel alignment of spins in these Fe pairs facilitates the sharing of the minority-spin electron. Upon cooling into the charge-ordered state, the minority spin localizes, these Fe pairs begin to couple antiferromagnetically, and the Fe^{2+} and Fe^{3+} spins alternate in all three directions (the G-type order illustrated in Figure 9.19).

Mössbauer spectroscopy is essential in studies of the local distribution of the iron valence and spin states. Analysis of $YBaFe_2O_5$ spectra reveals the evolution of the minority-spin electron's orbital occupancies from the doubly occupied d_{xz} in the charge-ordered state at low temperatures to shared occupancy at high temperatures by two d_{z^2} orbitals of the thus faintly attracted valence-mixing Fe atoms that face each other across the Y layer [7]. The valence mixing preserves the t_{2g} degeneracy to a large extent and occurs when the orbital-energy difference between the two sites in the charge-ordered state is small enough to be overcome by thermal energy.[4]

11.2.3 CaFeO$_3$ and SrFeO$_3$

The high-spin iron d^4 of the CaFeO$_3$ and SrFeO$_3$ perovskites contains one electron per two e_g orbitals in the octahedral ligand field. The system has three ways in which to respond to the need to lift the orbital degeneracy: One is Jahn–Teller distortion that splits the e_g levels by c-axial expansion of the Fe-coordination octahedron (Section 5.3.10). The second is disproportionation of d^4 into the more stable d^3 and d^5 high-spin half-filled orbital levels of Fe^V and Fe^{III}. The third is to make the single e_g electron itinerant by mixing with oxygen $2p$ bands. The latter two are found in CaFeO$_3$ and SrFeO$_3$, respectively.

CaFeO$_3$ has Fe^{IV} that disproportionates into Fe^V and Fe^{III} on cooling through 290 K (Figure 11.5). SrFeO$_3$ is metallic down to at least 4 K. The structural difference is that the relatively large Sr^{2+} makes SrFeO$_3$ a cubic perovskite, whereas in CaFeO$_3$ the octahedra tilt, lowering the symmetry. The subtle competition between electron delocalization and disproportionation depends upon the width of the conduction band. As discussed in Section 6.6.5, linear M–O–M bonds (as in cubic SrFeO$_3$) lead to a relatively wide conduction band. This favors delocalization of the e_g electrons. Bending the M–O–M bonds due to tilts in CaFeO$_3$ narrows the conduction band, destabilizing the metallic state and favoring disproportionation into a charge-ordered semiconductor[5] below 290 K.

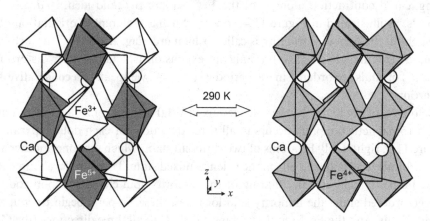

Figure 11.5 Charge disproportionation in CaFeO$_3$.

---

[4] In most general terms, such a difference will not be high when the orbitals about to share the mixing electron are weakly antibonding (practically nonbonding). Localization of the electron will cause only a small expansion of bonds, a small structural distortion; a shallow potential well that will not trap the electron unless the temperature is very low. A strong antibonding character would cause a larger lattice distortion, creating a deep potential well that could trap the electron even at high temperatures.

[5] Flat (narrow) bands have a small curvature, small second derivative, and this leads to a large effective mass (Section 10.2.5) that favors localization.

Box 11.1 Characterizations: Mössbauer spectroscopy

While resonant X-ray absorption creates excited states in the electron shell of atoms, resonant γ-ray absorption excites the nucleus. The complication is that the γ photon has such a high energy that its momentum recoils the atom it hits in a solution or gas. The energy loss then prevents resonant fluorescence by that atom; a radiative emission that would carry meaningful information. Rudolf Mössbauer realized that atoms in solids might allow absorption of the γ photon without recoil if the photon momentum brings less energy than that of a vibronic quantum.* The fraction of absorption events that proceeds in such a manner is called the recoilless fraction.

Only relatively low-energy γ radiation gives a high recoilless fraction. The most popular is the radiation associated with the transition of the metastable excited state $I = 3/2$ (of the nuclear-spin angular momentum I; mean lifetime 0.14 µs) to the stable state $I = 1/2$, in the ^{57}Fe isotope (about 2% of natural iron). Radioactive ^{57}Co is used as the γ source (the half-life is 272 days). About 90% of its decay path goes via the excited $I = 3/2$ state of ^{57}Fe that then de-excites via two main paths: In 89% of cases, it ejects an electron. In 11% of cases, it emits a γ photon of energy 14.412 keV. This photon is then resonantly absorbed in ^{57}Fe of the sample. The energy peak of this photon is extremely sharp; 14412 eV high yet only 5×10^{-9} eV wide at half its height. This extreme sharpness allows tuning the incoming γ energy over the narrow span of the γ-absorption peaks by the Doppler effect. All that is needed is to mechanically oscillate the source back and forth relative to the flat transmission sample. As an example, a velocity $v = 1$ mm/s changes the γ-photon energy by a fraction v/c, hence by 48×10^{-9} eV; about ten times the peak's natural width. With this resolution, the so-called hyperfine structure of the resonance absorption appears as shifts and splitting of absorption energies, as illustrated in the figure below. A disadvantage is that only samples containing ^{57}Fe, ^{119}Sn, ^{151}Eu, and ^{197}Au do not have to be cooled down in order to limit the recoil. The ultimate disadvantage is that some elements have no Mössbauer-suitable isotopes.

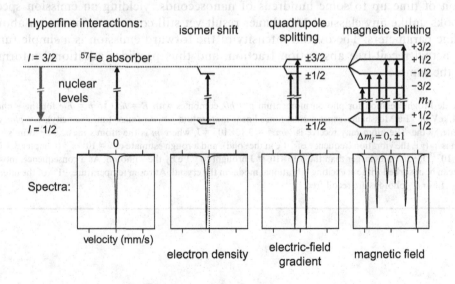

Box 11.1 (cont.)

The hyperfine features are isomer shift, quadrupole splitting, and magnetic splitting. The isomer shift reflects the chemical environment around the ^{57}Fe nucleus—oxidation state, covalency, coordination number—anything that affects the electron density of the nearby $3s$ and $4s$ electrons (only s electrons have non-zero probability at the nucleus). The quadrupole splitting is caused by the electric-field gradient that the ^{57}Fe nucleus experiences as a result of its own valence electrons and the surrounding ligands. It tells us something about the symmetry of its coordination environment. The magnetic splitting is caused by cooperative magnetism as seen at the nucleus. It is very sensitive to both local and long-range magnetic ordering. The isomer shift, the "electric" quadrupole splitting, and the magnetic splitting build on each other, and the mutual orientation of the electric and magnetic fields affects the shape of the sextet of peaks. The ^{57}Fe Mössbauer spectrum therefore provides an account of all local valence, coordination, and spin states of iron in the sample.

Two additional techniques take a different approach. One is conversion-electron Mössbauer spectroscopy, a reflection technique for thin-film samples (~100 nm) or very thick samples that can't be thinned. It utilizes the 89% probability of the sample's excited ^{57}Fe state not emitting a γ photon but ejecting an electron. The ejected electrons are detected in backscattering mode, together with secondary electrons emitted by the X-ray radiation released when higher-lying electrons fill the holes vacated by the ejected electrons. Simplified, the spectrum is the inverse of the absorption spectrum described above, but probes only a thin surface layer of the measured sample. The second method utilizes synchrotron γ radiation (6–30 keV) with high-intensity pulses. After exciting the ^{57}Fe atoms in the sample by an extremely short γ pulse (0.1 ns), a coherent (fluorescent) γ re-emission upon de-excitation of the atoms occurs in the forward direction (while the incoherent portion emits in all directions). It is detected as a function of time up to some hundreds of nanoseconds, yielding an emission spectrum that looks unlike any classical Mössbauer result yet still contains information about the hyperfine parameters. The overall intensity of the forward emission is a simple function of the actual recoil-free absorption fraction, and thus provides additional information about the sample.

---

* The de Broglie relation for photon momentum $p = h/\lambda$ combines with $E = hc/\lambda$ to $p = E/c$ for the γ photon of $E = 14.4$ keV in ^{57}Fe Mössbauer experiments. Since linear momentum $p = mv$ (mass times velocity), $v = p/m$, and the kinetic energy the ^{57}Fe atom may receive is $\frac{1}{2}mv^2 = 3.13\times10^{-22}$ J, where m is the atom's mass. The atom's vibronic quantum is $h\nu$ (ν is the vibration frequency of ^{57}Fe in the solid), and a rough estimate of $\nu = 10^{13}$ s^{-1} (Chapter 3.3.1) yields $h\nu = 66\times10^{-22}$ J; ~20 times larger than the 3.13×10^{-22} J brought to ^{57}Fe by the γ photon. As a consequence, most of the photons can be absorbed without exciting vibrational modes in the crystal. At room temperature, 91% of the interactions between ^{57}Fe and γ photons are recoil-free.

11.3 Charge and Orbital Ordering in Perovskite-Type Manganites

The instability of the d^4 configuration encountered in the previous section on $CaFeO_3$ and $SrFeO_3$ is an important ingredient of magnetotransport in the perovskite oxides containing trivalent and tetravalent manganese. Before we proceed by way of example to the series $La_{1-x}Ca_xMnO_3$, let's have a look at the pure parent phases.

11.3.1 Spin and Orbital Ordering in CaMnO₃ and LaMnO₃

$CaMnO_3$ is isotypical but not isoelectronic with $CaFeO_3$. The nearly linear d^3–d^3 super-exchange interaction in $CaMnO_3$ is mediated by an oxygen-anion p orbital linking the empty e_g orbitals of neighboring Mn^{4+} cations. The top of Figure 9.20 shows that this dictates alternating electron spins across the oxygen links. All superexchange interactions are anti-ferromagnetic, and the spin order is therefore G-type (see Figure 9.19, top left). The degenerate half-filled d^3 shell of pseudospherical symmetry does not permit any charge or orbital ordering in stoichiometric $CaMnO_3$.

Stoichiometric $LaMnO_3$ and $CaFeO_3$ are isotypical and their transition-metal ions are isoelectronic, yet the two compounds find different solutions to the d^4 instability. Mn^{3+} has a lower electronegativity than Fe^{4+}, hence reduced covalency that narrows the $Mn(e_g)$–$O(2p)$ conduction band and prevents itinerancy of the single e_g electron. Instead, the coordination octahedron removes the e_g degeneracy by uniaxial expansion (Jahn–Teller distortion) to a 4.36 Å apex-to-apex distance (versus 3.81 Å and 3.92 Å for the other two distances) [8]. This lowers the d_{z^2} orbital energy relative to $d_{x^2-y^2}$ and accommodates the single electron in d_{z^2}. The d_{z^2} orbitals follow the elongated axis of each tilted octahedron, an orbital ordering, see Figure 11.6.

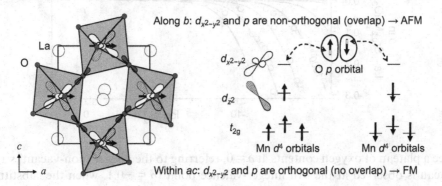

Figure 11.6 $LaMnO_3$: Ordering of half-filled Mn d_{z^2} orbitals (dashed, shape simplified), via uniaxial Jahn–Teller distortion of Mn coordination octahedra (*Pnma*, neutron diffraction at 9 K [8]), yields ferromagnetism (FM) in the ac plane and antiferromagnetic coupling (AFM) along b. See text for description of superexchange interactions.

Box 11.2 Synthetic Methods: Valence control via O_2 partial pressures

LaMnO$_{3+\delta}$ has a relatively wide oxygen nonstoichiometry range of about $-0.1 < \delta < 0.3$. Upon partial substitution of La^{3+} with suitable alkaline-earth cations, the range remains wide but shifts to lower values. How can one synthesize a material with a specific desired mixed valence of Mn?

From Chapter 3, we know that equilibrium is established between point defects of the oxide and O$_2$(g) at high temperatures. Let's start with LaMnO$_3$ and let's say we are not aware of the actual type of intrinsic defects in it, so we simply consider a plain oxygen deficit and excess. This defines anion-Frenkel defects (Figure 3.1) that act as compensators of the Mn redox equilibria we wish to control. Following Table 3.2, we write equations for oxidation, $O_{2(g)} = 2O_i'' + 4h^\bullet$; reduction, $2O_O^\times = 2v_O^{\bullet\bullet} + 4e' + O_2$(g); the intrinsic anion-Frenkel defect formation, $O_O^\times = v_O^{\bullet\bullet} + O_i''$; and the intrinsic ionization, nil $= e' + h^\bullet$. The respective mass-action terms are: $K_{ox} = [O_i'']^2[h^\bullet]^4 \cdot p_{O_2}^{-1}$, $K_{red} = [v_O^{\bullet\bullet}]^2[e']^4 \cdot p_{O_2}$, $K_F = [O_i''][v_O^{\bullet\bullet}]$, and $K_i = [e'][h^\bullet]$, where square brackets mean molar fractions. Let's now acceptor dope LaMnO$_3$ into La$_{1-x}$Sr$_x$MnO$_3$. The electroneutrality condition (Section 3.1.5) states $2[O_i''] + [e'] + [Sr_{La}'] = [h^\bullet] + 2[v_O^{\bullet\bullet}]$. From this, with the help of the mass-action terms, one evaluates the $f([O_i'']) = 0$ and $f([v_O^{\bullet\bullet}]) = 0$ polynomial functions for the oxygen defects, with p_{O_2} as the independent variable and three of the four equilibrium constants as parameters. Each of the equilibrium constants, K, is associated with a standard reaction-enthalpy and -entropy change ΔH and ΔS, as $K = \exp[(\Delta S/R) - (\Delta H/RT)]$. This determines the temperature dependence. The La$_{1-x}$Sr$_x$MnO$_{3+\delta}$ oxygen non-stoichiometry, $\delta = [O_i''] - [v_O^{\bullet\bullet}]$, is least-squares fitted to the $\delta = f(p_{O_2}, T)$ data obtained by experiment [9] and plotted over a range of temperatures for a given x:

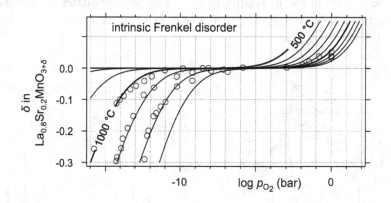

We see a plateau of oxygen contents at $\delta = 0$, referring to the integer (non-vacant) structure. No plateau is observed at the Mn integer-valence point ($\delta = -0.1$, when the substitution is compensated by oxygen vacancies). This implies that electronic defects strongly outnumber anion-Frenkel defects ($K_i > K_F$ more than $K_i > K_S$ in Figure 3.4).

Say, we now realize from the literature [10, 11] that the intrinsic defects in LaMnO$_{3+\delta}$ actually are of the Schottky type. We can approach it as an M$_2$O$_{3+\delta}$ oxide (M = La, Mn), in

Box 11.2 (cont.)

which metal and oxygen vacancies are the intrinsic defects. The new equation for oxidation is $O_2(g) = (4/3)v_M''' + 4h^\bullet + 2O_O^\times$, the Schottky-defect formation nil $= 2v_M''' + 3v_O^{\bullet\bullet}$, the other two equations remain the same. The corresponding new mass-action terms are: $K_{ox} = [v_M''']^{4/3}$ $[h^\bullet]^4 \cdot p_{O_2}^{-1}$ and $K_S = [v_M''']^2 [v_O^{\bullet\bullet}]^3$. The electroneutrality condition for the acceptor-doped phase $La_{1-x}Sr_xMnO_3$ is $3[v_M'''] + [e'] + [Sr_{La}'] = [h^\bullet] + 2[v_O^{\bullet\bullet}]$. From it, with the help of the mass-action terms, the $f([v_M''']) = 0$ and $f([v_O^{\bullet\bullet}]) = 0$ polynomial functions are evaluated, and the oxygen nonstoichiometry $\delta = (3/2)[v_M'''] - [v_O^{\bullet\bullet}]$ is fitted to the experimental $\delta = f(p_{O_2}, T)$ data; here again for $La_{0.8}Sr_{0.2}MnO_{3+\delta}$:

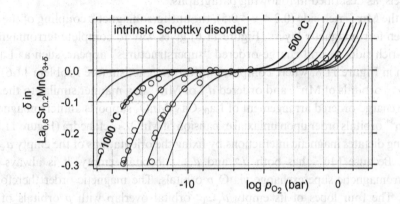

Because both Frenkel and Schottky defects involve oxygen vacancies, the reductive non-stoichiometry is the same in both plots, only the oxidative parts differ slightly. The message for $La_{0.8}Sr_{0.2}MnO_{3+\delta}$ synthesis is the same. Despite Sr doping and high temperatures both promoting O_2 release, some extra oxygen ($\delta > 0$) is still accommodated at all feasible temperatures under syntheses in air where $\log(p_{O_2}/bar) = -0.71$. Graphs like these help finding the right p_{O_2} and temperature to equilibrate the product before it is rapidly quenched to obtain the desired oxidation state of Mn.

Ordering of the half-filled d_{z^2} orbitals dictates the orientation of the empty $d_{x^2-y^2}$ orbitals that participate in superexchange interactions via oxygen atoms. Along the b axis, we get antiferromagnetic superexchange just as we did in all three directions in $CaMnO_3$. In the ac plane, the p orbital of each O overlaps on one side with the empty spin-up sub-band $d_{x^2-y^2}$ of Mn; on the other side it points against the half-occupied d_{z^2}, hence overlaps with its spin-down sub-band, making the ac-plane spins parallel (the coupling is analogous to the d^3–d^5 case shown in Figure 9.20). The bulk magnetic order is antiferromagnetic ($T_N = 140$ K), of the type A shown in Figure 9.19, top right.

11.3.2 The La$_{1-x}$Ca$_x$MnO$_3$ Phase Diagram

Although the most interesting magnetoresistance properties of La$_{1-x}$Ca$_x$MnO$_3$ occur in a limited range of compositions around $x \approx \frac{1}{3}$, their understanding is assisted by considering the wider structural and magnetic phase diagram of this solid solution. Let's recollect that the magnetic order of LaMnO$_3$ is antiferromagnetic A type whereas CaMnO$_3$ is G type. At high temperatures, solid LaMnO$_3$ and CaMnO$_3$ are completely miscible, paramagnetic, with Mn^{3+} and Mn^{4+} charges mixed. The phase diagram in Figure 11.7 shows that at the La$_{0.5}$Ca$_{0.5}$MnO$_3$ composition, a 1:1 ordering of the Mn charges occurs below the critical charge-ordering temperature, T_{co}. Simultaneously, the conflict of differing magnetic orders unravels, as described in following paragraphs.

On the Mn^{3+}-rich side ($0.2 < x < 0.5$), the antiferromagnetic coupling of the LaMnO$_3$ type between ferromagnetic layers (Figure 11.6) gives way to complete ferromagnetism. On the Mn^{4+}-rich side, several charge-ordered "superstructures" appear, such as La$_{1/3}$Ca$_{2/3}$MnO$_3$ shown in Figure 11.8, where complex antiferromagnetism (see Problem 11.6) occurs due to empty e_g orbitals of Mn^{4+} and ordered d_{z^2} of Mn^{3+} in a manner similar to the next example.

The charge-ordered arrangement of La$_{0.5}$Ca$_{0.5}$MnO$_3$ supports antiferromagnetism when the d_{z^2} Mn^{3+} orbitals order upon uniaxial expansion of MnO$_{6/2}$ octahedra (Figure 11.9). This orbital ordering dictates magnetic interactions by fixing the orientations of the empty $d_{x^2-y^2}$ orbitals of Mn^{3+}. Because Mn^{4+} has both d_{z^2} and $d_{x^2-y^2}$ orbitals empty, it is always able to make antiferromagnetic superexchange via O p orbitals. The magnetic order therefore depends on Mn^{3+}. The four lobes of its empty $d_{x^2-y^2}$ orbital overlap with p orbitals of four oxygens

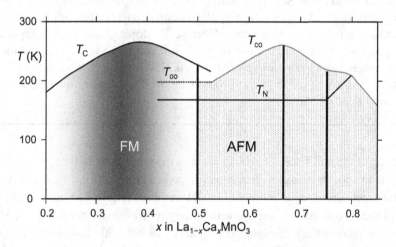

Figure 11.7 La$_{1-x}$Ca$_x$MnO$_3$ phase diagram. Ordering of spins (ferromagnetic below T_C, antiferromagnetic below T_N), of orbitals below T_{oo}, and of the Mn^{3+} and Mn^{4+} charges below T_{co}. Compiled from a variety of sources. Vertical lines mark compositions for which the actual ordered "superstructure" was determined experimentally. Stability regions of different phases change with magnetic field.

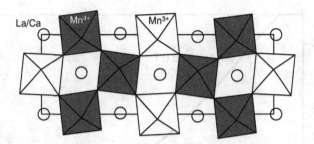

Figure 11.8 Charge ordering below T_{co} in $La_{1/3}Ca_{2/3}MnO_3$ [12].

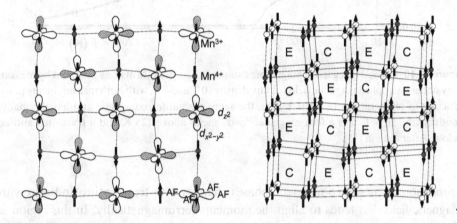

Figure 11.9 Magnetism in charge- and orbital-ordered $La_{0.5}Ca_{0.5}MnO_3$ (dark Mn^{4+}, white Mn^{3+}). Left: A plane with ordered half-filled d_{z^2} Mn^{3+} orbitals (hatched) that fit the structural distortions. The empty $d_{x^2-y^2}$ orbitals each make four antiferromagnetic Mn–O–Mn couplings via p orbitals of oxygen (not shown) with empty e_g orbitals (not drawn) of two Mn^{4+} (in plane) and two Mn^{3+} (out of plane) neighbors. Right: The resulting CE-type order of spins in a slightly tilted view for clarity. After [13,14].

mediating four antiferromagnetic couplings towards Mn neighbors. The other two couplings are ferromagnetic.[6] The total antiferromagnetic order combines the C and E types introduced in Figure 9.19, and is denoted CE.[7]

The Mn^{3+}-rich ferromagnetic phase and the CE-type antiferromagnetic phase compete when the composition parameter x increases towards the $x = 0.5$ line in Figure 11.7. The

---

[6] Note in Figure 11.9 that the half-filled d_{z^2} orbitals of Mn^{3+} expand the Mn^{4+}–O–Mn^{3+}–O–Mn^{4+} distance, along which the three Mn couple ferromagnetically.

[7] To compare the C (or E) cells identified in Figure 11.9 with the types of antiferromagnetic order in Figure 9.19, sketch any C (or E) cell from Figure 11.9 with a + and − on the corners for the two opposing moment orientations, do the same with the dotted cell in Figure 9.19, and rotate till they unite.

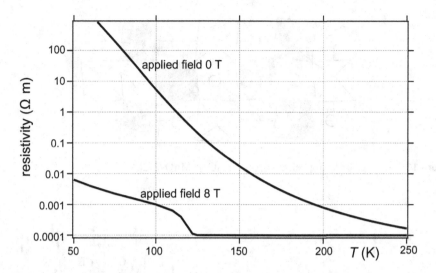

Figure 11.10 Colossal (negative) magnetoresistance in $La_{0.5}Ca_{0.5}MnO_3$, as observed by measuring resistivity under an applied magnetic field (of flux density) 0 T and 8 T, with both measurements performed as a function of temperature. In the 0 T field, the sample is antiferromagnetic and has thermally activated conductivity. In 8 T, it is a ferromagnetic "bad" metal above 125 K and a frustrated antiferromagnet below. After data in [15].

proportion of these two magnetic phases changes with temperature and also with applied magnetic field that tends to align the moments ferromagnetically. In this region where the phases compete, the increased filling of e_g electrons makes metallic delocalization the preferred method of removing the e_g degeneracy, further stabilizing the ferromagnetism by double-exchange interactions (Section 9.9.1). In contrast to this metallic delocalization, the competing antiferromagnetic phase has a thermally activated conductivity of localized electrons. As a consequence, up to a six-orders-of-magnitude decrease in electrical resistivity (Figure 11.10) may be observed in such samples upon exposure to a strong magnetic field. This negative[8] magnetoresistance observed in manganites is therefore termed **colossal magnetoresistance**.

11.3.3 Tuning the Colossal Magnetoresistance

Replacement of the large La^{3+} in $La_{0.5}Ca_{0.5}MnO_3$ by smaller trivalent lanthanide ions reduces the width of the $Mn(e_g)$–$O(p)$ conduction band by increasing the octahedral tilting that bends the Mn–O–Mn angles (Table 6.2). The narrow band favors electron localization (Section 10.2.5) and stabilizes the charge-ordered phase. This also means that the range over which the ferromagnetic and antiferromagnetic phases coexist in Figure 11.7 (controllable by

---

[8] It is called "negative" because the resistance decreases under the magnetic field.

the magnetic field) shifts to lower x values. Phases like $Nd_{0.7}Ca_{0.3}MnO_3$ show the typical two-phase instabilities and colossal magnetoresistance.

Replacement of Ca^{2+} with larger Sr^{2+} destabilizes the charge-ordered phase in favor of the metallic ferromagnetic phase because the reduced octahedral tilting increases the bandwidth, and this favors delocalized electrons. This general trend can be seen in Figure 11.11. While $La_{0.55}Sr_{0.45}MnO_3$ (on the right) is a metallic ferromagnet with high T_C, its calcium analogue $La_{0.55}Ca_{0.45}MnO_3$ (middle) has a much lower T_C, and replacement of La with a smaller lanthanide converts it into a charge-ordered antiferromagnet such as $Nd_{0.55}Ca_{0.45}MnO_3$ (left). As discussed below, two parameters are responsible for this complex behavior—the general size effect and the local variance effect of the solid solution at the A site.

From Chapter 1, we know how the size of the perovskite A-site cation determines the octahedral tilting through the tolerance factor. As discussed in Sections 10.4.3 and 6.6.5, the tilting in turn controls the width of the $Mn(e_g)$–$O(p)$ conduction band and thus the stability of the ferromagnetic metallic state. When the cations occupying the A site have similar size, the properties can be easily tuned by controlling the average radius. However, in most cases these cations differ in size, often considerably. Then we have to consider not only the average size, but also the mismatch in size of the A-site cations, which creates additional local distortions. A suitable parameter to evaluate the mismatch is the statistical variance σ^2 of ionic radii. Termed the **variance effect** [16], it is expressed as $\sigma^2 = \sum_i (x_i r_i^2 - r_A^2)$, where $r_A^2 = \sum_i x_i r_i^2$ is the weighted average of ionic-radii

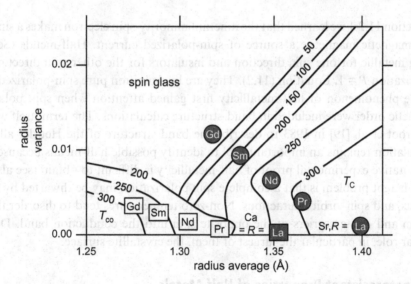

Figure 11.11 Magnetic behavior in a contour map of charge-ordering (T_{co}) and Curie-temperature (T_C) isotherms (in K), in the A-site ionic-radius [17] variance–average plot [18] for $R_{0.55}AE_{0.45}MnO_3$ (R = lanthanoid, $AE = Ca_{1-x}Sr_x$ and $Sr_{1-x}Ba_x$). Of many such compositions, two paths are marked, AE = Ca with squares, AE = Sr with circles (ferromagnetic shaded, antiferromagnetic not).

squares for the in total i atoms of fractional site occupancies x_i at A. With fine compositional scaling in terms of a three-atom solid solution at the A site, the effects of the size parameter and of the variance parameter can be discerned [17]. The result is shown in the contour plot of Figure 11.11 that maps the stability of the charge- and spin-ordered variants of $R_{0.55}AE_{0.45}MnO_3$ (R = lanthanoid, AE = $Ca_{1-x}Sr_x$ and $Sr_{1-x}Ba_x$) in terms of their ordering temperature.

The contour map of isotherms in Figure 11.11 explains all. Low variance (equal size) means low size-induced disorder at the A site, which favors either the ferromagnetic metallic phase or the charge-ordered antiferromagnetic phase, which in turn depend on the average radius at the A site as described above. Increasing the variance, i.e. the local size disorder, quickly disrupts the charge ordering (as seen on left) that facilitates the antiferromagnetic order, yet much less so the ferromagnetic spin ordering (seen on right). Ferromagnetism persists to moderate values of variance, but eventually a magnetically disordered spin-glass phase wins out.

In total, three compositional parameters tailor magnetotransport properties of these perovskite manganates: The doping level of the alkaline-earth ion(s), which controls the oxidation state of Mn. The average radius at the perovskite A site, which varies the stability of the antiferromagnetic charge-ordered versus ferromagnetic states. The A-site cations' size variance, which fine tunes the critical competition between these two.

11.4 Half-Metals and Spin-Polarized Transport

In Section 11.2.1 we learned that the itinerant minority-spin electron makes a single crystal of ferromagnetic magnetite a source of spin-polarized current. Half-metals (Section 9.9.1), being metallic for one spin direction and insulators for the other spin direction, have spin polarization $P = 1$, Equation (11.2). They are by definition pure spin-polarized conductors.

The phenomenon of half-metallicity first gained attention when spin polarization and magnetic order were included in band-structure calculations. The term itself was coined by de Groot et al. [19] in 1983 to describe the band structure of the Heusler alloy NiMnSb. Calculation remains an important tool to identify possible half-metals because specific and quantitative experimental proof of half-metallicity is difficult to obtain (see also Box 11.3). An inherent problem is that a complete spin polarization may be thwarted by temperature, defects, and spin–orbit interactions. Non-zero temperatures tend to disorder the ferromagnetism and excite carriers of the opposite spin into the conduction band. Defects play a similar role, in particular the largest of them, the crystallite surface.

11.4.1 Magnetoresistant Properties of Half-Metals

Tunneling magnetoresistance is one of the potentially interesting transport effects that can arise between two half-metallic electrodes separated by a thin tunneling barrier. Let's consider in more detail the previously mentioned Jullière experiment [1] (Section 11.1) with two *ferromagnetic* epitaxial films as the electrodes. The tunneling from one electrode (emitter, filled states) to the other electrode (collector, empty states) occurs in two independent channels, one for ↑ and one for ↓ electrons (spin conservation assumed). The two conductances, $G^\uparrow$ and $G^\downarrow$, are each assumed to be proportional to the product of the densities of states[9] at the Fermi level for the two ferromagnetic electrodes, for example $G^\downarrow = n_1^\downarrow \cdot n_2^\downarrow$ (the qualifier E_F is omitted for simplicity). Let's now consider varying directions of magnetization at the two electrodes. Two limiting cases emerge—G_{par} for the two ferromagnetic films of parallel and G_{apar} of antiparallel magnetization:

$$G_{par} = n_1^\uparrow n_2^\uparrow + n_1^\downarrow n_2^\downarrow \tag{11.3}$$

$$G_{apar} = n_1^\uparrow n_2^\downarrow + n_1^\downarrow n_2^\uparrow \tag{11.4}$$

In a magnetoresistance experiment, a strong enough applied field turns the moments parallel. The maximum tunneling magnetoresistance, TMR, is the normalized difference between G_{par} and G_{apar}. The normalization is performed against either value, giving two alternative expressions justifying the Jullière formula[10] in Equation (11.1),

$$TMR \equiv \frac{R_{apar} - R_{par}}{R_{apar}} = \frac{G_{par} - G_{apar}}{G_{par}} = \frac{(n_1^\uparrow n_2^\uparrow + n_1^\downarrow n_2^\downarrow) - (n_1^\uparrow n_2^\downarrow + n_1^\downarrow n_2^\uparrow)}{(n_1^\uparrow n_2^\uparrow + n_1^\downarrow n_2^\downarrow)}$$

$$= \frac{2P_1 P_2}{1 + P_1 P_2} \tag{11.5}$$

$$TMR \equiv \frac{R_{apar} - R_{par}}{R_{par}} = \frac{G_{par} - G_{apar}}{G_{apar}} = \frac{(n_1^\uparrow n_2^\uparrow + n_1^\downarrow n_2^\downarrow) - (n_1^\uparrow n_2^\downarrow + n_1^\downarrow n_2^\uparrow)}{(n_1^\uparrow n_2^\downarrow + n_1^\downarrow n_2^\uparrow)}$$

$$= \frac{2P_1 P_2}{1 - P_1 P_2} , \tag{11.6}$$

where spin polarizations P_i of the two electrodes 1 and 2 were introduced by substitution for $n_1^\uparrow$ and $n_2^\uparrow$ from Equation (11.2). The formula in Equation (11.5) has a maximum $TMR = 1$, whereas the maximum is ∞ for Equation (11.6). Both are used in the literature, along with R_{apar}/R_{par}.

As noted in Section 11.1, ferromagnetic metals have low P values. Even with half-metals, only single-crystalline films have P_1 and P_2 close to 1 that would yield high TMR for the two spin alignments at the two sides of the tunneling barrier [20]. Since the grain and domain structure of half-metallic films can be difficult to control, attention has been given to **powder magnetoresistance**, the change in conductance/resistance of half-metal powder compacts upon a change in the external magnetic field. Its advantage is the isotropy of powders that eliminates directional variations. The insulating surfaces of the individual powder grains are the tunneling barriers, as electrons have to percolate across many grains of random spin

---

[9] More precisely, to an effective value called tunneling density of states.

[10] The version with the + sign appears in Jullière's article [1] but now both versions are in use.

orientation. After Coey and Venkatesan [20], we simplify the general result of Guinea [21] to assume that the probability of magnetoresistant tunneling between two average half-metallic grains i and j of the mixture varies as $\cos^2(\theta_{ij}/2) = (1 + \cos\theta_{ij})/2$ of the average angle $0° < \theta_{ij} < 90°$ between any two grain spins in the mixture, and so does the conductance G:

$$G \propto \frac{1 + \cos\theta_{ij}}{2} \qquad (0° < \theta_{ij} < 90°) \tag{11.7}$$

The maximum observed TMR is the normalized difference in G (or in $R = 1/G$). Figure 11.12 illustrates the effect of the external magnetic field on the average angle. The $\theta_{ij} = 0°$ case is the strong-field limit for all spins aligned parallel, and resistance R_{par} is measured in Figure 11.12. The $\theta_{ij} = 90°$ is the average angle of particle pairs, randomly distributed between having spins 0° parallel and 180° antiparallel, for the demagnetized powder at the coercive field, and R_{random} is measured in Figure 11.12. The spin-aligned case means $\cos\theta_{ij} = 1$, say $G_{par} = 1$ after Equation (11.7). The random demagnetized powder has a $\cos\theta_{ij}$ average equal to 0, and $G_{random} = ½$. For the two types of normalization of the resistance difference given in Equations (11.5) and (11.6), the following maxima for the TMR in powder compacts of half-metals are obtained, respectively:

$$TMR \equiv \frac{R_{random} - R_{par}}{R_{random}} = \frac{G_{par} - G_{random}}{G_{par}} = \frac{1 - ½}{1} = ½ \tag{11.8}$$

$$TMR \equiv \frac{R_{random} - R_{par}}{R_{par}} = \frac{G_{par} - G_{random}}{G_{random}} = \frac{1 - ½}{½} = 1 \tag{11.9}$$

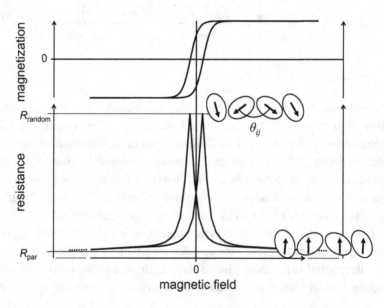

Figure 11.12 Magnetoresistance of a powder compact and magnetic configurations in a percolative conduction link at saturation and coercivity, after [22].

Given that half-metals at non-zero temperatures never have 100% spin polarization, similar spin polarizations can also be achieved for materials that are not strictly half-metallic: EuO (becomes ferromagnetic at low temperatures; Section 9.9.1) has a calculated band gap [23] of 3.4 eV for one spin direction and 2.5 eV for the other. The $La_{1-x}Ca_xMnO_3$-type perovskites have a small minority-spin density at the Fermi level, which makes them nearly half-metallic spin-polarized conductors and contributes to the exhibited variety of magneto-transport features. Similarly, magnetite at ambient conditions has a nearly zero band gap for minority-spin electrons and ~0.7 eV for majority-spin d electrons. The resulting spin polarization is sometimes termed **transport half-metallicity** or transport spin polarization. In some connotations, these phases are simply counted among half-metals as well.

Half-metals can be categorized in several ways. One of them considers the origin of the spin-polarized itinerant electron: (1) half-metals with integer-valence itinerant electrons, such as CrO_2 and NiMnSb; and (2) half-metals with valence-mixing itinerant electrons, such as Fe_3O_4 and Sr_2FeMoO_6. Coey and Venkatesan [20] list a more detailed categorization of half-metals that also includes the transport half-metals.

Box 11.3 Characterizations: Experimental proofs of half-metallicity

Magnetization of half-metals, in contrast to magnetic order in metals, leads to integer values of the saturated magnetic moments per magnetic atom because of the existence of the insulating spin direction in which the number of localized electrons per atom must be an integer dictating the number of electrons in the itinerant spin direction. The integer magnetic moment is a necessary but not sufficient condition of half-metallicity.

Positron annihilation is a selective but demanding test of half-metallicity. A half-metal has a Fermi surface for one spin direction only, not for both as in ordinary metals. Spin-polarized positrons radiating from a suitable beta source, such as ^{22}Na, would annihilate with electrons of opposite spin at the Fermi surface of a half-metallic thin film or single crystal. Each individual annihilation forms two γ photons, the directions of which conserve the linear momentum and energy that are detected and evaluated. One of a small number of such studies was performed on half-metals and concerned NiMnSb [24].

Andreev reflection [25] occurs for conduction across a metal–superconductor interface. Say, a $\downarrow$ electron crosses from the metal into a superconductor well below its T_c. As it does, it has to grab a $\uparrow$ electron from the metal it leaves because in the superconductor it will only move as a Cooper pair ($\downarrow\uparrow$). This leaves a backscattered or "reflected" hole in the $\uparrow$ band of the metal, doubling the conductance $G = dI/dV$ (for voltage $V \to 0$) against what would be achieved if the collector was non-superconducting. In half-metals, only the $\downarrow$ band is occupied, there is no $\uparrow$ electron to grab, and the Andreev reflection is suppressed [26]. One observes that for $V \to 0$, the conductance dI/dV has a minimum. The spin polarization P is given by the position of the actual conductance at $V \to 0$ on the scale between this minimum ($P = 1$) and the doubling maximum of the

---

Box 11.3 (cont.)

Andreev reflection ($P = 0$). Because the Andreev reflection requires very good interface contacts with little or no tunneling, its application on half-metals suffers from limitations for half-metallicity at the surface of the solid.

Spin-resolved photoemission directly yields the degree of spin polarization on oriented surfaces. The information is deduced from the spatial energy distribution of electrons emitted from thin films or single crystals of half-metals by polarized light. Analogously, spin-resolved inverse photoemission uses a low-energy polarized-electron beam and measures the distribution of the emitted photons. The disadvantage is that both methods only see a few atomic layers below the surface, and the obtained spin polarization depends on how well the surface of the given crystal structure reconstructs to carry the half-metallicity of the bulk. As a result, it is not easy to compare spin polarization values among different half-metals.

11.4.2 CrO$_2$

Half-metallicity of CrO$_2$ was identified by Schwarz [27] in 1986. Chromium dioxide at that time was a popular magnetic tape-recording material. The ionic model of CrO$_2$ implies a $3d^2$ electron configuration, and the integer spin-only ferromagnetic moment of 2.0 μ_B per Cr at 0 K suggests that the CrO$_2$ ferromagnetism mentioned in Section 9.9.1 is not one of itinerant electrons such as in Fe or Ni (Section 9.9.2). Actually, in a simple picture, superexchange interactions in its rutile-type structure could well make it an antiferromagnet, and the integer local moment would be as expected for a semiconductor. Yet CrO$_2$ is a ferromagnet and metallic conductor [28]. Two circumstances are essential for CrO$_2$ ferromagnetism: (1) The CrO$_6$ octahedra are slightly deformed into two short and four long Cr–O bonds [29]. The latter expand the xy plane of the octahedron, which lowers the d_{xy} orbital below the energy of d_{xz} and d_{yz}, so that the d_{xy} orbital accommodates one electron of the two. (2) The high Cr^{4+} electronegativity lends high covalency to the d_{xz} and d_{yz} interaction with the oxygen p orbitals. The final result is that while the electron in the d_{xy} orbital is localized, the other two orbitals accommodate the second Cr electron in a band with π^* character. The π^* electrons are itinerant and the ensuing double-exchange interaction favors ferromagnetism. Hund's rule favors parallel alignment of the two electrons, accounting for the saturation moment of 2.0 μ_B per Cr as the temperature approaches 0 K [28].

The half-metallicity of CrO$_2$ at 0 K is manifested by spin polarization of nearly 1.0 by Andreev reflection (Box 11.3) at freshly made point contacts at liquid-helium temperatures. Owing to the natural surface layer of antiferromagnetic Cr$_2$O$_3$, tunneling magnetoresistance of CrO$_2$ powder compacts at low temperatures also yields high levels of spin polarization. However, increasing temperature brings a strong contribution from hopping (not polarized) conductivity, and this effectively eliminates the tunneling contribution. At room temperature, the *TMR* is only ~3% [30].

11.4.3 Heusler Alloys

Half-metallicity occurs in some Heusler intermetallics such as NiMnSb or Co_2MnSb (Section 1.4.2). A simple picture of the electronic band structure that elucidates the half-metallic behavior is obtained by considering these alloys as Zintl phases (Section 1.5.6), where the more electronegative Sb atom achieves an octet, Sb^{3-}. In the following treatment, we assume these anions cause a crystal-field splitting of the transition-metal d orbitals that then form molecular orbitals and eventually bands.

Figure 11.13 shows the $[MnNi]^{3+}$ molecular orbitals in the half-Heusler alloy NiMnSb, formed by overlap (mixing, hybridization) of the Ni and Mn d orbitals that split in the tetrahedral field of Sb^{3-} into a doubly degenerate e set and a higher-lying triply degenerate t_2 set (Figures 7.2 and 5.23). Orbitals of the same symmetry overlap to form five bonding and five antibonding orbitals. These widen into bands, for which this approximate picture is valid close to the full-symmetry Γ point in the Brillouin zone. Note that the atomic orbitals of the more electronegative Ni are lower in energy than those of Mn. The properties of the phase then depend on the filling of these molecular orbitals.

Filling of the scheme in Figure 11.13 with the 22 valence electrons of NiMnSb is shown in Figure 11.14. The energy shift between the spin-up and spin-down d bands[11] is such that the electron-filling process, like pouring in water, places the Fermi level between the bonding and

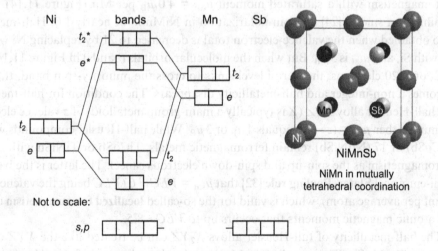

Figure 11.13 Molecular-orbital scheme for the half-Heusler phase NiMnSb, after [31].

[11]   The electron-pairing energy; a sum of the Hund-rule coupling and coulombic repulsion.

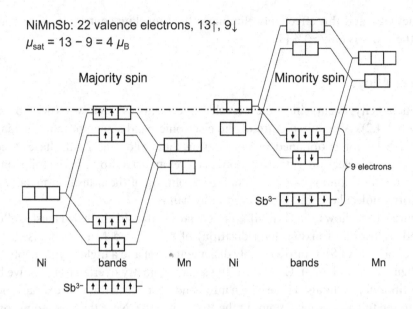

NiMnSb: 22 valence electrons, 13↑, 9↓
$\mu_{sat} = 13 - 9 = 4\,\mu_B$

Majority spin   Minority spin

9 electrons

Ni   bands   Mn   Ni   bands   Mn

Figure 11.14 Filling of molecular orbitals (bands) in NiMnSb by majority-spin and minority-spin electrons.

antibonding bands of the minority spin while crossing the incompletely filled antibonding t_2^* band of the majority spin. The majority spin is metallic; the minority spin has a gap. We see four uncompensated majority spin electrons of predominantly Mn parentage that give rise to ferromagnetism with a saturated moment $\mu_{sat} = 4.0\,\mu_B$ per Mn (Figure 11.14). Positron-annihilation studies [24] see spin polarization in NiMnSb of nearly 1.0. Half-metallicity is also obtained when the valence-electron total is decreased to 21 by replacing Ni with Co, or Sb with Si, and μ_{sat} is $3\,\mu_B$. But when the molecular orbitals (bands) in Figure 11.14 are filled with only 20 electrons, the Fermi level sinks to cross the minority-spin band, the moment becomes a non-integer and half-metallicity disappears. The condition for half-metallicity in the half-Heusler alloy XYZ (Z is typically a main-group metalloid) of z valence electrons per formula is then $\mu_{sat} = (z - 18)$ equals $4\,\mu_B$ or $3\,\mu_B$. While half-Heusler compounds with $z = 19$ (VCoSb) or 17 (NdNiSb) remain ferromagnetic metals, TiCoSb or TiNiSn with $z = 18$ lose ferromagnetism as the spin-up and spin-down electrons cancel. The latter is the basis for the semi-empirical **Slater–Pauling rule** [32] that $\mu_{sat} = VEC - 6$ (VEC being the valence-electron count per average atom), which is valid for the so-called **localized ferromagnetism** of individual atomic magnetic moments that occurs up to $VEC \approx 8.5$.[12]

The half-metallicity of full-Heusler alloys X_2YZ can be treated like the XYZ case. Let's consider Co_2MnSb. The formally $[Co_2Mn]^{3+}$ molecular orbitals are formed by overlap of the

---

[12] At higher VEC, itinerant ferromagnetism due to conduction electrons takes over, as discussed in Section 9.9.2 and shown in Figure 9.27 for the example of Ni.

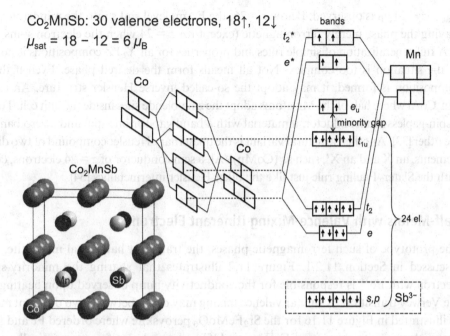

Figure 11.15 Formation of the minority gap in the $[Co_2Mn]^{3+}$ molecular orbitals of the Co_2MnSb full-Heusler alloy with 30 valence electrons per formula, after [31].

Mn atomic orbitals with the Co_2 molecular orbitals.[13] Note in Figure 11.15 that the Co sites alone would form a primitive cubic lattice of Co cubes filled alternately with Mn and Sb. The Co $3d$ orbitals split in the octahedral field of their six Co neighbors into the e_g and t_{2g} sets that overlap to form five bonding Co_2 molecular orbitals of e and t_2 symmetry. These orbitals can mix with the Mn orbitals. The five antibonding orbitals have a symmetry that does not mix with the e and t_2 orbitals of Mn. They therefore remain nonbonding with respect to Mn. The resulting orbital-energy scheme is shown in Figure 11.15.

The full-Heusler alloy becomes half-metallic when filling of the molecular-orbital scheme places the Fermi level between the minority-spin nonbonding e_u and t_{1u} orbitals. This happens for example in Co_2MnSb with 30 valence electrons. These are filled into the bands in Figure 11.15, placing 12 electrons into the minority bands and 18 in the majority bands. The spin of 6 electrons remains uncompensated. Of these, two occupy the e_u nonbonding orbitals of Co_2 parentage; one electron per Co. The remaining four will occupy the antibonding orbitals of Mn parentage. This corresponds to the actual location of the magnetic moments in the structure. Owing to ferromagnetic coupling, $\mu_{sat} = 6\,\mu_B$. Co_2MnSi (29 valence electrons) is also half-metallic, with $2\,\mu_B$ at the two Co atoms and $3\,\mu_B$ at Mn. The Co_2YZ Heusler alloys remain half-metallic down to 26 valence electrons per formula, and the integer

[13] The orbital approach is used here as a local precursor of bands present in these extended structures.

$\mu_{sat} = z - 24\,\mu_B$ is observed. Below $z = 26$, the Fermi level sinks into the minority-spin band, leaving the phase merely ferromagnetic (except for $z = 24$ when the electron spins cancel).

A full generalization of simple rules and properties for all X_2YZ compositions is not possible as the situation is too complex. Not all metals form the desired phase. Even if the correct composition is formed, it may adopt the so-called inverse Heusler structure. An example is Mn_2CoAl where half of the Mn atoms adopt the zinc-blende sites inside the unit cell. The result is a spin-gapless semiconductor, a material with a band gap for one spin and a zero band gap for the other [33]. Another structural variant is the quaternary Heusler compound of two different X elements, an X and an X', such as (CoMn)VAl, a semiconductor of $z = 24$ electrons, complying with the Slater–Pauling rule just like the other Heusler intermetallics [34].

11.4.4 Half-Metals with Valence-Mixing Itinerant Electrons

The prototype of such ferromagnetic phases, the transport half-metal magnetite, has been discussed in Section 11.2.1. Figure 11.2 illustrates that sharing the minority-spin Fe^{2+} electron with Fe^{3+} is responsible for the conductivity jump observed upon heating through the Verwey transition. A similar valence mixing may occur between two different metal ions, as illustrated in Figure 11.16 for the Sr_2FeMoO_6 perovskite where ordered Fe and Mo share the single minority electron of Fe^{2+} origin [35] and couple antiferromagnetically:

Sr_2FeMoO_6 gained attention when a relatively high $TMR = 0.1$, as defined in Equation (11.9), was reported [36] even at room temperature, compared with $TMR = 0.3$ at 4.2 K. This fueled hopes for a material with large magnetoresistance at room temperature. Unfortunately, Sr_2FeMoO_6 preparations tend to suffer from intersite disorder between Fe and Mo, which affects the observed magnetic moment as well as the magnetic ordering temperature. By extrapolation of neutron powder-diffraction data to the extremes of order and disorder, the saturated magnetic moment of the highly ordered phase is shown [37] to be approximately $4\,\mu_B$ at low temperatures and disappears at a T_N of 440 K. This is the normal moment of a high-spin Fe^{2+} in an octahedral field, but also that of $Fe^{2.5+}$ ($4.5\,\mu_B$) minus the opposite spin of $0.5\,\mu_B$ on Mo (both due to the ½ electron present at Fe and Mo), hence consistent with the picture of a ferrimagnet with antiferromagnetic coupling of spins between Fe and Mo. In contrast, the disordered samples form Fe–Fe and Mo–Mo antiferromagnetic clusters of higher T_N than the ordered phase, lowering the total measured magnetic moment.

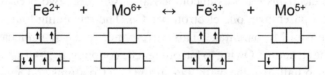

Figure 11.16 Valence mixing $Fe^{2+} + Mo^{6+} \rightarrow Fe^{2.5+} + Mo^{5.5+} \leftarrow Fe^{3+} + Mo^{5+}$ in the Sr_2FeMoO_6 perovskite. The minority-spin electron of Fe^{2+} origin is shared with the thus antiferromagnetically coupled Mo.

11.5 Problems

11.1 Tunneling magnetoresistance between two ferromagnetic films oriented parallel and antiparallel was measured to be 0.4, referring to Equation (11.1). Calculate the spin polarization P and the corresponding relative percentages $n^\uparrow(E_F)$ and $n^\downarrow(E_F)$ for the tunneling electrons.

11.2 All atomic magnetic moments in $YBaFe_2O_5$ lie along the y axis. Based on Figure 11.3 (in which two chemical unit cells are drawn on the left, four on the right) and information in the text, sketch with arrows the Fe magnetic moments in the smallest magnetic-unit-cell parallelepipeds of charge-ordered and valence-mixed $YBaFe_2O_5$.

11.3 $YBaMn_2O_5$ has a structure with two crystallographically different Mn sites:

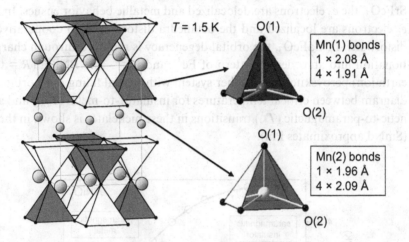

(a) Determine the oxidation state and orbital occupancies of Mn(1) and Mn(2) in charge-ordered $YBaMn_2O_5$. (b) The occupation of one of the d orbitals alternates between Mn(1) and Mn(2). Which orbital is involved in $YBaMn_2O_5$ orbital ordering?

11.4 For each of the following perovskites predict whether there will be a cooperative Jahn–Teller distortion: (a) $LaCrO_3$, (b) $LaTiO_3$, (c) low-spin $LaCoO_3$, (d) high-spin $NdMnO_3$.

11.5 You are asked to synthesize two samples of $La_{0.8}Sr_{0.2}MnO_{3+\delta}$, with $\delta = 0.1$ and 0. At which temperature would you equilibrate the $\delta = 0.1$ composition in ambient O_2 atmosphere prior to quenching? How would you equilibrate the $\delta = 0$ composition in commercial Ar gas of $p_{O_2} \approx 10^{-4}$ bar? Use the Schottky plot in Box 11.2, where isotherms are spaced evenly by 100 degrees.

11.6 Into the figure below, sketch the ordered d_{z^2} orbitals of Mn^{3+} such that their orientation explains the magnetic moments in this $La_{1/3}Ca_{2/3}MnO_3$ layer. Use the principle that follows from the magnetic order of $La_{0.5}Ca_{0.5}MnO_3$ in Figure 11.9.

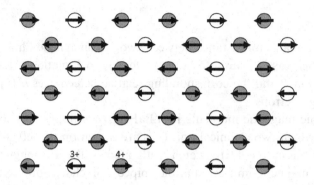

11.7 Perovskites with partially filled e_g orbitals at the octahedrally coordinated atom respond to this impending violation of orbital degeneracy in a variety of ways. In $SrFeO_3$, the e_g electrons are delocalized and metallic behavior ensues. In $LaMnO_3$, the e_g electrons are localized and the octahedra distort through cooperative Jahn–Teller distortions. In $CaFeO_3$, the orbital degeneracy is removed through charge disproportionation into an ordered pattern of Fe^{3+} and Fe^{5+}. The $RNiO_3$ (R = trivalent rare-earth ion) perovskites are another system with partial filling of e_g orbitals. The phase diagram between critical temperatures for insulator-to-metal (T_{IM}) and antiferromagnetic-to-paramagnetic (T_N) transitions in these nickelates is shown in the figure below (SmNd approximates Pm).

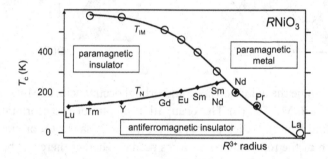

(a) What is the oxidation state, d-electron count and occupation of the t_{2g} and e_g orbitals for the low-spin nickel in these compounds? (b) Would you expect the nickel-centered octahedra in $LaNiO_3$ to be symmetric or to be distorted as a result of a cooperative Jahn–Teller distortion? (c) Whereas $LaNiO_3$ is a paramagnet and a metallic conductor at all temperatures, the lanthanoid $RNiO_3$ phases undergo a metal–insulator transition upon cooling. What stabilizes metallic $LaNiO_3$ even at low temperatures and why? (d) Why does the insulator-to-metal transition temperature T_{IM} increase along the rare-earth series? (e) Which changes in the crystal structure of $YNiO_3$ would you expect upon cooling through T_{IM}? How would the

Ni–O distances allow you to see whether the low-temperature paramagnetic insulating phase is stabilized by a cooperative Jahn–Teller distortion or by a charge disproportionation?

11.8 Sketch the Andreev reflection of an electron crossing from a metal into a superconductor in an electric circuit.

11.9 Calculate the total valence-electron content z per formula and predict the properties of the following half-Heusler alloys:

	PdMnSb	IrMnSb	TiNiSn	TiCoSb	MnCoSb
z					
Ferromagnet?					
μ_{sat} (in μ_B)					
Half-metal?					
Semiconductor?					

11.10 Sketch the spin-up (majority) and spin-down (minority) orbitals of Co_2MnSi around the Fermi energy so that half-metallicity is achieved.

11.11 Spin-resolved density of states were obtained by M. Kallmayer et al. [38] on Co_2MnSi thin films. The DOS of the minority spin was 24 units, of the majority spin 306 units. Calculate the degree of polarization and the optimal tunneling magnetoresistance between two epitaxial films with both alternative formulas.

11.6 Further Reading

S. Bandyopadhyay, M. Cahay, "*Introduction to Spintronics*" (2015) CRC Press.

C. Felser, A. Hirohata (editors), "*Heusler Alloys*" Volume 222 in Springer Series in Materials Science, (2016) Springer.

F. Hagelberg, "*Magnetism in Carbon Nanostructures*" (2017) Cambridge University Press, Part IV—Transport Phenomena".

11.7 References

[1] M. Jullière, "Tunneling between ferromagnetic films" *Phys. Lett. A* **54** (1975), 225–226.

[2] M.B. Robin, P. Day, "Mixed valence chemistry. A survey and classification" *Adv. Inorg. Chem. Radiochem.* **10** (1967), 247–422.

[3] E.J.W. Verwey, "Electronic conduction of magnetite and its transition point at low temperatures" *Nature (London)* **144** (1939), 327–328.

[4] Z. Kąkol, "Magnetic and transport properties of magnetite in the vicinity of the Verwey transition" *J. Solid State Chem.* **88** (1990), 104–114.

[5] M.S. Senn, J.P. Wright, J.P. Attfield, "Charge order and three-site distortions in the Verwey structure of magnetite" *Nature* **481** (2012), 173–176.

[6] P.M. Woodward, P. Karen, "Mixed valence in $YBaFe_2O_5$" *Inorg. Chem.* **42** (2003), 1121–1129.

[7] J. Lindén, F. Lindroos, P. Karen, "Orbital occupancy evolution across spin- and charge-ordering transitions in $YBaFe_2O_5$" *J. Solid State Chem.* **252** (2017), 119–128.

[8] B.C. Hauback, H. Fjellvåg, N. Sakai, "Effect of nonstoichiometry on properties of $La_{1-t}MnO_{3+\delta}$ III. Magnetic order studied by powder neutron diffraction" *J. Solid State Chem.* **124** (1996), 43–51.

[9] J.H. Kuo, H.U. Anderson, D.M. Sparlin, "Oxidation–reduction behavior of undoped and Sr-doped lanthanum manganese oxide $LaMnO_3$ nonstoichiometry and defect structure" *J. Solid State Chem.* **83** (1989), 52–60.

[10] B.C. Tofield, W.R. Scott, "Oxidative nonstoichiometry in perovskites, and experimental survey. Defect structure of an oxidized lanthanum manganite by powder neutron diffraction" *J. Solid State Chem.* **10** (1974), 183–194.

[11] J.A. Alonso, M.J. Martínez-Lope, M.T. Casais, J.L. MacManus-Driscoll, P.S.I.P.N. de Silva, L.F. Cohen, M.T. Fernandez-Diaz, "Non-stoichiometry, structural defects and properties of $LaMnO_{3+\delta}$ with high delta values $(0.11 < \delta < 0.29)$" *J. Mater. Chem.* **7** (1997), 2139–2144.

[12] P.G. Radaelli, D.E. Cox, L. Capogna, S.-W. Cheong, M. Marezio, "Wigner-crystal and bi-stripe models for the magnetic and crystallographic superstructures of $La_{0.333}Ca_{0.667}MnO_3$" *Phys. Rev. B* **59** (1999), 14440–14450.

[13] E.O. Wollan, W.C. Koehler, "Neutron-diffraction study of the magnetic properties of the series of perovskite-type compounds $La_{1-x}Ca_xMnO_3$" *Phys. Rev.* **100** (1955), 545–563.

[14] J.B. Goodenough, "Theory of the role of covalence in the perovskite-type manganites $[La,M^{(II)}]$ MnO_3" *Phys. Rev.* **100** (1955), 564–573.

[15] G.Q. Gong, C.L. Canedy, G. Xia, J.Z. Sun, A. Gupta, W.J. Gallagher, "Colossal magnetoresistance in the antiferromagnetic $La_{0.5}Ca_{0.5}MnO_3$ system" *J. Appl. Phys.* **79** (1996), 4538–4540.

[16] L.M. Rodríguez-Martínez, J.P. Attfield, "Cation disorder and size effects in magnetoresistive manganese oxide perovskites" *Phys. Rev. B* **54** (1996), R15622–R15625.

[17] R.D. Shannon, "Revised effective ionic radii and systematic studies of interatomic distances in halides and chalcogenides" *Acta Crystallogr. Sect. A* **232** (1976), 751–767.

[18] Y. Tomioka, Y. Tokura, "Global phase diagram of perovskite manganites in the plane of quenched disorder versus one-electron bandwidth" *Phys. Rev. B* **70** (2004), 014432/1–5.

[19] R.A. deGroot, F.M. Mueller, P.G. Van Engen, K.H.J. Buschow, "New class of materials: Half-metallic ferromagnets" *Phys. Rev. Lett.* **50** (1983), 2024–2027.

[20] J.M.D. Coey, M. Venkatesan, "Half-metallic ferromagnetism. Example of CrO_2" *J. Appl. Phys.* **91** (2002), 8345–8350.

[21] F. Guinea, "Spin-flip scattering in magnetic junctions" *Phys. Rev. B* **58** (1998), 9212–9216.

[22] J.M.D. Coey, "Powder magnetoresistance" *J. Appl. Phys.* **85** (1999), 5576–5581.

[23] Z. Szotek, W.M. Temmerman, A. Svane, L. Petit, P. Strange, G.M. Stocks, D. Koedderitzsch, W. Hergert, H. Winter, "Electronic structure of half-metallic ferromagnets and spinel ferromagnetic insulators" *J. Phys.: Cond. Matter* **16** (2004), S5587–S5600.

[24] K.E.H.M. Hanssen, P.E. Mijnarends, L.P.L.M. Rabou, K.H.J. Buschow, "Positron-annihilation study of the half-metallic ferromagnet nickel manganese antimonide (NiMnSb): Experiment" *Phys. Rev. B* **42** (1990), 1533–1540.

[25] A.F. Andreev, "Thermal conductivity of the intermediate state of superconductors" *Zh. Eksp. Teor. Fiz.* **46** (1964), 1823–1828. (*Sov. Phys. JETP* **19** (1964), 1228).

[26] R.J. Soulen, Jr., J.M. Byers, M.S. Osofsky, B. Nadgorny, T. Ambrose, S.F. Cheng, P.R. Broussard, C.T. Tanaka, J. Nowak, J.S. Moodera, A. Barry, J.M.D. Coey, "Measuring the spin polarization of a metal with a superconducting point contact" *Science* **282** (1998), 85–88.

[27] K. Schwarz, "Chromium dioxide predicted as a half-metallic ferromagnet" *J. Phys. F: Met. Phys.* **16** (1986), L211–L215.

[28] M.A. Korotin, V.I. Anisimov, D.I. Khomskii, G.A. Sawatzky, "CrO$_2$: A self-doped double exchange ferromagnet" *Phys. Rev. Lett.* **80** (1998), 4305–4308.

[29] J.K. Burdett, G.J. Miller, J.W. Richardson, Jr., J.V. Smith, "Low-temperature neutron powder diffraction study of chromium dioxide and the validity of the Jahn–Teller viewpoint" *J. Am. Chem. Soc.* **110** (1988), 8064–8071.

[30] S. Sundar Manoharan, D. Elefant, G. Reiss, J.B. Goodenough, "Extrinsic giant magnetoresistance in chromium(IV) oxide, CrO$_2$" *Appl. Phys. Lett.* **72** (1998), 984–986.

[31] I. Galanakis, P.H. Dederichs, "Half-metallicity and Slater–Pauling behavior in the ferromagnetic Heusler alloys" *Lecture Notes in Physics* **676** (2005), 1–39.

[32] L. Pauling, "The nature of the interatomic forces in metals" *Phys. Rev.* **54** (1938), 899–904.

[33] S. Ouardi, G.H. Fecher, C. Felser, "Realization of spin gapless semiconductors: The Heusler compound Mn$_2$CoAl" *Phys. Rev. Lett.* **110** (2013), 100401/1–5.

[34] T. Graf, C. Felser, S.S.P. Parkin, "Simple rules for the understanding of Heusler compounds" *Prog. Solid State Chem.* **39** (2011), 1–50.

[35] J. Lindén, T. Yamamoto, M. Karppinen, H. Yamauchi, T. Pietari, "Evidence for valence fluctuation of Fe in Sr$_2$FeMoO$_{6-w}$ double perovskite" *Appl. Phys. Lett.* **76** (2000), 2925–2927.

[36] K.-I. Kobayashi, T. Kimura, H. Sawada, K. Terakura, Y. Tokura, "Room-temperature magnetoresistance in an oxide material with an ordered double-perovskite structure" *Nature* **395** (1998), 677–680.

[37] D. Sánchez-Soria, J.A. Alonso, M. García-Hernández, M.J. Martínez-Lope, J.L. Martínez, A. Mellergård, "Neutron-diffraction magnetic scattering in ordered and disordered Sr$_2$FeMoO$_6$" *Appl. Phys. A* **74** (2002), S1752–S1754.

[38] M. Kallmayer, P. Klaer, H. Schneider, E. Arbelo Jorge, C. Herbort, G. Jakob, M. Jourdan, H.J. Elmers, "Spin-resolved unoccupied density of states in epitaxial Heusler-alloy films" *Phys. Rev. B* **80** (2009), 020406/1–4.

12 Superconductivity

Superconductivity is the phenomenon whereby a significant number of elements and many compounds can conduct electricity with zero resistance below a critical temperature, T_c, field, H_c and current, J_c. There are many technological applications for materials with this remarkable property and therefore a large global research and development effort in the area; around 7000 original research articles are published on superconductivity every year. In this chapter, we will look at the history of superconductivity and its physical origins in so-called BCS or conventional systems. We'll then focus on the solid state chemistry of five distinct families of superconducting materials: A_3C_{60} alkali-metal intercalates, molecular superconductors, $Ba(Pb,Bi)O_3$ perovskites, the cuprate- or "high-T_c" superconductors, and the LaOFeAs-related "iron" superconductors. These families will highlight several recurrent themes and show how chemistry is used to prepare and tune superconducting materials.

12.1 Overview of Superconductivity

The discovery of superconductivity is a wonderful example of how "blue skies" research can lead to completely unexpected discoveries. In 1908, Heike Kamerlingh Onnes, a Dutch physicist working at the University of Leiden in the Netherlands, succeeded in liquefying helium. Access to liquid He, which boils at 4.22 K, allowed him to perform physical measurements on materials at much lower temperatures than previously possible. At that time, little was known about what would happen to the electrical resistance (R) of metals at very low temperatures. There were three basic possibilities (Figure 12.1a): (1) Would the known decrease in R with T seen at higher temperatures continue such that a metal had zero resistance at low temperature? (2) Would R flatten off such that it approached a finite value? (3) Or would R rise to infinity at low T as mobile conduction electrons "froze" and became bound to individual metal atoms rather than delocalized? Initial measurements on Au and Pt suggested hypothesis (2), but experiments on Hg (Figure 12.1b) showed a very dramatic drop in R over just a few hundredths of a degree to extremely low

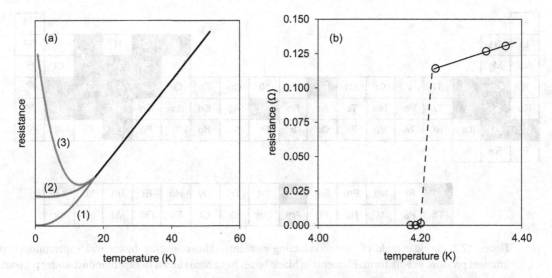

Figure 12.1 (a) Early postulated models for how a metal's resistance might vary with temperature as discussed in the text. (b) Onnes' original data on the resistance of Hg metal; below a critical temperature T_c of 4.2 K (the best modern value is 4.153 K), resistance falls to ~0 Ω.

values. Onnes' report [1] on a sample with a resistance of 172.7 Ω as a liquid at 0 °C stated that "at 4.3 K [resistance] had sunk to 0.084 Ω ... at 3 K the resistance was found to have fallen below 3×10^{-6} Ω, that is to one ten-millionth of the value it would have [in the solid state] at 0 °C. Mercury has passed into ... the superconductive state."

Superconductivity below a critical temperature T_c has since been identified as the ground state of around a quarter of the elements at ambient pressure and more than half at ambient and higher pressure or as thin films[1], as summarized in Figure 12.2. The element with the lowest discovered T_c at ambient pressure is Rh (0.000325 K); that with the highest Nb (9.25 K). Li superconducts at 0.004 K at ambient pressure yet has one of the highest elemental T_c values under pressure (20 K at ~30 GPa). Eu is the most recently discovered superconducting element (2009), with a T_c of 1.8 K at 80 GPa. For our later discussions, there are two interesting points to take from Figure 12.2. Firstly, none of the naturally magnetic elements (Fe, Co, Ni, Gd are ferromagnetic; Mn antiferromagnetic) superconduct at ambient pressure. Secondly, the elements with the highest normal metallic conductivity (Cu, Ag, Au have conductivities of around 6×10^7 S/m at room temperature, Chapter 10) do not superconduct.

In addition to elements, many compounds have been shown to superconduct, including some that are magnetically ordered. Until 1986, the alloy Nb_3Ge had the highest T_c known (23.2 K). Superconductivity was then discovered in a large family of copper-containing oxides at temperatures up to 138 K (~164 K at high pressure), which held the record for

---

[1] Thin films of elements and compounds can be significantly strained relative to the bulk. This influences local coordination geometries and thereby electronic properties.

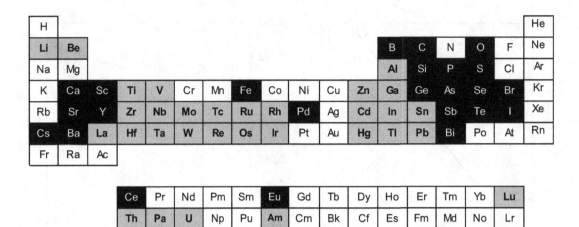

Figure 12.2 Periodic table of superconducting elements. Those in gray boxes will superconduct at ambient pressure in bulk form. Elements in black boxes have been shown to superconduct under pressure or as thin films.

the highest well-established value of T_c for many years. The Nb$_3$Ge T_c barrier has also been surpassed by A$_3$C$_{60}$ compounds, MgB$_2$, some nitride halide intercalates, and by a family of iron-containing materials related to LaOFeAs. A prediction in 1968 by Neil W. Ashcroft [2] that metallic hydrogen, if formed at high pressures, would be a high-temperature superconductor, is behind another pathway to high T_c—high-pressure accommodation of super-stoichiometric hydrogen around pinning points of suitable central atoms in a crystal structure. In 2015, researchers reported $T_c > 200$ K in samples of H$_2$S held at very high pressure (forming H$_3$S), and in 2018 a $T_c \approx 270$ K was reported in a superstoichiometric lanthanum hydride.[2] Table 12.1 gives a selection of superconducting materials chosen either because they led to historically significant breakthroughs, or because they are representative members of a larger family of materials. The Holy Grail in this area is, of course, ambient-pressure superconductivity at room temperature or above. There are periodic reports of higher-T_c materials than those in Table 12.1, but they prove hard or impossible to reproduce. These are often semi-jokingly referred to as USOs—unidentified superconducting objects.

12.2 Properties of Superconductors

We've stated above that superconductors have zero resistance below the critical temperature T_c. It's important to consider initially whether there is anything intrinsically special about the

---

[2] $T_c = 203$ K superconductivity has been reported in samples of H$_2$S pressurized to ~150 GPa [A.P. Drozdov et al., *Nature* **525** (2015), 73–76]. The superconducting phase has been suggested to be H$_3$S (created by decomposition of 3H$_2$S to 2H$_3$S + S) or [H$_3$S]$^+$[SH]$^-$. Various high-pressure metal-poor hydrides were predicted to have T_c close to or above room temperature, such as LaH$_{10}$ (~280 K at >200 GPa) [H. Liu et al., *PNAS* **114** (2017), 6990–6995], the first experimental support appeared in 2018.

Table 12.1 Selected superconductors and their critical temperatures, organized partly chronologically and partly by chemical type. TMTSF and ET are defined and drawn in Figure 12.12. T_c values are collated from references in [3].

Material	T_c (K)	Comment
Rh	0.0003	Lowest T_c of any superconducting element
Hg	4.153	First superconductor discovered in 1911
Pb	7.2	Highest-T_c type-I superconducting element
Nb	9.25	Highest-T_c type-II superconducting element
Li	0.004/20	Ambient/~30 GPa
Ca	21–25	Highest-T_c element ~220 GPa
Nb_3Sn	18	High-T_c intermetallic
Nb_3Ge	23.2	Highest-T_c intermetallic
NbO	1.5	First oxide superconductor, 1933
$SrTiO_{3-\delta}$	0.3	First perovskite superconductor
$BaPb_{1-x}Bi_xO_3$	13	First high-temperature superconductor of 1970s
$Ba_{0.6}K_{0.4}BiO_3$	34	Highest T_c of $BaBiO_3$-related systems
$La_{1.85}Ba_{0.15}CuO_4$	30	First cuprate superconductor, 1986
$La_{1.85}Sr_{0.15}CuO_4$ or $La_2CuO_{4.08}$	38	Highest-T_c "214" cuprate superconductor
$YBa_2Cu_3O_7$	93	First >77 K (N_2 boiling point) superconductor, 1987
$HgBa_2Ca_2Cu_3O_{8+x}$	134/164	Highest-T_c cuprate at ambient/elevated pressure
$Ba_{1-x}Sr_xCuO_2$	90	Infinite-layer cuprate
$Ca_{1-x}Sr_xCuO_2$	110	Highest-T_c ternary infinite-layer cuprate
$Nd_{2-x}Ce_xCuO_4$	~25	n-type superconducting cuprate
$Sr_{1-x}Nd_xCuO_2$	40	n-type superconducting infinite-layer compound
$RuSr_2(Gd,Eu,Sm)Cu_2O_8$	~58	Ferromagnetic cuprate superconductor
$Nd_{0.8}Sr_{0.2}NiO_2$	9–15	d^9 nickel oxide potential cuprate analogue
Sr_2RuO_4	1.5	Non-cuprate "214" superconductor
$Na_x(H_2O)_yCoO_2$	~5	Layered cobalt oxide
$LaO_{1-x}F_xFeAs$ ($x \approx 0.12$)	26	First superconducting iron oxide-pnictide
$SmO_{0.9}F_{0.1}FeAs$	55	Highest-T_c oxide pnictide
$Fe_{1+x}Se$	37	Under pressure
$PbMo_6S_8$	15	Chevrel phase
NbN	17	Widely used nitride superconductor
CaC_6	11.5	Graphite intercalate
K_3C_{60}	18	C_{60} intercalate
Cs_3C_{60}	38	Highest-T_c molecular system under pressure
$Li_x(THF)_yHfNCl$	25.5	Intercalate; highest-T_c nitride based superconductor
MgB_2	39	Highest-T_c conventional superconductor
YPd_2B_2C	23	Highest-T_c borocarbide
$MgCNi_3$	7–8	A non-oxide anti-perovskite
$CeCu_2Si_2$	1	First heavy-fermion superconductor

Table 12.1 (cont.)

Material	T_c (K)	Comment
UPd_2Al_3	2	Antiferromagnetic heavy-fermion superconductor
$PuCoGa_5$	18	High-T_c heavy-fermion superconductor
$(SN)_x$	0.33	Inorganic polymer
$(TMTSF)_2PF_6$	1	First organic superconductor at 12 kbar
$ET_2Cu[N(CN)_2]Br$	11.6	Charge-transfer salt
H_2S	203	At 150 GPa

superconducting state, or whether superconductors merely represent extreme examples of metallic behavior. If we compare resistivities for metals and superconductors, we find an excellent low-temperature metallic conductor like Ag has $\rho \approx 10^{-11}$ Ω m at 1 K, whereas the best estimates for superconductors are around 10^{-25} Ω m. The resistivity of superconductors is lower than even the best metal by many orders of magnitude, suggesting a qualitatively different electronic state.

Superconductors also display remarkable magnetic properties. When a superconductor is placed in a magnetic field, it displays the **Meissner** or **Meissner–Ochsenfeld effect**—it expels the magnetic field so that the magnetic induction B within the sample is zero (Figure 12.3).[3] From our Chapter 9 definitions of magnetic induction as $B = \mu_0(H + M)$ and magnetic susceptibility as $\chi = M/H$, it follows that, for a superconductor, $\mu_0(H + \chi H) = 0$ or $\chi = -1$.[4] A superconductor is thus a **perfect diamagnet**. However, the total field exclusion occurs only for certain sample geometries and below a critical value of the applied field. The exclusion arises because the supercurrents flowing in a thin layer at the surface of the superconductor generate a magnetic field that exactly opposes the external field.

This magnetic behavior is fundamentally different to that of a hypothetical "perfect metallic" conductor with resistivity $\rho = 0$. If one took either a superconductor below T_c or a perfect metal with $\rho = 0$ and applied a magnetic field, the magnetic field within both samples would be zero. If, however, one started with both samples in a magnetic field at temperatures where they had a finite resistance and then cooled them such that their resistivity fell to zero, the two samples would show different behavior. The superconductor would expel the magnetic field, whereas the perfect conductor would trap the field.

Superconductors only display the Meissner effect below certain applied field strengths. In a **type-I superconductor**, such as elemental Pb, perfect diamagnetism with $\chi = -1$ (i.e. $M = -H$) is only observed up to a critical field H_c. Above H_c, the sample reverts to a normal conducting state, and χ falls to the much smaller value of -1.8×10^{-5}, which is typical for a normal metal (Figure 12.3a). The magnitude of H_c is temperature dependent, being largest

---

[3] This is the origin of the famous magnet levitation demonstration using a superconductor.

[4] H is the magnetic-field strength, M is the magnetization, and μ_0 is the permeability of free space; see Section 9.2.2 for full definitions.

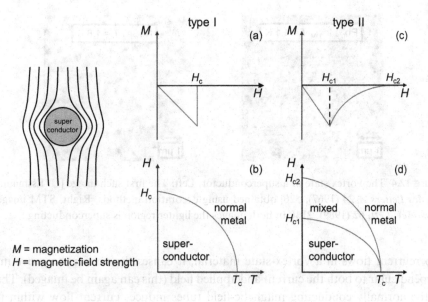

Figure 12.3 Magnetization at $T < T_c$ and schematic phase diagrams of superconductors in an applied magnetic field.

near $T = 0$ K and falling to 0 at $T = T_c$. This leads to the behavior shown in Figure 12.3b. An internal field created by current flow in a superconductor wire can also destroy superconductivity, leading to the related concept of a critical current, J_c.[5]

Other materials exhibit so-called **type-II superconductor** behavior (Figure 12.3c and d), in which they show perfect diamagnetism up to a critical field H_{c1} followed by a gradual decrease of the absolute value of the magnetization up to a critical field H_{c2}. Above H_{c2}, normal magnetic behavior is observed. Critical fields H_{c2} are typically much higher than H_c values for type-I superconductors. As an example, while pure Pb is a type-I superconductor with $T_c = 7.2$ K and $H_c \approx 550$ G at 4.2 K, alloying with In yields a type-II superconductor; at 20% In, perfect diamagnetism is only observed up to H_{c1} of ~150 G but H_{c2} is ~3700 G. Values of H_{c2} up to 110 T (1.1×10^6 G) have been reported in nanocrystalline $PbMo_6S_8$ [4].

In the range between H_{c1} and H_{c2}, a type-II superconductor exists in a mixed or vortex state where certain regions of the sample superconduct and expel the magnetic field, while other regions don't. The normal region is separated from the rest of the superconductor by a circulating supercurrent called a vortex. There is a magnetic repulsion between the magnetic-field "tubes" that penetrate the sample, which then arrange themselves to maximize their separation. This often leads to a regular hexagonal array that can be imaged (Figure 12.4) by placing fine magnetic particles on the specimen ("decoration") or by surface-sensitive techniques such as scanning tunneling microcopy (STM) and magnetic force microscopy. When a

---

[5] A high critical current is particularly important in superconducting cables.

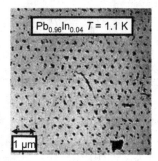 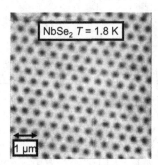

Figure 12.4 The vortex state of a superconductor. Left: The first such image [U. Essmann, H. Trauble, *Physics Letters A* **24** (1967), 526] obtained using decoration methods. Right: STM image [H.F. Hess, *Phys. Rev. Lett.* **62** (1989), 214]. In both images the lighter region is superconducting.

supercurrent flows in a vortex-state material, it causes the vortices to move in a direction perpendicular to both the current and applied field (this can again be imaged). The movement of the normally conducting magnetic-field tubes induces current flow within them, which dissipates energy and destroys superconductivity. If, however, the vortices can be spatially trapped or **pinned**[6] (e.g. by lattice defects in the sample), the sample will transport current with zero resistance via the fixed superconducting regions. Note that there is no structural or chemical difference between the normal and superconducting regions in the vortex state.

12.3 Origins of Superconductivity and BCS Theory

How does superconductivity arise? In this section, we will discuss theoretical models of the so-called **conventional superconductors**. Some texts that discuss the theory of **unconventional superconductors** are given in Further Reading.

We've already encountered several clues that will be important in understanding the origin of superconductivity. Firstly, we've seen that there is an inverse relationship between normal metallic conductivity and superconductivity—the best metals don't superconduct. Secondly, internal or external magnetic fields normally disrupt superconductivity.[7] We can gain more

---

[6] Pinning is also behind the fascinating levitation properties of type-II superconductors. If you cool a small, flat, superconductor disc with a small, very strong, magnet on its flat surface, a pinned vortex state will trap the magnetic field in specific places in the superconductor, and, below T_c, the magnet will start "hovering" there without being held. This is the Meissner effect. When you then gradually raise the magnet with bamboo tweezers, if the superconductor is light enough and has enough pinned vortices, it will hang below the magnet. This is not the Meissner effect, but an attraction due to the magnetized pinned vortex state. If the magnet is removed and then placed back on the superconductor, the "magnetic memory" of its original position stored in the vortex state returns it to its original position. With the magnet removed, the still-cooled superconductor will attract ferrous objects for the same reason.

[7] There are exceptions. For example, in alloys such as UGe$_2$ [Saxena et al., *Nature* **406** (2000), 587–592], URhGe [Aoki et al., *Nature* **413** (2001), 613–616] and UIr [Akazawa et al., *J. Phys. Condens. Matter* **16** (2004), L29–32], superconductivity can coexist with itinerant ferromagnetism. These are examples of unconventional superconductors.

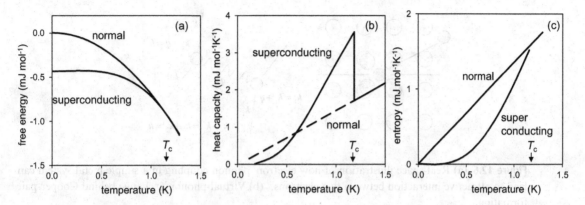

Figure 12.5 (a) Free energy, (b) heat capacity, and (c) entropy of Al in the superconducting and normal states of aluminum. Parts (a) and (c) after [5].

insight by considering the thermodynamics of the transition to the superconducting state. Figure 12.5 shows the temperature dependence of the free energy, heat capacity, and entropy of aluminum in the normal and superconducting state. These data were obtained by performing the measurements initially in a zero magnetic field and then repeating them in a field above H_c to suppress the superconducting transition. The free energy of the superconducting state is lower than that of the normal state at all temperatures below T_c (1.2 K), though by just 0.4 mJ/mol at 0 K. At T_c, the heat capacity shows a marked change, which, along with the lack of a discontinuity in the free energy, suggests a second-order transition (see Section 4.4.4). In addition, the electronic contribution to the specific heat at low temperature shows an $e^{-aT_c/T}$-like dependence, which suggests an energy gap is present with magnitude $\sim kT_c$; we'll discuss this gap later. Finally, the entropy is lower in the superconducting state, which suggests that the electrons must adopt a more ordered configuration.

Since there are no changes in a sample's cell parameter or structure at T_c, the thermodynamic data of Figure 12.5 led to the idea that there must be attractive electron–electron interactions in the superconducting state, which lower the free energy of the system despite the loss of entropy. How do these attractions occur? While it was originally thought that the structure of the sample played little role, in the 1950s it was discovered that the T_c values of superconducting elements exhibit an isotope effect whereby $T_c \propto M^{-\alpha}$, where M is the isotope mass. Isotopes of Hg, for example, have T_c values ranging from 4.161 K to 4.126 K as the mean atomic weight varies from 199.7 to 203.4 and $\alpha = 0.49(2)$ [6]. This dependence mirrors the dependence of vibrational frequencies on mass,[8] and suggests that lattice vibrations, or phonons (Section 4.4.6), give rise to the electron–electron attractions.

The simplest model for the electron–phonon (or **Fröhlich**) interaction is shown in Figure 12.6a. As an electron (e_1) moves right to left through a metal, its negative charge will attract

---

[8] A simple diatomic molecule, for example, has vibrational frequency (in Hz) of $(1/2\pi)\sqrt{k_{bond}/\mu}$, where μ is the reduced mass of the atoms involved ($1/\mu = 1/m_1 + 1/m_2$) in kg and k_{bond} the force constant in N/m.

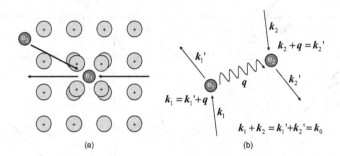

Figure 12.6 (a) Real-space illustration of how electron–phonon coupling in a simple metal system can lead to attractive interaction between two electrons.[9] (b) Virtual-phonon exchange behind Cooper-pair formation.

the positive metal ions around it, creating a temporary deformation of the structure and a local region with excess positive charge. This moving region of positive charge, or *polarization trail*, will then attract a second electron, forming a **Cooper pair**. We therefore have a situation in which two electrons attract to form a pair not via a direct interaction but via a first-electron-to-lattice, lattice-to-second-electron process[10] of phonon-mediated pairing.

Since electrons move much more rapidly than the heavier metal ions, there will be a time delay between the motion of the first electron that causes the lattice distortion, and the subsequent attraction of the second electron. To estimate the distance, d, the first electron has passed by the time the second is attracted, we need to know v_F (velocity of an electron at the Fermi level) and the Debye angular frequency, ω_D (a measure of the frequency at which the lattice can respond):

$$d = v_F \frac{2\pi}{\omega_D} \tag{12.1}$$

For aluminum, $v_F \approx 2\times10^6$ m/s (see Section 10.2.6) and $\omega_D \approx 5\times10^{13}$ s^{-1}, which gives $d \approx$ 2500 Å; i.e. the distance between electrons in pairs is large. At this distance, there will be effectively no Coulomb repulsion between the electrons to counteract the attraction, particularly given the screening by the positive charges on the metal ions.

A more precise way to describe this interaction is in terms of the exchange of a **virtual phonon** between the electrons. In this process, we imagine electron 1 with an initial wave vector,[11] k_1, emitting a phonon with wave vector, q, after which its wave vector becomes k_1' (Figure 12.6b). A second electron absorbs this phonon and is scattered from wave vector k_2 to k_2'. Conservation of momentum leads to the relationships:

---

9   Note that in BCS superconductors the distance between e_1 and e_2 is sufficiently large that this local sketch is a gross
    approximation; the distance is much smaller in the bismuth oxide and cuprate superconductors discussed later.

10 A common analogy is placing heavy bowling balls on a sprung mattress. One ball creates a depression that the
    second ball will roll into. The balls are attracted to each other by an indirect first-ball-to-mattress, mattress-to-
    second-ball interaction.

11 Remember from Chapter 6 that the wave vector just labels one of the crystal orbitals or an allowed electron state.

$$k_1 = k_1' + q \text{ and } k_2 + q = k_2' \tag{12.2}$$

Since momentum is also conserved for the overall process, Equation (12.2) leads to the relationship:

$$k_1 + k_2 = k_1' + k_2' = k_0 \tag{12.3}$$

The phonon is said to be virtual because it is reabsorbed during the process, and no net vibrational energy is transferred to the lattice to produce electrical resistance.

Is there any special relationship between the two electrons that will maximize the overall attractions (i.e. is there a special value of k_0)? We can best explore this in k space using the schematic of Figure 12.7. If the attraction is mediated by a phonon of energy $\hbar\omega_D$, then the Pauli exclusion principle means that only electron states within $\hbar\omega_D$ of E_F can be involved in the coupling,[12] corresponding to a shell of finite width δk in k space. The states involved can be represented graphically, as shown in Figure 12.7b, and their number is maximized, giving the largest number of electron–electron attractions, when $k_0 = 0$. Equation (12.3) then means that we have the condition on the electrons that $k_1 = -k_2$. If the two-electron wavefunction of the Cooper pair is spherically symmetric,[13] quantum mechanics requires a singlet spin state with $S = 0$, leading to the overall condition that the electrons in the Cooper pair have opposite momenta and spin: $k\uparrow$ and $-k\downarrow$. This is called **s-wave pairing** and occurs in conventional or BCS (see below) superconductors.[14] Note that the situation is a complex and dynamic one: Cooper pairs in a superconductor shouldn't be thought of in the same way as electron pairs in a chemical bond, but as in a dynamic state in which individual electrons are constantly exchanging phonons.

From the idea of Cooper pairs, John Bardeen, Leon Cooper, and John Schrieffer were able to build the first quantitative model for superconductivity, **BCS theory**, for which they were awarded the 1972 Nobel Prize in Physics. Their theory showed that, at $T < T_c$, many electrons are involved in Cooper pairs and that it is the cooperative properties of the ensemble of pairs that determines overall properties.[15] For example, the pair binding energy (usually called 2Δ), depends on how many other pairs are present. As the Cooper pairs combine two fermions (the electrons), they behave as a boson and will all have the

---

[12] Electrons must transition to empty states; at $T = 0$ K all states $< E_F$ are full.

[13] This is the case for conventional superconductors but not for unconventional superconductors such as cuprates or heavy-fermion materials.

[14] A d-wave pairing, where the two-electron wavefunction has nodes like those in d orbitals, also requires $S = 0$ and is thought to occur in cuprates. Spin triplet states ($S = 1$, see Section 7.3.2) are also possible but associated with p- or f-wave pairing. These have been proposed in heavy-fermion superconductors such as Sr_2RuO_4 and UPt_3. See, e.g., the text by Norman in Further Reading for more detail.

[15] The cooperative behavior is perhaps unsurprising given that the radius of a Cooper pair is $\sim 10^{-7}$–10^{-6} m and that $\sim 10^6$–10^7 other electrons pairs will be present in the corresponding volume.

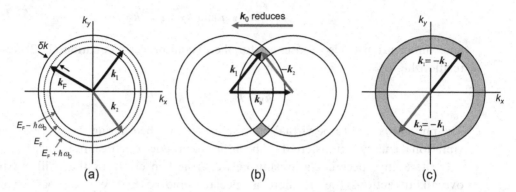

Figure 12.7 (a) A k-space sketch of a spherical shell of states within δk of Fermi level k_F. Electrons with arbitrary wave vectors k_1 and k_2 are shown. The change in k_1 on phonon exchange is restricted to a small area determined by δk. (b) The overall condition of $k_1 + k_2 = k_0$ for virtual phonon exchange can be represented geometrically by two such shells offset by k_0. For a given value of k_0 the overlap region (shaded) gives the states that can be involved in coupling. The maximum coupling will occur when the two shells overlap such that $k_0 = 0$ and $k_1 = -k_2$.

same quantum state at $T = 0$ K; i.e., they have undergone a transition similar[16] to a Bose–Einstein condensation.[17]

One of the important features of the BCS ground state is that its energy is lower than that of the normal free-electron state (recall Figure 12.5a). In particular, an energy gap, the **superconducting gap**, opens up between the superconducting state and the normal free-electron state.[18] We can understand this gap with reference to Figure 12.8. The exchange of virtual phonons between electrons close to E_F lowers the electrons' energy, which can be represented as a peaking in the density-of-states (DOS) plot in Figure 12.8. Note that this figure represents the single electron states and does not specifically show Cooper pairs. The typical energy, $\hbar\omega_D$, of a phonon giving rise to coupling is around 10^{-3} eV, whereas E_F is typically ~5 eV. This means that only a small fraction, around 10^{-3}, of the conducting electrons have their energy changed and explains why the energy difference between the normal and superconducting states is small. At $T = 0$ K, levels above the gap are empty. As T rises, electrons are excited across the gap to normal states. Since this destroys some Cooper pairs, the gap decreases but the sample remains superconducting. At $T = T_c$, the gap reduces to zero and the sample becomes a normal metal.

[16] "Similar" as Cooper pairs only exist in the overall cooperative state of a superconductor, the superconducting state wouldn't arise from a gas of independent electrons.

[17] A Bose–Einstein condensate is a state that forms when a very dilute gas of bosons is cooled to sufficiently low (nanokelvin) temperatures that a large fraction of them adopt the same quantum state and quantum-mechanically derived behavior becomes observable on a macroscopic scale. The 2011 Nobel Prize in Physics was awarded to Cornell, Ketterle, and Wiemann, who first formed Bose–Einstein condensates in Rb and Na gases.

[18] More exotic gapless superconductors are also possible, in which Cooper pairs exist without a gap.

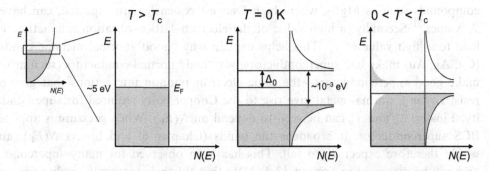

Figure 12.8 DOS plot of a metal close to E_F in its normal and superconducting states. In the superconducting state, peaks appear in $N(E_F)$ just below and above the gap. Away from E_F, the DOS of the superconductor is similar to a normal metal. At $T > 0$ K, there is some occupancy of normal states above the gap.

For an isotropic system with a 0 K gap defined as $2\Delta_0$,[19] BCS theory predicts that:

$$\Delta_0 = \frac{\hbar\omega_D}{\sinh\left[\dfrac{1}{N(E_F)\cdot V'}\right]} \approx 2\hbar\omega_D e^{-1/N(E_F)\cdot V'} \tag{12.4}$$

where $N(E_F)$ is the DOS at the Fermi level and V' the strength of the electron–lattice-vibration interaction (electron–phonon coupling).[20] Values for typical materials predict $\Delta_0 \approx 5\times10^{-4}$ eV, which matches experimental measurements of the gap obtained by infrared radiation adsorption. The theory predicts that the gap decreases continuously with increasing temperature, becoming 0 at T_c. The T_c is related to Δ_0 by:

$$2\Delta_0 = 3.52kT_c \tag{12.5}$$

and we see why conventional superconductivity only happens at low temperature. Combining Equations (12.4) and (12.5) gives:

$$T_c = 1.14\frac{\hbar\omega_D}{k}e^{-1/N(E_F)\cdot V'} \tag{12.6}$$

The form of Equation (12.6) helps us to rationalize some of the experimental observations on superconductivity. Firstly, as the Debye frequency, ω_D, will scale with $M^{-1/2}$ (Footnote 8 in this chapter), T_c will show the same dependence. This explains the experimentally observed isotope effect. We can also see why light-element-containing

---

[19] At $T = 0$ K, a BCS superconductor has Cooper pairs in the same quantum state. The first single-electron state is higher in energy by Δ_0 such that energy $2\Delta_0$ must be supplied to break a pair.

[20] The approximation only holds for weak coupling $N(E_F)\cdot V' \ll 1$.

compounds such as MgB_2, where high Debye frequencies are expected, can have high T_c values.[21] Secondly, a high value of the electron–lattice-vibration interaction V' will lead to a high value of T_c. This helps explain why "good" normal metallic conductors (Cu, Ag, Au) make bad superconductors, yet "bad" normal conductors (such as oxides) make good superconductors—the same electron–phonon interactions that give rise to resistivity in a normal metal give rise to the Cooper pairs required for superconductivity. Finally, Δ_0 and T_c can be seen to depend on $N(E_F)$. When pressure is applied to a BCS superconductor, it broadens the bands (Chapter 6) and lowers $N(E_F)$, and we would therefore expect T_c to fall. This has been observed for many superconductors and will be discussed in Section 12.4. Note that this behavior only applies for conventional BCS superconductors, and the effect of pressure on unconventional non-BCS systems can be very different.

A superconductor's signature property of zero resistance is a consequence of the cooperative nature of the ensemble of Cooper pairs, and stems from the fact that they are all in the same quantum state. In Chapter 10, we described conductivity as arising when individual electrons have their momenta shifted by the application of an electric field from k to $k+\Delta k$, and resistivity as arising due to defects and phonons scattering electrons back into lower momentum states. Why doesn't this occur in superconductors? When there is zero current in a superconductor, the net momentum of each Cooper pair is zero ($k\uparrow$ and $-k\downarrow$). When they are accelerated by an electric field, the entire Fermi surface is shifted and all Cooper pairs gain momentum (changing to $k+\Delta k\uparrow$ and $-k+\Delta k\downarrow$). Since the pairs have coupled momenta, the only way to change the overall momentum is to break up a Cooper pair by excitation across the superconducting gap, and this requires an energy of 2Δ. An elastic process such as impurity scattering clearly won't be able to do this and can't give rise to resistivity. What about inelastic processes such as phonon absorption or emission? As we've discussed, these processes happen continuously in a superconductor at $0 < T < T_c$ and create a dynamic equilibrium between normal electrons and Cooper pairs. However, the Cooper pairs that form in these processes must have the same wavefunction as the pre-existing pairs and hence the same momentum. This means that their creation and destruction won't cause any change in the current. The overall current is therefore only affected by factors that influence all Cooper pairs in the same way, such as changing the applied electric field. We can also understand the origin of the superconducting critical current J_c from the energy gap. When the applied electric field increases the pair's kinetic energy to be comparable with its binding energy, Cooper pairs are gradually destroyed until resistance re-emerges.

As a final point, we should consider which electrons are involved in carrying the supercurrent. Is it just the small fraction of electrons close to E_F involved in Cooper pairs, such that we can think of a normal current and supercurrent being present

---

[21] Note also the high T_c of Li (20 K) and H_2S-derived compositions (203 K) under pressure (Table 12.1).

simultaneously? It turns out that the situation is more complex. At 0 K, there are no excitations possible across the superconducting gap, and all the electrons behave as if they're carrying the supercurrent. At higher temperatures, where excitations are possible, the number of electrons involved in the supercurrent falls, reaching 0 at T_c. The presence of at least some electrons carrying the supercurrent is sufficient to give zero resistance.

Box 12.1 Materials Spotlight: A serendipitous superconductor MgB$_2$

In 1954, the synthesis and crystal structure of MgB$_2$ were reported in the *Journal of the American Chemical Society* [7]. Other researchers cited the paper just 12 times in the following 10 years, once in the 1970s and once in the 1990s. Almost 50 years later, Akimitsu set out to dope the ferromagnetic semiconductor CaB$_6$ with Mg. When the magnetic and electronic properties of the MgB$_2$ starting material were tested, it was found to superconduct at a remarkable 39 K. The discovery was reported in a wonderfully succinct paper in *Nature* in February 2001 [8], which was cited almost 300 times by the end of the calendar year.

The structure of MgB$_2$ (see below, left) is remarkably simple, containing Mg sandwiched between graphene-like hexagonal B layers. It's also strikingly similar to the graphite intercalation compounds we'll meet in Chapter 13. This structural relationship is unsurprising since electronic calculations show the material is essentially ionic, Mg^{2+}B$_2^{2-}$, making the B layers isoelectronic with graphite. Their band structure is also similar to that of graphene (Chapter 6), with one major difference; both the 2D sp^2-derived σ band and the more three dimensional p_z-derived π band cross the Fermi level. This gives rise to holes in the σ band (it would be full in graphite) and a corresponding number of electrons in the π band, creating a $\sigma \approx 10^5 \ \Omega^{-1} \ \mathrm{cm}^{-1}$ metallic conductivity at room temperature. Hall-effect measurements have shown the presence of both carriers.

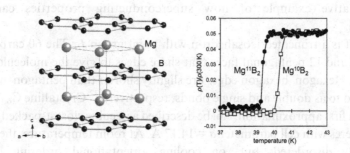

Despite the high T_c, MgB$_2$ appears to be a conventional *s*-wave superconductor. It shows a strong isotope effect, with T_c shifts of ~1 K on substitution of ^{10}B for ^{11}B (see right-hand side of the figure [9]) and ~0.1 K on Mg substitution. The α values in $T_c \propto M^{-\alpha}$ are 0.26–0.30 and 0.02,

Box 12.1 (cont.)

respectively, suggesting that phonons involving B are particularly important. The strong B–B bonding gives a stiff material with high-frequency phonons ($\omega_D \approx 3300$ cm^{-1}), and there is strong electron–lattice-vibration coupling ($V' = 0.82$). Equation (12.6) shows these factors will lead to a high T_c. One peculiar feature of MgB$_2$ is that the phonons couple differently with electrons in the σ and π bands, which leads to two distinct superconducting gaps (around 7.1 meV and 2.3 meV). These have been observed by a number of experimental techniques, most strikingly heat-capacity measurements that show a superposition of two distinct features of the type plotted for Al in Figure 12.5b.

After the discovery of superconductivity in pure MgB$_2$, many attempts were made at chemical doping to enhance T_c. To date only three substitutions are possible: Al and Mn for Mg and C for B. In each case, there is a decrease in T_c. Al and C are both electron dopants and the band structure shows a decrease in $N(E_F)$ on adding electrons, while the magnetic moment of Mn^{2+} is predicted to destroy Cooper pairs. Despite reducing T_c, carbon doping raises the $T = 0$ K critical field from $H_{c2} \approx 16$ T in MgB$_2$ to $H_{c2} \approx 36$ T in Mg(B$_{0.95}$C$_{0.05}$)$_2$. This high H_{c2} in a light and relatively cheap material has meant that MgB$_2$ is already being used in MRI magnets.

It's remarkable that an optimally doped conventional superconductor with the highest T_c of any binary compound remained overlooked for so long!

12.4 C$_{60}$-Derived Superconductors

In this section, we will discuss the superconducting alkali- and alkaline-earth A$_x$C$_{60}$ intercalates. These have been shown to superconduct up to a maximum T_c of 38 K, and currently hold the T_c record for ambient-pressure molecular systems. While the full detail of the electronic properties of these materials is complex [10], they provide an illustrative example of how superconducting properties can be chemically controlled.

C$_{60}$ itself is a truncated icosahedron with point group I_h. The 60 carbon atoms form 20 hexagonal and 12 pentagonal faces that share edges to give the molecule's familiar soccer-ball shape. Hexagon–hexagon edges are slightly shorter than pentagon–hexagon edges, and are referred to as double and single bonds, respectively.[22] Crystalline C$_{60}$ adopts a structure which, to a first approximation, can be described as a cubic closest-packed (ccp) arrangement of C$_{60}$ spheres with cell parameter $a = 14.17$ Å. At room temperature, the C$_{60}$ molecules are rotationally disordered, but, on cooling, orientational ordering occurs[23] and the

[22] Average bond lengths of the two sets are 1.391 Å and 1.455 Å; intermediate between typical carbon–carbon double- and single-bond lengths of 1.32 Å and 1.54 Å.

[23] In fact, the molecules remain dynamic down to ~100 K in the $Pa\bar{3}$ form but flip between discrete orientations.

incompatibility of the $m\bar{3}m$ site symmetry of the ccp arrangement of spheres and the icosohedral symmetry of the C$_{60}$ molecules lowers the space-group symmetry. If we ignore this subtlety, the ccp arrangement makes two tetrahedral and one octahedral holes per C$_{60}$ (Section 1.4.2). The tetrahedral and octahedral holes have a radius of 1.12 Å and 2.07 Å, respectively. C$_{60}$ has a relatively high electronegativity and is easily reduced down to C$_{60}^{6-}$. Since the interstitial holes are comparable in size to alkali and alkaline-earth ions (six-coordinate "CR radius" [11] $r = 0.9$ Å for Li^{+} and $r = 1.81$ Å for Cs^{+}), intercalation[24] compounds A$_x$C$_{60}$ form. Filling all the octahedral and tetrahedral holes leads to an A$_3$C$_{60}$ composition. In Na$_2$A$_x$C$_{60}$ (A = K, Rb, Cs), the smaller Na^{+} ions occupy tetrahedral holes and the larger A^{+} octahedral holes, and the symmetry remains $Pa\bar{3}$. In A$_3$C$_{60}$ (A = K, Rb, Cs), larger cations must occupy the tetrahedral holes and this forces the C$_{60}$ molecules to realign. They become disordered over two possible orientations such that the symmetry of the average structure becomes $Fm\bar{3}m$. Upon switching to body-centered cubic (bcc) packing, A$_6$C$_{60}$ compositions can be formed, and metal content up to $x = 12$ is possible in compounds such as Li$_{12}$C$_{60}$.

In 1991 it was reported that A$_3$C$_{60}$ compounds are metallic and superconduct at low temperature [12], with $T_c = 18$ K for K$_3$C$_{60}$ and 29 K for Rb$_3$C$_{60}$. The origin of the metallic behavior can be understood from the simplified molecular orbital (MO) diagram of Figure 12.9. In the C$_{60}$ molecule, the highest occupied molecular orbital (HOMO) is fivefold degenerate (h_u symmetry[25]), and separated by ~1.8 eV from a triply degenerate t_{1u} lowest unoccupied molecular orbital (LUMO) and by ~2.5 eV from the t_{1g} LUMO+1. In the solid state, these molecular orbitals broaden to form bands. For the A$_3$C$_{60}$ composition, the band derived from the t_{1u} level will be half filled, and one might anticipate metallic properties.

In fact, much of the interesting behavior of A$_3$C$_{60}$ materials arises because they lie close to a Mott–Hubbard metal-to-insulator boundary (see Section 10.4.1). In an isolated C$_{60}$ molecule, the electron–electron Coulomb repulsion is estimated to be around 3 eV and the Hubbard U in the extended solid around 1 eV. The bandwidth, W, is around 0.5 eV. Since $U > W$, we would predict the materials to be Mott–Hubbard insulators. In fact, when the triply degenerate nature of the LUMO is taken into account, the Mott–Hubbard boundary moves to $U > 2.5W$ for insulating behavior, and we predict metallic properties for A$_3$C$_{60}$. We'll see below that proximity to the metal–insulator boundary can lead to dramatic changes in conduction properties as the bandwidth W is changed.

One key question is whether these materials are BCS superconductors. Equation (12.6) showed that one prediction from BCS theory is that T_c depends on the DOS at the Fermi level $N(E_F)$, and $T_c \propto e^{-1/N(E_F)\cdot V'}$ implies that T_c will decrease as $N(E_F)$ is

---

[24] Intercalation chemistry is discussed more fully in Chapter 13.

[25] See Footnote 5.19 for symbolism; h implies fivefold degeneracy.

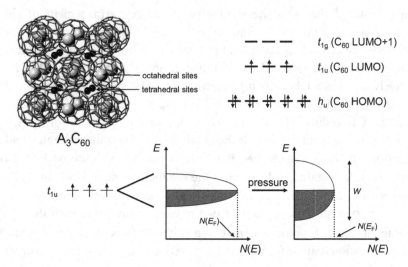

Figure 12.9 Top: Structure of A_3C_{60} and C_{60} MO diagram with electron filling appropriate for C_{60}^{3-}. Bottom: The t_{1u} becomes a half-filled band in solid K_3C_{60}; the bandwidth W and $N(E_F)$ depend on pressure.

lowered. Do A_3C_{60} compounds follow this behavior? The easiest way to change $N(E_F)$ would be to change the separation between the C_{60} molecules. As the molecules are moved closer together, we would expect the width W of the conduction band to increase as bonding interactions move to lower energy and antibonding interactions to higher energy. As the molecules are moved apart, the opposite would happen. In the limit of infinitely separated C_{60} molecules, one would have infinitely sharp bands, or the discrete energy levels of the molecule, Figure 12.9. Since the area of the t_{1u}-derived band must stay constant (it can always adopt up to six electrons per C_{60}), $N(E_F)$ will increase as molecular separation increases. Equation (12.6) then predicts that T_c will increase with the cell parameter a. The easiest chemical way to increase a is to increase the size, or average size, of the alkali metal. The size of the cation on the smaller tetrahedral site will be of particular importance.

Figure 12.10 shows how T_c changes as a increases for a large number of A_3C_{60} compounds,[26] and reveals that they broadly follow the predictions of BCS theory up to a volume per C_{60} of ~775 $Å^3$, or face-centered cubic (fcc) cell parameter of ~14.6 Å. As we discuss below, a Mott–Hubbard transition occurs beyond this value and the compounds become antiferromagnetic insulators at low temperature. This behavior is similar to that of non-BCS highly correlated electronic systems such as the cuprates and pnictides of Sections 12.7 and 12.8.

---

[26] This and related plots actually show the volume per C_{60} molecule, allowing direct comparison of different structure types.

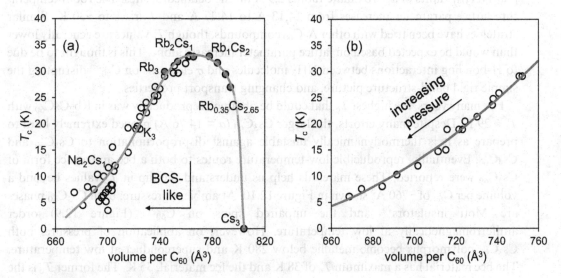

Figure 12.10 (a) Dependence of T_c on volume per C$_{60}$ molecule (adjusted to room temperature) for a variety of C$_{60}^{3-}$ containing materials. Open circles from ref. [13], shaded from ref. [14]. (b) Evolution of T_c as volume per C$_{60}$ changes by external pressure for Rb$_3$C$_{60}$. Solid lines are guides to the eye.

The dependence of T_c on $N(E_F)$ is also revealed by the effect of external pressure. As C$_{60}$ molecules are pushed together, orbital overlap will increase, leading to broader bands and lower $N(E_F)$. This would be expected to decrease the T_c of a BCS superconductor (though note that it can have the opposite effect on non-BCS materials), and this is again observed experimentally. Figure 12.10b shows data for Rb$_3$C$_{60}$. The reason why these A$_3$C$_{60}$ materials show this apparently simple behavior is that the two important parameters for influencing T_c, i.e. the electron–phonon coupling constant and the DOS at the Fermi level, can be influenced separately (at least to a first approximation). In fact, the phonons responsible for superconductivity are thought to be essentially intramolecular C$_{60}$ vibrations and not to involve the alkali metal. This is supported by isotope-effect values α in $T_c \propto M^{-\alpha}$, which are significant for carbon but zero for rubidium (carbon $\alpha = 0.30\pm0.05$ in K$_3$C$_{60}$ [15] and either 0.30 ± 0.05 or 0.21 ± 0.012 [16, 17] in Rb$_3$C$_{60}$, whereas rubidium $\alpha = -0.028\pm0.036$ [18]).

There have been considerable efforts to increase T_c in C$_{60}$-derived materials, including some spectacular high-profile claims that were later retracted.[27] A genuine example of how to increase the T_c of a parent A$_3$C$_{60}$ compound is via ammoniation of the 10.5 K superconductor Na$_2$CsC$_{60}$. This reaction leads to the formation of (NH$_3$)$_4$Na$_2$CsC$_{60}$, which contains

[27] See, for example, "*Plastic Fantastic: How the Biggest Fraud in Physics Shook the Scientific World*" by E.S. Reich (MacMillan Science).

$[Na(NH_3)_4]^+$ units of approximate radius 2.9 Å on the octahedral sites. The room-temperature lattice parameter increases from 14.13 Å to 14.47 Å and T_c rises to ~30 K. Similar strategies have been tried with other A_3C_{60} compounds, though T_c values are generally lower than would be expected based on lattice parameter increases alone. This is thought to be due to H-bonding interactions between NH_3 molecules and σ electrons on C_{60}^{3-} disrupting the simple rigid band-structure picture, and changing transport properties.

For many years, the highest T_c that could be obtained reproducibly was in Rb_2CsC_{60} with $T_c \approx 29$ K. Despite many efforts, the larger Cs_3C_{60} ($a = 14.76$ Å) proved extremely hard to prepare as it is thermodynamically unstable against disproportionation to Cs_1C_{60} and Cs_4C_{60}. Eventually, reproducible low-temperature routes to both a bcc and an fcc form of Cs_3C_{60} were reported. These materials help us understand the dip in T_c values beyond a volume per C_{60} of ~760 Å^3 shown in Figure 12.10. At ambient pressure, both Cs_3C_{60} phases are Mott insulators[28] and the unpaired spins on C_{60}^{3-} (Figure 12.9) order antiferromagnetically at low temperature. However, on application of pressure, both Cs_3C_{60} polymorphs become metallic below 280 K and superconduct at low temperature. The bcc material has a maximum T_c of 38 K and the fcc material, 35 K. The former T_c is the highest of any C_{60} compound, and the latter the highest for an fcc A_3C_{60} structure. T_c has a strong dependence on pressure leading to the "dome-like" superconducting-phase diagram of Figure 12.11.

The effect of pressure is to reduce the C_{60} separation, broaden the conduction bands and cause a transition to the metallic state. It's not surprising that similar behavior can be achieved chemically. For example, introduction of Rb to form $Rb_xCs_{3-x}C_{60}$ will also move C_{60} molecules closer together and broaden bands. Compounds in this family are found to be metallic for $x \gtrsim 0.35$ and T_c varies from 26.9 K ($x = 0.35$) to 32.9 K ($x = 1.0$; the highest T_c in this family) and back down to 31.8 K ($x = 2$) on increasing the Rb content. There is again a dome-like dependence of T_c (shown with the shaded circles in Figure 12.10) similar to Cs_3C_{60} under pressure. The three compositions are said to be respectively over-, optimally- and under-expanded. The under-expanded composition Rb_3C_{60} ($T_c = 29.3$ K) shows a value of $2\Delta_0/kT_c$ of 3.6(1), close to the value of 3.52 predicted by BCS theory in Equation (12.5), whereas the over-expanded samples show values up to 5.6(2) for $x = 0.35$, indicating non-BCS behavior.

We can understand the domes of the superconducting-phase diagrams as arising from two competing effects. If we start in the over-expanded region, contracting the cell destroys the Mott insulator as the bandwidth increases and the electronic states become less molecule-like. This initially gives rise to a metal with strong electron–electron correlations,[29] where the superconductivity is non-BCS in origin. In this region, decreasing the volume per C_{60}

---

[28] More formally they are Mott–Jahn–Teller insulators as the C_{60} molecules undergo a Jahn–Teller distortion, in which the degeneracy of the three t_{1u} orbitals is lost, giving a low-spin $S = \frac{1}{2}$ state. As discussed in the text, removing orbital degeneracy reduces the U/W ratio, above which insulating behavior occurs.

[29] The initial metallic state has been described as retaining Jahn–Teller distortions, suggesting the presence of both localized and extended states and has been called a Jahn–Teller metal.

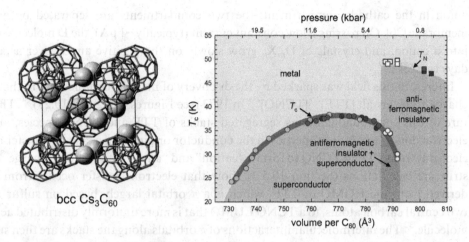

Figure 12.11 bcc Cs_3C_{60} undergoes an antiferromagnetic insulator to superconducting phase transition on application of pressure; figures reproduced from [19] with permission.

increases T_c. As contraction continues and the bandwidth increases, one reaches more conventional metallic behavior and BCS superconductivity occurs. In the BCS region, T_c decreases with increasing bandwidth, as $N(E_F)$ decreases. The maximum value of T_c appears to occur on the boundary of these two behaviors, highlighting the importance of both local molecule-like features and extended properties in controlling superconductivity.

Superconductivity has also been observed in compounds with a higher level of doping, such as Ba_4C_{60} that has a T_c of 6.7 K. This material has a body-centered arrangement of ordered C_{60} molecules with space group *Immm*. In this formally C_{60}^{8-} compound, the superconductivity comes from the t_{1g} LUMO+1-derived band (Figure 12.9). The t_{1g} band appears to simultaneously give rise to better metallic conductivity and lower T_c values. For example, the isostructural Ba_4C_{60} (electrons in t_{1g} band) and K_4C_{60} (electrons in t_{1u} band), are a $T_c = 6.7$ K metal and an insulator, respectively. It seems likely that mixing of Ba and C orbitals broadens the t_{1g} band.

12.5 Molecular Superconductors

In addition to the C_{60} superconductors described above, there has been huge interest in the development of so-called **molecular** or **organic superconductors**. Many of these are based on salts of organic electron-donor (D) and -acceptor (X) molecules, $D^{\delta+}_m X^{x-}_n$. What distinguishes these from the C_{60} intercalates is that they are prepared from molecular precursors in solution. This is often accomplished electrochemically by placing a solution of D and a salt of X^{x-} in the anode compartment of an electrochemical cell, and a solution containing the anion in the cathode compartment; the two compartments are separated by a porous membrane [20]. On passing a low constant current (typically ~1 μA), the D molecule oxidizes into a cation, and crystals of D_mX_n grow slowly on the positive anode over a period of days to weeks.

Interest in this field was sparked by the discovery of metallic conductivity in the organic charge-transfer salt $[TTF]^{\delta+}[TCNQ]^{\delta-}$ in 1972 (see Figures 12.12 and 10.29).[30] The structure of this compound contains segregated stacks of TTF and TCNQ species, and their electron donor/acceptor properties in the conductor originate in a charge transfer of ~0.59 electrons from TTF to TCNQ to form a "cation" and "anion" of fractional charge.[31] Band-structure calculations (Section 10.5.3) show that electron transfer occurs from a band derived from the HOMO of TTF, which is a π orbital largely based on sulfur and the two central carbon atoms, to a TCNQ LUMO that is more uniformly distributed across the molecule.[32] The intermolecular interactions of π orbitals along the stacks are then sufficient to give metallic conductivity in a band that contains contributions from both the donor HOMO and acceptor LUMO. Conductivity along the stack increases from ~4×10^4 S/m at room temperature to ~10^6 S/m at 60 K.[33] Below this temperature, a Peierls distortion[34] (or charge-density wave[35]) occurs, mainly on the TCNQ stacks, and the material becomes semiconducting.

The scientific insights from TTF–TCNQ led to the synthesis of many related salts with metallic conductivity containing TTF or one of the chemically related donors shown in Figure 12.12. The first organic superconductors (often called **Bechgaard salts** after their discoverer) have the formula $(TMTSF)_2X$, where X can be a variety of inorganic anions (PF_6^-, ClO_4^-, ReO_4^-, etc.). The anions lie between layers of $[TMTSF]^{0.5+}$ cations formed when two octet-obeying TMTSF molecules gave up 1 π electron to achieve partial aromatic stabilization. Despite its fixed oxidation state, the anion is key to controlling superconductivity as it determines the distances, hence the interactions, between donor molecules. It also

---

[30] It's common to use abbreviations for donor and acceptor molecules: TTF = tetrathiafulvalene; TCNQ = tetracyanoquinodimethane; TMTTF = tetramethyltetrathiafulvalene; TMTSF = tetramethyltetraselenafulvalene; BEDT-TTF (or ET) = bis(ethylenedithio)tetrathiafulvalene; $dmit^{2-}$ = 1,3-dithiole-2-thione-4,5-dithiolate.

[31] TTF in Figure 12.12 has two rings, each of them having seven σ electrons (two from each S, one from each C). According to the Hückel rule, losing one electron per ring (off the C=C central bond) by oxidation to a $[TTF]^{2+}$ cation would generate full aromatic stabilization. In the salt with organic acceptors, this tendency leads to a partial oxidation that is more precisely described in band-structure terms.

[32] Calculations suggest that the electron density of the TTF HOMO is around 61% S, 22% from the central C atoms and 17% from the four outer C [Fraxedas et al., *Phys. Rev. B.* **68** (2003), 195115].

[33] Cu has $\sigma \approx 10^8$ S/m at room temperature.   [34] Similar to that described for polyacetylene in Section 10.5.1.

[35] The terms charge-density wave (CDW) and spin-density wave (SDW) are frequently encountered when discussing complex electronic materials. If we have a charge-ordered metal oxide with alternating sites, such as Mn^{3+} and Mn^{4+} in $La_{0.5}Ca_{0.5}MnO_3$ (Chapter 11), we could imagine describing the charge distribution using a periodic (charge-density) wave propagating through the structure. Similarly, we could represent magnetic-moment direction and magnitude on different sites of an antiferromagnet by a wave propagating through the structure. Such waves could have wavelengths matching the lattice repeat and give rise to commensurate charge/spin ordering, or could have wavelengths that don't have a simple relationship to the lattice periodicity and give rise to incommensurate charge/spin ordering.

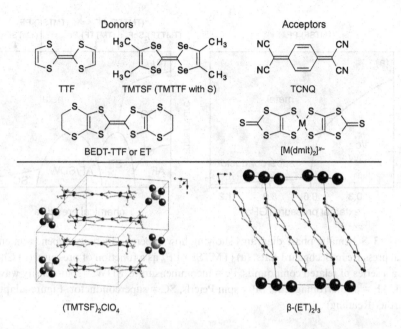

Figure 12.12 Top: Molecular structures of donor and acceptor molecules discussed in the text. Bottom: Structures of two molecular superconductors (H atoms omitted in β-(ET)$_2$I$_3$ for clarity).

influences subtle structural distortions of the structure, which can destroy metallic conductivity. In the first such superconductor discovered, (TMTSF)$_2$PF$_6$, a transition to an insulating state occurs on cooling below 12 K, in which the residual unpaired spins of the cations' HOMO couple antiferromagnetically (a spin-density wave). If this transition is suppressed by applied pressures of 0.6 GPa, superconductivity is observed below $T_c \approx 1.5$ K (Figure 12.13a).

At room temperature, the molecular anions in these materials are rotationally disordered, but this disorder freezes out on cooling. In the case of X = ClO$_4^-$, the freezing leads to doubling of the cell dimensions in the same manner as observed for a Peierls distortion, and a metal–insulator transition occurs. Under slow cooling, however, anion ordering occurs at 24 K in such a way that the periodicity is not doubled along the stacking axis, the metal–insulator transition is suppressed, and superconductivity occurs below $T_c = 1.4$ K at ambient pressure.

We saw in A$_3$C$_{60}$ compounds that both external and internal chemical pressure can influence superconductivity. Similar observations hold for molecular superconductors as illustrated in the schematic phase diagrams of Figure 12.13. As discussed above, Figure 12.13a shows how pressure tunes the properties of (TMTSF)$_2$PF$_6$ by influencing molecular overlap in the donor stacks, leading to superconductivity above 0.6 GPa. Figure 12.13b shows how the same effect can be achieved chemically. On the left-hand side of the diagram, we have (TMTTF)$_2$SbF$_6$ that has sulfur-containing organic electron donors and becomes

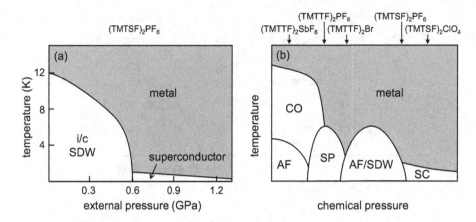

Figure 12.13 Schematic phase diagrams showing how superconductivity depends on either external or chemical pressure in Bechgaard salts. (a) (TMTSF)$_2$PF$_6$ as a function of pressure. (b) Chemical-pressure effect on a series of related compounds. i/c = incommensurate, SDW = spin-density wave, CO = charge ordered, AF = antiferromagnetic, SP = spin Peierls, SC = superconductor. Figure adapted from Brown (see Further Reading).

insulating on cooling at relatively high temperatures. If we change the anion to the smaller PF$_6^-$, donor–donor interactions increase and charge ordering is suppressed to lower temperatures. When we swap TMTTF for Se-containing TMTSF, even higher donor–donor interactions further suppress charge ordering, and the application of moderate pressure then induces superconductivity. The smaller ClO$_4^-$ salt will superconduct even at ambient pressure. It has also been shown that application of sufficiently high pressures will drive TMTTF salts into the superconducting state.

The origin of the pairing needed for superconductivity in these materials isn't known definitively, but there is strong evidence that intra-stack antiferromagnetic spin fluctuations are important; i.e. that a non-BCS mechanism is responsible. This is consistent with the observation that superconductivity in (TMTSF)$_2$PF$_6$ emerges just above the pressure at which the spin-density wave is destroyed, and that T_c has its highest value here.

Far higher superconducting temperatures are observed in (ET)$_2$X salts (e.g. Figure 12.12, lower right). These compounds again have structures in which layers of electron donors are separated by anions, with the donor packing leading to more 2D electronic properties. In the β-(ET)$_2$X (X = [I$_3$]$^-$, [IBr$_2$]$^-$, [AuI$_2$]$^-$, etc.) compounds, the donors have their molecular planes parallel (Figure 12.12). In the so-called κ-phase salts, e.g. κ-(ET)$_2$Cu[N(CN)$_2$]Br, the donor layers contain "dimers" of closely separated pairs of face-to-face donors, which then interact with other dimers that are oriented at approximately 90°. Conductivity is highly anisotropic; typically ~5000 S/m in plane but 1000 times lower perpendicular to the plane. The compound κ-(ET)$_2$Cu[N(CN)$_2$]Br has one of the highest ambient pressure T_c values (11.6 K) known for a molecular system. By contrast, the Cl analogue undergoes a metal-to-insulator transition at 40–50 K on cooling. However, a modest pressure (~0.3 kbar) suppresses this and a 12.8 K superconductor results.

Salts of metal complexes containing ligands with strong structural similarities to TTF have also shown superconductivity. TTF[Ni(dmit)₂]₂ contains TTF-like species as both the electron donor and acceptor, and superconducts at 1.6 K under pressure. Superconductivity supported purely by the metal-complex anions has also been observed at 5 K under pressure in NMe₄[Ni(dmit)₂]₂.

The range of molecular superconductors is large[36] and will undoubtedly continue to grow. One much-heralded advantage of molecular systems is the ability of the chemist to tailor the structure of the electron-donor and electron-acceptor species at will. Such changes can help suppress the transitions (Peierls and Mott–Hubbard) that compete with superconductivity. The disadvantage is that the large number of intermolecular interactions present in the crystal structure means that different structural arrangements can have similar energies, and polymorphism is common. It's therefore not surprising that even making subtle changes to the electron donors and acceptors can frequently lead to major changes in structure and properties. Such structural changes are extremely difficult to predict or control a priori.

12.6 BaBiO₃ Perovskite Superconductors

Perovskite-derived superconductors have been extensively studied since the discovery of superconductivity in reduced SrTiO₃₋δ ($T_c \approx 0.3$ K) in 1964 [21]. One early example, which paved the way for later work on cuprates, was found by Sleight and co-workers in the solid solution between two perovskite materials, BaPbO₃ and BaBiO₃ [22].

BaPbO₃ has a tilted version of the perovskite structure with orthorhombic symmetry *Ibmm* at room temperature and might be expected to be an electronic insulator like most other main-group perovskites. In fact, the energy of the Pb 6s orbital is such that the empty 6s conduction band partially overlaps the O 2p valence band, and BaPbO₃ has metallic conductivity with $\sigma \approx 3000$ S/cm at 300 K.

From simple size arguments, one might expect BaBiO₃ to be a cubic perovskite and that the single 6s electron of Bi^{4+} could also give rise to metallic conductivity. In reality, the electronic structure is more complex. We know from chemistry that 4+ oxidation states are unusual for group 15 elements, and in BaBiO₃ the nominal Bi^{4+} should disproportionate to Bi^{3+} and Bi^{5+}, giving $Ba_2Bi^{III}Bi^VO_6$, analogous to compounds like $Ba_2Bi^{III}Ta^VO_6$. As with the molecular systems of Section 12.5, this disproportionation can be described as a charge-density wave. Structural studies support this view and reveal two distinct octahedral sites,

---

[36] Over 100 superconducting phases are known with donors chemically related to BEDT-TTF (ET) and TMTSF.

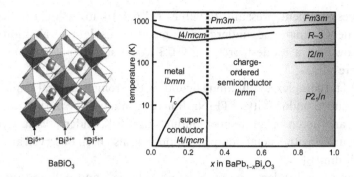

Figure 12.14 Structure of $BaBiO_3$ and schematic structural phase diagram of $BaPb_{1-x}Bi_xO_3$. Space groups of important phases are given. Issues with sample purity mean that it is hard to be definitive about the position of the metal-to-semiconductor boundary.

one (Bi^{3+}) with Bi–O bond distances around 2.28–2.30 Å and a second (Bi^{5+}) with shorter distances of 2.11–2.13 Å. The combined Bi-site ordering and tilting of octahedra gives $P2_1/n$ monoclinic symmetry at room temperature and below. Band-structure calculations show that it is a semiconductor. We can understand the conductivity of $BaBiO_3$ using a polaron model, in which a pair of electrons moves from Bi^{3+} to Bi^{5+}, turning it to Bi^{3+}, etc. This requires a significant change in the size of the $BiO_{6/2}$ octahedron.[37]

In $BaPb_{1-x}Bi_xO_3$ solid solutions, one can tune the electronic properties across a metal-to-semiconductor boundary at $x \approx 0.3$. On cooling, superconductivity is observed for samples with $0.05 < x < 0.30$, with the maximum T_c of 13 K at $x \approx 0.27$ (Figure 12.14), occurring very close to the metal-to-semiconductor boundary. The structural phase diagram of $BaPb_{1-x}Bi_xO_3$ is complex, and superconducting samples are often found to be a metastable mixture of the orthorhombic phase of space group $Ibmm$ and a tetragonal phase ($I4/mcm$) that is believed to be responsible for superconductivity. What electronic role does the seemingly innocent Pb^{4+} play in controlling the structure of these materials and therefore their superconductivity? If we start at the Bi-rich end of the series, increasing the Pb content disrupts the Bi^{3+}/Bi^{5+} charge order as the intermediate size and valence of Pb^{4+} [$r(Pb^{4+}) = 0.78$ Å; $r(Bi^{3+/5+}) = 1.03/0.76$ Å] breaks down the large/small ordered pattern of octahedra. The onset of superconductivity occurs close to the doping level at which charge ordering disappears. At this level, the material is metallic at high temperatures and the electrons are delocalized.

Superconductivity at a higher T_c of up to 34 K has also been created via A-site doping of $BaBiO_3$ to produce $Ba_{1-x}K_xBiO_3$, with superconductivity observed for $0.30 < x < 0.45$. The doping by the K^+ acceptor is compensated by oxidation of Bi^{3+}, which will again disrupt the

---

[37] As a pair of electrons moves, it swaps the identities of Bi^{3+} and Bi^{5+} sites. Oxygens will rearrange such that Bi^{3+} sites have longer Bi–O distances and Bi^{5+} short, as electrons move between different Bi–O bonding orbitals. We can imagine a "breathing" mode of the octahedra being involved.

pattern of charge ordering. The phase diagram is again complex, but increasing the K content initially destroys charge ordering, and the material superconducts once the low-temperature symmetry becomes $I4/mcm$ at $x > 0.30$ (i.e. the same symmetry as observed for $BaPb_{1-x}Bi_xO_3$ superconductors).[38] A 12 K superconductivity has been observed in $Sr_{0.4}K_{0.6}BiO_3$ that can be prepared at high pressure. Charge ordering is again evident in the undoped material and suppressed in the superconducting phase. Superconductivity has also been observed in the A-site-deficient $(K_{0.45}Na_{0.25})Ba_3Bi_4O_{12}$ below 27 K and in the related Sb phase $BaPb_{0.75}Sb_{0.25}O_3$ below 3.5 K.

The origin of superconductivity in these compounds is still debated, and detailed understanding is made difficult by the fact that two structurally similar phases are usually present in superconducting samples of $BaPb_{1-x}Bi_xO_3$. There is, however, good evidence that they have BCS-like properties with both $BaPb_{1-x}Bi_xO_3$ and $Ba_{0.6}K_{0.4}BiO_3$ having $^{16}O/^{18}O$ isotope effects ($T_c \propto M^{-\alpha}$), with α of 0.4–0.5. It's clear from their structural behavior that the significant electron–lattice-vibration interactions required by BCS theory exist, and that superconductivity occurs close to the metal–insulator boundary and close to charge-localizing phase transitions. It's worth concluding by emphasizing two features we'll see again in cuprates: electrons are in a band with significant oxygen and metal character, and there is a tendency for charge disproportionation of Bi over oxidation states ranging from Bi^{III} to Bi^V.

12.7 Cuprate Superconductors

One of the most significant breakthroughs in the field of superconductivity came in November 1986 when Georg Bednorz and Alex Müller at IBM Research Zurich reported superconductivity at 28 K in the La–Ba–Cu–O system [23]. By the end of that year, the superconducting phase had been identified as $La_{1.85}Ba_{0.15}CuO_4$, which has a T_c up to 30 K. This released a frenzy of world-wide research activity aimed at understanding and exploiting these materials. Superconductivity below 38 K was rapidly found in $La_{1.85}Sr_{0.15}CuO_4$, 93 K superconductivity (above the 77 K boiling point of readily available liquid nitrogen) in $YBa_2Cu_3O_7$, and eventually 134 K superconductivity (164 K under pressure) in $HgBa_2Ca_2Cu_3O_8$. The observation of such high T_c values overthrew predictions from BCS theory that the maximum T_c would be limited to ~30 K[39] and suggested that another pairing mechanism must be at play. We will look at the structural chemistry of key **cuprate superconductors** and the electronic features that lead to their remarkable properties.

---

[38] These samples are cubic at room temperature.

[39] More recent work on MgB_2 shows a higher BCS limit of ~39 K.

12.7.1 La$_2$CuO$_4$ "214" Materials

Pure La$_2$CuO$_4$ is an antiferromagnetic insulator with a Néel temperature of ~317 K. It has a slightly distorted version of the K$_2$NiF$_4$ structure, as shown in Figure 12.15. The K$_2$NiF$_4$ structure can be viewed as an intergrowth of the perovskite and rock-salt structures. If one separates 2D infinite slabs of corner-sharing MO$_2$O$_{4/2}$ octahedra from the perovskite structure (Figure 12.15a) and translates them by $a/2 + b/2$ alternately back and forth such that the axial oxygens lie directly above and below positions of A-site cations from neighboring layers, one forms the K$_2$NiF$_4$ structure. The A-site cation is nine-coordinate, with one oxygen directly above it (see the lowest A site in Figure 12.15c), four oxygens in its plane and four oxygens in a plane below. From the "waist down", this cation has the coordination environment found for the A site in perovskite; from the "waist up", it has the coordination of a cation in the rock-salt structure. The crucial structural feature as far as superconductivity is concerned is that there are infinite 2D layers of distorted corner-sharing CuO$_2$O$_{4/2}$ octahedra that have axial (often called apical in the cuprate literature) Cu–O bond lengths of ~2.4 Å and equatorial bond lengths of ~1.9 Å (ICSD 68381).

There are two distinct ways in which La$_2$CuO$_4$ can be doped to make it superconducting. Both rely on **hole doping** (which is a partial/fractional oxidation, here of Cu^{2+}, as opposed to electron doping that would be a partial reduction), and we'll discuss the electronic changes that occur in Section 12.7.4. One way is to partially replace La^{3+} with a divalent ion to form compounds such as La$_{2-x}$Sr$_x$CuO$_4$, with the highest T_c = 38 K at x = 0.15.[40] The average copper oxidation state is then Cu$^{2.15+}$. Alternatively, samples can be doped with excess oxygen to give La$_2$CuO$_{4+\delta}$. The highest-T_c material has $\delta \approx 0.08$, implying a similar average oxidation state of Cu$^{2.16+}$. The oxygen interstitial occupies a site midway between adjacent La layers in the structure such that it is surrounded by a tetrahedron of La atoms, as shown in Figure 12.15c. Significant local distortion is required to allow this, and nearby oxygen atoms move by up to 0.5 Å. These oxygen interstitials are often mobile in the structure, and the structural and electronic properties of these and related cuprate superconductors can therefore vary with time.

It is interesting to note that optimal superconducting properties occur at similar Cu oxidation states for both doping routes and occur at doping levels just sufficient to suppress low-temperature antiferromagnetic ordering of the Cu^{2+} spins. If such a material is underdoped, superconductivity is not observed; if overdoped, normal metallic behavior results without superconductivity. As with A$_3$C$_{60}$, superconductivity occurs near the metal-to-insulator boundary. A schematic phase diagram of cuprate superconductors illustrating these and other points is discussed later in Section 12.7.4.

---

[40] T_c up to 52 K can be achieved in related compositions under pressure ($\Delta T_c \approx 0.009$ K per GPa) or in deliberately strained thin films. Attempts to increase T_c by chemical rather than external pressure by replacing La^{3+} with the smaller Y^{3+} led to the identification of YBa$_2$Cu$_3$O$_7$.

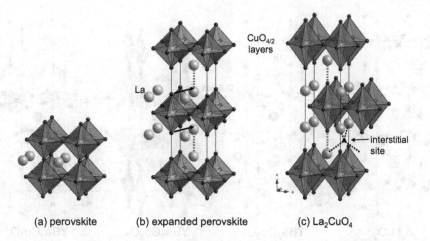

Figure 12.15 Relationship between the (a) perovskite and (c) La_2CuO_4 structures. The central layer of a hypothetical expanded perovskite (b) is displaced by $a/2$ and $b/2$ such that its apical oxygens lie above and below the La atoms of adjacent layers. A potential La_4O interstitial site, as found in $La_2CuO_{4+\delta}$, is labelled.

It's also worth noting that even rather subtle changes in the local structure of a material can influence superconducting properties. For example, it's found experimentally that for a variety of "optimally doped" $Ln^{3+}_{1.85} A^{2+}_{0.15}CuO_4$ materials, all with the same $Cu^{2.15+}$ oxidation state but with different combinations of Ln^{3+} and A^{2+} cations, there is a maximum T_c for a mean A-site nine-coordinated ionic radius of ~1.22 Å. In addition, Attfield et al. [24] have shown that not only the mean cation radius but the variance[41] of the cation radius is important—at a constant doping level, T_c is lowered if a large variety of small and large cations is used to produce a given average radius.

12.7.2 $YBa_2Cu_3O_{7-\delta}$ "YBCO" or "123" Materials

Perhaps the best known of the cuprate superconductors is the yttrium barium copper oxide $YBa_2Cu_3O_{7-\delta}$ first reported by Wu and Chu in 1987 [25]. It is frequently referred to as YBCO or 123 (after the cation ratios). $YBa_2Cu_3O_7$ has a T_c of 93 K.

The structure of $YBa_2Cu_3O_7$ can again be derived from perovskite. As shown in Figure 12.16, the prototype perovskite cell is tripled such that it contains $A_3M_3O_9$ with three $MO_{6/2}$ octahedra. The simplest way (Pauling's parsimony rule, Chapter 1) to remove two O atoms is to form $1MO_{4/2} + 2MO_{5/2}$. The square-planar $CuO_{4/2}$ link in chains, the corner-sharing bases of the $CuO_{5/2}$ square pyramids link in planes. This leads to the colloquial names for the two sites as *chains* and *planes*, respectively.

---

[41] The population variance, σ^2, is defined as $\sigma^2 = 1/N \sum_{i=1}^{N} (x_i - \mu)^2$, where there are N values x_i and μ is their mean. A range of different radii will give a large variance.

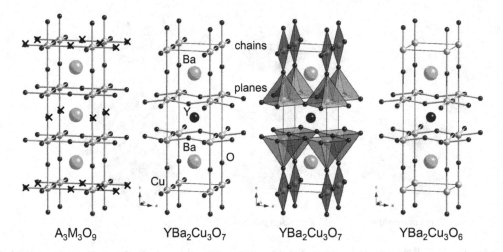

Figure 12.16 A hypothetical triple perovskite structure $A_3M_3O_9$. Deleting two oxygens per unit cell (crosses) leads to $YBa_2Cu_3O_7$, shown in ball-and-stick and polyhedral views. Deleting a further oxygen leads to $YBa_2Cu_3O_6$.

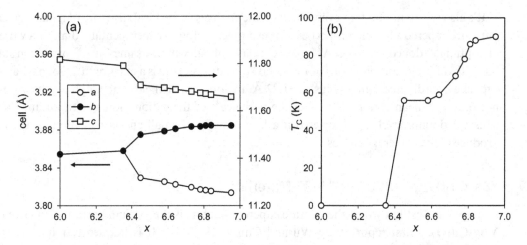

Figure 12.17 The dependence of (a) unit-cell parameters and (b) T_c on oxygen content for a series of $YBa_2Cu_3O_x$ samples. Data from [27].

The structural and superconducting properties of $YBa_2Cu_3O_{7-\delta}$ depend critically on oxygen content (Figure 12.17 shows data from an early publication). As the oxygen content is lowered below 7, oxygen is removed from $CuO_{4/2}$ at sites halfway along the b cell edge, until at $\delta = 1$ one reaches the $YBa_2Cu_3O_6$ structure shown in Figure 12.16. Below $YBa_2Cu_3O_{6.41}$, the material no longer superconducts. This loss of superconductivity is associated with a structural change from orthorhombic ($a \approx b \neq c$) to tetragonal ($a = b \neq c$)

Box 12.2 Synthetic Methods: Synthesis and tuning superconductivity in $YBa_2Cu_3O_{7-\delta}$

It's straightforward to synthesize a sample of $YBa_2Cu_3O_{7-\delta}$ that will show superconductivity. A simple procedure is to grind and mix a stoichiometric ratio of $0.5Y_2O_3$, $2BaCO_3$, and $3CuO$ and then heat it to 950 °C for a few hours in air. The mixture is then reground, pelletized, and reheated to 950 °C, then slow-cooled at around 1 °C a minute in air (or better in O_2) to room temperature. Such samples will show the Meissner effect by repelling a magnet when cooled by liquid N_2. However, we know that superconductivity in this system depends sensitively on the oxygen nonstoichiometry, with the highest T_c when δ approaches 0. How do we make an optimal sample? At what temperatures and in what specific atmosphere should it be prepared?

Experimental data for the dependence of δ in $YBa_2Cu_3O_{7-\delta}$ on temperature and oxygen partial pressure are shown below:

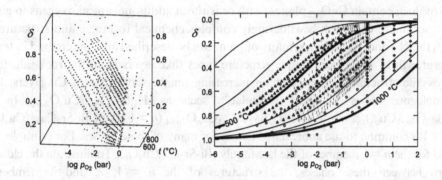

In Chapter 3, we simplified the defect equilibria in $YBa_2Cu_3O_{7-\delta}$ to a single pseudochemical oxidation of oxygen vacancies with the equilibrium constant $K_{vox} = p_{O_2}^{-1/2} \cdot 4(1-\delta)^2/\delta$; see Equation (3.5). How do we derive the temperature dependence for this equilibrium constant? The easiest approach is to express the Gibbs energy as $\Delta G = \Delta H - T\Delta S$ and to introduce $\Delta G = -RT\ln K$, which gives $K_{vox} = \exp[(\Delta S_{vox}/R) - (\Delta H_{vox}/RT)]$. A least-squares fitting of the experimental data gives $\Delta S_{vox} = -69.8(4)$ J/(mol K) and $\Delta H_{vox} = -79.8(3)$ kJ/mol as parameters. The calculated isotherms shown in the graph above agree best with the experiment in the middle range of nonstoichiometry where the concentration of the neglected defects (see Section 3.3) is lowest. We can see from this graph that to prepare samples with $\delta \approx 0$, one should anneal the sample at relatively low temperatures (<500 °C) in $p_{O_2} > 1$.

symmetry, related to oxygen atoms randomly occupying ½ 0 0 and 0 ½ 0 positions in the tetragonal unit cell; a short-range order occurs at the local level [26].

In $YBa_2Cu_3O_7$, the average Cu oxidation state +2.33 is distributed over one square-planar $CuO_{4/2}$ and two $CuO_{5/2}$ square pyramids. Since we know that d^8 ions favor square-planar coordination, a first approximation is that $YBa_2Cu_3O_7$ contains one square-planar Cu^{3+} and two square-pyramidal Cu^{2+}. $YBa_2Cu_3O_6$ then has the typical linear Cu^+ and two Cu^{2+}. As in La_2CuO_4, both compounds have their *planes* occupied by Cu^{2+} in this first approximation. What happens next in superconducting $YBa_2Cu_3O_7$ is that the mismatch of the two average Cu–O bond orders $3^+/4$ versus $2^+/5$ along the same length of the short unit-cell edge, and the highly oxidizing nature of Cu^{3+}, both make the *chains* partially hole dope (oxidize) the Cu^{2+} of the *planes*, as can be shown by bond-valence analysis. This **intrinsic hole doping** again leads to **p-type superconductivity** of this cuprate, suggesting the superconducting pairing derives from diluted holes in the *planes*. More in Section 12.7.4.

12.7.3 Other Cuprates

The remarkable properties of the 214 and 123 cuprates prompted the search and subsequent discovery of superconductivity in a wide variety of copper-containing mixed metal oxides. All of them contain $CuO_{4/2}$ planes (with or without additional apical oxygens to give five- or six-coordinate Cu), with bewilderingly complex chemical formulae and structures. Just as with the 214 and 123 families, many of them can be described and understood in terms of the intergrowth of slabs from simple structure types (like layers from the rock-salt, fluorite, or perovskite structures). Some contain increasing numbers of adjacent Cu layers, leading to homologous series of ideal formulae[42] such as $HgBa_2Ca_{n-1}Cu_nO_{2n+2}$ (n = 1–7), $TlBa_2Ca_{n-1}Cu_nO_{2n+3}$ (n = 1–4), $Tl_2Ba_2Ca_{n-1}Cu_nO_{2n+4}$ (n = 1–4), or $Bi_2Sr_2Ca_{n-1}Cu_nO_{2n+4}$ (n = 1–3). It's common to use abbreviated names for many of these phases. For example, Bi-2201 is used for the n = 1 member of the last family ($Bi_2Sr_2Ca_0Cu_1O_6$). To illustrate the close relationships between these phases, the structures of the n = 1, 2, and 3 members of the $HgBa_2Ca_{n-1}Cu_nO_{2n+2}$ family are shown in Figure 12.18. Note that CuO_2 layers separated by small Ca^{2+} cations don't contain apical O atoms. The n = 1 material has the highest T_c of any cuprate with a single CuO_2 plane (98 K). In general, it's found that T_c of such materials increases as the number of adjacent Cu planes increases from n = 1 to n = 3 but decreases thereafter.

These chemically complex materials are difficult to prepare as single phases, and they frequently have compositions that differ from ideal stoichiometric formulae. This can be caused by a variety of defects (e.g. in $TlBa_2Ca_{n-1}Cu_nO_{2n+3}$ by Tl/Ca substitution, Cu/Tl substitution, Tl/Ca vacancies, and/or oxygen nonstoichiometry), by loss of volatile species (e.g. CuO or Bi_2O_3) during synthesis, or by intergrowths of different structures between the cuprate planes. This leads to a variety of mechanisms for the hole doping responsible for their superconductivity. Many materials, in particular the Bi phases, have complex incommensurate structures due to the size mismatch between adjacent structural layers. Many other cuprate superconductors are known with species such as halide anions, CO_3^{2-}, BO_3^{2-}, or $(GaO_3^{3-})_\infty$ zig-zag chains of corner-sharing tetrahedra between Cu planes.

---

[42] As in YBCO, the exact oxygen content may differ from these formulae. For example, the Hg series is often expressed as $HgBa_2Ca_{n-1}Cu_nO_{2n+3-\delta}$.

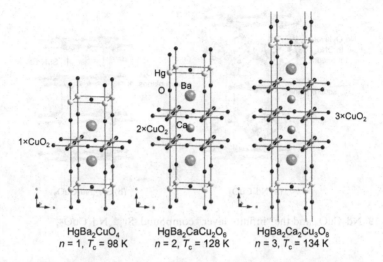

HgBa₂CuO₄   HgBa₂CaCu₂O₆   HgBa₂Ca₂Cu₃O₈
$n = 1$, $T_c = 98$ K $n = 2$, $T_c = 128$ K $n = 3$, $T_c = 134$ K

Figure 12.18 Structures of the $n = 1$, 2, and 3 members of the $HgBa_2Ca_{n-1}Cu_nO_{2n+2}$ series. The ½ ½ 0 oxygen positions would be occupied by O in the related $TlBa_2Ca_{n-1}Cu_nO_{2n+3}$ series. The $n = 3$ member has $T_c = 164$ K under pressure; the reason for the T_c enhancement is not understood.

In addition to hole doping leading to p-type superconductivity, some cuprates can be electron-doped to **n-type superconductivity**. These contain CuO_2 planes without apical oxygens. The first such family discovered was based on Nd_2CuO_4 doped with CeO_2, or with substitution of F for O, to give $Nd_{2-x}Ce_xCuO_4$ or $Nd_2CuO_{4-x}F_x$. A T_c up to ~25 K has been found. The structure of Nd_2CuO_4 is shown in Figure 12.19a.[43] The smaller radius of Nd^{3+} compared to La^{3+} leads to a different structure type containing slabs of fluorite-like $[Nd_2O_2]^{2+}$ separated by $[CuO_{4/2}]^{2-}$. The size of the Nd_2O_2 block is such that $CuO_{4/2}$ planes are under slight stretch tension. Electron doping puts electrons in Cu–O antibonding states (see below), relieving this tension. In these systems, it is again difficult to produce phase-pure samples, and the exact composition of the superconducting component is unsure. n-type superconductivity has also been achieved at $T = 40$ K in $Sr_{1-x}Nd_xCuO_2$ synthesized at high pressures. This material has the so-called **infinite-layer**[44] **superconductor structure** (Figure 12.19b), of alternating $CuO_{4/2}$ plane and cation layers, and can be thought of as the "parent structure" of all cuprate superconductors.[45]

12.7.4 Electronic Properties of Cuprates

What can we say about the origin of superconductivity in the cuprates? Currently there is still no universally accepted theory to explain their remarkable properties. One

[43] Colloquially known as the T′ phase versus La_2CuO_4 the T phase.

[44] "Infinite" refers to a hypothetical $n = \infty$ member of the series $HgBa_2Ca_{n-1}Cu_nO_{2n+2}$ in Figure 12.18, a $CaCuO_2$ with only the apex-free CuO_2 planes.

[45] $Ca_{0.84}Sr_{0.16}CuO_2$ appears to be the only phase with the infinite-layer structure accessible at ambient pressure.

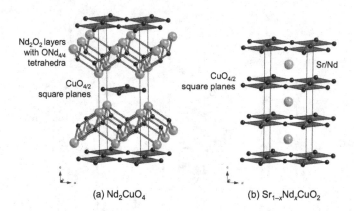

Nd$_2$O$_2$ layers
with ONd$_{4/4}$
tetrahedra

CuO$_{4/2}$
square planes

Sr/Nd

CuO$_{4/2}$
square planes

(a) Nd$_2$CuO$_4$ (b) Sr$_{1-x}$Nd$_x$CuO$_2$

Figure 12.19 Nd$_2$CuO$_4$ and the "infinite-layer" compound, Sr$_{1-x}$Nd$_x$CuO$_2$.

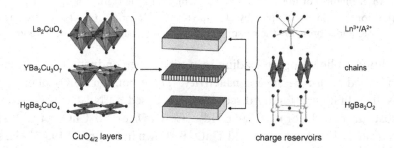

La$_2$CuO$_4$

YBa$_2$Cu$_3$O$_7$

HgBa$_2$CuO$_4$

CuO$_{4/2}$ layers

Ln^{3+}/A^{2+}

chains

HgBa$_2$O$_2$

charge reservoirs

Figure 12.20 Schematic representation of the key structural features of superconducting cuprates.

common feature of all the cuprates discussed is the presence of one or more CuO$_{4/2}$ planes. Controlling their electronic properties by doping is key to controlling super-conductivity. This leads to a generic description of cuprates (Figure 12.20) as electron-ically active CuO$_{4/2}$ planes separated by so-called **charge reservoirs** that dope the planes. The CuO$_{4/2}$ planes can contain purely corner-sharing square-planar copper or additional apical oxygens may be present to form square pyramids (as in YBa$_2$Cu$_3$O$_7$) or octahedra (as in La$_2$CuO$_4$).

As we've seen in Section 6.5.2, d^9 Cu^{2+} in a square-planar oxygen environment will have a d-orbital splitting pattern that leads to a half-filled band of energy levels derived from the antibonding overlap of p orbitals on oxygen and the $d_{x^2-y^2}$ orbital on Cu (Figure 12.21). Due to the similarity in energy of Cu $3d$ and O $2p$ orbitals, there is significant covalence, and the band has both Cu and O character. The width of the band is also such that we expect the material to be a Mott–Hubbard insulator rather than a metallic conductor. In the majority of cuprates, the role of the charge reservoir is to introduce holes in this band, which induces an

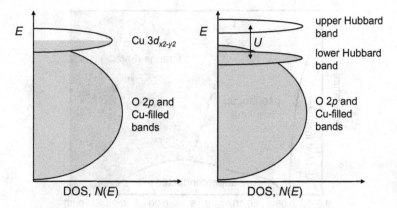

Figure 12.21 Left: Schematic density-of-states plots for CuO$_{4/2}$ planes; Right: Splitting of the $d_{x^2-y^2}$ into an upper and lower Hubbard band due to energy cost of electron pairing.

insulator–metal transition.[46] The mechanisms for doping vary. In the La$_2$CuO$_4$ series, the extrinsic hole doping by either La^{3+}/A^{2+} substitution or incorporation of excess oxygens produces a Cu oxidation state of +2.16. In YBa$_2$Cu$_3$O$_7$ the hole doping is intrinsic: Cu^{3+} of the chains partially oxidize the Cu^{2+} of planes to Cu$^{(2+x)+}$ and are themselves reduced to Cu$^{(3-2x)+}$. A variety of similar doping mechanisms occur in the chemically more complex systems.

Many cuprates have a T-versus-doping phase diagram similar to Figure 12.22, in which undoped materials are antiferromagnetic insulators at low temperature. Long-range antiferromagnetism is suppressed at ~3% hole doping and superconductivity emerges at ~5% doping (between these regions local phase segregation occurs). T_c is typically maximized at ~15–20% doping, and superconductivity destroyed above ~25% doping, leading to a superconducting dome similar to those we've seen in other materials. The regions prior to and after the maximum T_c are referred to as **under-** and **over-doped** regions, respectively.

The electronic properties of the cuprates above T_c turn out to be just as unusual as those below, and are still not properly understood. Initial doping of the Mott insulator causes a gradual transition to what's known as the **pseudogap** regime. In this region there is a gap in the electronic density of states for some values of k (i.e. some momentum directions of electrons) but not for others. There is also a strong tendency

[46] In 2019, superconductivity was discovered in thin films of Nd$_{0.8}$Sr$_{0.2}$NiO$_2$ [D. Li et al., *Nature* **572** (2019), 624–627]. The Ni oxidation state is around one unit lower than in the Cu superconductors, suggesting superconductivity again occurs by hole doping a d^9-derived half-filled band. The Ni possibility was discussed by Karen et al. [*J. Solid State Chem.* **97** (1992) 257–273] when they correlated high-T_c superconductivity with the presence of a well-separated nearly half-filled Cu-localized HOMO so that a dilute occurrence of a hole pair (if less than half filled) or an electron pair (if more than half filled) in that orbital was imaginable.

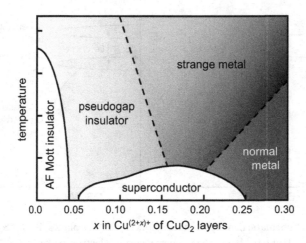

Figure 12.22 A generic phase diagram for hole-doped cuprate superconductors. Dashed lines separate regions where there are poorly defined changes in electronic behavior.

towards forming a variety of ordered states such as charge-density waves,[47] spin-density waves, and more exotic forms of electronic order. The pseudogap regime has several electronic features in common with the superconducting state, though there is debate about whether it is a precursor to superconductivity or a competing state. Beyond the pseudogap region, one enters a region where the material is often described as a "strange" or "bad" metal to reflect its differences to a normal metal. Examples of these differences include high-temperature resistivities two orders of magnitude higher than normal metals[48] and a linear dependence of resistivity with temperature up to the highest temperatures measured.[49] It is perhaps unsurprising that the electronic properties of cuprates in this region are unlike those of a normal metal. The reason for the insulating nature of the undoped materials (which have band gaps of ~2 eV) is the high Hubbard U that reflects strong electron–electron repulsions on moving an electron from one site to another that already has an electron (see Section 10.4.1). Some of these repulsions will still be present in doped cuprates, and we can imagine electrons as having to "get out of each other's way" to allow conductivity by moving in concert. As such, the mobile electrons will be strongly correlated, unlike those in a normal metal. As the hole-doping levels are increased, correlations reduce and the high-temperature properties gradually become more like those of a normal metal.

---

[47] One form of separate ordering of charge and spin leads to so-called "stripes" of electron density in some cuprates.

[48] The in-plane resistivities (per CuO_2 layer) are very similar for different families of optimally doped cuprates and typically around 5×10^{-4} Ω cm at 300 K. Out-of-plane resistivities are higher by a factor rising from ~30 for $YBa_2Cu_3O_{6.95}$ to ~1×10^6 for $Bi_2Sr_2Ca_2O_{8+\delta}$ just above T_c [N. Hussey, *J. Phys. Condens. Matter.* **20** (2008), 123201].

[49] In a normal metal, resistivity saturates at high temperatures when the mean free path becomes comparable to the de Broglie wavelength; see Chapter 15.

How do these arguments influence superconductivity? While it's clear that the cuprates obey the general rule of "bad-metal, good-superconductor", this doesn't explain Cooper-pair formation at temperatures far higher than those predicted by BCS theory for conventional superconductors. There's good evidence that the antiferromagnetic nature of the parent insulating phases survives into the superconducting regime in the form of short-range dynamic magnetic fluctuations, and that these are much stronger than in a normal metal. Most experts think that these are important for pair formation.

Despite over 30 years of intense scientific effort, it's still not possible to provide a complete explanation for superconductivity in the cuprates. We shouldn't feel too bad about this. Writing in 2015, Chu (co-discoverer of $YBa_2Cu_3O_7$) and co-workers concluded that it is still necessary to "let the experts resolve the different views" [28].

12.8 Iron Pnictides and Related Superconductors

In February 2008, Hosono and co-workers reported superconductivity at $T_c = 26$ K in $LaO_{1-x}F_xFeAs$ with $0.14 < x < 0.4$ [29]. The observation of such a high T_c in an Fe-based superconductor was a considerable surprise and rapidly led on to the discovery of ambient-pressure $T_c = 55$ K in $SmO_{1-x}F_xFeAs$, a temperature only surpassed by the cuprates. Superconductivity has since been discovered in a range of related materials, most notably those based on $BaFe_2As_2$ (T_c up to 30 K), $LiFeAs$ ($T_c = 18$ K), and even in the binary phase β-FeSe[50] ($T_c = 37$ K at 8.9 GPa). In contrast to other families, superconductivity in these

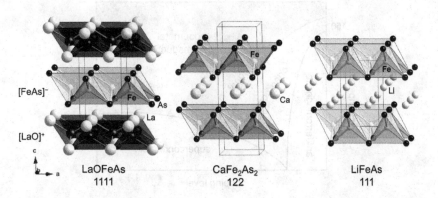

Figure 12.23 Pnictide superconductors. All contain layers of edge-sharing $FeAs_{4/4}$ tetrahedra.

---

[50] It is the tetragonal β-FeSe and not the hexagonal α form that superconducts; there is some confusion in the literature over the use of α/β.

materials can be remarkably tolerant to compositional changes, and chemical changes that can modify superconducting properties are typically possible on all sites.

As with the cuprates, these materials are often referred to by a shorthand notation: LaOFeAs as "1111", CaFe$_2$As$_2$ as "122", LiFeAs as "111" (Figure 12.23). All of them contain [FeAs]$^-$ layers made up of edge-sharing FeAs$_{4/4}$ tetrahedra. In the 1111 family, these layers alternate with [LaO]$^+$ layers of OLa$_{4/4}$ edge-sharing tetrahedra. The La ions "bridge" the oxide and pnictide portions of the structure and have a 4 + 4 square-antiprismatic coordination LaO$_4$Se$_4$. In the 122 family, A^{2+} replaces [La$_2$O$_2$]$^{2+}$ segments, giving a ThCr$_2$Si$_2$-type structure (Th → Ca, Cr → Fe, Si → As). LiFeAs has Li$^+$ ions in five-coordinate square-pyramidal sites between the [FeAs]$^-$ tetrahedral layers (Cu$_2$Sb structure type). Finally, FeSe adopts the α-PbO structure type and has the same FeSe$_{4/4}$ layers but with (in the ideal structure) no cations separating them.

All these pure/undoped compounds share common features. Each contains Fe^{2+} arranged on a square grid of edge 2.63–2.85 Å at ~200 K.[51] The Fe–Fe distance is much shorter than the Cu–Cu distances in the cuprates (~3.9 Å), and band-structure calculations show that the important electronic states near E_F are dominated by iron d–d overlap. Experimentally, these materials are metallic conductors at room temperature with resistivities of $\rho \approx 10^{-2}$ Ω cm for LaOFeAs and ~10^{-3} Ω cm for β-FeSe. This is in contrast to the Mott-insulating or bad metallic behavior of the cuprates.

The most straightforward way of generating superconductivity in the 1111 family is by electron doping. In LaO$_{1-x}$F$_x$FeAs, superconductivity is observed for 0.04 < x < 0.14 with a maximum T_c around $x \approx 0.11$. Doping has several effects. The room-temperature resistivity is approximately halved; the antiferromagnetic ordering, which occurs around 150 K in

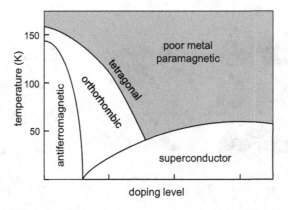

Figure 12.24 Generic phase diagram for LaOFeAs-derived superconductors.

---

[51] Small distortions from tetragonal to orthorhombic symmetry occur on cooling most systems.

LaOFeAs,[52] is shifted to lower temperature and eventually suppressed; and superconductivity is observed at low T. The schematic phase diagram of the pnictide superconductors (Figure 12.24) is therefore strikingly similar to that of the cuprates, with superconductivity emerging as the antiferromagnetic order is suppressed. Superconducting hole-doped compositions such as $La_{0.85}Sr_{0.15}OFeAs$ have also been prepared.

In the 122 family, superconductivity can again be achieved by both hole doping in materials such as $A_{1-x}B_xFe_2As_2$ (A = Ca, Sr, Ba; B = Na, K, Cs)[53] or electron doping in materials such as $AFe_{2-x}Co_xAs_2$. For FeSe, the superconducting phase is believed to be the stoichiometric β-$Fe_{1.01\pm0.02}Se$,[54] suggesting that electron doping is not required in this compound. The ambient-pressure T_c of 8.5 K can be raised to ~15 K by partial substitution of Te for Se, even though FeTe itself is non-superconducting. Under pressure (8.9 GPa), a T_c of 37 K has been reported in FeSe.

At the time of writing, the iron pnictide superconductors are being intensively researched, and the origin of their superconductivity is unknown. However, their superconducting phase diagram suggests that, like the cuprates, antiferromagnetic spin fluctuations may be important in electron pairing. One hope is that understanding their properties will help shed further light on the cuprates. The pnictides' metallic parent state and their relatively high values of critical field and critical current may also make them more suitable for certain applications.

12.9 Problems

12.1 Consider the main families of superconducting materials discussed in this chapter: elemental metals, intermetallic compounds, fullerenes, organic conductors, cuprates, and iron pnictides. (a) State the highest T_c (in K) observed in each of the different families. (b) Order the six families in ascending order of the year in which superconductivity was first discovered. (c) In which families does BCS theory provide a reasonable description for superconductivity?

12.2 State two properties that distinguish a superconductor from a normal metallic conductor.

12.3 Al has T_c = 1.14 K. Assuming it is in the BCS weak-coupling limit, calculate the superconducting gap $2\Delta_0$. Give your answer in J, eV, and cm^{-1}.

12.4 Al has a Debye frequency $\omega_D = 5\times10^{13}$ s^{-1} (~1670 cm^{-1}) and $N(E_F)\cdot V' = 0.168$. Estimate the superconducting gap $2\Delta_0$.

12.5 Sn and Pb have T_c = 3.7 K and 7.2 K respectively. When illuminated with far-IR radiation, they show a marked drop in reflectivity at frequencies of ~10 and ~20 cm^{-1}, respectively. Rationalize these observations.

---

[52] Magnetic ordering is often closely associated with the tetragonal to orthorhombic transition. While doping suppresses this transition, superconductivity is observed for both symmetries.

[53] For example $Ba_{0.6}K_{0.4}Fe_2As_2$ has T_c = 38 K.   [54] Early reports suggested $Fe_{1+x}Se$ was responsible.

12.6 The table below contains T_c values for different isotopically enriched samples of Sn and Hg. Assuming $T_c \propto M^{-\alpha}$, determine α in each case. Give a short explanation of why T_c depends on the isotopic mass. Sn data are from ref. [30]; Hg data are from ref. [6].

Sn samples, mean atomic mass	113.58	116.67	118.05	118.7	119.78	123.01
T_c (K)	3.8082	3.7708	3.7442	3.7419	3.7238	3.6651
Hg samples, mean atomic mass	199.7	200.7	202.0	203.4		
T_c (K)	4.161	4.150	4.143	4.126		

12.7 Draw a sketch of the fcc structure of C_{60}. Show the positions of octahedral and tetrahedral holes and state the number of each. Assuming that C_{60} has a radius of 5 Å, estimate the cubic cell parameter and size of these sites. Compare the diameters of the alkali-metal cations with the size of the octahedral holes. You can find ionic radii at webelements.com.

12.8 The table below contains data for the alkali-metal fullerides A_3C_{60}. (a) Give a brief explanation for the T_c changes observed in the 0 GPa data. (b) Give a brief explanation for the pressure dependence of T_c in Rb_3C_{60}. (c) Comment on differences in T_c between the two series.

Compound	P (GPa)	a (Å)	T_c (K)
K_3C_{60}	0	14.25	18
K_2RbC_{60}	0	14.30	22
$K_{1.5}Rb_{1.5}C_{60}$	0	14.35	24
Rb_2KC_{60}	0	14.36	26
Rb_3C_{60}	0	14.44	29
Rb_3C_{60}	0.113	14.40	28.1
Rb_3C_{60}	0.258	14.37	26.8
Rb_3C_{60}	0.434	14.32	25.4
Rb_3C_{60}	0.661	14.27	23.5

12.9 $Rb_xCs_{3-x}C_{60}$ compounds have T_c = 26.9 K, 32.9 K, and 31.8 K and room-temperature cell parameters a = 14.70 Å, 14.60 Å, and 14.45 Å for x = 0.35, 1.0, and 2.0, respectively. Predict the dependence of T_c on pressure for each of these compositions.

12.10 Perovskite-related $SrBiO_3$ and $Sr_{0.4}K_{0.6}BiO_3$ can both be prepared at high pressure. $SrBiO_3$ has two BiO_6 octahedral sites, one with bond distances of 2 × 2.281, 2 × 2.317, 2 × 2.321 Å and a second with bond distances of 2 × 2.140, 2 × 2.209, and 2 × 2.123 Å. $Sr_{0.4}K_{0.6}BiO_3$ has a single site with Bi–O distances of 2 × 2.110 and 4 × 2.114 Å. Discuss the origins of these structural differences and predict the differences in low-temperature conductivity you might expect for each material. Calculate the Bi bond valence sum for each site assuming $R_{Bi-O}^0 = 2.094$ Å.

12.11 Although the true structure of La_2CuO_4 is orthorhombic, it can be approximated as tetragonal with space group #139 *I4/mmm*, unit-cell parameters of $a = 3.814$ Å and $c = 13.15$ Å, and crystallographically non-equivalent atoms at the following coordinates (and Wyckoff sites): Cu 0 0 0 (2*a*); La 0 0 0.362 (4*e*); O1 0 ½ 0 (4*c*); O2 0 0 0.187 (4*e*). The symmetry generates equivalent positions: Rotational symmetry generates the additional coordinate ½ 0 0 for Wyckoff site 4*c* and 0 0 −*z* for Wyckoff site 4*e*. Translational symmetry generates equivalent positions at (½,½,½)+ for all coordinates of the four Wyckoff sites. (a) Sketch the structure. (b) State the coordination geometry around each metal ion. (c) Calculate the Cu–O bond distances.

12.12 Depending on the level of oxygen vacancies and temperature, $YBa_2Cu_3O_{7-\delta}$ can be either a (poor) metallic conductor, an antiferromagnetically ordered insulator, or a superconductor. A simplified phase diagram separating the different regions is shown below. (a) Give a brief description of the structure of $YBa_2Cu_3O_{7-\delta}$, emphasizing the features relevant for its superconducting properties. (b) Describe key changes in structural or physical properties of $YBa_2Cu_3O_{7-\delta}$ as you move along each of the arrows 1, 2, and 3 in the figure. (c) State which characterization techniques would give information on each of the arrowed transitions. (d) Describe how a material with optimum superconducting properties could be prepared.

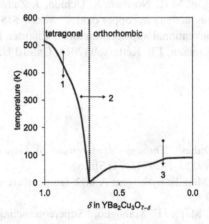

12.13 A black material **A** was prepared by heating 0.3978 g of CuO, 0.1480 g of $BaCO_3$, and 1.5069 g of La_2O_3 to 700 °C under N_2 and rapidly cooling to room temperature. **A** contains 15.79% O by mass and exhibited no weight loss on heating under a flow of He. A second black material **B** was prepared by heating La_2CuO_4 to 1070 °C for 17 hours in air and then slowly cooled under flow of O_2. When 20 mg of **B** was heated to 380 °C under flow of Ar, it lost 0.059 mg. Calculate the chemical formulae of **A** and **B**, and indicate how their electronic properties might be expected to differ from those of pure La_2CuO_4.

12.14 What is unusual about the superconducting phases $Nd_{2-x}Ce_xCuO_4$ and $Sr_{1-x}Nd_xCuO_2$?

12.10 Further Reading

W. Buckel, "*Superconductivity Fundamentals and Applications*" (1991) VCH.

S.J. Blundell, "*Superconductivity: A Very Short Introduction*" (2009) Oxford University Press.

J.E. Hirsch, M.B. Maple, F. Marsiglio (editors), "*Superconducting Materials: Conventional, Unconventional and Undetermined (Physica C, volume 514)*" (2015) Elsevier. This special edition contains reviews of many different superconducting families. The editorial by Hirsch, Maple, and Marsiglio gives an overview of the area

P. Batail, "Molecular conductors" *Chemical Reviews* **11** (2004), 4887–4890 (editorial); 4891–5737 (various authors).

S.E. Brown, "Organic superconductors: The Bechgaard salts and relatives" *Physica C* **514** (2015), 279–289.

A.W. Sleight, "Bismuthates: $BaBiO_3$ and related superconducting phases" *Physica C* **514** (2015), 152–165.

J.P. Attfield, "Chemistry and high temperature superconductivity" *J. Mater. Chem.* **21** (2011), 4756–4764.

C.W. Chu, L.Z. Deng, B. Lv, "Hole-doped cuprate high temperature superconductors", *Physica C* **514** (2015), 290–313.

B. Keimer, S.A. Kivelson, M.R. Norman, S. Uchida, J. Zaanen, "From quantum matter to high-temperature superconductivity in copper oxides" *Nature* **518** (2015), 179–186.

M.R. Norman, "Unconventional superconductors", Chapter 13 in "*Novel Superfluids, Volume 2*", Editors: K.H. Bennermann, J.B. Ketterson (2014) Oxford University Press.

12.11 References

[1] H. Kamerlingh Onnes, "*Through Measurement to Knowledge: The Selected Papers of Heike Kamerlingh Onnes 1853–1926*" (1991) Springer.

[2] N.W. Ashcroft, "Metallic hydrogen: A high-temperature superconductor?" *Phys. Rev. Lett.* **21** (1968), 1748–1749.

[3] J.E. Hirsch, M.B. Maple, F. Marsiglio, "Superconducting materials classes: Introduction and overview" *Physica C* **514** (2015), 1–8.

[4] H.J. Niu, D.P. Hampshire, "Critical parameters of disordered nanocrystalline superconducting Chevrel-phase $PbMo_6S_8$" *Phys. Rev B* **69** (2004), 175403.

[5] C. Kittel, "*Introduction to Solid State Physics*", 8th edition (2005) John Wiley and Sons.

[6] C. Reynolds, B. Serin, W. Wright, L. Nesbitt, "Superconductivity of isotopes of mercury" *Phys. Rev.* **78** (1950), 487.

[7] M.E. Jones, R.E. Marsh, "The preparation and structure of magnesium boride, MgB_2" *J. Am. Chem. Soc.* **76** (1954), 1434–6.

[8] J. Nagamatsu, N. Nakagawa, T. Muranaka, Yuji Zenitani, J. Akimitsu, "Superconductivity at 39 K in magnesium diboride" *Nature* **410** (2001), 63–64.

[9] S.L. Bud'ko, P.C. Canfield, "Superconductivity of magnesium diboride" *Physica C* **514** (2015), 142–151.

[10] O. Gunnarsson, "*Alkali Doped Fullerides*" (2004) World Scientific.

[11] R.D. Shannon, "Revised effective ionic radii and systematic studies of interatomic distances in halides and chalcogenides" *Acta. Crystallogr. Sect. A* **32** (1976), 751–767.

[12] A.F. Hebard, M.J. Rosseinsky, R.C. Haddon, D.W. Murphy, S.H. Glarum, T.T.M. Palstra, A.P. Ramirez, A.R. Kortan, "Superconductivity at 18 K in potassium-doped C_{60}" *Nature* **350** (1991), 600–601.

[13] S. Margadonna, Y. Iwasa, T. Takenobu, K. Prassides in "Fullerene based materials structure and property" *Structure and Bonding* **109** (2004), 127–164.

[14] R.H. Zadik, Y. Takabayashi, G. Klupp, R.H. Colman, A.Y. Ganin, A. Potocnik, P Jeglic, D. Arcon, P. Matus, K. Kamaras, Y. Kasahara, "Optimized unconventional superconductivity in a molecular Jahn–Teller metal" *Sci. Adv.* **1** (2015), e1500059.

[15] C.C. Chen, C.M. Lieber, "Synthesis of pure $^{13}C_{60}$ and determination of the isotope effect for fullerene superconductors" *J. Am. Chem. Soc.* **114** (1992), 3141–3142.

[16] C.C. Chen, C.M. Lieber, "Isotope effect and superconductivity in metal-doped C_{60}" *Science* **259** (1993), 655–658.

[17] M.S. Fuhrer, K. Cherrey, A. Zettl, M.L. Cohen, "Carbon isotope effect in single-crystal Rb_3C_{60}" *Phys. Rev. Lett.* **83** (1999), 404–407.

[18] B. Burk, V.H. Crespi, A. Zettl, M.L. Cohen, "Rubidium isotope effect in superconducting Rb_3C_{60}" *Phys. Rev. Lett.* **72** (1994), 3706–3709.

[19] Y. Takabayashi, A.Y. Ganin, P. Jeglic, D. Arcon, T. Takano, Y. Iwasa, Y. Ohishi, M. Takata, N. Takeshita, K. Prassides, M.J. Rosseinsky, "The disorder-free non-BCS superconductor Cs_3C_{60} emerges from an antiferromagnetic insulator parent state" *Science* **323** (2009), 1585–1590.

[20] P. Batail, K. Boubekeur, M. Fourmigué, J.-C.P. Gabriel, "Electrocrystallization, an invaluable tool for the construction of ordered, electroactive molecular solids" *Chem. Mater.* **10** (1998), 3005–3015.

[21] J.F. Schooley, W.R. Hosler, M. Cohen, "Superconductivity in semiconducting $SrTiO_3$" *Phys. Rev. Lett.* **12** (1964), 474–475.

[22] A.W. Sleight, J.L. Gillson, P.E. Bierstedt, "High-temperature superconductivity in the $BaPb_{1-x}Bi_xO_3$ system" *Solid State Commun.* **17** (1975), 27–28.

[23] J.G. Bednorz, K.A. Müller, "Possible high T_c superconductivity in the Ba–La–Cu–O system" *Z. Phys. B* **64** (1986), 189–193.

[24] J.P. Attfield, A.L. Kharlanov, J.A. McAllister, "Cation effects in doped La_2CuO_4 superconductors", *Nature* **34** (1998), 157–159.

[25] M.K. Wu, J.R. Ashburn, C.J. Torng, P.H. Hor, R.L. Meng, L. Gao, Z.J. Huang, Y.Q. Wang, C. W. Chu, "Superconductivity at 93 K in a new mixed-phase Y–Ba–Cu–O compound system at ambient pressure" *Phys. Rev. Lett.* **58** (1987), 908–910.

[26] M. v. Zimmermann, J.R. Schneider, T. Frello, N.H. Andersen, J. Madsen, M. Käll, H.F. Poulsen, R. Liang, P. Dosanjh, W.N. Hardy, "Oxygen-ordering in underdoped $YBa_2Cu_3O_{6+x}$ studied by hard x-ray diffraction" *Phys. Rev B* **68** (2003), 104515/1–13.

[27] R.J. Cava, A.W. Hewat, E.A. Hewat, B. Batlogg, M. Marezio, K.M. Rabe, J.J. Krajewski, W.F. Peck Jr, L.W. Rupp Jr., "Structural anomalies, oxygen ordering and superconductivity in oxygen deficient $Ba_2YCu_3O_x$" *Physica C* **165** (1990), 419–433.

[28] C.W. Chu, L.Z. Deng, B. Lv, "Hole-doped cuprate high temperature superconductors" *Physica C* **514** (2015), 290–313.

[29] Y. Kamihara, T. Watanabe, M. Hirano, H. Hosono. "Iron-based layered superconductor $LaO_{1-x}F_xFeAs$ ($x = 0.05$–0.12) with $T_c = 26$ K" *J. Am. Chem. Soc.* **130** (2008), 3296–3297.

[30] E. Maxwell, "Superconductivity of the isotopes of tin" *Phys. Rev.* **86** (1952), 235–242.

13 Energy Materials: Ionic Conductors, Mixed Conductors, and Intercalation Chemistry

In previous chapters we have concentrated on properties of materials that are largely derived from the electrons they contain. In this chapter and the next, the focus will change to materials in which the main applications derive from the transport of ions, atoms, or molecules through them. Long-range atomic, ionic, or molecular migration in a solid may initially come as a surprise, especially given the conventional view of solids as strongly bonded arrays, with atoms undergoing only small vibrations around their mean positions. Some solids do, however, allow high ionic migration in the solid state. We will see that they can have important applications as **electrolytes** in batteries, sensors, and fuel cells. We'll look at the general requirements for this functionality, then focus on conductors of metal cations, protons, and oxide anions. We will also consider **mixed ionic electronic conductors**: materials that can conduct both ions and electrons and are used as electrodes in many energy-related devices. Finally, we will discuss **intercalation chemistry** that involves the reversible insertion/removal of chemical species into a host structure. This chemistry is exploited in most lightweight rechargeable batteries.

13.1 Electrochemical Cells and Batteries

Since energy storage and production is the reason why many of the materials in this chapter are of interest, it's worth reminding ourselves of the components of an electrochemical cell, Figure 13.1. Such a cell consists of **electrodes** separated by an **electrolyte**. In some cases (such as the $Zn(s)|ZnSO_4(aq)||CuSO_4(aq)|Cu(s)$[1] cell shown), the electrode compartments might be physically separated by a salt bridge.

A **galvanic** cell is one in which a spontaneous reaction produces electricity, and an **electrolytic** cell is one in which a non-spontaneous reaction is driven by an external current source. The **anode** is always defined as the electrode at which oxidation occurs; the **cathode** as

[1] In this notation, a solid line | represents a phase boundary between cell components and a double line || a phase boundary such as an ideal salt bridge that doesn't contribute to the cell potential.

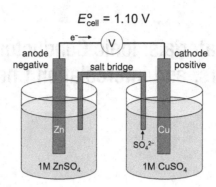

Figure 13.1 Example of a galvanic cell under standard conditions, with standard cell potential E°_{cell}. The salt bridge (typically KCl dissolved in agar) allows ion flow without excessive mixing of electrolytes.

where reduction occurs. In a galvanic cell such as the one shown in Figure 13.1, oxidation of Zn metal occurs at the anode, Zn^{2+} ions pass into solution, and the electrons left behind give the anode a negative charge. These electrons flow through the external circuit to the cathode. Here, Cu^{2+} ions being reduced from solution attract the electrons, giving the cathode a positive charge. The circuit is completed by the flow of ions through the salt bridge and the electrolytes.

In an electrolytic cell, the spontaneous direction of the reaction is reversed using an external current source. Reduction must now occur at the electrode where oxidation occurred originally. This can only be achieved by supplying electrons, i.e. giving that electrode a negative charge. Since the electrode at which reduction occurs is always defined as the cathode, the cathode is negative (and anode positive) in an electrolytic cell.

In any galvanic cell that has not reached its final state of chemical equilibrium, the movement of electrons through a potential difference can be used to do electrical work: the higher the potential difference and larger the number of electrons, the more work. The maximum amount of work occurs when the cell operates reversibly at a fixed chemical composition. This occurs when the electric current approaches zero (the resistance of the voltmeter approaches infinity) and the measured voltage is called the cell potential, E_{cell}.[2] The zero-current cell potential is related to the Gibbs free energy of the cell's chemical reaction by:

$$\Delta G = -nFE_{cell} \tag{13.1}$$

where n is the number of electrons in the cell half equations and F is Faraday's constant (96485 C/mol). A spontaneous reaction (ΔG negative) means a positive cell potential. The cell potential depends on the chemical state of the cell. When it is far from equilibrium, the cell potential will be high; when it is close to equilibrium, the cell potential will be close to zero.

---

[2] Or the zero-current potential or (obsolete) electromotive force (emf) of the cell.

For a cell such as $Zn(s)|ZnSO_4(aq)||CuSO_4(aq)|Cu(s)$ (Figure 13.1), we can use standard (reduction) electrode potentials[3], $E°$, to determine both the direction of spontaneous change and the cell potential under standard conditions. We write down half equations for the two electrodes and realize that the oxidizing agent will be in the half equation with higher standard (reduction) potential. In our cell, the oxidizing agent is Cu^{2+} because $E° = +0.34$ V for the $Cu^{2+}(aq) + 2e^- \rightarrow Cu(s)$ half equation and the reducing agent is Zn because $E° = -0.76$ V for $Zn^{2+}(aq) + 2e^- \rightarrow Zn(s)$. The total reaction is then $Cu^{2+}(aq) + Zn(s) \rightarrow Cu(s) + Zn^{2+}(aq)$, and the standard cell potential is the difference of the two $E°$ values: 1.10 V.

Under non-standard conditions, the electrode potential is given by the Nernst equation:

$$E = E° - \frac{RT}{nF}\ln(Q) \qquad (13.2)$$

where Q is the reaction quotient (a mass-action term) for the half equation. Since the mass-action term depends on activities or concentrations of species present, we can see how electrochemical cells can be used as **chemical sensors**—they will give a changing electrical signal as the chemical concentration of a particular metal or ion changes. Application of the Nernst equation is explored in the end-of-chapter problems.

These basic ideas have led to a number of different battery designs based on different chemical processes. A battery is, strictly speaking, an array of cells connected together, but the term is used loosely. **Primary batteries** are based on reactions that are irreversible under normal conditions and can only be used once. **Secondary batteries** have reversible reactions and can be recharged.

The standard disposable alkaline cell is a primary battery. It relies on the reaction between a Zn anode (connected to the flat base) and an MnO_2 cathode (connected to the raised button) with a KOH electrolyte (hence the alkaline cell name). The cathode is a cylinder of MnO_2 powder mixed with carbon black to improve conductivity, into which a central cylinder of a gel containing zinc anode powder and electrolyte is inserted. The surface shared by the two cylinders is electrically and mechanically isolated by an ion-conducting porous polymeric membrane.

The most familiar secondary battery is the lead–acid rechargeable car battery. It contains Pb and PbO_2 plates in a sulfuric acid electrolyte when charged, which both form $PbSO_4$ on discharge. Rechargeable "nickel–cadmium" batteries were used in many portable devices and rely on the reversible reaction of Ni^{III}-containing $NiO(OH)$ and Cd to produce $Ni(OH)_2$ and $Cd(OH)_2$. In many devices, these were replaced by "nickel–metal hydride" batteries. These again rely on an $NiO(OH)$ cathode that is reduced to $Ni(OH)_2$ as the cell is discharged. The hydride reducing agent is stored in the charged cell in a metal hydride anode based on the $LaNi_5$ alloys discussed in Chapter 2. The chemistry of these traditional batteries is explored further in the end-of-chapter problems.

---

[3] The potential versus the standard hydrogen electrode under unity activities. Standard *reduction* potential because it is a potential for reduction of the oxidized species in the redox couple.

13.2 Fuel Cells

A second energy-producing device that requires high-functionality materials is the **fuel cell**. Fuel cells again convert chemical energy into electrical energy but, in contrast to batteries, the fuel and oxidant are supplied from an external source. Fuel cells have high efficiency (60–70% or up to 90% if waste heat is also used), no moving parts, are quiet to operate, and can have zero or low polluting emissions at the point of use. There are many possible designs, and Figure 13.2 shows a configuration using H_2 as fuel. In this cell, H_2 releases electrons at the anode, which flow through an external circuit. The protons formed migrate through the electrolyte, and, at the cathode, combine with electrons and oxygen to form H_2O. The Gibbs free energy of $H_2(g) + \frac{1}{2}O_2(g) \rightarrow H_2O(l)$ ($\Delta G = -236$ kJ/mol) can, in theory, be used to do electrical work; the work is limited in practice by ohmic-heat losses and residual concentrations of the two reactants. At low temperatures, a Pt catalyst is typically employed at both electrodes to speed up reaction rates.

Fuel cells are used in a variety of large-scale applications, and several companies have demonstrated small devices that could replace rechargeable batteries in some portable appliances [1]. Other applications of similar technology include reversing the direction in which the fuel cell operates to produce an electrochemical H_2 pump capable of removing H_2 from a system. In a NEMCA (nonfaradaic electrochemical modification of catalytic activity) reactor, this has been applied to reactions such as the dehydrogenation of ethane or conversion of methane to other hydrocarbons. We will discuss the materials for different fuel-cell designs later in the chapter.

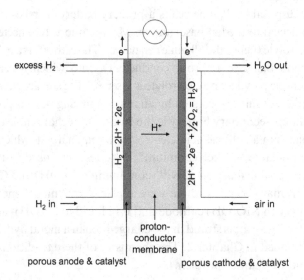

Figure 13.2 A schematic diagram of a hydrogen fuel cell with a proton-conductor electrolyte. 2–5-nm Pt nanoparticles are typically used as the catalyst.

13.3 Conductivity in Ionic Compounds

In Chapter 10, we defined conductivity as $\sigma = L/RA$ where L, R, and A are the sample length, resistance, and area. Values of conductivity vary over a huge range from $<10^{-20}$ S/cm for insulators to $>10^{25}$ S/cm for the superconductors discussed in Chapter 12.[4] Figure 13.3 reviews the typical ranges for different categories of conductors. Although ionic compounds such as NaCl are considered insulators at low temperatures, their conductivity rises dramatically on heating and can approach the values of semiconductors. Certain ionic materials such as AgI and PbF_2 have conductivities approaching those of metals even at relatively modest temperatures. How does this occur?

In Chapter 2, we saw that the structures of ionic materials always contain defects. NaCl, for example, will always have a certain level of cation vacancies due to minor aliovalent impurities of the type $Na_{1-2x}Ca_x \square_x Cl$ (extrinsic defects). In addition, at temperatures above 0 K, it is thermodynamically favorable to form intrinsic Schottky defects (which predominate in NaCl) or Frenkel defects (e.g. in AgCl). When NaCl is heated, Na^+ can hop to an adjacent vacant defect site as described in Section 3.3.1 and shown schematically in Figure 13.4. The position of the vacancy has then moved, and a different Na^+ can hop into the newly vacant site. Under an applied field, the cation vacancy (which has a formal negative charge; see Section 2.2) can migrate giving rise to a net current. The overall conductivity will depend on the number of vacant sites available, as we know from the relationship between conductivity (σ), the number of charge carriers (n), their charge (z), and their mobility (μ), developed in Chapter 10:

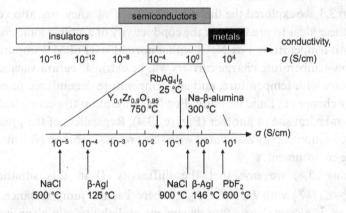

Figure 13.3 The range of conductivities σ, in S/cm, for different conductors.

---

[4] We use siemens per centimeter (S/cm) for conductivity in this chapter rather than siemens per meter (S/m) as these are the units most commonly used in the literature; 1 S/cm is 100 S/m.

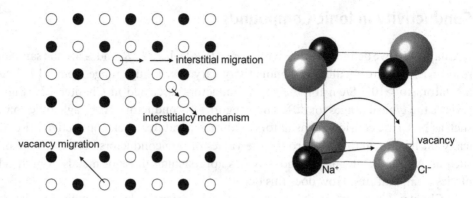

Figure 13.4 Left: Schematic mechanisms of vacancy and interstitial migration. Right: Na^+ migration from an occupied to a vacant site.

$$\sigma = nz\mu \tag{13.3}$$

For a compound where interstitial defects dominate, such as AgCl, one can imagine two mechanisms that allow charge movement. Given the large number of interstitial sites available, one possibility is that direct hopping occurs from an occupied interstitial to a vacant interstitial. A second possibility is that an Ag^+ cation hops from an interstitial site to a normal lattice site, provided that in so doing it drives the Ag^+ off this site and pushes it onto a new interstitial position: a "billiard ball" or interstitialcy-type mechanism. In AgCl, this latter mechanism is believed to predominate.

In Section 3.3, we explored the thermodynamics of vacancy migration quantitatively, and we can use these ideas to predict how the conductivity of a simple ionic compound like NaCl will vary with temperature. For NaCl with a normal level of M^{2+} impurities, the dominant source of low-temperature charge carriers will be extrinsic cation vacancies. Their number will not change with temperature, and any temperature dependence of conductivity will be governed by changes in ionic mobility. For Na^+ migration to occur, the Na^+ ion must pass from one octahedral site to another (Figure 13.4). Regardless of the precise pathway, there will be an activation energy associated with this process since the Na^+ must pass through an unfavorable environment.

In Section 3.3.8, we saw that the diffusivity D in this situation has the form $D = D_0 \exp(-E_A/kT)$ with $D_0 = \lambda^2 f_c\, p_{dir} v$, where λ is the jump distance, v is a vibrational frequency, p_{dir} is a crystal-structure-dependent probability of a given jump moving an ion forwards, and f_c is a correlation factor related to the probability of an ion returning to its original site. For a simple 1D case with each jump allowed, this simplifies to $D_0 = \lambda^2 v$. In Section 3.3.6, we derived the Nernst–Einstein equation that relates the conductivity of a species to its diffusivity via $\sigma/D = q^2 n/kT$, where we use n as the number of mobile ions per unit volume in place of c in Equation (3.23) and q is the charge. This relationship applies when the same random-walk mechanism applies for both conduction and diffusion. We can then express the conductivity as:

$$\sigma = \frac{\lambda^2 v q^2 n}{kT} \exp\left(\frac{-E_m}{kT}\right) \tag{13.4}$$

where E_m is the energy barrier height for vacancy (or interstitial) diffusion. A plot of $\ln(\sigma T)$ versus $1/T$ will therefore give the activation energy for vacancy migration from its slope of $-E_m/k$. For samples of NaCl with different dopant levels, provided the activation energy doesn't change significantly on doping, the conductivity will be determined largely by the number of dopants in the pre-exponential factor of Equation (13.4).[5] For different doping levels, plots of $\ln(\sigma T)$ against $1/T$ would therefore give a series of parallel lines (see Figure 13.5).

We saw in Chapter 2 that the number of intrinsic defects increases rapidly as temperature increases, such that we may have to consider the contribution of both extrinsic and intrinsic defects to conductivity. The intrinsic contribution will be particularly important for high-purity samples at high temperature. In Section 2.3, we derived the number of Schottky defects (ignoring vibrational contributions to entropy) as:

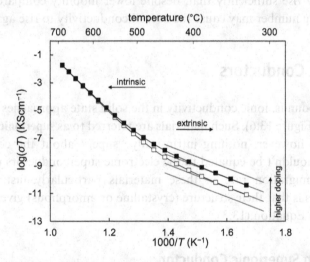

Figure 13.5 Plots of $\ln(\sigma T)$ against $1000/T$ for an ionic compound at two levels of extrinsic defects: high (■) and low (□) doping. Data are calculated with Equation (13.6) for an NaCl crystal with activation energies of $E_m = 0.70$ eV and $E_S = 2.37$ eV.

[5] Note the $1/T$ term in the pre-exponential factor arises in the presence of an applied field but isn't present in the equivalent expressions using diffusivity. If a plot of $\ln(\sigma)$ versus $1/T$ [as opposed to $\ln(\sigma T)$ versus $1/T$] is used to determine activation energies from conductivity measurements, values may be in error by ~10 kJ mol^{-1}.

$$n = N_0 \exp(-\Delta H_S / 2kT) \tag{13.5}$$

where N_0 is the number of sites per cm^3 of sample and ΔH_S is the formation enthalpy of a Schottky defect pair. This can be introduced into Equation (13.4) giving:

$$\sigma = \frac{\lambda^2 v q^2}{kT} N_0 \exp(-\Delta H_S / 2kT) \exp(-E_m / kT) \tag{13.6}$$

A plot of $\ln(\sigma T)$ against $1/T$ (Figure 13.5) will then give two broad regions: one where vacancies from extrinsic doping dominate conductivity with a slope of $-E_m/k$ (low T), and one where intrinsic vacancies dominate with slope $-(E_m + \Delta H_S/2)/k$ (high T). Between these two approximately linear regions, a curved region will occur where extrinsic and intrinsic contributions are comparable.

It's worth noting that while doping a compound like NaCl, where vacancy migration dominates, with a 2+ ion will increase conductivity, in AgCl, where interstitial migration dominates, the conductivity may decrease. This happens because doping will create additional vacancies on normal Ag^+ sites (v_{Ag}'), which, by simple equilibrium considerations, will decrease the number of Ag^+ interstitials (the product of the number of interstitials and the number of vacancies remains constant). At higher levels of doping, the number of vacancies may rise sufficiently that, despite lower mobility compared to interstitial sites, their increasing number may cause the overall conductivity to rise again.

13.4 Superionic Conductors

In some compounds, ionic conductivity in the solid state approaches or exceeds that of the molten state (Figure 13.6). Such materials are referred to as **superionic** or **fast ionic** conductors. There is, however, nothing intrinsically "super" about the conduction mechanism (certainly it shouldn't be equated to the electronic superconductors of Chapter 12), nor is the intrinsic migration rate in these materials particularly fast. The key feature of these materials is that their structure (crystalline or amorphous) gives a large overall value of $n \times z \times \mu$ in Equation (13.3).

13.4.1 AgI: A Cation Superionic Conductor

One material that illustrates many of the features required for high cationic conductivity is AgI. At room temperature, it adopts the wurtzite (β-AgI) or sphalerite (γ-AgI) structures (Figure 1.32) with Ag^+ occupying half the tetrahedral holes of a hexagonally or cubic closest packed (hcp or ccp) array of I^-, respectively. Upon warming to 420 K, there is a transition to α-AgI, in which iodide anions adopt a body-centered cubic (bcc) packing. This is accompanied by a ~10^4-fold increase in conductivity to values near 2 S/cm. The conductivity actually

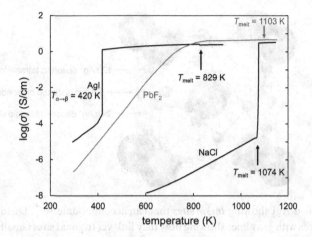

Figure 13.6 Conductivities of the archetypal fast cation conductor AgI and fast anion conductor PbF$_2$ compared to NaCl. After ref. [2].

drops by 10% when the material melts at 829 K. The electronic contribution to the overall conductivity is around 10^7 times lower than the ionic.

The bcc arrangement of two I$^-$ ions per α-AgI cell creates a large number of different sites that could be occupied by two Ag$^+$ ions: Wyckoff site 12*d* on the cell face, which has a distorted tetrahedral coordination environment of I$^-$; site 6*b* on the cell edge (a distorted octahedron); and 24*h* (distorted triangle); see Figure 13.7.[6] Neutron-scattering experiments and simulations suggest that Ag$^+$ cations spend around 75% of their time on tetrahedral sites, with hops between positions via trigonal sites being around six times as likely as those via octahedral sites. Calculations also suggest that after a given hop, there is a slight preference for the ion to migrate back to its starting position.[7] The diffuse orbitals of the large polarizable I$^-$ anion facilitate low-energy pathways for Ag$^+$ migration, and α-AgI has an extremely low activation energy for ionic migration of 0.03 eV (~3 kJ/mol). The Ag$^+$ ions in α-AgI are sufficiently disordered that they have been described as "liquid like" above 420 K, and the β/γ → α transition is frequently described as a "melting" of the cation sublattice. This view is supported by thermodynamic measurements, which show an entropy gain of ΔS = +14.5 J/(mol K) for the β → α transition and +11.3 J/(mol K) for melting of AgI at 829 K [3]. The sum of these two values is similar to the entropy of melting of a "normal" ionic solid such as NaCl of ~24 J/(mol K). The structural evidence, however, suggests that a picture of

---

[6] Bond-valence sums for Ag$^+$ at the three sites are 1.1, 1.2, and 1.4, respectively. The sites are highly distorted from ideal geometries: the "tetrahedral" site has bond angles of 101.5° and 126.9°; the triangular site has bond angles of 109.5° and 141.1°; and the "octahedral" site has bond distances of 2 × 2.5 Å and 4 × 3.6 Å.

[7] After an initial hop, we would statistically expect a 25% probability of Ag$^+$ hopping to each neighboring tetrahedral site. There is actually a ~40% chance of a hop back to the original position.

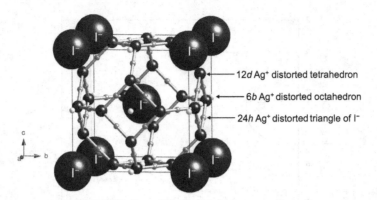

Figure 13.7 Cubic α-AgI showing *Im3̄m* sites that can accommodate Ag^+. Distorted tetrahedral sites are small black spheres with gray lines showing how they link via trigonal sites (small gray spheres). Distorted octahedral sites are the small white spheres at the cell edges.

frequent discrete hops between specific cation sites is a more appropriate description than a molten array.

α-AgI exemplifies many of the key requirements for a material to show high conductivity:

1. A large number of mobile ions to maximize n in $\sigma = nz\mu$.
2. An arrangement of oppositely charged ions that creates energetically favorable sites into which the mobile ion can hop, with low-energy migration routes between them.[8]
3. Polarizable immobile ions that can deform their electronic clouds to allow migration of mobile ions.
4. The absence of conditions that lead to trapping of mobile ions, such as attractions to the local charges of aliovalent dopants or vacancies; a process that leads to significant aging (deterioration of conduction over time) in some ionic conductors.
5. A low charge on mobile ions to minimize the activation energy for migration.
6. Other things being equal, small ions will have a lower activation energy for migration than large ones.

Many attempts have been made to further increase the ionic conductivity of AgI. Substitution on the cation site has led to materials such as $RbAg_4I_5$ (which has one of the highest room-temperature ionic conductivities known, $\sigma = 0.21$ S/cm), $Ag_{1-x}Cu_xI$, and many

---

[8] It is important to realize that conduction requires both mobile cations and vacant sites for them to hop to. In several materials, vacancies are introduced by doping, and one finds a maximum in conductivity when one has 50% vacancies and 50% ions on the site of interest. A good example occurs in $Li_{4-3x}Al_xSiO_4$. As Al is introduced, vacancies start to appear on one Li site that is completely empty by $x = 0.5$. When $x = 0.25$, the mobile Li site is 50% occupied and 50% vacant and maximum conductivity is observed [A. García, G. Torres-Trevino, A.R. West, *Solid State Ionics* **13** (1990), 40−41].

other materials. Substitutions on the anion sublattice leads to materials such as Ag_3SI and Ag_2S. At high temperatures, the structure of β-Ag_2S is similar to α-AgI but with twice as many of the tetrahedral sites occupied. Derivatives with oxoanions, such as $Ag_6I_4WO_4$, have also been investigated. Some have been successfully used as electrolytes in batteries.

For those fascinated by the ubiquity of the perovskite structure throughout materials chemistry, it's worth noting Ag_3SI. The α^*-Ag_3SI form has a high conductivity of ~0.3 S/cm at room temperature and a structure like that of α-AgI but with anions disordered over the bcc sites. The thermodynamically more stable β form is ordered and contains I^- at the center of the unit cell, S^{2-} at the corners, and has Ag^+ about halfway along each cell edge. As such, the structure is an anti-perovskite $ISAg_3$.

13.4.2 PbF₂: An Anionic Superionic Conductor

In the same way that AgI exemplifies many of the key features of cationic conductors, PbF_2 can be considered the archetypal anion conductor. Indeed, its unusual properties were reported as long ago as 1838 by Faraday. He noted that solid PbF_2 acted as an insulator but that on heating it became conducting long before it melted. The conductivity of PbF_2 as a function of temperature is shown in Figure 13.6. In contrast to AgI, it rises continuously from ~10^{-7} S/cm at ambient temperature to ~4 S/cm at 711 K before leveling off. The conductivity shows no significant change on melting at 1103 K. Despite the fact that there is no abrupt change in conductivity with temperature, evidence from heat-capacity and thermal-expansion data suggest that a transition to the superionic state can be defined as occurring at $T_c = 711$ K.

At low temperature, anion-Frenkel defects prevail in PbF_2. The interstitial F^- is thought to adopt a position at the center of the ccp cell of Figure 13.8 (marked as F4). This site is equivalent to the anion site in the rock-salt structure. A large number of these sites is available (one per formula unit), which favors high conductivity. As the material is heated and the number of Frenkel defects increases above the ~1% level, the proximity of neighboring defects is thought to destabilize this site such that it is not significantly occupied in the superionic state. Instead, the superionic state is thought to contain the interstitial ions predominantly at site F1 in Figure 13.8; midway between two anions but displaced towards the center of an empty anion site. A second anion site, F2, located approximately 1.5 Å from a cube corner along the body diagonal towards its center, is also observed. This location is thought to be due to anions close to F1 interstitials, relaxing away from their ideal positions. Sites F1 and F2 aren't occupied simultaneously in any local region, but appear in the average structure due to local structural clusters distributed throughout the material. One model that fits neutron-scattering data well has these local clusters containing vacancies, interstitial anions, and anions relaxed off their ideal position in a 3:1:2 ratio. The Frenkel defects in these clusters move dynamically through the crystal in the superionic state and have

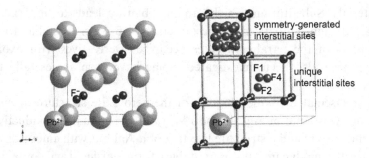

Figure 13.8 The fluorite structure of PbF_2 is a ccp array of Pb^{2+} with F^- in tetrahedral holes (left). It can also be described as an array of F^- cubes with centers alternately empty or filled by Pb^{2+}. Possible F^- interstitial sites in empty cubes shown on right as small dark gray spheres; labels are discussed in the text.

a lifetime of the order of 10^{-12} s. Similar conduction mechanisms are believed to occur in the fluorite-related oxygen-ion conductors of Section 13.7.1.

13.5 Cation Conductors

In the following sections we will discuss some of the important categories of cationic conductors, with a focus on materials that have found technological applications.

13.5.1 Sodium β-alumina

Sodium β-alumina is one member of a family of materials discovered in the 1960s, which have sufficiently high Na^+ conductivity to be used as electrolytes in high energy-density rechargeable sodium–sulfur batteries. Traditionally, the formula of these materials has been expressed as $Na_2O \cdot nAl_2O_3$, with n ranging from ~8 to ~11. To understand the structure, it's probably easier to express the formula as $NaAl_{11}O_{17}$ (the $n = 11$ member), while keeping in mind that the material can contain extra Na^+, charge-balanced by O^{2-}, up to a composition of $Na_{1.375}Al_{11}O_{17.185}$ ($n = 8$). These limiting compositions correspond to $Na_{0.18}Al_2O_{3.09}$ and $Na_{0.25}Al_2O_{3.125}$, respectively.

The ideal structure of $NaAl_{11}O_{17}$ is shown in Figure 13.9 and is related to that of spinel, $MgAl_2O_4$, which has a ccp arrangement of O^{2-} ions with Mg^{2+} filling one-eighth of the tetrahedral holes and Al^{3+} half of the octahedral holes. Blocks with cation site occupancies similar to those in spinel are present in sodium β-alumina, but in every fifth layer, three-quarters of the ccp O^{2-} ions are missing. The Na^+ ions are located in these anion-deficient layers, which contain a mirror plane such that the layer of oxygen atoms immediately above the plane is the same as that below. The occupied Na^+ sites (often referred to as Beevers–Ross,

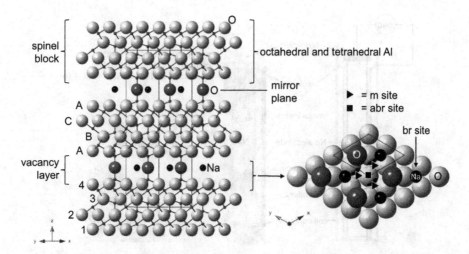

Figure 13.9 Sodium β-alumina. The side-on view (left) emphasizes the four spinel-like ccp layers of O^{2-} (large spheres), with Al^{3+} in tetrahedral and octahedral holes, separated by a single sodium/oxygen vacancy layer. A view perpendicular to the layers is shown on the right.

or br, sites) have coordination number nine (three O atoms below/above at 2.83 Å and three in plane at ~3.23 Å). The bond-valence sum for a Na^+ cation at the br site is 0.43, significantly lower than would normally be expected.

There are, in addition, other potential Na^+ sites possible. These are referred to as m (for midway, marked with a triangle in Figure 13.9) and abr (for anti-Beevers–Ross, marked with a square). These sites would be eight- and five-coordinate and have Na^+ bond-valence sums of 0.43 and 0.48, respectively. We therefore have a 2D network of possible Na^+ sites with similar bond-valence sums. As a result, the material has both a high conductivity ($\sigma = 0.1$ S/cm at 573 K) and a low activation energy (0.16 eV) for migration (see Figure 13.11). The mechanism for conductivity is thought to be of the billiard-ball type with Na^+ migrating from its original site via a pathway br → m → abr → m → br. When it reaches an m site adjacent to an occupied br site, that cation is expelled from its position.

The exceptional properties of sodium β-alumina led to the development of the sodium–sulfur rechargeable cell (Figure 13.10) that operates at around 300 °C. It comprises a liquid-sodium anode separated from a liquid-sulfur/carbon cathode[9] by a thin tube of sodium β-alumina. The use of liquid electrodes ensures good contact with the solid electrolyte. Upon discharge, Na from the anode releases electrons that flow through an external circuit, and the Na^+ ions formed migrate through the β-alumina electrolyte to the cathode where they combine with liquid sulfur and electrons from the external circuit to form Na_2S_x. The cell can be operated over a composition range of $Na_2S_{5.2}$ to $Na_2S_{2.7}$ and gives a voltage that varies from 2.07 V to 1.78 V.[10] The reaction $2Na + 3S \rightarrow Na_2S_3$ could theoretically yield 760 W h/kg, and batteries

[9] Carbon is added to increase conductivity.
[10] Note that Na_2S_5 contains two Na^+ cations and one polysulfide S_5^{2-} anion.

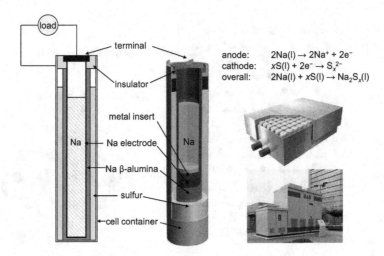

$$
\begin{aligned}
\text{anode:} &\quad 2\text{Na(l)} \rightarrow 2\text{Na}^+ + 2e^- \\
\text{cathode:} &\quad x\text{S(l)} + 2e^- \rightarrow \text{S}_x^{2-} \\
\text{overall:} &\quad 2\text{Na(l)} + x\text{S(l)} \rightarrow \text{Na}_2\text{S}_x\text{(l)}
\end{aligned}
$$

Figure 13.10 A cut-away diagram for one design of a sodium–sulfur rechargeable cell, of about 9×50 cm. Right: A battery of many cells within an electrically heated thermal enclosure with an output power of 50 kW and a storage capacity of 360 kW h. 3D images and photograph provided by NGK Insulators Ltd (www.ngk.co.jp).

with an energy density around 150 W h/kg have been built.[11] They have been shown to have around 86% efficiency on charge/discharge and can operate for thousands of cycles. Their main practical applications are as backup power supplies and load levelers. In the latter application, they help meet energy demands from power stations at peak times or smooth the output of wind- or solar-power facilities. For example, a 34 MW system containing around 250000 cells has been installed by NGK Insulators at a Japanese wind farm [4].

There are many materials closely related to sodium β-alumina with similar properties. The so-called β'' aluminas have a related but subtly different structure and occur at lower Na content. The β'' structure can also be stabilized with dopants such as Li^+ and Mg^{2+} and can incorporate a range of other cations such as K^+ and Ag^+.

13.5.2 Other Ceramic Cation Conductors

Many other materials have been examined for high cationic conductivity; the conductivities of some are compared with materials already discussed in Figure 13.11. One family that has undergone extensive investigation are the so-called NASICON phases (for Na Super Ionic Conductors). These are solid solutions of the type $\text{Na}_{1+x}\text{Zr}_2(\text{PO}_4)_{3-x}(\text{SiO}_4)_x$, in which a variety of Na^+ sites are occupied within a framework of corner-linked $\text{ZrO}_{6/2}$ octahedra and $(\text{P,Si})\text{O}_{4/2}$ tetrahedra. The two end members of the series are relatively poor ionic conductors, but the $x = 2$

---

[11] Compared to 20–40 W h kg^{-1} for lead–acid cells.

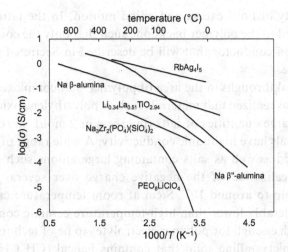

temperature (°C)

Figure 13.11 Electrical conductivities for selected cationic conductors. PEO = polyethylene oxide.

member $Na_3Zr_2(PO_4)(SiO_4)_2$, which has a mixture of full and empty Na^+ sites, has a high conductivity of ~0.2 S/cm at 573 K and activation energy of ~0.3 eV.

A number of ceramic phases showing high Li^+ conductivity have been investigated, but all have much lower conductivity than liquid electrolytes formed by dissolving salts such as $LiPF_6$ in non-aqueous solvents at low temperature. Examples include Li_2SO_4, Li_4SiO_4 and derivatives, and various sulfides. The name LISICON has been given to conductors based on solid solutions of the type $Li_{2+2x}Zn_{1-x}GeO_4$ ($-0.36 < x < 0.87$) with $Li_3Zn_{0.5}GeO_4$ having $\sigma = 0.125$ S/cm at 573 K. Perovskites such as $Li_{0.5-3x}La_{0.5+x}TiO_3$, also show high conductivity; these contain A-site layers that alternate between La^{3+} and a mixture of La^{3+}, Li^+, and vacancies, with Li^+ displaced from the ideal A-site position.

13.5.3 Polymeric Cation Conductors

In addition to ceramic ionic conductors, there is considerable interest in polymer electrolytes for many applications. These are potentially easier to process and more readily integrated into devices than ceramic materials. They also offer the benefit of intrinsic flexibility: as other components of devices change size due to temperature and/ or chemical changes, the polymer can adjust without cracking. Compared to liquid electrolytes, polymers eliminate problems related to leakage of flammable/toxic solvents from devices and have mechanical stabilities that remove the need for spacers to separate electrodes.

Polymer electrolytes can be divided into two main categories: **polymer–salt complexes** and **polyelectrolytes**. In the former, an ionic salt is dissolved in a polymer matrix leading to the possibility of both cation and anion motion. In the latter, charged groups are covalently bound to the polymer backbone such that only the counter ion is mobile. The NAFION proton conductor that will be described in Section 13.6.1 is one example of a polyelectrolyte.

The early breakthroughs in the area of polymer–salt complexes were made in the mid 1970s when it was realized that materials such as polyethylene oxide [PEO, $(CH_2CH_2O)_n$] could dissolve large quantities of salts (in excess of 2 mol/L in some cases) and that the resulting materials have high ionic conductivity. A wide range of simple ionic salts have been investigated as well as salts containing large anions such as $[(CF_3SO_2)_3C]^-$,[12] in which the delocalization of the negative charge over several atoms aids solubility. Conductivities up to around 10^{-4} S/cm at room temperature can be achieved, which, although considerably lower than high-temperature ceramic conductors or liquid electrolytes, are high enough for polymer electrolytes to be of technological importance.

PEO is a semicrystalline solid that contains helical $(CH_2CH_2O)_n$ chains, with two turns of the helix occurring every 19.3 Å. While the majority of the best PEO-derived conductors are amorphous, it has been possible to gain insight into the structures of their crystalline counterparts using powder-diffraction methods (single crystals are extremely difficult to prepare). For example, in $(PEO)_3 \cdot NaClO_4$, Na^+ ions are accommodated within helical polymer chains and coordinated by four ether oxygens and two oxygens from the ClO_4^- anions. In other compounds, such as $(PEO)_6 \cdot LiSbF_6$ (Figure 13.12), cations are fully coordinated by the polymer chains and completely isolated from the charge-balancing anions. The polymer chains provide a large number of donor O atoms, giving rise to an array of available cation sites along the chains. These pre-formed cation sites may be one reason why the crystalline form of this material has a higher conductivity than its amorphous counterpart, even above the glass-transition temperature of the latter [5].

For the majority of polymer–salt complexes, the conductivity of amorphous materials is several orders of magnitude higher than crystalline analogues. This is related to **segmental motion** contributing to ionic migration. In addition to local bond vibrations, so-called crankshaft torsional motions around C–C or C–O bonds occur in an amorphous polymer above its glass-transition temperature T_g. These motions are believed to promote the migration of ions by breaking bonds in one local coordination environment of Li^+ and simultaneously creating a more favorable nearby location for the ion to migrate to. This mechanism leads to an enhanced dependence of conductivity on temperature over the other materials we discuss, and polymer–salt electrolyte conductivity is often described using the Vogel–Tammann–Fulcher expression $\sigma = \sigma_0 exp\left(-B/(T-T_0)\right)$, which is discussed further in Section 15.8.

---

[12] Tris[(trifluoromethyl)sulfonyl]methanide(1−).

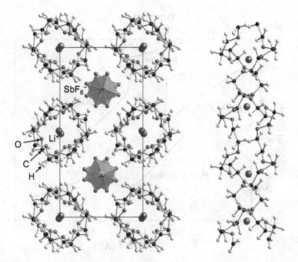

Figure 13.12 The structure of the crystalline polymer–salt complex $(PEO)_6 \cdot LiSbF_6$. Left: Axial view of helical polymer chains that surround Li^+ to create tubes interspersed with SbF_6^- octahedra. Right: Side-on view of a single helical tube.

13.6 Proton Conductors

Materials that display high proton conductivity are required for a number of applications. One is in the low-temperature fuel cell we discussed in Section 13.2, which relies on proton migration through a membrane separating fuel and oxidant to turn chemical into electrical energy. Unfortunately, the highest-conductivity materials only work well at low temperatures, where the only effective electrocatalyst is Pt that is both expensive and prone to deactivation by CO poisoning.[13] The challenge in this area is to identify a material that remains conducting at a temperature high enough to use cheap electrode materials and to either tolerate low-purity H_2 or to run directly on fuel sources such as methanol. A number of materials exhibiting high proton mobility, both organic and inorganic, have been discovered, and conductivity data of some are compared in Figure 13.13.

The mechanisms of proton conductivity are somewhat different to those encountered in other cationic conductors, due to the unique bonding requirements of the proton. In the electronically insulating materials required for pure proton conduction, hydrogen typically forms a short bond to oxygen (~0.95 Å) as well as an additional weaker interaction to a more distant oxygen; a **hydrogen bond**. The proton often lies in a double potential well, with a local minimum close to each oxygen. As O–O separations in the material decrease, the energy barrier for H migration from one oxygen to the other decreases until, for short O–O

---

[13] Consequently, the technological drive is to increase the operating temperatures of these polymer fuel cells, whereas it is to decrease those of the high-temperature solid-oxide fuel cells (based on either proton or oxide conducting solid electrolytes) from ~800–1000 °C to ~500–700 °C.

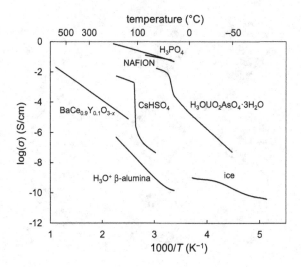

Figure 13.13 Conductivities of selected solid proton conductors. Data for liquid H_3PO_4 and water ice are included for comparison. Data are extracted from refs. [6, 7].

separations of ~2.4 Å, the potential has a single minimum with hydrogen equally bonded to both oxygens. Thus, although the strong proton–oxygen interaction means that protons don't diffuse freely, there is a low-energy mechanism for migration provided the proton motion is coupled to the motion (either local or long range) of the oxygen to which it is bound.

13.6.1 Water-Containing Proton Conductors

Many of the earliest proton conductors studied were materials containing structural water. The most widely used are the protonated forms of NAFION (for Na Fast ION) polymers (Figure 13.14) that were developed as sodium ion-exchange resins, but also show high H^+ conductivity when protonated and hydrated. They are often referred to as proton-exchange membranes or polymer electrolyte membranes (both **PEMs**).

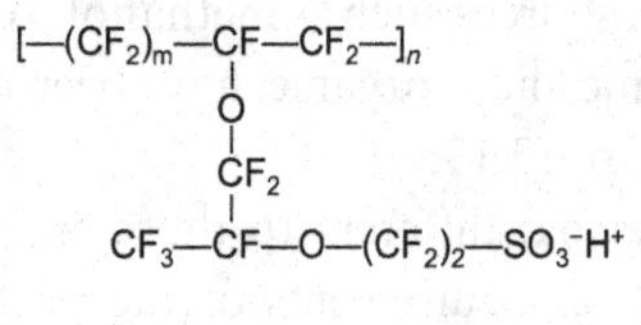

PEMs contain a stable, hydrophobic, fluorinated polymer backbone with pendant acidic $-SO_3H$ groups separated by around 6–9 Å. The high proton migration is believed to occur along low-dimensional water channels between polymer chains and to have a similar mechanism to that in water itself. The picture in pure water is complex, but excess protons are thought to exist predominantly in $H_5O_2^+$ or $H_9O_4^+$ clusters that occur with approximately equal probabilities, have lifetimes of the order of 10^{-13} s, and are hydrogen bonded to surrounding H_2O molecules. Proton migration

$$[-(CF_2)_m-CF-CF_2-]_n$$
$$|$$
$$O$$
$$|$$
$$CF_2$$
$$|$$
$$CF_3-CF-O-(CF_2)_2-SO_3^-H^+$$

Figure 13.14 The chemical structure of a typical NAFION-related polymeric proton conductor.

occurs via the largely uncorrelated shifting of protons within the hydrogen-bonded network and the dynamic destruction and creation of these clusters. As one H bond weakens prior to migration, others strengthen, meaning that the overall activation energy for proton mobility in pure water (~0.1 eV) is significantly less than the energy required to break an individual H bond (~1 eV). This mechanism is supported by the ratio of the proton-conduction diffusivity to the diffusivity of water, D_{proton}/D_{H_2O}, of around 4.5, which shows that proton movement greatly exceeds molecular migration.

In highly hydrated NAFIONs, proton mobility approaches that in water. Unfortunately, if the material is heated, it dehydrates and conductivity falls due to individual H bonds in the water-containing region becoming stronger, as well as due to trapping of protons by the anionic charge on $-SO_3^-$ groups. Hydrated NAFIONs can therefore only be used up to ~90 °C, which is their main technological limitation. In addition, membranes can be relatively permeable to fuel molecules, leading to the equivalent of a chemical short circuit in the fuel cell. Despite these shortcomings, fuel cells using hydrated NAFIONs and related polymers are widely used.

13.6.2 Acid Salts

A second category of proton conductors are acid salts such as $MHXO_4$ (e.g. M = NH_4, Cs, Rb, K; X = S, Se) and CsH_2PO_4. As they don't rely on water molecules for the conduction pathway, they can operate at higher temperatures. Perhaps the best understood of these materials is $CsHSO_4$. Its conductivity jumps from around 10^{-6} S/cm to 10^{-3} S/cm when it undergoes a first-order phase transition at ~413 K (Figure 13.13). This behavior is reminiscent of that of AgI, but, as with all proton conductors, the conductivity remains several orders of magnitude lower than the best cationic conductors.

The jump in conductivity at the phase transition is caused by the onset of disorder in the tetrahedral HSO_4^- groups, which tumble around their average site in the high-temperature phase much as they would in solution; such phases are frequently described as **plastic crystals**. As neighboring HSO_4^- groups tumble, adjacent oxygens will come into closer contact than usual (Figure 13.15). If a proton-carrying oxygen comes within ~2.4 Å of an oxygen on an adjacent tetrahedron, there will be little or no energy cost for a proton transfer between groups, leading to a low activation-energy pathway for migration. From an order–disorder point of view, the proton changes from being located on one specific S–O bond to being dynamically disordered over all four. Thermodynamic measurements support the high disorder of the high-temperature phase with ΔS values for the 413 K phase transition and eventual melting at 484 K of 13.2 J/(mol K) and 27.2 J/(mol K), respectively; i.e. a significant fraction of the entropy gain expected for melting occurs at the solid-solid phase transition. Phase-transition temperatures and resistivities in $MHXO_4$ generally follow the order $Cs^+ < Rb^+ < K^+$, such that the Cs-containing materials have the most favorable properties. This is

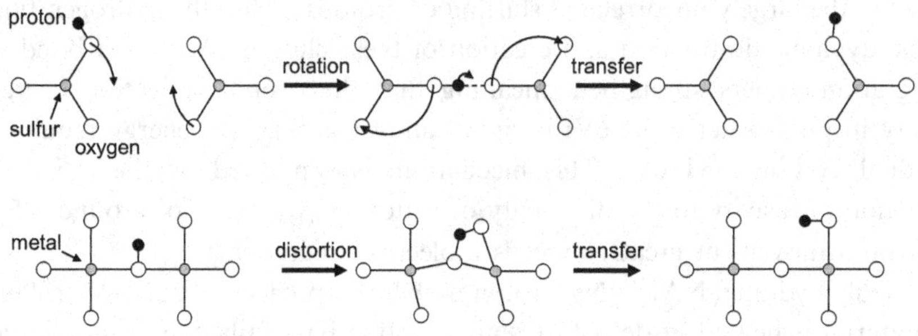

Figure 13.15 Proton migration from one sulfate tetrahedron to another in $CsHSO_4$ (top) and from one octahedron to another in the octahedral network of $BaH_xCe_{1-x}Y_xO_3$ (bottom).

consistent with weaker cation–anion bonding and lower activation energies for HXO_4^- reorientation with a larger, more polarizable counter ion. Working fuel cells based on $CsHSO_4$ have been demonstrated [8].

13.6.3 Perovskite Proton Conductors

High-temperature proton conductivity in perovskites is associated with OH^- groups, which can be stable up to very high temperature in some compounds. Hydroxide anions are formed when water molecules react with oxygen vacancies in a process represented in Kröger–Vink notation as:

$$H_2O(g) \; + v_O^{\bullet\bullet} + \; O_O^{\times} = 2OH_O^{\bullet\bullet} \tag{13.7}$$

(see Section 2.6 for a definition of terms). Essentially, the oxygen of a water molecule enters a vacant site in the structure and its protons form two OH groups. As the reaction involves deprotonation of H_2O, it is most favored for basic oxides. Perovskites that undergo this process include acceptor-doped phases compensated by oxygen vacancies, such as $BaCe_{1-x}Y_xO_{3-x/2}$ (with $BaCe_{0.8}Y_{0.2}O_{2.9}$ forming the best proton conductor) and $BaZr_{1-x}Y_xO_{3-x/2}$. Equation (13.7) is exothermic for both systems.

The mechanism for conductivity in these phases is related to that in the acid salts and involves proton jumps from OH^- to adjacent O^{2-} (see Figure 13.15). In a compound such as $BaCeO_3$, the average O–O separation (~3.2 Å) is too long for strong OH–O hydrogen bonding, unless there is a significant local distortion of the CeO_6 octahedron. Calculations suggest that the energy penalty for distorting the structure to produce O–O separations in the 2.5–3.0 Å range is almost perfectly offset by the gain in hydrogen-bond strength that follows, giving a range of structural configurations of similar energy. Locally, each OH group is thought to form transient H bonds to its eight nearest O sites such that there is a rotational diffusion of the H bond with an activation energy of around 0.1 eV. Proton transfer is believed to occur when local distortions lead to short instantaneous O–O distances of ~2.4 Å and has an activation energy of ~0.4–0.6 eV.

Note that even though there is an apparent net migration of OH groups through the structure, this occurs via proton transfer between adjacent oxide ions and not via the motion of an intact OH group. The concentration of oxygen vacancies is too low to allow significant OH migration. This can be proved by ^{18}O tracer diffusion experiments; for example, in $BaCeO_3$, ^{18}O diffusivities are three orders of magnitude too low to explain the proton diffusivity. The mechanism at work in these materials also implies that there will be an optimum temperature for proton migration. At too low a temperature, the H^+ mobility will be low; at too high a temperature, the equilibrium of Equation (13.7) will lie too far to the left, protons will be lost as water gas, and the number of carriers is low.

13.7 Oxide-Ion Conductors

While PbF_2 taught us many of the key features that lead to high anion conductivity, most technological applications require oxide-anion conductors. Two typical devices requiring oxide conductivity are shown in Figures 13.16 and 13.17. In the solid-oxide fuel cell (SOFC) shown in Figure 13.16, H_2 is oxidized to protons at the fuel–anode interface releasing electrons that flow through the external circuit to the cathode. At the cathode, these electrons reduce O_2 to O^{2-} ions that are sucked into the oxide-anion conducting electrolyte and

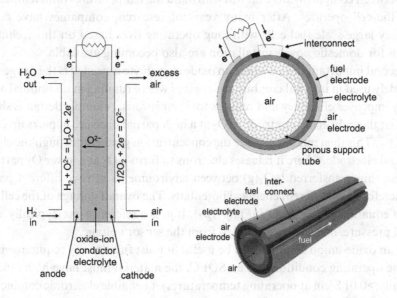

Figure 13.16 Left: Schematic design of a typical solid-oxide fuel cell. Right: Typical tubular design (3D image modified from ref. [9]).

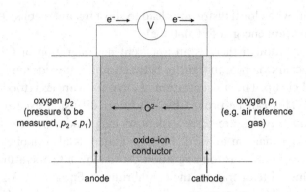

Figure 13.17 Schematic of a sensor to measure the partial pressure of O_2.

migrate along the strong concentration gradient to the anode where they react with the protons to form water. As for the proton-conductor fuel cell of Section 13.2, the free energy of the reaction $H_2(g) + \frac{1}{2}O_2(g) \rightarrow H_2O(g)$ can be used to do electrical work, though note that the direction of ion flow is opposite to that in Figure 13.2. The right-hand side of Figure 13.16 shows the design of a commercial tubular device in which air passes down the center of a hollow tube and fuel down its outside.

Much of the challenge in commercializing these devices lies in the difficulty of producing thin-enough components to keep the overall cell resistance low and in providing gas-tight seals between compartments that will withstand the temperatures (and temperature changes) when the cell operates. After many years of research, companies have developed high-efficiency large-scale fuel cells with long operating lives based on this technology. Smaller systems for domestic power installation are also becoming available.

A second important device where an oxide electrolyte is needed is the oxygen sensor. These are widely used in internal combustion engines where running at an optimal air-to-fuel ratio greatly improves efficiency and reduces toxic emissions. A simple design is shown in Figure 13.17. At the right-hand electrode, $\frac{1}{2}O_2$ at a high partial pressure p_1 picks up two electrons to form O^{2-}. This then migrates under the concentration gradient through the electrolyte to the left-hand electrode, where it releases electrons to form $\frac{1}{2}O_2$ at a lower O_2 partial pressure p_2. One has thus transferred $\frac{1}{2}O_2(g)$ between environments at two different partial pressures and therefore two different chemical potentials. The overall voltage of the cell is given by the Nernst equation as $E_{cell} = RT/nF \ln(p_1/p_2)$. If p_1 is a reference gas (typically air) at a known partial pressure, one can determine p_2 from the sensor voltage.

For an oxide-anion conductor to be useful, it must fulfill many requirements, often under extreme operating conditions. In an SOFC, the material must have high ionic conductivity (typically >0.01 S/cm at operating temperatures), negligible electronic conductivity, thermodynamic stability, low volatility of components, good mechanical properties, low reactivity with electrode materials, thermal expansion compatible with other components, as well as be cheap, environmentally benign, and easy to fabricate as dense thin layers of a few tens of microns. This is a demanding set of requirements! Many structural families have been investigated and shown to support oxide-ion conductivity (fluorites, perovskites, brownmillerites, Aurivillius phases, pyrochlores, melilites, apatites, and others). Conductivities of selected members of important families are shown in Table 13.1 and Figure 13.18; and the origins of their conductivities are discussed in the following sections.

Table 13.1 Ionic conductivities in S/cm at 1000 K of selected SOFC electrolytes in air. Data largely extracted from ref. [10].

Parent	Optimally doped electrolyte	Colloquial name	$\log(\sigma)$
ZrO_2	$Sc_{0.093}Zr_{0.907}O_{1.9535}$	ScSZ	−1.1
ZrO_2	$Y_{0.08}Zr_{0.92}O_{1.96}$	YSZ	−1.4
CeO_2	$Ce_{0.8}Gd_{0.2}O_{2-x}$	GDC	−1.3
Bi_2O_3	$Bi_{1.6}Er_{0.4}O_3$	ESB	−0.4
$Bi_2O_3{}^a$	$Bi_{0.913}V_{0.087}O_{1.587}$	BiVO	−1.4
$LaAlO_3{}^b$	$Sr_{0.1}La_{0.9}AlO_{3-x}$		−3.1
$LaGaO_3$	$Sr_{0.2}La_{0.8}Ga_{0.76}Mg_{0.19}Ca_{0.05}O_{3-x}$	LSGM	−1.1
$Ba_2In_2O_5{}^c$	$Ba_2In_2O_5$		−3.5
$Bi_4V_2O_{11}$	$Bi_2V_{0.9}Cu_{0.1}O_{5.35}$	bimevox	−0.8
$A_{10}(SiO_4)_6O_{2\pm x}$	$La_{9.75}Sr_{0.25}(SiO_4)_6O_{2.875}$	apatite	−1.6
$La_2Mo_2O_9$	$La_2Mo_2O_9$	lamox	−1.4

^{a}773 K, ^{b}under N_2, ^{c}at 10^{-6} atm where conductivity is predominantly ionic.

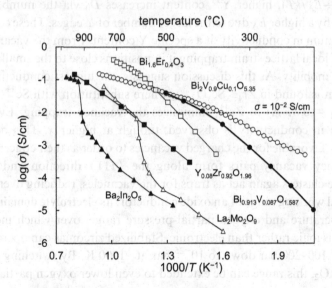

Figure 13.18 Conductivities of selected oxide-anion conductors.

13.7.1 Fluorite-Type Oxide-Ion Conductors

The most widely used oxide-anion conductors are derived from ZrO_2. ZrO_2 has a monoclinic structure at room temperature with seven-coordinate Zr. On heating, it undergoes phase transitions firstly to a tetragonal (~1370 K) and then (~2643 K) to a cubic fluorite-type form that remains stable up to the melting point of ~2988 K. The cubic structure can be stabilized at lower temperature by acceptor doping with Ca^{2+}, Sc^{3+}, or Y^{3+} to form solid solutions $A_xZr_{1-x}O_{2-x}$ (A^{2+}) or $R_xZr_{1-x}O_{2-x/2}$ (R^{3+}). For the A^{2+} doping, the dopant-oxide dissolution equation can be expressed in Kröger–Vink (Section 2.6) notation as:

$$AO = A_{Zr}'' + v_O^{\bullet\bullet} + O_O^{\times} \tag{13.8}$$

and we see that doping produces an oxide vacancy. With Ca^{2+}, the cubic material is stable for $0.15 < x < 0.28$ and with Y for $0.13 < x < 0.68$ at 1500 °C. These materials are known as calcia- and yttria-stabilized zirconias (CSZ and YSZ, respectively) and have high oxide-anion conductivities. YSZ is the most widely used material commercially, and its conductivity exceeds 10^{-2} S/cm above ~650 °C.

How do we optimize O^{2-} conductivity in these compounds? In a simple picture, we might expect conductivity to increase with vacancy concentration, reaching a maximum when 50% of the sites are vacant (if chemically possible). In practice, the conductivity peaks at much lower concentrations due to vacancy trapping; in YSZ this occurs at around 4% vacancies ($x = 0.16$). One contribution to the trapping comes from the different sizes [11] of Zr^{4+} (r_{VIII} = 0.84 Å) and Y^{3+} (r_{VIII} = 1.02 Å) cations. If we think of the fluorite structure in terms of edge-shared M_4O tetrahedra, the larger size of Y^{3+} means that the energy barrier for an O^{2-} to cross a Y–Y or Y–Zr tetrahedral edge is higher than a Zr–Zr edge. This reduces the mobility of anions close to a Y dopant. In terms of our expression for diffusivity of $D = D_0 \exp(-E_A/kT)$, higher Y^{3+} content increases D_0 via the number of vacancies, but this is offset by a higher E_A due to the larger number of Y edges. These competing effects lead to the maximum in conductivity at a specific Y content. From the vacancy's perspective, we can think of local lattice strain trapping it in positions close to the smaller Zr^{4+} ions, thereby reducing its mobility. As this discussion suggests, the highest conductivity in any zirconia-based system is found in $Zr_{1-x}Sc_xO_{2-x/2}$ where substitution with Sc^{3+}, which has a similar radius (r_{VIII} = 0.87 Å) to Zr^{4+}, minimizes the vacancy trapping. Even in this material, a maximum in conductivity is observed, though at higher x. This occurs as it becomes energetically favorable for the charged vacancies to order as their concentration is increased. Local vacancy–vacancy pairs form along the $\langle 111 \rangle$ direction and make longer-range clusters. The clusters again act as traps for the vacancies, reducing overall O^{2-} mobility.

The useful working range of an oxide conductor, its **electrolyte domain**, is typically defined as the temperature and oxygen partial-pressure range, over which more than 99% of its conduction is ionic rather than electronic. Stabilized zirconias can operate at oxygen partial pressures of 100–200 bar down to 10^{-25} bar at ~1000 K. By switching to materials such as Y-doped ThO_2, this range can be extended to even lower oxygen partial pressure.

A number of other fluorite-type materials have been investigated in the search for higher conductivities. Ceria-derived (CeO_2; Ce^{4+} $r_{VIII} = 0.97$ Å) materials have larger cubic cell parameters than zirconias, providing a more open metal–oxide framework that should make oxide-ion migration easier. Under oxidizing conditions, doped compositions such as $Gd_xCe_{1-x}O_{2-x/2}$ with $0.1 < x < 0.2$ do have higher conductivities, particularly at low temperature. Unfortunately, under the reducing conditions at the anode side of the fuel cell,[14] Ce^{4+} can be reduced to Ce^{3+}. This leads to n-type electronic conduction and a partial electronic short circuit in the cell. It also causes significant lattice expansion (Ce^{3+} $r_{VIII} = 1.14$ Å) that can lead to mechanical failure [12]. Despite these problems, working cells have been made with ceria electrolytes.

While ZrO_2 and CeO_2 must be acceptor doped to create vacancies, high vacancy content and ionic conductivity can be achieved in pure Bi_2O_3. At room temperature, α-Bi_2O_3 has a monoclinic structure, but at 1002 K it converts to a face-centered cubic δ form with a defect-fluorite structure, $BiO_{1.5}\square_{0.5}$, which is stable up to the melting point of 1103 K. The 25% of vacant oxygen sites leads to a conductivity of 1 S/cm at 1000 K, considerably higher than CSZ or YSZ. Doping is again required to stabilize the cubic structure at low temperature, and vacancy-trapping effects mean that the minimum level of dopant again gives the highest conductivity; materials such as $Bi_{2-x}Er_xO_3$ ($x \approx 0.4$) and $Bi_{2-x}Y_xO_3$ ($x \approx 0.46$ to 0.50) have the highest conductivities. The highest conductivity reported to date in a stable δ-Bi_2O_3 derivative is in $Bi_{1-x}V_xO_{1.5+x}$ with $x = 0.087$, which surpasses the $\sigma = 10^{-2}$ S/cm threshold at a mere 350 °C [13]. Its properties and stability have been related to three key structural features: variable coordination of V, tumbling motions of VO_n polyhedra (which help O^{2-} migration in a manner reminiscent of proton transfer in $CsHSO_4$), and adoption of a stable superstructure with ordered cation sites (to reduce aging). Low stability under reducing conditions and volatilization of Bi_2O_3 at relatively low temperatures have traditionally been seen as barriers to the practical use of Bi-containing electrolytes. New innovations in electrode and electrolyte designs and the possibility of operating at lower temperature, where Bi is less reducible, are helping to overcome these problems.

13.7.2 Perovskite, Aurivillius, Brownmillerite, and Other Oxide Conductors

A number of interesting oxide-ion conductors have been found in materials with perovskite-related structures. The earliest reports of ionic conductivity in pure perovskites centered on acceptor-doped $RAlO_3$ (R = rare earth), with materials such as $Sr_{0.1}La_{0.9}AlO_{3-x}$ having the highest conductivities. Later work focused on $LaGaO_3$-based materials that have higher conductivities than stabilized zirconias and operate at lower oxygen partial pressures. The best properties are achieved by acceptor doping at both La and Ga sites, and there is a trade-off between vacancy concentration and vacancy trapping. Optimal materials use size-matched substitutions, giving compositions $La_{1-x}Sr_xGa_{1-y}Mg_yO_{3-\delta}$, with $0.1 < x < 0.2$ and $0.15 < y < 0.20$. These materials don't absorb water, don't show significant aging at 800 °C, and have >98% oxide-anion conduction under the typical operating conditions of a fuel cell.

---

[14] Reducing conditions due to the excess of fuel at the anode side. Typical partial pressures might be 10^{-19} bar O_2.

At higher oxygen deficiency levels, some perovskite-related materials adopt structures in which the vacancies are ordered. Compositions ABO_{3-x} with $x = 0.5$ (i.e. 16.7% oxygen vacancies) can adopt the structure of the mineral brownmillerite, Ca_2FeAlO_5. One example of this is $Ba_2In_2O_5$ (Figure 13.19) that can be thought of as containing rows of oxygen vacancies parallel to [110] of the simple perovskite cell, running in every other equatorial plane of the corner-linked InO_6 octahedra. In doing this, one removes two of the six octahedral corners in every second layer, leading to alternate layers of tetrahedra and octahedra. Conductivity data for $Ba_2In_2O_5$ are included in Figure 13.18. At 1000 K and $p(O_2) = 10^{-6}$ atm, the conductivity is 10^{-3} S/cm, but jumps to 0.1 S/cm at 1200 K, when the material undergoes a first-order phase transition where vacancies disorder, ultimately reaching a disordered cubic perovskite structure at around 1300 K. Improvement in conductivity at lower temperature can be achieved in compositions such as $Ba_2In_{1.75}Ce_{0.25}O_{5.125}$, where the disorder is stabilized. The high number of vacancies means, as suggested by Equation (13.7), that $Ba_2In_2O_5$ is susceptible to water uptake and it is in fact a good proton conductor from 300 °C to 700 °C. Problems of water uptake at low temperature can be reduced by substitution to give materials such as $Ba_{1.2}La_{0.8}In_2O_{5.4}$.

Other structure types investigated include apatite-related $A_{10-x}(MO_4)_6O_{2\pm y}$ ($A = Ln^{3+}$/alkaline earth, M = Si/Ge), Aurivillius-type "bimevox" materials, and $La_2Mo_2O_9$-derived phases. The Aurivillius structure of Bi_2WO_6 is shown in Figure 13.19. It contains layers of corner-sharing $[WO_{4/2}O_2]^{2-}$ octahedra alternating with $[Bi_2O_2]^{2+}$ layers. The $[Bi_2O_2]^{2+}$ portion has a central layer of oxide ions tetrahedrally coordinated by Bi (similar to $[La_2O_2]^{2+}$ in the LaOFeAs superconductors of Chapter 12). The Bi^{3+} ions are coordinated by four of these oxygens (at 2.18–2.51 Å) and also by two oxygens of the WO_6 layers (2.44–2.58 Å) in an asymmetric coordination environment typical of a lone-pair cation. Replacement of W^{VI} by V^V leads to $Bi_2VO_{5.5}$ or $Bi_4V_2O_{11}$, which has a similar structure, but with oxygen vacancies predominantly in the octahedral layers. $Bi_4V_2O_{11}$ has a conductivity of 0.1–1.0 S/cm in its

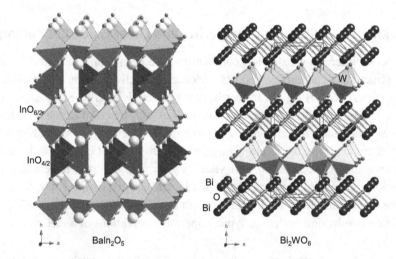

Figure 13.19 Brownmillerite-type $Ba_2In_2O_5$ and Aurivillius-type Bi_2WO_6.

high-temperature (> 843 K) γ form. The substituted "bicuvox" phase[15] $Bi_2V_{1-x}Cu_xO_{5.5-3x/2}$ with $x \approx 0.1$ retains a conductivity of 10^{-2} S/cm down to 623 K. $La_2Mo_2O_9$ has also been shown to exhibit high conductivity in its high-temperature structural form, and this has been related to the ability of Mo to adopt variable coordination numbers. This particular material highlights some of the complexities of functional oxide materials; its room-temperature structure contains a remarkable 312 crystallographically unique atoms!

13.7.3 SOFC Electrode Materials and Mixed Conductors

There are also significant challenges when considering materials to use as electrodes in SOFCs. The anode must catalyze the oxidation of the fuel, conduct O^{2-} ions from the electrolyte, and simultaneously have sufficiently high electronic conductivity (~100 S/cm or higher) to channel the electrons produced into the external circuit. It must also have thermal expansion compatible with the electrolyte. Traditionally, porous ceramic–metal composites or **cermets** have been used, such as Ni–YSZ composites. Having a fully connected (percolated) metal component ensures electrical conductivity, and fuel oxidation can occur at the three-phase boundary between metal, electrolyte, and fuel. This is, however, only a 1D intersection.[16] There is therefore significant benefit in using **mixed ionic and electronic conductors** (MIECs) where fuel oxidation can occur over the whole of the porous anode's high surface.[17] A variety of perovskite materials have been investigated for this purpose such as $(La_{0.75}Sr_{0.25})_{0.5}Cr_{0.5}Mn_{0.5}O_{3-x}$, Mn/Ga-doped $Sr_{1-x}La_x$ $TiO_{3+\delta}$,[18] and $PrBaMn_2O_{5+\delta}$ [14, 15, 16].

At the cathode, the key process is reduction of molecular O_2 and its transport to the electrolyte. Materials such as $La_{1-x}Sr_xMnO_{3-x}$ (see Chapter 11) are frequently used. A variety of perovskite-related materials such as $Ba_{0.5}Sr_{0.5}Co_{0.8}Fe_{0.2}O_3$ and $LnBaCo_2O_{5+\delta}$ [17] are also being investigated for this application.

13.8 Intercalation Chemistry and Its Applications

The last category of materials we will discuss where the migration of ionic species leads to energy-related applications are **intercalation compounds**. In everyday usage, the term intercalation refers to the insertion of extra days into the calendar, such as the introduction of February 29 in leap years to allow for the fact that the Earth takes 365.26 days to orbit the Sun. In solid state chemistry, intercalation is defined as a reversible insertion of **guest species** (atoms, ions, molecules, or molecular ions) into vacant sites of a **host structure**. It is a process that has been practiced inadvertently for millennia in the processing of clays by pottery industries around the world and is shown schematically in Figure 13.20.

---

[15] One of the wider family of "bimevox" materials.

[16] The use of Ni in cermets can also lead to significant coke formation if hydrocarbons are used as fuels.

[17] Mixed conductors have other areas of application, including as semi-permeable oxygen membranes.

[18] Typical composition $La_{0.33}Sr_{0.67}Ti_{0.92}Mn_{0.04}Ga_{0.04}O_{3.125}$.

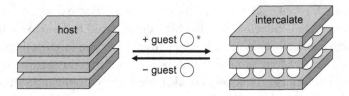

Figure 13.20 Schematic intercalation reaction of a layered 2D host.

13.8.1 Graphite Intercalation Chemistry

The first scientific report of intercalation chemistry is probably that of Schäuffatl, who described the reaction of graphite with concentrated sulfuric acid in the 1840s. Since then, a large number of graphite intercalation compounds have been prepared with guest species between the graphite layers that move apart along the c axis to accommodate them. One of the most controllable synthetic methods for combining the graphite and guest is by vapor-phase transport. This is done by heating the guest and solid graphite to slightly different temperatures in different compartments of an evacuated glass tube. Potassium intercalates $K_{0.125}C$ ($\equiv KC_8$) with K between all the graphite sheets (Figure 13.21) can be prepared at ~250 °C.[19] A bromine intercalate $(Br_2)_{0.0625}C$ can be obtained by holding the guest and graphite at ~20 °C. Intercalation can also be achieved using a solution of K in liquid ammonia[20] or Br_2 in CCl_4. Intercalates of a wide variety of metals and molecular guests can be prepared in similar ways.

Graphite is unusual as an intercalation host in that it will intercalate both electron donors like K and acceptors like Br_2. The key to this behavior lies in its semimetallic band structure discussed in Chapter 6. On intercalation, K is ionized to K^+ and electrons are donated to the empty conduction band. With an acceptor, on the other hand, electrons from the valence band partially reduce the guest, leaving holes at the top of the valence band. Both these processes lead to partially filled bands, and intercalation causes significant changes in conductivity, as shown in Table 13.2. As discussed in Chapter 6, graphite shows significant anisotropy in its conductivity. The anisotropy is decreased in the alkali-metal intercalates but increased in molecular intercalates due to poor overlap between the graphite p orbitals and molecular orbitals of the guest. It is interesting to note that the carrier mobility of graphite layers is sufficiently high that AsF_5-acceptor intercalates[21] have a conductivity comparable to Cu despite having only around a quarter of the density.[22]

In addition to compositions such as $K_{0.125}C$, which contains guest molecules between all host layers as shown on the left of Figure 13.21, it is also possible to prepare compounds containing fewer guests in which only every nth layer is occupied. This

---

[19] The graphite is held at a slightly higher temperature than the guest to prevent condensation of the guest on the graphite. The K content is limited to 0.125 due to the relative sizes of K and graphite layers (see Problem 13.13).

[20] Here ammonia co-intercalation to form $K_x(NH_3)_yC$ occurs.

[21] The intercalate formed from graphite and AsF_5 is believed to contain a mixture of molecular ions and neutral molecules between the layers formed via reactions such as $3AsF_5 + 2e^- \rightarrow 2AsF_6^- + AsF_3$.

[22] Graphite/AsF_5 has mobility of 1.3 m^2 per volt per second [$m^2/(V\ s)$] at room temperature. Cu has $\sigma = 5.8 \times 10^5$ S/cm and mobility 0.0035 $m^2/(V\ s)$.

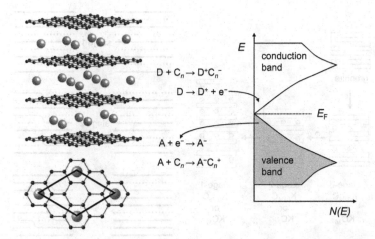

Figure 13.21 Left: A $K_{0.125}C$ intercalation compound of graphite viewed parallel and perpendicular to the layers. Right: DOS plot for pristine graphite with indication how a donor (D) partly fills the conduction band or an acceptor (A) partly empties the valence band.

Table 13.2 Room-temperature conductivity of graphite and various intercalation compounds in plane (σ_a) and out of plane (σ_c) [18].

Guest	Stage	σ_a (S/cm)	σ_c (S/cm)	σ_a/σ_c
Graphite (none)	–	2.5×10^4	8.3	3000
K	1	1.1×10^5	1.9×10^3	56
Li	1	2.4×10^5	1.8×10^4	14
Br_2	2	2.2×10^5	1.6	1.4×10^5
AsF_5	2	6.3×10^5	0.24	2.7×10^6

phenomenon is called **staging**. For the potassium–graphite intercalate, stage-2 compounds with every other layer full can again be prepared by the vapor-transport method, if the graphite is heated to ~375 °C rather than 250 °C to lower the driving force for intercalation. The simplest model for staging would be that shown on the left of Figure 13.22, with host layer gaps being either completely full or completely empty. However, the experimental observation that it is possible for different stages to evolve smoothly from n to $n-1$ as more guest is added is inconsistent with this model. For example, transforming from a stage-1 to a stage-2 compound would require the improbable removal of guest species entirely from one layer and their complete reintroduction in another. A more plausible model was suggested by Daumas and Herold [19], and is shown on the right of Figure 13.22. Here, staged compounds have pleated host layers with islands of guest between them. With this model, smooth transformation between stages is possible with only lateral movement of guests. A possible driving force for staging that is consistent with experimental observations is that it is due to

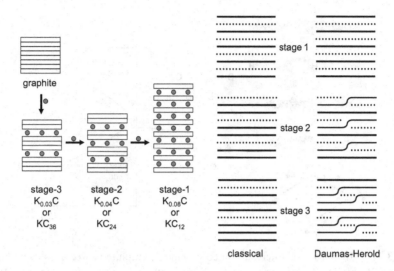

Figure 13.22 Staging in graphite intercalation compounds.

strain in host layers. As guests are intercalated, the addition or removal of electrons in the graphite layers changes the equilibrium C–C bond distance, creating local strain (C–C bond lengths increase on K intercalation as its electron is accommodated in graphene antibonding orbitals). The strain throughout the crystal can be minimized by packing guest species as closely as possible within one layer and maximizing the distance to the next intercalated layer. Stages can order over large distances, giving rise to sharp (00l) peaks in diffraction patterns, and $n = 8$ and higher stages are well documented in alkali-metal intercalates. Different staged compounds have different conductivities and different colors. In the case of K, stage-1 compounds are gold, stage-2 are blue, stage-3 are blue-black, and stage-4 and higher are black.

Box 13.1 Synthetic Methods: Chimie Douce

The reversibility of intercalation chemistry offers unusual opportunities to synthesize materials at low temperature that would not otherwise be stable. These are examples of "Chimie Douce", French for "Soft Chemistry". One of the earliest applications of this was the preparation of a metastable layered form of VS_2 that couldn't be prepared by other methods. When Li, V, and S are heated together in a sealed silica-glass tube, crystals of $LiVS_2$ form, with Li between VS_2 layers. If this compound is stirred with a solution of I_2 in acetonitrile, lithium is removed, leaving a metastable layered form of VS_2. The reactions are:

$$Li + V + 2S \rightarrow LiVS_2$$
$$LiVS_2 + \tfrac{1}{2}I_2 \rightarrow LiI + VS_2$$

Box 13.1 (cont.)

The low reaction temperature traps the metastable layered form of VS_2 [20]. Similar tricks have led to a metastable λ-MnO_2 by Li extraction from $LiMn_2O_4$ [21].

Chimie Douce has also been used to synthesise layered $LiMnO_2$ [22, 23], analogous to $LiCoO_2$. Due to size considerations, the layered material can't be prepared directly at high temperatures. However, its Na analogue can be made at 923 K and $LiMnO_2$ prepared by stirring $NaMnO_2$ with LiBr in hexanol at 433 K.

13.8.2 Lithium Intercalation Chemistry and Battery Electrodes

One of the main reasons for interest in intercalation chemistry is its application in the electrodes of rechargeable batteries. Lithium is the most electropositive metal ($E° = -3.04$ V) and has the lowest density ($\rho = 0.53$ g/cm^3), making it particularly suitable for application in high-energy, low-weight batteries. We will therefore look at intercalation reactions of Li with layered metal chalcogenides and metal oxides, and how they can be exploited in lithium-ion batteries.

A large number of MX_2 metal dichalcogenides adopt layered CdI_2 structures (Figure 1.28), where octahedral holes in every second layer of an hcp array of anions are occupied by cations (e.g. TiS_2, ZrS_2, SnS_2). Owing to the relatively weak van der Waals forces holding layers together, these materials are ideal intercalation hosts and take up a variety of guest species. A simple room-temperature reaction with n-butyllithium, for example, leads to the rapid formation of Li_xMS_2, with octane as the side product. The Li "atoms" formed from butyllithium reduce the MS_2 host, and the resulting Li^+ ions enter the empty octahedral sites between the host lattice layers; the electrons enter the conduction band of the host.

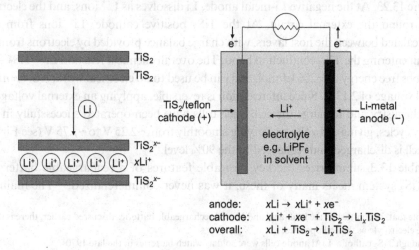

anode: $xLi \rightarrow xLi^+ + xe^-$
cathode: $xLi^+ + xe^- + TiS_2 \rightarrow Li_xTiS_2$
overall: $xLi + TiS_2 \rightarrow Li_xTiS_2$

Figure 13.23 Left: Schematic intercalation reaction. Right: Schematic design for a rechargeable Li cell. Typical solvents used are mixtures of dimethyl carbonate, diethyl carbonate, and ethylene carbonate.

Table 13.3 Requirements for intercalation-based electrode materials for rechargeable Li batteries.

Battery requirement	Chemistry to meet requirement
High cell voltage, E_{cell}	Requires a large ΔG for the cell reaction and will be governed by the change in chemical potential of Li in the charged and discharged states.
Approximately constant E_{cell} during discharge	Requires ΔG to change little as a function of x during the xLi + cathode $\rightarrow$ Li$_x$ [cathode] intercalation reaction.[25]
High capacity	The more Li ions per formula unit that can be reversibly intercalated/deintercalated, the higher the charge that can be stored. The capacity is usually expressed in A h/kg ($\equiv$ mA h/g).
Storage energy density	A battery needs high energy density either per unit mass or volume, depending on the application. The more exchangeable Li per electrode, the better. The overall energy density (ed) is the product of the electrode potential E and the charge per unit mass or volume Q: $ed = EQ$ in W h/kg or W h/L.
High current/power	Rapid intercalation is required to give high current flows and high power. This requires high mobility of both Li$^+$ and electrons through electrodes and is favored by minimal structural rearrangement during intercalation. It is helped by cell designs that minimize diffusion length.
Reversibility/capacity fade	Intercalation must be reversible over a wide composition range and allow many cycles of charge/discharge without damage to either the atomic, micro, or bulk structure of the electrodes or on-going reaction between electrodes and electrolyte.
Manufacturing and commercial considerations	The device must be easy to manufacture, cheap, safe to use, have minimal environmental impact, use abundant elements, have a long shelf life, be recyclable at the end of life, and outperform established technology sufficiently to warrant commercialization.

This reaction can also be performed in an electrochemical cell of the type shown in Figure 13.23. At the negative Li-metal anode, Li dissolves as Li$^+$ ions, and the electrons released pass round the external circuit. At the TiS$_2$ positive cathode, Li$^+$ ions from solution are intercalated between the host layers, with charge balance provided by electrons from the external circuit entering the TiS$_2$ conduction band. The overall chemical reaction Li + TiS$_2$ $\rightarrow$ LiTiS$_2$ has a Gibbs free energy of -206 kJ/mol, and can be used to do electrical work; $\Delta G = -nFE_{cell}$ implies a cell voltage of 2.13 V. Since intercalation is reversible, applying an external voltage can reverse the chemistry and recharge the cell. Such cells have been operated successfully in the lab over many cycles, giving cell voltages varying smoothly from ~2.25 V to ~1.75 V (see Figure 13.28) as the cell is discharged and recharged to the 90% level.[23]

Table 13.3 summarizes the key desirable features of a rechargeable battery. While the Li–TiS$_2$ system meets many of these, it was never commercialized.[24] The main reason was

[23] Note that, in contrast to the lead–acid and Ni–Cd rechargeable batteries discussed earlier, there is no reaction with the electrolyte as the cell operates.

[24] Though TiS$_2$ cathode–LiAl anode cells were sold as watch batteries in the late 1970s.

[25] Note that a small change in voltage during discharge can be advantageous as it then gives an indication of the battery charge. This is how the battery lifetime indicators on portable devices function.

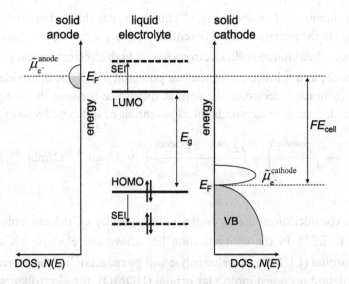

Figure 13.24 Electron energy levels in a typical Li cell in its charged state. As drawn, the formation of an SEI (see text) passivating layer gives kinetic stability such that electrolyte reduction doesn't occur despite $\tilde{\mu}_{e^-}^{anode}$ being higher than its LUMO. All levels are in joules per mole (J/mol) as F is the Faraday constant.

safety. As the cell is discharged, Li^+ ions are removed from the anode, then redeposited under charging, but not necessarily at the same point. Over many cycles, dendritic growth of Li metal at the anode surface is observed. The growing "strands" of Li have such a high surface area that reactions with the electrolyte become a problem. If they become too large, they can short-circuit the cell, leading to sudden temperature rises and the cell igniting.

13.8.3 Lithium-Ion Batteries with Oxide Cathodes

How can we design a better intercalation-based battery; one that meets more of the requirements in Table 13.3? We'll need to consider the chemistry of each component of the cell, the reactions that can occur at the interfaces between components, and the passage of both ions and electrons during its operation.

Let's consider first what determines E_{cell}. Since we no longer have a simple aqueous galvanic cell (like in Figure 13.1) operating under standard conditions, we can't just determine E_{cell} from the difference of standard reduction potentials of, say, Ti^{4+}/Ti^{3+} and Li^+/Li. Instead, we have to consider the energetics of adding or removing electrons at the Fermi levels, E_F, of the two crystalline electrodes. The important electron-energy levels are shown in Figure 13.24. When these energy levels are in units of eV[26] (units of voltage times the elementary charge) the E_F difference divided by 1 (electron) yields directly the cell voltage E_{cell}.

---

[26] E_F values are typically given in electronvolts (1 eV = $1.602176634 \times 10^{-19}$ J) and represent the energy released for each electron added to the solid. Multiplying by N_A converts 1 eV per added electron to 96485 J per added mole of electrons (the charge of 1 mole of electrons is 96485 C or the Faraday constant; and 1 J = 1 C × 1 V).

Since the chemistry of an operating cell changes as it is charged or discharged, it is useful to express E_{cell} via the electrochemical potentials $\tilde{\mu}_A = \mu_A + z_A F \phi$ per mole of electrons[27] at the two electrodes. In a charged cell, electrons have a high electrochemical potential $\tilde{\mu}_{e^-}^{anode}$ at the reduced anode, and a low electrochemical potential $\tilde{\mu}_{e^-}^{cathode}$ at the oxidized cathode. If we connect leads to each electrode, the voltage difference between them, E_{cell}, will be proportional to the difference in electrochemical potentials or in Fermi levels of the two electrodes:

$$E_{cell} = \left(\frac{\tilde{\mu}_{e^-}^{anode} - \tilde{\mu}_{e^-}^{cathode}}{F} \right) = \left(\frac{E_F^{anode} - E_F^{cathode}}{F} \right) \; [\text{V}, \text{J/mol}^{-1}/(\text{C/mol}^{-1}), \text{J/mol}^{-1}/(\text{C/mol}^{-1})]$$

$$(13.9)$$

Our first consideration in cell design is the stability of the electrolyte to oxidation or reduction. If $\tilde{\mu}_{e^-}^{anode}$ in the charged state lies above the electrolyte's lowest unoccupied molecular orbital (LUMO), the electrolyte will be reduced[28]; if the corresponding $\tilde{\mu}_{e^-}^{cathode}$ is below the highest occupied molecular orbital (HOMO), the electrolyte will be oxidized. As both processes could decompose the electrolyte, there is an **electrolyte stability window** that limits the cell potential to $E_{cell} \leq E_g/F$, where E_g is the electrolyte HOMO–LUMO gap in J/mol.[29] With aqueous electrolytes, this limits practical voltages to $\lesssim 1.5$ V.[30] Much larger E_{cell} values are possible with organic electrolytes, such as for $LiPF_6$ dissolved in a mixture of organic carbonates.[31] With a Li-metal anode, $\tilde{\mu}_{e^-}^{anode}$ lies above the LUMO of even these electrolytes. Fortunately, the electrolyte stability window can be expanded via formation of a thin amorphous passivating layer on the electrode surface, which is produced by reaction with the electrolyte during the initial charge. This is called the **solid electrolyte interphase** (SEI), and ethylene carbonate is particularly effective in this role. SEI formation reduces the battery capacity but, since it provides a kinetic barrier to further electrolyte reduction in use, it improves the lifetime.

Let's now consider how to choose optimum electrode materials for an intercalation-based Li cell. When we discharge the cell, we move Li from the anode (where it has a high chemical potential) to the cathode (where it is low) in the reaction Li(anode) → Li(cathode). Li^+ ions move through the electrolyte and electrons flow through the external circuit where they do work on the surroundings. If we break the circuit so that no current can flow, we create an "electrochemical

---

[27] The electrochemical potential of the electrons has two contributions. The first is the plain chemical potential μ_A of the electron, taken as a component A in a reacting mixture of A, B, C, ... and describes how the Gibbs energy of this "system" changes per unit change in the number of moles of the reaction component. At a given temperature and pressure and fixed amount n of other species $\mu_A = (\partial G/\partial n_A)_{T,p,n_B,n_C} ...$, and is typically expressed in J mol^{-1}. The second contribution is $z_A F \phi$, where z_A is the charge on A and ϕ the electrostatic potential in V, and reflects the energy required to bring a charged species to a specific location. Note that $\tilde{\mu}_A = \mu_A$ for a neutral species such as Li.

[28] As electrons move to lower energy if they transfer from the anode to the empty electrolyte LUMO.

[29] Or $E_{cell} \leq E_g$ for the HOMO–LUMO gap given in eV.

[30] Sum of standard potentials for $H^+ + e^- \rightarrow \frac{1}{2}H_2$ and $\frac{1}{2}O_2 + 2H^+ + 2e^- \rightarrow H_2O$ plus a small kinetic overpotential.

[31] Dimethyl carbonate, diethyl carbonate, and ethylene carbonate are commonly used.

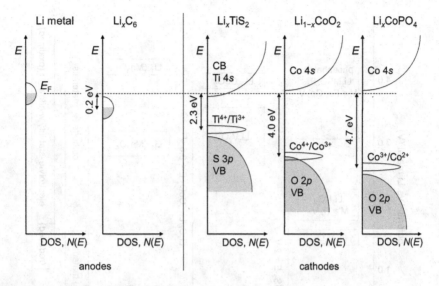

Figure 13.25 DOS plots showing relative positions of E_F (in eV = 1.602×10^{-19} J) in Li metal and Li_xC_6 anodes and approximate energies and characters of charged cathode-material orbitals that will be filled on Li intercalation. VB = valence band, CB = conduction band. After ref. [24].

equilibrium" state,[32] in which local equilibrium for the process $Li \rightleftarrows Li^+ + e^-$ at each electrode requires $\widetilde{\mu}_{Li^+} + \widetilde{\mu}_{e^-} = \mu_{Li}$.[33] Since Li^+ can diffuse freely through the electrolyte, under these conditions $\widetilde{\mu}_{Li^+}$ will be identical at both electrodes and it follows from Equation (13.9) that:

$$E_{cell} = \left(\frac{\widetilde{\mu}_{e^-}^{anode} - \widetilde{\mu}_{e^-}^{cathode}}{F} \right) = \left(\frac{\mu_{Li}^{anode} - \mu_{Li}^{cathode}}{F} \right) \quad (13.10)$$

We can see from this equation that we need to manipulate the chemical potential of Li in the anode and cathode environments to influence E_{cell}.

In commercial batteries, the most widely used trick at the anode is to use an intercalation compound such as Li_xC_6 instead of Li metal. Here, μ_{Li}^{anode} is lower than in Li metal (Li is more stable in Li_xC_6, Figure 13.25). This has the disadvantage of reducing E_{cell}, as can be seen from Equation (13.10), but two major advantages: Firstly, reduction of the electrolyte when the cell is charged becomes less likely. Secondly, the formation of Li metal dendrimers during cell charging is eliminated,[34] greatly improving battery safety.

How do we compensate for the loss of E_{cell} caused by using a graphite anode? The only way is to look for a cathode in which $\mu_{Li}^{cathode}$ is as low as possible. Since Li is invariably present in the

[32] Note that this is not a global chemical equilibrium state.
[33] At equilibrium, products and reactants must have equal electrochemical potentials.
[34] At least under slow charging, which approximates equilibrium conditions, when all Li will intercalate.

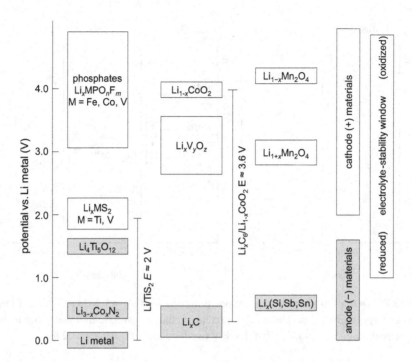

Figure 13.26 Voltage ranges for various cathode/anode combinations. The cell voltage is the difference between cathode and anode potentials. The electrolyte-stability window to reduction/oxidation for $LiPF_6$ in a 1:1 mixture of ethylene carbonate and diethyl carbonate is indicated. Batteries must be operated within this window.

cathode as Li^+ ions and electrons, it is convenient to consider the terms $\tilde{\mu}_{Li^+}^{cathode}$ and $\tilde{\mu}_{e^-}^{cathode}$ separately. A good cathode will therefore need low-energy sites to adopt Li^+ ions[35] and to have a low E_F ($\tilde{\mu}_{e^-}^{cathode}$), meaning low-energy orbitals for the electrons. It is therefore important to understand the density of states (DOS) of the cathode, using the concepts introduced in Chapter 6. DOS plots for some of the cathodes discussed in this chapter are included on the right of Figure 13.25. The first thing we can appreciate from Figure 13.25 is that the energy of the top of the cathode valence band will set an upper limit on the value of E_{cell} achievable with a given anode. The maximum E_{cell} achievable using the Co^{4+}/Co^{3+} couple in an oxide cathode such as $Li_{1-x}CoO_2$ will therefore be much larger than using Ti^{4+}/Ti^{3+} in a chalcogenide such as Li_xTiS_2. Even in an oxide, changing the counter ion for a given transition metal has a significant impact on E_{cell}. For example, E_{cell} is greater using the Co^{3+}/Co^{2+} couple in Li_xCoPO_4 than the Co^{4+}/Co^{3+} couple in $Li_{1-x}CoO_2$.[36] Figure 13.26 summarizes these ideas and shows the range of cell voltages achievable with different anode and cathode materials.

---

[35] We explore the influence of different Li^+ sites in the end-of-chapter problems where we see E_{cell} changing for a given cathode when Li^+ adopts different coordination sites.

[36] Another example is the observation that the position of the Fe^{3+}/Fe^{2+} couple can be moved by around 1 eV within a closely related series of $Li_xFe_2(XO_4)_3$ materials with X = As, P, Mo, or S (see ref. [20]).

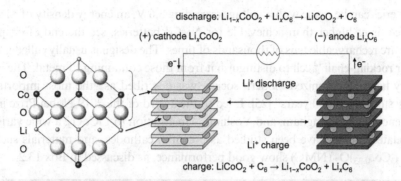

Figure 13.27 The simplified[38] structure of $LiCoO_2$ and a schematic representation of its function in a Li-ion cell. The direction of electron flow shown is that during discharge.

The first commercially successful rechargeable battery was introduced by Sony in the early 1990s and used all of these ideas. It employed a Li_xC_6 intercalation anode to prevent dendrimer formation and an oxide $Li_{1-x}CoO_2$ cathode based on the Co^{4+}/Co^{3+} couple. The structure of $LiCoO_2$, Figure 13.27, is related to that of NaCl, and can be described as a slightly distorted ccp of oxide ions with the octahedral holes in alternate layers occupied by Li^+ and Co^{3+}.[37]

The cell is manufactured in a discharged state with a $LiCoO_2$ cathode, carbon anode, and electrolyte of $LiPF_6$ dissolved in a mixture of alkyl carbonates,[39] with a porous polymer film separating the electrodes. There are various physical designs that try to maximize electrode surface areas to maximize Li^+ diffusion rates.[40] On the first charge, Li ions are partially deintercalated from $LiCoO_2$, forming $Li_{0.5}CoO_2$, and intercalated by the anode, forming LiC_6. When the cell discharges, the reverse occurs: Li is deintercalated from graphite and Li^+ ions from the electrolyte intercalate between cathode layers, with the charge-balancing electrons flowing through the external circuit. The overall cell reaction is $0.5LiC_6 + Li_{0.5}CoO_2 \rightarrow LiCoO_2 + 0.5C_6$, the free energy of which arises from moving Li ions and electrons from the carbon to the oxide material where their chemical potential is lower. An important operating feature of the $Li_{1-x}CoO_2$ cell is that only half the available Li ions are exploited in the cell reaction. Beyond this point, one approaches the stability limit of $Li_{1-x}CoO_2$ with respect to O_2 loss, and oxidation of the electrolyte becomes more likely.[41]

[37] Remember from Chapter 1 that the close-packed layers are perpendicular to the $\langle 111 \rangle$ (body-diagonal) directions of the cubic unit cell. Note the similarity to $LiTiS_2$ that has alternate Li^+/Ti^{3+} layers but in an hcp anion array.

[38] Li/Co ordering lowers the symmetry from cubic to rhombohedral in $LiCoO_2$.

[39] For example, dimethyl carbonate, diethyl carbonate, and ethylene carbonate (see Footnote 31). Ethylene carbonate is particularly important for anode passivation.

[40] The ionic current density in the electrolyte and through the SEI layer is lower than the electronic current density of the external circuit, requiring a thin electrolyte and high-surface-area electrodes.

[41] As shown in Figure 13.25, the Co^{4+}/Co^{3+} couple is sufficiently close to the O $2p$ valence band that electrons may be lost from valence-band orbitals with high oxygen character if more Li is removed on charging; i.e. O^{2-} oxidation occurs. Near the surface, peroxide formation can lead to the reaction $O_2^{2-} = O^{2-} + \frac{1}{2}O_2(g)$ and oxygen loss.

Batteries can be made with a voltage exceeding 3.6 V, an energy density of ~180 W h/kg (two to three times higher than achievable for Ni–Cd batteries, see the end-of-chapter problems), which are rechargeable tens of thousands of times. The design is usually called a **LION** (lithium ion) or **rocking-chair**[42] cell to distinguish it from those containing Li metal. The rechargeable Li battery has revolutionized modern society was described as "the most important advance in energy storage for 100 years" [25]. It led to the award of the 2019 Nobel Prize in Chemistry to Goodenough, Whittingham, and Yoshino.[43] In addition to $LiCoO_2$, a wide variety of substitution-related phases have been studied as potential cathodes, and materials such as $Li(Ni_{\sim0.33}Mn_{\sim0.33}Co_{\sim0.33})O_2$ (NMC) show good performance, as discussed in Box 13.2.

Box 13.2 Nanoscale Concepts: Cathode structure control from the micro- to nanoscale

A rechargeable battery places extreme demands on its constituent materials and on the interfaces between them. Electrode materials must undergo major yet rapid and reversible changes in their local structure and bonding, and must shuttle between highly reactive oxidation states. Going to the nanoscale offers the potential benefit of rapid reactions due to short diffusion lengths, but the high surface area of components can lead to significant side reactions. Properties therefore need to be optimized across the length scales from ångstroms (controlling atomic-level structure) to nanometers (morphology and composition of primary particles or domains) to microns (particles made up of several primary particles).

One challenge in improving the properties of Li_xMO_2 cathodes is to find cheaper materials that simultaneously improve capacity without compromising safety or lifetime. Compositions like $Li(Ni_{0.8}Mn_{0.1}Co_{0.1})O_2$ have higher capacity (up to 200 mA h/g versus 140 mA h/g for $LiCoO_2$) but have severe safety implications. On charging to low Li levels, they can undergo an exothermic transformation to spinel structures at around 200 °C ($3MO_2 \rightarrow M_3O_4 + O_2$). The heat and oxygen released can lead to the cell catching fire. In contrast, the higher Mn content in $Li(Ni_{0.5}Mn_{0.5})O_2$ stabilizes the layered structure such that decomposition doesn't occur until 300 °C, but leads to lower capacity since Mn^{4+} doesn't take part in the electrochemical reaction.

How might we get the best of both worlds? One possibility is to prepare **core-shell** cathode particles (see Figure B13.2.1) that contain a high-capacity Ni-rich core enclosed by a high-stability Mn-rich shell. Such particles can be synthesized by reacting aqueous transition-metal solutions with hydroxide to form particles of $(Ni_aMn_bCo_c)(OH)_2(s)$ of controlled and uniform size. By varying the concentration of metal ions in the feedstock over time it's possible, for example, to precipitate Ni-rich $(Ni_{0.8}Mn_{0.1}Co_{0.1})(OH)_2$ cores surrounded by $(Ni_{0.5}Mn_{0.5})(OH)_2$. On calcining these hydroxides with lithium carbonate, hydroxide, or nitrate, the spherical particle morphology and transition-metal distribution

---

[42] The widely used "rocking-chair" term seems to have been introduced to reflect the fact that the Li ions move reversibly to and fro between the cathode and anode.

[43] Stanley Whittingham for the Li/TiS_2 battery, John Goodenough for metal oxide cathodes, and Akiro Yoshino for introducing carbon-based anodes.

Box 13.2 (cont.)

is retained, leading to core-shell particles. The particles show good stability when used as cathodes, with just a small capacity reduction due to the shell layer.

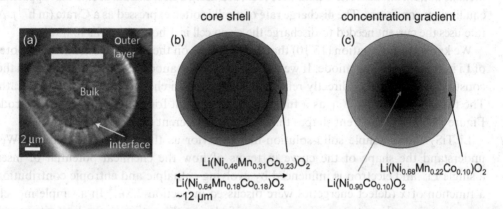

Figure B13.2.1 (a) Scanning electron microscope (SEM) image of a core-shell $LiMO_2$ cathode particle with compositions indicated in (b). (c) Concentration-gradient particle. SEM image from Sun et al., *Nat. Mater.*, **8** (2009), 320–324 with permission.

One problem with this design is that the core and shell show very different volume changes on delithiation (around 9–10% and 2–3%, respectively). The resulting stresses could ultimately lead to the protective shell cracking away from the core. One solution to this is to produce particles in which the shell has a composition that gradually becomes richer in the stabilizing manganese as one moves from the core to the surface. Figure B13.2.1a shows such a particle with overall composition $Li(Ni_{0.64}Mn_{0.18}Co_{0.18})O_2$ but with a Mn-rich and therefore stabilized surface of $Li(Ni_{0.46}Mn_{0.31}Co_{0.23})O_2$. These particles showed similar initial capacity to uniform particles, but with 95% capacity retention over 500 cycles in a cell compared to 70%.

An alternative approach is to produce full **concentration-gradient particles** in which the composition varies continuously from the core to the outer surface (Figure B13.2.1c). Particles of overall composition $Li[Ni_{0.75}Mn_{0.15}Co_{0.10})O_2$ could be prepared, in which Mn content varied from 0% at the core to 22% at the surface. By adjusting the OH^- concentration it was also possible to produce the primary particles as extended nanorods. Importantly, microspheres could be prepared with these nanorods aligned radially. This morphology maintains good diffusion pathways for Li^+ into the particle (10 times higher than a core-shell particle). These particles showed exceptional stability in working cells (e.g. 95% capacity retention over 1000 cycles at 25 °C). They also showed good thermal stability in their highly delithiated state, with an onset temperature for exothermic decomposition to spinel phases of 280 °C [26].

13.8.4 Electrochemical Characteristics of Lithium Batteries

Electrochemical measurements give considerable insight into the performance of battery electrodes and the details of the chemical changes that occur. Figure 13.28 shows voltage as a function of Li content for cells using two cathode materials, Li_xTiS_2 and Li_xFePO_4, against metallic lithium anodes. These data were recorded using a slow discharge rate that approximates equilibrium conditions. The **discharge rate** of a cell is often expressed as a C rate (in h^{-1}). A 1 C rate uses the current needed to discharge the given cell in 1 hour, 0.1 C in 10 hours.

We know from Equation (13.10) that E_{cell} depends on the difference in chemical potential of Li in the cathode and anode. If we use Li metal as the anode, its chemical potential there is constant, and the $-E_{cell}$ directly reflects the changing Li chemical potential at the cathode. The two plots showing E_{cell} as a function of Li content for the two different electrodes in Figure 13.28 have different shapes that reflect the different chemistries that occur.

Li_xTiS_2 shows simple solid-solution-like behavior as the Li content changes. We can understand the shape of the curve in terms of how the chemical potential of inserting a single Li^+ and electron is influenced by evolving enthalpic and entropic contributions as a function of x (defect energetics were discussed in Section 2.3).[44] In a simple model, the entropic term reflects the number of ways of placing the cation in the host sites available (configurational entropy) and is therefore particularly important near $x = 0$ and $x = 1$ where the increase in configurational entropy per Li is highest. The enthalpic term can be related to repulsive Li^+–Li^+ interactions that lead to an increase in $\mu_{Li}^{cathode}$ with increasing x, thus causing E_{cell} to decrease with x. The combination of these effects leads to the S-shaped cell voltage observed. The difference in voltage on charge and discharge is due to the internal

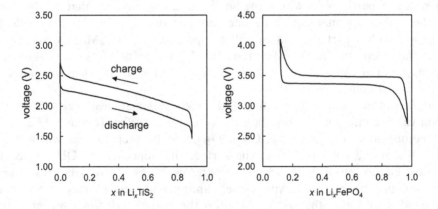

Figure 13.28 Voltage versus Li^+/Li for Li_xTiS_2 and Li_xFePO_4 at current density $10\,mA/cm^2$ and discharge rate 0.1 C after several charge/discharge cycles [27, 28].

---

[44] Since μ is the partial molar Gibbs energy at constant pressure it follows that $\mu = \left(\dfrac{\partial G}{\partial n}\right) = \left(\dfrac{\partial H}{\partial n}\right) - T\left(\dfrac{\partial S}{\partial n}\right)$.

resistance R of the cell to the ionic current I (which equals the external current). This reduces/increases the discharge/charge voltage by an Ohm's law $I \cdot R$ term.

In contrast, the electrochemical behavior of Li_xFePO_4 (Figure 13.28, right) is affected by the miscibility gap for $0.1 < x < 0.95$ that leads to a different E_{cell} dependence on x.[45] When the cell is discharged between $x = 0.1$ and $x = 0.95$, the chemical change that occurs is a direct conversion of $Li_{0.1}$[cathode] to $Li_{0.95}$[cathode], and there are two phases present. The cell voltage versus x therefore shows a plateau.

In practice, the intercalation chemistry and therefore electrochemical properties of cathodes can be much more complex due to changes in the Li-ordering patterns within layers as a function of x, or changing coordination requirements of the cathode transition metal on reduction. For example, Mn^{4+} in an MnO_2 cathode would become Jahn–Teller active on reduction to Mn^{3+} during discharge, changing local coordination geometry and influencing E_{cell}. Voltage steps might also occur when particularly favorable Li^+ sites in a material become fully occupied, or a particular host band becomes filled by electrons. A careful combination of electrochemical and structural studies (often *in situ*) is required to unravel this behavior.

13.8.5 Other Lithium Battery Electrode Materials

Despite its excellent properties, $LiCoO_2$ has a drawback of high cost.[46] In addition, the cell's internal resistance means that full charge and discharge rates cannot be quicker than ~1 hour in order to avoid overheating. In addition to the NMC materials mentioned above, there has also been considerable interest in materials with the olivine[47] structure such as $LiFePO_4$ [29]. The structure (Figure 13.29) contains edge- and corner-sharing FeO_6 octahedra and PO_4 tetrahedra and has chains of edge-sharing LiO_6 octahedra that allow a 1D Li diffusion pathway through

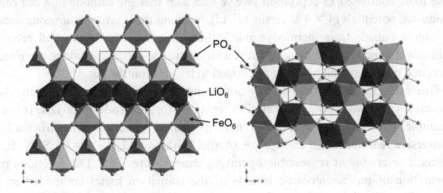

Figure 13.29 $LiFePO_4$ viewed down c and b axes.

[45] For example a material of nominal composition $Li_{0.525}$[cathode] would phase-separate to domains of $0.5Li_{0.1}$[cathode] and $0.5Li_{0.95}$[cathode].
[46] Co is relatively scarce, and projected usage in automotive batteries exceeds current mining production.
[47] Olivine is a mineral with composition $(Fe,Mg)_2SiO_4$.

the material. By coating individual crystallites with thin layers of carbon, it has been possible to overcome the low intrinsic electronic conductivity of $LiFePO_4$ ($\sim 10^{-9}$ S/cm) and produce cells with capacities of $\sim$160 A h/kg (close to the theoretical limit of 170 A h/kg), which can be charged and discharged 30000 times. As discussed in the previous section, a bulk Li_xFePO_4 cathode acts as a two-phase system with $x = 0.1$ and $x = 0.95$ at room temperature and a resultant flat E_{cell} versus x curve (Figure 13.28, right). A single-phase behavior, with the normal sloping E_{cell} versus x curve, only occurs above $\sim$450 °C or when working with metastable 40 nm particles. These small particles give good battery performance even without carbon coating. A range of related materials with Mn, Co, and Ni substitution, and mixed-anion materials such as $LiVPO_3F$ have been studied or already commercialized. These cathode materials are of particular interest for high-power applications.

In the discussion so far, we've focused on the cathode systems that have already been exploited commercially. There is, however, a huge research effort to improve the properties of all the components used in rechargeable batteries, how they can be prepared from cheaper materials, and how they can eventually be recycled. Alternative anodes to Li_xC_6 have been investigated, including Si and Sn or Sb-based alloys. These have higher capacities (the saturated stoichiometry is up to Li_4Si for pure Si) but undergo large volume changes (300% for Si) on charging, meaning that they must be prepared as sponges or nanoparticles surrounded by an Li^+ conducting medium. Oxides such as $Li_4Ti_5O_{12}$ have also been studied as potential anodes. They lead to lower cell voltages (Figure 13.26) but can be charged more rapidly without Li metal plating the anode, improving safety. Methods that could allow Li-metal anodes to be re-introduced are also important, as Li metal has a much higher gravimetric capacity than LiC_6 (3361 mA h/g versus 339 mA h/g). These could include using solid polymeric or ceramic electrolytes, changes to electrolyte solvents/additives, or more advanced thin-film cell designs to give more controlled Li deposition. We've also seen that the cathode in a cell can be highly oxidizing (potentials of > 4 V versus Li^+/Li), meaning that even non-aqueous electrolytes are operating outside their thermodynamic stability window of $\sim$3.5 V. Cell reversibility and ultimate lifetime is frequently associated with electrolyte decomposition at the electrode interface, and much research has been performed to try and control this.

One exciting idea is that it may not be necessary to rely on the relatively few host structures that will allow intercalation reactions for electrode materials. It is known, for example, that reaction kinetics of nanostructured materials are sufficiently fast that **conversion reactions** such as $M_aX_b + (b \cdot n)Li = aM + bLi_nX$ (X = O, S, N, P, F, H) can be used reversibly at reasonable operating temperatures [30]. The reactions proceed via formation of nanometer-scale islands of the transition metal embedded in Li_nX, and controlling the morphology of the reactants and products is crucial. There is also considerable interest in lithium–oxygen and lithium–sulfur batteries. The former operate on the reaction $2Li + O_2 = Li_2O_2$ and could offer capacities up to 1200 mA h/g, but there are huge materials challenges that need to be overcome before commercialization [31].

Box 13.3 Materials Spotlight: Supercapacitors

In addition to the batteries and fuel cells discussed in the main text, there is a third category of electrochemical devices used for energy storage: **electrochemical capacitors** or **supercapacitors**. The basic set-up of a supercapacitor is shown in Figure B13.3.1 and again contains two electrodes (usually called the anode and cathode even though redox reactions don't necessarily occur) in contact with a liquid electrolyte. At the electrode–electrolyte interface, ions accumulate, attracted by the charge on the electrode surface, forming an electrically charged double layer separated by 2–10 Å. This gives rise to a capacitance we can estimate from the expression $C = Q/V = \varepsilon_0 \varepsilon_r A/d$ as being of the order of 10 μF/cm^2, assuming ε_r of 10 for ions in water and a separation d = 9 Å. If an electrode such as a porous carbon is used, which can have a surface area up to 1000 m^2/g, one can achieve extremely high capacitances of ~100 F/g [32]. Since the device needs two electrodes, its overall capacitance is given by $C = 1/(1/C_1 + 1/C_2) = C_1/2$ for symmetric electrodes. The specific capacitance is then 25 F/g (given double the mass), much higher than conventional capacitors which are in the pF to μF range. The energy stored by such a supercapacitor is given by $1/2\,CV^2$. With organic electrolytes, which are stable to decomposition up to a few volts, energy densities of ~25 W h/kg are possible.

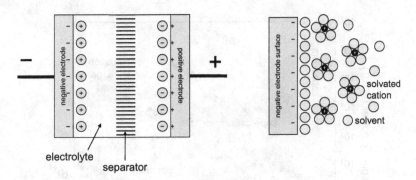

Figure 13.3.1 Schematic of a symmetric supercapacitor. Right-hand figure shows a close-up of the electrical double layer formed at the negative electrode. The overall capacitance is due to the charge from any specifically adsorbed ions and solvent and the charge in the outer layer.

One key difference between supercapacitors and batteries or fuel cells is the timescale of the processes responsible for energy storage. The double layer at the electrode surface of a supercapacitor forms rapidly (10^{-8} s) compared to the redox reactions in batteries and fuel cells (10^{-2}–10^{-4} s). This means that charging and discharging processes are extremely fast. There is also no structural change needed in these processes and there are fewer side reactions between electrodes and electrolytes. Supercapacitors can therefore be charged and discharged millions of times without capacity loss.

Significantly higher capacitance can be achieved by using electrochemically active electrodes in what are called **faradaic supercapacitors** or **pseudocapacitors**. In these devices, redox reactions take place at the electrode surface, with charge passing from the electrolyte to the electrode in a process similar to that in batteries [33]. Capacitances 10–100 times higher than for electrostatic double-layer electrodes are possible, though one pays the price of lower cycling stability and lower power as the (dis)charging reactions are slower. Hybrid supercapacitors use one electrode of each type to overcome some of these limitations. Electrodes using conducting polymers and metal oxides such as RuO_2, MnO_2, and Co_3O_4 have been extensively investigated. RuO_2 works via reactions such as $RuO_2 + xH^+ + xe^- = RuO_{2-x}(OH)_x$ ($0 < x < 2$) and has many attractive properties: a theoretical capacitance of up to 2000 F/g, reversible redox reactions, good electronic and proton conductivity, and high stability. It has the drawback of being very expensive.

The properties of the different energy devices we've considered are captured in the **Ragone plot** shown in Figure B13.3.2, which plots their specific power (in W/kg) and specific energy (W h/kg). Fuel cells are high-energy sources but have low power, supercapacitors have low energy but high power, and batteries lie somewhere in between. While none of these systems can approach the simultaneous high energy and power of a traditional combustion engine, hybrid systems offer this important option. For example, in an automotive application one could combine a high-energy fuel cell (for range) with batteries and supercapacitors to provide high power when needed for acceleration. The high charging rate of supercapacitors also means that they can be used for regenerative energy storage on braking. Supercapacitors are also widely used when a short-term energy supply is needed such as in power-supply backups.

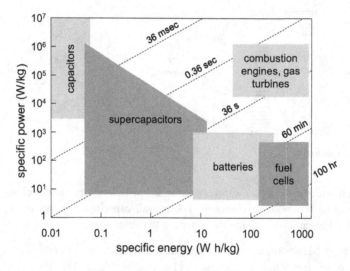

Figure B13.3.2 A Ragone plot comparing the power and energy density of different electrochemical storage devices. The sloped dashed lines indicate the relative times to store or extract energy.

13.9 Problems

13.1 Write out half equations, the cell reaction, and determine the standard potential of the following cells given standard electrode potentials $E°(Cu^{2+}/Cu) = +0.34$ V, $E°(Sn^{2+}/Sn) = -0.14$ V, $E°(Co^{3+}/Co^{2+}) = +1.81$ V, $E°(Sn^{4+}/Sn^{2+}) = +0.15$ V: (a) $Sn(s)|SnSO_4(aq)\|CuSO_4(aq)|Cu(s)$, (b) $Pt|Co^{3+}(aq),Co^{2+}(aq)\|Sn^{4+}(aq),Sn^{2+}(aq)|Pt$.

13.2 In its charged state, a lead–acid battery contains PbO_2 and Pb electrodes. The half equations that occur during discharge can be written as below. Give the overall cell equation and estimate $E°_{cell}$. State which electrode is the cathode and which the anode.

$$PbO_2(s) + 3H_3O^+(aq) + HSO_4^-(aq) + 2e^- \leftrightarrow PbSO_4(s) + 5H_2O(l); \quad E° = 1.685 \text{ V}$$
$$PbSO_4(s) + H_3O^+(aq) + 2e^- \leftrightarrow Pb(s) + HSO_4^-(aq) + H_2O(l); \quad E° = -0.356 \text{ V}$$

13.3 The non-rechargeable alkaline battery contains a Zn anode, MnO_2 cathode, and KOH electrolyte, with MnO_2 present slightly in excess of a 1:2 molar ratio for safety reasons. Propose a balanced equation for the cell reaction and give individual half equations.

13.4 Nickel–cadmium batteries rely on the reaction between NiO(OH) and Cd metal with an alkaline electrolyte, with both metals forming hydroxides. Propose a balanced equation for the cell reaction and give individual half equations.

13.5 Use the Nernst equation to estimate the potential of an electrochemical cell $Zn(s)|Zn^{2+}(c_1, aq)\|Zn^{2+}(c_2, aq)|Zn(s)$ in which Zn concentrations c_1 and c_2 are 0.3 M and 0.8 M, respectively.

13.6 Write down the half equations that occur at each electrode in the oxygen sensor of Figure 13.17 and state the overall cell equation. Oxygen sensors of this type are used in car engines to monitor the air to fuel ratio via changes in the partial pressure of O_2 in the exhaust gas relative to air. Calculate the voltage expected from the sensor for the following air to fuel (A:F) ratios: (a) a lean mixture of A:F = 10.5 and $p(O_2) = 1\times10^{-14}$ Pa, (b) a stoichiometric ratio of A:F = 14.7 and $p(O_2) = 5\times10^{-11}$ Pa, (c) a rich mixture of A:F = 18 and $p(O_2) = 4500$ Pa. Assume the partial pressure of O_2 in air is 21.2 kPa and the exhaust gases are at 1000 K.

13.7 The table below contains conductivity data for a crystal of NaCl. Comment on the shape of a plot of $\ln(\sigma T)$ versus $1/T$. Estimate the activation energy for vacancy migration and Schottky defect formation from your graph.

T (K)	550	600	650	700	750
σ (S/cm^{-1})	1.60×10^{-8}	5.06×10^{-8}	1.37×10^{-7}	3.51×10^{-7}	9.92×10^{-7}
T (K)	800	850	900	950	1000
σ (S/cm^{-1})	3.36×10^{-6}	1.25×10^{-5}	4.47×10^{-5}	1.47×10^{-4}	4.34×10^{-4}

13.8 AgI is often described as a fast ion conductor. Given that NaCl has around 10^{-4} carriers per formula unit immediately below its melting point and that conductivities are ≈ 1 S/cm (AgI) and $\approx 10^{-5}$ S/cm (NaCl), comment on the term *fast* ion conductor. Assume cell parameters of 5.84 Å for NaCl and 5.06 Å for AgI.

13.9 The table below contains conductivity data for two materials. The melting point of PbF_2 is 1103 K. (a) Describe the structure of each material and the origin of the relatively high conductivities. (b) Comment on the temperature dependence of conductivities. (c) For $Zr_{1-x}Y_xO_{2-x/2}$, estimate the activation energy for charge-carrier migration. (d) State two possible uses of $Zr_{1-x}Y_xO_{2-x/2}$.

T (K)	PbF_2 σ (S/cm)	$Zr_{1-x}Y_xO_{2-x/2}$ σ (S/cm)
670	0.1	0.0001
840	4.6	0.002
1250	4.6	0.1

13.10 At high temperature, $La_2Mo_2O_9$ has a structure closely related to cubic β-$SnWO_4$, which contains WO_4 tetrahedra and Sn in a distorted six-coordinate environment. On cooling through 580 °C, a $2a \times 3b \times 4c$ monoclinic superstructure is formed. Correlate these observations with the conductivity data in Figure 13.18, and suggest a mechanism for conductivity above 580 °C.

13.11 The table below lists the in-plane conductivity for graphite and $K_{0.125}C$ at two different temperatures. Comment on these values.

Substance	σ(90 K) (S/cm)	σ(300 K) (S/cm)
Graphite	1.3×10^4	2.5×10^4
$K_{0.125}C$	1.1×10^6	1.1×10^5

13.12 The table below lists d spacings of the first four strong diffraction peaks observed in the diffraction patterns of graphite and two Rb intercalation compounds. Note that the graphite interlayer spacing is 3.35 Å and that the preferred orientation of crystallites means that $(00l)$ reflections are typically strongest. Comment on these data.

Substance	d_{hkl} (Å)	d_{hkl} (Å)	d_{hkl} (Å)	d_{hkl} (Å)
Graphite	3.35	1.68	1.12	0.84
$Rb_{0.03}C$	12.34	6.17	4.11	3.08
$Rb_{0.125}C$	5.65	2.83	1.41	0.94

13.13 The structure of the stage-1 K_xC intercalate has been reported in space group *Fddd* with a pseudohexagonal unit cell of $a = 4.92$ Å, $b = 8.51$ Å, $c = 21.4$ Å with K at 0 0 0, C1 at $\frac{1}{4}\ \frac{1}{12}\ \frac{1}{8}$, C2 at $\frac{1}{4}\ \frac{1}{12}\ \frac{5}{8}$. With the assistance of a structure-drawing program: (a) Sketch the structure of a single layer and annotate a simple 2D unit cell. (b) Comment on the stacking sequence of layers relative to graphite. (c) State the value of x in K_xC and comment on the value given that bcc K has $a = 5.21$ Å. (d) Comment on the fact that Rb (Rb metal bcc $a = 5.63$ Å) and Cs ($a = 6.06$ Å) intercalates have similar structures.

13.14 The Li graphite intercalate has a limiting composition LiC_6. Sketch a possible unit cell in 2D and estimate the Li–Li distance assuming a C–C bond length of 1.42 Å.

13.15 Calculate the theoretical capacity in A h/kg for the electrode materials: (a) Li metal anode, (b) LiC_6 anode, (c) $LiTiS_2$ cathode, (d) $LiCoO_2$ cathode.

13.16 For LiC_6 calculate the theoretical capacity in A h/kg of carbon.

13.17 Assuming densities for $LiFePO_4$, carbon black, and Teflon® to be 3.6, 1.8, and 2.2 g/cm^3 respectively, calculate the reduction in volumetric density of a cathode comprising 10 wt% carbon and 5 wt% Teflon.

13.18 Calculate the energy density in W h/kg for the Li/TiS_2 cell described in Section 13.8.2. Assume an operating voltage of 2 V and that the electrolyte and packaging make up 50% of the mass.

13.19 Calculate the energy density in W h/kg for a battery based on a Li_xC_6 anode and a $Li_{1-x}CoO_2$ cathode, operating at an average voltage of 3.6 V. Assume that the electrolyte and casing make up 50% of the battery mass and that the cathode can only be charged to $Li_{0.5}CoO_2$. Repeat the calculation for a pure Li anode assuming the cell voltage is 0.6 V higher. Comment on these values.

13.20 The figure below shows E against x at 30 °C for a cell with a spinel-based $Li_xMn_2O_4$ cathode ($0 < x < 1.8$) and a Li anode. (a) Describe the structure you might expect for a $LiMn_2O_4$ spinel. (b) State what can you conclude from E_{cell} versus x in the regions $0 < x < 0.5$, $0.5 < x < 1$, and $1 < x < 1.8$. (c) State what could cause the marked drop in E_{cell} at $x = 1$.

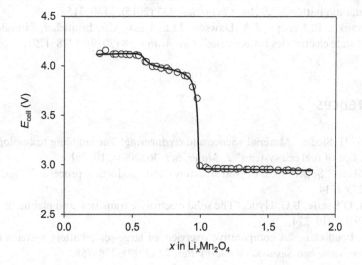

13.21 $Li_1Mn_2O_4$ (a = 8.2495 Å) is cubic, whereas $Li_2Mn_2O_4$ is tetragonal (a = 5.653 and c = 9.329 Å). Suggest a reason for lower symmetry of $Li_2Mn_2O_4$. Calculate the volume change upon its formation from cubic $Li_1Mn_2O_4$ and state how that might impact the long-term performance of a cathode cycling between these two compositions.

13.10 Further Reading

M. Winter, R.J. Brodd, "What are batteries, fuel cells and supercapacitors" *Chem. Rev.* **104** (2004), 4245–4269. The October 2004 issue of *Chemical Reviews* contains a number of articles on energy-related materials.

S. Hull, "Superionics: Crystal structures and conduction processes" *Rep. Prog. Phys.* **67** (2004), 1233–1314.

L. Malavasi, C.A.J. Fisher, M.S. Islam, "Oxide-ion and proton conducting electrolyte materials for clean energy applications: Structural and mechanistic features" *Chem. Soc. Rev.* **39** (2010), 4370–4387.

P.G. Bruce (editor), "*Solid State Electrochemistry*" (1995) Cambridge University Press.

J. Maier, "Thermodynamics of electrochemical lithium storage" *Angew. Chem. Int. Ed. Engl.* **52** (2013), 4998–5026.

M. Armand, J.-M. Tarascon, "Building better batteries" *Nature* **451** (2008), 652–657.

M.S. Whittingham, "Lithium batteries and cathode materials" *Chem. Rev.* **104** (2004), 4271–4301.

J.B. Goodenough, Y. Kim, "Challenges for rechargeable Li batteries" *Chem. Mater.* **22** (2010), 587–603.

J.B. Goodenough, "Evolution of strategies for modern rechargeable batteries" *Acc. Chem. Res.* **46** (2013), 1053–1061.

L. Croguennec, M. Rosa Palacin, "Recent achievements on inorganic electrode materials for lithium-ion batteries" *J. Am. Chem. Soc.* **137** (2015), 3140–3156.

T. Famprikis, P. Canepa, J.A. Dawson, M.S. Islam, C. Masquelier, "Fundamentals of inorganic solid-state electrolytes for batteries" *Nat. Mater.* **18** (2019), 1278–1291.

13.11 References

[1] B.C.H. Steele, "Material science and engineering: The enabling technology for the commercial-isation of fuel cell systems" *J. Mater. Sci.* **36** (2001), 1053–1068.

[2] S. Hull, "Superionics: Crystal structures and conduction processes" *Rep. Prog. Phys.* **67** (2004), 1233–1314.

[3] M. O'Keeffe, B.G. Hyde, "The solid electrolyte transition and melting in salts" *Philos. Mag.* **33** (1976), 219–224.

[4] A. Poullikkas, "A comparative overview of large-scale battery systems for electricity storage" *Renewable and Sustainable Energy Rev.* **27** (2013), 778–788.

[5] Z. Gadjourova, Y.G. Andreev, D.P. Tunstall, P.G. Bruce, "Ionic conductivity in crystalline polymer electrolytes" *Nature* **412** (2001), 520–523.

[6] K.D. Kreuer, "Proton conductivity: Materials and applications" *Chem. Mater.* **8** (1996), 610–641

[7] K.D. Kreuer, "Proton conducting oxides" *Annu. Rev. Mater. Res.* **33** (2003), 333–359.

[8] S.M. Haile, D.A. Boysen, C.R.I. Chisholm, R.B. Merle, "Solid acids as fuel cell electrolytes" *Nature* **410** (2001), 910–913.

[9] http://commons.wikimedia.org/wiki/File%3ATubular_sofc_de.png.

[10] V.V. Kharton, F.M.B. Marques, A. Atkinson, "Transport properties of solid oxide electrolyte ceramics: A brief review" *Solid State Ionics* **174** (2004), 135–149.

[11] R.D. Shannon, "Revised effective ionic radii and systematic studies of interatomic distances in halides and chalcogenides" *Acta Crystallogr. Sect. A* **232** (1976), 751–767.

[12] M. Mogensen, N.M. Sammes, G.A. Tompsett, "Physical, chemical and electrochemical properties of pure and doped ceria" *Solid State Ionics* **129** (2000), 63–94.

[13] X. Kuang, J.L. Payne, M.R. Johnson, I. Radosavljevic Evans, "Remarkably high oxide ion conductivity at low temperature in an ordered fluorite-type superstructure" *Angew. Chem. Int. Ed. Engl.* **51** (2012), 690–694.

[14] S. Tao, J.T.S. Irvine, "A redox-stable efficient anode for solid-oxide fuel cells" *Nat. Mater.* **2** (2003), 320–323.

[15] J.C. Ruiz-Morales, J. Canales-Vázquez, C. Savaniu, D. Marrero-López, W. Zhou, J.T.S. Irvine, "Disruption of extended defects in solid oxide fuel cell anodes for methane oxidation" *Nature* **439** (2006), 568–571.

[16] S. Sengodan, S. Choi, A. Jun, T.H. Shin, Y.-W. Ju, H.Y. Jeong, J. Shin, J.T.S. Irvine, G. Kim, "Layered oxygen-deficient double perovskite as an efficient and stable anode for direct hydrocarbon solid oxide fuel cells" *Nat. Mater.* **14** (2015), 205–209.

[17] S. Yoo, A. Jun, Y.-W. Ju, D. Odkhuu, J. Hyodo, H.Y. Jeong, N. Park, J. Shin, T. Ishihara, G. Kim, "Development of double-perovskite compounds as cathode materials for low-temperature solid oxide fuel cells" *Angew. Chem. Int. Ed. Engl.* **53** (2014), 13064–13067.

[18] M.S. Dresselhaus, G. Dresselhaus, "Intercalation compounds of graphite" *Adv. Phys.* **51** (2002), 1–186.

[19] N. Daumas, A. Herold, "Relations between phase concept and reaction mechanics in graphite insertion compounds" *C. R. Seances Acad. Sci., Ser. C* **268** (1969), 373.

[20] D.W. Murphy, C. Cros, F.J. DiSalvo, J.V. Waszczak, "Preparation and properties of Li_xVS_2 ($0 \leq x \leq 1$)" *Inorg. Chem.* **16** (1977), 3027–31

[21] J.C. Hunter, "Preparation of a new crystal form of manganese dioxide: λ-MnO_2" *J. Solid State Chem.* **39** (1981), 142–147.

[22] A.R. Armstrong, P.G. Bruce, "Synthesis of layered $LiMnO_2$ as an electrode for rechargeable lithium batteries" *Nature* **381** (1996), 499.

[23] F.C. Capitaine, P. Gravereau, C. Delmas, "A new variety of $LiMnO_2$ with a layered structure" *Solid State Ionics* **89** (1996), 197–202.

[24] J.B. Goodenough, Y. Kim, "Challenges for rechargeable Li batteries" *Chem. Mater.* **22** (2010), 587–603.

[25] J.M. Tarascon, S. Grugeon, M. Morcette, S. Laruelle, P. Rozier, P. Poizot, "New concepts for the search of better electrode materials for rechargeable lithium batteries" *C. R. Chim.* **8** (2005), 9–15.

[26] S-T. Myung, H-J. Noh, S-J. Yoon, E-J. Lee, Y-K. Sun, "Progress in high-capacity core–shell cathode materials for rechargeable lithium batteries" *J. Phys. Chem. Lett.* **5** (2014), 671–679.

[27] M.S. Whittingham, "Chemistry of intercalation compounds: Metal guests in chalcogenide hosts" *Prog. Solid State Chem.* **12** (1978), 41–99.

[28] J.M. Tarascon, M. Armand, "Issues and challenges facing rechargeable lithium batteries" *Nature* **414** (2001), 359–367.

[29] A.K. Padhi, K.S. Nanjundaswamy, J.B. Goodenough, "Phospho-olivines as positive electrode materials for lithium batteries" *J. Electrochem. Soc.* **144** (1997), 1188–1194.

[30] J. Cabana, L. Monconduit, D. Larcher, M. Rosa Palacín, "Beyond intercalation-based Li-ion batteries: The state of the art and challenges of electrode materials reacting through conversion reactions" *Adv. Mater.* **22** (2010), E170–E192.

[31] P.G. Bruce, S.A. Freunberger, L.J. Hardwick, J.M. Tarascon, "Li–O_2 and Li–S batteries with high energy storage" *Nat. Mater.* **11** (2012), 19–29.

[32] X. Zhao, B. Mendoza Sánchez, P.J. Dobson, P.S. Grant, "The role of nanomaterials in redox-based supercapacitors for next generation energy storage devices" *Nanoscale* **3** (2011), 839–855.

[33] G. Wang, L. Zhang, J. Zhang, "A review of electrode materials for electrochemical supercapacitors" *Chem. Soc. Rev.* **41** (2012), 797–828.

14 Zeolites and Other Porous Materials

In this chapter we will discuss the chemistry of porous materials, that is materials whose 3D structure contains pores or channels accessible by chemical species. IUPAC has defined three classes of porous materials—**microporous** (<2 nm), **mesoporous** (2–50 nm), and **macroporous** (>50 nm). We'll see that these pore sizes allow uptake of atoms and molecules leading to important technological applications. For example, crude-oil cracking using microporous zeolites underpins a multibillion US dollar industry. Materials with significantly larger pore sizes, such as the mesoporous MCM family of aluminosilicates and similar materials based on metal oxides and chalcogenides, have additional exciting applications. We will also discuss metal–organic frameworks (MOFs), where one can design materials with extremely high porosity and correspondingly large internal surface areas.

14.1 Zeolites

Naturally occurring zeolites were first identified in 1756 by the Swedish mineralogist Axel Cronstedt, who noted that certain minerals appeared to bubble and froth when heated. He named these minerals after the Greek words zeo (boil) and lithos (stone). We will see that this behavior is directly related to the structures and chemistry of these materials. Zeolites can be described as framework aluminosilicates constructed from corner-sharing $SiO_{4/2}$ and $AlO_{4/2}$ tetrahedra, which contain cations, anions, water, and/or other guest molecules in the framework pores. As such, we can represent the general formula of a zeolite as $G^{n+}_{x/n}[(AlO_2)_x(SiO_2)_{1-x}]\cdot mH_2O$.[1] Materials with similar structures but containing other elements on the tetrahedral sites are called **zeotypes**.

---

[1] We show cations only in this formula; several zeolites contain cations and anions in their pores.

579

We can understand many of the structural features of aluminosilicates by starting with a simple SiO_2 structure such as cubic β-cristobalite shown in Figure 1.39.[2] If we replace half the Si^{4+} sites in each unit cell with Al^{3+}, we will need some form of charge compensation. One possibility is to introduce Na^+ ions into 12-coordinate sites between the corner-linked tetrahedra to obtain $NaAlSiO_4$, the mineral carnegieite. A similar derivation can be performed for the different crystalline polymorphs of SiO_2, giving the so-called stuffed silica structures.

Zeolite structures are conceptually similar, and the large number of ways in which tetrahedra can be corner-linked gives a rich structural diversity. Each zeolite framework is built from different rings (see Section 1.4.3) forming cages that surround various "pores" and "channels" containing the charge-balancing ions. The important chemistry of zeolites is derived from their ability to exchange non-framework ions, absorb molecules into their pores, and catalyze chemical reactions, and is a direct consequence of their structure. The 3D porosity of zeolites gives them extremely high effective surface areas of up to 900 m^2/g.[3]

From the discussion above, the distinction between zeolites and other aluminosilicates may seem slightly arbitrary. In practice, it is based on the relative amounts of "pore" and "framework" present and quantified by the **framework density** (FD) or number of tetrahedron-center atoms (T) per nm^3. Corner-linked tetrahedral aluminosilicates typically have framework densities $\gtrsim 21$ T/nm^3, while zeolites have low framework densities ranging from ~10 to ~20 T/nm^3 (see Table 14.1).

Table 14.1 Characteristics of selected zeolites and related phosphates.

Name	Date[a]	Framework[b]	Ring	Channels (Å)	T/nm^3	vol%[c]	Code[d]	SBU[e]
Sodalite	1930	$Al_6Si_6O_{24}$	6	–	17.2	0	SOD	6–2
Zeolite A	1956	$Al_{12}Si_{12}O_{48}$	8	4.1 × 4.1	12.9	21.4	LTA	8
Faujasite	1958	$Al_{58}Si_{134}O_{384}$	12	7.4 × 7.4	12.7	27.4	FAU	6–6
ZSM-5	1978	$Al_nSi_{96-n}O_{192}$ $n<27$	10	5.5 × 5.1	17.9	9.8	MFI	5–1
Mordenite	1961	$Al_8Si_{40}O_{96}$	12	7.0 × 6.5	17.2	12.3	MOR	5–1
UTD-1	1996	$Si_{64}O_{128}$	14	8.1 × 8.2	17.2	15.6	DON	5–3
ITQ-33	2006	$Si_{20}Ge_{26}O_{92}$	18	12.3 × 12.3	12.3	25.1	ITT	6
VPI-5	1988	$Al_{18}P_{18}O_{72}$	18	12.7 × 12.7	14.2	25.4	VFI	6
Cloverite	1991	$Ga_{96}P_{96}O_{372}F_{24}(OH)_{24}$	20	4 × 13.2	11.1	33.8	-CLO	4–4

[a]Date of first reference in atlas of zeolite types. [b]Framework composition of species first used to establish the framework type. [c]Volume percent accessible by a 2.8 Å diameter spherical probe molecule (see text). [d]A three-letter code used to identify the framework (see Section 14.1.1); a hyphen precedes for frameworks where the T sites are not all 4-connected. [e]Secondary building units (SBUs, shown in Figure 14.4).

2 $Fd\bar{3}m$ (227, origin 1) cubic, $a = 7.133$ Å at 300 °C, Si at 0 0 0 (8a), O at ⅛ ⅛ ⅛ (16c) in the ideal structure.

3 The "soccer pitch" is a common non-SI unit of area. The most common championship pitch size is 7140 m^2.

14.1.1 Representative Structures of Zeolites

All zeolites have a 3D framework of corner-sharing tetrahedra. One common unit found in several important zeolites is the **sodalite** or **β cage**, an approximately spherical unit made up of 24 tetrahedra (Figure 1.57). There are various ways in which this unit can be represented, as shown in Figure 14.1. Figure 14.1a shows a ball-and-stick view, Figure 14.1b a polyhedral view, and Figure 14.1c a view in which oxygen atoms have been omitted and tetrahedral centers are linked by a single line. This third view simplifies the sodalite cage into a truncated cuboctahedron (24 vertices) and is frequently used when drawing structures of zeolites. It also emphasizes the two types of rings of tetrahedra present: 6T-rings[4] (of 6 Al/Si central atoms) and 4T-rings. Figure 14.1d gives an idea of the size of the ring openings. Taking a 1.35 Å Shannon [1] ionic radius for O^{2-} in ideal cage geometry, the O–O separations across 6T- and 4T-rings are around 2.4 Å and 1 Å respectively.

If we share each of the 4T-rings of a sodalite cage with another cage, we create the structure shown in Figure 14.2a, which we can think of as a sodalite cage occupying each corner of a cube. Interestingly, the truncated cuboctahedron of the sodalite cage tessellates in three dimensions such that the cavity we might expect to form at the center of our cube is itself a sodalite cage. This is the structure of the mineral sodalite,[5] which has a typical formula $Na_8Cl_2[(AlO_2)_6(SiO_2)_6] \cdot 8H_2O$[6] and a framework density of 17.2 T/nm³.

The space available inside different zeolite frameworks can be estimated by calculating the volume percent (vol%) of the unit cell that can be occupied by a spherical probe molecule of water (assumed to be a 2.8 Å diameter sphere). This is called the **occupiable volume**. However, there is a snag. For the sodalite framework, 10.3% of the volume can be occupied but cannot

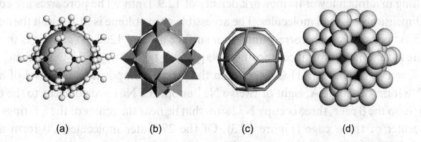

(a) (b) (c) (d)

Figure 14.1 Views of a sodalite cage showing (a) atoms involved, (b) $TO_{4/2}$ tetrahedra, (c) T centers linked by a line, and (d) with oxygens shown using a radius of 1.35 Å. The central sphere emphasizes the approximately spherical shape of the cage. In (a)–(c) it is drawn with a radius of 4.2 Å; in (d) a 3.1 Å radius sphere just contacts the O atoms.

---

[4] In the Liebau nomenclature of ordinary silicates (Appendix C), the 6-ring of tetrahedra would be termed a "sechser ring". This 6-ring is part of the structure's 6–2 secondary building units (defined in Figure 14.4).

[5] We note that sodalite is mineralogically classified as a feldspathoid (similar to feldspar), yet its structural features are similar to the other zeolites we discuss.

[6] The mineral sodalite contains Cl^- at the center of each cage tetrahedrally coordinated by Na^+.

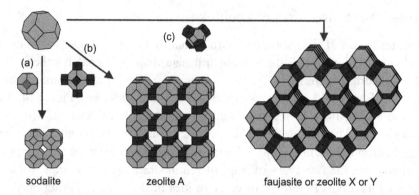

Figure 14.2 Schematic representation of how the structures of (a) sodalite, (b) zeolite A, and (c) faujasite can be derived by sharing the light gray sodalite cage via the dark gray linkers.

be reached once the sodalite has been synthesized. This volume lies inside the sodalite cages, but the size of the ring opening makes it inaccessible to a molecule. The **accessible volume** is therefore 0% and the corresponding **accessible surface area** 0%.

Instead of directly sharing 4T-rings between sodalite cages, the cages can be linked by a T–O–T bridge, as shown in Figure 14.2b. Remember that each line in this figure represents a T–O–T linkage (two tetrahedra sharing a vertex oxygen). The introduction of this extra linker unit (a 4–4 or double-4-ring, see Figure 14.4) leads to a far more open structure and creates a large 48-sided cavity (called an **α cage**) between the sodalite cages. The structure also now contains large 8T-ring windows, with a pore size of about 4.1 Å × 4.1 Å, which link the α cages, resulting in a much lower framework density of 12.9 T/nm^3. The pore sizes are comparable to the dimensions of small molecules. The accessible pore volume is 21.4% and the accessible area 281.5 Å^2 per cell, which corresponds to a surface area of 1205 m^2/g. This is the structure of zeolite A that has a typical composition $Na_{12}[(AlO_2)_{12}(SiO_2)_{12}]\cdot27H_2O$. The sites occupied by non-framework Na^+ and H_2O depend on the precise composition and level of hydration. In the Na form of zeolite A, eight of twelve Na^+ adopt the Na1 site adjacent to the 6T-rings that open onto the β cage, three occupy Na2 sites that lie near the center of the 8T-rings and one is in the center of the α cage (Figure 14.3). Of the 27 water molecules, 20 form a pentagonal dodecahedron within the α cage similar to that found in water clathrates. Approximately four water molecules lie in the β cage, and the remaining approximately three hydrate the Na^+ cation in the center of the α cage.

Instead of linking 4-rings via 4–4 units one can link 6T-rings via 6–6 units in a similar way. This leads to the faujasite structure (Figure 14.2c), in which the sodalite cages again adopt a cubic arrangement, but with positions analogous to those of carbon in diamond. The structure contains 12T-rings leading to larger pore sizes of 7.4 Å × 7.4 Å, the formation of a large supercage with an internal diameter ~14 Å, an overall framework density of 12.7 T/nm^3, accessible pore volume of 27.4%, and surface area of 1211 m^2/g. Faujasite itself is a naturally

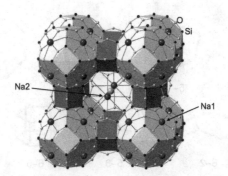

Figure 14.3 Approximate Na sites in zeolite A. Na1 lies at the center of the 6-ring, Na2 at the center of the 8-ring, and Na3 (colored white) at the center of the α cage.

occurring mineral $(Ca,Mg,Na_2)_{29}[(AlO_2)_{58}(SiO_2)_{134}]\cdot 240H_2O$. Synthetic analogues include zeolite X $Na_{86}[(AlO_2)_{86}(SiO_2)_{106}]\cdot 220H_2O$ and zeolite Y; X and Y differ only in their Si:Al ratio, X has a ratio around 1.0–1.4 and Y typically above ~2.5. The formula is written in this apparently complex way to reflect the fact that, due to the high space-group symmetry ($Fd\bar{3}m$), each T atom on a general position (192i Wyckoff site) generates 192 equivalent atoms in the unit cell. The faujasite structure again contains a number of potential cation sites, whose occupation depends on the degree of hydration and the precise cations present.

In addition to the sodalite-cage zeolites, many other topologies are possible. The unique framework types known are documented in the *Atlas of Zeolite Framework Types* [2]. The 2018 online atlas (www.iza-structure.org/databases/) listed 230 ordered frameworks, with each given a unique three-letter code identifier. It also contains coordinates, drawings, and interactive 3D views of all the framework types known.

There are several useful ways of categorizing different zeolite structures. One of the early systems used the **secondary building units** (**SBUs**) they contain [2]. These are units containing several tetrahedra (**basic building units**, **BBUs**) from which the entire framework can be built. The SBUs important for the zeolites in this chapter are shown in Figure 14.4 and included in Table 14.1. As more framework types were discovered, the zeolite community switched to discussing structures in terms of the common **composite building units** (**CBUs**) present [3] or the more mathematical language of natural tilings [4, 5]; information on both is listed for each framework in the online zeolite atlas [2]. To help summarize a given framework, the atlas lists two other terms—the **coordination sequence** and the **vertex symbol**. The combination of these is thought to be unique to a given framework and can be used to decide if a newly prepared material has a novel framework. The coordination sequence is evaluated for each crystallo-graphically unique[7] T (tetrahedron taken as a unit, or as if represented by just its central atom) in the structure, and it is obtained by counting the number of neighboring T that each such unique T is bonded to in increasing shells. For example, each T will link to $N_1 = 4$ neighboring T in the first shell, each of which will link to a maximum of 3 T in the second shell ($N_2 \le 4\times 3 =$

---

[7] Atoms generated from a given site by space-group symmetry are crystallographically equivalent. Only crystallo-graphically unique sites need to be specified when describing a crystal structure or considered when generating the coordination sequence.

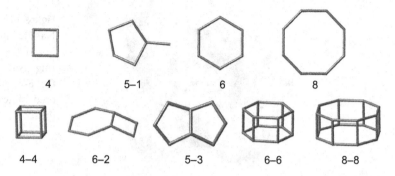

Figure 14.4 Secondary building units found in some of the zeolite structures discussed.

12), etc. Coordination sequences are listed up to N_{10}. The vertex symbol was introduced in Section 1.4.3[8] and gives the size of the smallest ring associated with each of the six tetrahedral angles of a T site, with rings from the three pairs of opposite angles placed adjacent to each other in the list. If more than one ring of a given size is found, the number of such rings is added as a subscript (e.g. 6_2 for two 6-rings at an angle). For zeolite A (of the three-letter code LTA for "Linde Type A"), the coordination sequence (CS) and vertex symbol (VS) of the single T are: CS 4 9 17 28 42 60 81 105 132 162 and VS 4·6·4·6·4·8, and we can see that the largest ring size is 8, consistent with Figure 14.3.

Computer algorithms have been developed to investigate possible zeolite structure types, and many thousands of hypothetical framework structures have been predicted [6, 7]. Some of these have subsequently been found experimentally. One such example, whose synthesis adopted many of the lessons learned from years of zeolite research, is the large-pore ITQ-33 [8].[9] The guiding principle used in its preparation was the observation that large-pore structures generally contain a significant proportion of 3- and 4-rings, which are favored by including Ge in the synthesis [9].[10] ITQ-33 was isolated and identified using automated high-throughput synthetic methods that allowed exploration of unusual synthetic conditions. It contains 18T cavities (~12.3 Å across) interconnected by perpendicular 10T pores (~6.1 Å × ~4.3 Å). These large pores lead to an unusually low framework density of 12.3 T atoms per 1000 A^3, and a density of only 0.37 g/cm^3. The material is more thermodynamically stable than other large-pore materials, and the presence of Al in the framework leads to catalytic properties (Section 14.1.3).

---

[8] We also discuss rings and coordination sequences in the context of amorphous materials in Chapter 15.

[9] Formula: $(hexamethonium)_{0.07}F_{0.07}(H_2O)_{0.37}[Si_{0.66}Al_{0.04}Ge_{0.30}O_{2.02}]$.

[10] The same synthetic tactic has also led to ITQ-40 that has the lowest framework density of any zeolite (10.1 T atoms per nm^3) and a framework composition of $Ge_{32}Si_{43.6}O_{150}(OH)_4$. The presence of OH groups means that not all T sites are 4-connected. Calculations show that Ge is stabilized in structures with small T–O–T angles [9].

Box 14.1 Synthetic Methods: Zeolites and hydrothermal synthesis

Modern synthetic routes to zeolites stem from the works of Barrer, Milton, and Breck in the 1940s and 1950s. For example, Barrer investigated the synthesis of mordenite by taking a "batch composition" of $Na_2O:Al_2O_3:8.2-12.3SiO_2$ and heating to ~290 °C in the presence of water [10]. Milton and Breck, working at Union Carbide, realized that adding hydroxide anions to reaction mixtures would greatly increase the solubility of Al via formation of $[Al(OH)_4]^-$, and were consequently able to lower reaction temperatures to ~100 °C. They successfully prepared many previously known natural zeolites as well as zeolite A, a material with no natural counterpart. The addition of organic cations as templates (see main text) to these hydrothermal preparations later led to the production of many other new zeolites.

In a typical modern zeolite **hydrothermal synthesis**, a silica source and an alumina source are combined with water, template molecules, and guest cations at high pH to produce an inhomogeneous amorphous **gel** (see Section 14.3 for definitions). Typical alumina sources include $NaAlO_2$, $Al(OH)_3$, and $Al_2(SO_4)_3$; typical silica sources are soluble silicates such as $Na_2SiO_3 \cdot nH_2O$, aqueous silica sols (typically 30% by weight SiO_2), and other forms of amorphous silica. The addition of F^- anions as mineralizers allows syntheses to proceed at lower pH, helps incorporation of heteroatoms, and can promote the growth of large single crystals. The gel is aged and then heated until the amorphous material transforms to a crystalline phase. This is often performed above 100 °C, under the autogenous pressure generated in a Teflon-lined steel autoclave (colloquially known as a "bomb"). Common lab-scale autoclaves can be used up to 250°C and 1800 psi (~12.5 MPa), and are designed to be "fail safe" in the event of unexpected pressure build-up.

The transformation from gel to final crystalline zeolite often occurs over an extended time period, during which different products may evolve. Frequently, Ostwald's rule of successive transformations applies, with initial tiny metastable particles evolving into more stable ones. High-density zeolites such as analcime (FD = 18.6 T/nm^3) and sodalite (FD = 17.2 T/nm^3) are therefore typically observed later in reactions.

There has been considerable debate over the precise roles the initial gel, species in solution, free cations, and template molecules play in the synthesis. Does the amorphous gel transform directly in the solid state to form the zeolite? Does it merely act as a nutrient source to produce soluble species from which the zeolite grows? The commercial importance of zeolites means that huge efforts have been made to understand the process [11]. The best picture is probably as summarized in Figure B14.1.1. The initial amorphous gel (a) will be in equilibrium with solutes such as $[Al(OH)_4]^-$, various oligomeric silicates, template molecules, and other cations (b). As species rearrange in solution and reprecipitate by making and breaking T–O–T bonds, guest cations or templates will energetically favor the formation of certain rings (c), increasing their likelihood and leading to areas of local order (d). Ordered nuclei of a critical size will then progress to form a fully ordered crystalline material (e). As apparent from this figure, traditional distinctions between solid- and solution-phase syntheses are probably not applicable. In fact, even apparently clear synthesis liquids that might imply a solution mechanism can

Box 14.1 (cont.)

actually be optically transparent colloidal suspensions. Crystallization within or on such colloidal particles has been observed experimentally.

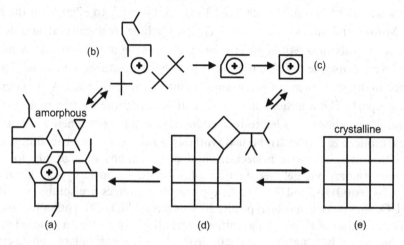

Figure B14.1.1 Schematic of processes involved in the synthesis of zeolites (after [11]).

14.1.2 Roles of Template Molecules in Zeolite Synthesis

One factor that led to an explosion in the number of synthetic zeolites was the inclusion of organic cations during synthesis (see Box 14.1). These cations play two important roles: they control the Si:Al ratio and they influence the specific framework topology formed. An illustration of how the Si:Al ratio can be controlled is given by the synthesis of compounds with the sodalite structure. In the sodalite structure, each of the 24 tetrahedra making up a sodalite cage is shared by four cages (when we remember that the void in the center of Figure 1.57 is itself a sodalite cage). There are therefore six T sites per cage. When sodalite is prepared in the presence of NaOH, the size of each cage is such (Figure 14.1) that it can adopt four Na^+ cations and an OH^- anion. These will charge-balance three Al^{3+} and three Si^{4+} giving $Na_8Al_6Si_6O_{24}(OH)_2 \cdot nH_2O$ [12] with an Si:Al ratio of 1:1 and $n = 3$–4.[11] If the synthesis is performed in the presence of $N(CH_3)_4OH$ [13], the larger size of the $[N(CH_3)_4]^+$ cation means that there's only space for one cation per cage and the Si:Al ratio is increased to 5:1 ($[N(CH_3)_4]_2Al_2Si_{10}O_{24}$). The final zeolite contains $[N(CH_3)_4]^+$ trapped inside a cage that it can't diffuse into or out of. This suggests a **templating** role where the cage crystallizes around the $[N(CH_3)_4]^+$ cation during synthesis.

---

[11] Similarly, Figure 1.57 shows that in the cubic mineral sodalite there are three Na^+ cations to charge-balance the framework along with an additional one Na^+ charge-compensated by one Cl^-.

The use of different organic templates can allow the Si:Al ratio to be varied continuously from ∞:1 down to 1:1 in some zeolites. At the high limit, the zeolite would essentially be a polymorph of crystalline SiO_2. The lower Si:Al limit of 1:1 is rationalized by **Loewenstein's rule**, which states that Al–O–Al linkages are not stable in the tetrahedral framework of zeolites.[12] Since each Al site must therefore be surrounded by four Si (each of which links on to three more T sites that may contain Al), one cannot have a Si:Al ratio lower than 1:1. This rule does not hold for less open networks of aluminosilicates produced at higher temperatures, which can have more Al than Si.

The Si:Al ratio has a significant impact on the properties and applications of zeolites. Firstly, when the Si:Al ratio increases, the internal surface changes from hydrophilic to hydrophobic as the amount of charge-compensating cations decreases; high-silica zeolites will therefore sorb nonpolar molecules. Secondly, if an Al-containing zeolite charge-compensated by NH_4^+ or $[N(CH_3)_4]^+$, or another protonated organic cation is calcined (heated to remove or burn the volatile template) in a final synthetic step, protons will be left behind to maintain charge balance. By increasing the Si:Al ratio, one reduces the number of such acid sites.

The exact role of organic cations in controlling the framework topology is more controversial, and the term **structure-directing agent** is perhaps more realistic than template. For example, zeolite ZSM-5[13] was first prepared in the presence of tetrapropylammonium cations, and the size and shape of this molecular ion was thought to be crucial for producing that specific framework. However, the same ZSM-5 has since been prepared in the presence of over 20 different organic species, in the presence of purely inorganic cations and, in 1997, it was found in nature as the rare mineral mutinaite. In fact, the organic cation plays multiple roles: It may simply be a "space filler", with non-specific van der Waals interactions helping stabilize the growing zeolite. It may work as a structure-directing agent that increases the likelihood of a certain framework topology under specific conditions. It can be a true template whereby specific electronic and steric properties uniquely direct formation of the desired framework. Perhaps the best example of pure templating is the use of "tri-quat", $[C_{18}H_{36}N_3]^{3+}$, in the synthesis of ZSM-18. The size- and shape-match between the guest and framework suggests templating (Figure 14.5), though subsequent theoretical studies have successfully predicted other molecules that also lead to ZSM-18 formation [14]. It has also been possible to design template molecules of specific chirality to produce enantiomerically enriched bulk zeolite samples [15]. In addition to using organic molecules to favor the formation of certain frameworks, the alternate approach of *blocking* formation of an undesired framework is also possible. One uses molecules that either wouldn't fit inside common frameworks, or that would bind to surfaces of unwanted frameworks inhibiting their growth.

---

[12] Al on a Si site carries a formal negative charge (Al_{Si}') that repels other Al sites.

[13] Zeolite Socony-Mobil (Socony = Standard Oil Company of New York) after the research labs where it was developed.

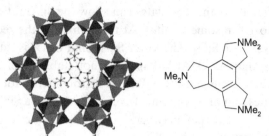

Figure 14.5 The structure of the MEI framework of ZSM-18 and the tri-quat molecule first used in its synthesis.

14.1.3 Zeolites in Catalysis

One extremely important application of zeolites is in heterogeneous catalysis, where they underpin multi-billion dollar sectors of the chemical industry. Many of the most important applications rely on their properties as Brønsted or Lewis **solid acids**.[14] We've discussed above that zeolites contain cations in their pores to balance framework charge. These cations can be exchanged for protons either by stirring in dilute acid (for high-silica zeolites such as ZSM-5), by heating an ammonium-exchanged form of the zeolite, or via calcination of organic-template-containing zeolites. Typical reactions for these procceses are:

1. $Na^+[\text{zeolite}] + H_3O^+ \xrightarrow[-Na^+]{\text{exchange}} H_3O^+[\text{zeolite}] \xrightarrow{\text{heat}} H^+[\text{zeolite}] + H_2O$

2. $Na^+[\text{zeolite}] + NH_4^+ \xrightarrow[-Na^+]{\text{exchange}} NH_4^+[\text{zeolite}] \xrightarrow{\text{heat}} H^+[\text{zeolite}] + NH_3$

3. $[\text{template}]^+[\text{zeolite}] \xrightarrow{\text{heat}} H^+[\text{zeolite}] + CO_2 + H_2O + \text{other products}$

The protons are present in the zeolite as Brønsted-acid sites with the H atom covalently bonded to a framework O adjacent to an Al site.

The true structures of commercially useful catalysts are often significantly more complex than a simple crystallographic picture would suggest. For example, a HY zeolite[15] with a Si:Al ratio of ~2.5 can be prepared by reaction 2 above, but is unfortunately unstable under the conditions needed in real catalytic applications.[16] Catalysts with a higher Si:Al ratio would be more stable, but are more expensive to prepare directly. It is, however, possible to turn the ammonium form of Y into so-called "ultrastable" HY (USY) by contacting it with steam at ~750 °C for around 24 hours. In this process, Al is lost from the framework, leaving

---

[14] IUPAC defines a Brønsted acid as a molecular entity capable of donating a proton to a base, whereas a Lewis acid is a molecular entity that is an electron-pair acceptor and able to react with a Lewis base (an electron-pair donor). The base in the specific case of zeolites is a sorbed molecule. Extra-framework Al and rare-earth cations (both discussed below) are believed to act as Lewis-acid sites in zeolites. The term solid acid refers to it remaining undissolved, solid, in the gas or liquid reaction medium.

[15] Here and later, the nomenclature term HY implies the H^+ form of zeolite Y, similarly NaA implies the Na^+ form of zeolite A.

[16] Particularly under the high-temperature conditions described later to regenerate catalysts.

defects behind and initially producing so-called extra-framework aluminum in the zeolite pores and cages.[17] Parts of the zeolite framework collapse and become amorphous as more Al is lost, but local recrystallization allows Si from these regions to "heal" nearby framework defects, effectively refilling them with Si. A portion of the Al migrates to the surface of the crystallites during the process, where amorphous alumina has been observed experimentally. Framework healing and Al loss occur at comparable rates and result in large mesoporous cavities forming inside the zeolite crystallites, which can make up 20–30% of the crystallite volume. The framework Si:Al ratio increases during the process and the HY formed is more stable.[18] The process has two additional benefits: The large mesopores can speed diffusion through the zeolite, leading to better catalytic activity, and the extra-framework aluminum species increase activity. Low-coordinate Al species can act as Lewis acids, and cationic species have been calculated to lower the activation energies of key steps of the catalyzed reactions. Overall, catalysts prepared in this way can have activities two orders of magnitude higher than otherwise.

The commercial importance of zeolite acid catalysis means that there has been considerable effort to try and quantify the acidity of the Brønsted-acid sites[19] present and how acidity influences reactivity. Acidity, referring to a general removal of the proton (no base), can be determined computationally[20] or inferred using infrared spectroscopy to measure framework O–H stretching frequencies. Alternatively, thermal analysis[21] or spectroscopic techniques can provide insight using different probe-molecule bases [16]. Perhaps the most direct insight comes from NMR studies using probe molecules [17]. These show, for example, that a molecule that is basic in an aqueous environment, such as p-fluoroaniline ($C_6H_4FNH_2$), is protonated in the pores of HZSM-5 and HY, whereas a less basic one such as p-fluoronitrobenzene ($C_6H_4FNO_2$) is not. The picture that emerges in this specific case is of a degree of base protonation comparable to what would be found in 70% H_2SO_4.[22]

In a given framework, acidity usually increases as the Si:Al ratio increases up to about 6–10. This can be related to the increasing number of Si atoms in the framework sites increasing

---

[17] A variety of different extra-framework Al species can be present in zeolites depending on the specific conditions (varying as a function of temperature, water content, Al content, etc.). ^{27}Al NMR suggests the presence of four-, five-, and six-coordinate species and theoretical studies show that Al^{3+} can coordinate to O^{2-} (framework and non-framework), OH^-, and H_2O depending on conditions. Single-Al species are mobile within the framework, and clusters containing a few Al^{3+} may form within the zeolite pores. As mesopores develop or Al migrates towards the surface, larger clusters can evolve towards nanoscale hydrated alumina.

[18] The increasing Si content of the framework is supported by ^{29}Si NMR measurements and a decrease in unit-cell parameter. A portion of the extra-framework Al can be removed from the material by acid washing or ion exchange.

[19] Many catalysts contain rare-earth ions as additional Lewis-acid sites.

[20] By calculating the energy required to remove a proton to an infinite distance from the resulting framework anion.

[21] For example, a material containing sorbed amines can be gently warmed to remove weakly bound molecules and then heated to higher temperatures to remove more strongly bound molecules. Acidity correlates with the temperature at which amines that are protonated in the pores desorb. The number of accessible acid sites correlates with the number of desorbing species.

[22] This comparison leads to zeolites often being referred to as strong acids. Acidity is the ability to protonate a Lewis base in a particular solvent or medium, and comparing strengths between different media is problematic. Note that early literature often stated zeolite acidity as being in the so-called super-acid range. Super acids would, for example, protonate benzene to form a benzenium ion, $C_6H_7^+$. This is not observed in zeolites.

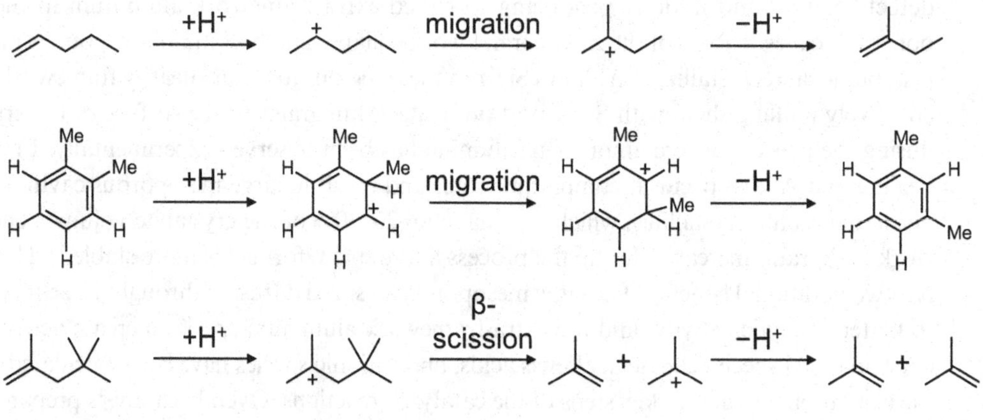

Figure 14.6 Acid-catalyzed transformations of simple organic molecules. The top two processes exemplify isomerization reactions and the lower process, cracking.

the framework electronegativity and hence its Brønsted acidity. After a plateau,[23] the acidity starts to fall as the number of protons present (to charge-balance Al) decreases. Whether a specific H-zeolite can act as an effective acid will also depend on the chemical transformation under consideration, as it will be influenced by how the protonated form of the molecule in question is stabilized when coordinated by the zeolite framework. These factors are probably the most important in determining the overall acidity of a specific zeolite.

Figure 14.6 shows examples of the acid-catalyzed transformations that can occur in zeolites. These are thought to proceed via carbenium-ion-like species.[24] For an alkene, protonation to produce a carbenium ion can be easily imagined: $H_2C=CH_2 + H^+ \rightarrow H_2C^+\text{--}CH_3$. For an alkane, one has to envisage either protonation to produce a five-coordinate carbocation followed by H_2 loss, or a direct hydride abstraction, both leading to the carbenium ion; either $H_3C\text{--}CH_3 + H^+ \rightarrow H_4C^+\text{--}CH_3 \rightarrow H_2C^+\text{--}CH_3 + H_2$ or $H_3C\text{--}CH_3 \rightarrow H_2C^+\text{--}CH_3 + H^-$.[25] Once formed, a carbenium ion can undergo a variety of processes, such as hydride or alkyl migration leading to **isomerization**, or β-scission leading to **cracking** (see Figure 14.6). Such a rearranged carbenium (cat)ion can then either deprotonate and desorb as an alkene or abstract hydride from another hydrocarbon and desorb as an alkane.[26]

---

[23] The major influence on acidity is the number of second-nearest-neighbor sites occupied by Al for the AlOH site in question (second neighbor, as Loewenstein's rule dictates that first neighbors are all Si). For high Si:Al ratios there is a high probability that these sites will all contain Si such that the local electronegativity will only change slightly as more Si is added.

[24] Carbenium ions have a carbon with three covalent bonds and a positive charge. In contrast, carbonium-ions contain C with five bonds (e.g. CH_5^+ or $C_2H_7^+$). Although protonated zeolites aren't generally sufficiently strong acids to directly protonate sorbed alkenes to form carbenium ions, the transition states involved in reactions are often thought to be carbenium-like. Some particularly stable carbenium ions have been observed experimentally in zeolite pores.

[25] More exotic non-classical carbonium ions involving two-electron, three-center C–H–C bridges formed by protonation of C–C bonds have also been proposed.

[26] The actual chemical processes occurring within zeolites can often be significantly more complex than the overall chemical transformation might suggest. For example, there is good experimental evidence that the transformation of methanol to short-chain alkenes is mediated by aromatic methyl benzenium cations and other cyclic species in the

Table 14.2 Relative cracking rates of various hexanes using HZSM-5 at 340 °C.

	n-Hexane	2-Methyl-pentane	3-Methyl-pentane	2,3-Dimethyl-butane	2,2-Dimethyl-butane
Relative rate	0.71	0.38	0.22	0.09	0.09

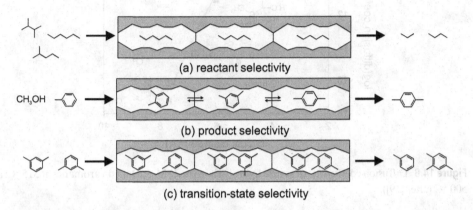

(a) reactant selectivity

(b) product selectivity

(c) transition-state selectivity

Figure 14.7 Shape-selective catalysis in zeolites.

The huge importance of these transformations to the chemical industry explains the enormous interest in acid catalysts. Solid zeolites have the advantage that they can be readily separated from products and recycled. They also have extremely high and easily accessible external and internal surface areas. Most importantly, the dimensions of their pores and cages can control the course of the catalyzed reaction, hence the pore size constrains the chemistry leading to a **shape-selective catalysis**. In this way, zeolites are analogous to enzymes since the local geometry fine-tunes the properties of the active center. Different pore and void sizes and shapes control **reactant-**, **product-**, or **transition-state selectivity**, as illustrated schematically in Figure 14.7.

In **reactant-size selectivity**, Figure 14.7a, access to the catalytically active sites of the zeolite is controlled by the size of the framework pores. An example of this is the observation of different rates of catalytic cracking for different isomers of C_6H_{14} by ZSM-5 as given in Table 14.2. Straight-chain n-hexane, which diffuses rapidly into the pores, cracks far more readily than branched isomers. Similarly, both n- and i-butanol are readily dehydrated by large-pore NaX zeolite at ~250 °C but only n-butanol by smaller-pore CaA zeolite. The size of the pores can be modified in a number of ways to influence reactant selectivity. In zeolite A, for example, alkali metals take up sufficient pore space so that only methane can penetrate inside. With divalent pore cations, only half the sites are occupied, and this allows straight-chain hydrocarbons to diffuse into the zeolite but not branched hydrocarbons. In general, molecules up to ~0.5 Å larger than the apparent pore size can diffuse through zeolites. This has been explained in terms of local vibrations increasing pore diameter, via the rigid-unit mode ideas discussed in Box 14.2.

framework via a so-called carbon-pool mechanism. Zeolite-bound alkoxide groups are long-lived intermediates in several processes. More details are given in refs. [16] and [17].

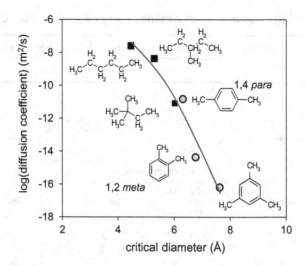

Figure 14.8 Diffusion coefficients of simple hydrocarbons in ZSM-5. Data on aromatics at 315 °C; aliphatics 500 °C (after [19]).

Product-size selectivity relies on the fact that different products of a reaction can have vastly different diffusion rates out of the zeolite (Figure 14.8). A wonderful example of this is provided by the disproportionation reaction of methylbenzene in ZSM-5 to produce benzene and dimethylbenzene; the 1,4-dimethylbenzene isomer is a valuable feedstock for the production of nylon [18]. Using unmodified or small crystallites of ZSM-5, one obtains an equilibrium mixture of isomers containing ~24% of the desired 1,4 (*para*) product. However, the diffusion coefficient of the *para* isomer is ~10^4 times higher than for the other isomers (Figure 14.8), such that increasing the average time spent in the catalyst should increase the proportion of the *para* isomer produced. This can be achieved by either blocking a fraction of zeolite pores with coke, or by increasing the crystallite size of the catalyst. At low conversion, 0.7 µm crystallites give ~60% selectivity for 1,4-dimethylbenzene while 3.6 µm crystallites give 80%; 97% selectivity can be achieved in coated or impregnated ZSM-5. In-situ solid state NMR studies on related reactions have supported this mechanism by showing that the isomer distribution within the pores of the zeolite is much closer to that expected from thermodynamic equilibrium than that observed in the extracted product mixture.

The final category of shape selectivity is **transition-state selectivity**, Figure 14.7c. In the example shown, the transfer of a methyl group from one dimethylbenzene molecule to another can produce two isomers of trimethylbenzene. The transition state required to produce 1,3,5-trimethylbenzene is sterically disfavored in the pores of mordenite, resulting in preferential production of 1,2,4-trimethylbenzene. Similarly, in the acid-catalyzed isomerization of dimethylbenzenes, no trimethylbenzene forms. This is presumably due to pore size disfavoring the bimolecular transition state needed to form the latter product. Another area in which transition-state selectivity is thought to be important is in preventing the build-up of coke in zeolite catalysts. For large-pore zeolites such as mordenite, coking can block pores, leading to a decrease in catalytic activity. In zeolites such as ZSM-5, the transition states leading to coke formation are believed to be disfavored by pore size, such that coke only forms on the zeolite surface. Transition-state selectivity can be distinguished from reactant and product selectivity in that there should be little dependence on crystallite size, aside from the indirect influence of surface sites. All the forms of selectivity we have discussed can be optimized by treatments (e.g. dealumination) to remove or block surface sites.

The largest catalytic application of zeolites in catalysis is in the cracking of crude oil to produce more useful fractions [20]. Over 99% of the world's petrol is produced in this way using >300000 tonnes of zeolites annually. In a typical system, preheated crude oil (~370 °C) is passed onto a fluidized bed of catalyst (**fluid catalytic cracking**, **FCC**) at ~500 °C, where cracking occurs. In a subsequent reactor, the catalyst is then treated with steam to remove sorbed products before flowing to a regeneration vessel, where it is heated in air at up to ~730 °C to remove coke before it returns to the catalyst bed. Most cracking is carried out using FAU zeolites (Figure 14.2c). Zeolites are also widely used in hydrocracking, in which low-grade oils react with hydrogen over a fixed catalyst bed at high temperature and pressure to produce higher-grade oils. Dual-function catalysts with both acidity and the capability to catalyze hydrogenation and dehydrogenation reactions are needed. This is particularly important to help promote the formation of carbenium-ion-like species when processing alkanes. These latter properties can be achieved by producing metal dispersions both on the surface and in the cages of the zeolites.[27] In all these applications, it is often the ease of catalyst regeneration and cost rather than its intrinsic properties that lead to commercial use. Large-pore materials, such as those in Section 14.1.6 and 14.2, are often unstable under regeneration conditions, and the exotic templates required to prepare some frameworks may make them too expensive for large-scale use.

Other catalytic applications involve modifying the chemistry of the framework to introduce active species. One example is the catalytic oxidation of simple organic species by H_2O_2 on Ti-containing zeolites such as TS-1 of the same structure as ZSM-5. This process works without toxic reagents and produces water as a byproduct. A fuller coverage of catalytic uses of zeolites both on industrial and small scales can be found in Further Reading.

14.1.4 Ion-Exchange Properties

Another extremely important aspect of zeolite chemistry is that the charge-balancing cations inside their frameworks can often be swapped for other cations in aqueous solution:

$$b A^{a+}(aq) + a B^{b+}[zeolite] \leftrightarrow b A^{a+}[zeolite] + a B^{b+}(aq)$$

The preference for forming a given zeolite–cation combination is determined by a number of factors: the number and size of sites available to coordinate cations (note that the type of site

---

[27] For example by ion exchange of the compensating cation with Ni^{2+} and reduction, or with $[(Pd,Pt)(NH_3)_4]^{2+}$ followed by decomposition and reduction. Protons charge-balance the final catalyst.

may change during the process as some sites may exchange preferentially); the cation charge, radius, and electronegativity; and the ability of water molecules to hydrate the cation and form hydrogen-bonding networks. It is also important to consider the entropy changes in both the zeolite and solution that occur on exchange.[28] It is common to report **selectivity**[29] via so-called **ion-exchange isotherms** that compare the amount of a given cation in the zeolite with the amount in solution at equilibrium. For zeolite A, for example, selectivities of alkali metals are $Na^+ > K^+ > Rb^+ > Li^+ > Cs^+$; alkaline earths are preferred over Na^+ with $Ca^{2+} \approx Sr^{2+} \approx Ba^{2+} > Mg^{2+}$. The higher Si:Al ratio of ZSM-5 increases the affinity for larger weakly hydrated cations: $Cs^+ > H^+ \approx NH_4^+ \approx Rb^+ > K^+ > Na^+ > Li^+$. Zeolite ion exchange will happen even with distilled water via $NaX + H_2O \leftrightarrow HX + NaOH$. This means that a zeolite such as NaX in contact with water gives an alkaline solution. Other zeolites undergo more complex slow hydrolysis reactions.

The maximum ion-exchange capacity of a given zeolite is determined by the Si:Al ratio that dictates the number of charge-balancing cations. In many cases, full ion exchange isn't possible. For example, Cs^+ ions are too large to fully replace all Na^+ ions in the Na^+ form of zeolite A (NaA); Rb^+ and Cs^+ are too large to ion-exchange with smaller cations in the β cages of zeolite X. Zeolite ion exchangers can therefore show size selectivity similar to the catalytic reactant-size selectivity discussed above.

The microporosity of zeolites also affects the kinetics of ion exchange. For example, while Mg^{2+} and Ca^{2+} will both undergo full ion exchange with NaA, the rate of exchange at room temperature is around 10 times slower for Mg^{2+}. This is believed to be due to the stronger hydration shell of Mg^{2+} that must be removed before the cation can enter the zeolite pores. At higher temperature, rates become comparable. The huge importance of ion exchange to industry has led to a large body of research in this area and more details can be found elsewhere [21, 22].

One major application area is in solid laundry detergents, where NaA (among other effects) softens water by removing Ca^{2+} and Mg^{2+}. Zeolites have the added benefit of reducing powder agglomeration during storage, allowing compact formulations. Up to 15% zeolite A might be present in a typical detergent.[30] The zeolite synthesis is carefully controlled to produce crystallites of 3–5 μm with beveled edges [23] so that they rinse from clothes more readily.

---

[28] Possible entropic contributions include: changes in the number of cations in the zeolite or solution for $a \neq b$, changing numbers of water in the framework for different cations, changes in the water coordination shell of cations in solution, etc.

[29] One way of depicting selectivity is through an ion-exchange isotherm that plots a quantity Z_A representing the equilibrium amount of cation A in the zeolite against S_A representing the amount in solution: $Z_A = \dfrac{a \cdot m_{A_z}}{a \cdot m_{A_z} + b \cdot m_{B_z}}$ and $S_A = \dfrac{a \cdot M_{A_s}}{a \cdot M_{A_s} + b \cdot M_{B_s}}$ with m_{A_z} the molality of A^{a+} in the zeolite in mol/kg and M_{A_s} the molarity of A^{a+} in solution in mol/L. A straight-line plot with gradient 1 shows no preference for cation A over B. A high initial slope shows a selectivity for A over B and a lower slope, the opposite. More complex shapes can reflect exchange occurring at different cation sites.

[30] In the detergent industry, the species used to control Ca^{2+} and Mg^{2+} levels are traditionally called builders. Various species called chelates or co-builders (e.g. citrate) are added to control other metal-ion concentrations. A typical solid formulation will also include sodium carbonate and silicates to produce the alkaline conditions needed for the wash, surfactant molecules, dispersing agents (e.g. polycarboxylates) to avoid reprecipitation of soil particles, as well as bleaches, perfumes, and optical brighteners.

This usage declined somewhat as the industry moved to liquid formulations and as surfactant systems were improved, but zeolites are still present in over 30% of laundry products (around 70000 tonnes of zeolite are used annually in a ~$40 billion market).

Zeolites are also used in water purification. Natural clinoptilolite[31] is used to remove NH_4^+ from waste water at sewage plants, and radioactive ions such as ^{137}Cs and ^{90}Sr after nuclear spills [21, 22]. Clinoptilolite was even added to the feedstock of sheep that had grazed on ^{137}Cs-contaminated land following the 1986 explosion at the Chernobyl nuclear power plant! Zeolites also provide a potential method for the medium-to-long-term storage of radioactive waste from the nuclear industry.

14.1.5 Drying Agents, Molecular Sieving, and Sorption

All chemical laboratories will have a bottle of 3A, 4A, or 5A zeolite beads that are used as general-purpose drying agents. The zeolites are heated in a vacuum to drive off water that coordinates the extra-framework cations, after which they have a high affinity for water uptake. The number in the symbol corresponds to the nominal size (in Å) of molecules the zeolite will absorb. Zeolite 4A is the dehydrated Na form of zeolite A, $Na_{12}[(AlO_2)_{12}(SiO_2)_{12}]$. Zeolite 3A is the K form, and the larger K^+ ions reduce the accessible pore size. Zeolite 5A is $(Ca_4Na_4)[(AlO_2)_{12}(SiO_2)_{12}]$. In 5A, the cations only adopt site 1 in the structure of Figure 14.3, effectively leaving the 8-rings "open" and creating a larger accessible pore. Zeolites 4A and 5A can be used to remove water from most alcohols, except methanol for which 3A is used because methanol itself would enter the larger pores of 4A and 5A. Zeolite 3A is widely used for drying natural gas, whereas 4A might be used if there is a need to also remove small amounts of the larger CO_2 molecule. Zeolite 5A is commonly used in the separation of straight-chain and branched hydrocarbons. For example, there are several industrial processes that isolate linear C_{10} to C_{18} hydrocarbons from gas streams containing branched and cyclic species by their sorption in the pores of 5A.

The separation of O_2 from air is another important application of zeolites. Historically, oxygen was produced by fractional distillation of liquid air, relying on the different boiling points of N_2 and O_2 (77 and 90 K, respectively). However, the need for complex liquefaction apparatus can be avoided by using zeolites such as Li-chabazite or a $(Li^+, M^{2+})X$ zeolite, which both operate at room temperature [24]. Since neither N_2 or O_2 have a permanent dipole moment, separation relies on the fact that N_2 has both a larger quadrupole moment and higher polarizability[32] than O_2 [25], leading to stronger electrostatic interaction with

---

[31] Natural clinoptilolite (framework code HEU) has typical composition $A_6Al_6Si_{30}O_{72} \cdot \sim 20H_2O$, A = (K, Na, $Ca_{0.5}$, $Sr_{0.5}$, $Ba_{0.5}$, $Mg_{0.5}$) with higher Na + K content than alkaline earths.

[32] The simplest electric dipole is a vector of a + − pair of point charges. The simplest quadrupole moment can be thought of as alternating point charges at the corners of a square (+ − + − clockwise round the square). Homonuclear diatomics have no permanent dipole but, since they are not spherically symmetric, have a quadrupole moment +⁻ +. Quadrupole moments are −1.115 and −0.225 atomic units (equal to ea_0^2, where e is the elementary charge and $a_0 = 0.529$ Å is the radius of the first Bohr orbit) and polarizabilities are 11.74 and 10.61 in units of a_0^3 for N_2 and O_2, respectively [25].

extra-framework cations. Under typical conditions this leads to a ~4–10-fold higher uptake of N_2 than O_2. Oxygen can thus be produced by a vacuum-swing-adsorption method whereby gases are loaded at ~1.2 bar and preferential N_2 adsorption leaves behind an O_2-rich gas. When the adsorbent is saturated, a vacuum is applied (~0.35 bar) and the zeolite sorbent regenerated. Multiple sorbent beds can be used to give a constant flow of O_2.

While the processes above all work via an absorb–desorb cycle, there is also considerable interest in producing extended membranes of zeolites to enable continuous sieve-like separation; incorporating catalytic activity in the membranes makes them more exciting still. It has, for example, been possible to produce extremely thin highly oriented (so pores are aligned) membranes of MFI zeolites that allow high fluxes of molecules and can separate pairs of species such as 1,4-dimethylbenzene/1,2-dimethylbenzene or benzene/cyclohexane [26].

14.1.6 AlPOs and Related Materials

Given the structural chemistry of zeolites, it is not surprising that similar materials can be prepared containing other tetrahedral metal cations. For example, starting from SiO_2, site ordering with Al^{III} and P^V (the idea shown in Figure 1.38) leads to dense aluminum phosphates, isostructural to quartz, tridymite, or cristobalite polymorphs of SiO_2. By heating mixtures of Al_2O_3:P_2O_5:amine-template:40–400·H_2O under hydrothermal conditions, porous aluminophosphates (AlPOs) similar to zeolites have been prepared. It is worth noting that the site ordering of Al and P sites in such AlPO frameworks constrains them to those containing even-numbered rings. The main difference to zeolite synthesis is that reactions are typically performed at low pH where it is easier to include elements in addition to Al and P in the framework during synthesis.[33] A wide variety of other tetrahedral $MO_{4/2}$ groups (commonly M = Ga, Ge, As, B, Be, Li, Co, Zn) can be introduced.

One interesting feature of AlPOs is that materials with large pores (>12 T) can be synthesized (see Table 14.1 and Figure 14.9). VPI-5, for example, can be synthesized hydrothermally at 150 °C from orthophosphoric acid and hydrated alumina in the presence of tetrabutylammonium hydroxide. It contains 12 Å-wide channels and has a 25% accessible volume. The gallophosphate cloverite $(C_7H_{14}N)_{24}[F_{24}Ga_{96}P_{96}O_{372}(OH)_{24}]$ (Figure 14.9, right) contains 20-membered rings and has four terminal OH groups that protrude into the pores (reducing their size slightly), giving the characteristic clover-leaf pore shape after which the material was named. In cloverite, the F is located in 4-4-rings and increases the coordination of each Ga from four to five (GaO_4F).

Large-pore AlPOs differ from the zeolites in four main ways: They may contain non-tetrahedral metal atoms (e.g. Al in octahedral coordination). They may contain terminal OH groups on metals. They may contain non-tetrahedral species (e.g. OH, H_2O, F) as part of the framework. Finally, they tend to be less stable than zeolites under thermal or hydrothermal conditions, which limits their potential applications.

---

[33] Under the high-pH conditions of zeolite synthesis, many transition elements are insoluble.

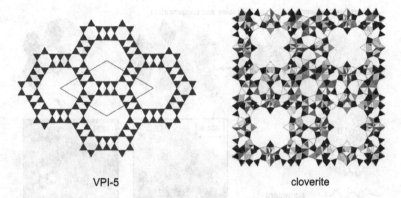

VPI-5   cloverite

Figure 14.9 Schematic views of the structures of VPI-5 and cloverite. In cloverite GaO_4F polyhedra shaded in light gray, PO_4 in dark gray.

14.2   Mesoporous Aluminosilicates

The crystalline zeolites and related frameworks we've discussed so far contain pores that have regular shape, but are limited in size to $\lesssim 12$ Å. Before the early 1990s, materials containing mesopores (2–50 nm) were restricted to amorphous silicas (Section 14.3) or modified layered materials in which a distribution of pore sizes are present. Since then, a variety of synthetic methods have been developed for producing mesoporous ceramics with uniform pore sizes up to several hundred ångstrøms.

The most famous of these are the MCM-41[34] aluminosilicates first described in detail by workers at Mobil [27]. They performed hydrothermal syntheses under basic conditions, using various silica and alumina sources in the presence of long-chain alkylammonium cations such as hexadecyltrimethylammonium, $[C_{16}H_{33}(CH_3)_3N]^+$.[35] These molecules are well-known cationic surfactants with charged hydrophilic headgroups and neutral hydrophobic hydrocarbon tails. Above a critical concentration in solution (the **critical micelle concentration, CMC**), they self-assemble into spherical or rod-like micelles, more complex 3D surfactant networks, or lamellar structures. Under the concentration, temperature, and pressure conditions used to prepare MCM-41, long rod-like micelles are formed (Figure 14.10, top), which cluster into ordered arrays with their headgroups coordinated by silicate anions of the initial mixture. The remaining anions are protonated by water and slowly condense during heating onto these pre-arranged anions, forming a solid polyacid aluminosilica network. This allows the base

---

[34] MCM after Mobil Composition of Matter.

[35] A typical preparation used 2 g of "Catapal alumina" [$AlO(OH)$ and $Al(OH)_3$ in ~1:1 mixture], 25 g of "HiSil" (a precipitated form of SiO_2), and 100 g of a tetramethylammonium silicate solution containing 10 wt% SiO_2 along with 200 g of a 29 wt% surfactant solution of $[C_{16}H_{33}(CH_3)_3N]^+[OH/Cl]^-$ (overall Si:surfactant ratio of ≤1). The mixture was heated to 150 °C in an autoclave for 48 hours followed by heating to 540 °C to remove the template.

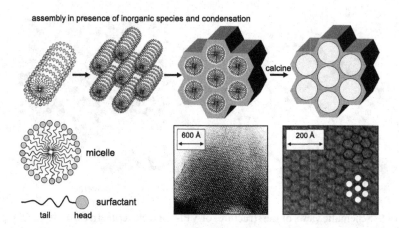

Figure 14.10 The formation of MCM-41. Surfactant molecules assemble into rod-like micelles around which inorganic walls form. Transmission electron microscope (TEM) images (bottom center/right) of typical samples looking along the pore direction. Dark regions are walls and lighter regions pores. Center image is a sample prepared with $[C_{12}H_{25}(CH_3)_3N]^+$. The material on the right has a pore size of 40 Å and was prepared with $[C_{16}H_{33}(CH_3)_3N]^+$. Images adapted from ref. [28] with permission.

present to dissolve a new portion of the amorphous silica into silicate. The process continues until the final templated aluminosilicate particles precipitate. Calcination then removes the template molecules leading to a porous material. This process is shown schematically in Figure 14.10. Electron microscopy reveals the formation of a regular array of ~40 Å pores running in one dimension, showing that the micelle morphology is retained in the solid. The walls themselves are amorphous, typically ~10–15 Å thick and contain a variety of silica species (Q^4, Q^3, and Q^2).[36]

The size of the pores can be controlled using surfactant templates with different alkyl-chain lengths, or by adding other organic molecules that can dissolve in the hydrophobic core of the micelles and increase their size during synthesis. For example, using a shorter-chain surfactant like $[C_{12}H_{25}(CH_3)_3N]^+$ in the synthesis gives ~30 Å pores in the final product. Materials with pores up to ~100 Å can be prepared by adding 1,3,5-trimethylbenzene to a $[C_{16}H_{33}(CH_3)_3N]^+$ synthesis.

Similar materials can also be prepared using non-ionic surfactant molecules. For example, high-stability large-pore structures can be prepared using block copolymers[37] of poly(ethylene oxide) and poly(propylene oxide) of the form $(EO)_n(PO)_m(EO)_n$. These again contain more hydrophilic (EO) and more hydrophobic (PO) regions and form

---

[36] The Q^n symbol is used to represent a SiO_4 tetrahedron linked to n other tetrahedra. In quartz, for example, all Si sites are Q^4 (Niggli formula $SiO_{4/2}$). An otherwise linked tetrahedron with a single Si–OH group would be Q^3. Different Q^n species can be identified and quantified by ^{29}Si NMR.

[37] Block copolymers contain regular sequences of different monomers and are known to undergo ordering on the mesoscale due to the differing physical properties (e.g. hydrophobic or hydrophilic) of different blocks.

various micellar structures in a solution, with EO surfaces and PO cores. With these surfactants, the mesoporous silicate synthesis is performed under acidic conditions where the silicate and surfactant species will be protonated. Anions from the added acid maintain charge balance in the initial products. Materials with pore sizes up to ~300 Å and thick (~30–65 Å) walls can be prepared using an $(EO)_{20}(PO)_{70}(EO)_{20}$ surfactant with 1,3,5-trimethylbenzene added to swell the micelle size. One was quoted as being stable up to 500 °C or in boiling water for 24 hours, had a surface area of 910 m^2/g, an average pore size of 260 Å, and a pore volume of 2.2 cm^3/g [29]. The expensive polymer template can be removed by solvent extraction and reused in subsequent syntheses. These materials often exhibit microporosity in the silica walls in addition to the microporosity of the large channels, though with an irregular pore distribution reflecting the amorphous nature of the walls.

By changing the surfactant or reaction conditions, it is also possible to make materials with 3D pore structures, or with spherical pores arranged in hexagonal- or cubic-closest-packed-like patterns. A variety of related porous metal oxides have also been produced, including ZrO_2, TiO_2, Nb_2O_5, Ta_2O_5, and Mn_xO_y examples.

Applications of mesoporous materials generally take advantage of their large-pore structures. Gas-sorption studies show that they have a high available pore volume and surface area, though they exhibit so-called type-IV[38] isotherms with a capillary conden-sation step such that the gaseous sorbate must be at relatively high pressure. While MCM materials don't show the high acidity of zeolites required for catalytic cracking,[39] their uniform large-pore structure leads to exciting opportunities for supporting mol-ecules usually used for homogeneous catalysis within a controlled and uniform environ-ment. The resulting heterogeneous catalyst is more easily separable from reaction mixtures for re-use, and the local pore environment can lead to shape-selective control of products.

In one early demonstration example, a chiral[40] Pd-containing catalytic molecule was attached to the walls of MCM-41.[41] The resulting catalyst was shown to catalyze the following reaction (Ac is acetyl, CH_3CO) of two alternative products:

Ph⟍⟍OAc + PhCH$_2$NH$_2$ ⟶ x Ph⟍*⟍ (NHCH$_2$Ph) + (1−x) Ph⟍⟍NHCH$_2$Ph + HOAc

---

[38] See e.g. Barton et al. in Further Reading for a discussion of isotherms in porous materials.

[39] Acidity appears to correlate with Al–O–T angles: the larger the angle, the higher the acidity. In the amorphous walls of MCMs, there are no structural features that force large angles.

[40] Chiral molecules have non-superimposable mirror images called enantiomers. They are discussed in Box 14.3. An enantioselective synthesis is one that produces more of one enantiomer than the other; an enantiomeric excess (ee). A chiral carbon in an organic molecule has four different groups attached and is marked with *.

[41] A ligand derived from 1,1′-bis(diphenylphosphino)ferrocene (dppf) was first attached to the walls of MCM-41 by the reaction of a chiral-amine side arm on one ring of the ferrocene with a 3-bromopropyltrichlorosilane molecule grafted onto the MCM-41 surface. The catalyst formed on subsequent coordination with Pd^{2+}.

and to be more active than an equivalent soluble Pd-containing catalyst. The confined geometry of the pore yields 50% of the branched isomer ($x = 0.5$), compared to just 2% ($x = 0.02$) for a catalyst supported on the surface of non-porous silica [30]. More importantly, the branched product had a 95% enantiomeric excess; much higher than the 43% found with the unconfined catalyst. The confined catalyst therefore shows both **regioselectivity** (the place where the double bond is attacked) and **enantioselectivity** over other forms.

There are numerous other applications of MCM materials. For example, mesoporous oxidation catalysts similar to the TS-1 zeolites can be prepared either by including $Ti(OC_2H_5)_4$ during synthesis or via a post-synthetic grafting of active organometallic species such as $Ti(\eta\text{-}C_5H_5)_2Cl_2$ and heat treatment. Supported metal-nanoparticle catalysts can also be prepared. More details are available in Further Reading, and we will encounter many similar ideas when we discuss applications of MOFs in Section 14.4.

It has also been possible to combine mesoporous (>2 nm) and microporous (<2 nm) cavities in materials [31]. For example, when some high-silica zeolites (e.g. ZSM-5 or Y) are treated with surfactants such as $[C_{16}H_{33}(CH_3)_3N]^+Br^-$ in the presence of a weak base, they appear to regrow around surfactant micelle species that become encapsulated in the zeolite crystallite. The surfactants can subsequently be calcined off to give zeolite crystallites with essentially unchanged shape but containing ~45–50 Å mesopores in addition to the normal framework micropores. The exact mechanism by which this occurs isn't fully understood, but it probably involves local recrystallization of the zeolites via a process similar to USY formation discussed in Section 14.1.3. These mesostructured zeolites have been prepared on a multi-tonne scale and show high activity in oil-cracking applications. In particular, the mesopores allow rapid diffusion of large molecules into the zeolite for cracking, and rapid diffusion of small molecules out of the zeolite, reducing the "over-cracking" that produces coke.

14.3 Other Porous Oxide Materials

There are a number of other ways to produce highly porous oxide materials with pore sizes controllable over a range of length scales. One widely used synthetic approach is the **sol–gel** method. A typical example would be the formation of amorphous porous silicas either by acidifying an aqueous solution of sodium silicate, or by the controlled hydrolysis of silicon alkoxides in a non-aqueous (typically alcohol) solvent, via the following summary reactions:

$$SiO_3^{2-} + 2H^+ \rightarrow SiO_2 + H_2O \quad \text{or}$$
$$Si(OR)_4 + 2H_2O \rightarrow SiO_2 + 4ROH$$

In this process, soluble molecular species gradually condense to form a colloidal solution of high-molecular-weight polysilicate particles (the **sol**). These particles then link together to form a 3D network (the **gel**[42]) with interparticle pores filled by solvent:

$$-Si-OH + HO-Si- \longrightarrow -Si-O-Si- + H_2O$$

On heating to remove solvent and any other organic species added during the preparation, porous **xerogels**[43] are formed. In contrast to the zeolites and mesoporous materials discussed above, neither the pores nor constituent silica particles have long-range order. Other porous metal oxides can be prepared using suitable metal-alkoxide precursors in place of the Si source.

Xerogels can be either microporous or mesoporous depending on the specific synthetic conditions, with the porosity governed by the sizes of particles formed during the early stages of polymerization. Porosity control can also be achieved by using pre-formed building blocks such as $[Si_8O_{12}](OCH_3)_8$ or $[Ti_{16}O_{16}](OC_2H_5)_{32}$ molecular precursors. Sol–gel synthesis can also be used to prepare materials with macroscopic porosity. In the preparation of so-called **inverse opals**, the interstices in a highly ordered close-packed array of polymer spheres (typically ~1 μm diameter polystyrene) are filled with a desired silica sol. Following gelation, the polystyrene spheres can be removed by careful calcination or dissolved in solvent to produce silica with ordered interconnected micron-sized spherical cavities.

Aerogels are closely related materials that typically have >50% free volume and can have extremely low densities down to 1 mg/cm^3. They are prepared by replacing the solvent in a sol–gel synthesis by a supercritical gas such as CO_2. Silica aerogels can have extremely low thermal conductivities as their open structures inhibit both conduction and convection modes of heat transport. They are used for thermal insulation, particularly in space applications due to their low weight.

Another conceptually simple but elegant approach to making porous solids is the use of selective leaching of a two-phase sample, as shown schematically in Figure 14.11 [32]. Here, one forms an intimate composite of two phases, one of which can be selectively removed by dissolving it in an appropriate solvent, either directly or after suitable chemical transformation (e.g. selectively reducing one oxide to a metal followed by acid leaching). The resulting porous material can itself be chemically altered in several ways: an oxide could be reduced to form a porous metal; the surface could be coated with other species; or an additional chemical reaction could render the walls microporous, creating a hierarchy of pore sizes.

---

[42] In a gel, the network of particles spans the entire volume of the container.   [43] Greek xeros means dry.

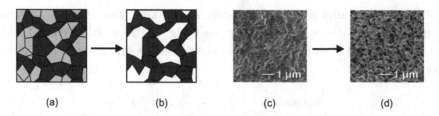

(a)   (b)   (c)   (d)

Figure 14.11 The production of porous materials by selective leaching. (a), (b) Schematics of the approach. (c), (d) Scanning electron microscope (SEM) images of how leaching of ZnO from a dense NiO/ZnO composite leads to a porous material. Reproduced from *Chem. Commun.* (2006), 3159–3165.

Box 14.2 Materials Spotlight: Fascinating flexible frameworks—shrinkage on heating and expansion on squeezing

The 3D structures of framework materials can give rise to some fascinating counterintuitive physical properties, several of which are technologically exploitable. For example, we expect that materials expand on heating. Linear coefficients of thermal expansion, $\alpha_\ell = (\ell_{T_2} - \ell_{T_1})/(\ell_{T_1}[T_2 - T_1])$, are typically around +2.5 for Si, +8 for Al_2O_3, and +17 for Cu (all $\times 10^{-6}\,K^{-1}$) near room temperature. This expansion can ultimately be traced back to the asymmetric nature of interatomic potentials (the shape is similar to the lattice-formation energy curve in Figure 5.2). As a bond is heated, higher vibrational levels are populated, which corresponds to larger mean atomic separations causing thermal expansion. Some framework materials, however, show the opposite behavior and contract on heating. Perhaps the most famous of these is cubic ZrW_2O_8, which shows isotropic contraction (**negative thermal expansion**, **NTE**) from 0.3 K to 1050 K and has $\alpha_\ell = -9 \times 10^{-6}\,K^{-1}$ between 2 K and 300 K [33]. The origin of this effect can be related to its structure, which contains O in two-coordinate Zr–O–W linkages between corner-sharing ZrO_6 octahedra and WO_4 tetrahedra (see Figure B14.2.1). The metal–oxygen bonds themselves are relatively strong and show low thermal expansion. At a local level, however, the O atom undergoes a transverse or sideways vibration on heating. The larger the magnitude of this dynamic displacement, the closer Zr and W polyhedral centers are pulled together, and the smaller the volume V.* This is called the tension effect.

To fully understand thermal expansion in a solid, we have to consider phonons that affect the whole structure; we cannot just focus on the local picture. Interestingly, the 3D network of ZrW_2O_8 has a set of phonon modes that keep local coordination polyhedra (nearly) undistorted. These low-energy, low-frequency states are called rigid-unit modes (RUMs), and can dominate thermal expansion. Many RUMs in ZrW_2O_8 involve transverse motions of O and tend to contract the structure. Similar behavior is found in some zeolites, some cyanide networks, and some MOFs [34].

Box 14.2 (cont.)

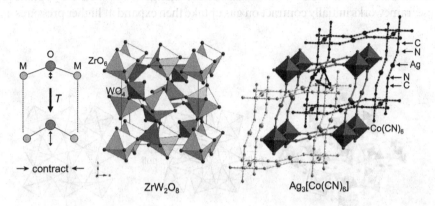

Figure B14.2.1 The structure of ZrW_2O_8 contains corner-sharing octahedra and tetrahedra. $Ag_3[Co(CN)_6]$ contains $[Co(CN)_6]^{3-}$ octahedra linked by Ag^+ into three interpenetrating pseudo-cubic networks (shown with different bond shadings). Exemplar short (~3.5 Å) Ag–Ag interactions between these networks are shown with double-headed arrows.

The framework material $Ag_3[Co(CN)_6]$ has an exceptionally anisotropic thermal expansion and contraction [35]. In its trigonal structure ($P\bar{3}1m$), large Ag^+ ions link $[Co(CN)_6]^{3-}$ octahedra into three interpenetrating 3D networks and themselves form densest-packed planes parallel with *ab* of the hexagonal cell. The geometry of the material at any temperature is governed by these weak Ag–Ag interactions. On heating, Ag–Ag distances expand profoundly in the *ab* plane, and the stronger network bonds force the structure to contract along *c*. The situation is a 3D equivalent of stretching a piece of garden lattice fencing or a folding wine rack. Values of $\alpha_\ell = -120\times10^{-6}$ K^{-1} along *c* are a truly colossal uniaxial NTE.

Coupled RUM rotations of rigid tetrahedra are also important in zeolites. Many zeolites have "windows of flexibility", which are cell-parameter ranges over which the structure can distort without individual TO_4 tetrahedra distorting [36]. Most zeolites with a pure SiO_2 composition adopt the highest cell volume that is possible without tetrahedral distortion. In some ways, this goes against traditional ideas of solids trying to maximize their density, but makes sense if one considers the driving force as maximizing the distance between O atoms of different tetrahedra to minimize Coulomb repulsion. We can also consider RUM-like distortions in zeolites as a low-energy mechanism whereby the framework structure can expand locally to allow the passage of diffusing guest molecules, and for understanding their displacive phase transitions.

Another counterintuitive property of zeolites is that under special circumstances some can appear to expand at the unit-cell level when put under pressure. The explanation for this apparent paradox is that when the sample is squeezed in a water-containing pressure medium, the pores take up more water allowing the overall volume to reduce under pressure despite the zeolite volume increasing in the process. Natrolite, for example, transforms from $Na_{16}Al_{16}Si_{24}O_{80}\cdot16H_2O$ to $Na_{16}Al_{16}Si_{24}O_{80}\cdot32H_2O$ under pressure and the super-hydrated form has a 4.5% larger volume than the normal phase [37, 38].

Box 14.2 (cont.)

Remarkable changes in cell volume on the uptake or loss of solvents and gases are also found in some MOFs, with materials such as MIL-88 showing a volume change of 100% on CO_2 uptake [39, 40, 41]. Some of these frameworks initially contract on gas uptake then expand at higher pressures as more gas is adsorbed.

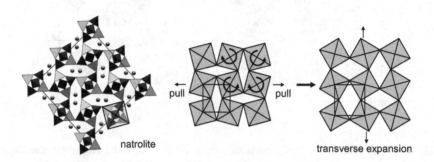

natrolite   transverse expansion

Figure B14.2.2 The structure of natrolite and a simple mechanism by which corner-hinged squares can give a negative Poisson's ratio by expanding in a transverse direction when pulled. Similar coupled rotations can occur on heating, giving rise to RUMs and NTE.

Finally, let's consider the **Poisson's ratio** of a material. This measures how much a material contracts in a direction perpendicular (transverse) to the direction in which it is pulled by a mechanical force (longitudinal) and is defined as $PR = -\varepsilon_{\text{transverse}}/\varepsilon_{\text{longitudinal}}$, where ε is a strain (deformation expressed as a fractional length increase). Most materials such as rubber contract in the transverse direction when stretched and therefore have a positive Poisson's ratio. Cork is an interesting material in that its Poisson's ratio is very close to zero. This has huge practical importance. It means a cork can be pushed into (and, more importantly, later pulled out of) a wine bottle's neck to produce a tight seal; a bulging rubber bung couldn't be! There are rare examples of materials, **auxetic materials**, which have a negative Poisson's ratio and become fatter when stretched or thinner when compressed [42]. One simple geometric way of achieving this is shown in Figure B14.2.2. If we have a 2D network of squares hinged at the corners, pulling the system in one direction will cause the squares to rotate cooperatively and expand in the other direction. The zeolite natrolite has been shown to display auxetic behavior [43]. Similar behavior can be engineered in macroscale objects, has been shown in foams processed to contain re-entrant pores [44], and leads to a number of exciting applications.

* These types of vibrations have a negative **Grüneisen parameter**, γ, which relates vibrational frequency ν to volume via $\gamma = -d(\ln\nu)/d(\ln V)$. The volume thermal expansion of a material depends on γ via the relationship $\alpha_V = \bar{\gamma} C_V K/V_{\text{mol}}$ (see, for example, ref. [34]; C_V is the specific heat per mole at constant volume, K is the isothermal compressibility, $\bar{\gamma}$ averaged over all modes). Since C_V and K are both positive, negative Grüneisen parameter modes lead to a negative α_V and contraction.

14.4 Metal–Organic Frameworks (MOFs)

14.4.1 MOF Structures

While the microporous zeolites, mesoporous silicas, and porous oxides discussed above take much of their inspiration from mineralogy and oxide chemistry, our final family of porous materials, **metal–organic frameworks (MOFs)** (also known as **coordination polymers** or **hybrid porous materials**), makes use of ideas from coordination chemistry. The key design concept behind these materials is to take metal ions or small metal-containing clusters as nodes and to link them with multidentate ligands to produce framework structures.

The simplest illustration of this idea is given by Prussian-blue-related materials (Section 7.4.2). If one mixes Cd^{2+} and $[Pd(CN)_6]^{2-}$ ions in solution, a precipitate is formed, in which the N atoms of six different $[Pd(CN)_6]^{2-}$ octahedra coordinate Cd^{2+} to produce an infinite network of linked octahedral centers (Figure 14.12).[44] The simplest analogy is to the network of corner-linked octahedra of ReO_3, but with the single O linking atom replaced by the CN group (a network expansion) and Re by Pd/Cd (site ordering); an alternative analogy is to the net of B atoms in CaB_6 (vertex decoration).[45]

Similarly, when $Zn(CN)_2$ is precipitated from solution, the preference of Zn for tetrahedral coordination leads to a structure containing a diamond-like network of linked $Zn(CN)_{4/2}$ tetrahedra.[46] The large distance between the tetrahedral Zn centers (Zn–CN–Zn is ~5.11 Å) would, however, lead to a structure with too high a porosity to be stable. $Zn(CN)_2$ overcomes

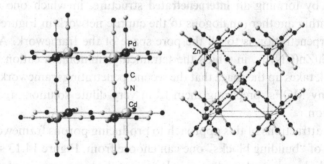

Figure 14.12 The structures of $Cd[Pd(CN)_6]$ (left) and $Zn(CN)_2$ (right). C/N sites are disordered in $Zn(CN)_2$. One network is drawn with dashed bonds.

[44] Prussian blue itself can be made by adding $K_4[Fe^{II}(CN)_6]$ solution into an excess of Fe^{3+}(aq) producing a blue precipitate $Fe_4[Fe(CN)_6]_3 \cdot xH_2O$ with $14 \leq x \leq 16$. Although the structure is based on a network of linked octahedral Fe centers, there are significant defects on the Fe^{II} and CN sites that are occupied by water; Fe^{III} is coordinated by an average of 4.5 N atoms and 1.5 O atoms.

[45] The structure of CaB_6 is shown in Figure 1.40. A B_6^{2-} octahedral anion decorates the Te^{2-} sites of the CsCl-structured CaTe, with short B–B distances to neighboring octahedra.

[46] CN groups show CN/NC disorder.

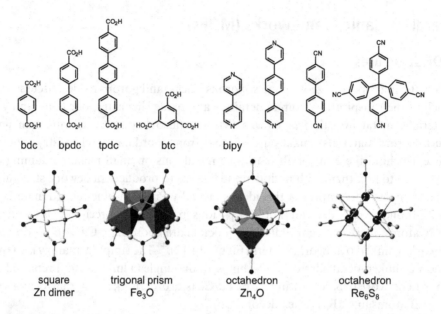

Figure 14.13 Top: Linkers commonly used in the preparation of MOFs; abbreviations are given for those discussed in text. Bottom: Nodes of specified geometrical connectivity with an increasing number of metal atoms (linkers' functional groups are drawn to indicate the connectivity).

this problem by forming an **interpenetrated** structure, in which one diamond network is interleaved within another (analogous to the cuprite network in Figure 1.42). A simple way to avoid interpenetration is to fill the pore space of the framework. An example of this is $[N(CH_3)_4]^+[CuZn(CN)_4]^-$, in which the tetramethylammonium cation that charge-balances the framework takes up the space that the second penetrating framework would have adopted [45]. For many MOFs, preparing them from very dilute solutions reduces the chances of interpenetration.

One of the attractions of this approach to producing porous frameworks is the essentially unlimited set of "building blocks" one can choose from. Figure 14.13 shows some common multidentate linking ligands and a tiny selection of the huge array of metal nodes possible. In addition, it is possible to complete the coordination environment of metals at nodes with non-linking ligands. This leads to an enormous range of potential structures and the possibility of designing materials with specific pore shapes and controllable chemical and physical properties. Note that with MOFs the structure's topology is controlled by the connectivity enforced by the nodes employed, rather than by the size/shape of a template, as in zeolite syntheses.

The natural language for describing MOF structures is that of nets [46]. In many cases, it is found that structures adopt one of a relatively small number of **default nets** that represent the simple high-symmetry ways of linking together nodes of different connectivity. Many of these are familiar from the structures in Chapter 1 and the discussion in Section 1.4.3. As shown in Table 1.6, for six-coordinate octahedral nodes we expect the α-Po or NaCl net; for four-coordinate tetrahedral nodes we expect nets related to diamond, quartz, cristobalite, or the zeolites; with equal numbers of square planar and tetrahedral nodes the PtS net; and for octahedral and trigonal centers in equal proportion the TiO_2 rutile net. For three-coordinate nodes a common arrangement is the $10_5 10_5 10_5$ network drawn in Figure 1.36, which is often referred to as the (10,3)-a network in the literature.[47] This network is chiral and contains fourfold helices of the same handedness parallel to each of the cubic axes [47, 48].

To describe and systematize MOF structures, their nets are often given simple codes of the form **abc-d**. For example, a primitive cubic net is called **pcu**. When a vertex is placed halfway along the edge, the expanded net is called **pcu-e**. This net is the O net in ReO_3 and sufficiently common that it is given its own name—**reo** (i.e. **pcu-e ≡ reo**). These symbols can also be used for zeolites. For example, the net of four-coordinated T vertices of the zeolite faujasite has been given the symbol **fau**. When considering the O atoms of faujasite, the expanded structure is called **fau-e**. There are excellent reviews available [49] and various online databases and software packages available for analyzing nets.[48]

The discipline of targeting a MOF with a specific topology has been termed **reticular chemistry**.[49] The MOF-5 family of materials illustrates this approach: MOF-5 (Figure 14.14) can be crystallized from Zn salts and benzenedicarboxylic acid (bdc) in N,N'-diethylformamide at 85–105 °C. These experimental conditions favor the in situ

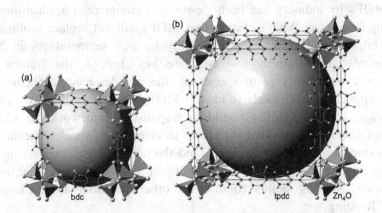

Figure 14.14 Structures of two MOF-5-related materials that form part of an isoreticular series. Spheres of radii 8 Å (a) and 11.8 Å (b) are drawn inside the pores to emphasize porosity.

---

[47] Here the (n,m) nomenclature refers to a network made of nodes of connectivity m, in which the shortest closed loop has n nodes; the attached "a" says it has the highest symmetry of several arrangements possible.

[48] For example, the reticular chemistry structure resource at http://rcsr.anu.edu.au/, the EPINET (Euclidean Patterns in Non-Euclidean Tilings) resource at http://epinet.anu.edu.au/, and ToposPro at http://topospro.com.

[49] From the Latin *reticulum*, meaning little net.

formation of $Zn_4O(CO_2R)_6$ clusters made up of four ZnO_4 tetrahedra that share a common corner (Figure 14.13). The six carboxylate groups, which complete the Zn coordination, emanate octahedrally from the cluster and link to other clusters in a CaB_6-type network. The framework of MOF-5 remains intact on heating to remove solvent,[50] leading to a material that is truly porous and stable up to around 400 °C (for the related IRMOF-6). The multidentate binding of the carboxylate anion linker and the inherent rigidity of the Zn_4O building block are thought to impart this stability. It is also possible to replace the short bdc linker by a variety of larger linkers, including bpdc and tpdc of Figure 14.13, to produce more open frameworks. This is the process of **expansion** that gives rise to structures that are members of an **isoreticular series** (maintain the node connectivity).[51] The tpdc material (Figure 14.14b) has a cell parameter of 21.49 Å, a percentage free volume of 91.1%, and a remarkably low density of just 0.21 g/cm^3.[52] A sphere of diameter 28.8 Å will fit in the cavity without overlapping any of the framework atoms. The available pore space in these and later generations of MOFs is far higher than in any zeolite and only exceeded in amorphous xerogels and aerogels. Surface areas up to 10000 m^2/g are possible. However, as with the cyanidometallates discussed above, the likelihood of forming interpenetrated networks increases with linker length.

14.4.2 Some Applications of MOFs

While zeolites and other porous oxides have had enormous industrial impact, the take-up of MOFs by industry has been slower and commercial applications are only just beginning to emerge. What's clear is that MOFs will not replace zeolites in their large-scale catalytic applications, where stability at high temperatures in harsh chemical environments and relatively low cost are key. Indeed, the framework linkers in MOFs are exactly the type of species zeolites are designed to chemically transform. Instead, applications are likely to exploit MOFs' high porosities for molecular storage and release, the possibility of building responsive frameworks in which bonds can break and reform dynamically, the ability to incorporate specific chemical functionality either during or after MOF synthesis, and the possibility of engineering multifunctionality in a single material. We will highlight just some of the potential uses of MOFs in this section. Starting points to discover other potential applications are given in Further Reading.

There has been significant interest in exploiting the large internal surface area of MOFs for the physisorption of gases such as CH_4, H_2, CO_2, CO, and NO. In the case of H_2, the

---

[50] This is not the case for many MOFs.   [51] An isoreticular series of structures have the same topology or net.

[52] When discovered, this was the lowest density known for a crystalline material. Still lower densities have since been found. For example the covalent organic framework COF-108 has a density of 0.17 g/cm^3 [El-Kaderi et al., *Science* **316** (2007), 268–272] and MOF-339 a density of 0.13 g/cm^3 and void volume of 94% [Furukawa et al., *Inorg. Chem.* **50** (2011), 9147–9152].

possibility of using MOFs for on-vehicle storage stimulated large amounts of research. MOF-5, for example, stores up to 10 wt% H_2 at 77 K and 100 bar, corresponding to a volumetric density of 66 g H_2 per liter, more than twice that of pure H_2 under the same conditions (31 g/L). Unfortunately, at 298 K the adsorption capacity decreases significantly to around 8.9 g/L, so it is little better than the pure gas (8.7 g/L). Similar behavior occurs with most hydrogen-storage MOFs and the best materials have gravimetric capacities of $\lesssim 0.5$–1 wt% at room temperature. The capacity decrease with temperature arises from relatively low isosteric heats of adsorption[53] of 5–12 kJ/mol; values of 15–25 kJ/mol are probably needed for realistic applications. The best room-temperature storage is achieved in MOFs that contain metal nodes with open coordination sites, but volumetric capacities remain limited to $\lesssim 12$ g/L.

A second fuel-storage application that was researched extensively targets methane-powered vehicles. MOF-loaded tanks capable of storing two to three times more CH_4 than regular tanks at comparable pressures and temperatures have been successfully cycled thousands of times, though have not yet been commercialized. Another fascinating potential application is the use of MOFs to capture water from the air. For example, Kim et al. [50] have shown that $Zr_4O_4(OH)_4(fumarate)_6$ (MOF-801) will take up water from air of a mere 20% humidity at 25 °C in the dark. Sunlight can then be used to heat the MOF and release the water. Devices capable of producing 2.8 L of water per kg of MOF have been described.

The gas-storage properties of MOFs are particularly dramatic when specific host–guest interactions can be built into the framework. For example, $Cu_2(pzdc)_2pyr$ (pzdc = pyrazine-2,3-dicarboxylate, pyr = pyrazine) contains 1D channels in which selective adsorption of acetylene over the similarly sized CO_2 occurs due to strong hydrogen bonding between acetylene and a carboxylate oxygen. The amount of acetylene stored is 0.43 g/cm^3 and would correspond to a hypothetical acetylene pressure of 41 MPa at room temperature, 200 times the pressure at which it would normally explode [51]. There is also interest in MOFs that can selectively trap CO_2 from mixed-gas streams and even transform it chemically [52]. Tricks such as decorating the interior of pores with amine groups to enhance the interaction with CO_2 have been particularly successful. At the time of writing, the first smaller-scale MOF-based products, based on gas capture/ release properties, are starting to be produced commercially [53]. These include: MOF-loaded gas cylinders to allow the storage of toxic gases at sub-atmospheric pressures, removing the risk of leakage; MOF sachets that release the fruit-ripening inhibitor 1-methylcyclopropane; and antimicrobial-release devices.

A second interesting aspect of MOF chemistry is the potential for coordination "flexibility". In the zeolites we discussed in the first part of this chapter, strong Si–O and Al–O bonds mean that their frameworks remain intact during most sorption and

---

[53] Isosteric heat of absorption is a measure of the strength of interaction between the gas and MOF surface. It's defined as the difference between the partial molar enthalpies of a species in the bulk fluid and adsorbed phases.

catalytic applications. With MOFs, it is possible to use weakly binding groups such that framework linkages can break and reform reversibly. For example, the structure obtained on dehydrating $Ni_2(4,4'\text{-bipy})_3(NO_3)_4 \cdot 4H_2O$ contains cavities large enough to hold a toluene molecule, but has connecting windows that are only ~50% of the size required for toluene to diffuse through the framework to reach them. Nonetheless toluene can be absorbed into the structure suggesting that local flexibility imparted by breaking weak C–H⋯O hydrogen bonds allows the windows to open, permitting passage of the guest, then close, to "trap" the guest in the structure. Guest uptake is therefore controlled by the size of the cavity itself rather than that of the window leading to it.

A third attractive feature of MOFs is the way in which their structures can be modified and exploited to facilitate useful chemistry. We'll exemplify this by touching on their possible use in heterogeneous catalysis. We can conceptually think of five different loci of MOF-catalysis centers, each of which can make use of the shape selectivities discussed in Sections 14.1.3 and 14.2 [54]. Metal-centered catalysis can occur either at (1) the nodes (particularly if they have vacant coordination sites), (2) at linkers containing catalytically active metal centers incorporated during MOF synthesis, or (3) at linkers modified post-synthesis to incorporate metals. Similarly, an active organic catalyst can be used as (4) a linker during MOF synthesis or (5) created by modifying the linker post-synthesis. In addition, one can prepare MOFs from chiral linkers. This leads to the possibility of performing enantioselective catalysis as explored in Box 14.3.

The advantages of a MOF-based catalyst over a homogeneous catalyst combine the advantages of zeolites and porous oxides discussed in earlier sections with additional factors: Their heterogeneous nature means that expensive catalysts can be readily separated from liquid products and re-used. This also helps reduce the contamination of products with potentially toxic metals. The ability to anchor and isolate catalytic species in pores can increase the catalyst lifetime as any deactivating processes involving more than one catalytic center (e.g. clustering or the attack of the fragile ligand set on one catalytic molecule by another) are decreased. MOFs also have a particular advantage in the tunability of their shape selectivity as pore sizes can be systematically controlled through reticular chemistry. One would expect, and often finds experimentally, that decreasing the pore size will increase the product selectivity, though often at the expense of the reaction rate. Conversely, increasing the pore size typically increases reaction rates as reagents and products can diffuse more rapidly to and away from active centers. Rates and selectivities are expected to approach those of homogeneous catalysts as pore sizes increase, and this has again been verified experimentally. One can even imagine dual MOF catalysts where tandem or sequential processes occur at different active centers, or using photochemistry at a metal center to promote organic catalysis at a linker. This range of possibilities and post-synthetic control sets MOFs apart from zeolites and mesoporous materials.

Box 14.3 Materials Spotlight: Asymmetric catalysis using MOFs

There is enormous interest in the synthesis of chiral molecules, that is molecules whose object and mirror image are non-superimposable (**enantiomers**). Enantiomers often have significantly different biological activity, such that **enantioselective synthesis** or catalysis is enormously important to the pharmaceutical industry. MOFs offer exciting opportunities in this arena since their design-and-synthesis strategy allows much of the learning from homogeneous asymmetric catalysis to be reapplied in heterogeneous catalysts, with the added benefits of easy catalyst capture and recycling and pore-size control. Enantioselective catalysis using MOFs has been demonstrated using each of the five strategies to produce catalytic sites discussed in the main text and in Figure B14.3.1.

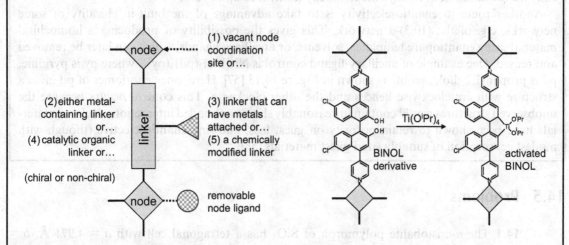

Figure B14.3.1 Types of catalytic center possible in MOFs and schematic example of post-synthesis linker functionalization.

The first demonstration of MOF-based asymmetric catalysis used an enantiopure Zn MOF called POST-1[54] and relied on organocatalysis using the linker [55]. POST-1 was synthesized using an enantiopure linker containing one carboxylate and one pyridyl binding site, producing large chiral 1D channels. Half the pyridyl nitrogens coordinate to Zn clusters at the nodes and half point into the framework channels. Its chemical tunability was demonstrated by showing that channel nitrogens could undergo chemical reactions post-synthesis, and the chiral properties of the framework by demonstrating enantioselective uptake of metal complexes from solution. Most importantly, the framework pyridyl groups were shown to catalyze an organic reaction[55] with a

---

[54] POST-1 is [Zn₃(μ-O)(LH)₆]·2H₃O·12H₂O, where linker L contains one carboxylate and one pyridyl group and can be prepared enantiopure from tartaric acid.

[55] Specifically a transesterification reaction with an ee of 8%.

Box 14.3 (cont.)

small enantiomeric excess (ee). The relatively low ee was ascribed to the active centers being remote from the full influence of the chiral pore walls.

Similar issues of low ee might be expected if one relied on using chiral framework linkers to induce enantioselective catalysis at metal nodes. Despite this, early work demonstrated that a framework made from Zn and a chiral triangular linker[56] could achieve a 40% ee for a Lewis-acid-catalyzed reaction.[57]

The first example of linker–metal-based asymmetric catalysis in a crystalline MOF was achieved in a Cd MOF built with a pyridyl-functionalized BINOL[58] linker (see Figure B14.3.1) [56]. After synthesis, the two free OH groups on the linker could be reacted with $Ti(O^iPr)_4$ to give a $[MOF]Ti(BINOL)(O^iPr)_2$ material. This mimics many solution-based catalysts and was shown to catalyze the addition of diethylzinc to aldehydes with ee up to 93%.

Another route to enantioselectivity is to take advantage of the built-in chirality of some networks, e.g. of the (10,3)-a network. This gives the possibility of producing a homochiral material using enantiopure templates, solvents, or ancilliary ligands, which can later be removed and reused. One example of ancillary-ligand control is $Ni_3(btc)_2(pd)_3(py)_6$, where py is pyridine, pd is propan-1,2-diol, and btc is shown in Figure 14.13 [57]. Here, one enantiomer of pd gives a structure with anticlockwise helices and the other clockwise. This control occurs because the unobserved structures would contain unreasonably short nonbonded interactions. Related materials have been shown to retain porosity on guest loss, allowing enantioselective (though with modest ee) sorption of suitably sized guest materials [58].

14.5 Problems

14.1 The α-cristobalite polymorph of SiO_2 has a tetragonal cell with $a = 4.971$ Å, $b = 6.922$ Å, with Si on $2a$ and O on $4f$ Wyckoff sites. The Si–O bond length is 1.60 Å and the average O–O distance within tetrahedra is 2.62 Å. The high-pressure stishovite polymorph of SiO_2 has the rutile structure (Figure 1.45) with a tetragonal cell of dimensions $a = 4.177$ Å and $c = 2.665$ Å, and occupied Wyckoff sites of $2a$ (Si) and $4f$ (O). Data for a zeolite A polymorph of SiO_2 (LTA) are given in the table accompanying Problem 14.4. (a) Calculate the theoretical density of each polymorph. (b) Assuming that Si and O can be treated as hard spheres and that O atoms are in contact in cristobalite, estimate hard-sphere radii for Si and O. (c) Estimate the volume percent filled by atoms in each polymorph. Comment on your answers.

14.2 Calculate the surface area of (a) a 1 g single crystal of cristobalite (density = 2.3 g/cm³, assume the crystal takes a spherical shape), (b) a uniform powder of spherical crystals

[56] The material was $Zn_3(ChirBTB-1)_2$ where BTB-1 is [4,4′,4″-benzene-1,3,5-triyl-tribenzoate] with a chiral oxazoline group at the ortho position of each benzoate.

[57] Specifically conversion of 1-napthylaldehyde in the Mukaiyama aldol reaction.

[58] BINOL is the colloquial name of [1,1′-binapthelene]-2,2′-diol. It is widely used as a ligand in homogeneous catalysis as its two enantiomers are easily separable and stable to racemization.

each with a diameter of 10 μm, and (c) uniform spherical particles with a diameter of 10 Å. Compare your values to the surface area of a typical zeolite of 900 m²/g.

14.3 Zeolite A can be described using a 11.9 Å primitive cubic cell with the origin at a β-cage center in Figure 14.3. From the figure, make a visual estimate of the size of a sphere that would fit inside the β and α cages. By using the information in the table accompanying Problem 14.4, estimate the surface area of these spheres in a 1 g sample of zeolite A. Compare your answer to the accessible area of 1205 m²/g quoted in the text.

14.4 Calculate the number of tetrahedral atoms per nm³ for each zeolite framework type listed in the table below.

Name	Space group	Cell (Å)	Site	x	y	z	Wyckoff
SOD	$Im\bar{3}m$	8.9650	T1	0.2500	0.5000	0.0000	12d
			O1	0.1467	0.5000	0.8533	24h
LTA	$Pm\bar{3}m$	11.919	T1	0.0000	0.1823	0.3684	24k
			O1	0.0000	0.2122	0.5000	12h
			O2	0.1103	0.1103	0.3384	24m
			O3	0.0000	0.2967	0.2967	12i
FAU	$Fd\bar{3}m$	24.34	T1	0.9469	0.1251	0.0364	192i
			O1	0.8958	0.1042	0.0000	96h
			O2	0.9669	0.0762	0.0762	96g
			O3	0.9966	0.1429	0.9966	96g
			O4	0.9283	0.1770	0.0730	96g
ANA	$Ia\bar{3}d$	13.567	T1	0.6616	0.5884	0.1250	48g
			O1	0.6417	0.4716	0.1180	96h
RHO	$Im\bar{3}m$	14.919	T1	0.2500	0.1037	0.3963	48i
			O1	0.2754	0.1193	0.5000	48j
			O2	0.3344	0.1298	0.3344	48k

14.5 The table above contains ideal framework coordinates for different zeolites. Use a package such as vesta (http://jp-minerals.org/vesta/en/) to produce a 3D drawing of each structure. For each example state: (a) which SBUs are present (see Figure 14.4); (b) the vertex symbol of the T site; (c) the coordination sequence (out to the level N3).

14.6 Assuming that O atoms are in contact and have a radius of 1.35 Å, estimate the pore opening formed by a planar six-, eight-, ten-, and twelve-membered $SiO_{4/2}$ ring. Note that if the rings are nonplanar, pore sizes can be considerably reduced. Hint: Consider the internal size of a circle drawn inside a ring of circles.

14.7 In NaA, the total void space accessible is approximately 926 Å³. Sorption studies suggest H_2O fills a volume ~833 Å³, whereas N_2 only fills ~755 Å³. Comment on these observations.

14.8 Under certain conditions, a zeolite with the ZSM-5 framework and formula $NaAlSi_{23}O_{48}$ (**A**) can be synthesized. In the presence of $(CH_3CH_2CH_2)_4NOH$ it is possible to prepare a material (**B**) with an extremely similar X-ray-diffraction pattern and the elemental composition N 0.86%, C 8.86%, H 1.73%, Al 1.66%, Si 39.70%, O 47.19% by weight. On heating to 500°C **B** loses 11.39% of its mass to form **C**. (a) Calculate the empirical formula of **B** from the elemental analysis. (b) Given the similarity in the X-ray-diffraction patterns of **A** and **B**, suggest a sensible chemical formula for **B**. (c) Calculate the empirical formula of **C**. (d) State how a Na-containing zeolite such as **A** can be converted to an acidic form. (e) State the roles played by organic cations such as $[(CH_3CH_2CH_2)_4N]^+$ in zeolite synthesis.

14.9 A mixture of $Si(OCH_3)_4$, $(CH_3)_4NOH(aq)$, and $Al(OCH_2CH_2CH_3)_3$ was heated to drive off alcohol, resulting in a clear solution. This solution was heated at 130 °C in a sealed autoclave for 8 days. The resulting white precipitate was washed with distilled water and dried at 100 °C, yielding a white powder **A** whose X-ray-diffraction pattern was very similar to that of the mineral gismondine ($CaAl_2Si_2O_8 \cdot 4H_2O$). Chemical analysis of **A** gave weight percentages N 4.2%, C 14.5%, H 4.2%, Al 8.1%, Si 25.4%, O 43.5%. (a) Calculate the empirical formula of **A**. (b) Given that the X-ray-diffraction pattern is similar to gismondine, suggest a sensible molecular formula for **A**. (c) In light of your answer to (a) and (b), discuss how template molecules may be used to influence the hydrophilicity of zeolites.

14.10 Haag and co-workers measured the cracking rate of *n*-hexane by HZSM-5 relative to a high-surface-area catalyst under identical experimental conditions [59]. The dependence of the rate on the Al:Si ratio and on the intensity of the signal due to tetrahedral Al sites in the ^{27}Al solid state NMR are shown in the figures below. State what you can conclude from the linear dependence of activity on both the Al:Si ratio and the NMR signal from tetrahedral sites.

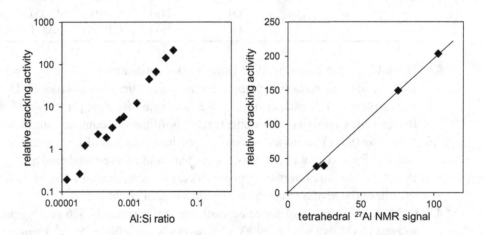

14.11 When dimethylbenzenes (xylenes) are passed over acidic zeolites (HZSM-5 that contains 10-rings and HY that contains larger 12-rings), two processes can occur: isomerization to different mixtures of *ortho*, *meta*, and *para* isomers; or disproportionation to toluene (methylbenzene) and trimethylbenzenes, as shown in the table. State which of the shape-selective catalysis effects shown in Figure 14.7 is responsible for the enhanced *para-* to *ortho*-selectivity of HZSM-5 and for favoring the isomerization reaction over disproportionation in HZSM-5.

Zeolite	*para-/ortho*-selectivity	Isomerization/disproportionation
HZSM-5	2.9	33
HY	1.0	1.5

14.12 The first peak in the powder diffraction pattern of a mesoporous MCM material with 30 Å pores recorded with a wavelength λ of 1.54 Å is at 2.2° 2θ. Estimate the thickness of the silica walls. A hexagonal material has $a = b \neq c$, $\alpha = \beta = 90$, $\gamma = 120$; Bragg's law states $\lambda = 2d_{hkl}\sin\theta$.

14.13 Crystallographers have a rough "rule of thumb" that non-H atoms occupy around 18 Å^3 in many crystal structures. Based on this assumption, estimate the percent pore space in the tpdc MOF on the right of Figure 14.14.

14.14 Consider the PtS net listed in Table 1.6. The dehydrated form of MOF-11 has composition $Cu_2(ATC)$ where ATC is 1,3,5,7-adamantane tetracarboxylate and contains a Cu paddle-wheel unit analogous to the Zn dimer shown in Figure 14.13. Describe the structure of PtS. Based on the information given and your chemical knowledge, suggest possible structures for MOF-11 and $ZnPt(CN)_4$.

14.15 Consider a heterogeneous enantioselective catalyst based on a catalytically active metal center attached to a MOF linker. (a) How would you expect the reaction rate to compare to the equivalent catalyst in solution? How might reaction rate vary as the pore size varied? (b) How might the enantiomeric excess compare to the equivalent catalyst in solution? (c) How might catalyst stability and turn-over compare to the equivalent catalyst in solution? (d) How would you prove that catalysis is truly heterogeneous and not caused by catalyst leaching from the MOF into solution? (e) How would you prove that catalysis occurs in the pores of the MOF and is not restricted to surface sites?

14.6 Further Reading

P.A. Wright, "Microporous Framework Solids" *RSC Materials Monographs* (2008) Cambridge.

M.E. Davis, "Ordered porous materials for emerging applications" *Nature* 417 (2002), 813–821.

T.J. Barton, L.M. Bull, W.G. Klemperer, D.A. Loy, B. McEnaney, M. Misono, P.A. Monson, G. Pez, G. W. Scherer, J.C. Vartuli, O.M. Yaghi, "Tailored porous materials" *Chem. Mat.* 11 (1999), 2633–2656.

F. Schüth, "Engineered porous catalytic materials" *Ann. Rev. Mater. Res.* 35 (2005), 209–238.

M. Eddaoudi, D.B. Moler, H. Li, B. Chen, T.M. Reineke, M. O'Keeffe, O.M. Yaghi, "Modular chemistry: Secondary building units as a basis for the design of highly porous and robust metal–organic frameworks" *Acc. Chem. Res.* 34 (2001), 319–330.

S. Kitagawa, R. Kitaura, S. Noro, "Functional porous coordination polymers" *Angew. Chem., Int. Ed.* **43** (2004), 2334–2375.

S. Kitagawa, K. Uemara, "Dynamic porous properties of coordination polymers inspired by hydrogen bonds" *Chem. Soc. Rev.* **34** (2005), 109–119.

There are many other excellent reviews of MOFs available. For example, the February 2012 issue of *Chemical Reviews* (**112**, 673–1268) contains 17 review articles on different aspects of their chemistry and applications. The August 2014 issue (**43**, 5403–6176) of *Chemical Society Reviews* contains 30 further reviews.

14.7 References

[1] R.D. Shannon, "Revised effective ionic radii and systematic studies of interatomic distances in halides and chalcogenides" *Acta Crystallogr. Ser. A* **32** (1976), 751–767.

[2] Ch. Baerlocher, L.B. McClusker, D.H. Olson, "*Atlas of Zeolite Framework Types*" 6th edition (2007) Elsevier; www.iza-structure.org/databases/.

[3] L.B. McCusker, F. Liebau, G. Engelhardt, "Nomenclature of structural and compositional characteristics of ordered microporous and mesoporous materials with inorganic hosts" *Pure Appl. Chem.* **73** (2001), 381–394.

[4] V.A. Blatov, O. Delgado-Friedrichs, M. O'Keeffe, D.M. Proserpio, "Three-periodic nets and tilings: Natural tilings for nets" *Acta Crystallogr. Ser. A* **63** (2007), 418–425.

[5] N.A. Anurova, V.A. Blatov, G.D. Ilyusin, D.M. Prosperio, "Natural tilings for zeolite-type frameworks" *J. Phys. Chem. C* **114** (2010), 10160–10170.

[6] M.M.J. Treacey, I. Rivin, E. Balkowsky, K.H. Randall, M.D. Foster, "Enumeration of periodic tetrahedral frameworks, II. Polynodal graphs" *Micropor. Mesopor. Mater.* **74** (2004), 121–132.

[7] O. Delgado-Friedrichs, A.W.M. Dress, D.H. Huson, J. Klinowski, A.L. Mackay, "Systematic enumeration of crystalline networks" *Nature* **400** (1999), 644–647.

[8] A. Corma, M.J. Díaz-Cabañas, J.L. Jordá, C. Martinez, M. Moliner, "High-throughput synthesis and catalytic properties of a molecular sieve with 18- and 10-member rings" *Nature* **443** (2006), 842–845.

[9] G. Sastre, A. Corma, "Predicting structural feasibility of silica and germania zeolites" *J. Phys. Chem. C* **114** (2010), 1667–1673.

[10] R.M. Barrer, "Syntheses and reactions of mordenite" *J. Chem. Soc.* (1948), 2158–2163.

[11] C.S. Cundy, P.A. Cox, "The hydrothermal synthesis of zeolites: Precursors, intermediates and reaction mechanisms" *Microporous and Mesoporous Materials* **82** (2005), 1–78.

[12] J. Felsche, S. Lugher, C. Baerlocher, "Crystal structure of the hydro-sodalite $Na_6[AlSiO_4]_6 \cdot 8H_2O$ and of the anhydrous sodalite $Na_6[AlSiO_4]_6$" *Zeolites* **6** (1986), 367–372.

[13] C. Baerlocher, W.M. Meier, "Synthese und Kristallstruktur von Tetramethylammonium-sodalith" *Helv. Chim. Acta.* **52** (1969), 1853–1860.

[14] K.D. Schmitt, G.J. Kennedy, "Toward the rational design of zeolite synthesis: The synthesis of zeolite ZSM-18" *Zeolites* **14** (1994), 635–642.

[15] S.K. Brand, J.E. Schmidt, M.W. Deem, F. Daeyaert, Y. Ma, O. Terasaki, M. Orazov, M.E. Davis, "Enantiomerically enriched, polycrystalline molecular sieves" *Proc. Natl. Acad. Sci. USA* (2017), 201704638.

[16] M. Boronat, A. Corma, "Factors controlling the acidity of zeolites" *Catal. Lett.* **145** (2015), 162–172.

[17] J.F. Haw, "Zeolite acid strength and reaction mechanism in catalysis" *Phys. Chem. Chem. Phys.* **4** (2002), 5431–5441.

[18] D.H. Olson, W.O. Haag, "Structure-selectivity relationship in xylene isomerization and selective toluene disproportionation" *ACS Symp. Ser.* **248** (1984), 275–307.

[19] N.Y Chen, W.O. Haag, "Hydrogen transfer in catalysis on zeolites" in "*Hydrogen Effects in Catalysis*" Editors: Z. Paal, P.G. Mena (1988) Marcel Dekker, 695–722.

[20] A. Primo, H. Garcia, "Zeolites as catalysts in oil refining" *Chem. Soc. Rev.* **43** (2014), 7548–7561.

[21] A. Dyer, "Ion exchange properties of zeolites and related materials" in "*Handbook of Porous Solids*" Editors: F. Schüth, K.S.W. Sing, J. Weitkamp (2002) Wiley, 525–554.

[22] W. Schmidt, "Applications of microporous materials as ion-exchangers" in "*Handbook of Porous Solids*" Editors: F. Schüth, K.S.W. Sing, J. Weitkamp (2002) Wiley, 1058–1097.

[23] A. Dyer, "*An Introduction to Zeolite Molecular Sieves*" (1988) John Wiley and Sons.

[24] T.R. Gaffeny, "Porous solids for air separation" *Curr. Opin. Mater. Sci.* **1** (1996), 69–75.

[25] M. Bartolomei, E. Carmona-Novillo, M.I. Hernandez, J. Campos-Martinez, R. Hernandez-Lamoneda, "Long-range interaction for dimers of atmospheric interest: Dispersion, induction and eletrostatic contributions for O_2–O_2, N_2–N_2 and O_2–N_2, *J. Comp. Chem.* **32** (2010), 279–290.

[26] Z.P. Lai, G. Bonilla, I. Diaz, J.G. Nery, K. Sujaoti, M.A. Amat, E. Kokkoli, O. Terasaki, R.W. Thompson, M. Tsapatsis, D.G. Vlachos, "Microstructural optimization of a zeolite membrane for organic vapor separation" *Science* **300** (2003), 456–460.

[27] C.T. Kresge, M.E. Loeonowicz, W.J. Roth, J.C. Vartuli, J.S. Beck, "Ordered mesoporous molecular sieves synthesized by a liquid-crystal mechanism" *Nature* **359** (1992), 710–712.

[28] J.S. Beck, J.C. Vartuli, W.J. Roth, M.E. Leonowicz, C.T. Kresge, K.D. Schmitt, C.T-W. Chu, D. H. Olson, E.W. Sheppard, S.B. McCullen, "A new family of mesoporous molecular sieves prepared with liquid crystal templates" *J. Am. Chem. Soc.* **114** (1992), 10834–10843.

[29] D. Zhao, J. Feng, Q. Huo, N. Melosh, G.H. Fredrickson, B.F. Chmelka, G.D. Stucky, "Triblock copolymer syntheses of mesoporous silica with periodic 50 to 300 ångstrom pores" *Science* **279** (1998), 548–552.

[30] J.M. Thomas, R. Raja "Exploiting nanospace for asymmetric catalysis: Confinement of immobilized, single-site chiral catalysts enhances enantioselectivity" *Acc. Chem. Res.* **41** (2008), 708–720.

[31] T. Prasomsri, W. Jiao, S.Z. Weng, J. Garcia Martinez, "Mesostructured zeolites: Bridging the gap between zeolites and MCM-41" *Chem. Commun.* **51** (2015), 8900–8911.

[32] E.S. Toberer, R. Sheshadri, "Template-free routes to porous inorganic materials" *Chem. Commun.* (2006), 3159–3165.

[33] T.A. Mary, J.S.O. Evans, T. Vogt, A.W. Sleight, "Negative thermal expansion from 0.3 to 1050 kelvin in ZrW_2O_8" *Science* **272** (1996), 90–92.

[34] M. Dove, H. Fang, "Lattice dynamics, negative thermal expansion, and associated anomalous physical properties of materials" *Rep. Progr. Phys.* (2016), 066503.

[35] A.L. Goodwin, M. Calleja, M.J. Conterio, M.T. Dove, J.S.O. Evans, D.A. Keen, L. Peters, M.G. Tucker, "Colossal positive and negative thermal expansion in the framework material $Ag_3[Co(CN)_6]$" *Science* **319** (2008), 794–797.

[36] A. Sartbaeva, S.A. Wells, M.M.J. Treacy, M.F. Thorpe, "The flexibility window in zeolites" *Nature Mater.* **5** (2006), 962–965.

[37] Y, Lee, J.A. Hriljac, T. Vogt, J.B. Parise, G.J. Artioli, "First structural investigation of a super-hydrated zeolite" *J. Am. Chem. Soc.* **123** (2001), 12732–12733.

[38] M. Colligan, Y. Lee, T. Vogt, A.J. Celestian, J.B. Parise, W.G. Marshall, J.A. Hriljac, "High-pressure neutron diffraction study of superhydrated natrolite" *J. Phys. Chem. B* **109** (2005), 18223–18225.

[39] F. Millange, C. Serre, N. Guillou, G. Férey, R.I. Walton, "Structural effects of solvents on the breathing of metal-organic frameworks: An in-situ diffraction study" *Angew. Chemie Int. Ed. Engl.* **47** (2008), 4100–4105.

[40] C. Mellot-Draznieks, C. Serre, S. Srublé, N. Audebrand, G. Férey, "Very large swelling in hybrid frameworks: A combined computational and powder diffraction study" *J. Am. Chem. Soc.* **127** (2005), 16273–16278.

[41] S. Bourrelly, P.L. Llewellyn, C. Serre, F. Millange, T. Loiseau, G. Férey, "Different adsorption behaviors of methane and carbon dioxide in the isotypic nanoporous metal terephthalates MIL-53 and MIL-47" *J. Am. Chem. Soc.* **127** (2005), 13519–13521.

[42] K.E. Evans, A. Alderson, "Auxetic materials: Functional materials and structures from lateral thinking!" *Adv. Mat.* **12** (2000), 617–628.

[43] J.N. Grima, R. Gatt, V. Zammit, J.J. Williams, K.E. Evans, A. Alderson, R.I. Walton, "Natrolite: A zeolite with negative Poisson's ratios" *J. Appl. Phys.* **101** (2007), 086102.

[44] R. Lakes, "Foam structures with a negative Poisson's ratio" *Science* **235** (1987), 1038–1040.

[45] B.F. Hoskins, R. Robson, "Design and construction of a new class of scaffolding-like materials comprising infinite polymeric frameworks of 3D-linked molecular rods. A reappraisal of the zinc cyanide and cadmium cyanide structures and the synthesis and structure of the diamond-related frameworks $[N(CH_3)_4][Cu^IZn^{II}(CN)_4]$ and $Cu^I[4,4',4'',4'''$-tetracyanotetraphenylmethane$]$ $BF_4 \cdot xC_6H_5NO_2$" *J. Am. Chem. Soc.* **112** (1990), 1546–1554.

[46] A.F. Wells, "*Three Dimensional Nets and Polyhedra*" (1977) Wiley-Interscience.

[47] R. Robson, "A net-based approach to coordination polymers" *J. Chem. Soc., Dalton Trans.* (2000), 3735–3744.

[48] O. Yaghi, M. O'Keeffe, N.W. Ockwig, H.K. Chae, M. Eddaoudi, J. Kim, "Recticular synthesis and the design of new materials" *Nature* **423** (2003), 705–714.

[49] M. O'Keeffe, O.M. Yaghi, "Deconstructing the crystal structures of metal organic frameworks and related materials into their underlying nets" *Chem. Rev.* **112** (2012), 675–702.

[50] H. Kim, S. Yang, S.R. Rao, S. Narayanan, E.A. Kapustin, H. Furukawa, A.S. Umans, O.M. Yaghi, E.N. Wang. "Water harvesting from air with metal–organic frameworks powered by natural sunlight" *Science* **356** (2017), 430–434.

[51] R. Matsuda, R. Kitaura, S. Kitagawa, Y. Kubota, R.V. Belosudov, T.C. Kobayashi, H. Sakomoto, T. Chiba, M. Takata, Y. Kawazoe, Y. Mita, "Highly controlled acetylene accommodation in a metal–organic microporous material" *Nature* **236** (2005), 238–241.

[52] C.A. Trickett, A. Helal, B. Al-Maythalony, Z.H. Yamani, K.E. Cordoval, O.M. Yaghi, "The chemistry of metal–organic frameworks for CO_2 capture, regeneration and conversion" *Nature Reviews Materials* **2** (2017), 17045.

[53] N. Notman, "MOFs find a use" *Chem. World* (May 2017), 44–47.

[54] M. Yoon, R. Srirambalaji, K. Kim, "Homochiral metal–organic frameworks for asymmetric heterogeneous catalysis" *Chem. Rev.* **112** (2012), 1196–1231.

[55] J.S. Seo, D. Whang, H. Lee, S.I. Jun, J. Oh, Y.J. Jeon, K. Kim, "A homochiral metal–organic porous material for enantioselective separation and catalysis" *Nature* **404** (2000), 982–986.

[56] C. Wu, A. Hu, L. Zhang, W. Lin, "A homochiral porous metal–organic framework for highly enantioselective heterogeneous asymmetric catalysis" *J. Am. Chem. Soc.* **127** (2005), 8940–8941.

[57] C.J. Kepert, T.J. Prior, M.J. Rosseinsky, "A versatile family of interconvertible microporous chiral molecular frameworks: The first example of ligand control of network chirality" *J. Am. Chem. Soc.* **122** (2000), 5158–5168.

[58] D. Bradshaw, T.J. Prior, E.J. Cussen, J.B. Claridge, M.J. Rosseinsky, "Permanent microporosity and enantioselective sorption in a chiral open framework" *J. Am. Chem. Soc.* **126** (2004), 6106–6114.

[59] W.O. Haag, R.M. Lago, P.B. Weisz, "The active site of acidic aluminosilicate catalysts" *Nature* **309** (1984), 589–591.

15 Amorphous and Disordered Materials

While the majority of this book has focused on crystalline materials with long-range translational order and symmetry, many of the oldest technological materials are amorphous. Glazed Egyptian artifacts have been dated back to 4000 BCE and glass beads used as currency to around 3000 BCE. Glasses are used in a wide variety of scientific instruments such as optical microscopes, telescopes, thermometers, barometers, and the plethora of glassware that chemists use. Many major scientific advances have relied on the ability to engineer a certain type of glass. Without glass, Hooke, van Leeuwenhoek, Pasteur, and Koch would not have been able to investigate microorganisms under a microscope. Without glass lenses in his telescope, Galileo would not have made the celestial observations that changed our view of the universe. Three hundred years after Galileo, the Hooker telescope at the Mount Wilson Institute was used by Edwin Hubble to devise his theory of an expanding Universe. Over 2 tonnes of glass had to be slow-cooled over a year-long period in the first stage of manufacturing its 2.5-m-diameter mirror.

In this chapter, we cover structural and dynamical aspects of glasses, amorphous, and disordered materials; terms that are often used interchangeably in this field. We discuss their formation and the concept of the glass transition, of which P.W. Anderson wrote 25 years ago: "The deepest and most interesting unsolved problem in solid state theory is probably the theory of the nature of glass and the glass transition" [1]. We highlight some of the unique physical properties of glasses, in particular their low-temperature dynamics and conductivity. We will focus mainly on inorganic glasses, but many of the concepts apply equally to organic glasses, amorphous polymers, and biopolymers (e.g. plastics, RNA, DNA, proteins), which can form randomly entangled and intermeshed chains and constitute an important group of amorphous materials called random coil glasses. We point to some of the most important technological uses of glasses. At various points in the chapter, we touch on general structural concepts regarding topology and space filling, which apply to amorphous, quasicrystalline, and crystalline materials.

15.1 The Atomic Structure of Glasses

Glasses differ from crystalline materials in that they don't have a lattice. Atoms positioned on lattices have **long-range order** and form a regular array described by translational symmetry as discussed in Chapter 1. While long-range order is absent in glasses, the distribution of atoms is not completely random since the presence of chemical bonds imposes similar interatomic distances throughout the solid. This local similarity is called **short-range order** and is a property of all solids. The absence of long-range order yields diffraction data (see Box 9.1) with no or very broad Bragg peaks, and it has important implications for the mechanical, electronic, and vibrational properties of glasses. The inorganic glasses we'll focus on can be classified into two large families—**covalent glasses** such as amorphous silica, and **random-packed glasses** such as metallic glasses.

Short-range order in glasses arises due to excessive bonding constraints on the connectivity of local coordination polyhedra. Too many, too strong, or too incompatible bonding constraints[1] destroy lattice periodicity rather than just decreasing the crystal symmetry. This may also happen when a system that would otherwise crystallize is not given enough time to equilibrate.

The nature of the short-range order in glasses is determined by local polyhedra like $BO_{3/2}$ triangles or $SiO_{4/2}$ tetrahedra in a covalent glass, or specific metal polyhedra in random-packed metallic glasses. Glassy SiO_2 is built of corner-sharing SiO_4 tetrahedra with Si–O bond lengths ~1.6 Å and O–Si–O bond angles near 109.5°. The range of distances encountered experimentally is shown in Figure 15.1. As one enlarges the range beyond a single tetrahedron, certain **structural configurations** with similar distances and angles tend to reappear, but not with lattice periodicity. The loss of periodicity is due to differing angles between tetrahedra. This calls for a new way to represent glass structures.

One of the most common ways to describe the structure of a glass is the **pair distribution function (PDF)**, $G(r)$, which describes the distribution (a statistics of occurrences) of *all* pairs of *all* atoms as a function of their interatomic distances, r, throughout the structure. An example of a PDF for silica glass is depicted in Figure 15.1. The function $G(r)$ gives the probability of finding *any* two atoms at a distance r, with the area of each peak corresponding to the number of contributing pairs scaled by the sensitivity of the experimental probe to the specific atoms forming the pair (see ref. [2] for more details). We can see that with increasing r, the peaks increase in width and decrease in amplitude. Above ~5 Å, the large number of different pair distances present in a glass means that only broad, weak peaks are observed in this representation.[2] The first peak near 1.6 Å is due to the four Si–O distances in the SiO_4 tetrahedron.[3] The next peak, due to six O–O

---

[1] An example is B_2O_3 that has to accommodate $BO_{3/2}$ vertex-sharing triangles in a 3D structure.

[2] In contrast, the PDF of a crystalline material shows sharp peaks out to very large r and is only limited by experimental constraints, not by the long-range order of the sample.

[3] The small peaks below 1.6 Å are due to termination errors of the Fourier transformation used to calculate $G(r)$ and arise from the finite experimental data range.

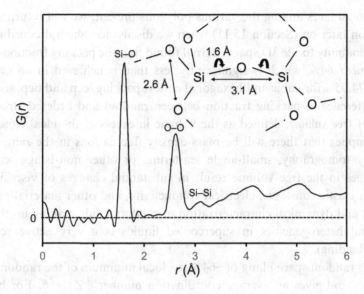

Figure 15.1 PDF describing the distribution of interatomic distances r in amorphous SiO_2. Adapted from ref. [3].

distances within each SiO_4 tetrahedron, appears near 2.6 Å. The increase in width of the second peak over the first one shows the variations in the O–Si–O angles around the ideal tetrahedral value of 109.5°. The average inter-tetrahedral Si–O–Si angle of ~135° can only be reliably extracted using a larger segment of structural configurations. The peak at 3.1 Å is caused by four Si–Si distances, and, as evidenced by its width, it is impacted by the distribution of the angles within the relatively large set of structural configurations. The peaks at 4 Å and 5 Å correspond to the O–O distances between two tetrahedra. These distances will vary upon rotating either tetrahedron around the bond to their shared vertex (which can be described with a torsion angle).

Many glasses form when a melt is cooled sufficiently quickly through a certain temperature, called the **glass-forming temperature**, T_g (more detail in Section 15.7). This approach, called **quenching**, does not allow enough time for an ordered crystalline phase to nucleate, and instead a solid snapshot of a liquid-like structure is formed. We can get some insight into the type of structures that form by considering a **random close packing of spheres**. If one pours spheres into a container and gently taps them down, they do not form the periodic closest-packed arrangement described in Chapter 1.[4] J.D. Bernal explored this type of random packed structure by making physical models using ball bearings and pouring wax over them so he could determine the positions of individual spheres [4, 5]. He observed a predominance

---

[4] Gentle tapping traps the system in a metastable state while more rigorous shaking would result in closer packing. Random close packing is not an intrinsic property of a given material but depends on the packing procedure. These ideas are further explored in Section 15.7.

of pentagonal faces among the various polygons present; we will return to this important observation later on (Section 15.11) when we discuss icosahedral coordination polyhedra and their inability to tile 3D space. Bernal found that the packing fraction of these arrangements is near 64%, which is significantly less than is achieved in an ordered crystalline packing (74.05% for cubic and hexagonal closest packing, ccp and hcp; see Chapter 1).

The difference in packing fraction between random and ordered structures led to the concept of **free volume**, defined as the volume in excess of the ideal closest packing. Free volume implies that there will be mass-density fluctuations in the sample, which can be probed by tomography, small-angle scattering, or other non-Bragg scattering methods [2]. Changes in the free volume result in substantial changes of viscosity and flow and also affect elastic constants, electrical conductivity, and other material properties [6]. The structural and dynamical characterization of random voids making up the free volume in glasses and heterogeneities in supercooled liquids is a very active research area (see Further Reading).

Bernal's random space filling of ~64% is a local minimum of the random packing density for spheres and gives an average coordination number $\langle Z \rangle \approx 6$. For both stretched or flattened spheres with gradually increasing aspect ratios (more prolate and more oblate[5]), the packing volume fraction initially increases above 70% of $\langle Z \rangle \approx 10$ and then falls off again as the aspect ratio grows further [7].

15.2 Topology and the Structure of Glasses

The **topology of glasses** describes the connectivity of the constituent objects rather than distances between them. Both covalent and random-packed glasses have distinctive topologies that can be described using graphs. A **graph** in this context is a three-dimensional object made up of vertices and edges. The graph vertices can be atoms and the edges can be bonds. The graph can therefore correspond to a real structure.[6] Atoms in glasses often have a fixed chemical coordination number, hence the graph vertices have fixed connectivity (e.g. 4 for Si and 2 for O in amorphous SiO_2 or 4 for amorphous Si[7]). Such glasses are called **continuous random networks** (CRN). Models of CRNs were initially built using simple rules: (1) atoms are fully coordinated, (2) there is no long-range order hence no translational symmetry, and (3) there are no coordination defects or voids. Due to their importance in understanding the properties of amorphous materials, advanced CRN-generating algorithms now also take

---

[5] Rotating an ellipse about its major or minor axis leads to a prolate (like a rugby ball) or oblate (like an M&M's chocolate drop) shape, respectively.

[6] This would be typical for covalent glasses. For random-packed glasses, the graph is a so-called Voronoi network where the graph edges are those of polyhedra describing the volume of influence of an individual atom.

[7] This is an idealization for amorphous Si that is known to have macroscopic voids into which some uncoordinated electron density of Si atoms can project. These "dangling bonds" are intrinsic defects occurring at a concentration of about one unpaired electron per 10^3 atoms, and they introduce electron states at the Fermi energy within the band gap. See Section 15.10 for more details.

defects and voids into account. The generated CRN models are used to calculate structural, vibrational, and electronic properties that can be compared to experiments [8].

In Section 1.4.3, we defined a ring as the smallest loop (cycle) that together with other such rings, the same or different, completes an "infinite" structural network. Such **rings** are an important topological feature found in both glasses and crystalline materials. The search for the smallest set of the smallest rings in liquids, crystals, and amorphous materials is also done using sophisticated computer algorithms [9].[8]

A fixed-connectivity network free of rings would give us an ever-growing tree-like topology called a **Bethe lattice**, with the number of nodes in the nth shell being $N_n = Z(Z-1)^{n-1}$, where Z is the vertex coordination number.[9] In the Bethe lattice of SiO_2, an $SiO_{4/2}$ tetrahedron that is the nth neighbor of an arbitrary original "seed tetrahedron" will share one vertex with its $(n-1)$th shell neighbor and three vertices with its $(n+1)$th shell neighbors giving $4 \times 3^{n-1}$ tetrahedra in the nth shell. This Bethe lattice with vertex coordination number 4 is shown in Figure 15.2 up to the third shell as a surface. In this network, we can form rings by pruning branches from the "seed tetrahedron" outward and connecting the resulting underbonded atoms with other atoms in close proximity.

These ideas are also useful in crystalline materials. 3-rings are the smallest ring we can form by "pruning two tetrahedral branches" that grow from the first shell. As a result, the seed tetrahedron is still surrounded by four nearest-neighbor tetrahedra but only by 10 and not 12 second-neighbor tetrahedra [$N_2 = (4 \times 3^1) - 2$]. The formation of 4-rings by pruning just one tetrahedral branch in the first shell is observed in the

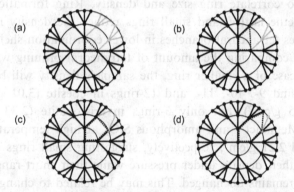

Figure 15.2 The "pruning of tetrahedral branches" to form rings in a Bethe lattice of 4-connected vertices (each represents one tetrahedron, say, of $SiO_{4/2}$ Niggli formula) out to the 3rd shell. The formation of 3-, 4-, 5- and 6-rings of closed paths is depicted in (a), (b), (c), and (d), respectively.

[8] Ring statistics are, for example, included in the program suite ISAACS (Interactive Structure Analysis of Amorphous and Crystalline Systems) found at http://isaacs.sourceforge.net/over.html.
[9] Dendrimers, first made by Vögtle in 1978, are repetitively branched molecules forming a Bethe lattice.

crystalline SiO_2 framework structure of coesite,[10] where the seed tetrahedron is surrounded by four nearest-neighbor tetrahedra but only by 11 and not 12 second-neighbor tetrahedra [$N_2 = (4 \times 3^1) - 1$]. 5-rings are the result of pruning two tetrahedral branches in the second shell and are found in keatite [10].[10] 6-rings are made by pruning only one tetrahedral branch in the second shell, while higher-membered rings result from pruning in more distant shells. 6-rings are observed in the tridymite and cristobalite forms of SiO_2. The cristobalite structure can be described as 1 seed tetrahedron + 4 first-neighbor tetrahedra + 12 second-neighbor tetrahedra + 24 third-neighbor tetrahedra; in total 41 tetrahedra going out to the third shell. In a Bethe lattice (without forming rings) there would be 52 tetrahedra, $N_{1,2,3} = (4 \times 3^0) + (4 \times 3^1) + (4 \times 3^2)$, out to the third shell as shown in Figure 15.2. The structure of tridymite differs. It is built up of 1 seed tetrahedron + 4 first-neighbor tetrahedra + 12 second-neighbor tetrahedra + 25 third-neighbor tetrahedra; in total 42 tetrahedra out to the third shell. It is thus only in the third neighbor shell that one can distinguish the topology of these two SiO_2 polymorphs from the Bethe lattice. Depicting this in either a polyhedral or network diagram results in very complex figures. This topological analysis provides us an explanation of why tridymite cannot transform into cristobalite by a low-energy distortion of the network but requires heating to break chemical bonds in a *reconstructive phase transition*: cristobalite has 24 third-neighbor tetrahedra and tridymite has 25. Note that both tridymite and cristobalite can undergo topology-preserving *displacive phase transitions* between their α and β forms at 114 °C and 270 °C, respectively, where no bonds are broken.

One can also correlate ring size and density. Ring formation always reduces the density of a Bethe lattice, and small rings reduce the density more than larger ones since one prunes tetrahedral branches in lower coordination shells. This can be seen in Figure 15.2 by comparing the amount of tetrahedral pruning when forming a 3- or a 6-ring. In the case of a smaller ring, the sample's density will be lower. In crystalline materials, we find 9-, 10-, 11-, and 12-rings in coesite (3.01 g/cm^3), 6- and 8-rings in quartz (2.65 g/cm^3), but only 6-rings in cristobalite (2.33 g/cm^3) and tridymite (2.26 g/cm^3). Metamict[11] and amorphous SiO_2 at high temperatures have densities of 2.26 g/cm^3 and 2.21 g/cm^3, respectively, suggesting small rings are present here. SiO_2 melts increase their density under pressure while their short-range order, as shown by PDF analysis, remains unchanged. This may be related to changing ring sizes. We will see later that the presence of rings in glasses might also be used to explain vibrational dynamics observed at very low temperatures (Section 15.9).

---

[10] Besides 4-rings, coesite also contains 6-, 8-, 9-, 10-, 11-, and 12-rings. See ref. [8]. Keatite also contains 7- and 8-rings.

[11] Metamictization describes the process where radiation damage results in a gradual amorphization of materials. Natural examples are zircon ($ZrSiO_4$), where U and Th atoms sit on Zr sites and their α-radiation destroys the mineral's crystallinity. Metamict materials form also by radiation in nuclear reactors and particle accelerators. Space travel requires materials resistant to metamictization.

15.3   Oxide Glasses

The best-known example of a technologically important amorphous material is silica-based window glass, which is made by fusing silica sand (SiO_2), soda ash (Na_2CO_3), and limestone ($CaCO_3$) in an approximate molar ratio of 0.7:0.15:0.15. After mixing, the raw materials are heated to 1500 °C for chemical homogenization. The temperature is then lowered to 1200 °C and the melt is poured onto molten tin, forming a floating ribbon with a smooth surface. During this step, the temperature is lowered to about 600 °C, and the glass is transferred to rollers and passed through annealing furnaces while continuing to cool. After exiting the annealing furnace, the glass is cut into its desired dimensions.

In general, oxide glasses are made from **network-forming** inorganic oxides SiO_2, B_2O_3, or P_2O_5, to which **network-modifying** oxides such as alkali- and alkaline-earth-metal oxides are added. Empirical rules (**Zachariasen rules**) for oxide-glass formation are: (1) the coordination number of the network-former atom is 3 or 4, (2) no oxygen atom links more than two polyhedra, (3) coordination polyhedra share only corners and not edges, and (4) for 3D networks at least three polyhedral corners are shared. Rules (2) and (3) facilitate construction of a network without long-range order; the corner-sharing allows variable angles of the M–O–M bonds, and these linkages can be randomly converted into terminal anionic oxygens that bond to cations originating from the network-modifying oxide.

In the following, we will focus on SiO_2-based glasses that, being based on a 3D network of corner-sharing SiO_4 tetrahedra, obey Zachariasen's rules. While pure SiO_2 glass only forms on rapid cooling, adding alkali-metal or alkaline-earth network modifiers in the form of carbonates or nitrates[12] increases the likelihood of glass formation. Cations, such as Li^+, Na^+, K^+, Sr^{2+}, Ca^{2+}, decrease the viscosity of the melt by orders of magnitude and reduce the melting temperature and the glass-forming temperature, T_g. The density, electrical conductivity, coefficient of thermal expansion, and refractive index of the glass are also influenced. We will touch on all of these properties in this chapter. Network modifiers also influence the mechanical strength of glasses, as shown in Box 15.1. The coordination requirements of alkali-metal and alkaline-earth cations underlie their role as network modifiers. They alter the tetrahedral $SiO_{4/2}$ network such that the Si–O–Si **bridging oxygen** is converted into $R_3Si–O^-$ units with **non-bridging oxygen** via $R_3Si–O–SiR_3 + O^{2-} \rightarrow 2R_3Si–O^-$ (where R stands for tetrahedral silicate units that make up the rest of the network and O^{2-} comes from, for example, Na_2O). The negative charge is compensated by the cations.

15.4   Optical Properties and Refractive Index

The technologically most important uses of glasses are, of course, derived from their optical properties. These allow us to build microscope and telescope mirrors, lenses, optical fibers,

---

[12] These decompose to oxides at the high temperature of the melt.

Box 15.1 Materials Spotlight: Strengthening glass—the serendipitous road to Corning's Gorilla® Glass

In 1952, Don Stookey at Corning Glass Works made an error when controlling the temperature of a furnace, which resulted in the serendipitous discovery of the first synthetic lithium-silicate glass ceramic [11]. It was lighter than aluminum metal, harder than high-carbon steel, and many times stronger than regular window glass. It was later named Pyroceram® and used for chemical glassware and cookware for microwave ovens. Another unexpected property was that this glass bounced without breaking when dropped on the floor. Corning subsequently launched "Project Muscle", exploring ways to strengthen glass. This led to a material called Chemcor®, which could be bent and twisted considerably before it broke. It could also withstand pressures of 690 MPa—over 10 times as much as normal glass.

One way of strengthening common glass is by careful thermal treatment known as **tempering**. The glass is initially poured into a form of desired shape and thickness. It is then reheated until it softens and rapidly cooled by forced air so that its surface cools much faster than the interior. Glass forms first in the cool surface region. Its faster thermal contraction compared to the interior means it forms under tensile stress due to the counterforce of the warmer, higher-volume interior still above T_g. At this stage, the interior itself is under compressive stress. When the interior of the sample eventually also cools below T_g and contracts, it will experience a tensile stress from the frozen outer layers, which themselves then experience compressive stress. The final frozen-in-stress profile of a tempered glass leaves the surface in compressive and the core in tensile stress. Compressive surface stresses make tempered glass much more mechanically resilient than normal glass. A mechanically induced defect can be contained in the compressed outer layer by blocking paths along which cracks could otherwise propagate; the sample only breaks once this layer is penetrated.[13] Tempered glass should have a minimum compressive stress near 70 MPa, while in safety glass it should exceed 100 MPa.

However, it was **chemical strengthening** that gave Chemcor® its superior performance. The glass is dipped into a 450 °C bath containing potassium salts, typically KNO_3. By diffusion, one partially replaces Na^+ (Shannon [12] ionic radius 99 pm) by larger K^+ (137 pm) in the near-surface region, forming a layer with inner compressive forces close to 700 MPa, significantly stronger than the ones created by thermal tempering.

Applications for car and airplane windscreens, phone booths, prison windows, and eye-glasses were envisioned, but these products never came to market due to too high a price. Corning discontinued the production of Chemcor®, and "Project Muscle" was shelved in 1971. However, about 35 years later, global demand for smart-phone screens led to chemical strengthening becoming the foundation for Corning's "Gorilla Glass®". While relying on a similar strategy for chemical strengthening, Gorilla Glass® has a different composition than Chemcor®, allowing its viscosity to be tailored for the specific manufacturing process needed to produce very thin glass sheets. In this process, called fusion-draw, molten glass is poured into a trough and allowed to overflow on both sides, rejoin underneath and be drawn down by rollers

---

[13] This can be observed in the mechanical stability found in Prince Rupert's drops made by dropping molten glass into ice water. Videos can be found online.

Box 15.1 (cont.)

to continuous sheets. The faster the rollers draw the melt, the thinner the glass. The specifications for Apple's iPhone® released in February 2007 demanded massive amounts of chemically strengthened glass with a thickness of only 1.33 mm. Corning was able to meet the specifications by May of the same year! Gorilla Glass® 3 was the result of a compositional modeling approach and resulted in improved damage resistance [13]. Gorilla Glass® 6, which is only 0.4 mm thick, survives on average 15 drops from a 1 m height. It is almost certain that you have touched Gorilla Glass®.

---

eyewear, and many other components of optical devices. One of the key features controlling the optical properties of a material is its **refractive index**, n, which relates the velocity of light inside a material, v, to that in vacuum, c (see Appendix J for the value of c):

$$n = c/v \tag{15.1}$$

While n for a vacuum is 1, most materials have $n > 1$, and typical optical glasses have $n \approx 1.5$, meaning that light travels at about two-thirds of its speed in vacuum. The refractive index depends on the wavelength of light, but it is conventionally specified at 589 nm.[14]

Let's consider what happens when light passes from one medium to another (Figure 15.3). The angles of incidence and refraction are measured with respect to the normal of the interface plane of the two media, 1 and 2. The ratio of the angles in medium 1 and medium 2 is given by Snell's law:

$$\frac{\sin\theta_1}{\sin\theta_2} = \frac{v_1}{v_2} = \frac{n_2}{n_1} \tag{15.2}$$

In the cases depicted in Figure 15.3, $n_2 > n_1$ and therefore $v_2 < v_1$ and $\theta_2 < \theta_1$. This means that light rays in medium 2 of higher refractive index n travel closer to the normal of the interface. An example would be light passing from air to water of higher refractive index, as shown in Figure 15.3a. This **refraction** explains why a straight rod appears broken when dipped in water.[15] In the opposite case, shown in Figure 15.3b, light is transmitted from water to air of lower refractive index, and Snell's law no longer holds for all angles. The larger angle of refraction (θ_1 in Figure 15.3b) reaches 90° at the critical incidence angle $\theta_{critical} = \arcsin(n_2/n_1)$, beyond which the light is no longer refracted but reflected. This phenomenon, called **total internal reflection**, is illustrated in Figure 15.3c. In this case, light

---

[14] The well-known yellow doublet in the sodium emission spectrum.

[15] For the same reason the Global Positioning System (GPS) has to take into account the delay of the radio signal due to the refractive index of the Earth's atmosphere to achieve high accuracy. The refractive index of air at 0 °C and 1 atm pressure is 1.000293.

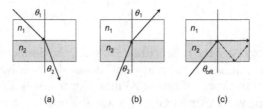

Figure 15.3 The refraction of light passing between two media. Medium 1 has a lower refractive index than medium 2. The angles are defined towards the normal of the media interface. Above a critical angle, passing from medium 2 to 1 results in total internal reflection (indicated by dashed lines).

propagates only in the medium with the higher refractive index. At a water–air interface, this critical incidence angle is 48.6°. This effect is particularly important for the transport of laser light through optical fibers (see Section 15.5 and Figure 15.4). A very special case occurs for X-rays with wavelengths near an atom's electronic absorption edge, where materials can have $n < 1$. This means that total reflection of the X-ray beam can occur outside such a material, a **total external reflection**. A negative n can even be created in some hybrid materials, leading to the fascinating possibility of "cloaking devices" that hide objects [14].

To understand these effects better, we have to know something about how light interacts with matter. In the expression for refractive index of Equation (15.1), v is actually the phase velocity of the light in the material, $v = c/n$,[16] which describes how fast the crests of the wave move. As an electromagnetic wave enters a material, its electrical field will interact with electrons. The strength of this interaction depends on the electric susceptibility χ_e (Section 8.1.1), which describes the coupling to the bulk polarization P (the average dipole moment per unit volume, Section 8.1.2). The electromagnetic wave causes the dipoles in the material to emit an electromagnetic wave with the same frequency as the original, but a different phase φ. The overall wave traveling in the material becomes a superposition of the original and emitted wave. This is an important concept in nonlinear optical materials and was discussed in detail in Section 8.7.

The refractive index is properly described as a complex quantity $n = n_r + i\kappa$, where the real part of the refractive index, n_r, refers to the actual phase velocity and the imaginary part, κ, describes the attenuation of the light's intensity. Depending on the phase difference $\Delta\varphi$ between the original and emitted wave, different physical effects are observed. In most materials, $\Delta\varphi$ is between 90° and 180°, and the overall wave will travel slower than the original, yielding "normal refraction" with $n > 1$ (e.g. the case in Figure 15.3). If $\Delta\varphi$ is 180°, a destructive interference of the waves leads to an imaginary refractive index that describes light absorption in opaque materials. There are two less common but interesting cases to consider: (1) If $\Delta\varphi$ is 270°, the wave will travel faster than the original. This "anomalous refraction" is possible as the phase velocity can be faster than the speed of light in a vacuum without violating the theory of relativity. An example of this is given by the total external

---

[16] The wavelength in a material is then $\lambda = \lambda_0/n$ with λ_0 being the wavelength in a vacuum. The light frequency in the material, $v = v/\lambda$, remains therefore as in a vacuum and is not dependent on the refractive index.

reflection discussed above: 0.4 Å wavelength X-rays have a refractive index of 0.99999974 (i.e. $n < 1$) in water. (2) If $\Delta\varphi$ is 0°, we observe constructive wave interferences that will amplify the light intensity. This occurs in lasers during stimulated emission.

We can link the refractive index of a material back to chemical composition, as discussed in Chapter 8, where we related the optical dielectric constant ε_{opt} to the refractive index n through Equation (8.14):

$$n^2 = \varepsilon_{opt} \tag{15.3}$$

The refractive index n and polarizability α are related through the Lorenz–Lorentz relationship we encountered in Equation (8.13) and can now write as:

$$\alpha_{LL} = \frac{3}{4\pi} V_a \left(\frac{n^2 - 1}{n^2 + 2}\right) \tag{15.4}$$

in the CGSes unit system with α_{LL} in Å^3 referring to the volume V_a in Å^3 per atom (for elements) or formula unit (for compounds), as determined by dividing the unit-cell volume by the number of formula units it contains. One can rearrange the Lorenz–Lorentz equation to calculate n from α_{LL}. In this approach, the overall polarizability of the material, α_{LL}, is estimated by assuming polarizabilities of the individual ions to be additive [see Equation (8.19) for the additivity rule]. Cation polarizabilities, which are assumed to be transferrable from one compound to another, were evaluated by Shannon and Fischer [15]. In contrast, anion polarizabilities show considerable variability from one compound to another. They depend on both coordination number and cation–anion distances and correlate strongly with the volume they occupy. Defining V_{an} as the volume of the formula unit divided by the number of anions, Shannon and Fischer [15] parameterized anion polarizabilities as:[17]

$$\alpha_{an} = \alpha^0 10^{-N_0/(V_{an})^{1.2}} \tag{15.5}$$

where α^0 (the polarizability of the free, non-coordinated anion) and N_0[18] are empirical parameters that were obtained by fitting a regression model to α_{an} from ~2600 refractive-index measurements of minerals and synthetic compounds [15]. Selected pairs for α^0 and N_0 are 1.79 Å^3 and 1.776 Å^3 for O^{2-}, 0.82 Å^3 and 3.00 Å^3 for F^-, 3.99 Å^3 and 1.800 Å^3 for Cl^-, and 1.62 Å^3 and 0.00 Å^3 for bound H_2O in hydrated materials.

The accuracy of refractive-index calculations with Equation (15.4) depends on the degree of covalent character in the ionic bonds. To parametrize this effect, we can correlate α and n using:

---

[17] In a 2006 paper (referenced in [15]), Shannon and Fischer used ⅔ as exponent. However, 1.2 yields smaller deviations between observed and calculated total polarizabilities and is therefore used in [15].

[18] These calculations are for polarizabilities at an infinite wavelength. For details, see [15].

$$\alpha = \frac{(n^2 - 1)V_a}{4\pi + \left(\frac{4\pi}{3} - Z_{eo}\right)(n^2 - 1)} \tag{15.6}$$

where Z_{eo} is an adjustable parameter with values between 0 for a hypothetical purely ionic case described by the Lorenz–Lorentz equation (15.4) and $4\pi/3$ for the limiting covalent case that yields the Drude equation, $\alpha_{Dr} = (n^2 - 1)V_a/4\pi$. With $Z_{eo} = 2.26$, Equation (15.6) is called the Anderson–Eggleton equation [16, 17]. It works well for network oxides and silicates, but less so for compounds containing individual poly-atomic anions such as carbonates, nitrates, sulfates, and perchlorates. By rearranging Equation (15.6) with $Z_{eo} = 2.26$, the Anderson–Eggleton refractive index, n_{AE}, can be calculated:[19]

$$n_{AE} = \sqrt{\frac{4\pi\alpha_{AE}}{\left(2.26 - \frac{4\pi}{3}\right)\alpha_{AE} + V_a} + 1} \tag{15.7}$$

Using this approach, predictions with errors of 1–2% can be made for oxides. For compounds containing edge- and corner-sharing transition-metal coordination poly-hedra, predicted values are typically up to 3% smaller than experimental ones. Furthermore, cations with a lone pair of electrons (e.g. Tl^+, Sn^{2+}, Pb^{2+}, Bi^{3+}) do not permit simple addition of their polarizabilities. The refractive index of the mineral orthoclase ($KAlSi_3O_8$) is explored in Problem 15.9. When attempting to use a similar approach for glasses, we are faced with the challenge of determining the volume per formula unit, V_a, of a material that has no unit cell. Given the density, ρ, and empirical formula of a glass, one can calculate the number of formula units in a bulk sample and then divide the volume by that number to obtain V_a. An example is given in Problem 15.11.

The Gladstone–Dale relationship, originally used to relate the density ρ in g/cm^3 of a liquid to its refractive index n_{GD}, is yet another approach to estimate the refractive index of a glass:

$$n_{GD} = 1 + \rho \sum_i k_i p_i \tag{15.8}$$

where k_i (listed in Table 15.1) is the chemical refractivity[20] (in cm^3/g) and p_i is the mass fraction of each oxide component of the glass [18]. See Problem 15.10 for an example of how Equation (15.8) can be used to estimate the refractive index.

---

[19] This approach is one of the methods used in the program POLARIO to calculate refractive indices; another is based on a modern adaptation of the approach of Gladstone and Dale.

[20] These are sometimes also called specific refractive energies of the components.

Table 15.1 Selected chemical refractivities k_i in cm^3/g for common oxides used in glasses. Values are from Table 3 in ref. [18].

Li$_2$O	0.307	MgO	0.200	B$_2$O$_3$	0.215
Na$_2$O	0.190	CaO	0.210	Al$_2$O$_3$	0.207
K$_2$O	0.196	SrO	0.145	P$_2$O$_5$	0.176
Rb$_2$O	0.128	BaO	0.128	As$_2$O$_5$	0.340
Cs$_2$O	0.119	PbO	0.133	SiO$_2$	0.208

15.5 Optical Fibers

The insulating network-forming compounds SiO$_2$, B$_2$O$_3$, and P$_2$O$_5$ have large electronic band gaps (Chapter 6), and this makes them transparent to visible and near-ultraviolet (UV) light (Chapter 7). When network-modifying oxides such as Na$_2$O are added to form a glass, different local bonding configurations and electronic energy levels are created. For example, a bridging Si–O–Si unit will have lower-energy bonding orbitals than a non-bridging Si–O unit, and the energy of these orbitals will be further influenced by their proximity to other non-bridging or bridging units. As a result of this variability, a range of different electronic energy levels will be present in glasses. This energy distribution leads to an increase in UV absorption as more Na$_2$O is added, and is referred to as a **UV tail**.[21] At low energies, an **infrared (IR) absorption band** peaks around 1385 nm due to hydroxyl groups bonded to the silicon.[22] These originate from the presence of minute quantities of water during glass manufacturing. The tails of these two absorption maxima bracket a window of low optical absorption between ~650 nm and ~1350 nm. It is also important to consider attenuation due to light scattering by inhomogeneities (impurities or density fluctuations) in a sample. Another important contribution to attenuation is Rayleigh scattering. It depends on λ^{-4} and on Δn_i, the latter being the refractive-index change induced by doping silica with a dopant i. For fiber-optics communication, we therefore use IR light at 850 nm and 1300 nm within the low-loss window or at 1550 nm beyond the strongest IR band (Figure 15.4).[23] These wavelengths allow the transmission of laser beams in glass fibers over thousands of kilometers with minimal losses by the process of internal reflection (Figure 15.3). This concept was the technological foundation of the rapid expansion of telecommunications at the end of the twentieth century. In 2009, Charles Kao received the Nobel Prize in Physics for his work on fiber optics.

Transmission under internal reflection above the critical angle (see Figure 15.3) is made possible by cladding a glass with lower refractive index around a core fiber with a higher

---

[21] These localized tails in the energy gap between the valence and conduction band are called Urbach tails and are discussed in Section 15.10.

[22] The "water peak" typically shows five main absorption bands at 1247 nm, 1352 nm, 1381 nm, 1391 nm, and 1412 nm.

[23] Multimode graded fibers are optimized for 850 nm and 1300 nm, while single-mode fibers are optimized for 1300 nm and 1550 nm. Transmitters for these wavelengths are lasers and light-emitting diodes.

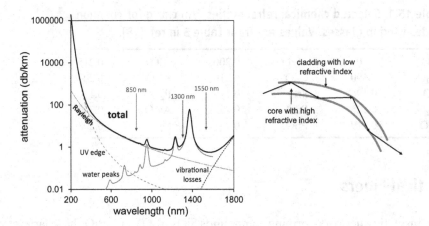

Figure 15.4 Contributions to attenuation in a typical glass fiber. Shape of water peaks taken from ref. [19]. Light with wavelengths of 850 nm, 1300 nm, and 1550 nm is transmitted with low attenuation.

refractive index.[24] Ultra-pure amorphous SiO_2 ($n = 1.458$) is used as the cladding material. The glass of the core needs to have a refractive index higher than SiO_2, and this is achieved by mixing SiO_2 with GeO_2, P_2O_5, and sometimes Al_2O_3 of refractive indices 1.609, 1.490, and 1.768, respectively [20]. Additionally, a gradient of the refractive index is introduced in the core to increase the amount of data that can be transmitted by overlapping a number of optical signals in one optical fiber using different wavelengths of laser light (**multiplexing**). This also allows communication in both directions of a fiber.

Attenuation α in optical fibers is measured in decibels (dB) per kilometer as $\alpha = (10/L)\log(P_{in}/P_{out})$, where P_{in} (in mW) is the electrical power of the signal needed on the input so that the light transmits $P_{out} = 1$ mW at the output distance L in kilometers. The initial target for a material useful for telecommunications was 20 dB/km, even though this means that just 1% of the input optical power has been transmitted. Modern fibers have much lower optical losses (near 0.1 dB/km), allowing the use of extremely long fiber cables [21]. On land, uninterrupted fiber lengths are of the order of 10 km. "Repeater" stations are placed between individual fiber segments that amplify and recondition the light signal. Initially, this was done by converting the optical signal into an electrical one and back. Nowadays, erbium-doped fiber amplifiers are used as a gain medium and "pumped" by a laser with a wavelength of either 980 nm or 1480 nm to produce stimulated emission near 1550 nm.

Low attenuation demands extreme purity. Over a 1 km length, 90% of the transmission can be lost by a metal ion impurity concentration of just 0.01 ppm. A second source of attenuation is due to small but unavoidable surface roughness and inhomogeneity, which reflects light in random directions. The production of optical fibers is therefore an extreme challenge. A hollow silica-glass tube (a "preform") is made and coated inside with high-purity oxide

---

[24] In practice, refractive index differences are about 1%.

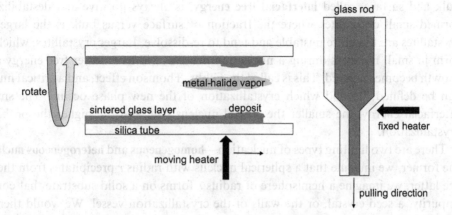

Figure 15.5 Optical-fiber growth using CVD in a tubular SiO$_2$ glass preform that is continuously rotated while layers with varying refractive index are deposited. An optical fiber is pulled vertically in a drawing tower.

glass using chemical vapor deposition (CVD) from a moderate-temperature flowing gas of SiCl$_4$ mixed with GeCl$_4$ or AlCl$_3$ or POCl$_3$, as well as fluorine sources, SiF$_4$ or SF$_6$, and oxygen. In single-clad fibers, the core has a higher refractive index than the silica cladding, and this is achieved by Ge-, P-, and Al-doping in the core. F-incorporation decreases the refractive index, which is needed in more complex fibers. P-containing glass has a lower viscosity, and this becomes important during fiber pulling. Subsequent layer-by-layer depositions with varying gas compositions allow a precise control of the refractive index and viscosity. Then an external torch heats the tube up to near 1900 K, sintering the oxide glass on its inner walls and collapsing the tube into a rod. Afterwards, the glass rod is either jacketed by an additional silica tube or directly heated locally in a drawing tower to near 2200 K where the tip of the preform melts and is rapidly stretched into a cooler area, forming an optical fiber with a typical diameter of 125 µm (Figure 15.5).

15.6 Nucleation and Growth

Before discussing the topic of glass formation, it is useful to consider some of the concepts in the alternative process of crystallization. The two key concepts are nucleation and growth. We can understand nucleation as a first-order phase transition, in which a new phase with a lower free energy emerges inside a large volume of the initial phase after having overcome an energy barrier associated with the formation of small crystallites. Growth describes the evolution of these nuclei to a bulk phase.

Let's consider crystallization from a melt. While the bulk of the new solid phase that emerges during crystallization will have a lower free energy than the liquid phase, its surface region, where atoms are underbonded, will be less stable. This energy difference between the bulk and surface, called **interfacial free energy**, is always positive and destabilizes newly formed small crystallites where the fraction of surface versus bulk is the largest. Small crystallites are therefore unstable and tend to re-dissolve. Larger crystallites, which initially form in small numbers during a nucleation process, will have lower free energy and their growth becomes favored. This is called the Gibbs–Thomson effect, and a critical nucleus size can be defined, beyond which crystallization of the new phase occurs. The smaller the interfacial energy, the smaller the critical nucleus size, and the higher the probability of crystallization.

There are two limiting types of nucleation—**homogeneous** and **heterogeneous nucleation**. In the former, we imagine that a spherical nucleus with radius r precipitates from the melt. In the latter, we imagine a hemisphere of radius r forms on a solid substrate that could be an impurity, a seed crystal, or the walls of the crystallization vessel. We would then have to consider two additional interfacial energies, one between the substrate and crystallite, and one between the liquid and substrate.

Normally, a melt will crystallize at its melting point, T_m, either spontaneously through homogeneous nucleation or via heterogeneous nucleation. Under very clean conditions, the absence of nucleation sites may allow the melt to be cooled considerably below T_m in a process called **supercooling**. At T_m, the free energies of the liquid melt and the solid are equal. At $T > T_m$, the melt is stable; at $T < T_m$, the solid. A supercooled liquid is therefore a metastable state that is out of chemical equilibrium. Water exemplifies this behavior. Pristine liquid water of $T_m = 273$ K can be supercooled to 232 K in the absence of any crystallization centers. Under the slightest disturbance, or on further cooling, the supercooled water crystallizes rapidly.

The goal in glass formation is to avoid crystallization, that is to cool the sample in such a way that nucleation and growth don't occur. The rate of both these crystallization steps is strongly temperature dependent. For example, close to T_m, the critical size for the nucleus to be stable is rather large, and most of the nuclei that form remelt. At lower temperatures, the critical radius is small and remelting is less likely. However, as the temperature is lowered, the viscosity of the liquid increases, and this slows down atomic transport and therefore nucleation. Similar considerations apply to growth. Glasses therefore form if a melt is cooled rapidly to a sufficiently low temperature so that neither nucleation or growth can occur.

15.7 The Glass Transition

In this section, we consider how kinetics and thermodynamics impact the formation of glasses, and we will see close analogies to the supercooled liquids just discussed. The most direct way to understand the glass transition is by comparing it with its alternative—crystallization (Figure 15.6). In the initial high-temperature liquid, the atomic arrangement at any point is in thermal equilibrium. Crystallization is a first-order phase transition (Chapter 4) manifested by a discontinuous decrease in the entropy of the crystallizing

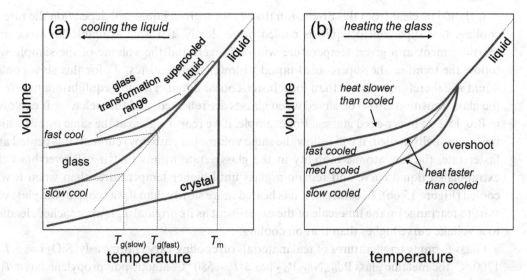

Figure 15.6 (a) Volume changes upon cooling a melt as a function of temperature and time at constant pressure. T_g is the glass-transition temperature at which a supercooled melt becomes a glass at a given rate of cooling. Glass formed at $T_{g(fast)}$ was cooled faster than glass formed at $T_{g(slow)}$. (b) Volume pathways on heating and cooling differ.

substance upon cooling through its T_m, which releases heat $H_m = T_m S_m$ that is absorbed by the cooler surroundings. This is accompanied by a change in volume (as shown in Figure 15.6) and enthalpy. If crystallization is avoided, the system changes continuously into the supercooled liquid regime with the viscosity gradually increasing on cooling (Section 15.8). At some point, the atoms no longer have sufficient time to fully rearrange to the (metastable) equilibrium liquid configuration for their temperature at the cooling rate used. When even lower temperatures are reached, the structure becomes frozen on the timescale of the experiment and a glass is formed. As this process occurs, the gradient of volume (or enthalpy) versus temperature changes gradually until at low temperature it becomes determined by the thermal expansion (or heat capacity) of the glass.[25] The temperature range over which the changes occur is called the **glass-transformation range**. It is often useful to quote a single representative temperature for this process. One way is to define the **glass-forming temperature**, T_g, from the intersection of the low- and high-temperature extrapolated behaviors measured at a specified cooling rate.[26]

[25] We therefore see corresponding changes in heat capacity and thermal expansion at the glass transition. There are similarities to the second-order transitions of Chapter 4, but the system is not at equilibrium.

[26] A second temperature related to glass formation, the fictive temperature, T_f, is often encountered in the literature. It is obtained just like T_g but upon heating the glass. If a sample is cooled then heated at identical rates, T_f and T_g will be essentially identical. If a glass is thermally annealed below T_m, the fictive temperature will change as discussed in Box 15.2. Note that different communities use slightly different definitions for T_f and T_g.

It should be clear from this discussion that T_g for a given liquid will depend on the rate of cooling. If a supercooled liquid is cooled more slowly, the time available for structural rearrangement at a given temperature will be longer, and the volume of the sample will follow the trend of the supercooled liquid to lower temperatures. T_g for this slow-cooled liquid will therefore be lower than for a faster-cooled liquid. The non-equilibrium nature of the glass transition is also observed when glasses are reheated, and this behavior is explored in Box 15.2 on hyper-aged glasses. For example, if we reheat a glass at the same rate at which it was originally cooled, it will follow the same volume (or enthalpy) curve. If it is heated at a faster rate, the low atomic mobility in the glassy state means that it will overshoot the extrapolated liquid curve and remain a glass until higher temperatures than when it was cooled (Figure 15.6b). In contrast, if it is heated more slowly than it was cooled, the glass can start to rearrange on the timescale of the experiment as its original T_g is approached, leading to a volume curve higher than that on cooling.

Glass-forming temperatures of real materials on cooling vary enormously: SiO_2 has a $T_g \approx$ 1200 °C, the metallic glass $Pd_{0.4}Ni_{0.4}P_{0.2}$ has a $T_g \approx 580$ °C, atactic polypropylene has a $T_g \approx$ −20 °C, and low-density polyethylene has a $T_g \approx -125$ °C. Good glass formers (e.g. SiO_2, GeO_2, Se) can be made by melt quenching and typically obey the Kauzmann rule, $T_g = 0.7T_m$. Poor glass formers ($T_g < 0.7T_m$), like metal alloys, have to be cooled very quickly to prevent crystallization.

To complicate matters further, deciding whether an actual sample is a glass, supercooled liquid, or liquid depends on the experimental probe used. In rheology, these properties are distinguished by the Deborah number, $D_e = \tau_c/\tau_p$, where τ_c is the timescale of the viscosity (time needed for a reference amount of viscous flow) and τ_p is the timescale of the experiment used to probe it. Whether a material responds like a liquid (low D_e) or a solid (high D_e) depends on the experimental timescale as well as the intrinsic properties of the material. As an example, one can happily swim slowly through water without bodily damage; the picture is very different if you hit water at high speed – τ_p, and therefore D_e, is different in these two "experiments". For similar reasons, a high-frequency (10^{15} Hz) technique like NMR will "see" the glass transition at a different temperature than a lower frequency technique such as calorimetry.

At $D_e \gg 1$, the system has insufficient time to explore all possible structural configurations (Section 15.1), as the probing timescale τ_p is too small. Glass is such a system and is called non-ergodic. In contrast, a melt is an ergodic system (same homogeneous behavior across the system over time) with $D_e \ll 1$, as it has sufficient time to equilibrate and probe all possible structural configurations. In an ergodic system, the configurational- and time-averages of properties such as viscosity are the same. In non-ergodic glasses, frozen-in configurations might average out over time if one probes at a longer timescale. At the glass-forming transition, ergodicity is lost, and the system can only experience a subset of configurations.

To help understand the interplay between structural configurations and their stability, it is useful to introduce the **potential-energy surface (PES)** or **energy landscape**. This is a

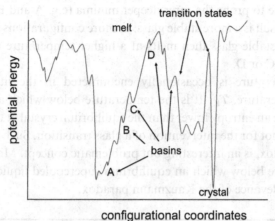

Figure 15.7 The PES of an N-body system separating the melt from various glass states, as well as the crystalline state as the lowest-energy configuration. The glass states A, B, C and D are discussed in Materials Spotlight 15.2.

representation of how the energy of an N-body system depends on its structural configurations. Any atomic structural configuration can be described by a multidimensional vector that contains the coordinates defining interatomic distances and angles of its atoms. For illustrative simplicity, we collapse all these **configurational coordinates**[27] onto one dimension of the horizontal axis in Figure 15.7, with the potential energy separating melt from glass or crystal on the vertical axis.

The PES captures the potential-energy states a material can access depending on its thermal history. At high temperatures, the system has sufficient internal energy to explore the whole energy landscape, and the dynamics resembles free atomic diffusion. Such an ergodic system with $D_e \ll 1$ has sufficient time to equilibrate. At lower temperatures, the system no longer has the energy to overcome the highest energy barriers and is then forced to adopt a minimum close to its current configuration. Its accessible configurational space is reduced or constrained. A system trapped in local low-energy arrangements (e.g. points C or D in Figure 15.7) is therefore unable to reach the thermodynamically more favorable global structural configurations of the glass or crystal because the diffusion is too slow. The kinetics of structural relaxation changes with temperature, and the probability that a certain activation barrier can be overcome becomes smaller with decreasing temperature; a regime called super-Arrhenius (Section 15.8). The PES also provides us with an intuitive visualization of why cooling rates are so important in glass transitions. Fast cooling rates will not allow the system to adequately probe many configurations before leaving the liquid state and forming a glass. Aspects of the liquid's structure then appear frozen in the glass. Slower cooling rates

[27] Whereas in PES diagrams and chemical reactions one uses 1D collective configurational coordinates describing the complete chemical system, in Section 7.8.2 we used configurational coordinates to describe the differences between electronic ground and excited states of localized areas near emission activators.

allow enough time to probe the rarer deeper minima (e.g. A and B) in the potential-energy landscape. The result is a more stable glass, as more configurations are explored, and a lower T_g.[28] That more stable glass then melts at a higher temperature than the less stable glass formed at points C or D.

Another temperature is occasionally encountered in the literature on glasses—the Kauzmann temperature, T_K. It is the temperature below which a supercooled metastable liquid would have an entropy lower than the equilibrium crystal at the same temperature and pressure, were it not for the intervention of a glass transition. Such a situation, known as the Kauzmann paradox, is an interesting but problematic concept. Modern theories [22] postulate a temperature below which an equilibrium supercooled liquid no longer exists, which invalidates the relevance of the Kauzmann paradox.

Box 15.2 Materials Spotlight: Amber—thermal history of a hyper-aged organic glass

Plant resins can polymerize over hundreds of millions of years and form organic glasses called amber, which often have a beautiful yellow-brown color leading to their use in jewelry [23]. Amber is different to other materials discussed in this chapter as the glass forms due to chemical and not physical (e.g. by cooling or applying pressure) processes. Ancient amber presents us with a unique opportunity to investigate a glass that has aged significantly beyond what is possible in a laboratory setting. Ambers are **hyper-aged organic glasses**.

In Figure 15.7, we showed a PES with distinct local minima for different glassy states denoted A, B, C, and D. In Section 15.7, glass formation is discussed as a process of the melt exploring different parts of the PES as we cool. Slow cooling allows more of the PES to be explored, leading to a more stable glass. With the same PES, we can also understand the behavior of amber upon heating, even though it formed by aging over time rather than by cooling. A possible location for the potential energy of hyper-aged (110-million-year-old) amber would be in a very deep basin such as A. If this amber is heated above its glass-melting temperature $^{heating}T_g$ of 438 K during a heat capacity (C_p) measurement, an endothermic peak is observed [24]. This represents the heat required to dislodge structural entities from the glassy environments in which they settled after a hundred million years of structural relaxation close to ambient conditions. In the PES diagram, a sample that's heated above $^{heating}T_g$ and cooled down again on a normal laboratory timescale might find itself located in basin D. One can quantify the stability difference between A and D by measuring the endotherm on melting amber D, which will be much smaller. Furthermore, one can investigate what happens to the size of the melting endotherm when amber A is annealed at different temperatures below $^{heating}T_g$, resulting in the formation of glasses located in basins B and C of the PES. If one gradually increases the annealing temperature towards $^{heating}T_g$, the amber becomes progressively less stable and the endothermic peak in the C_p measurement smaller and smaller. The materials located in basins B and C are referred to as partially rejuvenated ambers.

---

[28] T_g typically changes up to 5 °C as the cooling rate is changed by an order of magnitude.

Box 15.2 (cont.)

Heating amber tells us about the evolution of such a hyper-aged glass as it escapes the constraints of its local PES landscape and transitions towards free diffusion in the melt.

Heating amber thus erases the history of hundreds of millions of years of structural relaxation. But not all is erased. Remarkably, the study also showed that the specific heat at temperatures below 1 K is the same in hyper-aged, partially rejuvenated, and fully annealed amber; they fall on the same curve and show that C_p is proportional to $T^{1.27}$. This supports the conclusion that the tunneling two-level system we will discuss in Section 15.9 is an intrinsic and universal low-temperature dynamical property of glasses irrespective of their history.

15.8 Strong and Fragile Behavior of Liquids and Melts

According to Cohen and Turnbull [25], viscosity in glasses can be understood as a material flow into the voids or "free volume" caused by random packing (Section 15.1). An alternative purely empirical definition of T_g, based on the flow properties of a cooling melt, is to define T_g as the point at which the shear viscosity η reaches 10^{12} Pa s.[29] Shear viscosity is a measure of the resistance of a liquid to a deformation under shear stress.[30] At ambient conditions, liquids such as acetone, methanol, and ethanol have viscosities in the range of 10^{-3} Pa s, while the value for pitch is about 10^8 Pa s.[31] One of the remarkable features of the glass transition is that by cooling over a few hundred degrees, the shear viscosity can increase by 15 orders of magnitude. Just above T_g, the temperature dependence of viscosity for some melts like SiO_2 can be approximated using an Arrhenius-law expression, $\eta = \eta_0 \exp(-E_A/RT)$, with the pre-exponential factor η_0 and activation energy E_A being material-specific parameters. Plotting $\ln(\eta)$ versus $1/T$ allows us to identify $-E_A/R$ as the slope and $\ln(\eta_0)$ as the y intercept at infinitely high T. Over a wider temperature range, however, the viscosity of many liquids and most polymers and organic glass formers shows a significant deviation from this simple Arrhenius-type behavior and is better described using the Vogel–Fulcher–Tammann[32] equation:

$$\eta = \eta_0 \exp[DT_0/(T - T_0)] \tag{15.9}$$

---

[29] This value is close to the annealing point of glasses where they are already hard enough to break when dropped on the ground but still soft enough to relax internal strains by microscopic flow when modeled at the glassblowers' pipe. At higher viscosities, microscopic flow essentially stops.

[30] Shear stress is caused by a force parallel to the surface of a sample.

[31] One of the longest ever scientific experiments, the pitch-drop experiment, was started in 1927 at the University of Queensland by Parnell. He poured heated pitch into a sealed funnel, cut the seal in 1930, and monitored the falling drops. The ninth drop was observed in April 2014. There is now a webcam on the experiment and the tenth drop is expected in the 2020s.

[32] In polymer science, the equivalent equation is also known as the Williams–Landel–Ferry equation.

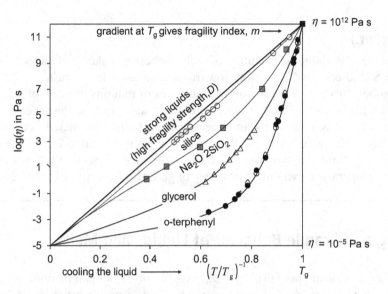

Figure 15.8 Viscosity of strong and fragile liquids versus inverse normalized temperature (an Angell plot). Arrhenius-like behavior would lead to a straight line. Cooling occurs from left to right on the $(T/T_g)^{-1}$ scale. Fragility index m is the slope of the curve at T_g.

with η_0 being the viscosity at infinitely high temperatures, D a material-dependent fit parameter for the liquid, called the **fragility strength**[33] (meaning strength *against* the liquid's fragility to easily crystallize as opposed to form glass; there is no relation to the brittleness of the material), and T_0 the Vogel–Fulcher–Tammann temperature where kinetic barriers rise to stop the flow.[34] We can see that Equation (15.9) approaches Arrhenius-type behavior as the temperature goes to infinity. This simple equation is capable of describing viscosity data over 15 orders of magnitude.

Experimental viscosities can be evaluated in a so-called Angell plot of $\log(\eta)$ versus the inverse normalized temperature, $(T/T_g)^{-1} = T_g/T$, as shown in Figure 15.8. The glass transition is then formally defined to occur at 10^{12} Pa s, and all curves are extrapolated to an infinite-temperature viscosity of 10^{-5} Pa s (a low viscosity).[35] Angell defined a **fragility index**, m, as the slope of the logarithm of the viscosity curve at T_g. Figure 15.8 shows that materials with a high fragility strength D in Equation (15.9) have a low fragility index and vice versa. Typical values are $2 < D < 100$. GeO_2 and SiO_2 are both strong liquids (strong glass formers) with D of 113 and 63, respectively, whereas toluene is a fragile liquid (easy to crystallize, poor glass former) with $D \approx 5$.

---

[33] The D equals B/T_0 of the conventional Vogel–Fulcher–Tammann equation, $\eta = \eta_0 \exp[B/(T - T_0)]$.

[34] T_0 is always less than T_g. As T_0 approaches zero, Arrhenius behavior is approached.

[35] The value 10^{-5} Pa s corresponds to relaxation times of 10^{-14} s, significantly shorter than the typical timescale (10^{-12} s to 10^{-13} s) of one atomic vibration. Typical liquids have viscosities near 10^{-3} Pa s. Extrapolation to 10^{-5} Pa s at infinite temperature is an arbitrary but reasonable choice.

Silica is an example of a strong liquid (a strong glass former; $D \approx 63$), and its shear viscosity is well approximated by the Arrhenius equation. Its activation energy is almost constant over a wide temperature range (constant slope in Figure 15.8), which indicates that the breaking and forming of Si–O bonds controls the temperature dependence of the viscosity. Another way to understand a strong liquid is via its PES (Figure 15.7), which is dominated by a single large basin modulated by smaller local energy variations due to the presence of different $SiO_{4/2}$ tetrahedral rings and clusters. A strong liquid has a close structural relationship to its glass and consequently has a high tendency for glass formation. Melts leading to bulk metallic glasses (see Section 15.11) are also typically strong liquids. They have high viscosities η, large fragility strengths D, show near-Arrhenius behavior, and have good glass-forming abilities.

Many organic molecules such as toluene or chlorobenzene are examples of fragile liquids (poor glass formers). Their viscosity can increase 10^{15} times as one cools the already supercooled liquid (Figure 15.8) towards the glass transition. The PES (Figure 15.7) of these fragile liquids is more complex than that of strong glasses and shows multiple relatively large basins modulated at the bottom by local energy variations. The structural rearrangements required to transition between different basins are complex and characterized by a broad spectrum of different bonding and nonbonding interactions with different relaxation times. In these molecular systems, it is the multitude of collective intermolecular interactions rather than chemical bonds that gives rise to the more complex multi-basin PES. The flow behavior is better accounted for by the Vogel–Tamman–Fulcher equation, as it isn't dominated by a single chemical bond strength. This has consequences for ionic conductivity in polymers (see Section 13.5).

High fragility (easy crystallization), manifested by a strongly non-Arrhenius behavior of $\eta(T)$, is an essential property of chalcogenide-based phase-change materials and the cause of their rapid crystallization that is exploited in phase-change memory. For example, if an alloy of Ge, Sb, and Te such as $Ge_2Sb_2Te_5$ is heated above its melting point ($T_m \approx 600$ °C) and cooled rapidly, it forms an amorphous low-conductivity phase (see Section 15.10) that is kinetically stable at room temperature. If this phase is subsequently heated to temperatures above T_g but below T_m, it rapidly crystallizes to a high-conductivity crystalline phase. These two states can be easily read out as 0 and 1 in an electronic device, leading to a non-volatile phase-change memory [26]. It is even possible to produce intermediate conductivity states so that each memory element can store more than one bit of information. The accompanying change in optical properties from a low-reflectivity amorphous to a high-reflectivity crystal-line state has also been exploited in optical storage devices such as Blu-ray disks. Research has led to the optimization of an alloy $Sc_{0.2}Sb_2Te_3$ with a recrystallization speed below 1 ns that reduces the rate-limiting step of the read/write operation [27]. It appears that the role of Sc is to seed heterogeneous nucleation, thereby accelerating crystallization.

To conclude our discussion of viscosity in amorphous materials, we will address the popular myth that the ambient-temperature viscous flow of glass can be observed in the windows of medieval cathedrals, since the panes are often thicker at the bottom than at the top. Work using the glass composition found in Westminster Abbey (dated 1268), revealed that, while the thermal history of the glass can alter the viscosity between 10^{24} Pa s and 10^{25} Pa s, the maximum flow under a gravitational force amounts to a thickness increase of 1 nm in 10^9 years. The actual observed thickness variation is due to the medieval manufacturing methods and the way glaziers chose to mount individual panes [28].

15.9 Low-Temperature Dynamics of Amorphous Materials

Glasses have different thermal properties than crystalline materials. Heat-capacity and thermal-conductivity measurements of glasses at temperatures below 1 K reveal they all show the same physical behavior (see Box 15.2). This can be related to the existence of dynamical excitations, which occur regardless of structure and composition and are therefore called *universal* low-temperature states.

In crystalline materials, we use the concept of phonons to understand the collective thermal motion of atoms. Phonons are wave-like dynamical displacements of atoms that occur in structures with translational symmetry and are the extended-structure analog of the vibrational modes of isolated molecules (Section 4.4.6). Atoms collectively and concertedly vibrate about their equilibrium positions, and the pattern of displacements can be described by long-range waves that propagate through the entire crystal. In a perfect harmonic crystal, the phonons are plane waves[36] with infinite lifetime, and, since phonons are responsible for heat conduction, one should observe an infinite thermal conductivity.[37] In reality, crystals are not perfect, and phonons are anharmonic, resulting in finite phonon lifetimes and finite thermal conductivity. A good approximation for thermal conductivity is:

$$\kappa = \frac{1}{3} C_V \, v_{\mathrm{s}} \, \ell_{\mathrm{p}} \tag{15.10}$$

where v_{s} is the average sound velocity in m/s, C_V the molar heat capacity at constant volume in J/(mol K), and ℓ_{p} is a parameter called the phonon mean free path in m, the average distance between two consecutive phonon-scattering events. Impurities and lattice defects reduce ℓ_{p}, suggesting that we might expect lower thermal conductivity in disordered materials.

Our model for phonons is based on crystalline materials, but, since glasses have no unit cell, no translational symmetry, and no long-range order, do they have phonons? The answer is, yes; collective excitations we call phonons exist, but they are no longer plane waves at the nanometer scale and, due to disorder, their lifetimes are shorter than in crystals.

Debye established [29] that for insulating crystalline materials (thus lacking mobile conduction electrons), the heat capacity C_p at constant pressure and the thermal conductivity κ should be proportional to T^3 at low temperatures. Zeller and Pohl [30] found that the

---

[36] A plane wave propagates with constant frequency perpendicular to 2D planes in which the wave fronts comprising points with the same phase are located (infinite parallel planes of constant amplitude).

[37] Superfluids, such as ^{3}He below 2 K, have infinite thermal conductivity, meaning that any volume of a sample will always have a uniform temperature; there is no resistance to heat transport.

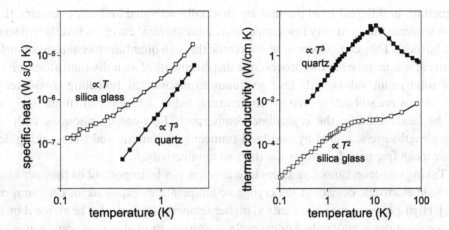

Figure 15.9 Comparison of specific heat and thermal conductivity of silica glass (open squares) and crystalline quartz (filled squares) where thermal conductivity was measured along the *c* axis. Data from ref. [26].

low-temperature C_p and κ of glasses are strikingly different; below 1 K, C_p is approximately proportional to T (not T^3) and κ is approximately proportional to T^2 (not T^3). We encountered this behavior in Box 15.2, where we learned that C_p of all ambers is proportional to $T^{1.27}$ below 1 K. The low-temperature thermal properties of crystalline α-quartz and amorphous silica are compared in Figure 15.9 and reveal that C_p is larger, whereas κ is lower in the glass.[38] Furthermore, a plateau is observed between 3 K and 10 K in $\kappa(T)$ for silica glass, which is due to the presence of an excess of low-frequency modes in the energy distribution of vibrations over the predictions of the Debye model. This plateau is referred to as the boson peak and is a universal feature of glasses (phonons are quasi-particles with zero or integer spin and therefore bosons). These excess vibrational modes are believed to scatter phonons, causing a precipitous reduction of their mean free path ℓ_p and a concomitant temporary saturation of $\kappa(T)$.[39]

One explanation for the universal ultra-low temperature dynamical behavior of amorphous materials is given by the **two-level tunneling** model developed by Anderson, Halperin, and Varma [31] and independently by Phillips [32]. In this model, localized excitations distinct from phonons are introduced as the lowest-energy excitations of glasses. The model assumes the presence of many local groups of atoms with nearly the same energy in the PES. In Section 15.7, we have seen that at high temperatures, atomic relaxations between

---

[38] The low-temperature peak in $\kappa(T)$ can be understood from the temperature dependence of the three terms in Equation (15.10). To a first approximation, ℓ_p increases as temperature decreases and phonon–phonon scattering is reduced but saturates at low T due to residual scattering by defects, v is independent of T, and C_V falls from a constant value to zero. The combination of these effects leads to the observed peak.

[39] Phonons are only well-defined quasi-particles when their mean free-path length ℓ_p is larger than the phonon half wavelength, $\lambda/2$. This is the Ioffe–Regel criterion for phonons that explains the saturation of thermal conductivity. In Section 15.10, it is discussed in more depth to explain saturation of electrical conductivity when the electron free-path length approaches the interatomic distances.

structures in different local minima are thermally activated and can overcome the barriers that separate them. At very low temperatures, the thermal energy is insufficient to overcome the barriers. They can, however, be overcome through quantum-mechanical tunneling. With many atoms involved in this process, the displacements of an individual atom can be tiny yet the total result substantial. Due to quantum-mechanical tunneling, two-level-tunneling excitations are still active even at temperatures below 1 K and contribute to C_p, causing it to be larger than in the crystalline counterpart. Low-energy phonons with very large wavelengths are scattered by two-level-tunneling excitations and the resulting decrease of their mean free path ℓ_p reduces the thermal conductivity.

Taking two-level tunneling states into account can be important as they act as sources of noise in electronic devices at the cryogenic temperatures explored for quantum computing [33]. High-precision measurements at higher temperatures can also be affected by two-level-tunneling states in materials. For example, the observation of gravitational waves is based on laser interferometry using reflective coatings of multiple alternating layers of amorphous silica glass and TiO_2-doped Ta_2O_5 glass[40] as mirrors. To detect metric changes of 10^{-18} m, two laser beams, split at 90° to each other, are sent down 4 km vacuum tubes and reflected by mirrors. Interference occurs when they are recombined in the presence of a gravitational wave. The main experimental obstacle to measuring miniscule interferometric signals is the absorption loss caused by the coating's thermal noise. This results in part from external friction, as the force resisting deformation in a material as stress (force per unit area) is never completely converted into strain/deformation without losses. Two-level-tunneling states are a significant source of internal friction and thermal noise in amorphous oxides. They have been included in modeling internal friction to help search for better coating materials for more precise interferometric gravitational-wave detectors [34].

In summary, the low-temperature structural and dynamic ground states that crystalline and amorphous materials display are quite different. Crystalline materials have structures described by unit cells and translational symmetry. Dynamically, they display a complex range of excitations described by phonons. Glasses have complex atomic-network structures with no translational symmetry and no unique structural ground state. In contrast to crystals, their dynamics at ultra-low T is described by universal two-level tunneling.

15.10 Electronic Properties: Anderson Localization

Another fundamental difference between amorphous and crystalline materials concerns electrical conductivity. We discussed in Chapter 10 how the conductivity of crystalline metals can be understood in terms of the mean free path ℓ travelled by an electron accelerated in an external electric field. Electrons are scattered by defects such as vacancies, interstitials,

---

[40] Besides reducing loss due to absorption, doping Ta_2O_5 with TiO_2 increases the refractive-index contrast versus SiO_2, enhancing the coating's reflectivity.

dislocations, and impurities, or by instantaneous local irregularities caused by phonons. In most metals, this scattering leads to a random walk of electrons, reducing ℓ (see Figure 10.2), and constrains conductivity to a diffusion process. The room-temperature value of ℓ for a metal such as Al is ~29 nm, about two orders of magnitude larger than its lattice parameter. In high-purity Al, ℓ increases to ~700 μm at low temperatures when phonon scattering is greatly reduced (see Chapter 10).

It is worth considering two extremes of conduction-electron behavior. In the case of a perfect crystal, scattering will be rare, and ℓ might exceed the size of the sample, particularly at low temperatures. This leads to so-called **ballistic conduction**, discussed for carbon nanotubes in Chapter 10, where the only scattering interactions are with the "imperfections" caused by the surfaces of the sample. At the other extreme, when there is a large amount of electron scattering due to phonons, disorder, and impurities, the path length ℓ between two collisions can approach the interatomic distance. From Equation (10.19), we define electrical resistivity as:

$$\rho = \frac{1}{\sigma} = \frac{m^* v_F}{ne^2} \frac{1}{\ell} \tag{15.11}$$

Since an electron can at most be scattered by every atom[41] as ℓ approaches the interatomic distance, we expect a limiting high value of ρ_{sat}, the saturation resistivity. This is called the Ioffe–Regel condition.[42]

The relevant question in this chapter is how the inherent high disorder of amorphous and glassy materials impacts electrical conductivity. Does the increasing disorder in a metallic conductor lower the mean free path ℓ and hence conductivity smoothly? P.W. Anderson provided the answer to this question. Surprisingly, he predicted a precipitous drop in electrical conductivity and localization of all the electrons after a critical amount of disorder is present. This **disorder-induced metal-to-insulator transition** is known as **Anderson localization**.[43]

Before we can understand Anderson localization, we have to introduce some results from quantum mechanics. The conductivity of a material is related to the probability of an electron moving from a point I to a point II. In a real material, there will be many different ways in which an electron can get from I to II, two of which are shown in Figure 15.10. Electrons have wave character described by a wavefunction with an amplitude and a phase. In our case, an electron moving as a wave along either path a or b has the (vector) amplitude A or B and phase φ_a or φ_b, respectively. We learned in Section 6.1.2 that the square of the wavefunction ψ^2 is called the **probability density**; the probability of finding an electron in a given region is equal to ψ^2 divided by the volume of the region. To calculate the probability of getting from I to II, quantum mechanics requires us to sum up the amplitudes of the two wavefunctions taking different paths, including an **interference term** that takes into account the phase difference, $\varphi =$

---

[41] In semi-classical theories like the Drude theory, the uncertainty of the k vector of an electron wavefunction becomes comparable to the size of the Brillouin zone when $\ell \approx$ the interatomic distance. See Section 6.4.

[42] Alternatively named Mott–Ioffe–Regel limit.

[43] Anderson localization explains early experiments in George Feher's group at Bell labs where phosphorous-doped Si at 2 K had long electron-spin relaxation times (>1000 s), consistent with electron localization.

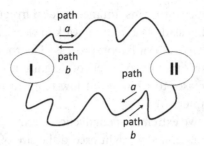

Figure 15.10 Two paths, *a* and *b*, for the wavefunction of an electron to propagate from point I to point II and back to I, which will result in constructive interference at point I.

$\varphi_a - \varphi_b$, of the two wavefunctions. If crests meet crests at II, we have constructive interference, and this occurs when there is zero phase difference ($\varphi = 0$), hence $\cos \varphi = 1$. Our probability is given by $|A + B|^2 = |A|^2 + |B|^2 + 2|A||B| \cos\varphi$. One might expect that in materials with random disorder, the phase difference φ (in the interval 0–180°) from many different pathways would average to 90°, and the interference term could be ignored. However, this is not the case. Mathematically, we can see that paths with the same phase ($\varphi = 0$) have the largest interference term ($\cos \varphi = 1$) and give the highest probability as they involve constructive interference of the two wavefunctions. One situation in which this is guaranteed to occur is when two waves take the same path but travel in opposite directions, as shown in Figure 15.10. The interference term in this situation is always 1, which increases the probability that electrons will "circle back on themselves" and become localized. This is called **weak backscattering**.

So how do these considerations impact the electronic structure of amorphous materials? In Section 6.1.6, we introduced electronic density-of-states (DOS) plots and showed that the valence and conduction bands in a semiconductor like Si are separated by an energy gap, and that conductivity occurs when either of them is partially occupied. This is depicted in Figure 15.11 and compared with the DOS of amorphous Si. The latter has two new features: defects in the middle of the gap due to "dangling bonds" in the macroscopic voids of the network (see Footnote 7 in this chapter) and regions of localized electron states at the band edges that extend into the gap, called **Urbach tails**.[44] The states lying near the middle of the gap are highly localized, not unlike localized energy levels of doped activator ions that were discussed in Section 7.8. States nearer the band edges are more extended but not over the entire sample, as they would in a crystalline material. Electrons and holes occupying these localized states do not contribute to the conductivity. The interface between extended and localized wavefunctions within a band is called the **mobility edge**, a concept introduced by Mott. The energy separation between the two mobility edges at the top and bottom of a single band is called the **mobility gap**.

---

[44] Exponential absorption edges are a universal property of all elemental and compound semiconductors and were described first in 1953 by Franz Urbach.

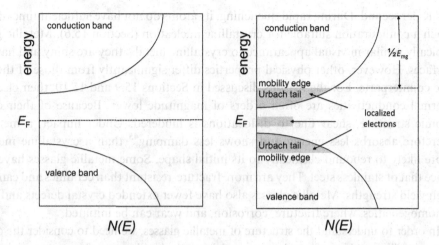

Figure 15.11 DOS of crystalline (left) and amorphous (right) semiconductors. Gray shadowing indicates localized electronic states including those at E_F which are due to "dangling bonds". As disorder increases, the mobility edges will grow within the bands and the mobility gap E_{mg} between the mobility edges at the top and bottom of a band will narrow, eventually leading to complete electron localization.

As disorder increases and backscattering becomes dominant, more electron states become localized, and the mobility gap decreases. Above a critical disorder, *all* electron states in a band will localize in what is called a Fermi glass. Such electron localization has been demonstrated in single crystals of $Li_xFe_7Se_8$, where Li above a critical concentration of $x = 0.53$ creates a sufficiently strong disorder to induce a metal–insulator transition and destroy both the crystallographic and magnetic order of Fe. Specific heat, electrical conductivity, as well as magnetic and optical measurements suggest that Anderson localization is taking place [35]. Such strong Anderson localization into a Fermi glass is a general quantum effect found also for light, sound, and microwaves that can be stopped in media; a high level of random disorder traps the quantum particle (or quasi particle) and prevents its random-walk migration. In 1977, Anderson and Mott received the Nobel Prize in Physics for their work on the conductivity of disordered materials.

15.11 Metallic Glasses

When metals or alloys cool down from the melt, they will generally solidify into the lowest energy state, usually crystalline. This typically occurs within microseconds and leads to polycrystalline samples with grains separated by grain boundaries and with line and planar defects. These areas with less than optimal atomic packing easily slip past each other under stress, leading to plastic deformation.

Although metals are generally "bad glass formers", in 1960 Klement et al. [36] discovered that the molten binary alloy $Au_{75}Si_{25}$ solidifies as a **metallic glass** if cooled at rates of at least

10^6 K per second. During rapid quenching, its atoms do not have sufficient time or energy to reach a configuration suitable for crystalline nucleation (Section 15.6). Metallic glasses are typically similar in visual appearance to crystalline metals; they are shiny and have smooth surfaces. However, other physical properties differ significantly from those of their crystalline counterparts. For the reasons discussed in Sections 15.9 and 15.10, their electrical and thermal conductivities are often orders of magnitude lower. Because of their disordered atomic structure, shear due to dislocations is hindered. Under impact, a metallic glass therefore absorbs less energy and shows less damping[45] than a crystalline metal and is more likely to rebound elastically to its initial shape. Some metallic glasses have strengths twice that of stainless steel. They are more fracture-resistant than ceramics and can have very high yield strengths. Metallic glasses also have fewer extended crystal defects and structural inhomogeneities, where fracture, corrosion, and wear can be initiated.

In order to understand the structure of metallic glasses, we need to consider the options for non-crystalline packing of metal atoms, and, in particular, the type of units that might be formed when a melt is rapidly cooled. Metallic bonding is much less directional than covalent bonding, and metal atoms can be approximated as spheres. These spheres will try to pack together efficiently so as to maximize the metallic bonding. The smallest 3D element comprising touching spheres is a tetrahedron. As more spheres are brought together they will touch in various directions, forming tetrahedra in different orientations. How can these regular tetrahedra be packed in three dimensions? The important observation is that tetrahedra can't be packed to fill space 100% efficiently, that is to say they don't **tessellate** 3D space. This is illustrated in Figure 15.12; five regular tetrahedra each trying to share two faces form a body similar to pentagonal bipyramid, but a gap is left of 7.4°. No such gap is possible in a real structure, so the tetrahedra must deform slightly in order to form the pentagonal bipyramid. This inability to tessellate space with a regular object is an example of **geometric frustration**. When the top five triangular faces of the pentagonal pyramid are each "capped" with one atom to complete new tetrahedra, and

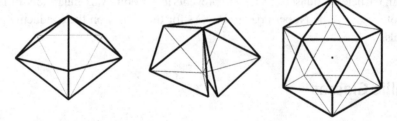

Figure 15.12 Regular pentagonal bipyramid (left) compared with a cluster of five regular tetrahedra that leaves a gap of ~7.4° (center). An icosahedron (right) of 13 atoms is built of 20 tetrahedra having their bases at its 20 triangular faces (icosa = 20 in Greek, hedros = bottom) and vertex at the center atom marked with a square dot.

---

[45] The process by which a material subjected to oscillatory deformations converts mechanical into thermal energy that is dissipated.

when the pentagon that these five new tetrahedra form is itself capped by one atom, an icosahedron is obtained. It is composed of 20 slightly deformed tetrahedra, the triangular bases of which form its 20 faces, and their common vertex is at center of the icosahedron. Analogous capping of the icosahedron's faces into subsequent layers is possible, and this creates series of larger and larger icosahedra, the **Mackay icosahedra**. Note that the icosahedron contains fivefold axes that aren't found in normal crystal structures.

This type of packing of spheres into random geometrically frustrated structures of deformed tetrahedral units is what Bernal found (Section 15.1) in his wax and ball models. In addition to being the principle behind packing of metallic glasses, it has another interesting consequence— the formation of **quasicrystals.** Quasicrystals of icosahedral or fivefold symmetry were discovered by Dan Shechtman in 1982. Fivefold symmetry is not one of the crystallographic lattice symmetries (introduced in Chapter 1), and a single equilateral pentagon will not tile 2D space. However, two tiles can be designed such that they tile 2D space in an aperiodic way called Penrose tiling, which has a fivefold and tenfold rotational symmetry.

In the context of metallic glasses, an interesting quasicrystal is one found for a $Ti_xZr_yNi_z$ alloy that normally forms a stable crystalline Laves phase [37]. As a droplet of its $Ti_xZr_yNi_z$ melt is cooled in a differential scanning calorimeter (see Box 4.1), *two* phase transitions occur. The first one corresponds to the formation of the metastable quasicrystal that, like the melt, has local icosahedral order. The second one is the crystallization of the Laves phase (with some intersite disorder of atoms), in which all icosahedral order is removed. This sequence complies with the Ostwald step rule that, in general, metastable products crystallize first, upon exploring/probing high-energy states (like in Section 15.7; see also Box 1.3). It suggests icosahedral symmetry of the local order in metallic melts, and this has been confirmed experimentally by electron diffraction [38].

One challenge in utilizing all properties of metallic glasses stems from the need for rapid cooling despite their low thermal conductivity. In early examples, the required cooling rates could only be achieved in thin films formed by casting the melt onto a spinning wheel with a very cold surface ("splat cooling"). It was not possible to cast samples in a mold, and this precluded many applications. To overcome this challenge, metallic glasses with slower rates of crystallization were needed.

Early experiments led to the development of empirical rules to compositionally precondition alloys to slow crystallization, thereby increasing the probability of glass formation. These include the following: (1) Use of multi-component alloys with three or more elements and high structural complexity to reduce the formation enthalpy of the undesirable crystal.[46] (2) Use of elements with an atomic radius mismatch greater than 12%, which leads to filling voids, higher packing density, and smaller free volume in the molten state [39], thus requiring an energetically unfavorable volume expansion during crystallization. (3) Use of elements with negative enthalpy of mixing that form weaker bonds to other elements in the glass than

---

[46] This is called the principle of confusion as the more complex chemical composition translates into a greater number of compounds that could nucleate. This competition frustrates crystal nucleation and growth, which will not occur during sufficiently rapid cooling.

to themselves, which increases the energy barrier at the liquid–solid interface, reduces atomic diffusivity, inhibits crystallization, and thereby extends the temperature range over which the supercooled liquid is stable. (4) Use of alloys with low-temperature eutectics to give a stable liquid state closer to ambient conditions.

In 1969, Chen and Turnbull [40] used these ideas to make metallic-glass spheres of ternary $Pd_xM_ySi_z$ alloys (M = Ag, Cu, Au) at cooling rates of only 100 K/s. This made it possible to prepare particles 0.5 mm in diameter. In 1974, the critical casting thickness was increased to about 1 mm in $Pt_xM_yP_z$ systems (M = Ni, Co, Fe) [41], and, in the early 1980s, glassy ingots of $Au_{55}Pb_{22.5}Sb_{22.5}$ with diameters of 5 mm were made [42]. Inoue et al. [43] and Peker and Johnson [44] later discovered metallic glasses based on La-, Mg-, Zr-, Pd-, Fe-, and Ti-containing alloys with even lower critical cooling rates (1–100 K/s), which were amenable to conventional molding.

Metallic glasses are found in a wide variety of commercial applications. Initially, they were used in sporting equipment such as golf-club heads, tennis rackets, and baseball bats. Golf-club heads using $Zr_{41.2}Ti_{13.8}Cu_{12.5}Ni_{10.0}Be_{22.5}$ ("Vitreloy") are twice as hard and four times as elastic as ones based on Ti metal, resulting in the transformation of 99% of the impact energy to the ball, compared to only 70% for Ti-based heads. Other applications exploit magnetic properties of metallic glasses, in particular those made of B, Si, P and Fe, Co, Ni, which are soft magnets (see Section 9.9 and Figure 9.23). Ferromagnetic metallic glasses can have high magnetic susceptibilities, low coercivity, and high electrical resistivity. When subjected to alternating magnetic fields, their high resistance limits eddy currents.[47] This, together with their low coercivity, is utilized in low-loss cores in power-distribution transformers. Bulk metallic glasses have also found use in optical mirrors, precision gears for micromotors, diaphragms for pressure sensors, and structural parts for aircrafts [45]. Furthermore, prototypes of efficient electrical motors demonstrated that permanent magnets made of metallic glasses can replace rare-earth magnets that have a potentially limited supply.

A more recent and related class of disordered materials are **high-entropy alloys**. Their mixing entropies increase by virtue of the high number of elemental components, and this lowers the Gibbs energy of the solid solution. In contrast to common alloys containing two or three elements in various proportions, high-entropy alloys are located near the middle of their multidimensional phase diagrams yet keep the simple face-centered-cubic, body-centered-cubic, or hexagonal closest packing due to averaged intersite disorder [46]. High-entropy alloys typically have high hardness, low ductility, and enhanced wear resistance compared to conventional metallic alloys [47].

---

[47] An alternating magnetic field will induce eddy currents in a conductor (Faraday's law of induction), which are inversely proportional to the resistivity of the material.

15.12 Problems

15.1 Using Zacharisen's rules, explain why SiO_2 forms a glass and MgO does not.

15.2 Explain why Zachariasen's rules suggest that Al_2O_3 will not form a glass but alumino-silicates with typical compositions of 11–16 mole percent (mol%) Al_2O_3, 52–60 mol% SiO_2, and 9–11 mol% K_2O will.

15.3 Consider a glass of composition $(M_2O)_x(SiO_2)_{1-x}$, where M_2O is an alkali-metal oxide. State the fraction of non-bridging oxygen and the ratio of non-bridging oxygen to silicon.

15.4 A sodium-silicate glass contains 25 mol% Na_2O. Calculate its O:Si ratio and derive a formula for the composition of a glass with the same O:Si ratio yet with Na_2O and CaO as the network modifiers (CaO results in a more durable glass).

15.5 The glass core of an optical fiber has an index of refraction 1.64. The index of refraction of the cladding is 1.50. (a) What is the maximum angle a light ray can make with the wall of the core if it is to remain inside the fiber? (b) Would this optical fiber work without cladding?

15.6 Calculate the maximum angle a light ray can make in internal reflection with the wall of the core in an optical fiber made with SiO_2 cladding ($n_2 = 1.609$) and a core with a 1% higher refractive index.

15.7 Yellow light of the Na doublet with a wavelength of 589.3±0.3 nm and frequency of 5.09×10^{14} Hz in vacuum enters Fe_2O_3 that has a refractive index of 3.00. Calculate the speed, wavelength, and frequency of the light in Fe_2O_3 and indicate which of the three change when the light passes from the vacuum to the Fe_2O_3.

15.8 Prove that Equation (15.6) reduces to the Lorenz–Lorentz equation (15.4) in the case of purely ionic bonding.

15.9 Use the Anderson–Eggleton relationship to calculate the refractive index of the mineral orthoclase ($KAlSi_3O_8$) with a unit-cell volume 720.4 $Å^3$ containing four formula units. The cation polarizabilities are 1.35 $Å^3$, 0.533 $Å^3$, and 0.284 $Å^3$, for K^+, Al^{3+}, and Si^{4+}, respectively. The free-ion polarizability α^0 for oxygen is 1.79 $Å^3$ and N_0 is 1.776 $Å^3$ (all values from Tables 4 and 5 in ref. [15]).

15.10 Calculate the refractive index using the Gladstone–Dale approach for orthoclase of Problem 15.9 using the chemical refractivities from Table 15.1. Compare the result with that of Problem 15.9.

15.11 Given the density of 2.655 g/cm^3 of an SiO_2 glass, estimate V_a in $Å^3$ per formula unit. Use the refractive index of 1.547 to calculate the polarization with the Lorentz–Lorenz equation (15.4).

15.12 Calculate the percentage transmission for a 0.1 dB/km and 0.001 dB/km optic fiber after 1 km.

15.13 Further Reading

J.L. Finney, L.V. Woodstock, "Renaissance of Bernal's random close packing and hypercritical line in the theory of liquids" *J. Phys. Condens. Matter* **26** (2014), 463102.

E. Le Bourhuis "*Glass: Mechanics and Technology*" (2008) Wiley.

J.C. Mauro, R.J. Loucks, A.K. Varshneya, P.B Gupta, "Enthalpy landscapes and the glass transitions" *Sci. Model. Simul.* **15** (2008), 241–281.

C.A. Angell, "The old problems of glass and the glass transition, and the many new twists" *Proc. Natl. Acad. Sci. USA* **92** (1995), 6675–6682.

M.D. Ediger, C.A. Angell, S.R. Nagel, "Supercooled liquids and glasses" *J. Phys. Chem.* **100** (1996), 13200–13212.

S.F. Swallen, K.L. Kearns, M.K. Mapes, Y.S. Kim, R.J. McMahon, M.D. Ediger, T. Wu, L. Yu, S. Satija, "Organic glasses with exceptional thermodynamic and kinetic stability" *Science* **315** (2007), 353–356.

J.F. Sadoc, R. Mosseri, "*Geometrical Frustration*" (2007) Cambridge University Press.

V. Lubchenko, P.G. Wolynes "The microscopic quantum theory of low temperature amorphous solids" *Adv. Chem. Phys.* **136** (2007), 95–296.

T. Egami "Magnetic amorphous alloys: Physics and technological applications" *Rep. Prog. Phys.* **47** (1984), 1601–1725.

15.14 References

[1] P.W. Anderson, "Through the glass lightly" *Science* **267** (1995), 1615–1616.

[2] T. Egami, S. Billinge, "*Underneath the Bragg Peaks: Structural Analysis of Complex Materials*" 2nd edition (2012) Pergamon Press.

[3] S. Kohara, K. Suzuya, "Intermediate-range order in vitreous SiO_2 and GeO_2" *J. Phys.: Condens. Matter* **17** (2005), S77–S86.

[4] J.D. Bernal, "A geometrical approach to the structure of liquids" *Nature* **183** (1959), 141–147.

[5] J.D. Bernal, J. Mason, "Packing of spheres: Coordination of randomly packed spheres" *Nature* **188** (1960), 910–911.

[6] D.B. Miracle, T. Egami, K.M. Flores, K.F. Kelton, "Structural aspects of metallic glasses" *MRS Bull.* **32** (2007), 629–634.

[7] A. Donev, I. Cisse, D. Sachs, E.A. Variano, F.H. Stillinger, R. Connelly, S. Torquato, P.M. Chaikin, "Improving the density of jammed disordered packings using ellipsoids" *Science* **303** (2004), 990–993.

[8] N. Mousseau, G.T. Barkema, S.M. Nakhmanson, "Recent developments in the study of random continuous random networks" *Phil. Mag. B* **82** (2002), 171–183.

[9] L.W. Hobbs, C.E. Jesurum, V. Pulim, B. Berger, "Local topology of silica networks" *Phil. Mag. A* **78** (1998), 679–711.

[10] J. Shropshire, P.P. Keat, P.A. Vaughan, "The crystal structure of keatite, a new form of silica" *Z. Kristallogr. Cryst. Mater.* **112** (1959), 409–412.

[11] G.H. Beall, "Dr. S. Donald (Don) Stookey (1915–2004): Pioneering researcher and adventurer" *Front. Mater.* **3** (2016), article 37.

[12] R.D. Shannon, "Revised effective ionic radii and systematic studies of interatomic distances in halides and chalcogenides" *Acta. Crystallogr. Sect. A* **32** (1976), 751–767.

[13] J.C. Mauro, A. Tandia, K.D. Vargheese, Y.Z. Mauro, M.M. Smedskjær, "Accelerating the design of functional glasses through modeling" *Chem. Mater.* **28** (2016), 4267–4277.

[14] J.B. Pendry, D. Schurig, D.R. Smith, "Controlling electromagnetic fields" *Science* **312** (2016), 1780–1782.

[15] R.D. Shannon, R.X. Fischer, "Empirical electronic polarizabilities of ions for the prediction and interpretation of refractive indices: Oxides and oxysalts" *Am. Mineral.* **101** (2016), 2288–2300.

[16] O.L. Anderson, "Optical properties of rock-forming minerals from atomic properties" *Fortschr. Mineral.* **52** (1975), 611–629.

[17] R.A. Eggleton, "Gladstone–Dale constants for the major elements in silicates: Coordination number, polarizability and the Lorentz–Lorenz relationship" *Can. Mineral.* **29** (1991), 525–532.

[18] J. A. Mandarino, "The Gladstone–Dale compatibility of minerals and its use in selected mineral species for further study" *Can. Mineral.* **45** (2007), 1307–1324.

[19] D.B. Keck, R.D. Maurer, P.C. Schultz, "On the ultimate lower limit of attenuation in glass optical waveguides" *Appl. Phys. Lett.* **22** (1972), 307–308.

[20] C.R. Hammond, S.R. Norman, "Silica based binary glass systems: Refractive index behavior and composition in optical fibers" *Opt. Quantum Electron.* **9** (1977), 399–409.

[21] K. Nagayama, M. Kakui, M. Matsui, T. Saitoh, Y. Chigusa, "Ultra low loss (0.1484 dB/km) pure silica core fiber and extension of transmission distance" *Electron. Lett.* **38** (2002), 1168–1169.

[22] H. Tanaka, "Possible resolution of the Kauzmann paradox in supercooled liquids" *Phys. Rev. E* **68** (2003), 011505.

[23] J.B. Lambert, G.O. Poinar, Jr., "Amber: The organic gemstone" *Acc. Chem. Res.* **35** (2002), 628–636.

[24] T. Pérez-Castañeda, R.J. Jiménez-Riobóo, M.A. Ramos, "Two-level systems and boson peak remain stable in 110-million-year-old amber glass" *Phys. Rev. Lett.* **112** (2014), 165901/1–5.

[25] M.H. Cohen, P. Turnbull, "Molecular transport in liquids and glasses" *J. Chem. Phys.* **31** (1959), 1164–1169.

[26] H.-S. Philip Wong, S. Raoux, S. Kim, J. Liang, J.P. Reifenberg, B. Rajendran, M. Asheghi, K.E. Goodson, "Phase change memory" *Proc. IEEE* **98** (2010), 2201–2227.

[27] F. Rao, K. Ding, Y. Zhou, Y. Zheng, M. Xia, S. Lv, Z. Song, S. Feng, I. Ronneberger, R. Mazzarello, W. Zhang, "Reducing the stochasticity of crystal nucleation to enable subnanosecond writing" *Science* **358** (2017), 1423–1427.

[28] O. Gulbiten, J.C. Mauro, X. Guo, O.N. Boratav, "Viscous flow of medieval cathedral glass" *J. Am. Ceram. Soc.* **101** (2018), 5–11.

[29] P. Debye, "Zur Theorie der spezifischen Wärme" *Ann. Phys.* **39** (1912), 789–839.

[30] R.C. Zeller, R.O. Pohl, "Thermal conductivity and specific heat of noncrystalline solids" *Phys. Rev. B* **4** (1971), 2029–2041.

[31] P.W. Anderson, B.L. Halperin, C.M. Varma, "Anomalous low-temperature thermal properties of glasses and spin-glasses" *Philos. Mag.* **25** (1972), 1–9.

[32] W.A. Phillips, "Tunneling states in amorphous solids" *J. Low Temp. Phys.* **7** (1972), 351–360.

[33] C. Müller, J.H. Cole, L. Lisenfeld, "Towards understanding two-level systems in amorphous solids: Insights from quantum circuits" *Rep. Prog. Phys.* **82** (2019), 12450.

[34] C.R. Billman, J.P. Trinastic, D.J. Davis, R. Hamdan, H.-P. Chen, "Origin of the second peak in the mechanical loss function of amorphous silica" *Phys. Rev. B* **95** (2017), 014109.

[35] T. Ying, Y. Gu, X. Chen, X. Wang, S. Jin, L. Zhao, W. Zhang, X. Chen, "Anderson localization of electrons in single crystals: $Li_xFe_7Se_8$" *Sci. Adv.* **2** (2016), e1501283.

[36] W. Klement, Jr., R.H. Willens, P. Duwez, "Non-crystalline structure in solidified gold–silicon alloys" *Nature* **187** (1960), 869–870.

[37] K.F. Kelton, G.W. Lee, A.K. Gangopadhyay, R.W. Hyers, T.J. Rathz, J.R. Rogers, M.B. Robinson, D.S. Robinson, "First X-ray scattering studies on electrostatically levitated metallic liquids: Demonstrated influence of local icosahedral order on the nucleation barrier" *Phys. Rev. Lett.* **90** (2003), 195504/1–4.

[38] A. Hirata, L.J. Kang, T. Fujita, B. Klumov, K. Matsue, M. Kotani, A.R. Yavari, M.W. Chen, "Geometric frustration of icosahedron in metallic glasses" *Science* **341** (2013), 376–379.

[39] A. Inoue "Stabilization of metallic supercooled liquid and bulk amorphous alloys" *Acta Mater.* **48** (2000), 279–306.

[40] H.S. Chen, D. Turnbull, "Formation, stability and structure of palladium–silicon alloy based glasses" *Acta Metall.* **17** (1969), 1021–1031.

[41] H.S. Chen, D. Turnbull, "Thermodynamic considerations on the formation and stability of metallic glasses" *Acta Metall.* **22** (1974), 1505–1511.

[42] M.C. Lee, J. M. Kendall, W. L. Johnson, "Spheres of the metallic glass $Au_{55}Pb_{22.5}Sb_{22.5}$ and their surface characteristics" *Appl. Phys. Lett.* **40** (1982), 382–384.

[43] A. Inoue, A. Kato, T.G.K.S. Zhang, T. Masumoto, "Mg–Cu–Y amorphous alloys with high mechanical strengths produced by a metallic mold casting" *Mater. Trans. JIM* **32** (1991), 609–616.

[44] A. Peker, W.L. Johnson, "A highly processable metallic glass: $Zr_{41.2}Ti_{13.8}Cu_{12.5}Ni_{10.0}Be_{22.5}$" *Appl. Phys. Lett.* **63** (1993), 2342–2344.

[45] A. Inoue, N. Nishiyama, "New bulk metallic glasses for applications as magnetic-sensing, chemical, and structural materials" *MRS Bull.* **32** (2007), 651–658.

[46] B. Cantor, "Multicomponent and high entropy alloys" *Entropy* **16** (2014), 4749–4768.

[47] M.-H. Tsai, J.-W. Yeh, "High-entropy alloys: A critical review" *Mater. Res. Lett.* **2** (2014), 107–123.

Appendices

APPENDIX A

Crystallographic Point Groups in Schönflies Symbolism

Throughout the book, the so-called Hermann–Maugin symbolism has been used for notation of the symmetry elements of crystallographic point groups because it has traditionally been associated with the analysis of long-range periodic structures. When the experimental method focuses on the local coordinations, as spectroscopic methods do, another symbolism is often used, named after Arthur Moritz Schönflies. The Schönflies symbolism differs from the Hermann–Maugin symbolism not only in symbols but also in the choice of some symmetry elements; instead of the Hermann–Maugin rotoinversion axes, rotoreflection axes are used. Instead of listing the generating symmetry elements in significant directions, the Schönflies symbolism uses a shorthand for the visual form or symmetry of the object. In Table A.1, we add the Schönflies notation to the Hermann–Maugin-style Table 1.1.

Table A.1 Classification of 32 crystallographic point groups into seven crystal systems.

Crystal system	Minimum symmetry		Higher-symmetry point groups	
	H.–M.	Schön.	Hermann–Maugin	Schönflies
Triclinic	1	C_1	$\bar{1}$	C_i
Monoclinic	2, m	C_2, C_s	$2/m$	C_{2h}
Orthorhombic	222	D_2	$mm2, mmm$	C_{2v}, D_{2h}
Tetragonal	4, $\bar{4}$	C_4, S_4	$4/m, 422, 4mm, \bar{4}2m, 4/mmm$	$C_{4h}, D_4, C_{4v}, D_{2d}, D_{4h}$
Hexagonal	6, $\bar{6}$	C_6, C_{3h}	$6/m, 622, 6mm, \bar{6}m2, 6/mmm$	$C_{6h}, D_6, C_{6v}, D_{3h}, D_{6h}$
Trigonal	3, $\bar{3}$	C_3, C_{3i}	$32, 3m, \bar{3}m$	D_3, C_{3v}, D_{3d}
Cubic	23	T	$m\bar{3}, \bar{4}3m, 432, m\bar{3}m$	T_h, T_d, O, O_h

APPENDIX B

International Tables for Crystallography

The International Tables for Crystallography, Volume A: Space-group symmetry is the first in a series of authoritative volumes published by the International Union of Crystallography (the first online edition, 2006, http://dx.doi.org/10.1107/97809553602060000100 or http://it .iucr.org/Ab/, and the second online edition, 2016, http://dx.doi.org/10.1107/9780955360206000 0114 or http://it.iucr.org/Ac/). Volume A lists symmetry properties of the 230 space groups used for analysis and description of crystal structures. The symmetry information is presented symbolically, graphically, and mathematically. As an example, the entry for space group *Pnma* is explained in this appendix.

International Tables contain a minimum of two pages for each space group. The first of these two pages for space group *Pnma* is shown in Figure B.1. The upper line of text contains the space-group symbol *Pnma*, the symbols of the point group in Schönflies (D_{2h}^{16}) and Hermann–Maugin (*mmm*) notations and the crystal system (Orthorhombic). The next line gives the number (62) of the space group (from 1 to 230; in order of ascending symmetry), the full Hermann–Maugin space-group symbol ($P\,2_1/n\,2_1/m\,2_1/a$), and the symbol of Patterson symmetry (*Pmmm*) of an electron-density map used to determine crystal structures from diffraction that adds $\bar{1}$. The symmetry elements in the full space-group symbol are oriented along the main symmetry directions listed in Table 1.3.

The **space-group diagram** is given below these lines and shows the location of the symmetry elements in the unit cell. For an orthorhombic space group, the figure in the top left is the standard projection/view of the unit cell against its *c* axis, with the *a* axis down the projection plane (the page), *b* horizontal, *c* coming out of the plane towards the reader, and the origin in the upper left corner. The second orthographic projection of the unit cell is on the right (*a* down, *c* horizontal), and the third (*c* down, *b* horizontal) is below the standard projection. The standard-projection symbol is identical with the full space-group symbol. Its corresponding point diagram at the bottom right of Figure B.1 shows how the initial point *x*,*y*,*z* represented by "O+" is repeated by the symmetry elements. The *z* coordinate given as the "+" or "½+" means +*z* or ½ + *z*, as we are looking down against *c*. A comma in the point's circle signifies handedness relative to the initial point. Handedness of two points behind each other would be symbolized by the circle divided into two halves, one of them with a comma.

The point diagram can be understood if one considers the **graphical symbols for symmetry elements** in the standard projection on top left: Here the *n*, *m*, and *a* planes of the space-group symbol are represented by, respectively, the dashed dotted line ⋅–⋅–⋅– for the diagonal glide, by the full line ——— for the mirror plane and by ⌐ for the glide of translation in the direction *a* of the arrow. This glide is at a height of *z* = ¼. In the other two projections, however, these same symmetry elements are drawn with symbols that correspond to their new orientation against the

International Tables for Crystallography (2006). Vol. A, Space group 62, pp. 298–299.

$Pnma$ D_{2h}^{16} mmm Orthorhombic

No. 62 $P\,2_1/n\,2_1/m\,2_1/a$ Patterson symmetry $Pmmm$

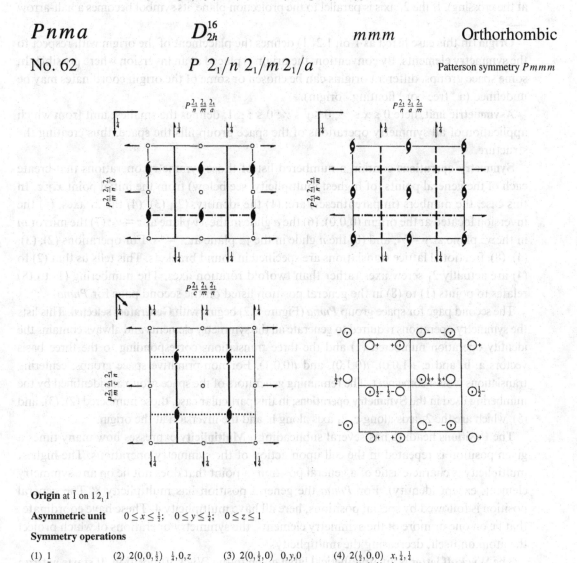

Origin at $\bar{1}$ on $1\,2_1\,1$

Asymmetric unit $0 \le x \le \tfrac{1}{2}$; $0 \le y \le \tfrac{1}{4}$; $0 \le z \le 1$

Symmetry operations

(1) 1 (2) $2(0,0,\tfrac{1}{2})$ $\tfrac{1}{4},0,z$ (3) $2(0,\tfrac{1}{2},0)$ $0,y,0$ (4) $2(\tfrac{1}{2},0,0)$ $x,\tfrac{1}{4},\tfrac{1}{4}$
(5) $\bar{1}$ $0,0,0$ (6) a $x,y,\tfrac{1}{4}$ (7) m $x,\tfrac{1}{4},z$ (8) $n(0,\tfrac{1}{2},\tfrac{1}{2})$ $\tfrac{1}{4},y,z$

Figure B.1 Page 298 of International Tables for Crystallography, Volume A (2006).

plane of the page. In the diagram on the top right (projection onto the *ac* plane), the mirror plane is now ¼ above this plane and has the symbol ⌐. In the lower left diagram (projection onto *bc* plane), the *n* glide is shown as ⌐, and the symbol ·········· is the original *a* glide viewed side on (implying that the direction of translation is now out of page, against the viewer). The inversions $\bar{1}$ are marked with small circles ∘, the twofold screw axes 2_1 have symbol ⟆. If this symbol has a white circle in its center, it means a 2_1 with a perpendicular m ($2_1/m$; the combination generates a center of inversion at the crossing). If the 2_1 axis is parallel to the projection plane, its symbol becomes a half-arrow ⟶.

Origin (in this case listed as $\bar{1}$ on $1\ 2_1\ 1$) defines the placement of the origin with respect to the symmetry elements. By convention, the origin is placed at an inversion where possible. In some space groups, different origins can be chosen or some of the origin coordinates may be undefined (a "free" or "floating" origin).

Asymmetric unit, here $0 \leq x \leq \frac{1}{2}$; $0 \leq y \leq \frac{1}{4}$; $0 \leq z \leq 1$, defines the smallest unit from which application of all symmetry operations of the space group fills the space, thus creating the structure.

Symmetry operations contain a numbered list of those symmetry operations that create each of the general points (of highest multiplicity, see below) from the initial point x,y,z. In this case, the numbers (in parentheses) are: (1) the identity; (2), (3), (4) the 2_1 axes; (5) the inversion located at the origin (0,0,0); (6) the a glide in the xy plane at $z = \frac{1}{4}$; (7) the mirror m in the xz plane at $y = \frac{1}{4}$; and (8) the n glide in the yz plane at $x = \frac{1}{4}$. For operations (2), (3), (4), (8), fractional lattice translations are specified in round brackets. This tells us that (2) to (4) are actually 2_1 screw axes rather than twofold rotation axes. The numbering (1) to (8) relates to points (1) to (8) in the general position listed on the second page for *Pnma*.

The second page for space group *Pnma* (Figure B.2) begins with **Generators selected**. This lists the symmetry operations required to generate all the symmetry elements and always contains the identity operation numbered (1) and the three translations corresponding to the three basis vectors **a**, **b**, and **c**; $t(1,0,0)$, $t(0,1,0)$, and $t(0,0,1)$. For non-primitive space groups, centering translations would come next. The remaining generators of the space group are identified by the numbering used in the **Symmetry operations**; in this particular case, those numbered (2), (3), and (5), which are the 2_1 axis along **c**, 2_1 axis along **b**, and the inversion at the origin.

The **Positions** heading has several subheadings. **Multiplicity** expresses how many times a given position is repeated in the cell upon action of the symmetry operations. The highest multiplicity is characteristic of a general position (a point that does not lie on any symmetry element, except identity). For *Pnma* the general position has multiplicity 8. The general position is followed by special positions, here all have multiplicity 4. These have coordinates that lie on one or more of the symmetry elements, the symmetry operations of which project the atom on itself, decreasing the multiplicity.

The **Wyckoff letter** is an alphabetical label given to the Wyckoff position. It starts with an "a" for one of the lowest-multiplicity sites at the bottom. The **Site symmetry** gives the point symmetry for the site. The **Coordinates** show how the original point x,y,z representing coordinates of an atom is transformed after the symmetry operations have been performed (eight such operations in Figure B.1), and the number of resulting coordinates corresponds to the multiplicity of the position. **Reflection conditions** are conditions for the appearance of *hkl* Bragg reflections when an atom is placed in that given Wyckoff position. Bragg reflections not fulfilling the listed condition are **absent** or **extinct**. When several special Wyckoff positions are occupied, their reflection conditions combine towards those listed for the general position that generates the maximum number of reflections.

CONTINUED   No. 62   *Pnma*

Generators selected   (1); $t(1,0,0)$; $t(0,1,0)$; $t(0,0,1)$; (2); (3); (5)

Positions

Multiplicity, Wyckoff letter, Site symmetry		Coordinates			Reflection conditions
					General:
8 d 1	(1) x,y,z	(2) $\bar{x}+\frac{1}{2},\bar{y},z+\frac{1}{2}$	(3) $\bar{x},y+\frac{1}{2},\bar{z}$	(4) $x+\frac{1}{2},\bar{y}+\frac{1}{2},\bar{z}+\frac{1}{2}$	$0kl: k+l=2n$
	(5) $\bar{x},\bar{y},\bar{z}$	(6) $x+\frac{1}{2},y,\bar{z}+\frac{1}{2}$	(7) $x,\bar{y}+\frac{1}{2},z$	(8) $\bar{x}+\frac{1}{2},y+\frac{1}{2},z+\frac{1}{2}$	$hk0: h=2n$
					$h00: h=2n$
					$0k0: k=2n$
					$00l: l=2n$

Special: as above, plus

4 c .m.	$x,\frac{1}{4},z$	$\bar{x}+\frac{1}{2},\frac{3}{4},z+\frac{1}{2}$	$\bar{x},\frac{3}{4},\bar{z}$	$x+\frac{1}{2},\frac{1}{4},\bar{z}+\frac{1}{2}$

no extra conditions

4 b $\bar{1}$	$0,0,\frac{1}{2}$	$\frac{1}{2},0,0$	$0,\frac{1}{2},\frac{1}{2}$	$\frac{1}{2},\frac{1}{2},0$	$hkl: h+l,k=2n$
4 a $\bar{1}$	$0,0,0$	$\frac{1}{2},0,\frac{1}{2}$	$0,\frac{1}{2},0$	$\frac{1}{2},\frac{1}{2},\frac{1}{2}$	$hkl: h+l,k=2n$

Symmetry of special projections

Along [001] $p2gm$	Along [100] $c2mm$	Along [010] $p2gg$
$\mathbf{a}'=\frac{1}{2}\mathbf{a}$ $\mathbf{b}'=\mathbf{b}$	$\mathbf{a}'=\mathbf{b}$ $\mathbf{b}'=\mathbf{c}$	$\mathbf{a}'=\mathbf{c}$ $\mathbf{b}'=\mathbf{a}$
Origin at $0,0,z$	Origin at $x,\frac{1}{4},\frac{1}{4}$	Origin at $0,y,0$

Maximal non-isomorphic subgroups

I	[2] $Pn2_1a$ ($Pna2_1$, 33)	1; 3; 6; 8
	[2] $Pnm2_1$ ($Pmn2_1$, 31)	1; 2; 7; 8
	[2] $P2_1ma$ ($Pmc2_1$, 26)	1; 4; 6; 7
	[2] $P2_12_12_1$ (19)	1; 2; 3; 4
	[2] $P112_1/a$ ($P2_1/c$, 14)	1; 2; 5; 6
	[2] $P2_1/n11$ ($P2_1/c$, 14)	1; 4; 5; 8
	[2] $P12_1/m1$ ($P2_1/m$, 11)	1; 3; 5; 7

IIa   none

IIb   none

Maximal isomorphic subgroups of lowest index

IIc   [3] $Pnma$ ($\mathbf{a}'=3\mathbf{a}$) (62); [3] $Pnma$ ($\mathbf{b}'=3\mathbf{b}$) (62); [3] $Pnma$ ($\mathbf{c}'=3\mathbf{c}$) (62)

Minimal non-isomorphic supergroups

I   none

II   [2] $Amma$ ($Cmcm$, 63); [2] $Bbmm$ ($Cmcm$, 63); [2] $Ccme$ ($Cmce$, 64); [2] $Imma$ (74); [2] $Pcma$ ($\mathbf{b}'=\frac{1}{2}\mathbf{b}$) ($Pbam$, 55); [2] $Pbma$ ($\mathbf{c}'=\frac{1}{2}\mathbf{c}$) ($Pbcm$, 57); [2] $Pnmm$ ($\mathbf{a}'=\frac{1}{2}\mathbf{a}$) ($Pmmn$, 59)

Figure B.2 International Tables for Crystallography, Volume A (2006) p. 299.

Subgroups and supergroups[1] are related by their order. The order of a group is the number of the symmetry operations it contains (as listed earlier), and it also corresponds to the multiplicity of the general point x,y,z. A group S is called a (proper) **subgroup** to the group G

---

[1] We skip the symmetry of special projections entry as it involves plane groups not used in this text.

if all symmetry elements of S are also present in G and the order of G divided by the order of S is an integer. Knowledge of group–subgroup relationships gives a clue on how a gradual removal of symmetry via deformations or via ordering of atoms or charges or magnetic moments can change the given parent structure into the less symmetric structure of its "child" subgroup. As we see in Chapters 2 and 4, these considerations may be used to investigate structural defects such as twinning and antiphase boundaries, and also to describe structural phase transitions.

The **Maximal non-isomorphic subgroups** are those closest to the parent group.[2] There is no other group in the hierarchy between the group and its maximal subgroup. The accompanied dilution of the symmetry operations is expressed by the so-called **index**, an integer number placed in square brackets that precedes the symbol of the subgroup. According to whether the group-to-subgroup transition involves the loss of some rotational or translational symmetry, two types of maximal non-isomorphic subgroups are distinguished: **Translationengleiche[3] subgroups** (under heading **I**) are those that keep the translations (all lattice points are kept), and the index tells us how many times the number of (rotational) **Symmetry operations** per lattice point was decreased upon the transition into the subgroup. **Klassengleiche[4] subgroups** are those that keep the crystallographic point-group symmetry, the "crystal class", while some translations (lattice points) are lost via dilutions such as loss of centering or a cell-edge multiplying. Those klassengleiche subgroups that retain the same conventional cell are listed under **IIa** and arise by removal of centering. Those klassengleiche subgroups that acquire a larger conventional cell than the parent group are listed under **IIb**. In both cases, the index in the square bracket is the dilution factor of lattice points upon the transition into the subgroup. For our *Pnma* in Figure B.2, only type **I** subgroups exist. One of them forms when **Symmetry operations** (5) to (8) are lost, lowering the space-group symmetry to $P2_12_12_1$. Another when **Symmetry operations** (2), (4), (5), (7) are lost—the translationengleiche subgroup $Pn2_1a$ and so on.

The category **IIc** lists the **Maximal isomorphic subgroups of the lowest index**, where the term isomorphic indicates that the space group remains the same (only the unit cell multiplies). Since the number of such enlargements is infinite, the condition of lowest index (here the dilution factor of lattice points) is applied to ensure that only "nearest" supercells are listed. **Minimal non-isomorphic supergroups** are the last category listed on the page. These relate to the situation in which the present group is a subgroup of a supergroup and let one investigate the higher possible symmetries. The subdivisions into **I** translationengleiche and **II** klassengleiche subgroups have the same meaning as explained above.

---

[2] Non-isomorphic; any two groups except for 11 enantiomorphic pairs of space groups: $P4_1$–$P4_3$ and 10 others (an inverted structure cannot be rotated to align with the original structure).

[3] German for "equal in terms of translations". Shorthand t-subgroups is also in use.

[4] German for "equal in terms of classes". Shorthand k-subgroups is also in use.

APPENDIX C
Nomenclature of Silicates

A systematic nomenclature of silicate anions has been developed by Friedrich Liebau. The newest version [1] places the formula of the silicate anion in square brackets and all geometry information in the preceding curly brackets, as $\{B,P,MD\}[Si_mO_n^{a-}]$. In this scheme, B stands for branchedness of the silicate anion, P for the periodicity of the tetrahedral chain, and M for multiplicity with D for dimensionality come as one combined symbol. The **branchedness parameter** B describes branching along a chain or ring; uB stands for unbranched, oB open branched, lB loop branched. The **multiplicity parameter** M describes how many parallel strings of tetrahedra the chain has. As an example, the tremolite chain in Figure 1.56 is a double chain. The **dimensionality parameter** D of the silicate anion is a "t" for a terminated anion, r for ring, $\frac{1}{\infty}$ for infinite chain, $\frac{2}{\infty}$ for a layer, $\frac{3}{\infty}$ for a framework. The **periodicity of the chain** P is the number of tetrahedra in the repeat unit along the chain direction (along the ring for ring chains), excluding any branches. When a name is formed for the silicate anion from the formula, this number is pronounced as a German enumerative. A chain with a repeat unit of 1 tetrahedron is called einer, 2 zweier, 3 dreier, 4 vierer, 5 fünfer, 6 sechser, 7 siebener, 8 achter, etc. As an example, the stokesite chain in Figure 1.56 is an unbranched sechser, the tremolite chain is an unbranched zweier double chain. More complex silicate chains are shown in Figure C.1 using examples of minerals that also contain other anions.

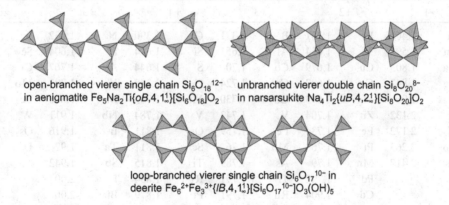

open-branched vierer single chain $Si_6O_{18}^{12-}$ in aenigmatite $Fe_5Na_2Ti\{oB,4,1^1_\infty\}[Si_6O_{18}]O_2$

unbranched vierer double chain $Si_8O_{20}^{8-}$ in narsarsukite $Na_4Ti_2\{uB,4,2^1_\infty\}[Si_8O_{20}]O_2$

loop-branched vierer single chain $Si_6O_{17}^{10-}$ in deerite $Fe_6^{2+}Fe_3^{3+}\{lB,4,1^1_\infty\}[Si_6O_{17}^{10-}]O_3(OH)_5$

Figure C.1 Examples of infinite silicate chains, their formulae, and nomenclature.

APPENDIX D

Bond-Valence Parameters in Solids

Equation (5.16) contains empirical parameters R_{ij}^0 and B for the bonded pair of ions ij. Each $R_{ij}^0 = B\ln(v_i/\sum e^{-d_{ij}/B})$ is calculated from from i–j bond distances d_{ij} in the coordination polyhedron and from the value v_i of the cation's oxidation state. The values for a select set of crystal structures with the given i solely coordinated by j are then averaged. Brown and Altermatt [2] used this procedure to compile a table of bond-valence parameters for use primarily in oxides. A similar approach was used by Brese and O'Keeffe [3]. Selected values of R_{ij}^0 for a variety of atoms bonded to oxygen are compiled in Table D.1. A more extensive set is at www.iucr.org/resources/data/datasets/bond-valence-parameters. For ions with s^2 lone pairs and those that may adopt one of several spin states, the values need to be applied with caution.

In other approaches, O'Keeffe and Brese [4] derived a formula of two tabulated parameters based on size and electronegativity, from which single-bond lengths R_{ij}^0 can be calculated for all combinations of 75 elements. More recently, Gagné and Hawthorn [5] evaluated both R_{ij}^0 and B independently for bonds to oxygen, with some improvement in standard deviations.

Table D.1 Selected bond-valence parameters R_{M-O}^0 (in Å) for oxides, sorted by oxidation states of M as column headers. The data are based on $B = 0.37$ Å and are compiled from refs. [2] and [3].

+1		+2		+3		+4		+5		+6	
Li	1.466	Be	1.381	B	1.371	C	1.390	N	1.432	S	1.624
Cu	1.593	Ni	1.654	Al	1.651	Si	1.624	P	1.604	Se	1.788
Na	1.803	Cu	1.679	Co	1.70	S	1.644	As	1.767	Cr	1.794
Ag	1.842	Co	1.692	Cr	1.724	Ge	1.748	V	1.803	Mo	1.907
Hg	1.90	Mg	1.693	Ga	1.730	Mn	1.753	Br	1.840	Te	1.917
K	2.132	Zn	1.704	V	1.743	V	1.784	Nb	1.911	W	1.921
Tl	2.172	Fe	1.734	Fe	1.759	Os	1.811	Ir	1.916	Os	2.03
Rb	2.263	Pt	1.768	Mn	1.76	Se	1.811	Ta	1.92	U	2.075
Cs	2.417	Mn	1.79	As	1.789	Ti	1.815	Sb	1.942		
		Pd	1.792	Ti	1.791	Ru	1.834	I	2.003		
		Cd	1.904	Rh	1.791	Pt	1.879	Bi	2.06		

Table D.1 (cont.)

	Ca	1.967	Au	1.833	Sn	1.905	U	2.075
	Hg	1.972	Sc	1.849	Hf	1.923		
	Sn	1.984	In	1.902	Zr	1.937		
	Pb	2.112	Lu	1.971	Te	1.977		
	Sr	2.118	Sb	1.973	Ce	2.028		
	Eu	2.147	Yb	1.985	Pb	2.042		
	Ba	2.285	Tm	2.000	U	2.112		
			Tl	2.003	Th	2.167		
			Er	2.010				
			Y	2.014				
			Ho	2.025				
			Dy	2.036				
			Tb	2.049				
			Gd	2.065				
			Eu	2.074				
			Sm	2.088				
			Bi	2.094				
			Nd	2.117				
			Pr	2.138				
			Ce	2.151				
			La	2.172				
			Ac	2.24				

APPENDIX E
The Effect of a Magnetic Field on a Moving Charge

A homogeneous magnetic field exerts a force on a wire carrying electric current, whether it is straight or in a loop. This short recapitulation from physics may help you to grasp some terms and units used in magnetism.

The left-hand side of Figure E.1 shows a straight segment of wire carrying current I. The force F exerted along a length l of the wire perpendicular to magnetic field of magnetic induction B is perpendicular to both the field and the wire, and its magnitude is $F = IBl$. This can be used to derive the force and moment exerted on a current loop of area A at an angle φ to a homogeneous field B. In order to do so, we draw a rectangular current loop of sides a and b. Forces F_1, F_2, F_3, and F_4 act on the four edges of the loop. Of these, F_3 and F_4 cancel. The moment of force acting on the loop is then $M = F_1 p = F_1 b \sin\varphi = Iab\sin\varphi$. As ab is the area A

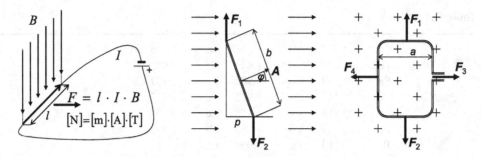

Figure E.1 Left: Force F exerted on a straight length l of a wire carrying current I in a homogeneous magnetic field of magnetic induction B. Middle and right: Force exerted on a current loop in magnetic field (two projections).

of the loop, the vector M of the moment is the cross product $M = I(A \times B)$ where A is the loop-area vector and B is the vector of the magnetic-field induction.

APPENDIX F

Coupling *j*–*j*

When spin–orbit coupling is relatively weak, it can be approximated by summing the spin momenta and orbital momenta separately and combining them afterwards—the Russell–Saunders coupling. For elements such as the actinoids, where spin–orbit coupling is strong, this scheme breaks down and one must use j–j coupling. For each electron, its orbital angular momentum is summed with its spin angular momentum to give the total angular momenta $j_1 = \ell - \frac{1}{2}$ and $j_2 = \ell + \frac{1}{2}$. The new quantum numbers for these states are j and m_j. The j values are then combined to form the total angular momentum J.

Let's use an example. Consider a Pb atom which has two unpaired electrons in its p shell, $[Xe]6s^2 6p^2$. For a configuration with x unpaired electrons in an orbital of an angular-momentum quantum number ℓ, the number of microstates is given by $N_\ell!/[x!(N_\ell - x)!]$ where $N_\ell = 2\ell(\ell + 1)$. Thus for our p^2 configuration we expect $6!/[2!4!] = 15$ microstates. Since $\ell = 1$ for p orbitals, and $|s| = \frac{1}{2}$ for an unpaired electron, addition of the electron's orbital momentum and spin momentum gives $j = \frac{3}{2}$ or $\frac{1}{2}$ with three possible combinations of j_1 and j_2 for the two electrons; $(\frac{3}{2},\frac{3}{2})$, $(\frac{3}{2},\frac{1}{2})$ and $(\frac{1}{2},\frac{1}{2})$. When $j = \frac{3}{2}$, the m_j values possible are $-\frac{3}{2}, -\frac{1}{2}, \frac{1}{2}, \frac{3}{2}$, whereas for $j = \frac{1}{2}$, m_j may be $-\frac{1}{2}$ or $\frac{1}{2}$ (see Figure 9.8). There are only six allowed pairs of m_{j1} and m_{j2} for the j_1,j_2 combination $\frac{3}{2},\frac{3}{2}$ because the Pauli exclusion principle states that if j_1 equals j_2, m_{j1} and m_{j2} must not be equal. The six m_{j1}, m_{j2} pairs are: ½,½ [$m_{j1} + m_{j2} = M_J = +2$]; ½,–½ [$M_J = +1$]; ½,–½ [$M_J = 0$]; ½,–½ [$M_J = 0$]; ½,–½ [$M_J = -1$]; –½,–½ [$M_J = -2$], corresponding to $J = 2, 1, 0$. For the second j_1, j_2 combination ½,½, there are eight pairs because m_{j1} and m_{j2} may be equal; the two unpaired electrons of Pb in this state are not equivalent, and this also means that ½,–½ and –½,½ combinations of m_{j1} and m_{j2} belong to two different sub-states. Again, $J = 2, 1, 0$ are the results of these combinations. For the third j_1, j_2 combination ½,½, the Pauli exclusion principle allows one possible state: $m_{j1} = ½$ and $m_{j2} = -½$ [$M_J = 0$] hence $J = 0$. Only the third Hund rule applies for the choice of the ground state. Because the shell is less than half full, the minimum J state, $J = 0$, is the ground state of the isolated Pb atom and lead is diamagnetic.

APPENDIX G
The Langevin Function

The Langevin function is an important expression that helps us understand the properties of paramagnets. It can be derived in a semi-classical treatment by considering the competition between thermal randomization and magnetic alignment. Let's assume that we have a magnetic moment free to point in any direction of space (i.e. we are ignoring the effects of quantization as if $J = \infty$). We can therefore imagine the moment lying anywhere on the surface of a unit sphere as in Figure G.1.

The probability of the moment *randomly* pointing into the interval between angle α and $\alpha + d\alpha$ towards the field direction z is the area of this interval, $2\pi \sin\alpha \cdot d\alpha$, divided by the entire unit-sphere surface area 4π, hence $(\sin\alpha \cdot d\alpha)/2$. The moments at angle α have a relative potential energy $U_m = -B\mu\cos\alpha$. The probability of pointing into the interval α to $\alpha + d\alpha$ under an applied field at a temperature T is then equal to the product of the random probability $(\sin\alpha \cdot d\alpha)/2$ and the Boltzmann probability $\exp(-U_m/kT) = \exp(B\mu\cos\alpha/kT)$ for the angle α. If we substitute $\cos\alpha = y$ and $B\mu/kT = x$, the average

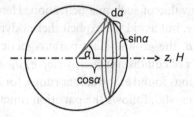

Figure G.1 The unit sphere for derivation of the Langevin function.

atomic moment in the direction of the field can be calculated by integrating over the top half of the unit sphere:

$$\bar{\mu}_z = \frac{\displaystyle\int_\pi^0 \mu \cos\alpha \cdot \exp\left(\frac{B\mu\cos\alpha}{kT}\right) \cdot \frac{1}{2}\sin\alpha \cdot d\alpha}{\displaystyle\int_\pi^0 \exp\left(\frac{B\mu\cos\alpha}{kT}\right) \cdot \frac{1}{2}\sin\alpha \cdot d\alpha} = \mu\,\frac{\displaystyle\int_{-1}^1 y\exp(xy)dy}{\displaystyle\int_{-1}^1 \exp(xy)dy} = \mu\left(\coth(x) - \frac{1}{x}\right)$$

The result is called the **Langevin function**. In terms of magnetization, it is written as $M/M_{sat} = \coth(x) - 1/x$, where M_{sat} is the saturation magnetization. At very low T and at high field, the function is markedly nonlinear. Under weak-field conditions or at higher temperatures, M versus B is essentially linear (Figure 9.16).

To obtain the Curie law of the paramagnet [Equation (9.17)], two approximations are made: $B \approx \mu_0 H$ [because M in Equation (9.7) is small compared to H]; and $\coth x \approx 1/x + x/3$ (because $x \ll 1$), yielding $M/M_{sat} = x/3$ where $x = \mu_0 H\mu/kT$. Since M_{sat} in the Langevin approximation is equal to $N\mu$, where N is the number of magnetic atoms, expressing M/H [Equation (9.8) defining χ] gives the Curie law in the form of Equation (9.19).

APPENDIX H
The Brillouin Function

The Brillouin function describes the relative magnetization, M/M_{sat} versus temperature as a function of the applied field of induction B. One way to derive it is to replace the integrals in the derivation of the Langevin function with summations over the discrete energy states allowed by J, but this is a task hard to get right. Another way is to obtain the partition function, Z, for the magnetic energy levels defined by the quantum numbers S or J, convert it to the Helmholtz free energy $F = -kT\ln Z$, express the magnetization thermodynamically as $M = -(\partial F/\partial B)T$ and relate it to the maximum value of such magnetization. Here, as in most textbooks, we shall follow a similar procedure, but avoid the explicit thermodynamics.

In the magnetic field B, the ground-state paramagnetic J level splits into an $M_J = -J$, $-J+1, -J+2, \ldots +J$ ladder of equidistant energy step $g_J\mu_B \cdot B$ (Figure 9.10).[5] The distribution (partition) of the populations found at the ladder rungs, for example among a huge collection of equivalent atoms in a crystal, follows the partition function Z,

---

[5] Notice that increasing B stretches the ladder.

$$Z = \sum_{M_J=-J}^{+J} e^{-\frac{g_J\mu_B M_J \cdot B}{kT}} \tag{H.1}$$

which represents the sum of the individual populations that are controlled by the ratio of the magnetic and thermal energy at each rung. The probability that an atom has one specific M_J rung of the ladder populated (or the fraction of such atoms in the large collection) is:

$$p_{M_J} = \frac{e^{-\frac{g_J\mu_B M_J \cdot B}{kT}}}{Z} \tag{H.2}$$

The mean projection $\bar{\mu}_z$ of the collection's moments onto the axis of the applied field is the sum, over all rungs, of the ladder-rung probability times that rung's magnetic moment:

$$\bar{\mu}_z = \sum_{M_J=-J}^{+J} \mu_z p_{M_J} = \frac{1}{Z}\sum_{M_J=-J}^{+J} g_J\mu_B M_J \cdot e^{-\frac{g_J\mu_B M_J B}{kT}} \tag{H.3}$$

Let's differentiate the Z in Equation (H.1) with respect to B, as it will prove useful,

$$\left(\frac{\partial Z}{\partial B}\right)_T = \frac{1}{kT}\sum_{M_J=-J}^{+J} g_J\mu_B M_J \cdot e^{-\frac{g_J\mu_B M_J B}{kT}} \tag{H.4}$$

because, by comparison of this result with Equation (H.3), we realize that:

$$\bar{\mu}_z = \frac{kT}{Z}\left(\frac{\partial Z}{\partial B}\right)_T = kT\left(\frac{\partial \ln Z}{\partial B}\right)_T \tag{H.5}$$

Let's now simplify Equation (H.1) for Z by introducing the negatively taken ratio of the ladder's energy step and the thermal energy,

$$\xi = -g_J\mu_B B/kT \tag{H.6}$$

to obtain

$$Z = \sum_{M_J=-J}^{+J} e^{\xi \cdot M_J} \tag{H.7}$$

in which further substitution $x = e^{\xi}$ provides us with an array (in parentheses)

$$Z = \sum_{M_J=-J}^{+J} x^{M_J} = x^{-J} + x^{-J+1} + \ldots + x^{J-1} + x^J = x^{-J}(1 + x + \ldots x^{2J-1} + x^{2J}) \tag{H.8}$$

that has a defined sum, $(1 - x^{2J+1})/(1 - x)$. Then we can write Z as:

$$Z = e^{-J\xi}\left(\frac{1 - e^{(2J+1)\xi}}{1 - e^{\xi}}\right) = \frac{e^{-J\xi} - e^{(J+1)\xi}}{1 - e^{\xi}} \tag{H.9}$$

The fraction, when expanded by $e^{-\xi/2}$, provides a result in which we recognize the hyperbolical sinus functions:

$$Z = \frac{e^{-(J+1/2)\xi} - e^{(J+1/2)\xi}}{e^{-\xi/2} - e^{\xi/2}} = \frac{\sinh[(J + 1/2)\xi]}{\sinh(\xi/2)} \tag{H.10}$$

Now we can do the differentiation suggested in Equation (H.5), but, because B appears in the functional parameter ξ inside a function, the differentiation looks like this:

$$\bar{\mu}_z = kT\left(\frac{\partial \ln Z}{\partial B}\right)_T = kT\left(\frac{\partial \ln Z}{\partial \xi}\right)_T\left(\frac{\partial \xi}{\partial B}\right) \tag{H.11}$$

Plugging in the derivative of Equation (H.6) with respect to B gives

$$\bar{\mu}_z = g_J\mu_B\left(\frac{\partial \ln Z}{\partial \xi}\right)_T \tag{H.12}$$

where plugging in Z from Equation (H.10) and differentiating yields:

$$\bar{\mu}_z = g_J\mu_B\left(\frac{(J + 1/2)\cosh[(J + 1/2)\xi]}{\sinh[(J + 1/2)\xi]} - \frac{1}{2} \cdot \frac{\cosh(\xi/2)}{\sinh(\xi/2)}\right) \tag{H.13}$$

Because $g_J\mu_B \cdot B$ is the saturated paramagnetic moment μ_{sat}, we multiply the large parenthesis with J/J in order to express the ratio $\bar{\mu}_z/\mu_{sat}$ of the average and saturated moments:

$$\frac{\bar{\mu}_z}{\mu_{sat}} = M_r = \frac{1}{J}\left\{\left(J + \frac{1}{2}\right)\coth\left[\left(J + \frac{1}{2}\right)\xi\right] - \frac{1}{2}\coth\left(\frac{1}{2}\xi\right)\right\} = B_J(\xi) \tag{H.14}$$

This is the **Brillouin function** $B_J(\xi)$ for the relative magnetization $M_r = M/M_{sat}$ versus ξ of Equation (H.6) and J.[6] In the spin-only case, J is replaced by S (also in ξ).

In a weak field, we expect to obtain the Curie law. Because ξ is small in a weak field, $\coth \xi \approx 1/\xi + \xi/3$ and Equation (H.14) becomes:

[6] In contrast to the effective moment, saturated or z-projection moments are summed directly into total magnetization.

$$\frac{\overline{\mu}_z}{\mu_{\text{sat}}} = \frac{J+1}{3}\xi \tag{H.15}$$

The saturation moment was a strong-field extrapolation, in which only the lowest level $M_J = -J$ is occupied, and the magnitude $\mu_{\text{sat}} = g_J\mu_{\text{B}}{\cdot}J$ in analogy with Equation (9.13):

$$\overline{\mu}_z = \frac{J+1}{3}J \cdot \frac{g_J^2\mu_{\text{B}}^2 B}{kT} \tag{H.16}$$

Next, we approximate B as $\mu_0 H$ because the paramagnetic magnetization M in Equation (9.7) is much smaller than the field that caused it. We sum the moments for N atoms and divide by H to get the magnetic susceptibility χ_N per that amount of atoms. Finally, we realize from Equation (9.14) that $(J+1)J{\cdot}g_J^2\mu_{\text{B}}^2$ is the square of the effective magnetic moment, μ_{eff}^2, consistent with the weak-field premise of the Curie law. The result is

$$\chi_N = N\frac{\overline{\mu}_z}{H} = N \cdot \frac{\mu_{\text{eff}}^2\mu_0}{3kT} \tag{H.17}$$

the Curie law of Equation (9.19) for the paramagnet of N isolated magnetic atoms.

When combined with the Weiss concept of internal field, the Brillouin function can also be used to describe the relative magnetization M_r of ferromagnets. One approach uses T_C to express the strength of the ferromagnetic interactions instead of the experimentally less accessible Weiss-field coefficient, λ. The idea is that at T_C the magnetization becomes low, so the approximation of Equation (H.15) for M_r applies and can be equated with a Weiss internal-field-based expression of M_r for a ferromagnet at $T = T_C$. The derivation of such a Weiss-field expression for M_r starts with μ/μ_{sat}, where $\mu_{\text{sat}} = g_J\mu_{\text{B}}{\cdot}J$ (per atom), as used above, and $\mu = H_W/\lambda$ (per atom) by analogy with Equation (9.22):

$$M_r = \frac{H_W}{g_J\mu_{\text{B}}J\lambda} \tag{H.18}$$

Now we need to express the Weiss internal field intensity H_W as a function of temperature. That is dictated by the ratio of the magnetic and thermal energy [Equation (H.6)] in which the field B now includes the Weiss field, $\mu_0(H_W + H)$. We neglect H as it is much weaker than H_W:

$$H_W = \frac{kT}{g_J\mu_{\text{B}}\mu_0} \cdot \xi \tag{H.19}$$

Combination of Equation (H.18) and Equation (H.19) gives the internal-field-based expression for the relative magnetization M_r of a ferromagnet, in which we replace T with T_C according to the premise:

$$M_r = \frac{kT_C}{g_J^2\mu_{\text{B}}^2\mu_0 J\lambda} \cdot \xi \tag{H.20}$$

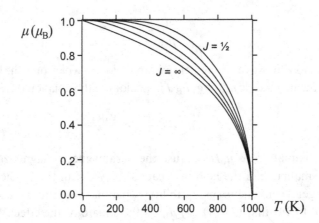

Figure H.1 Thermal decay of an ordered moment of 1 μ_B and $T_c = 1000$ K, for $J = \frac{1}{2}, \frac{3}{2}, \frac{7}{2}, 9$ and ∞.

At this temperature, it can be equated with the approximate M_r of Equation (H.15), which allows us to express λ (per atom) and plug it back into Equation (H.20). This yields the desired term for M_r, in which we expressed the Weiss field via the magnitude of T_C:

$$M_r = \frac{J+1}{3} \cdot \frac{T}{T_C} \cdot \xi \qquad (H.21)$$

In the ferromagnetic range, Equation (H.21) must be satisfied simultaneously with the Brillouin function for M_r of Equation (H.14). Unfortunately, plugging ξ from Equation (H.21) into Equation (H.14) yields no analytical solution for M_r, and these two equations must be solved numerically. The result, illustrated in Figure H.1, is valid for thermal disordering of any individual ordered atomic moment with a critical temperature T_c.

APPENDIX I

Measuring and Analyzing Magnetic Properties

Historically, most magnetic measurements were made using the Gouy or Faraday methods whose basic operating principle is shown in Figure 9.4. Nowadays most measurements are made using either a vibrating-sample or a SQUID (superconducting quantum interference device) magnetometer. In a SQUID magnetometer, the sample is moved through a detection coil where it induces a current. The coil is connected to a SQUID sensor which is a superconducting ring containing a weak link or Josephson junction and acts as an extremely sensitive current-to-voltage converter. The output voltage of the SQUID is proportional to the magnetic moment of the sample, and the instrument can be calibrated using a known material. Modern commercial systems are remarkable instruments with automated field and temperature control and can routinely operate from 0 T to 7 T over temperatures from 2 K to at least 400 K with sensitivities down to 10^{-8} emu. Specialized attachments allow work under non-ambient pressure, as the sample is laser irradiated, or as other physical properties are monitored.

Samples are mounted in holders of low or known diamagnetic contribution. One common method is to place the sample in a gelatin drug capsule, then glue (with varnish) the capsule in a suitably sized drinking straw which is in turn mounted on the magnetometer drive assembly. Alternatively, high-purity silica-glass holders can be used.

In an experiment to test low-temperature magnetic ordering, one might initially measure moment as a function of field at room temperature. A paramagnet should give a straight line passing through the origin. A small, non-zero, intercept might indicate the presence of a minor ferromagnetic impurity (such as Fe or Ni from a spatula). The magnetization of such an impurity would saturate at low field, leading to an approximately constant offset. A suitable field for measurements can be assessed from this plot. If there is a non-zero offset, a field large enough for this to be irrelevant should be chosen, or one could measure at two fields on the linear portion of M versus H and subtract data.

Common commercial instruments produce data files with columns labeled as Temperature, Field (Oe), and Moment (emu). The first stage of the analysis is to correct the measured moment for effects due the sample holder, if any are seen in a blank experiment, in order to obtain the moment due to the sample, μ_{sample}. Molar susceptibility is $\chi_{\text{mol}}^{\text{measured}} = \mu_{\text{sample}}/[\text{Field} \times (\text{mass/molar mass})]$ and should be corrected at this stage for diamagnetic contributions due to core electrons in the sample to give $\chi_{\text{mol}}^{\text{para}}$. Since $\chi_{\text{mol}}^{\text{measured}} = \chi_{\text{mol}}^{\text{para}} + \chi_{\text{mol}}^{\text{dia}}$, one must subtract $\chi_{\text{mol}}^{\text{dia}}$. Because $\chi_{\text{mol}}^{\text{dia}}$ is negative, $\chi_{\text{mol}}^{\text{para}}$ will be larger than $\chi_{\text{mol}}^{\text{measured}}$. For simple compounds, the temperature-independent contribution of $\chi_{\text{mol}}^{\text{dia}}$ can be estimated from tables of Pascal's constants (see e.g. ref. [6] for a compilation). Let's take $CuSO_4 \cdot 5H_2O$ as an example. The tabulated $\chi_{\text{dia}}(Cu^{2+})$, $\chi_{\text{dia}}(SO_4^{2-})$, $\chi_{\text{dia}}(H_2O)$ are -11, -40.1, and -13, in units of 10^{-6} emu/mol, giving $\chi_{\text{dia}}(CuSO_4 \cdot 5H_2O)$ as -116.1×10^{-6} emu/mol to be used as the diamagnetic correction. Note that its absolute value is significantly smaller than the $\chi_{\text{mol}}^{\text{para}}$ of about $+1300 \times 10^{-6}$ emu/mol at room temperature. For a Curie–Weiss paramagnet, a plot of $1/\chi_{\text{mol}}^{\text{para}}$ versus T should give a straight line as in Figure 9.15 (or Figure 9.18). The Curie constant C is given by the inverse gradient of this line, and the Weiss constant θ is the temperature where this line crosses the T axis. The effective moment μ_{eff} in Bohr magnetons per formula unit mol/N_A can be determined from this C obtained from *molar* susceptibility:[7]

$$\mu_{\text{eff}} = 2.828\sqrt{C} \text{ (CGSem)} \qquad \mu_{\text{eff}} = 797.7\sqrt{C} \text{ (SI)} \tag{I.1}$$

A second check of the paramagnetic moment is to plot μ_{eff} as a function of T after setting $C = \chi_{\text{mol}}T$ into Equation (I.1), which should give a clearly recognizable straight horizontal line (if

---

[7] The numerical factor is $[3k/(N_A\mu_B^2\mu_0)]^{0.5}$ in SI, derived from Equation (9.17) and Equation (9.19) after replacing N with N_A and expressing the moment in Bohr magnetons.

it does not, think). All these procedures can be readily automated by least-squares fitting in a spreadsheet package, and, with this method, it may be appropriate to fit all diamagnetic contributions and/or temperature-independent paramagnetic contributions as a single, temperature-independent contribution to $\chi_{mol}^{measured}$. For the most careful work, one should remember to weight fits according to experimental uncertainties in measured moments and propagate errors correctly.

If the sample is ferro- or antiferromagnetic, plots of reciprocal susceptibility versus temperature will become linear a little above the magnetic ordering temperature. Here the Weiss constant (temperature) and effective moment can be obtained, as for the plain paramagnet. T_N of antiferromagnetic powder can be determined from the peak in the $\chi_{mol}^{measured}$ versus T curve. A ferromagnet's T_C can be found from extrapolation of the sudden drop in magnetization upon heating towards this temperature.

APPENDIX J
Fundamental Constants of Exact Value

After major revision[8] in 2019, all seven SI base units are defined by exact numerical values for the fundamental physical constants[9] to which they relate. As some other constants used in chemistry and physics are composed solely of these fundamental physical constants, their values are also exact. A list relevant for this book follows:

Avogadro constant, $N_A = 6.02214076 \times 10^{23}$ mol^{-1}
Boltzmann constant, $k = 1.380649 \times 10^{-23}$ J/K
elementary charge, $e = 1.602176634 \times 10^{-19}$ C
Planck constant, $h = 6.62607015 \times 10^{-34}$ J s
speed of light in vacuum, $c = 299792458$ m/s
standard atmosphere: 101325 Pa (760 torr or mm Hg)
standard-state pressure: 100000 Pa (1 bar)
molar gas constant ($R = kN_A$): $R = 8.314462618\ldots$ J/(mol K)
molar volume ($V_m = RT/P$) **of ideal gas** at $P = 101325$ Pa and $T = 273.15$ K: $V_m =$
    $22.41396954\ldots$ L/mol (1 L $= 10^{-3}$ m^3)

---

[8] See for example https://en.wikipedia.org/wiki/2019_redefinition_of_the_SI_base_units.
[9] See for example https://physics.nist.gov/cgi-bin/cuu/Category?view=html&Adopted+values.x=103&Adopted+values.y=10.

References for Appendices

[1] H.-J. Klein, F. Liebau, "Computerized crystal-chemical classification of silicates and related materials with CRYSTANA and formula notation for classified structures" *J. Solid State Chem.* **181** (2008), 2412–2417, and references therein.

[2] I.D. Brown, D. Altermatt, "Bond valence parameters obtained from a systematic analysis of the Inorganic Crystal Structure Database" *Acta Crystallogr. Sect. B* **41** (1985), 244–247.

[3] N.E. Brese, M. O'Keeffe, "Bond valence parameters for solids" *Acta Crystallogr. Sect. B* **47** (1991), 192–197.

[4] M. O'Keeffe, N.E. Brese, "Atom sizes and bond lengths in molecules and crystals" *J. Am. Chem. Soc.* **113** (1991), 3226–3229.

[5] O.C. Gagné, F.C. Hawthorn, "Comprehensive derivation of bond-valence parameters for ion pairs involving oxygen" *Acta Crystallogr. Sect. B* **51** (2015), 562–568.

[6] G.A. Bain, J.F. Berry, "Diamagnetic corrections and Pascal's constants" *J. Chem. Educ.* **85** (2008), 532–536.

Index

111 superconductors 522
1111 superconductors 522
122 superconductors 522
123 superconductors, see YBa$_2$Cu$_3$O$_{7-\delta}$
214 superconductors 512
8−N rule **17**

A$_3$C$_{60}$ superconductors 501
  (NH$_3$)$_4$Na$_2$CsC$_{60}$ 503
  Cs$_3$C$_{60}$ 504
  K$_3$C$_{60}$ 501
  Rb$_3$C$_{60}$ 501
α-Bi$_2$O$_3$ 553
acceptor doping **62**, **93**, 416
acceptor ionization energy **419**
acceptors **416**, 419
accessible surface area **582**
accessible volume **582**
acid salts (MHXO$_4$) 547
activator **268**, 274, 279
additivity rule **310**
aerogel **601**
Ag$_3$[Co(CN)$_6$] 603
Ag$_3$SI 539
AgI 536
air separation 595
aliovalent doping 416
aliovalent substitution **62**
alizarin (C$_{14}$H$_8$O$_4$) 265
allochromatic **245**
alloy superconductors 487
alloy 55
alpha cage **582**
ALPO materials 596
alternative battery anodes 570
aluminosilicates 579
aluminum 398, 409, 410, 411–412
ambipolar diffusion **108**
ammonia borane (BH$_3$NH$_3$) 57
Anderson localization **645**, 647
Anderson–Eggleton equation 630
Anderson–Eggleton refractive index 630

Angell plot 640
angular momentum quantum numbers in one-electron
    atom **358**
anion conductors 539
anion polarizabilities 629
anode **529**
antibonding molecular orbital **170**
antiferroelectric **319**
antiferromagnet **356**
antiferromagnetic ordering 373
antiferromagnetism 472
anti-Frenkel intrinsic defect pair, see Figure 3.1
anti-perovskite 539
antiphase boundary **72**
anti-Schottky intrinsic defect pair, see Figure 3.1
Arrhenius equation **111**
asymmetric unit **12**
attenuation (in glass fibers) 631
Aurivillius structure 554
autoclave 585
auxetic materials **604**
Avogadro constant in SI **672**
axial glide plane, see glide plane, axial

β-BaB$_2$O$_4$ 340, 342
β-(ET)$_2$X 508
Ba$_{1-x}$K$_x$BiO$_3$ 510
Ba$_{1-x}$V$_x$O$_{1.5+x}$ 553
Ba$_2$In$_2$O$_5$ 554
BaBiO$_3$ 509
BaCe$_{1-x}$Y$_x$O$_{3-x/2}$ 548
BaCeO$_3$ 548
bad metal 520
BaFe$_2$As$_2$ 521
ballistic conduction **645**
ballistic transport **447**
band (of crystal orbitals) **207**
band emitters **274**, 277
band gap **213**, 214, 262–264
band insulator 434
band width 501
band-structure diagram **207**, 208

bandwidth **207**, 221, 431
BaPb$_{1-x}$Bi$_x$O$_3$ 510
BaPbO$_3$ (barium lead oxide) 509
bar magnet 349
barium bismuth oxide, see BaBiO$_3$
barium lead oxide, see BaPbO$_3$
barium titanate, see BaTiO$_3$
basic building units (BBUs) **583**
basis set **202**, 210, 226
BaTiO$_3$ 139, 314–318, 321, 326
BaZr$_{1-x}$Y$_x$O$_{3-x/2}$ 548
Bardeen–Cooper–Schrieffer theory **495**, 505
BCS theory, see Bardeen–Cooper–Schrieffer theory
Bechgaard salt 445, 506
    [TMTSF]$_2$X (Bechgaard salt) 507
Bednorz and Müller 511
Beer's law 255
Beevers–Ross sites 540
beta cage **581**
Bethe lattice **623–624**
BH$_3$NH$_3$, see ammonia borane
Bi$_2$Sr$_2$Ca$_{n-1}$Cu$_n$O$_{2n+4}$ 516
Bi$_2$WO$_6$ 554
Bi$_4$V$_2$O$_{11}$ 554
bicuvox 555
bimevox 554
binary phase diagram **123**
binodal networks (*N,M*-connected nets) **30**
binomial distribution 104
bipolaron **440**
birefringent crystals 332, 334
bismuth oxide, see Bi$_2$O$_3$
bismuth 228, 397
Bloch function **202**, 203, 208, 220, 226
Bloch walls **378**
block copolymers 598
Bohr magneton 352
Boltzmann constant in SI **672**
bond graph **15**
bond order 173
bonding molecular orbital **170**
bond-valence method **192**
bond valence 662
    parameter, single-bond length, $R_{ij}°$ 662
Born–Haber cycle 158, 159
Born–Mayer equation **157**, 158
Bose–Einstein condensate 496
boson peak 643
boundary surfaces **164**, 168
Bravais lattices 5, 7
    non-standard settings 6
    standard settings 7
Brillouin function 370, **668**
Brouwer diagrams **92**
brownmillerite (Ca$_2$FeAlO$_5$) 554

buckminsterfullerene, see C$_{60}$
buffer pO_2 97
bulk diffusion **101**
Burns temperature **320**

C$_{60}$ superconductors 500
C$_{60}$ 500
Ca$_2$FeAlO$_5$, see brownmillerite
CaFeO$_3$ 462
calcination 587
CaMnO$_3$ 465
carbenium ions 590
carbon nanotubes
    chiral angle 449
    circumferential vector 448
    multi-walled carbon nanotubes (MWCNT) 447,
        451
    single-walled carbon nanotubes (SWCNT) 447–451
    translational vector 448
CaSi Zintl phase 48
cathode materials 563
cathode 529
cation conductors 536, 538
cation polarizabilities 629
ccp, see cubic closest packing
cell potential 530
centered Bravais lattices 6
centering 6
centroid shift **277–279**, 294
cermet **555**
charge compensation 62
charge-density wave 506
charge disproportionation 462
charge order 459, 510
charge-transfer excitation 234, 256, 257, 258
    ligand-to-metal charge transfer (LMCT) **258**, 259, 273,
        275, 281
    metal-to-ligand charge transfer (MLCT) **258**
    metal-to-metal charge transfer (MMCT) **259**, 261
charge-transfer salts 443–445
chemical refractivity 630
chemical sensors **531**
chemical strengthening of glasses **626**
chemical-diffusion coefficient **100**
chemisorption 56
child structure 149
chimie douce 558
chiral molecules 611
chiral zeolites 587
CIF (crystallographic information file) 13
Clausius–Mossotti equation **305**, 309, 327
Clebsch–Gordon series **250**
closest packing, hexagonal-, cubic- **19**
cloverite 596
coercive field (electrical), $E_{coercive}$ 316

coercive field (magnetic), H_{ci} **379**
coercivity, see coercive field
color 244–245
color center **58**
    F center 58
    M center 58
    R center 58
color rendering index (CRI) **292**, 293
color temperature **291**
colossal contraction 603
colossal magnetoresistance **470**
    compositional control 470
    variance effect, of ionic radii **471**
commensurately modulated structure **79**
component 120, **121**
composite building units or CBUs **583**
compound semiconductor 261–264
concentration gradient electrodes 567
concentration quenching **284**
conductance 457
conducting polymers 289, 438–441, 572
conduction bands **213**
conductivity 213, **397**, 398, 401, 409, 421
configuration, structural **620**
configurational coordinate model **270**
configurational coordinates **637**
configurational entropy 58
congruent melting **126**
conjugated π bonding 181, 339, 437, 438, 441, 445
conjugated π network, see conjugated π bonding
conjugated π system, see conjugated π bonding
continuous phase transition **137**, 141
continuous random networks **622**
conventional superconductors **492**
converse piezoelectric effect **321**
conversion reactions **570**
Cooper pair **494**
cooperative Jahn–Teller distortion **461**, 465
coordination polyhedra 32
coordination polymers **605**
coordination sequence **583**
coordinations 34
copper 397, 398, 409
core-shell electrodes 566
correlation diagram **252**, 253–255
coulomb integral 171
coulombic potential energy **154–155**, 156, 157
cracking **590**, 593
critical current 491, 498
critical exponent **143**
critical field 490
critical micelle concentration (CMC) **597**
critical phase matching 333
critical point **122**
critical pressure **122**

critical temperature **122**, **135**, 487
cross-relaxation **284**
crystal classes **5**
crystal momentum **207**, 407
crystal orbitals **202**, 205–207, 219, 226
crystal systems **5**, 655
crystal-chemical formula **14**
    balance of bond valence **15**
    balance of connectivity **15**
    balance of electroneutrality **15**
crystal-field theory **247**, 279
crystallographic databases 13
crystallographic point groups **5**, 655
    Hermann–Maugin and Schönflies symbols 655
crystallographic shear **74**
crystal structure figures
    a perovskite, cubic (+data) (Figure 1.49) 40
    AgI (Figure 13.7) 538
    Al_2O_3 corundum (+data) (Figure 1.30) 26
    $Ba_2In_2O_5$ (Figure 13.19) 554
    $BaMnO_3$ (+data) (Figure 1.54) 44
    $BaZnF_4$ cis-layers (Figure 1.44) 36
    Bi_2WO_6 (Figure 13.19) 554
    BiI_3 (+data) (Figure 1.29) 26
    CaB_6 (+data) (Figure 1.40) 32
    CaC_2 (Figure 1.18) 18
    CaF_2 (fluorite) (+data) (Figure 1.31) 27
    $CaSi_2$ (Figure 1.36) 29
    CaTe (+data) (Figure 1.40) 32
    $CdCl_2$ (+data) (Figure 1.29) 26
    CdI_2 (+data) (Figure 1.29) 26
    CdSb (Figure 1.18) 18
    Cu (+data) (Figure 1.21) 20
    Cu_2O (+data) (Figure 1.42) 33
    diamond (+data) (Figure 1.37) 30
    diamond (Figure 1.16) 17
    GaSe (Figure 1.17) 18
    Heusler alloys, full-, half- (+data) (Figure 1.34) 28
    $HgBa_2Ca_{n-1}Cu_nO_{2n+2}$ superconductors (Figure 12.18) 517
    iodine, I_2 (Figure 1.16) 17
    IrF_4 network (Figure 1.44) 36
    iron, alpha (+data) (Figure 1.23) 21
    iron-pnictide superconductors (Figure 12.23) 521
    La_2CuO_4 (Figure 12.15) 513
    $LaFeO_3$ (+data) (Figure 1.52) 43
    $LaNiO_3$ (+data) (Figure 1.52) 43
    lanthanum, alpha (+data) (Figure 1.22) 21
    $LiFePO_4$ (Figure 13.29) 569
    $LiNbO_3$ (Figure 8.22) 338
    lonsdaleite (+data) (Figure 1.37) 30
    metallic elements (Figure 1.24) 22
    Mg (+data) (Figure 1.21) 20
    $Mg_2Al_2Si_3O_{12}$ garnet (+data) (Figure 1.48) 40
    $MgAl_2O_4$ spinel (+data) (Figure 1.47) 39

NaCl (+data) (Figure 1.26) 24
Nd_2CuO_4 (Figure 12.19) 518
NiAs (+data) (Figure 1.13) 13
NiAs (+data) (Figure 1.26) 24
phosphorus, white (Figure 1.16) 17
polonium, alpha (+data) (Figure 1.23) 21
$(SnS)_{1.17}NbS_2$ (Figure 2.14) 79
ReO_3 (Figure 1.39) 32
SiO_2 cristobalite, beta (Figure 1.39) 32
SiO_2 tridymite, beta (Figure 1.39) 32
$SmCo_5$ (+data) (Figure 1.46) 38
$SnCl_2$ (Figure 1.17) 18
SnF_4 trans-layers (Figure 1.44) 36
sodalite (+data) (Figure 1.57) 47
sodium β-alumina (Figure 13.9) 541
$SrSi_2$ (Figure 1.36) 29
tellurium (Figure 1.16) 17
TiF_4 triple chains (Figure 1.44) 36
TiO_2 anatase (+data) (Figure 1.45) 36
TiO_2 brookite (+data) (Figure 1.45) 36
TiO_2 rutile (+data) (Figure 1.45) 36
$Y_5O_5F_7$ (Figure 2.16) 81
$YBa_2Cu_3O_{7-\delta}$ (Figure 3.6 and 12.16) 98, 514
$YBaFe_2O_5$ (Figure 11.3) 460
YCl_3 (+data) (Figure 1.29) 26
zeolites A and X (Figure 14.2) 582
$Zn_4O(O_2CC_6H_4CO_2)_3$ (a MOF structure) (Figure 1.41) 33
ZnS sphalerite (+data) (Figure 1.32) 27
ZnS wurtzite (+data) (Figure 1.32) 27
CsCl-type structure **31**
$CsHSO_4$ 547
cubic closest packing (ccp) **19**
cubic stabilized zirconia CSZ 552
cuprate charge reservoir 518
cuprate superconductors **511**
Curie constant, C 368
Curie law **368**
Curie temperature (electrical) 314, 319
Curie temperature (magnetic) T_C **356, 377**
Curie–Weiss law **369**
current density **396**
Czochralski method **342**

databases in crystallography, see crystallographic databases
Daumas Herold staging 557
δ-Bi_2O_3 553
de Broglie relation 410
Deborah number 636
Debye frequency 494, 497
Debye model 308
decay time **268**
defect concentration 59
defect equilibria 88, 515

defect equilibria, behind oxygen nonstoichiometry (redox) 89
defect ordering, see defect clustering
defect thermodynamics 58
degrees of freedom 120, **122**
densest packing of spheres, cubic, see cubic closest packing
densest packing of spheres, hexagonal, see hexagonal closest packing
density of states **209**, 223, 405, 497, 503, 564, 646, 647
Dexter electron transfer **283**, 284
diagonal glide plane, see glide plane, diagonal
diamagnet **353**
diamagnetic correction 671
    Pascal's constants 671
diamagnetic materials **356**
diamagnetism 367
    Landau diamagnetism **367**
    Lenz's law of induction **367**
    perfect diamagnet **367**, **490**
diamond 228–230, 303, 306
diamond glide plane, see glide plane, diamond
Dieke diagrams 275
dielectric loss **308**, 312
differential scanning calorimetry (DSC) 138
diffusion **99**
diffusion coefficient **100**
diffusionless transition 71
diffusivity **100**, 106
diffusivity and redox equilibria of defects 111
diffusivity due to point defects, temperature dependence 111
dimensionless susceptibility, see magnetic susceptibility, dimensionless
Dirac cones, see Dirac points
Dirac points 446, 451
direct-gap semiconductor **214**, 263, 287
directions in the lattice **3**
discharge rate 568
discontinuous phase transition **137**, 144
disorder-induced metal-to-insulator transition **645**
displacive phase transition **135**
distortion theorem **191**, 327
domain boundary, see twin boundary
domain formation **378**
domain structure **377**
domains (ferroelectric) 315, 320, 338
donor doping 62, 94, 416
donor ionization energy 417, 420
donors **416**, 419
doping **416**, 419, 439
Doppler effect, in Mössbauer spectroscopy 463
double exchange **382**
double glide plane, see glide plane, double

double perovskites 73, 273
down-conversion photoluminescence **268**, 291
drift velocity **400**
Drude equation 630
Drude model 398–401, 403, 407, 409
drying agents 595
   zeolite 3A 595
   zeolite 4A 595
   zeolite 5A 595
dye **245**, 265

easy axes **378**
edge dislocation **66**
effective mass **403**, 406, 407
effective moment, μ_{eff} **361**
effective nuclear charge **166**, 175, 260
Ehrenfest classification 136
electric displacement **304**
electric solenoid 350
electric susceptibility **303**, 332, 628
electrical conductivity (σ_i) due to charged defect **107**
electric-field intensity **396**
electrochemical capacitors **571**
electrochemical cells 529
electrochemical characterization 568
electrochemical potential 562
electrode **529**
electrolyte **529**
electrolyte domain **552**
electrolyte stability window **562**
electromagnetic radiation 243
electron mobility **401**, 406
electron–electron interactions 172, 186, 249, 429, 434
electroneutrality condition **90**
electron–hole pair 290, 426
electronic conductors **396**
electron–phonon coupling **270**, 271, 272, 286
electron–phonon interaction, see electron–phonon
   coupling
electron-spin g factor **358**
electron-transport layer 289
elemental semiconductors 230–231, 263
elemental superconductors 487
elementary charge in SI **672**
enantiomeric excess 612
enantiomers **611**
enantioselective sorption 612
enantioselective synthesis 610, **611**
enantioselectivity **600**
energy landscape, see potential energy surface
energy product, of ferromagnets, see maximum energy
   product
energy transfer 281–283, 284, 292
energy, potential of the magnetic moment **355**
enhanced Pauli paramagnetism 383

ergodic system 636
Euler strut 137
eutectic point **124**
excess free energy, entropy, enthalpy 143, 146
exciton **290**
extended defect **66**
extraordinary beam **332**
extrinsic defects **61**
extrinsic ionic conductivity 534

faradaic supercapacitors **571**
fast ion conductors **536**
faujasite 582
$Fe_{1-x}O$ 77
FePt 73, 139
Fermi energy **210**, 497, 503
Fermi level, see Fermi energy
Fermi velocity **409**, 410, 447
Fermi–Dirac distribution function **404**, 415
ferrimagnetic materials **356**
ferrimagnetism 385
ferroelectric ferromagnets 388
ferroelectric 313, **314**, 314–319, 327, 330
ferroelectric perovskite 139
ferromagnetic materials **356**
   half-metals 381
   insulators 381
   metals 381
ferromagnetism 377
ferromagnets of spin-polarized bands 381
FeSe, see iron selenide
$FeTiO_3$, see ilmenite
Fick's first law **100**
Fick's second law **100**
filling holes 22
first Brillouin zone **204**, 216, 217, 219
first-order Jahn–Teller distortion **189**, 280
first-order phase transition **136**, 144
fluid catalytic cracking FCC 593
fluorescence **268**
flux **100**, 125
Förster resonant-energy transfer (FRET) **282**
fractional coordinates **2**
fragility index **640**
fragility strength **640**
framework density **580**
Franck–Condon principle **271**
free energy, interfacial, see interfacial free energy
free volume **622**, 639
free-electron model 402–403, 405, 407, 412, 446
freeze-out regime **420**
Frenkel defects **58**, 60, 533, 539
Frenkel intrinsic defect pair, see Figure 3.1
Fröhlich interaction **493**
frontier orbitals **180**, 328

frustration, in cooperative magnetism **387**
fuel cells **532**
fullerene superconductors, see C_{60} superconductors
fundamental SI constants 672

gallium nitride (GaN) 288, 314
galvanic cells **529**
garnet 39
gate dielectric 427
Gaussian error function (erf) 113
gel **585, 601**
gemstone 245, 256
generalized 8−N rule **17**
geometric frustration **648**
germanium 230, 397, 418, 422
Gibbs phase rule, see phase rule
Gibbs-energy barrier for migration 104, 111
Gibbs–Thomson effect 634
Gladstone–Dale relationship 630
glass forming range **635**
glass transition 634
glass transition temperature 544
glasses
    covalent **620**
    random-packed **620**
glass-forming temperature 621, **635**
Glazer tilt (classification) **41**
glide plane **9**
    axial **9**
    diagonal **10**
    diamond **10**
    double glide **10**
Goldschmidt tolerance factor **40, 44**
graph **622**
graphene 221–223, 266, 445–447, 448, 450
graphite intercalation chemistry **556**
Gruneisen parameter **604**

H_2S superconductor, see hydride superconductor
half-metallic CrO_2 476
half-metallicity
    by Andreev reflection **475**
    by positron annihilation **475**
    by spin-resolved photoemission **476**
    with spin-polarized transport 472
half-metals of itinerant electrons 480
hamiltonian operator 161
hard (ferro)magnets **379**
hcp, see hexagonal closest packing
heat capacity 138, 144, 146, 642
Heusler alloys 477
hexagonal closest packing (hcp) **19**
hexagonal perovskites **44**
HfO_2 428
$HgBa_2Ca_{n-1}Cu_nO_{2n+2}$ 511, 516

high-entropy alloys **650**
highest-energy occupied molecular orbital **180**, 222
high-spin (HS) configuration **187**, 254, 256
high-spin state, see high-spin (HS) configuration
high-T_c cuprate superconductors 223
hole doping 512, **516**, 518
hole filling, see filling holes
holes **407**
hole-transport layer 289
homeotypism 30, 45, 46
HOMO, see highest-energy occupied molecular orbital
hopping in electric field 106
hopping under concentration gradient 105
hopping, random 103
hopping/migration, in chemical potential gradient
    105
hopping/migration, in electric potential gradient 106
hopping/migration, under driving force 104
host **268**
Huang–Rhys parameter **272**
Hubbard model **429–431**
Hubbard U 429, 437
Hund's rules 360, 367
    Hund's first rule 171, **251**
    Hund's second rule **251**
    Hund's third rule **251**
hybrid porous materials **605**
hydride superconductors 488
hydrogen bond **545**
hydrogen fuel cell 532
hydrogen storage 56
hydrothermal synthesis **585**
hyper-aged organic glasses **638**
hysteresis **136**, 356
hysteresis loop (ferroelectric) **316**, 319
hysteresis loop (magnetic) 378, 379

identity **4**
idiochromatic **245**
ilmenite ($FeTiO_3$) 192, 261
immiscibility dome **130**
incipient ferroelectric 312, **327**
incommensurately modulated structure **78, 79**, 149
incongruent melting **126**, 342
indigo ($C_{15}H_9N_2O$) 265
indirect-gap semiconductor **214**, 263, 287
infinite-layer structure **517**
infinitely adaptive structure **80**
infrared absorption band (in glass fibers) **631**
intensity of the magnetic field, H **352**
interaction integral 171
intercalation compounds **555**
    guest species **555**
    host structure **555**
interdiffusivity 106

interfacial free energy **634**
interference term **645**
internal conversion **268**
internal field, H_W **379**
International Tables for Crystallography 656
internal friction 644
interpenetrated network **606**
interstitial **54**, 534
interstitialcy mechanism 534
intersystem crossing **268**, 279, 291
intrinsic defect pairs, Schottky, anti-Schottky, Frenkel,
       anti-Frenkel, see Figure 3.1
intrinsic defect **55**
intrinsic ionic conductivity 535
intrinsic regime **420**
invariant point **122**, 125
inverse opal **601**
inverse spinel **38**
inversion center **4**
Ioffe–Regel condition 645
ion exchange 593
ion-exchange isotherms **594**
ionic conductivity 533
ionic conductors **396**
ionic radii 193, 194
ionic-radii variance effect, see variance effect
iron monoxide (wüstite), see $Fe_{1-x}O$
iron platinum, see FePt
iron selenide 521
iron pnictide superconductors 521
isomerization **590**
isoreticular **608**
isotope effect 493, 503, 511
isovalent substitution **61**
itinerant ferromagnetism **382**
ITQ-33 584

Jagodzinski–Wyckoff notation **19**
Jahn–Teller distortion, see first-order Jahn–Teller
       distortion
Jahn–Teller theorem **188**
j–j coupling **249**, 275, 664
jump and thermal energy kT 101
jump balance **103**
jump directional probability, p_{dir} 102
jump frequency, $p_B v$ 101
jump Gibbs-energy barrier, $\Delta^{\ddagger}G_m$ 101
jump probability, p_B 101
jump progression rate, $r_{progression}$ 102
jump site availability, p_{avail} 102

K_2NiF_4 structure 512
Kamerlingh Onnes, Heike 486
Kauzmann rule 636
Kauzmann temperature 638

KDP, see KH_2PO_4
KGe Zintl phase 48
KH_2PO_4 (KDP) 336
klassengleiche subgroup 73, **660**
Koch cluster 77
Kondo effect **387**
Kröger–Vink notation 65, 87
$KTiOPO_4$ (KTP) 336
KTP, see $KTiOPO_4$
K_xC 556

$La_{1/2}Ca_{1/2}MnO_3$ charge and orbital ordering 469
$La_{1-x}Ca_xMnO_3$ phase diagram 468
$La_{1-x}Sr_xGa_{1-y}Mg_yO_{3-\delta}$ 553
La_2CuO_4 512, 513
$La_2Mo_2O_9$ 555
$La_2CuO_{4+\delta}$ 512
$La_{2-x}Sr_xCuO_4$ 511, 512, 513
$LaGaO_3$ 553
$LaMnO_3$ 465
$LaMnO_{3+\delta}$ nonstoichiometry control 466
Landau theory **140**
Landé factor **363**
Langevin function 370, **666**
$LaNi_5$ 56
$LaO_{1-x}F_xFeAs$ 521
Laporte selection rule **256**
Larmor precession 363
lattice **2**
lattice diffusion **101**
lattice energy **156**, 158
lattice metrics, see metrics of lattices
lattice modes **147**
lattice parameters **2**, 8
lattice planes, equidistant set **3**
lattice-formation enthalpy 158
LCAO, see linear combination of atomic orbitals
lead acid battery 531
lead fluoride, see PbF_2
lever rule **124**
$Li(Ni_{\sim0.33}Mn_{\sim0.33}Co_{\sim0.33})O_2$, see NMC
$Li_4Ti_5O_{12}$ 570
LiB_3O_5 340, 341, 342
$LiBH_4$, see lithium borohydride
$LiCoO_2$ (lithium cobalt oxide) 565
LiFeAs 521
$LiFePO_4$ **569**
ligand-field splitting **184**, 186, 234, **247**, 248, 277
ligand-field theory **247**
light-emitting diodes (LEDs) **287–288**, 291, 422, 426
$LiNbO_3$ 324, 338
line defect **66**
line emitters **274**, 284
linear combination of atomic orbitals **170**, 200, 202
liquidus **124**

LISICON 543
lithium air battery 570
lithium borohydride (LiBH$_4$) 57
lithium cobalt oxide, see LiCoO$_2$
lithium intercalation 559
lithium-ion batteries 559, 566
lithium iron phosphate, see LiFePO$_4$
lithium–sulfur battery 570
lithium–TiS$_2$ battery 560, 568
LiVPO$_3$F 570
Loewenstein's rule **587**
longitudinal mode 148
Lorenz–Lorentz relationship 629, 630
lowest-energy unoccupied occupied molecular orbital **180**, 222
low-spin (LS) configuration **187**
low-spin state, see low-spin (LS) configuration
luminescence 168, **267**, 268
    bioluminescence **267**
    cathodoluminescence **267**
    chemiluminescence **267**
    electroluminescence **267**, 287
    photoluminescence **267**, 268
    piezoluminescence **267**
    sonoluminescence **267**
    thermoluminescence **267**, 268
    triboluminescence **267**
luminous efficacy **291**
luminous flux **291**
LUMO, see lowest-energy unoccupied occupied
    molecular orbital

Mackay icosahedra **649**
macroporous **579**, 601
Madelung constant **155**, 156
Magnéli phases **74**
magnesium diboride, (MgB$_2$) 499
magnesium hydride (MgH$_2$) 56
magnetic (dipole) moment, μ **350**
magnetic constant 353
magnetic field strength, see intensity of the
    magnetic field
magnetic induction, B **352**
magnetic measurements 670
magnetic moment, absolute, μ_{abs} 358
magnetic moment, ground state 359
    for actinoids, j–j coupling 360
    in free d-metal atom, under Russell–Saunders coupling
        scheme 360
    in free f-metal atom, under Russell–Saunders coupling
        scheme 367
magnetic moment
    spin-only 361
    z-projection, μ_z 358
    of 3d ions 363

of 4f ions 366
magnetic ordering, temperature **356**
magnetic permeability, relative, μ_r **354**
magnetic quantum number **162**
magnetic structure 373, 377
magnetic susceptibility
    dimensionless 354, 355
    mass, χ_m **354**
    molar, χ_{mol} **354**
    volume, χ_v **353**
magnetic unit cell 374
magnetic-flux density, see magnetic induction
magnetic-moment vector 351
magnetism, electron movements and their quantization 357
magnetism, unit systems in use 355
magnetite 458
magnetization of half-metals **475**
magnetization, M **353**
magnetizing field, see intensity of the magnetic field
magnetocrystalline energy **378**
magnetoelectric multiferroics 388
magnetoresistant device 457
magnetostatic energy **378**
magnetostriction **378**
magnetostrictive energy **378**
magnetotransport 457
manganite perovskites, contour map of ordering
    isotherms 471
martensitic transformations **71**
mass susceptibility, see magnetic susceptibility, mass
maximum energy product $(BH)_{max}$ **379**
MCM-41 597
mean free path **400**, 409
Meissner or Meissner–Ochsenfeld effect **490**, 492
mercury superconductivity 487
mesoporous materials **46**, **579**, 597
mesoporous zeolites 589
mesostructured zeolites 600
metal–organic framework (MOF) 31, **605**
metallic glasses **647**
metallic glasses, ferromagnetic 650
metal-oxide-semiconductor field-effect transistor
    (MOSFET) **426**
metrics of lattices 8
MgB$_2$, see magnesium diboride
MgH$_2$, see magnesium hydride
micelles 597
microporous materials **46**, **579**
microstates **250**, 275, 365
microwave resonator 311
migration/hopping, see hopping/migration
mirror plane 4
miscibility gap 569
mixed anion 80
mixed ionic and electronic conductor (MIEC) **555**

MO diagram, see molecular-orbital diagram
MO, see molecular orbital
mobility **108**
mobility edge **646**
mobility gap **646**
MOF applications
  asymmetric catalysis 611
  CO_2 capture and storage 609
  dynamic frameworks 610
  heterogeneous catalysis 610
  hydrogen storage 609
  physisorption 608
  supported catalysis 610
  water capture 609
MOF, see metal–organic framework
molar gas constant in SI **672**
molar susceptibility, see magnetic susceptibility, molar
molar volume of ideal gas in SI **672**
molecular magnets 389
molecular-orbital diagram 169–186, 211, 325, 328
molecular orbital **169**
molecular sieves 596
molecular superconductors **505**
monotectic point **130**
morphotropic phase boundary (MPB) **323**
Mössbauer spectroscopy 463
motif **2**, 222, 225
Mott–Hubbard insulator **431**, 434, 435, 436, 518
Mott–Hubbard transition 501, 502
multi-phonon emission **273**
multiplexing (in glass fibers) **632**
multiplicity **250**

$NaAl_{11}O_{17}$ 540
$NaAlH_4$, see sodium aluminum hydride
NaCl conductivity 535
NAFION 544, 546
NASICON 542
NaTl Zintl phase 47, 48
natrolite 603
Nb_3Ge superconductor 487
Nd_2CuO_4 **517**
Néel temperature, T_N **356**, **372**
negative thermal expansion **602**
NEMCA reactor 532
Nernst equation 531, 550
Nernst–Einstein equation **107**, 534
nets **606**
network augmenting 31
network expansion 31, **608**
network similarity (site ordering), see similarity (network-based)
network-forming oxides **625**
network-modifying oxides **625**
networks binodal (N,M-connected nets), see binodal networks (N,M-connected nets)

neutron diffraction on ordered magnetic moments 376
nickel–cadmium battery 531
Niggli formula **34**
nitinol 71
NMC 566
nodal planes **164**, 176, 182, 265, 442
node **164**, 201
non-centrosymmetric crystals 313, 324, 332
noncritical phase matching **334**, 341
non-ergodic system 636
non-radiative recombination **426**
non-steady-state diffusion **100**
non-steady-state diffusion, of point defects 112
nonstoichiometry
  narrow 87
  oxidative 88
  reductive 88
normal spinel **38**
n-type cuprate superconductors **517**
n-type semiconductor **419**, 422
nucleation
nucleation, heterogeneous **634**
nucleation, homogeneous **634**

occupiable volume **581**
octahedral holes **22**
  fractional filling 23
  full filling 23
octahedral ligand-field splitting (Δ), see ligand-field splitting
octahedral tilting **41**, 194, 236
olivine 569
operation of symmetry, see symmetry operations
optic axis **333**
optical dielectric constant **307**
optical fibers 631–633
orbital angular-momentum quantum number **162**, 250
orbital ordering **461**
orbital-degeneracy removal 462
  by charge disproportionation 462
  by electron itinerancy 462
  by Jahn–Teller distortion 462
order parameter **139**, 140
order, long-range **620**, 642
order, short-range **620**
order–disorder phase transition **135**
ordinary beam **332**
organic light-emitting diodes (OLEDs) **287**, 289–291
organic magnets 390
organic superconductors **505**
orientational disorder **136**
Ostwald ripening **37**
Ostwald step rule **37**, 649
overlap integral 171, 173, 174–175, 177
oxidation catalyst 593

oxide fluoride 80
oxide-ion conductors 549
oxyfluoride, see oxide fluoride
oxygen control 515
oxygen nonstoichiometry (redox defect equilibria behind it) 89
oxygen sensor 550
oxygen, bridging **625**
oxygen, non-bridging **625**
oxygen-nonstoichiometry control in oxides 97

packing of spheres 19
pair distribution function 318, **620**, 621, 624
pairing energy 495
pairing-$s/p/d/f$-wave, see $s/p/d/f$-wave pairing
paraelectric **314**, 318, 320, 323
parallel plate capacitor **302**
paramagnet **353**
paramagnetic materials **356**
paramagnetism 367
    Curie paramagnets **368**
    Curie–Weiss paramagnets **368**
    Pauli paramagnets **368**
parent structure 149
partial pressure of oxygen, pO_2 91, 97
Pauli exclusion principle 162, 171
Pauli paramagnetism, enhanced **371**
Pauli paramagnetism, temperature independent **371**
$Pb(Mg_{1/3}Nb_{2/3})O_3$ 320
PbF_2 (lead fluoride) 539
$PbZr_{1-x}Ti_xO_3$ (PZT) 323
Peierls' distortion **212**, 228, 439, 507
Peierls' theorem **212**
PEM 546
penetration depth **113**
pentacene 266, 441–443
PEO, see polyethylene oxide
perfect conductor 490
peritectic point **126**
peritectic reaction **126**
permeability of free space, see magnetic constant
perovskites 40, 233–236, 320, 434–437
    microwave dielectrics 312
    proton conductors 548
    superconductors 509
persistent phosphors **268**
phase diagram 120
phase matching 332–334
phase rule 97, **120**, 123
phase separation 569
phase transition thermodynamics 142
phase transition **135**
phase velocity 148
phase 120

phase-change materials 641
phonons 147, **149**, 642
phosphor-converted light emitting diode 293–294
phosphorescence **268**, 275, 280, 281, 290
phosphors **267**, 268, 273, 276, 284, 292–293
photoionization 268
photon **243**
photovoltaic cell **426**
physisorption 56
piezoelectric effect **321**
piezoelectricity 313, 321
pigment **245**, 257–258, 259, 265
pinning 66
planar defects **67**
Planck constant in SI **672**
plastic crystals **547**
p–n junction 287, **422–425**, 427
    built-in potential **424**
    depletion region **424**, 425
    forward bias **425**
    reverse bias **425**
point defect **54**
point defects in a pure oxide 87
point group **3**
point of integer structure **92**, 96, 98
point of integer structure, see Figures 3.2, 3.4, 3.5
point of integer valence **92**, 96, 98
point of integer valence, see Figures 3.2, 3.4, 3.5
point-defect compensations upon oxidation and reduction, see Figure 3.1
point-defect movements 101
point-group order **4**
point-symmetry elements **4**
Poisson's ratio **604**
polar axis 313, 337
polar materials 313, 314, 321
polar nanoregions (PNR) **320**
polarizability 278, **304**, 305, 309–311, 629
    dipolar polarizability **305**
    electronic polarizability **305**
    ionic polarizability **305**
polarization 160, **303**, 304, 305, 314, 331, 628
polaron **440**, 510
polonium 227–228
polyacetylene 212, 397, 438–441, 442
polycyclic aromatic hydrocarbons 441
polyelectrolytes **543**
polyethylene oxide 544
polyhedral connectivities 34
    edge sharing 35
    face sharing 35
    vertex sharing 34
polymer-electolyte membrane, see PEM
polymeric cation conductors **543**
polymer-salt complexes **543**

polymorphism of nanocrystals 37
polymorphism **68**
polytypes **68**
porous oxides 599
porphyrin 266
position, see special positions or general positions
potential energy surface (PES) **636**, 637, 638, 641
primary batteries **531**
primitive Bravais lattices **6**
principal quantum number **162**, 164, 418
probability density **164**, 202, **645**
proton conductors 545
proton-exchange membranes, see PEM
Prussian blue ($Fe_4(Fe(CN)_6)_3 \cdot xH_2O$) 257, 261, 605
pseudocapacitors **571**
pseudoelasticity **72**
pseudogap regime **519**
p-type cuprate superconductors **516**
p-type semiconductor **419**, 422
pyroelectricity 313, **314**, 330
PZT, see $PbZr_{1-x}Ti_xO_3$

quality factor **312**
quantum efficiency **268**
quantum harmonic oscillator 271
quasicrystals **649**
quasi-phase matching **334**, 338
quenching **621**, 648

radial distribution function **164**, 169
radial node **164**
radiative recombination 287, **425**
Ragone plot **572**
Ramsdell symbol **19**
random walk (electrons) 645
random walk **103**
random-close packing of spheres **621**
Rayleigh scattering 631
$RbAg_4I_5$ 538
reciprocal-space lattice **204**, 216, 217, 218, 219
reconstructive phase transition **135**
rectification **424**
redox compensation **62**
refraction **627**
    anomalous 628
    normal 628
refractive index 307, 311, **330**, 332, 333, **627**, 628
regioselectivity **600**
relative coordinates, see fractional coordinates
relative dielectric permittivity **303**, 417, 418
relaxation time **400**, 409
relaxor ferroelectric 320
remanence, see remanent magnetization
remanent magnetization, $M_{remanent}$ **379**
remanent polarization 316

ReO_3 231–233, 234, 235, 397, 605
repulsive potential energy 156, 157
residual resistivity **411**
resistivity **396**, 645
reticular chemistry **607**
RGB model 244
rhenium trioxide, see ReO_3
rigid unit modes (RUMs) 602
ring **623**
rocking chair battery **566**
rotation axes **4**
rotational axes, see rotation axes
rotational symmetry 3
rotational-symmetry elements **5**
rotoinversion axis **4**
rubrene (5,6,11,12-tetraphenyltetracene, $C_{42}H_{28}$) 443
rule of parsimony **34**
Russell–Saunders coupling **249**, 275

$s/p/d/f$-wave pairing 495
SALC, see symmetry-adapted linear combinations
sapphire (Al_2O_3:Fe^{2+},Ti^{4+}) 260, 288
saturation moment, μ_{sat} **361**
saturation polarization 316
saturation regime **420**
saturation resistivity 645
Schottky defect **55**, 60, 533
Schottky intrinsic defect pair, see Figure 3.1
screw axis **9**
screw dislocation **66**
secondary batteries **531**
secondary building units (SBUs) **583**
second-harmonic generation 313, **330**, 332, 333, 334
second-order Jahn–Teller (SOJT) distortion **189**, 280, 325–330
second-order phase transition **136**
segmental motion **544**
Seitz operator **9**
Seitz symbol 9
selectivity **594**
self diffusion 104, 106
self-activating phosphors **269**, 284
semimetal **223**, 446
sensitizer **268**, 282, 283, 285, 292
shape-memory alloys **71**
shape-selective catalysis **591**
    product-selective catalysis **592**
    reactant-selective catalysis **591**
    transition-state-selective catalysis **592**
shear plane **74**, 75
shell **162**
Si:Al ratio 587
silicates, oligo-, cyclo-, catena-, phyllo-, tecto- **45**
silicate nomenclature 661
    branchedness **661**

multiplicity **661**
  dimensionality **661**
  periodicity **661**
silicon 230, 306, **397**, 415, 416, 418, 422, 424, 427
silver cobalt cyanide see $Ag_3[Co(CN)_6]$
silver iodide, see AgI
silver 397, 398, 414
similarity (network-based) 30
single-molecule magnets **391**
SiO_2 cristobalite 135
SiO_2 quartz 135
SiO_2 tryidymite **135**
SiO_2 121
site ordering **30**, 72
  diamond–sphalerite, see Figure 1.38
  lonsdaleite–wurtzite, see Figure 1.38
  polonium–sodium chloride, see Figure 1.38
site, see Wyckoff site
Slater–Pauling rule of localized ferromagnetism **478**
slip plane **66**
$SmCo_5$ structure type, see Figure 1.46
$SmO_{1-x}F_xFeAs$ 521
$Sn_{1.17}NbS_{3.17}$ 78
Snell's law 627
sodalite cage 47, **581**
sodalite 257, 581
sodium aluminum hydride ($NaAlH_4$) 56
sodium β-alumina 540
sodium sulfur battery 541
soft magnets 379
soft mode **149**
sol **601**
sol-gel synthesis **600**
solid acids **588**, 589
solid-oxide fuel cell (SOFC) 549
solid solubility 94
solid solution 55, **63**, **121**, 128, 568
  solid-solution limit **65**
solid electrolyte interphase (SEI) **562**
solidus **124**
soliton **440**
solvus **129**
space groups **8**, 656
  absent or extinct Bragg reflections **658**
  asymmetric unit **658**
  coordinates **658**
  diagram **656**
  generators selected **658**
  multiplicity **658**
  non-symmorphic **8**
  order 659
  origin placement **658**
  positions **658**
  reflection conditions **658**
  site symmetry **658**

subgroups and supergroups **659**
subgroups, klassengleiche **660**
subgroups, maximal isomorphic, of the lowest index **660**
subgroups, maximal non-isomorphic **660**
subgroups, proper **659**
subgroups, translationengleiche **660**
subgroups' dilution index **660**
supergroups, minimal non-isomorphic **660**
symmetry elements, graphical symbols **656**
symmetry operations **658**
symmorphic **8**
  Wyckoff letter **658**
space-group symbols 11
  standard 11, 12
spectroscopic term, see term symbol
speed of light in vacuum in SI **672**
sphere packing, see packing of spheres
spin-density wave 506
spin flip, in antiferromagnet 373
spin glass **357**, **387**
spin glass, reentrant **388**
spin polarization, P **458**
spin quantum number **162**, 250
spin selection rule **256**
spin valve 457
spinel 38
spin-only moment, strong-field limit **362**
spin-only moment, weak-field limit **362**
spin–orbit coupling 249, 291, **359**
  strong-field limit **363**
  weak-field limit **362**
spin-pairing energy **186**, 375
spintronics 457
spontaneous magnetization **377**
SQUID magnetometer 670
$Sr_{0.1}La_{0.9}AlO_3$ 553
$Sr_{0.4}K_{0.6}BiO_3$ 511
$Sr_{0.73}CuO_2$ 80
$SrFeO_3$ 462
$SrGa_2$ Zintl phase 48
$SrTiO_3$ 147, 149, 194, 233, 236, 303, 312, 325, 326, 327, 434
stacking faults **67**, **68**
stacking sequence 19
staging **557**
  stage-1 and stage-2 557
standard atmosphere in SI **672**
standard-state pressure in SI **672**
static dielectric constant **307**
steady-state diffusion **100**
Stokes shift **270**, 271, 272, 277, 280
strain 140
strongly correlated materials **429**, 520
structure directing agent **587**
structures of metals 22

stuffed silica 580
subshell **162**, 167, 168, 274, 360
substitutional disorder **55**, 61, **136**
supercapacitors **571**
supercell **31**
superconducting dome 504
superconducting gap 493, **496**, 497
superconducting levitation 492
superconducting transition 498
superconductivity 486
supercooled state 122
supercooling **634**
supercritical fluid **122**
superelasticity **72**
superexchange **374**
superferromagnets **384**
superionic conductors see fast ion conductors
superstructure **31**
supported catalysts 599
surface structure (versus bulk structure) 22, 37
switching field **316**
symmetry breaking, see symmetry lowering
symmetry elements rotational, see rotational-symmetry
    elements
symmetry elements translational, see Bravais lattices
symmetry forbidden **175**, 189, 226
symmetry lowering **137**, 138
symmetry operations 3
symmetry-adapted linear combinations **178–179**, 183,
    184, 202
system **120**

Tanabe–Sugano diagram **281**
temperature coefficient of resonant frequency **312**
tempering of glasses **626**
template molecules **586**, 587
tension effect 602
term symbol **250**, 252
ternary phase diagram **131**
tessellation (in 3D space) **648**
tetracyanoquinodimethane (TCNQ) 444
tetrahedral holes **22**
    fractional filling, see Figure 1.5
    full filling, see Figure 1.5
tetrathiofulvalene (TTF) 444
thermal conductivity 642
thermal expansion 602
thermal quenching temperature **273**
thermal quenching **272–274**, 277
thermal stimulation of luminescence, see
    thermoluminescence
thermal vibration 101
$ThrCr_2Si_2$ structure 522
tight-binding methods **202**
tiling **583**

time-independent Schrödinger equation 161, 402
TiO_x 75, 76
TiS_2 559
titanium disulfide, see TiS_2
$TlBa_2Ca_{n-1}Cu_nO_{2n+3}$ 516
$TlBa_2Ca_{n-1}Cu_nO_{2n+4}$ 516
TMTTF 508
top seeded solution growth **342**
topology of glasses **622**
total angular-momentum quantum number **251**
total external reflection **628**
total internal reflection **627**, 628, 631
transference number (t_i) of a point defect **107**
transition temperature **135**
translational symmetry 2
translationengleiche subgroup 70, **660**
transport half-metallicity **475**
transverse mode 147
triangle rule **132**
tricritical transition 147
triple point **122**
TS-1 593, 600
$TTF[Ni(dmit)_2]_2$ 509
TTF–TCNQ 444, 506
tungsten trioxide, see WO_3
tunneling magnetoresistance **472**
    between two separated epitaxial films 473
    in a powder compact **473**
turbostratic disorder **68**
twinning 68
    deformation twins **69**
    growth twins **69**
    transformation twins **70**
    twin boundary **69**
    twin categories **69**
    twin component **68**
    twin domain **68**
    twin law **68**
    twin variants **70**
two-level tunneling model **643**
type-I superconductor **490**
type-II superconductor **491**

ultrastable zeolite Y 588
unconventional superconductors **492**
under- and over-doping 519
uninodal network **28**
unit cell **2**
unit conversions in magnetism 355, 356
unit-cell parameters **2**
universal low temperature states 639, 642
up-conversion photoluminescence **268**, 285
upper consolute temperature **130**
Urbach tails 646
UV tail (in glass fibers) **631**

vacancy clustering **58**, 66, 77
vacancy trapping 552
vacancy **54**
vacuum permeability, see magnetic constant
valence bands **213**
valence mixing, preservation of degeneracy 461
valence-mixed state 459
valence-sum rule **191**, 192
van der Waals forces **159**, 447, 559
vanadium disulfide, see VS_2
variance, see degrees of freedom
VEC_A (valence-electron count per atom A) **17**
Vegard's law **63**
    positive deviation **64**
    negative deviation **64**
vertex augmenting, see vertex decoration
vertex decoration **31**
vertex symbol **28**, **583**
Verwey transition **459**
vibration (thermal), see thermal vibration
vibronic coupling, see electron–phonon coupling
virtual phonon 494, 495
Vogel–Fulcher–Tammann equation 544, 639, 641
Vogel–Fulcher–Tammann temperature 640
volume susceptibility, see magnetic susceptibility, volume
vortex pinning **492**
vortex state 491, 492
VPI-5 596
VS_2 558

Wadsley defect 74
water softening 594
wave vector 148, **217**, 219, 408, 494
wavefunction 161–166, 169, 170, 171, 202, 203, 204, 217
weak backscattering 646
Weiss constant, θ **369**
wide nonstoichiometry 98
WO_3 74, 234, 235, 326

WO_{3-x} 74
Wyckoff site **12**
    general position **12**
    special positions **12**

xerogel **601**
X-ray absorption fine structure (EXAFS) 318

$Y_3Al_5O_{12}$, see yttrium aluminum garnet
$YBa_2Cu_3O_7$ synthesis 515
$YBa_2Cu_3O_{7-\delta}$ 98, 108, 112, 134, 511, 513
$YBaFe_2O_5$, charge ordered 460
$YBaFe_2O_5$, valence mixed 460
YBCO superconductors, see $YBa_2Cu_3O_{7-\delta}$
$YO_{1-m}F_{1+2m}$ 80
yttria-stabilized zirconia (YSZ) 552
yttrium aluminum garnet ($Y_3Al_5O_{12}$) 293, 310
yttrium barium copper oxide, see $YBa_2Cu_3O_{7-\delta}$

Zachariasen rules **625**
Zeeman effect **358**
Zeeman splitting, see Zeeman effect
zeolite catalysts 588
zeolite A 582
zeolites 46, 579
zeolite X 583
zeolite Y 583
zeotypes **579**
zero resistance 486, 498
zero-field splitting **365**
zero-point energy **160**
Zintl phases 47
Zintl–Klemm concept **47**
zirconia (ZrO_2) 552
zirconium tungstate (ZrW_2O_8) 130, 602
ZrO_2, see zirconia
ZrW_2O_8, see zirconium tungstate
ZSM-18 587

Printed in the United States
by Baker & Taylor Publisher Services